Behavioural Mechanisms of Food Selection

NATO ASI Series

Advanced Science Institutes Series

A series presenting the results of activities sponsored by the NATO Science Committee, which aims at the dissemination of advanced scientific and technological knowledge, with a view to strengthening links between scientific communities.

The Series is published by an international board of publishers in conjunction with the NATO Scientific Affairs Division

A Life Sciences B Physics	Plenum Publishing Corporation London and New York
C Mathematical and Physical Sciences D Behavioural and Social Sciences E Applied Sciences	Kluwer Academic Publishers Dordrecht, Boston and London
F Computer and Systems Sciences G Ecological Sciences H Cell Biology	Springer-Verlag Berlin Heidelberg New York London Paris Tokyo Hong Kong

Series G: Ecological Sciences Vol. 20

Behavioural Mechanisms of Food Selection

Edited by

Roger N. Hughes

Unit of Functional and Evolutionary Biology
School of Biological Sciences, University College of North Wales
Bangor, Gwynedd LL57 2UW, United Kingdom

Springer-Verlag Berlin Heidelberg New York
London Paris Tokyo Hong Kong
Published in cooperation with NATO Scientific Affairs Division

Proceedings of the NATO Advanced Research Workshop on Behavioural Mechanisms of Food Selection held at Gregynog, Wales, United Kingdom, July 17–21, 1989

ISBN-13:978-3-642-75120-2 e-ISBN-13:978-3-642-75118-9
DOI: 10.1007/978-3-642-75118-9

Library of Congress Cataloging-in-Publication Data. NATO Advanced Research Workshop on Behavioural Mechanisms of Food Selection (1989 : University of Wales Conference Centre) Behavioural mechanisms of food selection / edited by Roger N. Hughes. p. cm.—(NATO ASI series. Series G, Ecological sciences ; vol. 20) "Published in cooperation with NATO Scientific Affairs Division." "Proceedings of the NATO Advanced Research Workshop on Behavioural Mechanisms of Food Selection, held at [University of Wales Conference Centre] Gregynog, Wales, United Kingdom, July 17–21, 1989"—T.p. verso. Includes bibliographical references.
ISBN-13:978-3-642-75120-2 (U.S.)
1. Animals—Food—Congresses. 2. Food preferences—Congresses. I. Hughes, R. N. II. North Atlantic Treaty Organization. Scientific Affairs Division. III. Title. IV. Series: NATO ASI series. Series G, Ecological sciences ; vol. 20. QL756.5.N38 1989 591.53—dc20 89-26334

Softcover reprint of the hardcover 1st edition 1990

2131/3140-543210 – Printed on acid-free-paper

PREFACE

Twenty four years ago, MacArthur & Pianka (1966) and Emlen (1966), sought an evolutionary explanation for observed patterns in foraging behaviour. As a way forward, they assumed that an animal would promote its fitness by maximizing the net rate of energy intake, or more precisely the ratio of energetic gain to cost while foraging. So began a major development in behavioural ecology, soon to become known as Optimal Foraging Theory. Having grown steadily throughout its first decade, OFT exploded in popularity during the mid to late seventies. By the early eighties, this popularity had begun to decline (Pyke, 1984). Yet almost a decade later, OFT is still with us, attracting theoreticians, guiding empiricists and inciting sceptics.

The Energy Maximization Premise (EMP) has proved to be a powerful general principle, underpinning theoretical exploration of diverse aspects of foraging behaviour, often classified under the broad headings of searching, patch use and diet selection (Stephens & Krebs 1986). But whereas students of macro–carnivores have followed the course of classical OFT, theoreticians concerned with grazers have adopted the modelling procedures of range management and animal husbandry (Westoby 1974), while those involved with particle–feeders have independently developed their own approach (Lehman 1976, Lam & Frost 1976, Taghon *et al* 1978).

Energy maximization, of course, is not the only contibutor to fitness. Animals must derive specific nutrients from their food, react to competitors, avoid predators, shelter from environmental hazards and allocate time to other essential functions. Early models excluded such "constraints" on foraging behaviour. Indeed, the simplicity of these models and of their predictions was the key to their popularity: clear, simple predictions invite experimental testing if suitable laboratory subjects can be found. Initially, the simple models enjoyed great success, not only in predicting diet selection under controlled conditions, but also in guiding the formulation of experimental programmes. Nevertheless, explaining the inaccuracies of prediction became a preoccupation of Optimal Foraging theorists. In general, animals took energetically inferior items more often than predicted by the EMP. Sub–models were introduced, dealing with most types of constraint, and new insights were developed. As with the accumulation of terms in a polynomial regression, however, there was a danger of turning a simple heuristic model, albeit of limited applicability, into a model that has generality but is merely descriptive.

More recently, OFT experienced a renaissance. Acknowledging the complexity of nature, theorists began to formulate models in terms of optimal solutions to competing demands on the animal's time and energy budget (McNamara & Houston 1986, Mangel & Clark 1986). At this pivotal stage, it seemed appropriate to review the state–of–the–art across the board, looking at concepts old and new. Does OFT still have an important role to play in explaining diet selection? Are models developed for one trophic group applicable to others and if so, could they generate new insights ? Can events at the population or community level tractably be included in models and experiments investigating the foraging behaviour of individuals ? Does OFT have anything useful to say at the population or community level ?

In order to answer these and other questions, some fifty guests were invited to participate in an Advanced Research Workshop, sponsored by NATO and held on 16 – 21 July, 1989 at the University of Wales Conference Centre, Gregynog, near Newtown in mid–Wales. A calculated risk was taken in choosing participants from fields as diverse as ethology, population ecology, animal husbandry, range management, fisheries biology, oceanography and theoretical ecology. Too much discrepancy might have discouraged interaction, but the risk paid off. During four intensive days of presentations and discussion, refreshed by episodes of bowls, croquet and jogging in the beautiful grounds of Gregynog, enthusiasm ran high and interactions were prolific. Answers to all our questions did not emerge, nor could they have been expected to. The following papers, together with records of discussion, document the outcome of the Workshop. To facilitate accession, papers are arranged according to subject matter and do not follow the sequence of the original programme, which was organized around major topics for discussion. We hope to have achieved an honest appraisal of OFT, as applied to diet selection, and that readers will derive interest and perhaps some inspiration from this Volume.

Roger N. Hughes

REFERENCES

Emlen JM (1966) Am. Nat. 100:611–617

Hughes RN (1980) Oceanogr Mar Biol Ann Rev 18:423–481

Lam RK & Frost BW (1976) Limnol Oceanogr 21:490–500

Lehmann JT (1976) Limnol Oceanogr 21:501–516

MacArthur RH & Pianka ER (1966) Am Nat 100:603–609

Mangel M & Clark CW (1986) Ecology 67: 1127–1138

McNamara JM & Houston AI (1986) Am Nat 127: 358–378

Pyke GH (1984) Ann Rev Ecol Syst 15:523–575

Stephens DW & Krebs JR (1986) Foraging Theory. Princeton University Press, Princeton, NJ

Taghon GL, Self RFL & Jumars PA (1978) Limnol Oceanogr 23:752–759

Westoby M (1974) Am Nat 108: 290–304

ACKNOWLEDGEMENTS

The organizing committee wishes to thank the NATO Scientific Affairs Division for support. Thanks are extended to the staff of Gregynog for helping things to run so smoothly, the Llanfair Caereinion Harpists and Singers for local entertainment, to the Midland Bank for providing stationery and folders, and to serendipity for providing glorious weather when we least expected it.

Roger Hughes

Jean–Guy Godin

Bill DeMott

Paul Hart

NATO–ARW GREGYNOG JULY 1989

Left to right, standing: (1)Lennart Persson, (2)Frietson Galis, (3)Phil Hockey, (4)John Ball, (5)Gary Belovsky, (6)Alan Ansell, (7)Beth Okamura, (8)Tom Hobbs, (9)Roy Caldwell, (10)John McNamara, (11)Pete Jumars, (12)Jim Gilliam, (13)Jean–Guy Godin, (14)Iain Gordon, (15)Bernt Saether, (16)Hank Vanderploeg, (17)Pete Peterson, (18)Les Real, (19)Marion Croy, (20)Bob Vadas, (21)Andy Sih, (22)Barker Jørgensen, (23)Werner Schuler.

Left to right, crouching:(1)Fred Provenza, (2)Andrew Illius, (3)Dawn Bazely, (4)Bob Jaeger, (5)Bill DeMott, (6)Roger Hughes, (7)George Turner, (8)Alex Kacelnik, (9)Jeff Lucas

Contents

PREDATOR SWITCHING AND THE INTERPRETATION OF ANIMAL CHOICE BEHAVIOR: THE CASE FOR CONSTRAINED OPTIMIZATION.

LESLIE A. REAL
Department of Zoology
North Carolina State University
Raleigh, NC 27695-7617
USA

The past decade has seen considerable debate over the choice of methodology appropriate for the analysis of animal behavior. Unquestionably, the dominant approach uses a variety of optimization methods as a basis for modeling behavior. The optimization approach posits the existence of an objective function that is either maximized or minimized subject to a set of constraints, i.e.

$$\max f(x_1, \ldots, x_n, t)$$

subject to

$$\begin{aligned} g_1(x_1, \ldots, x_n, t) &\geq b_1(t) \\ &\vdots \\ g_m(x_1, \ldots, x_n, t) &\geq b_m(t) \end{aligned}$$

where $(x_1, \ldots, x_n)$ comprise a set of state variables and where the objective function and constraints may be a function of time t.

Optimization techniques have a venerable history of application in physics, economics, psychology, and physiology, and their extension to animal behavior is neither unique nor arbitrary. Optimization methods can generally be employed to describe any system that is *tending toward* some equilibrium or final state; for example, systems subject to directional selection. There are, of course, systems where no objective function can adequately summarize the dynamics since no equilibrium final state is achieved, as in systems subject to frequency-dependent selection where limit cycles are likely to emerge. But many natural systems are directional "hill-climbing" systems, and for these, optimization methods can be quite useful. The objective function simply states mathematically what is produced as a result of selection on the system as it tends toward the final state.

For example, simple optimization models using the calculus of variations show that a balloon is round because that shape minimizes the potential energy of the surface.

NATO ASI Series, Vol. G 20
Behavioural Mechanisms of Food Selection
Edited by R. N. Hughes

The Newtonian dynamical description relies on a set of equations describing the stresses at each point on the surface. The optimization formulation replaces that point by point description with an integral equation that summarizes the net result of the dynamical system (Lanczos 1966). The two approaches are complimentary and reflect different aspects of a single problem. Optimization arguments also account for why apples fall in straight lines, hanging telephone wires form catenary arches and why sheets of soap bubbles form hexagonal arrays when pressed - like the wax comb in a honeybee colony.

It should be clear from these examples that optimization models do not *ipso facto* imply conscious purposeful action. The charge that optimization models are inherently wrong because they are teleological is not very compelling. They are not teleological in any deep theological sense of the word. They do not represent some outside intention. They can imply directed change, but with no greater implication than the directed change inherent in a falling apple. Optimization techniques can be used to make very precise quantitative predictions when the investigator believes he or she can capture the appropriate dynamics in a summary mathematical statement.

While there is nothing inherently wrong with using optimization models, there is nothing inherently right about using them either. Other modeling approaches could prove to be of equal or greater value for analyzing animal behavior. Several alternatives to optimization have been suggested: satisficing (Simon 1956), statistical decision rules (e.g. "Best-of-n"), and *ad hoc* decision rules (i.e. "rules-of-thumb"), for example. Each of these alternatives attempts to represent behavior more accurately and truly in that they appear to be similar to simple mental processes, rather than more complicated mathematical calculations. The resemblance to what we "actually do" gives them their appeal.

Each of these alternatives, however, can fall short on several fronts. First, when given a particular fitness measure or analog of fitness (such as expected energy gain over time), and a particular probability distribution of outcomes (such as prey types), constrained optimization models can be constructed whose solutions generate higher expected fitness compared to the expected fitness resulting from the alternative models. Janetos (1980) and Janetos and Cole (1981) were the first to systematically compare the performances of alternative decision strategies. Janetos (1980) calculated the expected fitness resulting from choices over a fixed distribution of alternatives using four different decision rules. In his paper, alternatives were mates of different qualities, but the analysis is equally applicable to choices over different prey types. Each of the four different rules represents a different basic methodology. The four models were

(a) Random choice - accept any mate (prey) encountered.

(b) Fixed threshold - accept any mate (prey) above some critical threshold. This model corresponds to Simon's model of satisficing. It is important to recognize that in both the Simon (1956) and the Janetos (1980) papers, the critical threshold is arbitrarily defined. I will return to this point later.

(c) Sequential search ("optimal one-step") - accept any mate (prey) above a critical threshold where the threshold is derived through a dynamic programming algorithm. This model represents an optimization approach where the critical threshold is non-arbitrary.

(d) Best-of-n rule - accept the best mate (prey) after sampling n encounters. This rule constitutes the simplest statistical decision-rule based solely on sampling from the underlying distribution of prey qualities. The difference between rule (c) and rule (d) is that in (c) the decision variable is the critical threshold, while in (d) the decision variable is the sample size n.

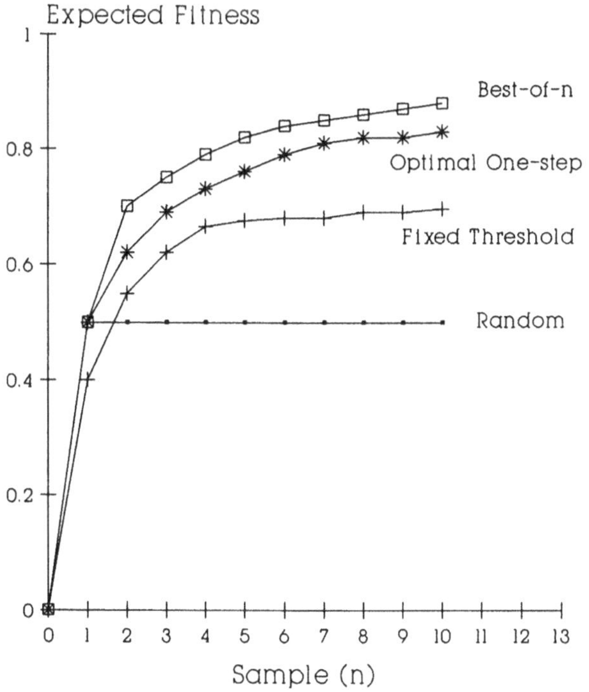

Figure 1. Comparative performance of the four decision rules suggested by Janetos (1980). The best-of-n rule dominates all other decision rules with respect to expected fitness gains for a specified distribution of prey qualities. The dominance of the best-of-n rule depends on the assumption of costless search by predators. Sequential search (optimal one step) dominates when there is a cost to searching (Real 1989).

Janetos (1980) then compared the expected fitness resulting from each of these rules over a range of sample sizes. He concluded that the Best-of n strategy dominates all other strategies and therefore should be adopted by searching individuals (Figure 1).

The analyses presented by Janetos (1980) and Janetos and Cole (1981) are flawed in two respects: their technical mathematical analysis is incorrect in several important respects (Houston and McNamara 1984), and they neglect to include any costs in the decision process (Real 1989). Nonetheless, their basic intention to compare the performance of alternative rules is still valid. After correcting the mathematics and including search costs in the models, we find that rule (c), the optimization rule employing sequential search and dynamic programming, dominates (Real 1989). In contrast, Janetos (1980) concludes that the statistical decision rule (d) dominates all other rules. Reformulation improves the optimization approach through increasing the sensitivity to the stream of encounters and adjustment of critical thresholds as a function of search and sampling costs.

While the reanalysis of Janetos (1980) and Janetos and Cole (1982) by Houston and McNamara (1984) and Real (1989) suggests the superiority of sequential search over the three alternatives listed above, no one has considered the relative performance of *ad hoc* processes (rules-of-thumb). Consider two potential rule-of-thumb decision processes:

(e) engage in at least two searches, and stop if a prey item encountered is lower in quality than the previous one,

(f) continue searching and sampling prey as long as successive prey encountered are higher in quality by at least the cost of the search.

Rule (e) has been used to model mate choice behavior in sculpins (Brown 1981) and seems to generate a fair approximation of the animal's behavior. Rule (f), to my knowledge, has not been used to model animal choice behavior but is interesting in that it explicitly includes the cost of search and sampling. However, neither rule (e) nor rule (d) demonstrate superior performance when compared to the optimal sequential search rules established through dynamic programming (McKenna 1985, Hey 1981, Schotter and Braunstein 1980).

In conclusion, when alternative models of choice are compared, dynamic optimization models generally can be constructed that lead to higher expected benefits. Since most of our modeling is based on the assumption that observed behavior is the product of Darwinian evolution (at least to some degree), then it seems appropriate to concentrate on those models and rules that systematically out-perform competing models.

Some investigators might raise the following objection to this argument: while it is true that optimization models out-perform alternatives, the behavioral assumptions underlying such models are beyond the capabilities of most organisms. Consequently, organisms can not follow some optimization principle even if it is in their evolutionary interests to do so. Many of the more sophisticated optimization models currently being explored, however, make no more assumptions about behavior then the "simpler" alternatives. A comparison with Simon's (1956) bounded rationality model suffices to illustrate this point.

Simon attempted to relax the assumption of global knowledge inherent in most optimization models used up to that time in economics. Instead, the deciding agent confronts a stream or sequence of choice situations where the agent must decide to accept or reject an action. The structure of this model is identical to the fixed threshold model of Janetos (1980). Simon's example uses a homeowner that must decide to accept or reject an offer to buy at some predetermined fixed value. Bounded rationality posits the existence of a critical threshold above which offers are accepted and below which offers are rejected. In the body of Simon's article no attempt is made to describe how the critical threshold is established. In fact, Simon argues (and reasonably so) that the critical threshold will change over time as the homeowner encounters offers and acquires information. However, without an independent theory to account for the dynamics of the threshold, no predictive theory can be constructed. Economists have recognized this for some time, and on this basis, have rejected satisficing as a model of behavior.

The lack of predictive power in the model was not lost on Simon! In the appendix to his now classic paper – where Simon actually tries to derive a specific quantitative model – he reverts to an optimization argument to derive the critical threshold. Homeowners (his example) are assumed to maximize the expected profit based on a knowledge of the *distribution* of potential offers. It seems that Simon trades one kind of global knowledge for another - instead of knowing the exact sequence of encounters, one knows the probability distribution of offers. Without at least this kind of knowledge, satisfying makes no specific predictions.

This kind of model, however, is no different from other forms of sequential decision theory based on optimization principles. For instance, consider the following formulation of the diet selection problem for predators selecting a single prey item during each decision period:

The energetic quality W of encountered prey is a random variable with probability density function f(w). What is the critical value of $W = W_{crit}$ that induces a predator to stop searching and to pursue an encountered prey item? What value W_{crit} determines the rule:

$$\begin{cases} \text{accept if } W \geq W_{crit} \\ \text{reject if } W < W_{crit} \end{cases} \tag{1}$$

The maximum expected fitness gain from time t onward to a searching predator is designated V_t. At time t+1, such an individual will encounter an additional prey after incurring the cost c of one period's additional search. If the predator encounters a prey of quality W then there are two options: either accept the prey, in which case the predator's fitness gain is W, or continue searching and begin with a new value of search V_{t+1}

$$V_t = \begin{cases} W - c & \text{if accepted} \\ V_{t+1} - c & \text{if rejected} \end{cases} \tag{2}$$

The predator chooses the greater of the two options, so $V_t = \max [\, W, V_{t+1}] - c$. The cost of search c is incurred whether or not the prey is accepted or rejected which is why it can be left out of the bracketed pair. If the prey is accepted, the predator gets (W − c), if rejected the predator continues search and starts with a new value V_{t+1} but minus the cost of search. Since W is a random variable, the condition (2) must be computed over the expectation of W, to give

$$V_t = E \max [\, W, V_{t+1} \,] - c. \tag{3}$$

For simplicity at the onset, I will assume that there is no time limit on the periods over which the predator can make choices, i.e. the simplest model has infinite time horizon. This is actually a common assumption, although not commonly recognized as one, in most traditional foraging models, e.g. Pyke et al 1977. The principle points can be made using this simplifying assumption.

If the organism is behaving optimally, it will not stop search unless the fitness gain from the encountered prey is greater than or equal to the maximum expected fitness gain from future search V_{t+1}. The calculation of future fitness gain in the infinite time horizon model is the same for any time t. Consequently, $V_t = V_{t+1} = V$. Equation (3) can then be rewritten in the simpler form $V = E \max [W,V] - c$. But from equation (2) the value V must correspond to the critical value W_{crit} that terminates search. Therefore the optimization criterion (3) can be rephrased according to the threshold acceptance fitness W_{crit} that causes search to stop:

$$W_{crit} \;=\; E \max [\, W, W_{crit} \,] - c. \tag{4}$$

Condition (4) is used to establish the critical value W_{crit}, and the method for explicitly determining its value is straightforward (see Real 1989). Condition (4) is the fundamental relation for sequential choice using dynamic programming and has been derived numerous times in the economics literature (Lippman and McCall 1976, McCall 1970), and has recently been extended into behavioral ecology (see Mangel and Clark (1988) for a thorough discussion of dynamic programming in behavioral ecology). Houston and Macnamara (1985) construct similar models for the central place foraging problem, and I have used this model elsewhere to examine patch choice and habitat selection (Real 1983) and mate choice (1989). The model can be extended to any kind of critical threshold model of choice behavior over time.

This kind of sequential search model based upon dynamic programming has all the salient features of the Simon's satisficing model. Decisions are based on the stream of encounters, and the only assumption is that the organism knows the distribution of potential prey qualities. Unlike the Simon model, however, the threshold is determined by an optimization criterion and can be predicted *a prior*. The sequential search model dominates all fixed threshold models where the fixed threshold is some value other than W_{crit} and therefore constitues a dominating strategy (Real 1989). This kind of model is every bit as "behavioral" as the original Simon model with the advantage of increased predictive power and performance. Simon's original intention to provide an alternative to optimization must be abandoned in the light of its inability to make concrete predictions and our ability to construct an alternative that does without losing the major features of the satisficing idea.

Statistical decision rules and rules-of-thumb also incorporate some form of optimization criterion. They are not predictive models until some objective function is attached to their structure. The appeal of the alternatives is not so much that they abandon optimality arguments as that they capture what we feel are more likely the actual processes of decision making. For instance, rules-of-thumb often capture the underlying dynamics that can generate optimal behavior according to a more global model. In the theory of risk-sensitive foraging (Stephens and Krebs 1986, Real and Caraco 1986, McNamara and Houston 1986, Ellner and Real 1989), organisms are assumed sensitive to the mean and the variance in reward distribution or prey profitability distribution. Stephens and Krebs (1986) suggest that the risk-sensitivie response can be accounted for by the simple rule-of-thumb:

(g) For a given mean reward, choose the alternative with the lowest variance, and for a given variance choose the reward with the highest mean.

Ironically, this rule-of-thumb is the exact computational algorithm used by Markovitz (1956, 1959) to solve the classic quadratic programming problem. The variance-discount model of risk-sensitive foraging is one special form of the quadratic programming problem, as has been pointed out by Regelman (1985). Consequently, the rule-of-thumb (g) does not *replace* the optimization problem but specifies the computational algorithm for its solution.

Optimization models appear to be superior both in performance and predictive power and provide a global description of the end product of microbehavior that may be associated with particular rules-of-thumb. The superiority of optimization approaches has only been judged against particular alternatives. We are not guaranteed that the optimization model is the "correct" model, only that we can generally construct a better optimization model when compared with a particular alternative model. Future research should be directed toward making the connection between the actual dynamics of behavior and the implicit maximization or minimization process inherent in those dynamics. To that end we must identify and describe the appropriate constraints on the dynamics.

Behavioral Constraints

Every optimization problem is explicitly or implicitly subject to a variety of constraints. Without a clear elucidation of the nature of these constraints, quantitative predictions will only loosely approximate observed behavior. The existence of constraints does not change the status of the optimization problem. The change in solution subject to constraint does not change the behavior from "optimal" to "suboptimal" as is often erroneously asserted. To use a physical rather than biological example, imagine a chain of fixed length suspended between two angled poles like a telephone wire. What shape will the chain assume at equilibrium?

This is a classic problem in the calculus of variations, and the solution is determined by using Euler's equation (Sagan 1969). The equilibrium solution, however, depends on the particular set of constraints operating. First, imagine the chain fixed at two points A and B on each of the respective poles. The chain is free to hang in any curve of length L subject to this two-fixed-endpoint condition. Empirically, we would see that the curve formed is a catenary arch (Figure 2). The catenary arch is that curve which minimizes the total potential energy in the suspended chain. Now, if we change the constraint set to include only a single fixed-endpoint at A, what curve will result? The chain will now hang in a vertical straight line. A different constraint set changes the qualitative behavior of the system from non-linear to linear. But in what way can we say that one solution is optimal while the other is suboptimal? Is the vertical straight line the optimal solution? Does adding the fixed endpoint constraint at B change the solution to a suboptimal one? These questions don't seem to make sense. The different solutions simply reflect different systems and interactions.

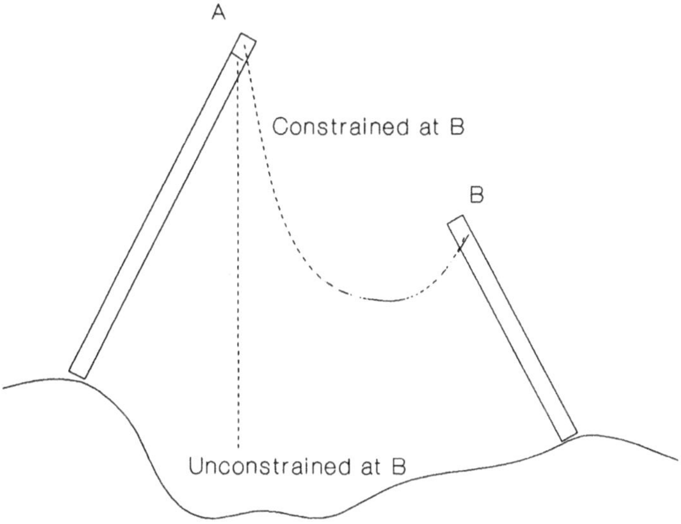

Figure 2. A chain suspended from two fixed end-points A and B will form a catenary arch. Removing the constraint at point B results in a new equilibrium, the hanging straight chain. In both systems potential energy is minimized at the equilibrium configuration.

To suggest suboptimality is to suggest that there is a final state that truly is the intended state of nature. Our emphasis should be on describing and modeling the full dynamics of the system as it is, rather than interpreting the system as a deviation from some idealized state. Viewed in this way, it makes no sense to imply that learning and/or memory constraints, for example, lead to "suboptimal" performance and behavior.

In order to characterize the organism's decision making, we must begin to catalog the set of constraints that potentially influence choice. Current research has focused on at least six broad categories of constraint: time, learning and memory, perceptual, genetic/developmental, and nutritional constraints.

There are many recent reviews of most of these topics, and several of the papers in this volume examine the role of one or another of these six types of constraint: for learning see Kacelnik and Krebs (1985), Stephens and Krebs (1986); for memory, Sherry and Schacter (1987); for perception of time, Kacelnik (this volume); for developmental constraints (Galis, this volume), and for genetic constraints and diet selection in herbivores, see Gould (1988).

Though these types of constraint are explored frequently, few investigators attempt to actually add quantitative characterizations of the constraints to mathematical optimization models. Sometimes this can be difficult, but most often the constraints can be added quantitatively. For example, nutrient constraints can be added to form a linear programming problem (Belovsky 1978, Pullium 1975), or some form of Bayesian learning can be added to sequential decision theory to account for information acquisition or loss (Mangel and Clark 1988, Green 1980, 1984, Iwasa et al 1981). The incoporation of constraints on the assignment of probabilities provides another example of how constraints can be added to traditional optimization models of choice behavior.

Models for foraging in stochastic environments (i.e., risk-sensitive foraging theory) can all be thought of as one form or another of Bernoulli expected fitness/utility maximization (Ellner and Real 1989). That is, the organism's decision process is influenced by two factors, a non-linear fitness function $f(x_1, x_2, \ldots, x_n)$ of alternative reward states x_i and a probability set $(p_1, \ldots p_n)$ where p_i represents the probability of the occurrence of the reward states x_i. The organism is assumed to maximize

$$E\ f(x_1, \ldots x_n) = \sum_{i=1}^{n} P_i\ f(x_i) \qquad (5)$$

To date, the primary focus has been to empirically derive the quantitative shape of the underlying fitness function $f(x_1, \ldots, x_n)$, (e.g., Caraco et al 1980, Real et al 1982, Caraco and Lima 1985, Stephens and Paton 1985, Wunderle and Cotto-Navarro 1988). Little attention has been paid to the probability set and to the organism's perception of likelihood in the process of decision making. We know that humans are quite bad at assessing probabilities and show a systematic bias in perception of likelihoods. Low probability events are over-estimated, while high probability events are under-estimated (Kahneman and Tversky 1982). Probabilities are essentially homogenized with the tails of the distribution being pulled into the center. Such a bias in perception of likelihood acts as a constraint on the decision making process.

In Real (1987), I explore the implications of subjective bias in probability assignment for interpreting experiments on risk-sensitive foraging. We can postulate that should animals show the same kinds of bias as humans, their subjective probabilities $p'=h(p)$ may be expressed as

$$p' = h(p) = p^b / [p^b + (1-p)^b] \tag{6}$$

This expression translates the objective probability p into the subjectively perceived probability p'. The parameter *b* influences the magnitude of curvature in the subjective transformation and determines the deviation between objective and subjective probability (Figure 3). For *b*>1 low probability events are over-estimated; while for *b*<1 low probability events are under-estimated. Equation (6) acts as a constraint on choice and can be added to modify the traditional expected fitness model to form a new subjectively weighted form of stochastic foraging:

$$V = \sum_{i=1}^{n} \frac{h(p_i)}{\sum_j h(p_j)} f(x_i) \tag{7}$$

which is now non-linear in the p's (Ellner and Real 1989). Equation (6) is only one suggested form of perceptual constraint on estimating likelihoods. There are many other possibilities, and the exact quantitative relationship between objective probabilities and subjective perception of probabilities remains an open empirical question.

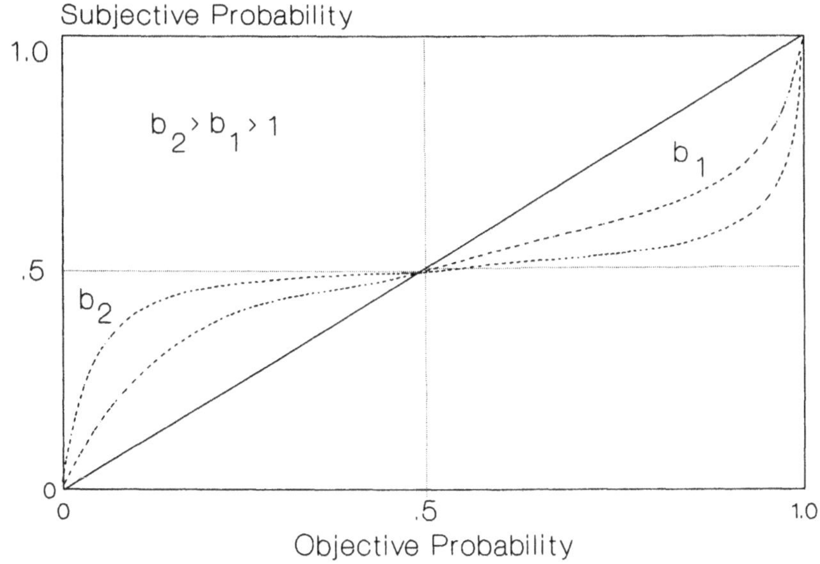

Figure 3. Graph of the relation $p' = p^b / \{p^b + (1-p)^b\}$ representing the shift in subjective probability away from the objective probability p. Parameter *b* influences the curvature of the relationship. As *b*>1 increases there is greater deviation from objectively established probabilities.

Waddington (1989) has recently extended the concept of subjective perception to include the subjective estimate of overall profitability in accounts of risk-sensitive foraging. Waddington suggests that foragers evaluate overall profitability non-linearly, and that subjective profitability shows diminishing returns. Individuals maximizing the expected subjective profitability would then behave in a risk-averse manner.

In the same way that we attempt to explicitly quantify perception of likelihood and add that constraint to existing models, we should include explicit treatment of other types of behavior constraints in current models. This has most successfully been accomplished in foraging/search models with learning constraints (Mangel and Clark 1988) and in diet selection models with nutrient constraints (Pullium 1975, Belovsky 1978).

Switching

At times it may be difficult to determine if observed behavior is a consequence of behavioral constraints or changes in the postulated objective function. Consider, for example, explanations for predator switching. Switching occurs when prey that are relatively rare in the environment are disproportionately under-exploited and common prey are over-exploited (Figure 4).

The first explanations offered to account for the observed lag in feeding rate relied on behavioral constraints. Holling (1959) suggested that below some critical food density, animals could not remember the prey and could not exploit it. Tinbergen (1960) suggested an alternative form of perceptual limitation and postulated the existence of a "specific searching image." Below the critical encounter rate, animals could not maintain a visual image of the prey and therefore could not exploit the prey. Evidence for a "specific search image" or for memory constraints as a basis for switching is quite controversial (Krebs 1973, Guillford and Dawkins 1987, 1989, Lawrence 1989, Lawrence and Allen 1983). There is some convincing evidence of the existence of this kind of perceptual constraint on foraging from laboratory studies using birds (Dawkins 1971a,b, Pietrewicz and Kamil 1981), but the exact perceptual mechanism involved is still not well understood.

Alternative explanations of the lag rely on energetic considerations. Rapport (1981), for instance, provides some empirical support for the claim that organisms are perfectly cognizant of alternate food types but choose not to exploit sparse populations of prey since it is energetically inefficient to pursue rare items. Only as the prey increases in abundance does it become profitable to exploit this resource. These two explanations (perceptual constraint versus energetic efficiency) may not be mutually exclusive and, at times, may operate simultaneously. For example, switching between resources may involve some learning that increases the efficiency of energy extraction, as in starfish (Fisher-Piette 1935, cited in Emlen 1973), as well as behavior influenced by shifts in attention and the processing of sensory data.

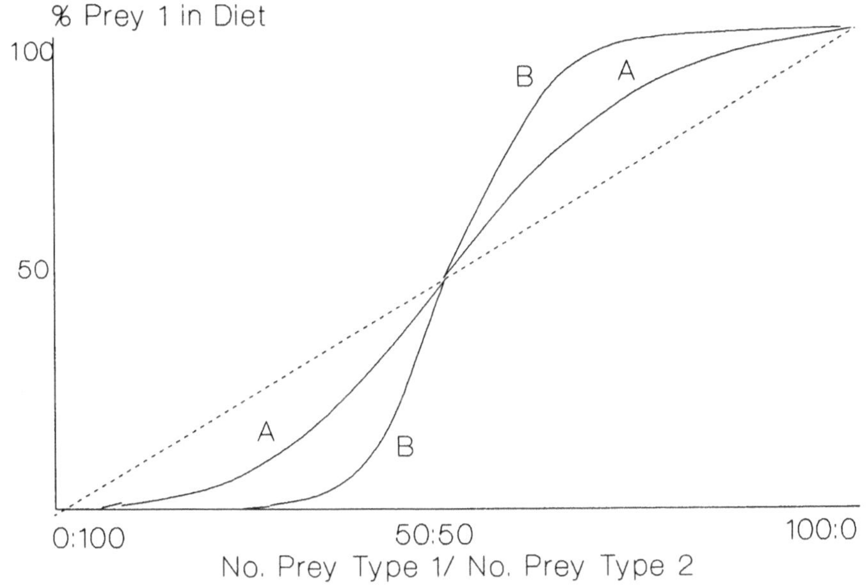

Figure 4. Idealized switching curves for a predator consuming two prey types as a function of the prey types' relative abundance. For predators showing no intrinsic preference for one type of prey at equal representation in the environment (100:100) prey type 1 should comprise 50 % of the diet. As prey type 1 becomes relatively scarce, it is disproportionately under-represented in the diet. When it is relatively common, it is disproportionately over-represented in the diet. The steepness of the curve (measured by the slope at the 50:50 point) determines the magnitude of switching. Curve A responses demonstrate less switching than curve B type responses.

To demonstrate the existence of the perceptual constraint, Dawkins (1971a) and Krebs (1973) emphasize that we must exclude six alternative scenarios (listed in Lawrence and Allen 1983). Predators may (a) learn to search in a particular place for food, (b) learn to distinquish a particular *type of place* for food, (c) alter their search path so as to increase encounters with prey, (d) learn to handle prey more efficiently, (e) avoid prey over others based on factors other than perceptual cues and (f) learn specialized hunting techniques. Excluding these alternatives should not be a difficult task if the appropriate experiments are carried out.

I have performed floral choice experiments using bumblebees that may help illustrate the dual operation of perceptual constraints and energetic efficiency in the switching behavior of pollinators. A colony of bumblebees, *Bombus pennsylvanicus*, was housed in an enclosure and provided with an artificial patch of flowers filled with varying quantities of nectar. The methods are similar to those used in Real (1981), Real et al., (1982) and Ott et al., (1985). The artificial patch of flowers consisted of a randomly distributed set of 200 blue and/or yellow plastic squares affixed to a plexiglass sheet with small wells drilled into the center of each artificial flower. Nectar was then dispensed into the flower, and individual foraging worker bees were allowed to visit the patch. To examine switching behavior, I varied the ratio of blue to yellow flowers present at any given trial where the total number of flowers in the patch always equalled 200. In all trials of the first experiment, every flower contained 2μl of nectar. Individual workers were allowed to forage from the artificial patch, and the percent visits to yellow flowers was monitored as a function of its relative abundance. The derived curve represents the responses of several different foragers to different ratios of blue and yellow flowers when the energetic qualities of blue and yellow flowers are constant across ratios. The experiment clearly demonstrates switching behavior, and yellow flowers are under-exploited when they are relatively rare (Figure 5).

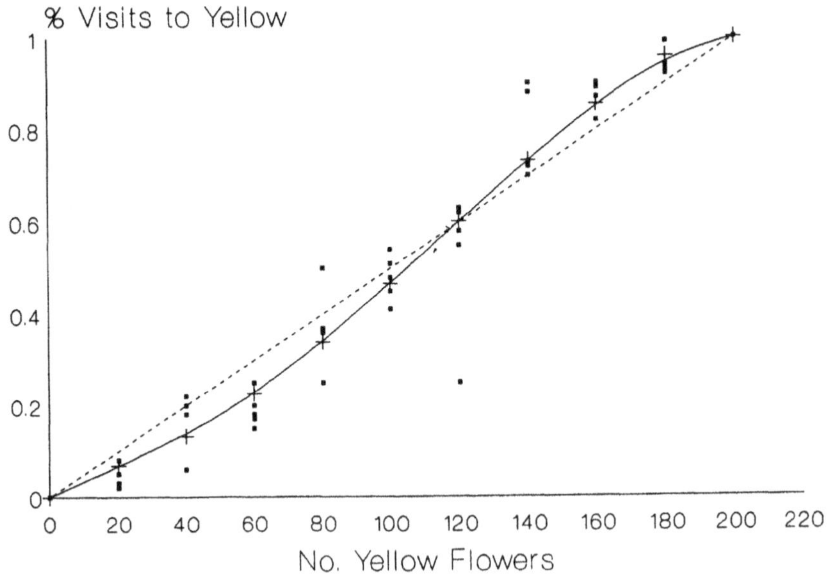

Figure 5. Switching response of individual worker bumblebees choosing among artificial blue and yellow flowers randomly spaced and intermixed within an enclosure. All flowers contained 2μl of nectar. The response shows a significant non-linear component and the best fit cubic regression is given by $Y = .042 + 3.06 \times 10^{-4}X + 5.57 \times 10^{-5} X^2 - 1.62 \times 10^{-7}X^3$ ($R^2 = .94$, df=44, $p<.001$). The dashed line represents the predicted percentage in the diet based on preference when both flower types are of equal abundance (100:100). The slope $Y'(100) = .006 \pm .008$.

In this experiment it is difficult to see how any of the six alternate hypotheses to perceptual constraint can operate. The location of flowers are randomly intermixed and equally accessible so foragers cannot learn location or type of location for rewarding flowers. All flowers are identical in shape and energetic quality, and learning to handle one type of flower is the same as learning how to handle the other type of flower. Flowers are randomly intermixed, and altering the search path will not improve foraging since blue and yellow flowers are equally likely to be nearest neighbors.

I hesitate to suggest that these bees are forming search images since most researchers in this field restrict this behavior to detecting cryptic prey, and it is hard to imagine that the yellow flower is more cryptic than the blue flower. Rather, it appears that something akin to Dawkins' (1971b) "shifting attention" mechanism may be operating.

The foraging and switching response of individual bees is not governed entirely by perceptual constraints; energetic efficiency also enters into the formation of preferences. The above experiment was repeated; however, yellow flowers were of variable quality with every third yellow flower containing 6μl of nectar and the remaining flowers empty. Blue contained 2μl in each flower. In each new trial the location of yellow flowers containing the 6μl was randomized so bees could not learn location differences. Bees once again demonstrate disproportionate under exploitation of the yellow flowers when they are rare (Figure 6). However, when both flowers types are equal in abundance the bees no longer show equal exploitation. Rather they continue to under-exploit the variable yellow flowers. This response is predicted from the already well documented risk-sensitive foraging behavior of bumblebees, and can be explained on simple energetic grounds (Harder and Real 1987). What cannot be explained simply on energetic grounds is the continued under-exploitation of yellow flowers after the risk-sensitive response has been factored into the analysis.

When presented in equal frequencies, the preference for yellow flowers averages 21 percent. Based strictly on energetic grounds, bees should visit yellow flowers in proportions determined by the straight line through this value and the origin. This predicted pattern of behavior assumes that bees are making decisions on the basis of the qualities of individual flowers, rather than the relative frequency of distribution types. If there were some form of non-linear effect on the bees' energy acquisition with changes in relative frequency of distribution type then we might not expect a linear response to changes in relative abundance when the two type of flowers have different distributional characteristics. Fortunately there is some empirical evidence supporting the claim that bees are making their decisions on a flower-by-flower basis (Carter and Dill 1989, Harder and Real 1987). The response should be linear with slope determined by the preference at the 50:50 point for relative abundance. The bumblebees, however, show a significant non-linear response, under-exploiting yellow when rare. While energetic efficiency has produced the overall decline in exploitation of the variable yellow flowers, there remains some form of perceptual constraint.

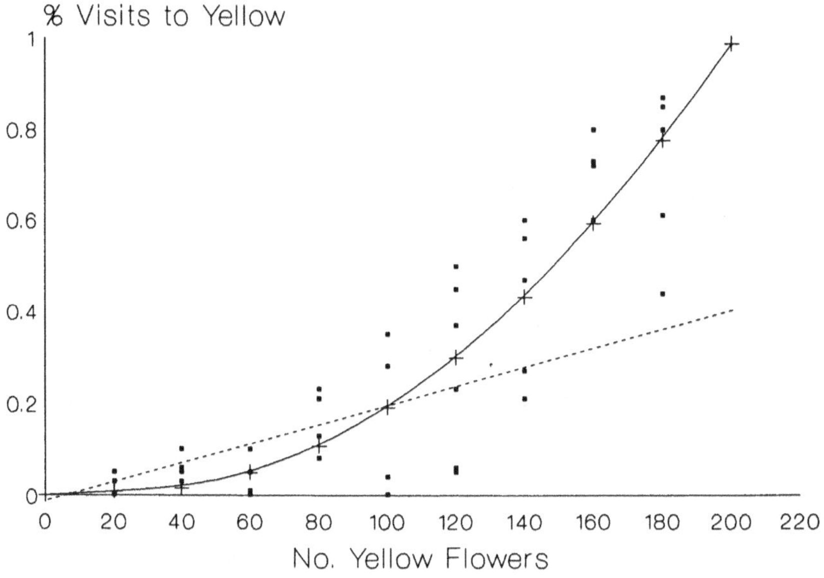

Figure 6. Switching response of individual worker bumblebees choosing among artificial blue and yellow flowers when yellow flowers are variable in quality and blue is constant. Blue flowers contained 2μl in every flower. Yellow flowers had 1/3 with 6μl and 2/3 with no nectar. The average qualities in the two flowers were the same. The response shows a significant non-linear component and is best represented by the quadratic regression $Y = .027 - .002\ X + 3.15 \times 10^{-5} X^2$ ($R^2 = .87$, df = 45, $p<.001$). The slope $Y'(100) = .0043 \pm .003$.

The increased preference for blue flowers is not strictly a response to variance in reward as opposed to constant rewards. In a third experiment, nectar was variable in all flowers - both blue and yellow - with 6 μl in a third and no nectar in the remaining flowers. Thus, the flower types were identical in distribution and of the same average quality as in the first experiment where all flowers had 2μl of nectar. The bees showed switching, and the best fit cubic regression through the responses is statistically indistinquishable from the curve for experiment one (Figure 7).

The magnitude of switching in predator response can be estimated by computing the slope of the switching function at the 50:50 point (Y' evaluated at 100 in the fitted regression equations). This point represents the change from being rare to being common. Given two curves A and B as in Figure 4, curve B seems to entail greater switching than curve A. This increased propensity toward switching is reflected in the slope at 50:50 ratios of relative abundance. Interestingly, the slopes at the 50:50 point of the empirically derived switching functions in the three bumblebee experiments are statistically indistinquishable (Z=.04, NS). The magnitude of switching seems constant across different energetic environments! The two components of the observed behavior – perceptual constraint and energetic efficiency – can be partitioned into effects influencing the magnitude of relative preference at the 50:50 point and the slope of the response curve at the 50:50 point. Each of these components could, in principle, take on independent values revealing different aspects of the animal's behavior.

These experiments indicate the need for detailed and simultaneous experiments on the interaction between energetic optimization criteria and perceptual constraints, as well as the need for mathematical models that quantify specific constraints and incorporate them as side conditions. Modeling efforts incorporating signal detection theory may be a step in that direction (Getty 1985, Getty and Krebs 1985).

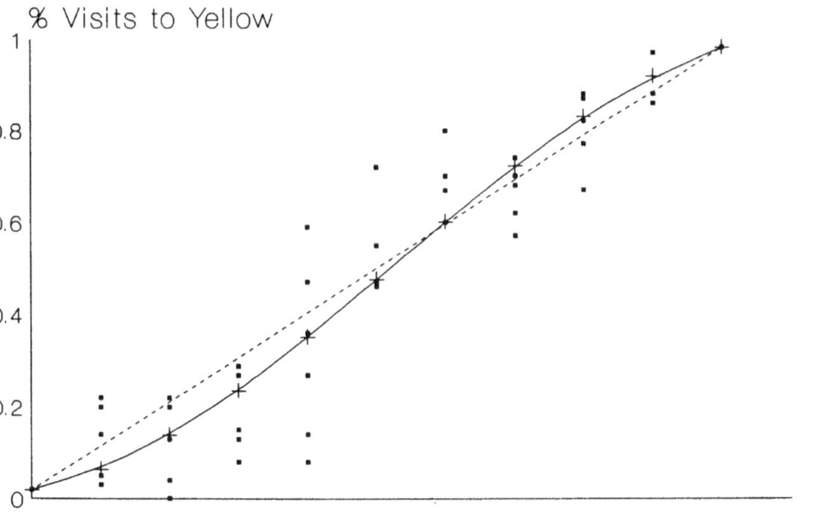

Figure 7. Switching response of individual worker bumblebees to changes in relative floral abundance when both blue and yellow floral types varied in nectar quantities. The distribution was the same as in the yellow variable case - 1/3 with 6μl and 2/3 with no nectar. The best fit cubic regression is $Y = .149 - 4.23 \times 10^{-3}X + 1.11 \times 10^{-4}X^2 - 3.62 \times 10^{-7}X^3$ ($R^2 = .87$, df= 45, $p<.001$) and is statistically indistinquishable from the regression established in experiment one. The slope $Y'(100) = .0071 \pm .008$.

Conclusion

The analysis and modeling of animal behavior should explore the entire panoply of mathematical devices which constitute the wide ranging field of decision theory – most of which is based on optimization in form or another. Any restricted adherence to a single methodogy would be foolish. Nonetheless, there are compelling reasons to believe that optimization approaches can be particularly enlightening and useful. Optimization models make specific quantitative predictions that are amenable to experimental verification. Results of optimization analyses can be compared with results based on dynamical models of behavioral microprocesses, and one hopes there will be some clear connection between these two levels of analysis. We should recognize the superiority of certain forms of optimization modeling with respect to overall evolutionary performance, and the fact that natural selection will drive the underlying dynamics into a configuration describable through some form of optimization criterion.

To fully understand the connection between behavioral microprocesses and their global equilibrium manifestations, we must carefully, thoroughly and *quantitatively* catalog the set of behavioral constraints likely to impinge on the performance of the animal. Often changes in behavioral constraints will be confounded with changes in behavior attributable to changes in objective function or its solution. Teasing apart the influences of behavioral constraints from features of the objective function can be difficult at times. Nonetheless, a partitioning of effects should be possible under the appropriate experimental designs. The switching behavior of predators is in many respects an ideal case for detailed study. Predators responding to changes in relative abundance in prey clearly have the objective of energetic efficiency and are subject to a variety of perceptual and learning constraints. The experiments using bumblebees illustrate, at least tentatively, how a partitioning of these effects may be possible.

Acknowledgments

I wish to thank James R. Ott and Eva Silverfine for their assistance in carrying out the experiments on the switching behavior of bumblebees. The manuscript was improved through suggestions and comments by Steve Ellner, Damienne Real, and Reuven Dukas. This research was supported through National Science Foundation grants BSR 8214599 and BNS 8719292.

Literature Cited

Belovsky GE (1978) Diet optimization in a generalist herbivore: the moose. Theor Pop Biol 14:105-134.

Brown L (1981) Patterns of female choice in mottled sculpins (Cottidae, Teleostei). Anim Behav 29:375-382.
Caraco T, Lima S (1985) Foraging juncos: interaction of reward mean and variability. Anim Behav 33:216-224.
Caraco T, Martindale S, Whitham TS (1980) An empirical demonstration of risk-sensitive foraging preferences. Anim Behav 2:820-830.
Carter R, Dill L (1989) Why are bumblebees risk-sensitive foragers. Behav Ecol Sociobiol (in press).
Dawkins M (1971) Perceptual changes in chicks: another look at the 'search image' concept. Anim Behav 19:566-574.
Dawkins M (1971) Shifts of 'attention' in chicks during feeding. Anim Behav 19:575-582.
Ellner S, Real L (1989) Optimal foraging models for stochastic environments: are we missing the point? Comments Theor Biol (in press).
Emlen JM (1973) Ecology: an evolutionary approach. Addison-Wesley, New York.
Fisher-Piette E (1935). Histoire d'une mouliere. Bull Biol 69:154-180.
Getty T (1985) Discriminability and the sigmoid functional response: how optimal foragers could stabilize model-mimic complexes. Am Nat 125:239-256.
Getty T, Krebs JR (1985) Lagging partial preferences for cryptic prey: a signal detection analysis of great tit foraging. Am Nat 125:39-60.
Gould F (1988) Genetics of pairwise and multispecies plant-herbivore coevolution. In Spencer K (ed) Chemical Mediation of Coevolution. Pages 13-55. Academic Press New York.
Green RF (1980) Bayesian birds: a simple example of Oaten's stochastic model of optimal foraging. Theor Pop Biol 18:244-256.
Green RF (1984) Stopping rules for optimal foragers. Am Nat 12:30-40.
Guilford T, Dawkins MS (1987) Search images not proven: a reappraisal of recent evidence. Anim Behav 35:1838-1845.
Guilford T, Dawkins MS (1989) Search image versus search rate: a reply to Lawrence. Anim Behav 37:160-162.
Guilford T, Dawkins MS (1989) Search image versus search rate: two different ways to enhance prey capture. Anim Behav 37:163-165.
Harder LD, Real LA (1987) Why are bumblees risk averse? Ecology 68:1104-1108.
Hey JD (1981) Economics in disequilibrium. New York University Press New York.
Holling CS (1959) The components of predation as revealed by a study of small-mammal predation of the European pine sawfly. Canad Entomol 91:293-320.
Hollis KL (1989) In search of a hypothetical construct: a reply to Guilford and Dawkins. Anim Behav 37:162-163.
Houston AI, McNamara JM (1984) Imperfectly optimal animals. Behav Ecol Sociobiol 15:61-64.
Houston AI, McNamara JM (1985) A general theory of central place foraging for single-prey loaders. Theor Pop Biol 28:233-262.
Iwasa Y, Higashi M, Yamamura N (1981) Prey distribution as a factor determining the choice of optimal foraging strategy. Am Nat 117:710-723.
Janetos AC (1980) Strategies of female choice: a theoretical analysis. Behav Ecol Sociobiol 7:107-112.
Janetos AC, Cole BJ (1981) Imperfectly optimal animals. Behav Ecol Sociobiol 9:203-209.
Kacelnik A, Krebs JR (1985) Learning to exploit patchily distributed foods. In Sibley RM, Smith RH (eds) Behavioral Ecology. Pages 189-205. Blackwells Oxford.
Kahneman D, Tversky A (1982) The psychology of preferences. Sci Amer 246:160-173.
Krebs JR (1974) Behavioral aspects of predation. In Bateson PB, Klopfer P (eds) Perspectives in Ethology. Pages 73-111. Plenum New York.
Lanczos C (1966) The variational principles of mechanics. University of Toronto Press, Toronto.

Lawrence ES (1989) Why blackbirds overlook cryptic prey: search rate or search image? Anim Beh 37:157-160.
Lawrence ES, Allen JA (1983) On the term 'search image.' Oikos 40:313-314.
Lippman SA, McCall JJ (1976) The econonmics of job search: a survey. Econ Inquiry 14:155-189.
Mangel M, Clark CW (1988) Dynamic modeling in behavioral ecology. Princeton University Press Princeton.
Markovitz H (1956) The optimization of quadratic function subject to linear constraint. Nav Res Logistics Quart 3:111-134.
Markovitz H (1959) Portfolio selection. Wiley New York.
McCall JJ (1970) Economics of information and job search. Q J Econ 84:113-126.
McKenna CJ (1985) Uncertainty and the labor market. St. Martin's Press New York.
McNamara JM, Houston AI (1986) The common currency for behavioral decisions. Am Nat 127: 358-378.
Ott J, Real LA, Silverfine E (1985) The effect of nectar variance on bumblebee patterns of movement and potential gene dispersal. Oikos 46:333-340.
Pietrewicz AT, Kamil AC (1981) Search images and the detection of cryptic prey: an operant approach. In Kamil AC, Sargent TD (eds) Foraging Behavior. Pages 311-331. New York Garland Press.
Pullium HR (1975) Diet optimization with nutrient constraints. Am Nat 109:765-768.
Pyke GH, Pullium HR, Charnov EL (1977) Optimal foraging: a selective review of theory and tests. Q Rev Biol 52:137-154.
Rapport DJ (1981) Foraging behavior of *Stentor coeruleus*: a microeconomic interpretation. In Kamil AC, Sargent TD (eds) Foraging Behavior. Pages 77-83. Garland New York.
Real LA (1981) Uncertainty and pollinator-plant interactions: the foraging behavior of bees and wasps on artifical flowers. Ecology 62:20-26.
Real LA (1983) Microbehavior and macrostructure in pollinator-plant interactions. In Real LA (ed) Pollination biology. Pages 287-304. Academic Press New York.
Real LA (1987) Objective benefit versus subjective perception in the theory of risk-sensitive foraging. Am Nat 130:399-411.
Real LA (1989) Sequential search theory and mate choice I. Models of single sex discrimination. Am Nat (in press).
Real LA, Caraco T (1986) Risk and foraging in stochastic environments. Ann Rev Ecol Syst 17:371-390.
Real LA, Ott JR, Silverfine E (1982) On the trade-off between the mean and variance in foraging: effect of spatial distribution and color preference. Ecology 63:1617-1623.
Regelman K (1985) A remark on the theory of risk-sensitive foraging. J Theor Biol 110:217-222.
Sagan H (1969) Introduction to the calculus of variations. McGraw-Hill New York.
Schotter A, Braunstein YM (1980). Economic search: an experimental study. Econ Inquiry 19:1-25.
Sherry D, Schacter DL (1987) The evolution of multiple memory systems. Psych Rev 94:439-454.
Simon HA (1956) Rational choice and the structure of the environment. Pscyhol Rev 63:129-138.
Stephens DW, Krebs JR (1986) Foraging theory. Princeton University Press Princeton.
Stephens DW, Paton SR (1986) How constant is the constant of risk-aversion? Anim Behav 34:1659-1667.
Tinbergen L (1960) The natural control of insects in pine woods I. Factors influencing the intensity of predation by songbirds. Arch Neerland Zool 13:265-343.
Waddington KD (1989) Are bumblebees risk-sensitive? It's a matter of choice. (manuscript).
Wunderle JM, Cotto-Navarro Z (1988) Constant vs. variable risk-aversion in foraging bananaquits. Ecology 69:1434-1438.

ORAL DISCUSSION – L. Real

Q. (Galis): Rules of thumb (ROT) are not the optimal solutions.
A. Rules of thumb are algorithms that give rise to global optimality. For example, a ROT for risk sensitive foragers would be: for a given mean, maximize the variance, and for a given variance, maximize the mean. This is the standard algorithm used to solve the quadratic programming optimization problem.

Q. (Kacelnik): If you assume an animal uses a ROT, this could change the optimality problem, rather than the ROT being a way to solve the problem.
A. One path is to use the ROT as a descriptive mechanism that generates an optimal principle. Another is to use an optimality principle to get at the ROT. If you start with the ROT, you often generate an optimization problem that does better than the ROT.

Q. (Peterson): Knowledge of the parameters (distribution of prey, etc.) is an important assumption, e.g. switching could be constrained by perception.
A. For the models to work, an organism must have some subjective distribution that it acts on. But the subjective distribution may be biased by memory, learning constraints, etc. In this case, ROT may be more accurate because they make no assumptions about the underlying distribution.

FORAGING IN THE CONTEXT OF LIFE-HISTORY : GENERAL PRINCIPLES AND SPECIFIC MODELS.

Alasdair I Houston
NERC Unit of Behavioural Ecology
Department of Zoology
University of Oxford

Introduction

This chapter attempts to explore some aspects of the tension between general principles of foraging and the prediction of specific models. There are three main sections. The first describes some models based on maximizing the long-term rate of energetic gain. It is shown that the predictions of the standard model of prey choice do not necessarily hold in other contexts. The second section outlines a more general framework for modelling behaviour. It has the advantage of allowing decisions to depend on the animal's state. It also puts foraging into the context of the animal's life-history. The final section discusses the issue of simple versus complex models.

Rate Maximization

Schoener (1971) made it clear that the basis of optimal foraging theory (OFT) was the relationship between food and fitness. Unfortunately, this general principle has tended to be ignored in favour of various simple slogans. One is that OFT requires that animals maximize their mean net rate of energetic gain while foraging. I will refer to this as rate maximizing. Within the context of rate maximizing, two paradigms have exerted an unhealthy fascination on empiricists. These paradigms are patch use (eg Charnov 1976a) and prey choice (eg Charnov 1976b). The original models of foraging in these contexts are well known (see Stephens and Krebs 1986 for a review), so I will give only a very brief account of them.

Patch use

Charnov (1976a) considered a foraging animal that encounters well-defined patches of food. Within a patch the forager obtains energy at a smoothly decreasing rate r(t). If G(t) is the amount of energy obtained if a time t is spent in the patch, then the rate of gain $\gamma(t)$ is given by

$$\gamma(t) = G(t) / (t + \tau),$$

NATO ASI Series, Vol. G 20
Behavioural Mechanisms of Food Selection
Edited by R. N. Hughes

where τ is the time taken to travel between patches. The marginal value theorem (MVT) states that γ is maximized by leaving a patch when $\gamma(t)$ falls to γ^*, where γ^* is the maximum rate that can be achieved.

Although this rule may appear to be circular, it is well-defined. It can be shown (McNamara and Houston 1985) that a naive animal can learn γ^* by using a simple rule.

One way in which the MVT has been misapplied is to use it to analyse the partial consumption of prey items. In such cases, the predator may be able to encounter new items while it is in the process of handling an item. This makes the standard patch use model inappropriate. Lucas and Grafen (1985) discuss this point and present a model that allows for simultaneous handling and search.

Prey choice.

The standard prey choice model is based on a forager that can search simultaneously for various types of item. An item of type i has an energetic content e_i and a handling time h_i . Encounters with each type constitute an independent Poisson process with encounter rate λ_i for type i . Searching and handling are mutually exclusive activities. This notation will be used repeatedly throughout this paper. The simplest characterization of the optimal prey choice policy is that prey type i should be included in the diet if and only if

$$\gamma^* < e_i / h_i ,$$

where γ^* is the maximum mean rate of energetic gain that can be achieved (cf the rule for leaving a patch). γ^* can be found by ranking types in terms of their profitability (e/h) and adding types to the diet in order of decreasing profitability until the resulting rate goes down. A given type should either always be included in the diet, or always rejected, i.e. there should be no "partial preferences". McNamara and Houston (1987a) review this topic.

To give a simple example of prey choice, let there be two prey types, with $e_1/h_1 > e_2/h_2$. Type 1 is then said to be more profitable. It should always be included in the diet, giving a rate of $e_1/(1/\lambda_1 + h_1)$. Then type 2 is taken if and only if

$$e_2 / h_2 > e_1 / (1/\lambda_1 + h_1) .$$

For further discussion, see Pyke (1984) Schoener (1987) and Stephens and Krebs (1986). A feature of this condition that is often regarded as surprising is the absence of λ_2 , the encounter rate with type 2 items.

It is sometimes claimed that rate-maximization models are very robust. I do not agree with this view. McNair (1979) showed that generalizing the prey choice model produced optimal diets in which prey types are not necessarily added to the diet in order of decreasing profitability, and in which the criterion for accepting a prey type depends on its encounter rate. I now give some further examples in which the results of the standard model do not hold.

Concurrent chains.

On the concurrent chains procedure, an animal is presented with two alternatives to which it can respond. Each alternative consists of an initial link and a terminal link. Denoting the alternatives by the subscripts 1 and 2, the initial links are typically variable interval (VI) schedules with mean intervals $1/\lambda_1$ and $1/\lambda_2$. Each VI gives access to its terminal link, which consists of an amount of food e_i and a time delay h_i ($i = 1,2$). On constant probability VIs, the distribution of intervals on alternative i is negative exponential with parameter i, so that while the animal is responding to one side (say side 1), access to the terminal link is a Poisson process with a rate λ_1 . During the time that the animal is responding to this initial link, the timer for the VI in the other side is running, so that if the animal switches to side 2 after a time away a then the probability of immediate access to a terminal link is $1 - \exp(-\lambda_2 a)$.

In experiments involving the concurrent chains procedure, attention is focused on the relative allocation of responding to the two initial links as a function of their values and the parameters of the terminal links.

There is clearly some resemblance between the terminal links of the concurrent chains and the food items envisaged by foraging theory. Furthermore, responding to the initial links can be though of as searching, and I have chosen my notation for the chains procedure so as to make this correspondence obvious.

Fantino and Abarca (1985) suggest that the prey choice model may not explain all aspects of concurrent chain behaviour. For example, with equal reward magnitudes let h_1 be less than h_2, so that the terminal link on side one is better than that of side 2 (higher profitability in terms of e/h). The data shows that λ_2 will have an effect on behaviour, whereas the standard prey choice result is that λ_2 has no effect. The trouble with this argument is that the prey choice model is concerned with whether or not to accept an item once it has been found. A further problem is that the concurrent chains procedure involves separate searching (ie initial link responding) for each prey type. If we work from first principles to find the pattern of responding on the initial links that maximizes the overall rate of energetic gain, then it turns out that λ_2 does influence optimal choice (Houston, Sumida and McNamara 1987). An example is given in Table 1.

TABLE 1

The relative allocation p* to the initial link on side 1 that maximizes overall rate on the concurrent chains procedure. $e_1 = e_2 = 1$, $h_1 = 5$, $h_2 = 20$. Time required for switching between initial links = 4, Heyman's switching index = 3 . See Houston et al. 1987 for further details.

$1 / \lambda_1$	$1 / \lambda_2$	p*
60	60	0.60
60	120	0.80
120	60	0.32
60	30	0.37
30	60	0.87

Central-place foraging (CPF)

Orians and Pearson (1979) introduced the term central-place foraging (CPF) to describe cases in which the foraging animal brings food back to a fixed location (the "central place"). A parent bird bringing food back to its young in a nest is an obvious example. Orians and Pearson defined two sorts of central-place forager. Single prey loaders (SPLs) return with only one item, whereas multiple-prey loaders (MPLs) return with more than one item. A foraging cycle for central-place forager consists of a journey from the central-place to the foraging area a time spent collecting food and a journey back to the central-place. The total travel time will be denoted by τ. Given some sort of decrease in the rate of collecting food (perhaps as a result of local depletion or an increase in handling time as a result of food already collected), the case of a MPL can be thought of as a patch-use problem, with the optimal behaviour being specified by the MVT (Orians and Pearson 1979).

Lessells and Stephens (1983) emphasise the distinction between MPL behaviour as choice of load size and SPL behaviour as prey choice. It must not be forgotten, however, that choice of prey will determine the expected size of the load. Consider a

SPL that searches simultaneously for two prey types, denoted by the subscripts 1 and 2. Using the usual notation for prey choice, (see above), we can evaluate the mean rate for each of three prey choice rules: take back the first type 1 item encountered, take back the first item encountered, and take back the first type 2 item encountered. Denoting these rates by R(1), R(G) and R(2) respectively we have

$$R\,(1) = e_1 \,/\, (\, 1/\lambda_1 + h_1 + \tau\,)$$

$$R\,(G) = (\, pe_1 + qe_2\,) \,/\, (\, 1/\lambda + ph_1 + qh_2 + \tau\,)$$

$$R\,(2) = e_2 \,/\, (\, 1/\lambda_2 + h_2 + \tau\,)\,,$$

where $\lambda = \lambda_1 + \lambda_2$ and $p = \lambda_1 / \lambda$ and $q = \lambda_2 / \lambda$.

If $e_2 > e_1$ but $e_1 / h_1 > e_2 / h_2$, then it is possible for the highest rate to be given first by R(1), then by R(G) and then by R(2) as τ is increased. The corresponding expected energetic contents of the load are e_1 , $pe_1 + qe_2$, and e_2 , so that as Stephens and Krebs (1986) point out, the expected load increases with travel time. The change in prey selection with τ can be understood by noting that the problem is equivalent to a standard prey choice model in which the handling times are $h_1 + \tau$ and $h_2 + \tau$ (Orians and Pearson 1979). We can therefore define a critical travel time τ_c at which

$$e_1 \,/\, (\, h_1 + \tau_c\,) \;=\; e_2 \,/\, (\, h_2 + \tau_c\,)\,.$$

Below τ_c , $e_1 / (h_1 + \tau) > e_2 / (h_2 + \tau)$, and so type 1 is always taken and the issue is whether type 2 should be taken. Similarly above τ_c , $e_2 / (h_2 + \tau) > e_1 / (h_1 + \tau)$ and so type 2 is always taken and the issue is whether or not to take type 1 items.

Houston and McNamara (1985a) generalise the model of CPF by SPLs to include a cost of being away from the central place. In doing so they point out that the model applies to an animal that hunts underwater but must return to the surface to eat each item, and to breathe. In other words, the surface is now the central place. One of the models assumes that the diver dies if it remains under water for longer than some maximum time duration T. Houston and McNamara show that it can be optimal to switch from rejecting a type to accepting a type as a function of how much time is left until T. To give an example, let there be two prey types with $e_1 > e_2$ and $h_1 = h_2$, and let γ^* be the maximum possible mean rate of energetic gain. Then if

$$e_2 < e_1 - \gamma^* / \lambda_1$$

then when there is more than some critical time S* left till T it is optimal to reject type 2 items, whereas when there is less than S* left it is optimal to accept them. If T < S*, then the animal should accept the first item that it encounters, whereas if T >

S*, type 2 items are rejected until a time T - S* has elapsed. If no type 1 item has been found by this time, then the next item that is found is accepted. The critical time left S* depends on the encounter rate with both prey types. For further models based on a limited amount of time for foraging see Lucas (1985) and this volume.

Although the classification of the behaviour of MPLs as involving patch time (or equivalently load size) rather than prey choice is convenient, it is not clear that it will always be biologically accurate. Some species of bird bring back loads that consist of a mixture of prey types (eg Brooke 1983 Turner 1982). Houston (1985) presents two models that explore the consequences of prey choice for MPLs. The "swallow" model was motivated by the work of Turner (1982) on the swallow (*Hirundo rustica*). Turner's data suggested that the encounter rate with the less profitable prey type influenced prey selection, in apparent contradiction to the predictions of the standard prey choice model. The swallow model assumes that there are two types, large items with mass m_L and small items with mass m_S. The swallow collects a ball or bolus of items in its throat and brings it back to the young. The handling time of an item is assumed to depend on the mass B of the bolus at the time that the item is caught. The particular form of dependence used for calculations was

$$H_L(B) = h_L + k_L B$$

and

$$H_S(B) = h_S + k_S B \quad .$$

It is assumed that items are small compared to the typical final mass of the bolus and are found relatively frequently, so that the bolus accumulates as a smooth deterministic flow. The optimal bolus B* depends on τ, but the form of the optimal policy can be obtained without knowing this dependence. There are two critical bolus sizes, B_1 and B_2, such that

(i) if $B^* < B_1$ then only large items are collected

(ii) if $B_1 < B^* < B_2$ then only large items are collected until $B = B_1$, after which both types are collected

(iii) if $B^* > B_2$ then only large items are collected while $B < B_1$, both types are collected while $B_1 < B < B_2$, and only small items collected when $B > B_2$.

The critical values are given by the equations

$$B_1 = (m_L h_S - m_S (1 / \lambda_L + h_L)) / (k_L m_S - k_S m_L)$$

and

$$B_2 = (m_L (1 / \lambda_S + h_S) - m_S h_L) / (k_L m_S - k_S m_L),$$

from which it can be seen that a necessary condition for a change in policy is that

$$k_L m_S > k_S m_L .$$

Because B* increases with τ , we may see a pattern of policy (i), policy (ii) and policy (iii) as τ increases. There is an obvious similarity between this pattern and the pattern of first taking a type 1 item, then taking either type, then taking a type 2 item that emerged as τ increased in the SPL model. It must be noted, however, that in the SPL model the modified profitability depends on τ, and in the final policy the animal is taking only type 2 items. In the swallow model, the profitabilities depend on bolus size, and in the final policy the animal goes through a phase of collecting only large items and a phase of collecting both sorts of item before entering the phase in which it collects only small items.

Can the standard prey choice model be applied to a central-place forger?

Lifjeld and Slagsvold (1988) removed some wing and tail feathers from female pied flycatchers (*Ficedula hypoleuca*). Because these birds sit while searching for a prey item and then fly to catch it, Lifjeld and Slagsvold argue that the feather removal would increase handling cost more than search cost. To predict the effect of such a change they used the standard prey choice model, and showed that the optimal diet becomes narrower as handling cost increases and becomes broader as searching cost increases. Lifjeld and Slagsvold argue that standard diet choice results can be applied directly to CPF by MPLs. They say that there is a well-defined function, the loading curve, that relates the time spent foraging to the accumulated energy. The optimal time spent foraging (and hence the load) will depend on travel time, but

" .... it should be stressed that travel time will not affect the shape of the loading curve; the loading curve is assumed to be fixed for a particular patch, irrespective of the distance to the CP. Thus travel time is not a factor that influences prey choice for a multiple-prey loader, and therefore the optimal diet model can be tested directly on such foragers without regard to travel time. The only complicating factor, as compared with a non-CPF situation, is that the actual prey choice may change during loading (cf Houston 1985). This is because the handling time of prey types may be affected differently by variation in load size." (Lifjeld and Slagsvold 1988 p114).

I am not completely happy with this argument. As long as the animal has a choice of prey types, it must be able to vary the loading curve by varying the prey that it loads. (Lifjeld (1988) makes this point). I will consider a weaker version of this claim after the effects of prey choice for MPLs have been discussed in a bit more detail.Lifjeld and Slagsvold found that females with feathers removed ("handicapped females") delivered food to their young at lower rates than females that had not been handicapped ("control females"). Lifjeld and Slagsvold were unable to measure load size directly, but data on feeding frequency suggested that handicapped females brought back bigger loads than control females in one of the two years studies (1983). There was also evidence that handicapped females brought back a narrower range of items than control females in this year. In discussing possible reasons for this pattern, Lifjeld and Slagsvold mention the results of Houston (1985). They say "if we assume that the increase in handling time is greater for large than for small prey then the diet should become broader with load size (cf the swallow model of Houston (1985)." One objection to this point is that its not clear that my model is relevant to the case that Lifjeld and Slagsvold are considering. I was analyzing load size and prey selection as a function of travel time, not as a function of energetic costs. There is, however, another study of the pied flycatcher in which the results of Houston (1985) may be relevant. Lifjeld (1988) investigated the effect of the number of nestlings and their hunger on the food that adults brought to them. Adults showed a marked increase in the frequency with which they delivered food when they were given five hungry nestlings to feed. Lifjeld suggests that this increase might be brought about by reducing collecting time, thus bringing back a smaller load. He then goes on to say that in the two models proposed by Houston (1985), the forager should take more small prey as load size increases. The swallow model does show a switch from taking only large items to taking both sorts of items if τ is big enough for B^* to be greater than B_1 , but it also shows a switch from taking both types to taking only small items if τ is big enough for B^* to be greater than B_2 . The other model in the Houston (1985) is the wheatear model, in which it is assumed that only two items can be brought back. There are two types, with parameters denoted by the subscripts 1 and 2. In contrast to the swallow model, in which handling time depends only on the total mass of the load, in the wheatear model the handling time of the second item depends on the type of item that is already on the load. The handling type for the first item is denoted by h_i ($i = 1, 2$), and the handling time for a type j item, given that the first item is of type i is denoted by h_{ij}. An interesting feature that emerged from this model is that the highest rate is sometimes achieved by a "sequential " or "contingent" policy in which behaviour depends on which type of item is encountered first. For example it may be optimal to return with the first item if it is a type 1 item but to load the first item if it is a type 2 item and then load the second item, whatever type it is. This policy is denoted by 10/2G, and the resulting rate is denoted by R(10/2G). The "G" stands for "Generalise". A similar policy is 10/21, in which the animal returns with the first item if it is type 1 but loads it if it is type 2 and then waits to load a type 1 item. Table 2 gives the optimal policy at four values of τ. The policy G1 involves loading the first item that is encountered and then loading the first type 1 item that is encountered, and the policy 11 involves loading the first and second type 1 items encountered. The table also gives the resulting mean load size, mean energetic content of the load, and the proportion of type 1 items in the load. It can be seen that as τ increases, the mean amount of energy brought back increases, and there is a trend towards specialising on type 1 items. This contradicts Lifjeld's claim.

TABLE 2

Results form the wheatear model (Houston 1985) with $e_1 = 9$, $e_2 = 1_1$ $\lambda_1 = 1.2$ $\lambda_2 = 6.0$ $h_1 = 0.333$ $h_2 = 0.08$, $h_{11} = 0.7$, $h_{12} = 0.26$, $h_{21} = 0.35$, $h_{22} = 0.082$. The table gives, for four values of τ, the optimal policy together with the resulting mean number of items in the load, $\bar{N}$, the mean energetic content of the load, $\bar{E}$, and the proportion of type 1 items in the load, Prop 1. See text for further details.

τ	Opt.Pol	$\bar{N}$	$\bar{E}$	Prop 1
0.1	10/2G	1.83	4.28	0.167
0.3	10/21	1.83	9.83	0.545
0.5	G1	2.0	11.33	0.583
0.7	11	2.0	18.0	1.0

I now return to the issue of using standard prey choice theory in the context of CPF. Lifjeld and Slagsvold claimed that travel time does not influence prey choice for a MPL. They admitted the counter-examples found in Houston (1985), but implied that in the absence of changes in handling time with load, their argument would stand. As a rejoinder to such a view, consider a CPF that can take at most N items, regardless of type. In the absence of either increases in handling time with load or decreases in encounter rate as a result of depletion it will always be optimal to take N items. Ignoring any contingent policies, this problem is essentially equivalent to that of a SPL with a travel time given by τ/N, so the optimal policy will depend on τ .

I have given so much space to CPF for a variety of reasons. The example of parents feeding their young is of obvious biological importance and provides a context in which data is relatively easy to collect. The models provide simple and realistic settings in which the conclusions of standard prey choice models do not hold. More fundamentally, they provide a schematic representation of the "edge effects" that might occur when an animal is limited by time or the amount of food that it can handle. The time limits might arise because of interruptions in the foraging process (eg Barnard et al. 1985, Lucas 1985 and this volume, McNamara and Houston 1987b) or because of the advantages of performing other activities (eg McNamara et al. 1987).

The Common Currency

The essence of rate-maximization is the assumption that a unit increase in energy always has the same value in terms of the forager's fitness (Pyke 1984, McNamara and Houston 1987b). The assumption is implausible, and is also at variance with two classes of observations.

(a) rate-maximization requires that all options that yield the same mean rate should be treated as equivalent. It is clear, however, that this is not the case. Animals are sensitive to the variance about a given mean. Such "risk-sensitive" foraging is reviewed by Real and Caraco (1986) and McNamara and Houston 1987b).

(b) An animal's foraging behaviour depends on its state. There is no obvious way to incorporate this effect in rate-maximization models. More generally, there is no way in which to compare foraging with other activities.
The obvious way to compare different activities is in terms of their contribution to fitness. It is not quite so obvious how to do this. Working independently, McNamara and Houston (1986) and Mangel and Clark (1986) proposed the same approach. It is described in detail by Mangel and Clark (1988) and is reviewed by Houston et al. (1988) so I will only summarise it here. The approach involves an explicit representation of the animal's state and the way in which its behaviour changes its state. The state might include the animal's size, its level of energy reserves stored as body fat, or the number of eggs that it has. The state of the animal is denoted by $\underline{x}$. The animal has a set of actions, and performs a given action at discrete times 1, 2 ... A policy is a rule for choosing behavioural actions as a function of $\underline{x}$ and time. The action that an animal

performs may make a contribution to the animal's reproductive success, and it will also typically change the animal's state. This change may depend, in a probabilistic way, on the animal's environment. Assuming that natural selection evaluates policies in terms of the resulting expected future reproductive success (efrs), we introduce

$\Psi^* (\underline{x},t)$ = efrs of an animal in state $\underline{x}$ at time t that follows the optimal policy.

It is now possible to compare actions at time t by means of

H (a, $\underline{x}$, t) = efrs of an animal in state $\underline{x}$ at time t that performs action a and then follows the optimal policy from time t + 1.

The H's constitute a common currency for comparing actions. The optimal action a* is the one with the largest H, and this H establishes Ψ^* at this time, ie.

$$\Psi^* (\underline{x}, t-1) = \max H (a, \underline{x}, t).$$

Thus from Ψ^* at t we can find Ψ^* at time t - 1 . This backward procedure can be repeated to find Ψ^* at time t - 2 etc. The technique is known as dynamic programming.

The calculations yield not only the optimal policy but also a measure of the costs of deviating from it. McNamara and Houston (1986) define the canonical cost of performing action a in state $\underline{x}$ at time t as H (a, $\underline{x}$, t) - H (a*, $\underline{x}$, t) .

We have now produced a general principle for evaluating behaviour, including foraging behaviour. It is not clear, however, that very much progress has been made, because it has been assumed that Ψ^* is known. How can this function be found? I will outline two possibilities.

(i) We can consider the whole of the animal's remaining life. From this perspective, traditional life-history theory becomes a special case.
Furthermore, phenotypic plasticity (eg Stearns and Koella 1986) emerges automatically because decisions depend on the animal's state.
This method can be used when insects are being studied (eg Mangel 1987) but it is unlikely to be feasible in the case of relatively long-lived animals, especially when a detailed investigation of a relatively short period of their life is desired. In such cases, it is necessary to relate this short period to the rest of the animals life (McNamara and Houston 1982).

(ii) We consider a biological relevant period of the animals life, such as a breeding

season or a winter. This period ends at some time T. We define a terminal reward function R ($\underline{x}$) that specified the efrs of an animal as a function of its state at T. The terminal reward function provides the link to the rest of the animal's life. It is now possible to work back from this function to investigate decisions in the times before T. This still leaves us the problem of finding R ($\underline{x}$). One possibility is to take the animals life after T as fixed and use measurements of reproductive success to estimate R ($\underline{x}$). For example, the work of Fritz and Morse (1985) shows how the efrs of a female crab spider (*Misumena vatia*) depends on its mass. This function could be used as a terminal reward at the end of the foraging period to assess foraging decisions.
If suitable data are not available, then the model can be used to explore the consequences of various forms of the terminal reward. An example of this approach is provided by McNamara in this volume.

Maximising survival probability

John McNamara and I have used the common currency to investigate the survival of an animal during a period in which it does not reproduce. We take the animal's state to be its level of energy reserves, x, and assume that the animal dies of starvation if x falls to zero. Houston and McNamara (1985b) analyzed prey choice for an animal that could only forage during the day and dies during the night if its reserves at dusk were less than a critical level x_c . It was found that survival probability is not necessarily maximized by maximizing γ . For example, when there are two prey types, it is possible that taking only type 1 items gives the highest rate, but also has a higher variance than taking both types. Under such circumstances there can be a wedge shaped region in the space of x v time of day in which it is optimal to take both sorts of items. The shape of this region depends on (amongst other things) the encounter rate with type 2 items. Although the exact shape of the region has to be found by computation, its upper boundary can be found by a diffusion approximation (McNamara 1984). Let a and b denote two foraging options, with means μ_a and μ_b and variances σ_a and σ_b (where $\sigma_b < \sigma_a$). McNamara shows that there is a boundary z (t) given by

$$z(t) = x_c - k(T - t)$$

where

$$k = (\mu_b\sigma_a - \mu_a\sigma_b) / (\sigma_a - \sigma_b)$$

such that it is optimal to prefer the high variance option if x (t) is below z (t) and the low variance option if x (t) is above z (t) .

$$\text{When } \mu_a = \mu_b = \mu \text{ ,}$$

$$z\ (t) = x_c - \mu\ (T - t)\ .$$

This condition might be mistaken for the expected daily energy budget rule (Stephens 1981). The fundamental difference is that Stephen's analysis allows for only one decision at the start of the foraging period, whereas McNamara's analysis is based on a series of decisions that depend on the animal's state. This issue is discussed in more detail by Houston and McNamara (1982, 1988).

The expected daily energy budget rule seems to be well on the way to becoming a dogma, more frequently cited than understood. It is based on a particular model in which the animal has to acquire enough energy during a foraging period to enable it to survive a following non-foraging period. Although Barnard and Brown (1981) mention the rule in the context of shrews, it is not clear that it is directly applicable in this case. Barnard et al. (1985) developed a model for shrews based on random interruptions of foraging, with the interruptions having a random duration. The expected daily energy budget rule does not hold in this model; in fact the mean proportion of time for which the animal can forage does not suffice to determine the optimal behaviour.

When the forager is never interrupted (i.e. foraging is continuous), the results are simple when the means are equal. If the animal has a positive mean net gain it should prefer low variance whereas if the animal has a negative mean net gain it should prefer high variance if reserves are low and low variance if reserves are high. (See McNamara and Houston (in press) for further results and discussion.)

Why Make Models?

A good model should be more simple than the phenomena that it models, but should not be so simple as to loose essential elements of reality. There is no guaranteed prescription for achieving this balance. As an illustration of some of the points that must be taken into consideration, I will discuss two cases.

(1) Central place foraging

In my discussion of CPF, I used rate maximization models. A reasonably realistic model might involve the state of each nestling and the state of each parent (including the composition of the load and perhaps the information that each parent has about the other parent). It would also be necessary to consider game-theoretic interactions between the birds (cf. Houston and Davies 1985, Parker and McNair 1978, 1979). This is a daunting list, and in the absence of specific information, it seems advisable to start by analyzing various simplified aspects of the problem to obtain some understanding of which features are likely to be important.

(2) Foraging behaviour of crab spiders

Morse (1979) claims that the foraging behaviour of the crab spider *Misumena vatia* can be used to make a quantitative test of the standard prey choice model. There are various reasons why such a view might be questioned.

(i) There are marked changes in encounter rates with prey during a day. For example, bumblebees make aboug 14 visits to the rose *Rosa carolina* between 9.00 and 10.30 but make less than 2 visits per hour between noon and 13.30 (Morse 1979).

(ii) Handling times are very long (mean of 5.48h for bumblebees, 1.51h for syrphid flies, Morse 1979), so that encounter rates may change significantly while the spider is handling an item.

(iii) It may be possible for a spider to detect a visitor to a flower while it is handling an item, so that search and handling are not mutually exclusive. Given that spiders adjust their handling times as a function of prey availability (Morse 1981), the model presented by Lucas and Grafen (1985) might be relevant.

(iv) Satiation seems to be important (Morse 1981).

These points invalidate the standard prey choice model, but even a relevant rate-maximization model would be inappropriate because the relationship between fitness and female size found by Fritz and Morse (1985) is actually incompatible with rate maximization.

Conclusions

I have shown how in the context of rate maximization the predictions of the standard prey choice model may change if various aspects of the biology are changed. One sort of effect that is not found in the standard model is the influence of chance, as illustrated by the "contingent" or "sequential" policies in the wheatear model.

The next section describes a more general approach than rate maximization. It is able to allow decisions to depend on the animal's state. This is clearly more realistic than the rate-maximization approach, but it also tends to require more information. The final section discusses the issue of simple versus complex models. It is argued that there may be circumstances in which it is useful to construct a rate-maximization model even though it does not include all the relevant aspects of the biology. Although such a model will not be able to predict exact details, it may provide an indication of which features are important. There will be circumstances in which the biology is incompatible with the assumptions of rate-maximization. If enough information is available, specific models can be constructed in such cases.

Acknowledgements

I thank John McNamara for comments on an earlier version of this chapter. I was supported by the Natural Environment Research Council.

References

Barnard CJ, Brown CAJ (1985) Risk-sensitive foraging in common shrews (*Sorex araneus* L). Behav Ecol Sociobiol 16:161-164

Barnard CJ, Brown CAJ, Houston AI, McNamara JM (1985) Risk-sensitive foraging in common shrews: an interruption model and the effects of mean and variance in reward rate. Behav Ecol Sociobiol 18:139-146

Brooke MdeL (1983) Wheatears, leatherjackets and a comment on central place foraging. Anim Behav 31:304-305

Charnov EL (1976a) Optimal foraging: the marginal value theorem. Theor Pop Biol 9:129-136

Charnov EL (1976b) Optimal foraging: attack strategy of a mantid. Am Nat 110:141-151

Fantino E, Abarca N (1985) Choice, optimal foraging and the delay-reduction hypothesis. Behav Brain Sc 8:315-330

Fritz RS, Morse DH (1985) Reproductive success and foraging of the crab spider *Misamena vatia*. Oecologia 65:194-200

Houston AI (1985) Central-place foraging: some aspects of prey choice for multiple-prey loaders. Am Nat 125:811-826

Houston AI, Clark CW, McNamara JM, Mangel M (1988) Dynamic models in behavioural and evolutionary ecology. Nature 332:29-34

Houston AI, Davies NB (1985) The evolution of cooperation and life history in the dunnock *Prunella modularis*. In: Sibly RM, Smith RH (eds) Behavioural Ecology, Blackwell Scientific Publications, Oxford, p 471

Houston AI, McNamara JM (1982) A sequential approach to risk-taking. Am Behav 30:1260-1261

Houston AI, McNamara JM (1985a) A general theory of central-place foraging for single-prey loaders. Theor Pop Biol 28:233-262

Houston AI, McNamara JM (1985b) The choice of two prey types that minimises the probability of starvation. Behav Ecol Sociobiol 17:135-141

Houston AI, McNamara JM (1988) A framework for the functional analysis of behavior. Behav Brain Sci 11:117-154

Houston AI, Sumida BH, McNamara JM (1987) The maximization of overall reinforcement rate on concurrent chains. J Exp Anal Behav 48:133-143

Lessells CM, Stephens DW (1983) Central place foraging: single-prey loaders again. Anim Behav 31:111-118

Lifjeld JT (1988) Prey choice and nestling hunger: an experiment with pied flycatchers, *Ficedula hypoleuca* . Anim Behav 36:134-139

Lifjeld JT, Slagsvold T (1988) Effects of energy costs on the optimal diet: an experiment with Pied Flycatchers *Ficedula hypoleuca* feeding nestlings. Ornis Scand 19:111-118

Lucas JR (1985) Time constraints and diet choice: different predictions from different constraints. Am Nat 126:680-705

Lucas JR, Grafen A (1985) Partial prey consumption by ambush predators. J theor Biol 113:455-473

Mangel M (1987) Oviposition site selection and clutch size in insects. J Math Biol 25: 1-22

Mangel M, Clark CW (1986) Towards a unified foraging theory. Ecology 67:1127-1138

Mangel M, Clark CW (1988) Dynamic modeling in behavioral ecology. Princeton University Press, Princeton, NJ

McNair JN (1979) A generalized model of optimal diets. Theor Pop Biol 15:159-170

McNamara JM (1984) Control of a diffusion by switching between two drift-diffusion coefficient pairs. SIAM J Cont Opt 22:87-94

McNamara JM, Houston AI (1982) Short-term behaviour and life-time fitness. In: McFarland DJ (ed) Functional Ontogeny, Pitman, p 60-87
McNamara JM, Houston AI (1985) Optimal foraging and learning. J theor Biol 117:231-249
McNamara JM, Houston AI (1986) The common currency for behavioral decisions. Am Nat 127:358-378
McNamara JM, Houston AI (1987a) Partial preferences and foraging. Anim Behav 35:1084-1099
McNamara JM, Houston AI (1987b) A general framework for understanding the effects of variability and interruptions on foraging behaviour. Acta Biotheor 36:3-22
McNamara JM, Houston AI Starvation and Predation in a patchy environment. In: Shorrocks B, Swingland I (eds) Living in a patchy environment
McNamara JM, Mace RH, Houston AI (1987) Optimal daily routines of singing and foraging in a bird singing to attract a mate. Behav Ecol Sociobiol 20:399-405
Morse DH (1979) Prey capture by the crab spider *Misumena calycina* (Araneae: Thomisidae) Oecologia 39:309-319
Morse DH (1981) Prey capture by the crab spider *Misumena vatia* (L) (Thomisidae) on three common native flowers. Am Midl Nat 105:358-367
Orians GH, Pearson NE (1979) On the theory of central place foraging. In: Horn DJ, Mitchell R, Stair GR (eds) Analysis of Ecological Systems, Ohio State University Press, Columbus, p 155
Parker GA, MacNair MR (1978) Models of parent-offspring conflict I Monogamy. Anim Behav 26:97-110
Parker GA, MacNair MR (1979) Models of parent-offspring conflict II Suppression: evolutionary retaliation of the parent. Anim Behav 27:1210-1235
Pyke GH (1984) Optimal foraging theory: a critical review. Ann Rev Ecol Syst 15:523-575
Real LA, Caraco T (1986) Risk and foraging in stochastic environments: theory and evidence. Ann Rev Ecol Syst 17:371-390
Schoener TW (1971) Theory of feeding strategies. Ann Rev Ecol Syst 2:369-404
Schoener TW (1987) A brief history of optimal foraging theory. In: Kamil AC, Krebs JR, Pulliam HR (eds) Foraging Behavior, Plenum Press, New York, p 5
Stearns SC, Koella JC (1986) The evolution of phenotypic plasticity in life-history traits: predictions of reaction norms for age and size at maturity. Evol 40:893-913
Stephens DW (1981) The logic of risk-sensitive foraging preferences. Anim Behav 29: 628-629
Stephens DW, Krebs JR (1986) Foraging theory. Princeton University Press, Princeton
Turner AK (1982) Optimal foraging by the swallow (*Hirundo rustica* L): prey size selection. Anim Behav 30:862-872

ORAL DISCUSSION - A. Houston

In answer to the question of how a simple energy maximization model may sometimes resemble a more complex formulation, Houston pointed out that often the complex model has a range of energy values over which fitness is essentially a linear function of energy, so that within those limits the two approaches concur: it is outside those limits where different predictions arise. Houston also responded to a question about the meaning of partial preference by noting that in nature, animals almost always ate some of the prey types that the model predicted should be ignored. One approach to this phenomenon is to incorporate stochastic choice as a constraint. For further discussion see McNamara & Houston (1987a).

THE STARVATION-PREDATION TRADE-OFF AND SOME BEHAVIOURAL AND ECOLOGICAL CONSEQUENCES

John M. McNamara,
University of Bristol

Introduction

There are a variety of ways in which an animal may increase the amount of food it obtains at the expense of an increase in predation risk. I give some examples:

(a) The most obvious way to increase the amount of food gained is to increase time spent foraging. Since foraging will typically be a more risky activity than, say resting. An increase in time spent foraging will increase the probability of being taken by a predator.

(b) Some areas of food give a higher yield but are more risky (Sih, 1980, 1982).

(c) Decreasing the level of vigilance will increase the rate of gain but increase predation risk (Pulliam, Pyke & Caraco, 1982; Milinski & Heller, 1978).

(d) A bird can decrease its probability of starvation by carrying extra fat reserves. Lima (1986) and McNamara & Houston (in press, a) consider two different mechanisms by which increasing fat reserves, and hence body mass, increases predation risk. Increasing body mass results in increased metabolic costs, especially for flight. This means that more food must be obtained to maintain body weight and hence more time must be spent foraging. Thus the bird is exposed to predators for a greater time each day (cf (a)). Increasing body mass may also decrease a bird's manoeverability and hence increase the bird's chances of being taken by a predator.

It is natural to ask about an animal's optimal response when faced with a trade-off between food and predation. For example, one might seek the maximum predation risk an animal should accept in order to obtain a food item. To answer this sort of question one must know the value of the food items to the animal; that is, one must know how much the item contributes to future reproductive success. The value of an item will depend on the animal's hunger state and the availability of other food sources in its

NATO ASI Series, Vol. G 20
Behavioural Mechanisms of Food Selection
Edited by R. N. Hughes

environment. It will also depend on how the energy obtained from the food item is used. The animal may use the item to build up its reserves before the breeding season or before migration, or might use the item to help it survive a lean period. Alternatively, it may not eat the item itself but use it to feed young or to attract a mate.

Thus, to examine optimal behaviour when faced with a trade-off, we must first specify the biological setting - different settings can lead to very different results. In this paper I restrict attention to the case of an animal attempting to survive an extended period of time. I will introduce a simple schematic model of the feeding options available to the animal and the associated predation risk. In this model the animal dies (of starvation) if its energy reserves fall to zero or if it is taken by a predator. We will be concerned with the policy which maximises the animal's chances of surviving the period. The model will be used to illustrate four main issues.

The form of the optimal survival policy. How does optimal behaviour change with the animal's level of energy reserves?

Modelling behaviour and robustness. I will use the survival model to illustrate problems in modelling behaviour and in the design of experiments.

The selection pressure on state dependent choices. Does allowing behaviour to depend on energy reserves give an animal a significant advantage over an animal whose behaviour is not state dependent?

Starvation and predation levels. Can observed population levels of mortality from starvation and predation be used to make inferences about which source of mortality is more important to the population?

Alasdair Houston and I have been concerned with these issues previously, and have discussed aspects of them in a number of published works. Here I use the simple survival model to give a unified account of them.

The Survival Model

Here I present a simple model of the decision problem faced by an animal attempting to survive an extended non-reproductive period. The period is assumed to be sufficiently long so that the animal would die in a small fraction of it if it received no food. This assumption would certainly hold for a small mammal or bird attempting to survive the winter.

Whilst the animal is alive its energy reserves, x, satisfy $x > 0$. To model an animal's finite capacity to carry fat we suppose there is a limit L such that the animal forages while $0 < x \leq L$, but must rest when $x > L$. Reserves decrease continuously at rate d due to metabolic loss and increase if a food item is eaten. Death can be due to predation, or from starvation if the animal's energy reserves fall to zero.

The animal is assumed to have n behavioural options. An option might be, for example, a possible area to search, or a level of vigilance to employ. At each of the times $t = 0,1,2,\ldots$ a foraging animal chooses one of these options for the next unit of time. Under every option the animal finds either 0 or 1 food item between times t and $t + 1$. If it chooses Option i the probability it finds an item is p_i and the probability it is killed by a predator is z_i. All food items have the same mean net energy content E. Thus the mean net rate of energy gain under option i is $\mu_i = p_iE - d$.

All computations presented here are for an upper limit $L = 100$ and metabolic loss $d = 4$. Food items have mean energy content $E = 8$ but are taken to be slightly variable, so that items found have energy content 7,8 or 9 with probabilities 0.25, 0.5 and 0.25 respectively.

I will be interested in the behavioural policy for the animal which maximises its long term survival probability. I will give a brief description of what is meant by such a policy and how it can be found. The expected future reproductive success of an animal at the end of our extended non-reproductive period will depend on the animal's energy reserves. The "terminal reward function" specifies this dependence. Given a terminal reward function one can find the animal's optimal policy over the whole period by dynamic programming (see, for example, McNamara and Houston, 1986). At a given time in the period an animal's best choice of option will depend on its energy reserves. If the time chosen is near the end of the period the best choice will also depend on time to go and the form of the terminal reward function. As time to go increases the dependence on time to go and the terminal reward function disappear, and the optimal policy tends to a stationary limit in which the animal's best choice of action depends only on its energy reserves. It is this limiting policy which I will be concerned with in this paper. The policy can be thought of as maximising the animal's long term survival probability. Mathematical details proving convergence of policies to a stationary limit are given in McNamara (in press).

I will use the following example to illustrate a number of points throughout the paper.

Example 1. The animal can choose between the two behavioural options whose parameters are given in Table 1. Option 1 involves no predation risk but has a mean net rate of gain of zero. Option 2 is risky in terms of predators but has a higher net rate of gain. The policy which maximises long term survival probability is to choose Option 2 when reserves are 49 or less and to choose the safe Option 1 when reserves are 50 or above.

Table 1. An example of the optimal policy and the resultant percentage of time spent taking each option. There are two options. For Option i, p_i gives the probability of a reward per unit time, μ_i the mean net rate of gain and z_i the predation risk per unit time. Range gives the values of reserves x for which the option is optimal. % time gives the percentage of time an animal following the optimal policy will spend on each option. The animal must rest if reserves exceed L = 100. Metabilic loss is d = 4 units of energy per unit time. Food items give E = 8 units of energy.

	p_i	μ_i	z_i	range	%time
Option 1	0.5	0	0	$50 \leq x \leq 100$	81.8
Option 2	0.6	0.8	0.002	$1 \leq x \leq 49$	14.8
[Rest	0	-4	0	$x \geq 101$	3.4]

An animal following a stationary policy will change the option it chooses as its reserves change. If we follow the behaviour of a large population of animals which independently follow the stationary policy, the proportion choosing a particular option at a given time will settle down to a limiting value. Table 1 gives the limiting proportions of animals performing Options 1 and 2 and resting, expressed as a percentage, for the case of Example 1.

In addition to calculating proportions one can also calculate limiting mortalities from starvation and predation. If we consider those members of a large population that are still alive at some time t, then these mortalities give the proportion dying from starvation and predation respectively within the next unit of time. For Example 1 the mortality from starvation is $S = 0.70 \times 10^{-4}$ and the mortality from predation is $P = 2.96 \times 10^{-4}$.

General Form Of The Optimal Policy

In the model of the previous section options differed in the probability of a food item, p_i, and the predation risk, z_i. McNamara (in press) considers a more general problem in which food items may also differ in size and in variability. A less technical account of the results obtained there can be found in McNamara and Houston (in press, b). McNamara and Houston (in press, b) gives the general form of the policy which maximises long term survival in two special cases. In the first case there is no predation but options differ in the mean and variability of the amount of energy obtained. The second case considered is the one we are considering here.

When the animal has a range of options differing in p_i and z_i McNamara and Houston (in press, b) show that certain options should never be used. Of those used it is shown that both p_i and z_i should decreased as energy reserves increase. In other words the animal should take less risks in terms of predators as its reserves increase. This results in the animal getting less food as its reserves increase. This form of the optimal policy has already been illustrated in Table 1.

A second illustration is given in Table 2 where there are five options. These options have the property that p_i can only be increased at the expense of increasing z_i. Again it can be seen that the optimal policy results in p_i and z_i decreasing as reserves increase.

Although the general form of the optimal policy is intuitively obvious, I would like to emphasis that the optimal policy might have a different form if we were concerned with a biological situation other than long term survival.

Table 2. An example of the optimal policy and the resultant percentage of time spent taking each option when there are five options. See Table 1 for terminology. L = 100, d = 4, E = 8.

	p_i	μ_i	z_i	range	%time
Option 1	0.75	2.0	0.05	$1 \leq x \leq 8$	0.2
Option 2	0.60	0.8	0.01	$9 \leq x \leq 16$	1.1
Option 3	0.55	0.4	0.0025	$17 \leq x \leq 34$	7.5
Option 4	0.52	0.16	0.0005	$35 \leq x \leq 66$	34.1
Option 5	0.50	0	0	$67 \leq x \leq 100$	53.7
[Rest	0	-4	0	$x \geq 101$	3.3]

Modelling Behaviour And Model Robustness

When one makes a quantitative model of behaviour it is often assumed that the animal under study knows certain parameter values. This is especially true in laboratory experiments in which there is a pre-experimental training period designed explicitly so that the animal can know the quality of the available sources of food. For example, if one is performing an experiment to examine risk sensitivity in which the animal has a choice between two sources of food which have the same mean gain but different variances (Caraco, Martindale and Whittam, 1980), then it is obviously important that the animal "knows" that the means are indeed the same. Two obvious general questions are:

(i) does the animal make an accurate estimate of a parameter?

and (ii) if it makes an estimate which is not accurate will behaviour be very different from that based on the true value of the parameter?

Accurate estimation.

It may be difficult for an animal to make an accurate estimate of a parameter. For example, if a feeder delivers a food item with probability p per trial it needs a lot of trials before even a statistician can accurately estimate p. Estimating predation risk is liable to be far more difficult given the rare nature of attacks by predators.

There are also good biological reasons why animals in the laboratory may never make accurate estimates of certain parameters. I give some obvious examples.

(a) The food supply of an animal may never be constant over time in its natural environment. In a laboratory experiment in which it receives a reward with constant probability p per trial it may thus be impossible for the animal to learn that p is constant. The animal is likely to continually interpret statistical fluctuations in the supply of food as a change in the underlying food availability (ie p) (McNamara and Houston, 1980).

(b) An animal may be often interrupted while foraging in the wild. Interruptions can come from bad weather, or from the competitive activities of conspecifics or from potential predators. For such an animal it is important to consume food quickly. In the laboratory it may continue to behave as if its food source may suddenly be taken away even though this is not the case (Houston and McNamara, 1989). This may lead to a preference for immediate rewards (Houston, Kacelnik and McNamara, 1982; Kagel, Green and Caraco, 1986; McNamara and Houston, 1987a).

(c) Given the difficulty of accurately estimating predation risk an animal may never respond to a laboratory environment as if it is completely predator free even if it is so (Houston and McNamara, 1989).

A consequence of these problems is that it may not be fruitful to make a completely accurate representation of the laboratory set up in a model. The model may be a good predictor of the animal's behaviour only if it contains elements, such as interruptions in the food supply, which are not actually present. This type of problem is discussed in Houston and McNamara (1989).

Robustness

From the above discussion it is clear that there are really two different questions concerning the robustness of a model. One concerns whether the predictions of a model are significantly qualitatively altered by small changes in parameter values. The second concerns whether predictions are qualitatively altered by the inclusion of factor or effects which are not in the original model.

One general approach to the problem of robustness uses canonical costs (McNamara and Houston, 1986). These are defined to be the loss in expected reproductive success that an animal incurs by using a suboptimal action. If these costs are high for all suboptimal actions then one might expect the predictions of the model to be robust in both the above senses. Canonical costs are also important in another sense. Even if the predictions of an optimality model are robust it may be that the fitness of an animal which follows a suboptimal action is almost as high as an optimising animal. Under these circumstances the selection pressure to adopt the exact optimal rather than a slightly suboptimal action will be weak. The canonical cost measures the strength of this selection pressure.

Rather than use canonical costs here, I will use the model of an animal attempting to survive an extended period to look directly at the effects of parameter and model changes on optimal behaviour. To look at the effect of predation risk I first find the maximum predation risk an animal should accept in order to obtain a food item.

Consider an animal at a time well before the end of the period we are considering. Clearly the probability that the animal will survive till the end of the period increases with the animal's energy reserves. To measure the advantage of having high reserves we introduce the function ϕ where $\phi(x)$ is the ratio of the probability that an animal with reserves x will survive the winter to the probability that an animal with reserves L (above which the animal rests) will survive the winter[1]. Figure 1(a) illustrates ϕ for the case of Example 1. It can be seen that ϕ increases rapidly for small reserves and then decelerates. To measure the relative value of a food item of energetic value e we introduce the function Δ_e where $\Delta_e(x) = \phi(x + e) - \phi(x)$. Thus $\Delta_e(x)$ is the (relative) increase in survival value which results from consuming the item. It is approximately equal to $e\phi'(x)$. Figure 1(b) illustrates Δ_e for Example 1. It can be seen that $\Delta_e(x)$ decreases very rapidly with reserves x.

Now suppose that an animal is offered a food item of energetic content e. To obtain the item the animal exposes itself to predators and has probability λ of being killed. Metabolic and time costs of obtaining the item are assumed to be negligible. It can then be shown that the animal should take the item if and only if λ is less than the critical value

$$\lambda_c = \frac{\Delta_e(x)}{\phi(x + e)}.$$

[1]It can be shown that ϕ does not depend on the time to go or on the final terminal reward at the end of winter, provided the time to go is large (McNamara, in press).

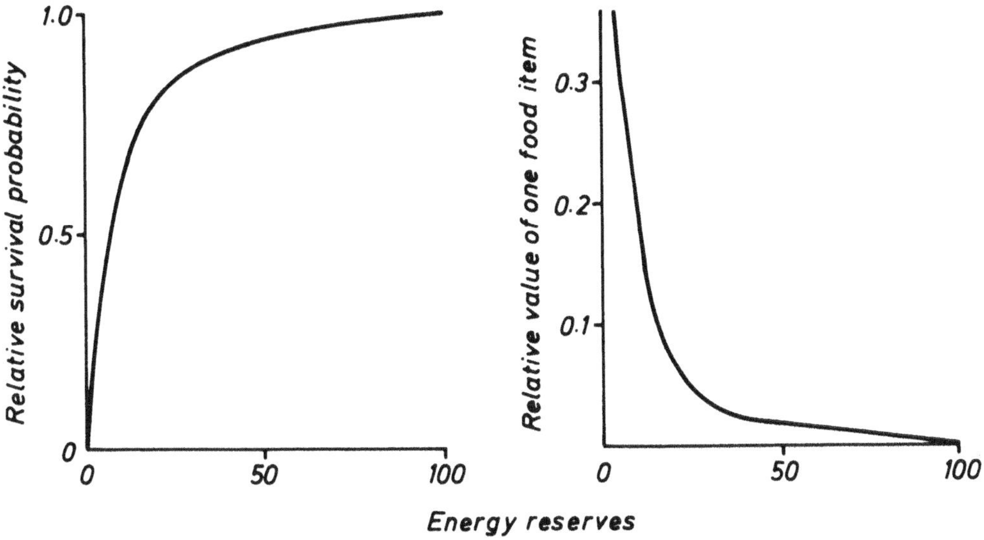

Figure 1. (a) relative survival probability, and (b) relative value of a food item of energy content e = 8, as a function of energy reserves. The background options available to the animal and other parameters are as for table 1.

One can understand this formula intuitively by observing that ϕ represents the (relative) loss in reproductive success on being killed by a predator and Δ_e represents the (relative) gain in reproductive success as a result of obtaining the food item. Not surprisingly, in weighing food against predation risk one looks at the ratio of gains to losses. A similar formula is given in, for example, Houston and McNamara (1988a).

Table 3 shows two examples of the dependence of λ_c on energy reserves. In (a) parameters are as in Example 1. In (b) the environment provides the animal with a better source of food than in (a). The table shows that the animal should accept a very high predation risk when its reserves are low, but that the maximum acceptable risk can be very small for high reserves, especially if the animal's environment provides a reasonable source of food.

These results show that the optimal behaviour of an animal with high energy reserves can be extremely sensitive to predation risk. In particular in a laboratory experiment in which there is no real risk, a very small perceived risk can alter an animal's optimal action.

Table 3. The maximum predation risk, λ_c, that an animal should accept in order to obtain a food item worth e = 8 units. Two cases are shown. In (a) the background environment consists of two food sources with parameters $p_1 = 0.5$, $z_1 = 0$ and $p_2 = 0.6$, $z_2 = 0.002$ (c.f. Table 1). In (b) the background environment has one food source with $p_1 = 0.6$, $z_1 = 0$. L = 100, d = 4, E = 8 throughout.

		energy reserves				
		10	30	50	70	90
λ_c	(a)	0.2023	0.0366	0.0180	0.0104	0.0030
	(b)	0.1981	0.0243	0.0032	0.00037	0.000035

When reserves are low behaviour is fairly insensitive to predation risk but extremely sensitive to food availability. To illustrate this suppose that an animal in an environment described by Example 1 is given a third option which yields a food reward with probability p_3 but at predation risk $z_3 = 0.01$. It can then be shown that Option 3 should be preferred to Option 2 where reserves are x = 12 if $p_3>0.602$. Thus although Option 3 is five times as risky as Option 2 it should be preferred if the probability of a food item is 0.002 greater than under Option 2. This sensitivity to food quality rapidly disappears as reserves increase. For x = 52 Option 2 should be preferred over Option 3 even if $p_3 = 1$.

The Selection Pressure On State Dependent Choices

The optimal policies presented in Tables 1 and 2 are state dependent: an animal's best option depends on its energy reserves. But how much of an advantage does an animal following the optimal policy have over an animal which adopts a good non-state dependent rule? State dependent rules are more complex to implement and one would expect them to be found in animal populations only if they confer a significant selective advantage on population members.

Alasdair Houston and I have been concerned with this issue previously. For example, one can consider an animal which has a choice between two sources of food which give the same mean gain but differ in variance in gain. Both choices involve no predation risk. When there is overnight energy loss it is known that the policy which maximises long term survival probability depends on both state and time of day (McNamara and Houston, 1986). One can investigate the strength of selection pressure for risk prone or risk averse decisions in this context.

Houston and McNamara (1986) consider the case where there is significant overnight energy loss, but the common mean net rate of gain under the foraging options is sufficient to give a positive 24 hour expected energy balance. It is shown that the selection pressure to adopt the high variance option when it is optimal to do so is much less strong than the selection pressure to adopt the low variance option when that option is optimal.

McNamara and Houston (in press, b) look at the case where there is no day night cycle. It is then never optimal to choose the high variance option when the common mean net rate of energy gain is positive. When the

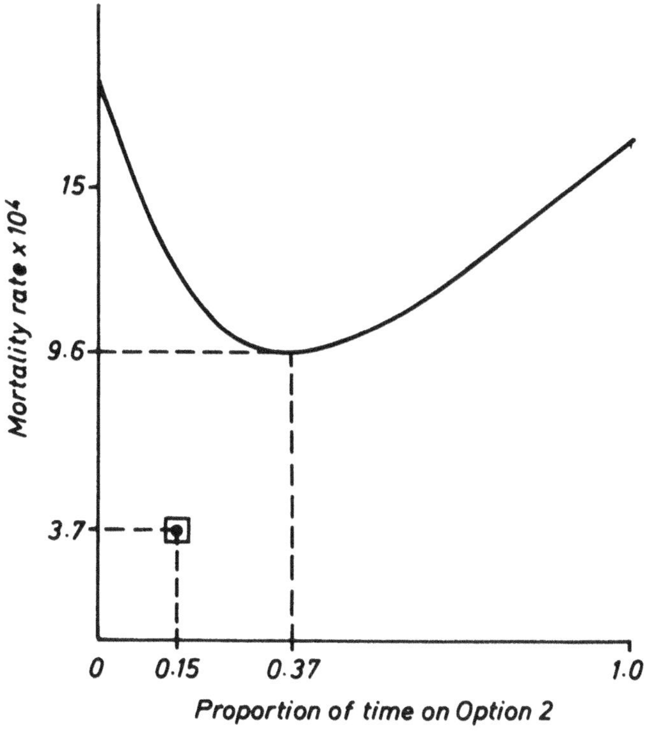

Figure 2. The solid curve gives the rate of total mortality as a function of the non-state dependent policy used. For comparison the mortality rate under the optimal (state-dependent) policy is shown. The options available to the animal and other parameters are as for table 1.

mean net gain is negative it can be optimal to choose the high variance option at low reserves, and the strength of selection pressure on such a choice can be high.

To investigate selection pressure in the present setting we allow an animal two behavioural options which differ in p_i and z_i. We then compare the rate of mortality under the optimal policy with that under a class of non-state dependent policies. For given p with $0 \leq p \leq 1$ we consider the policy which (for every state) chooses Option 1 with probability 1 - p and choose Option 2 with probability p. Thus if p = 0 Option 1 is chosen exclusively and if p = 1 Option 2 is chosen exclusively. Figure 2 plots total rate of mortality as a function of p in the case of Example 1. It can be seen that the best proportion of time to spend choosing Option 2 is 0.37. The mortality rate under this policy is 9.6×10^{-4}. This can be compared with the optimal policy under which the animal spends a proportion 0.15 of its time choosing Option 2 and suffers a mortality rate of 3.7×10^{-4}. Thus the mortality under the optimal policy is much lower than that under the best non-state dependent policy.

To obtain some idea on how rapidly state dependent behaviour might evolve, let us suppose that the extended non-reproductive period we are considering is a winter. We can consider a large population of animals independently following a behavioural policy, and look at the percentage surviving at the end of winter. Table 4 shows this percentage for four behavioural policies as a function of the length of winter. It is clear that the selection pressure to adopt state dependent behaviour can be enormous, even over just one year.

Table 4. Percentage of a large population surviving the winter as a function of the length of winter. Four policies are compared: use Option 1 alone; use Option 2 alone; the best non-state dependent policy (see figure 2); and the optimal state dependent policy. Option parameters as for Table 1.

N° of intervals in winter	percentage survival			
	Option 1	Option 2	best mix	optimal
500	39.4	43.4	61.9	83.3
1000	15.5	18.9	38.3	69.3
2000	2.4	3.6	14.6	48.1
3000	0.06	0.13	2.1	23.1

Other examples illustrating selection pressure are given in McNamara and Houston (in press, b). It is clear that it really does matter, not just how often an option is chosen, but in which state the option is chosen.

Starvation And Predation Levels

Starvation and predation are both factors which could potentially limit population size. In this section I argue that one can not make inferences about their relative importance as limiting factors from relative mortality levels. More strongly, I would like to argue that it is in general not meaningful to talk about relative importance. This is because the two sources of mortality are coupled together; animals may take risks in terms of predators to obtain food, or may forgo food to avoid predators. Because of this coupling one can not isolate starvation from predation.

Although this point has been made before (Lack, 1954), it can be seen most clearly by using simple models. I will present a schematic general model which can be used to analyse the starvation predation trade-off. I will then use the survival model developed above to illustrate a variety of effects that can be produced by the trade-off. Some of these effects are at first sight counter-intuitive.

The optimal trade-off

McNamara and Houston (1987b) present a simple model of the starvation-predation trade-off for an animal which must build up its reserves in order to survive a period without food. One can put the essence of their argument in a more general setting by supposing that the animal can control its behaviour by its choice of a single control variable u. The choice of u determines the rate of mortality from starvation S(u) and the rate of mortality from predation P(u). To capture the idea of a trade-off we assume that S(u) decreases and P(u) increases with increasing u. I give some examples of u based on the trade-offs given in the Introduction.

(a) If food gained and predation risk both increase with time spent foraging, one could take u to be the proportion of a day allocated to foraging.

(b) Suppose an animal has a source of food near cover which is safe from predators, and a better source of food away from cover which is risky. Then under an optimal policy the animal should choose the safe option when its reserves are high and switch to the risky option when its reserves fall below some critical threshold u^*. In this case the control variable u can be taken to be the level below which the animal switches to the risky option.

(c) Let 1 - u be the proportion of time a foraging animal spends looking around for predators.

(d) Let u be the body weight which a small bird attempts to maintain.

We can write the total mortality rate as

$$M(u) = S(u) + P(u).$$

The best choice u^* of u minimises total mortality. Thus

$$-S'(u^*) = P'(u^*).$$

The levels of starvation and predation at this optimal trade-off are $S(u^*)$ and $P(u^*)$ respectively. From the above equation it can be seen that there is no *a priori* reason why these quantities should be comparable.

Alasdair Houston and I have investigated relative starvation and predation levels at the optimal trade-off for a number of different models. McNamara and Houston (1987b) present a simple model of the decision problem facing an animal which must build up its energy reserves to survive a lean period. McNamara and Houston (in press, a) model the optimal body mass for an over wintering 19g passerine. In both of these cases the parameter values used are such that S(u) decreases rapidly and P(u) increases slowly with increasing u. Consequently, starvation levels are much lower than predation levels at the optimal trade-off (ie $S(u^*) \ll P(u^*)$). Other ranges of parameter values could reverse this result.

I will now investigate mortality levels for the model of an animal surviving an extended non-reproductive period developed in this paper. Suppose that the animal has two options. Option 1 yields energy at mean net rate μ_1 and has no predation risk ($z_1 = 0$). Option 2 yields energy at mean net rate μ_2 and has a predation risk of $z_2 = 0.002$. We look at the effect of increasing μ_2. Figure 3 shows the resulting levels of starvation

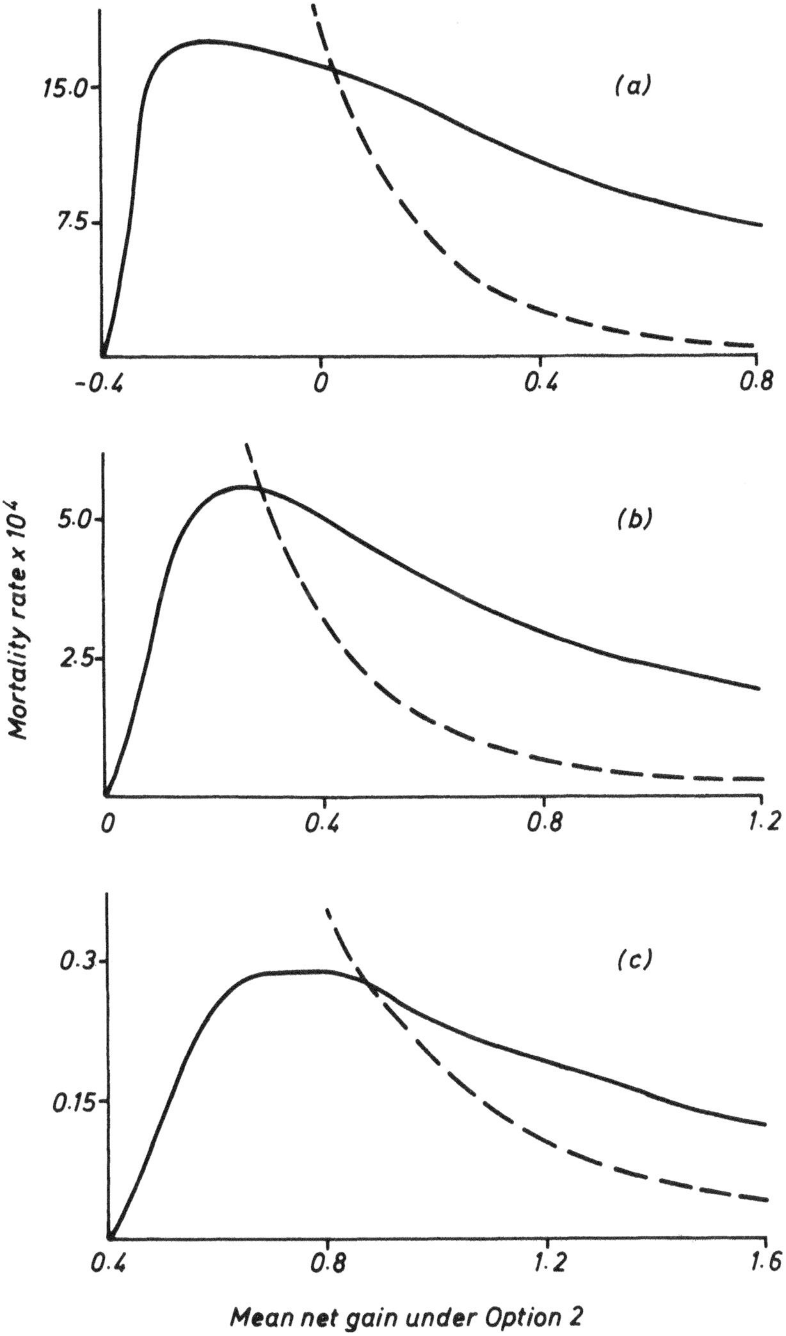

Figure 3. Starvation rates (dashed lines) and predation rates (solid lines) as a function of the mean net rate of gain under Option 2, μ_2. There are two options. Predation risks are $z_1 = 0$ and $z_2 = 0.002$. The mean net rate of gain under Option 1 is (a) $\mu_1 = -0.4$, (b) $\mu_1 = 0$, and (c) $\mu_1 = 0.4$. (L = 100, d = 4, E = 8.)

and predation for three values of μ_1. Clearly Option 2 should never be used when $\mu_2 < \mu_1$. In each case shown in the figure we look at values of μ_2 greater than μ_1. When $\mu_2 > \mu_1$ Option 2 should be used when reserves are sufficiently low. For small $\mu_2 - \mu_1$ starvation rate is much higher than predation rate. As $\mu_2 - \mu_1$ increases starvation rate drops well below predation rate in each case. Although absolute levels of mortality are very different for the three cases it can be see that relative levels, ie $S(u^*)/P(u^*)$, are comparable for given $\mu_2 - \mu_1$. These relative levels are also remarkably robust under changes in the predation level under Option 2. For example when $\mu_1 = 0$ and $\mu_2 = 1.2$ $S(u^*)/P(u^*) = 0.124$ when $z_2 = 0.002$ (figure 3(b)). Increasing z_2 to 0.01 changes the ratio to 0.162. Decreasing z_2 to 0.0002 changes the ratio to 0.134.

This type of example shows that one can not make inferences about the relative importance of starvation and predation from observed mortality levels. Drawing inferences from changing levels of mortality is also a dubious procedure. In the model of McNamara and Houston (1987b) an animal must build up its reserves to a level G by a given time if it is to survive. It is shown that increasing G can lead to a decrease in the starvation level $S(u^*)$ at the optimal trade-off.

To illustrate similar effects for the survival model developed here, suppose that an animal has two options. Option 1 has mean net rate of gain μ_1 and no predation risk ($z_1 = 0$). Option 2 has mean net rate of gain $\mu_2 = 0.8$ and a non-zero predation risk z_2. Figure 4 shows the dependence of starvation rate on μ_1 for three value of z_2. It can be seen that as μ_1 increases, and hence food in the environment increases, starvation initially increases. This effect can be understood as follows. When $\mu_1 = -0.8$ Option 1 is so poor that animals spend almost all their time using Option 2. This option gives a good supply of food and hence the rate of starvation is low, although the rate of predation is high. As μ_1 increases, to say $\mu_1 = 0$, an animal spends a much greater proportion of its time taking the predator-free Option 1. Since the mean net gain on this option is less than under Option 2, starvation increases. Of course, the total mortality $M(u^*) = S(u^*) + P(u^*)$ decreases with increasing μ_1.

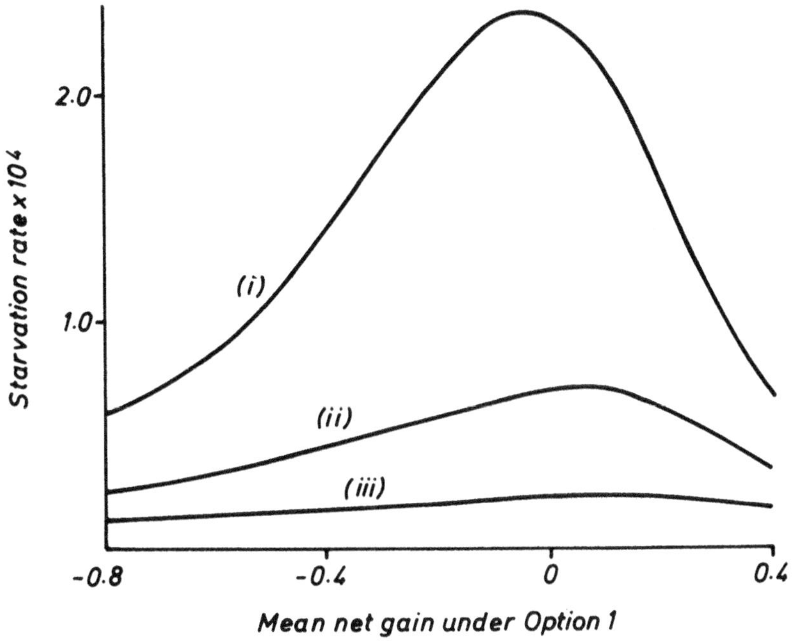

Figure 4. Effect of the mean net rate of gain under the safe alternative on the rate of starvation. There are two options. Option 1 is safe ($z_1 = 0$). Option 2 has mean net rate of gain $\mu_2 = 0.8$. The predation risk under this option is (i) $z_2 = 0.01$, (ii) $z_2 = 0.002$, and (iii) $z_2 = 0.0004$. ($L = 100$, $d = 4$, $E = 8$).

In figure 5 the animal again has the choice between two options. Option 1 yields a mean net rate of energy gain of $\mu_1 = 0$ and has no predation risk ($z_1 = 0$). The mean net gain under Option 2, μ_2, is positive. The figure shows how the level of predation changes as the predation risk under Option 2 increases. Increasing z_2 initially results in a greater level of mortality due to predation because of the increased risk. As z_2 increases further there is a rapid decline in the proportion of time an animal spends performing Option 2. Consequently predation levels drop, although starvation levels and total mortality increase. Since it is always optimal to choose Option 2 for very low reserves, increasing z_2 still further increases predation levels again. This can only be seen in curve (i) here.

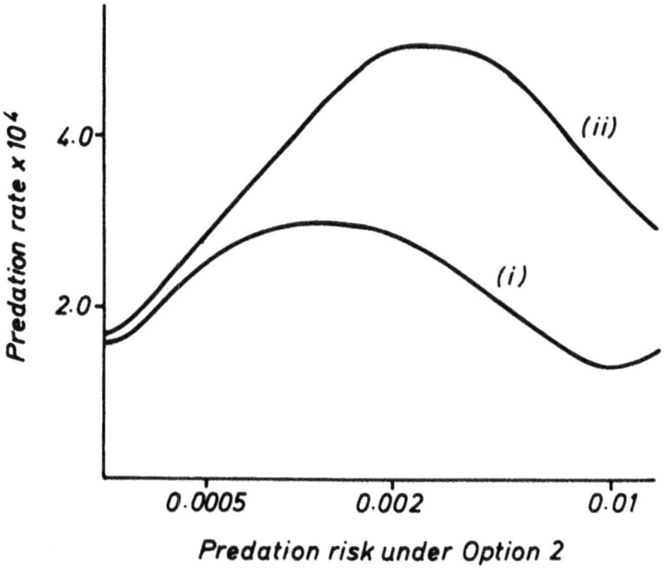

Figure 5. Effect of the predation risk under the risky alternative on the rate of predation. There are two options. Option 1 has $\mu_1 = 0$ and $z_1 = 0$. The mean net rate of gain under Option 2 is (i) $\mu_2 = 0.08$, and (ii) $\mu_2 = 0.16$. ($L = 100$, $d = 4$, $E = 8.$).

Further Remarks

The models which have been presented deal with an animal whose optimal behaviour does not depend on the behaviour of other members of the same population. Once there is a dependence one must seek a game theoretical solution rather than a simple optimum. McNamara and Houston (in press, c) consider a population in which each member has a choice of foraging area, where the food obtained and predation risk in any area depend on the number of other animals using the area. The game theoretical solution is now an evolutionary stable dynamic strategy. (See Houston and McNamara, 1987, 1988b). Following the strategy results in a state-dependent ideal free distribution for the population.

Once density dependent effects are introduced into the model of surviving an extended period one can use it to look at population dynamic questions. The model explicitly includes predation, and the relative levels of mortality from starvation and predation depend crucially on behaviour. It is therefore a natural setting in which to examine the influence of individual behaviour on the dynamics of a predator-prey system. Alasdair Houston and I are currently analysing this influence.

Acknowledgement

I should like to thank Alasdair Houston for his comments on this paper.

References

Caraco T, Martindale S, Whittam TS (1980) An empirical demonstration of risk sensitive foraging preferences. Anim Behav **28**: 820-830.

Houston AI, Kacelnik A, McNamara JM (1982) Some learning rules for acquiring information. In: *Functional Ontogeny*, ed McFarland DJ. Pitman, London.

Houston AI, McNamara JM (1986) Evaluating the selection pressure on foraging decisions. In: *Relevance of models and theories in ethology*, ed Campan R, Zayan R. Privat, Toulouse.

Houston AI, McNamara JM (1987) Singing to attract a mate - a stochastic dynamic game. J Theor Biol **129**: 57-68.

Houston AI, McNamara JM (1988a) A framework for the functional analysis of behaviour. Behav and Brain Sci **11**: 117-163.

Houston AI, McNamara JM (1988b) Fighting for food: a dynamic version of the Hawk-Dove game. Evolutionary Ecology **2**: 51-64.

Houston AI, McNamara JM (1989) The value of food: effects of open and closed economies. Anim Behav **37**: 546-562.

Kagel JH, Green L, Caraco T (1986) When foragers discount the future: Constraint or adaptation ? Anim Behav **34**: 271-283.

Lack D (1954) *The Regulation of Animal Numbers*. Clarendon Press, Oxford.

Lima SL (1986) Predation risk and unpredictable feeding conditions: determinants of body mass in birds. Ecology **67**: 377-385.

McNamara JM (in press) The policy which maximises long term survival of an animal faced with the risks of starvation and predation. J Appl Prob.

McNamara JM, Houston AI (1980) The application of statistical decision theory to animal behaviour. J Theor Biol **85**: 673-690.

McNamara JM, Houston AI (1986) The common currency for behavioural decisions. Am Nat **127**: 358-378.

McNamara JM, Houston AI (1987a) A general framework for understanding the effects of variability and interruptions on foraging behaviour. Acta Biotheoretica **36**: 3-22.

McNamara JM, Houston AI (1987b) Starvation and predation as factors limiting population size. Ecology **68**, 1515-1519.

McNamara JM, Houston AI (in press, a) The value of fat reserves and the trade off between starvation and predation. Acta Biotheoretica.

McNamara JM, Houston AI (in press, b) Starvation and predation in a patchy environment. In: *Living in Patchy Environment*, ed Shorrocks B, Swingland I. Oxford University Press, Oxford.

McNamara JM, Houston AI (in press, c) State dependent ideal free distributions.

Milinski M, Heller R (1978) Influence of a predator on the optimal foraging behaviour of sticklebacks (*Gasterosteus aculeatus*) Nature **275**: 642-644.

Pulliam HR, Pyke GH, Caraco T (1982) The scanning behaviour of juncos: a game-theoretical approach. J Theor Biol **95**: 89-103.

Sih A (1980) Optimal behaviour: can foragers balance two conflicting demands ? Science **210**: 1041-1043.

Sih A (1982) Foraging strategies and the avoidance of predation by an aquatic insect, *Notonecta hoffmanni*. Ecology **63**: 786-796.

ORAL DISCUSSION – J. McNamara

Q. (Kacelnik): If you increase the mean food supply in the safe patch, starvation increases; does the mean fitness of the population go up ?
A. Total mortality goes down because predation plummets, but starvation goes up.

Q. (Real): Is it fair to experimentally presume that a model is correct ? This also pertains to subjective estimates of probability, and use of the observed patterns of behaviour to try to construct the subjective estimate of predation risk. *You* know there is no risk, but the animal thinks there must be some risk. How do we measure this subjective estimate of risk ? Do we have any alternative to estimating lambda from its behaviour ?
A. It is dubious procedure to assume things are optimal and then make inferences from it. But if one does that, it would be good to have a method of cross checking, i.e. by maintaining subjective predation risk but altering other parameters, say food, and see what the effect is. In the laboratory, animals don't only worry about predation, they worry about lots of other things which you may not be able to characterise. It would be therefore difficult to measure subjective predation risk because you would also be measuring subjective estimates of many other things.

Q. (Sih): Could you compare natural populations, which you know have different levels of predation risk, and see if they have different levels of risk they are willing to take in the lab ? How can you measure predation ? It is very difficult in the field.
A. The take–home–message may be not to try to quantify subjective levels of predation risk, but just to be aware of situations in which they might be important i.e. if the animal has low levels of reserves then predation risk may not be important, but it may be very important where the animal has a high level of resources.

TIMING MECHANISMS IN OPTIMAL FORAGING: SOME APPLICATIONS OF SCALAR EXPECTANCY THEORY

Alejandro Kacelnik, Dani Brunner[1] and John Gibbon[2]
King's College Research Centre
King's College
Cambridge CB2 1ST
UK

INTRODUCTION

Uncertainty and its effects on optimal foraging predictions have been extensively discussed (for a review see Stephens & Krebs 1986). Uncertainty has usually been assumed to result from stochastic properties of the environment (Caraco 1980; McNamara & Houston 1980) or from transient incomplete knowledge while learning proceeds (Krebs, Kacelnik & Taylor 1978; McNamara & Houston 1985; Shettleworth, Krebs, Stephens & Gibbon 1988). Nevertheless, uncertainty in foraging may in addition originate from intrinsic properties of the predator's information-processing capabilities. Here we consider foraging problems of ambiguity in the psychological system responsible for the perception, storage and retrieval of information about the duration of time intervals.

As an illustration, we consider a foraging problem discussed by Davies (1977), and show that assuming "cognitive" ambiguity (we use the term cognitive in a loose sense to refer to information processing by the forager) leads to qualitatively new optimal foraging predictions. It is not our intention to belabour this particular foraging example, but rather to use it as a springboard to link a natural foraging problem to the psychology of timing. Other aspects of the same example have been related to timing by Roberts (1983).

The spotted flycatcher

Davies (1977) studied foraging by spotted flycatchers (*Muscicapa striata*) hunting for flying insects from fixed perching sites. Typically, the bird waits until an insect is within range, then makes a foray to catch the insect in mid-air and returns to the same or a nearby perch to start a new cycle. Occasionally the flycatcher leaves the site to travel to a new perch where it resumes the same hunting

[1]Department of Experimental Psychology, Cambridge University, Cambridge, UK
[2]N.Y. State Psychiatric University and Columbia University, NY, USA

NATO ASI Series, Vol. G 20
Behavioural Mechanisms of Food Selection
Edited by R.N. Hughes

strategy. Among other problems, Davies examined the factors involved in the change of foraging site. His main observations were: a) number of prey captured from each site before departure had an approximately exponential distribution, indicating that departure occurred with roughly equal probability after various numbers of prey had been obtained; b) average inter-catch interval, S(S ≈ 20 seconds) was independent of the number of prey already captured in a site, indicating that there was no gradual resource depression (Charnov, Orians & Hyatt 1976; Stephens & Krebs 1986); c) unsuccessful waiting time before abandonment of a perch (the giving up time, GUT) was longer than the average inter-capture interval (mean GUT ≈ 30 seconds). As a first approximation, it makes sense to expect the GUT to be longer than S, as this allows detection of a site's exhaustion, but a number of interesting issues arise. In his description of the field data Davies reported that

> "On average the bird would wait about 1.5 x the inter-catch interval before it left for another perch. This suggests that the flycatcher has a time expectancy and leaves after a certain time if a suitable prey has not come along (see also Krebs 1974).".

Since only one mean inter-catch interval was observed, one cannot accept the results as anything else than a suggestion for the existence of a time expectancy[1]. Waiting time may be independent of inter-capture interval and inter-site travel time. Further, even if the existence of a time expectancy is accepted, while the description of GUT as being "1.5 x the inter-catch interval" is accurate for this single value of inter-catch interval, the suggested proportionality cannot be assessed with a single habitat type. Giving up times are pervasive in foraging theory and thus the existence of timing ability and the generality of the proportional relation between GUT and inter-catch interval deserve attention.

An idealized foraging problem

In order to stress the importance of cognitive ambiguity, we now examine an abstraction of the flycatcher problem. We assume that inter-capture interval is fixed, rather than randomly distributed around its mean as in the natural example. Our discussion is intuitive rather than absolutely rigorous, as a more complete discussion will be presented elsewhere. We shall illustrate our ideas with data obtained from starlings (*Sturnus vulgaris*) foraging in the laboratory.

The following properties define the idealized foraging environment:

[1] Expectancy is a controversial and complex concept. Here we use the colloquial meaning implied by Davies as indicating that flycatcher's behaviour is adjusted to the experienced inter-catch interval. For a review of the concept of expectancy in animal psychology see Dickinson (in Press).

a. Fixed inter-prey intervals, S. This embodies two properties: lack of gradual resource depression (i.e. no progressive increase in inter-capture interval, Stephens & Krebs 1986) and deterministic intervals between encounters.

b. Stochastic abrupt depletion, with constant probability (p) of sudden patch exhaustion after each prey encounter or after arrival in the patch.

c. Habitats may differ in S and in between-patch travel times.

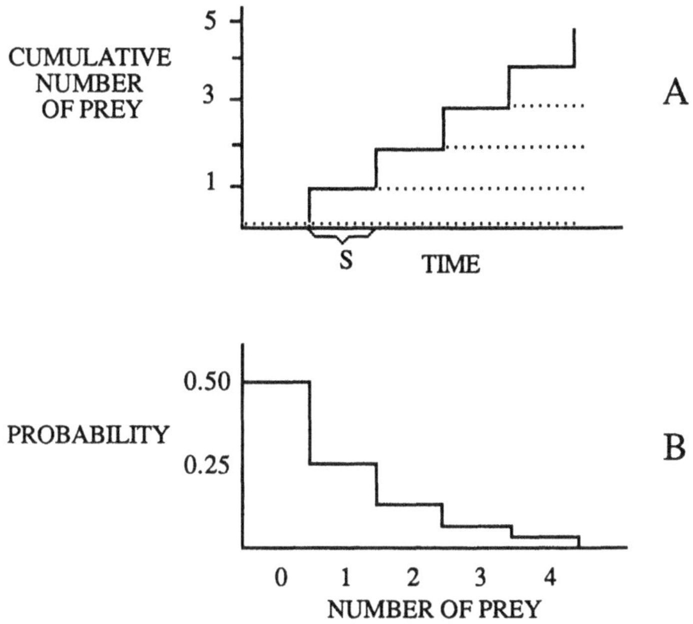

Figure 1: Idealized foraging environment. A: Cumulative gain as a function of time in the patch. The solid line shows an unusually long-lived patch. Dotted lines indicate the possibility that a patch gets exhausted after any number of prey. Inter-prey interval S is fixed. B: Frequency distribution of patches containing various numbers of prey. As probability of sudden patch depletion is constant after 0,1,2,...,m prey, this distribution is geometric. The expected number of prey to be obtained after each capture (n) is the mean of this distribution.

Figure 1 summarizes these properties. According to the marginal value theorem (MVT, Charnov 1976; McNamara 1982; Stephens & Krebs 1986) overall rate of energy gain in an environment with depleting patches is maximized if patches are left when the expected instantaneous rate in the current patch falls below the overall rate of gain for the environment. Instantaneous rate of gain is a problematic variable when prey come in discrete lumps, but the principle can be formulated using "local" expected rate in the current patch at time t, R(t), instead of instantaneous rate. R(t) is defined as the expected number of forthcoming captures in the current patch divided by the interval between t and the expected time at which the patch will be abandoned. In patches with renewal after each capture, t is the time since reaching the patch or since the last capture, whichever is shorter.

Unambiguous timing.

If there were no ambiguities in time estimates, the optimal departure rule would be to abandon the patch as soon as depletion is detected, namely when a prey fails to be encountered after waiting for S seconds. Local expected gain at the time of entering a patch or just after catching a prey (t=0) is:

$$R(o) = \frac{n}{(n+1)\,S} \qquad (1)$$

n: expected number of prey per patch; S: inter-prey-interval. The denominator is the time taken to collect the average number of prey plus the minimum time to detect patch exhaustion, namely a full interval S.

As the animal waits for food, expected local rate in the current patch is given by:

$$R(t) = \frac{n\,P(t)}{(S-t) + n\,S} \qquad (2)$$

where P(t) is a function that incorporates the sudden abrupt depletion; $P(t) = 1$ for $t < S$ and $P(t) = 0$ for $t \geq S$. The time in the denominator is now decomposed in two terms: the expected interval between the current instant and the next capture (S-t) and the expected time between the forthcoming capture and patch departure (nS). Equation 2 shows that because we consider fixed interval patches, expected rate in the current patch grows with t as S is approached and drops abruptly to zero when t = S because of the step transition in P(t). This is shown in figure 2.

Expected pay off in the environment (γ) evaluated at the time of giving up one patch, before any travel has been done is:

$$\gamma = \frac{n}{(n+1)S + \text{Travel}} \qquad (3)$$

Optimal departure time (GUT*) is the time at which local rate R(t) becomes lower than global rate γ. This implies that GUT* = S, for all values of γ. Notice that GUT* is independent of travel time.

Ambiguous timing

Now consider the same problem when there is ambiguity about the estimate of S. For the moment, we gloss over potential sources of this ambiguity, and simply assume that the animal must base its foraging decisions on a continuous subjective function f(t) that describes the probability that the waiting time t equals the environmental fixed interval S. The problem is depicted in figure 3A.

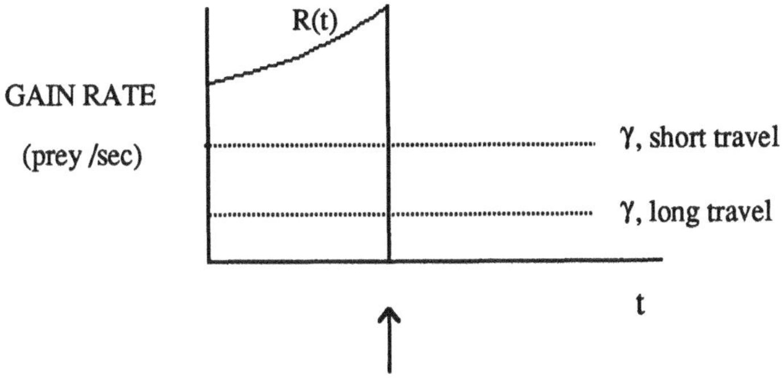

Figure 2: Unambiguous timing. Expected rates of gain as a function of waiting time t since arrival at the patch or since the last capture, whichever is shorter. Solid line is the expected rate in the current patch R(t). Dotted lines show expected rate of gain in the environment as a whole for two travel times. The hyperbolic increase in local expectation for t < S is due to the fixed S. Expected local rate falls abruptly at t = S, and thus it becomes lower than expected overall gain at a point that is independent of travel time.

In this case, because animals are uncertain about the exact moment at which t = S, instead of using the function P(t) they may estimate the probability that patch exhaustion has occurred by:

$$F(t) = \int_0^t f(t)\, dt \qquad (4)$$

The estimated probability that t has not yet exceeded the standard interval S is given 1-F(t) (Fig 3B).

The re-formulation of equation 2 to calculate local rate for a forager with time ambiguity ($R_S(t)$) offers some difficulties. The numerator of equation 2 can be transformed by replacing P(t) for 1-F(t), and n for n', defined as the average number of captures per patch experienced by the ambiguous timer. The transformation of the denominator, or expected time until departure, is more complex. Expected time until the potential forthcoming reinforcer must be a decreasing, positive function of t, with shape that depends on the process by which f(t) is sampled by the subject. The giving up time GUT will not equal S as before, as the subject will show variability and GUT will be an average with potential bias above or below S. This results in:

$$R_s(t) = \frac{n' (1-F(t))}{d(t) + (n'-1)S + GUT} \qquad (5)$$

where d(t) is the expected delay between t and the forthcoming capture. A rigorous expression for d(t) requires explicit assumptions for the process of sampling from f(t), but in order to show graphically the consequences of ambiguity we approximate d(t) by assuming that d(t) = S-t for t<S and d(t) = 0 after that. With this approximation, beyond the point when t = S the equation gives expected local rate assuming that the next prey will arrive "right now" if it arrives at all. We are further assuming that the time left in the patch beyond the forthcoming capture is assessed, on average, without error.

$R_s(t)$, the subjective local rate, is continuous rather than showing an abrupt fall to zero at t=S as in equation 3. This is shown schematically in figure 3C.

Optimal departure, as usual, occurs when $R_s(t) < \gamma$, but because $R_s(t)$ drops gradually, GUT* depends on γ, which in turn depends on travel time (Figure 3c). The presence of a hump in $R_s(t)$ is not a function of our approximation for d(t), but its exact shape is. The hump would not be present if inter-capture intervals were completely random rather than fixed[1], but the gradual decline would still be preserved.

Including subjective ambiguity about the inter-catch intervals leads to the qualitatively new prediction that in the absence of gradual resource depression and with deterministic intervals between captures, longer giving up times should be observed in habitats with longer travel times. Further, optimal departure time is a function of the shape of f(t), a subjective entity. As we will show later, this subjective distribution is also relevant to the question of the putative proportionality between GUT and S. We take this dependency of the optimal policy on f(t) as the main justification for looking into mechanisms of information processing with the prospect of refining optimal foraging hypotheses.

In the following section we revise some psychological ideas about behavioural timing which are relevant to foraging and present a preliminary report of foraging experiments using starlings in the laboratory that aim at testing these concepts. We restrict our review to Scalar Expectancy Theory (SET), (Gibbon 1977; Gibbon & Allan 1984; Gibbon, Church, Fairhurst & Kacelnik 1988). Ours is a sketchy, foraging-oriented outline and readers with a serious interest in the psychology of timing are referred to the original papers cited in the text.

---

[1] Notice that local and global rates are a partial function of GUT while the choice of optimal GUT depends on the comparison between these rates. This apparent circularity can be solved by iterative processes of adjustment of the value of γ and GUT. Similar problems are discussed by McNamara and Houston (1985).

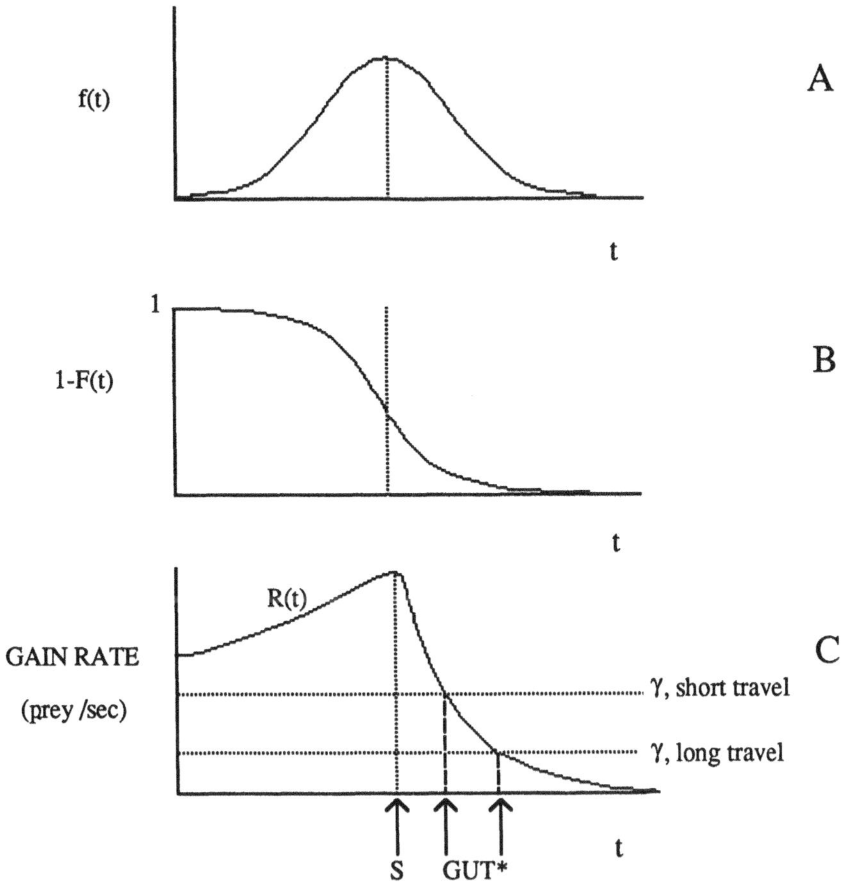

Figure 3: Ambiguous timing. A: estimated probability of t = S. B: Probability that the time of patch exhaustion has not been exceeded. C: expected rates of gain. The solid line is the local estimated rate and the two dotted lines are overall rates of gain for two environments which differ in travel time. See text for details. In contrast with Figure 2, the times at which local rate falls below environmental gain (GUT*) depends on travel time.

SCALAR EXPECTANCY THEORY

SET was initially formulated in terms of intervening motivational variables (local and background reward expectancies) and a decision rule (Gibbon 1977). Later, Gibbon and co-workers (Gibbon , Church & Meck 1984) incorporated structural components such as clock and memory processes which had been used in pre-existent psychophysical literature (Treisman 1963) and for which there is independent empirical evidence. Extensive work has been done in identifying the properties of these components (Roberts 1983; Church & Meck 1988). Of special interest for foraging theory is the analysis of sources of informational ambiguity and its effect on choice between alternative actions (Gibbon & Church 1984; Gibbon, Church & Meck 1984; Gibbon et al. 1988). Some essential features of the theory will now be described with reference to fixed interval (FI) schedules of reinforcement, which have some similarity with the foraging example described above.

In FI schedules, an event signals the initiation of a trial and a food reinforcement is contingent on the first response (such as a peck at an illuminated key) after a fixed interval. We call this fixed interval S to emphasize the similarity with the foraging example. Although responses executed before waiting for S seconds have no programmed consequence, subjects normally respond *before* S has elapsed, but not immediately after the initiating signal. Usually subjects wait for some time (often responding at a low background rate) shorter than S before starting to respond at a high rate (the FI "scallop"). Waiting time before the onset of fast responding, called the response breakpoint, is approximately proportional to S (Ferster & Skinner 1957; Schneider 1969; Mackintosh 1974, Staddon & Ettinger 1989). This dependency of waiting on S is evidence for some time discrimination.

SET was first applied to the timing of the response breakpoint in FI schedules. According to Gibbon (1977), the -subjective- instantaneous value of an event such as obtaining a food item, termed its expectancy, equals the event's total motivational value, say H, spread over the estimated time until its occurrence. Although as we have pointed out expectancy can be used with different meanings, we follow this notation from now on. In an FI, assuming that the total value of a food reward is H, that the subject has an accurate estimate of the value of S and that a time t has elapsed since the initiating signal, then instantaneous (local) expectancy h(t) is:

$$h(t) = \frac{H}{(S-t)}. \qquad (6)$$

Gibbon postulated that active responding starts when local expectancy h(t) exceeds a threshold and that this threshold is a function of the background expectancy of reinforcement in the environment or the expectancy when the trial starts, $(h(o) = \frac{H}{S})$. The similarity between equation 6 and equations 2 and 3 is apparent, and it is tempting to relate the hypothetical subjective variable expectancy to expected rate of energy gain. Local expectancy h(t) could be related to expected rate in the current patch (Eq. 2) and background expectancy h(o) to average rate in the environment, γ (Eq. 3). To establish an analogy between h(o) and γ requires including the inter-trial interval as an equivalent of travel time in the estimate of h(o). Since the MVT gives optimal patch departure time as a function of local and overall rates, by making this translation it is possible to compare the normative predictions of MVT with the predictions of SET based on the relation between local and global expectancies.

Experimental results in fixed interval and other schedules indicate that if the values of t and S in equation 6 are close to the real values (as opposed for example to some logarithmic transformation), then a ratio comparison rather than an absolute difference between expectancies must be involved in determining the breakpoint. Under a ratio comparison, in FI schedules the breakpoint should occur when

$$\frac{h(t)}{h(o)} > b \qquad (7)$$

b: the proportional improvement in local over background expectancy for a response "right now" to be subjectively worthwhile. b may be considered as random variable which fluctuates somewhat between trials. Although response cost is not usually mentioned in timing studies, some non-zero subjective cost is implicit in this formulation, as the subject needs an improvement over background reward probability to respond (i.e. b>1). The magnitude of the required improvement is a natural question for functional analysis, but in psychological studies it is usually treated as a free parameter that is adjusted to individual data.

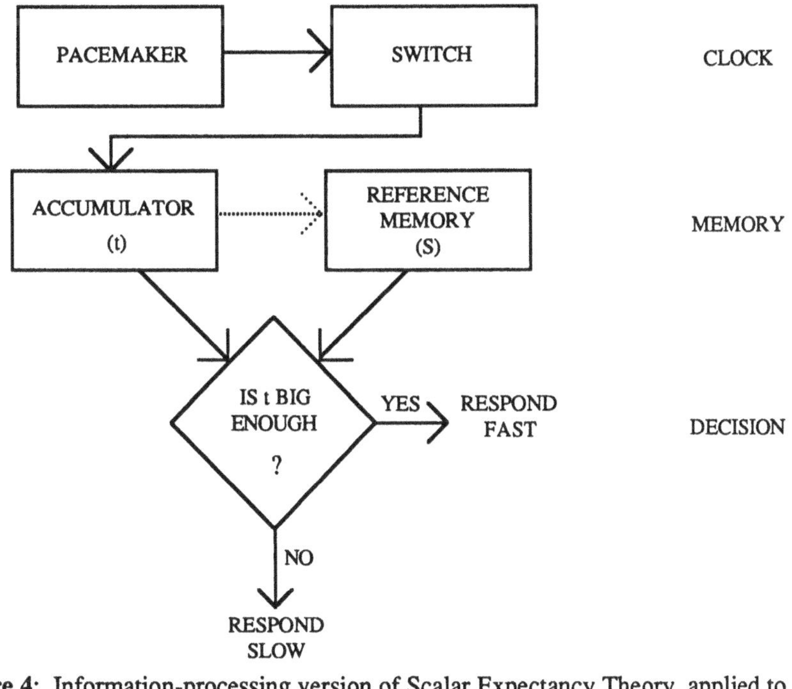

Figure 4: Information-processing version of Scalar Expectancy Theory, applied to the timing of onset of fast responding (the behavioural breakpoint) in fixed interval schedules of reinforcement. Re-drawn from Gibbon et al. 1984.

In order to use SET to study uncertainty and its sources, it is best to examine the theory in an information-processing formulation, instead of using motivational variables such as expectancy. This formulation is shown diagrammatically in figure 4 (Gibbon and Church 1984; Gibbon, Church & Meck 1984).

The model contains three subsystems:

A clock system allows for the measurement of elapsed times, using a pacemaker that delivers pulses with mean rate Λ and a switch that allows the clock output to be entered into an accumulator when appropriate.

The memory section of the model includes two subsystems. An accumulator stores the pulses passing through the switch and is reset to zero when a new trial starts, providing a running value for the estimated waiting time, which we call t' to distinguish from the real waiting time t. A reference memory holds information about average waiting time during previous cycles. The value contained in the accumulator before resetting is passed to reference memory when food is obtained. Reference memory contains a probability density function with mean S' formed by the storing of successive values of t' at the time of reinforcement. This function is all the information about the real value of S available to the subject.

Behaviour is generated by a decision process that involves comparing the values in both memory subsystems. When t' is sufficiently close to an estimate of S obtained from reference memory, active responding starts.

Notice that in this model comparisons are made between times and not between expectancies, i.e. the motivational value of the reinforcer does not appear in the decision process. When there is a single reward type and there is no inter-trial interval, after replacing h(t) and h(o) in equation 7 by their expressions as given in equation 6, the value of the reinforcer (H) drops out of the comparison ratio. In individual trials S is estimated by the subject using a sample s' from its reference memory. The breakpoint occurs when

$$\frac{1}{1 - \frac{t'}{s'}} = b \qquad (8)$$

If reference memory is formed without non-linear distortions in the transfer from the accumulator, then average breakpoint in real time will follow the relation

$$\frac{1}{1 - \frac{T}{S}} = B \qquad (9)$$

Where T is the real-time average value of the response breakpoint and B is the mean of the distribution of b, the decision threshold used in individual trials. Re-arranging,

$$T = S\,(1 - \frac{1}{B}) \qquad (10)$$

which shows that the assumptions embodied in the model generate a proportional relation between S and mean breakpoint (Gibbon 1977, Eq. 2). This proportionality depends on the assumed linearity between t' and S' and their respective real time variables. If there is more than one reward type and inter-trial interval is included, then h(o) must be based on average motivational value of all reward types in the environment and on average inter-reward time in the environment as a whole, taking into account inter-trial intervals. A decision based exclusively on the values of t' and S' could not show adjustments to reward types and inter-trial or travel intervals, unless the threshold itself incorporates some dependency on these factors.

The original formulation as a function of expectancies (Eq. 7) is perhaps closer to the intuition of foraging researchers than this purely time-based analysis, as it allows for examination of the value of the reinforcer and travel time as well as mapping to rate currencies more directly.

Sources of behavioural variability

When the breakpoint for fast responding in a single session of an FI experiment is compared between trials, some variability is observed. When variability is studied in experiments with various values of S, between-trial variability follows an orderly pattern: the frequency distribution of breakpoints becomes broader as S increases. Further, when breakpoints are plotted in relative time, that is when the frequency of breakpoints is plotted against the ratio t/S (Schneider 1969; Schneider & Neuringer 1972; Gibbon 1977) the frequency distributions obtained for different S have similar shapes. This is called the superposition property. In relative time breakpoints appear almost normally distributed, with a slight skew towards longer times. The coefficient of variation of breakpoints in absolute time in within-individual comparisons is nearly constant for different values of S, except for very small values. Gibbon, Church & Meck (1984) used the information processing model in figure 4 to seek forms of noise in the components that could account for the variability observed in experimental data.

The observed behavioural variability is compatible with assuming that the values stored in reference memory are normally distributed around S', that S' is linearly related to S and that the breakpoint in each trial is determined by picking a single random sample from the distribution of S'. This normal distribution in memory could be formed if there was a) random drift between trials in ticking rate of the pacemaker ; b) random drift between trials in closing and opening latencies of the switch and c) a noisy transfer process from working to reference memory. Given these properties, the observed results are also compatible with random variation between trials in the comparison threshold b. Other putative sources of variance can be eliminated. Perhaps surprisingly, the data are not compatible with a Poisson pacemaker within trials if the rest of the model is maintained. This is because if the pacemaker were Poisson with constant mean rate between trials, then variance in memory should increase more slowly with increasing interval length, and relative timing accuracy would then increase with S. This is not observed.

In addition to the previous conditions, for the model in figure 4 to account for the observed behavioural variability in fixed intervals, the contents of reference memory must be scalar, namely have constant coefficient of variation when S is changed. The scalar property of reference memory for time intervals is shown in figure 5. If this structure of the components of the model is maintained, the data on variability are compatible with a ratio comparison between t' and S' (equations 8 and 9), but not with a decision rule based on the absolute difference between t' and S'. Other possibilities for the properties of the system which would make absolute difference rules acceptable cannot be excluded (Killeen & Weiss 1987; Gibbon & Church 1981) but a full review of these options exceeds the scope of this report.

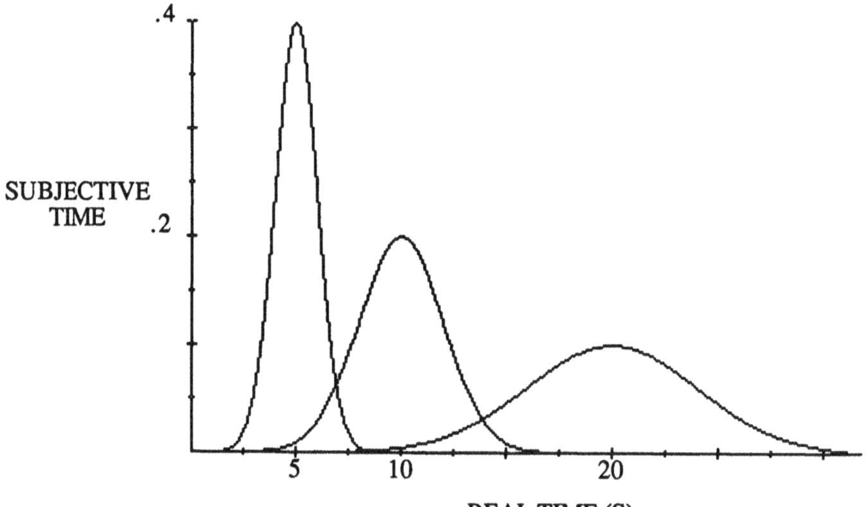

Figure 5: Postulated representation in reference memory of intervals of various duration. Subjective representation of time is assumed to be normally distributed around the real time interval, with constant coefficient of variation. This is the definition of the scalar property in memory for time.

Beyond Fixed intervals

SET has been applied to a variety of timing problems. Of special interest with respect to foraging are the accounts of choice between variable and fixed intervals of same average length. Gibbon et al. (1988) discuss a variety of experiments in which SET successfully accounts for choices between delayed reinforcers with or without variability. Their approach is normative but constrained by explicit information processing mechanisms, in contrast with the normative and unconstrained functional analyses in the risk-sensitivity literature (Caraco 1980; Caraco, Martindale & Whitham 1980). Gibbon and co-workers found that in a variety of experimental procedures animals chose between options as if they remembered a range of distribution of delays associated with the memory for each alternative and responded by picking the minimum delay in comparisons between just one or a small number of samples from each alternative. The crucial variable was thus not variability per se, but the pattern of remembered delays to reinforcement as caused by the third moment of the experienced distributions. Gibbon et al. discussed open but not closed economy experiments and variability in time but not in reward size, and thus the full implications of their findings for risk-sensitivity foraging cannot yet be assessed.

Is SET applicable to the problem of patch departure?

Our short review of SET addressed the onset of responding in fixed intervals. In contrast, the timing problem in the foraging situation with which we started this paper is about when to give up a given patch. We believe that if the information processing system identified in psychological research proved to have generality, important consequences may follow for optimal foraging theorizing. This is because animals processing information according to a system similar to the one depicted in figure 4 are effectively facing a different problem than a subject with perfect perception of elapsed intervals.

In the following section we present a preliminary report of a study of giving up times in starlings in the laboratory. Full details of this work will be presented elsewhere (Brunner, PhD dissertation, in prep). Our aim is to test if the behaviour of starlings foraging in a simulation of a patchy environment in the laboratory follows the predictions of SET. This would be evidence in favour of the generality of this account of timing and it would suggest that further applications of this mechanistic theory might be important for foraging studies.

Giving up times and Scalar Expectancy Theory: a laboratory study with starlings

The features of the abstraction of the flycatcher problem described in the Introduction were simulated in an experimental set up in the laboratory. Six wild caught starlings were trained to obtain food in a closed economy experiment by pecking at a key in a panel. The foraging environment was as follows. Upon "reaching a patch" (an event simulated by the lighting up of the pecking key) the first peck triggered a timer, and in half of the occasions 7 seconds of access to a feeder were allowed at the first peck after an interval S had elapsed. When a reward was obtained, pecking again at the key restarted the timer and a new interval started. In the remaining half of the occasions no rewards were given and the key stayed illuminated until the bird started to "travel" to a new patch. New patches could be sought at any time by completing 20 flights between two perches placed approximately 1m apart. On completion of a travel, the key was illuminated again and a new patch cycle could be started. An important difference between this environment and Davies' spotted flycatchers is that in the laboratory the interval between captures was fixed within each patch, instead of randomly distributed as in the field. In addition, in the laboratory the birds could not detect patch exhaustion other than by timing (no prey being offered after S had elapsed), while in the field the flycatchers may have used additional visual cues as well as waiting time to detect movement of the insect "batches". Uncertainty in the laboratory was only due to timing ability of the animal, as required to distinguish between the unambiguous and ambiguous timing models presented in figures 2 and 3. Because food could only be obtained by pecking at the key when food was due, giving up can be expressed in two forms: the *giving in time* (GIT), marked by the last attempt to get food by pecking in a patch, and the *moving on time* (MOT), namely the time of starting travel to a new patch. GIT is probably a better indicator of time expectancy, while MOT reflects more directly motivation towards seeking new patches.

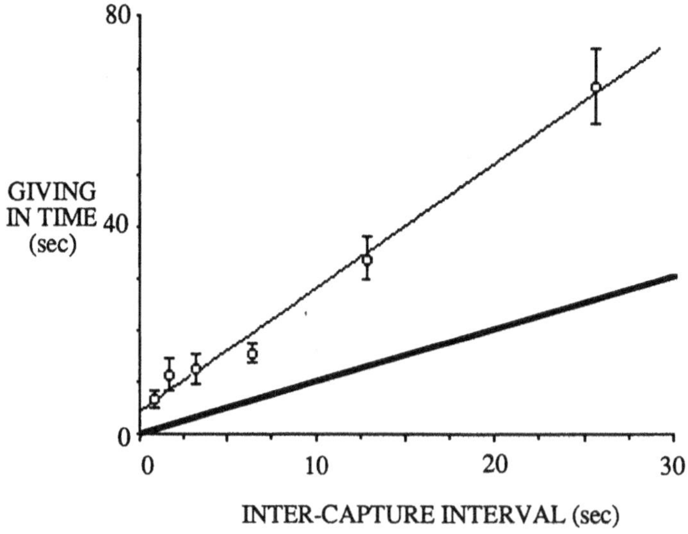

Figure 6: Average giving up time as a function of inter capture interval. GUT is measured as the last attempt to obtain food in a given patch (GIT). Each point is the mean of the individual means of six starlings. Error bars are the standard errors between individual means. The solid line is the line of equality, or the predicted GUT under unambiguous timing.

Figure 6 shows that when S was changed in the range between 0.8 and 25.6 seconds, GIT was linearly related to S (regression of the means of the six individual means for each S: GIT = 2.38 S + 4.46, $r^2 = 0.989$). Figure 7 shows that variability in GIT increased with S, but the frequency distributions appeared very similar when the data were plotted in relative time, namely when GITs were standardized with respect to S (Figure 8). The superposition is specially marked for values of inter prey intervals longer than 3.2 seconds.

The linearity of giving up times and the increased variability with superposition in relative time is what one might expect if the principles of scalar timing initially formulated in fixed interval situations applied to giving up times.

In a second experiment with four starlings in the same laboratory set up, we tested whether the qualitative predictions on the effect of travel time under timing ambiguity were present under our conditions. In this case, inter capture interval S was kept constant at 12.8 seconds, but there were three travel times, short, medium and long, depending on whether the number of flights to complete a travel was 10, 20 or 40 respectively. Average travel time varied monotonically with the number of flights required, but the relation was not strictly proportional. GUT should be independent of travel time if there is no timing ambiguity (Figure 2), but should be an increasing function of travel if the subject is constrained by its own variability (Figure 3).

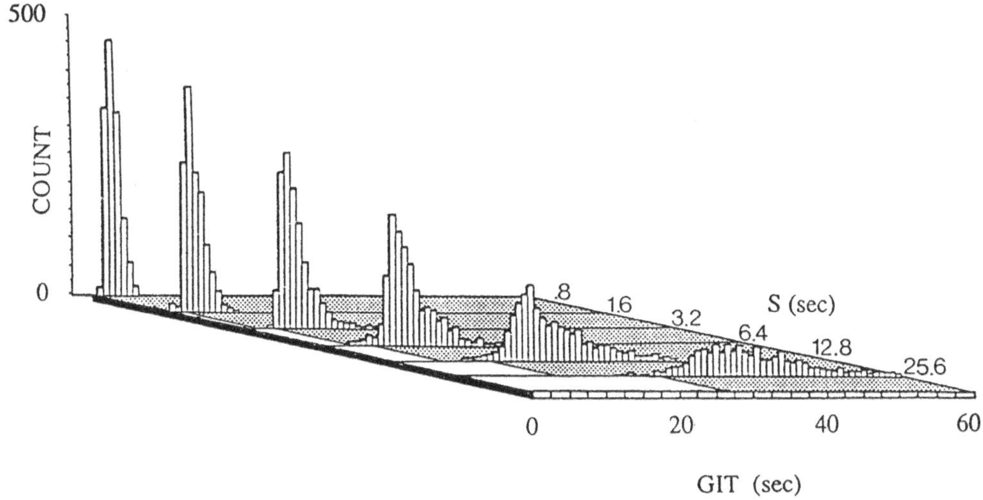

Figure 7: Variability in GIT as a function of inter capture interval. The frequency distributions include pooled data from all subjects. Distributions become broader as S increases.

Full analysis of these experiments will be presented elsewhere, but a preliminary view of the results is shown in figure 9. Both measures of giving up, GIT and MOT, increased with travel. The effects were more extreme in terms of MOT. A series of linear regressions showed that GIT increased significantly with travel for 3 out of 4 subjects while MOT increased significantly for all 4 subjects.

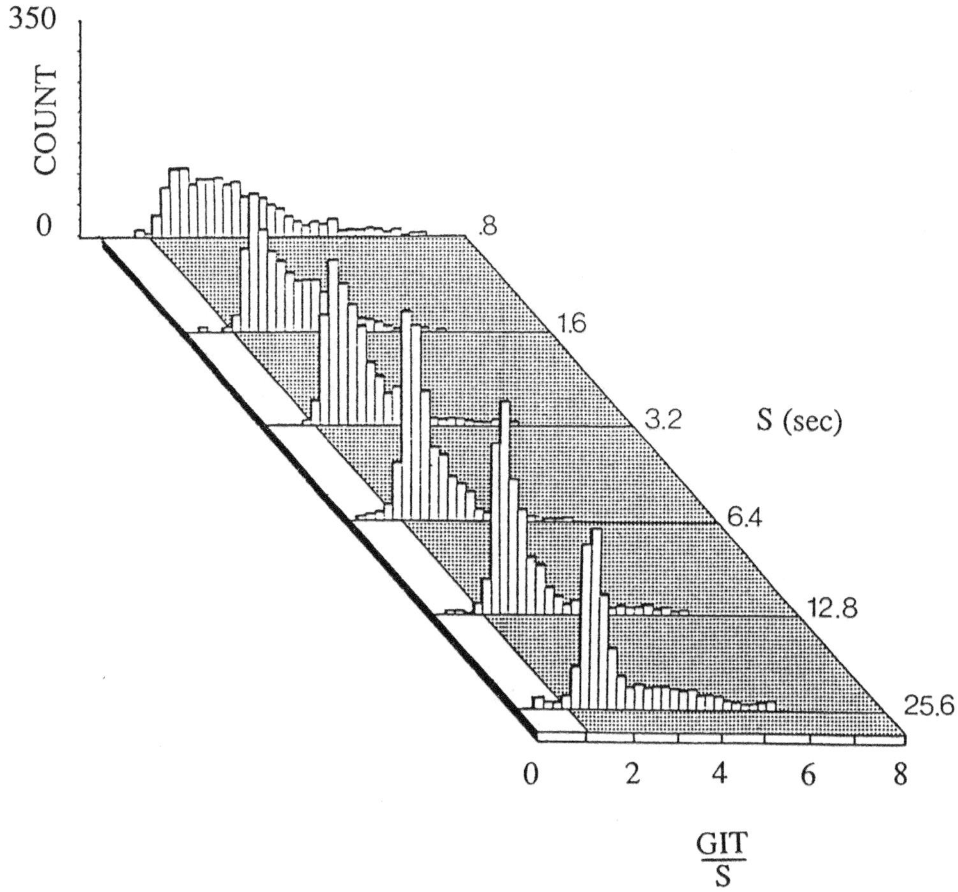

Figure 8: Variability in GIT expressed in relative time. The data in Figure 7 are now plotted after standardization as a fraction of S. The white line in the horizontal plane indicates equality. Notice that for short intervals the distributions are now flatter, but beyond 3.2 seconds the distributions have the same shape and are similarly placed with respect to S. This is equivalent to the superposition phenomenon in fixed interval experiments.

Summarizing the results of these two experiments:

a) Giving up time was a linear function of the inter-capture interval.

b) Variability in giving up time increased with inter capture interval.

c) Frequency distributions of giving up times obtained for S > 3.2 seconds superimposed when plotted in relative time.

d) Giving up time was a positive function of travel between patches.

All four features of the results are compatible with rate maximizing under the mechanistic account of scalar expectancy theory, and could not have been predicted from considerations of unconstrained rate maximization, that is by solving the optimality task of maximizing rate of intake with reference to the environment but excluding subjects' capabilities for information processing.

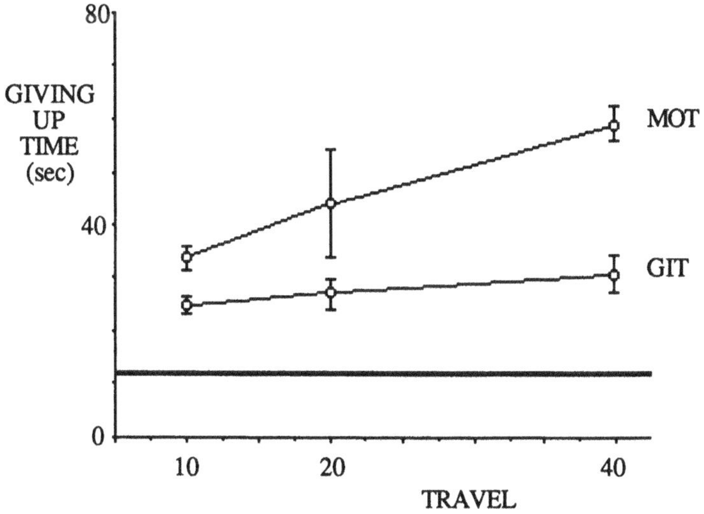

Figure 9: Effect of travel on giving up time. Although inter capture interval is constant (the solid horizontal line), two measures of giving up time increase significantly with travel ratio. The effect is more marked for the moving on threshold (MOT) than for the giving in threshold (GIT).

GENERAL CONCLUSIONS

This communication is a limited attempt to reappraise the relation between mechanistic and functional analyses in ethology and behavioural ecology (Tinbergen 1963). This problem has concerned us for some time (Cheverton, Kacelnik & Krebs 1985; Houston, Kacelnik & McNamara 1982; Kacelnik & Cuthill 1987). It is our view that during the development of optimal foraging theory psychological mechanisms have played an undeserved subordinate role. Researchers (including ourselves) usually begin by describing the general behaviour of the animal, then formalize the problem without reference to psychological processes and go on to predict the behavioural patterns which are expected to maximize fitness under the environmental circumstances considered in the formal model. As animals are not expected to go through the same calculations as researchers in computing the optimal behaviour but some solving algorithm is necessary, mechanistic hypotheses are formulated in the form of rules of thumb, namely blind adjustments of behaviour to a limited set of environmental features. These rules must be capable of generating behaviour close to the optimum.

Typical of this approach is the field example discussed earlier (Davies 1977). First, the general behaviour of the flycatcher is characterized by observation (fly-catching from fixed perches with occasional movement to another site). Then, various foraging problems are identified within that general behaviour and that environment (which prey are worth an attack, when should the bird return to the same perch after an attack, when should it move to another perch after an unsuccessful waiting time in a perch) and optimal solutions for each of these problems are found with the aid of theoretical models. The next step is to propose potential decision rules that would approximate those solutions. For example, in relation to the flycatcher problem of when to move to a new perch after unsuccessful waiting, Davies suggested that flycatchers form a time expectancy and abandon perches when they wait for longer than a certain time without detecting any prey worth attacking. He also surmised that this waiting time may be proportional to the average inter-catch interval. His discussion assumed that all variability was resident in the environmental side of the problem. In the model developed in collaboration with Richard Cowie in the same paper, the effectiveness of various GUTs, expressed without variability, is examined in a Montecarlo simulation where prey appear following a pattern that mimics observed inter-attack intervals. In contrast, to demonstrate the importance of mechanistic considerations theoretically and experimentally we constructed a situation in which randomness of inter-catch intervals was eliminated from the environment and instead the animal was assumed to be unable to discriminate or produce time intervals without errors. In our approach the secondary proposal of rules of thumb is replaced by seeking a foundation on the established literature on the psychology of timing.

We found that several features of the behaviour of starlings in such a situation are consistent with previous accounts of a theory of timing, Scalar Expectancy Theory. Interestingly, we also found that Davies' intuition of the formation of a time expectation with rough proportionality to inter-catch intervals was correct for starlings in our conditions - we found a linear rather than a proportional relation, and the slope was 2.4 rather than 1.5 but these are of course details considering the many differences in the two studies. To our knowledge there is no established optimality justification for this proportionality, as opposed to simple postulates that giving up times should exceed inter-catch intervals. The answer, we believe, may emerge from the nature of the information processing mechanism involved and we are currently working on more rigorous proposals for optimal choice of giving up thresholds within the framework of the psychology of time discrimination. Our hypothesis is that the linear relation will be found to be rather general and that this linearity minimizes the cost of deviations from the unconstrained optimal GUT.

The experimental results also highlight an aspect of behaviour for which we have no full interpretation as yet. In rate maximizing models the giving up time is basically a moving-on threshold (Charnov & Orians 1973; see discussion in Stephens & Krebs 1986, p.31). Movement of predators between patches is governed by the principle of lost opportunity, which states that a rate maximizer should move when its expected pay off in the current site is less than the expected pay off elsewhere. In our experiments with starlings, we found that they "give in" attempting to get food in a given patch some measurable time *before* they "move on" to seek another patch. It is as if after some waiting time (the GIT) the starlings are convinced that a patch is exhausted but they still delay the beginning of traveling to other patches. The gap between GIT and MOT is more pronounced for longer travel times (Figure 9). In psychological terms derived from open economy analyses, a possible interpretation is that travel is a punishing task and that its negative motivational value is greater for longer travels, thus delaying the start of new travels when they are expected to be longer. Since in the experiments travel was arranged according to a ratio schedule, this gap may be equivalent to the post-reinforcement pause in fixed ratio schedules of reinforcement (Ferster & Skinner 1957). Nevertheless, when considering rate maximization in a closed economy this account does not make sense without the addition of post-hoc modifications such as the need to perform other behaviours. This observation suggests new questions for field research. One may ask, for example, if flycatchers are attentive (equally responsive) to passing prey until the time of their departure from each perch, or whether, as the captive starlings, they cease to attack for some time before switching perches. Notice that this is not a trivial point for the optimal foraging analysis, because rates of encounter with prey in the field is often estimated from observed attacks and all quantitative predictions depend on estimated rates. This difficulty has been encountered in other foraging studies. Killeen, Smith & Hanson (1981), for example, found that although they gave rats a non-depressing patch simulation, the rats generated longer intervals between successive prey by pressing a lever at a lower rate as they accumulated pellets and they seemed to adjust their behaviour to this experienced decelerating gain curve instead of the programmed linear gain.

We refer to the psychological properties of animals as constraints for optimal foraging strategies. This is only a matter of research tactics. One may just as well discuss from an optimality perspective why do animals have those psychological properties. In doing that, ecological, developmental and neurobiological factors would be brought in as constraints, and one would be asking why, given all those factors, animals use, say, proportional timing, instead of logarithmic timing or some other transformation. Although animals evolve as integrated systems where individual properties interact to establish the range of possibilities for phenotypic adaptation, the process of study is analytical and requires temporary reification of specific features such as giving up time or diet breath. With a reversed perspective, we hope that the integration of foraging theory with the psychology of timing and other aspects of memory and learning will lead to new insights into the functional implications and perhaps the very structure of these psychological systems. Today's constraints are tomorrow's optimal strategies.

In conclusion, our main message is a call for including psychological properties of maximizing agents in the formulation of optimal foraging predictions. The common approach of postulating simple rules of thumb purely derived from examining the environment is not an alternative to the study of psychological mechanisms involved in foraging. Psychological properties such as the structure of memory and the ability to perform discriminations should be incorporated in the description of the problem from the start. This, we believe, can result in a productive integration of causal and functional analysis following a methodology within the spirit of Tinbergen's seminal distinction between levels of ethological enquiry.

SUMMARY

Optimal foraging models usually acknowledge environmental stochasticity, by assuming that the consequences of actions can only be predicted in probabilistic terms. Some models also take into account that incomplete information about the environment may add transient uncertainty that the optimizing subject must take into account while it learns.

One additional source of randomness which can be studied empirically and included in optimality models of decision making is the psychological system that processes information about time intervals. Time measurement is important in optimal foraging because optimal strategies depend on expected values of the times for handling, pursuing and consuming prey, the times for traveling between patches, the probability of interruption per unit of foraging time or other time-dependent events. Animals must use strategies that depend on estimates of these intervals. Psychological research on subjective representation of time shows that animals do not have unlimited accuracy in tasks which require interval identification or production. We show that capabilities to measure, remember and retrieve time intervals can be included in defining strategies for optimal patch exploitation. The main feature of our approach is that we explicitly identify psychological sources of uncertainty and incorporate current hypotheses on aspects of subjective representation of time to optimal decision making.

ACKNOWLEDGEMENTS

We are grateful to Tony Dickinson and Nick Davies for comments on the manuscript and to the other members of this Symposium, specially Frietson Galis, Les Real and John McNamara for enjoyable discussions on its content. AK was supported by King's College and by a research grant from the Wellcome Trust; DB by an external scholarship of the Argentinian Research Council-CONICET and JG by NIMH Grant MH 41649 to NYS Psychiatric Institute.

REFERENCES

Brunner D. Time discrimination and optimal foraging in starlings *Sturnus vulgaris*. Unpublished PhD dissertation , Cambridge University

Caraco T (1980) On foraging time allocation in a stochastic environment. Ecol 61:119-128

Caraco T, Martindale S, Whitham TS (1980) An empirical demonstration of risk-sensitive foraging

Charnov EL (1976) Optimal foraging: the marginal value theorem. Theor Popul Biol 9:129-136

Charnov EL, Orians GH (1973) Optimal foraging: some theoretical explorations. Unpublished manuscript

Charnov EL, Orians GH, Hyatt K (1976) The ecological implications of resource depression. Am Nat 110:247-259

Cheverton J, Kacelnik A, Krebs JR (1985) Optimal Foraging: constraints and currencies. In: Holldobler B, Lindauer M (eds) Experimental Behavioural Ecology and Sociobiology. Fortschritte der Zoologie, 31:109-126. G Fischer Verlag, NY

Church RM, Meck WH (1988) Biological Basis of the Remembered Time of Reinforcement. In: Commons ML, Church RM, Stellar JR, Wagner A (eds) Quantitative Analysis of Behavior: Biological Determinants of Reinforcement. Vol VII. Erlbaum, Hillsdale, NJ

Davies NB (1977) Prey selection and the search strategy of the spotted flycatcher (*Muscicapa striata*): a field study of optimal foraging. Anim Behav 25:1016-1033

Dickinson, A (to be published) Expectancy theory in animal conditioning. In: Klein SB, Mowrer RR (eds) Contemporary learning theories Chapter 9. Erlbaum, Hillsdale, NJ

Ferster CB, Skinner BF (1957) Schedules of reinforcement. Appleton-Century-Crofts, NY

Gibbon J (1977) Scalar Expectancy Theory and Weber's Law in Animal Timing. Psychol Rev 84:279-325

Gibbon J, Allan L (eds) (1984) Timing and Time Perception. The New York Academy of Sciences, NY

Gibbon J, Church RM (1981) Time Left: Linear Versus Logarithmic Subjective Time. J exp Psychol. Anim Behav Proc. 7:87-108

Gibbon J, Church RM (1984) Sources of variance in an information processing theory of timing. In: Roitblat HL, Bever TG, Terrace HS (eds) Animal Cognition. Erlbaum, Hillsdale, NJ

Gibbon J, Church RM, Fairhurst S, Kacelnik A (1988) Scalar expectancy theory and choice between delayed rewards. Psychol Rev 95:102-114

Gibbon J, Church RM, Meck WH (1984) Scalar Timing in Memory. In:Gibbon J, Allan L (eds) (1984) Timing and Time Perception. The New York Academy of Sciences, NY

Houston AJ, Kacelnik A, McNamara JM (1982) Some learning rules for acquiring information. In: McFarland DJ (ed) Functional Ontogeny. Pitman, London pp.140-191
Kacelnik A, Cuthill I (1987) Starlings and optimal foraging theory: modelling in a fractal world. In: Kamil AC, Krebs JR, Pulliam HR (eds) Foraging Behavior. Plenum, NY, pp 303-333
Killeen PR, Smith JP, Hanson SJ (1981) Central place foraging in *Rattus norvegicus* . Anim Behav 29:64-70
Killeen PR, Weiss NA (1987) Optimal Timing and the Weber Function. Psychol Rev 94:455-468.
Krebs JR, Kacelnik A, Taylor P (1978) Test of optimal sampling by foraging great tits. Nature 275:27-31
Krebs J (1974) Colonial nesting and social feeding as strategies for exploiting food resources in the Great Blue Heron, *Ardea herodias*. Behaviour 51:99-134
Mackintosh NJ (1974) The psychology of animal learning. Academic Press, London
McNamara JM (1982) Optimal patch use in a stochastic environment. Theor Popul Biol 21:269-288
McNamara JM, Houston A. (1980) The application of statistical decision theory to animal behaviour. J theor Biol 85:673-690
McNamara JM, Houston A. (1985) Optimal Foraging and Learning. J theor Biol 117:231-249
Roberts S (1983) Properties and Function of an internal clock. In: Mellgren RL(Ed) Animal Cognition and Behavior. North-Holland Publishing Co, Amsterdam
Schneider BA (1969) A two-state analysis of fixed-interval responding in the pigeon. J exp Anal Behav 12:677-688
Schneider BA, Neuringer AJ (1972) Responding under discrete-trial fixed-interval schedules of reinforcement. J exp Anal Behav 18:187-199
Shettleworth SJ, Krebs JR, Stephens DW, Gibbons J (1988) Tracking a fluctuating environment: a study of sampling. Anim Behav 36:87-105
Staddon JER, Ettinger RH (1989) Learning. An Introduction to the principles of adaptive behavior. Jovanovich, San Diego
Stephens DW, Krebs JR (1986) Foraging theory. Princeton University Press, Princeton, NJ
Tinbergen N (1963) On aims and methods of ethology. Z Tierpsychol 20:410-433
Treisman M (1963) Temporal discrimination and the indifference interval: implications for a model of the "Internal Clock". Psychol Monogr 77:1-31

ON EVALUATION OF FORAGING STRATEGIES THROUGH ESTIMATES OF REPRODUCTIVE SUCCESS

Robert G. Jaeger
Department of Biology
University of Southwestern Louisiana
Lafayette, Louisiana 70504, USA

Jeffrey Lucas
Department of Biological Sciences
Purdue University
West Lafayette, Indiana 47907, USA

INTRODUCTION

Foraging theories make predictions about how animals should behave so as to maximize energy gains, or gains in some other currency, given certain constraints. One assumption of such theories is that an animal that forages according to predictions will acquire more resources, and thus will have higher "fitness", than a conspecific that fails to forage by the theoretical rules. This assumption is seldom tested in any way directly related to reproductive success of individuals, and thus foraging theories may be accused of a certain amount of circular reasoning. For example, say that theory predicts that under certain conditions, a forager should ingest all encountered items of prey type 1 and should "ignore" encountered items of prey type 2; the assumption is that such behavior will eventually contribute to the fitness (defined vaguely here as reproductive success) of the forager. The theoretical prediction is tested empirically, and the resulting data closely fit the prediction. The inference drawn from the test is that the forager must have higher potential fitness than foragers that might have performed in other ways (such as eating prey type 2 and ignoring prey type 1).

The fitness assumption in foraging theories is often unstated as an assumption and is often only implicitly inferred from the results. Yet the fitness assumption forms the very basis for all foraging theories. The critical question remains, do animals that forage according to theoretical predictions differ in fitness, on average, from animals that forage otherwise? Empirical studies of foraging theories have seldom utilized designs (experimental or otherwise) that might test for differential fitnesses among conspecifics that forage in different ways (but see Morse, 1988, for an exceptional study). Indeed, Stephens and Krebs (1986), in their extensive treatment of foraging theory, hardly mentioned the connection between foraging behavior and measures of ultimate fitness of individuals. This reluctance to confront empirically the

NATO ASI Series, Vol. G 20
Behavioural Mechanisms of Food Selection
Edited by R. N. Hughes

fitness consequences of foraging behavior is surprising when compared to certain other areas of behavioral research. For example, a large recent literature has accumulated in the area of "mate choice" by animals, and particularly "female choice" of mates. It is clear that females of many species are choosy in selection of mates, but the primary issue today is, does a choosy female (on average) have any higher fitness than a conspecific female that mates randomly? There is presently considerable research ongoing concerning adaptive hypotheses of mate choice (e.g., Boake, 1986), and researchers in foraging behavior might profit from pursuing this literature.

Our goal is to explore several empirical ways in which the fitness assumption of foraging theories might be tested. We do not present specific experimental designs, because such designs will vary according to the type of animal tested. It is not even likely that any of the following methods can be applied to every particular species of interest, but a careful choice of species may lead to critical tests of differential reproductive success by foragers.

CLASSICAL APPROACHES

Animals make innumerable foraging decisions during their lifetimes. The more subtle the decisions are, the more difficult it will be to relate them directly to fitness. For example, decisions between a nearly continuous range of prey sizes that change with prey density are more subtle than decisions between gross prey categories that exhibit a strong genetic (i.e., fixed) component. Because foraging decisions can be quite complicated, researchers have used more short-termed currencies or goal functions as a tool to understand the adaptiveness of foraging behaviors. If these currencies are well chosen, maximization of the currency and maximization of reproductive success will be closely correlated. The trick is to find a way to choose the appropriate currency.

Early studies treated energy/time (E/T) as a "universal" currency that was applied to virtually all systems (but see Pulliam 1975). Unfortunately, E/T was chosen as much for ease of measurement as for the importance of this currency as an index of fitness. Obviously, the ideal would be to analyze currencies from first principles and to ask exactly which currencies are most likely to result ultimately in an increase in fitness. Several authors have provided important examples which suggest ways in which this could be done.

For example, foraging efficiency (energetic benefit/cost) can be shown to be better than E/T as a criterion for maximizing colony fitness of social bees

(Houston et al. 1988), and the behavior of worker bees is consistent with this prediction (Schmid-Hempel & Wolf 1988; Schmid-Hempel et al. 1985). For small birds in winter, survival rate can be shown to be a better predictor of long-term fitness than E/T (e.g., McNamara & Houston 1982). Carolina Chickadees exhibit caching strategies that do not maximize E/T but instead appear to maximize net survival rates (Lucas, this volume). Lastly, Gilliam and co-workers have shown that minimizing the ratio of predation rate to growth rate will maximize population growth rates in certain age-structured populations (Werner & Gilliam 1984; Gilliam and Fraser 1987), and the habitat choice of juvenile fish has been shown to change in a manner consistent with this prediction (Werner et al. 1983).

In each of these examples, theory has shown us that E/T is a poor predictor of fitness. More importantly, the theory has suggested alternative currencies that give us a more realistic understanding of the selective pressures acting upon the evolution of foraging decisions. As a result, these alternative currencies produce more robust predictions of foraging behavior than predictions derived from models using E/T. However, these currencies are system-dependent. Thus there is no panacea, no single currency that can be useful under all circumstances. In addition, the currencies still give indirect measures of fitness. An alternative approach to a theoretical analysis of currencies is to estimate empirically the relationship between foraging decisions and fitness components.

Two common ways of statistically testing the consequences of behavioral acts are through experimental designs (with experimental and control conditions) and through regression designs. We shall take these in order.

The experimental design would compare the reproductive success of animals that are allowed to forage normally (the control condition) with that of animals manipulated in their foraging choices (the experimental condition). Consider the prediction that foragers should choose prey type 1 but not prey type 2. One group of animals would be allowed to forage at will while the experimental group would be "fooled": i.e., for every prey item selected by the forager, the opposite prey item would be forced upon the animal at the point of ingestion. Obviously, forcing an animal to ingest a prey item that it previously avoided presents severe technical problems for the researcher, for many species of foragers; thus, this experimental approach may have poor utility for most studies. Still, if it could be implemented for particular species, the experiment might yield important evidence of relative fitnesses of animals in the control and experimental conditions. An interesting way to approach this would be to utilize some species in which females lay clutches of eggs and then must forage before the next clutch can be laid. Subjecting females to control or experimental conditions between the two clutching periods would allow assessment of their differential reproductive successes (e.g., number of eggs, sizes of eggs, mass of clutch). Alternatively, the forager could be fed a variety of diets, and growth rate or reproductive success would then be estimated for each. These data could then be compared to the forager's behavior under choice conditions. The data could also be compared to theoretical analyses of the system to give some insight into how closely the theory matches empirically derived estimates of the benefit of alternative strategies available to the forager.

The regression approach takes advantage of naturally differential foraging behavior of individuals. It is reasonable to assume that individuals in a population differ among themselves genetically and thus might differ in the genetic expression of foraging behavior. A theoretical prediction of foraging might be tested by subjecting numerous conspecifics to identical treatments and monitoring differences in reproductive success. Foraging behavior (relative to the prediction) would be the independent variable while reproductive output would be the dependent variable. Again, females that multiply clutch and forage between clutches would be an excellent system in which to test statistically for significant regression.

ALTERNATIVE APPROACHES

The above approaches look for indirect effects of foraging on subsequent fitness. The hypothesis is that animals that forage according to theoretical foraging predictions will somehow be "healthier" than those that forage otherwise, and that such "healthiness" will be visible to the researcher through differential reproductive investment by individuals. We now consider alternative approaches that might provide easier empirical ways for testing the fitness assumption in foraging theories. These alternative approaches deal with the circumstances where foraging decisions are made during some phase of the reproductive cycle, and therefore directly affect reproductive success. We shall focus on three phases: mate choice, reproductive output, and reproductive success associated with parental care.

(1) Mating Success. Hamilton and Zuk (1982) proposed that females may be able to detect healthy males by some "truth in advertising"

characteristic of those males, such as intensity of body color. Although Hamilton and Zuk posed this concept in terms of males that differ in parasitic loads (for which they might have varying levels of genetic resistance), the same concept could be posed in foraging studies. As in the regression approach above, consider males that differ in their foraging behavior relative to some theoretical prediction. The hypothesis would be that those foraging as predicted would be "healthier" than those not doing so, and this might be expressed phenotypically in characteristics detectable by females. Thus, the mating success of different males could be statistically compared to their differential foraging behavior. Mating success could be taken to mean the number of females attracted and mated.

Several systems have been described where male mating success may be directly related to short-term foraging decisions. For example, male balloon flies (Kessel 1955) and scorpion flies (Thornhill 1981) offered females prey which the females fed on during copulation. Thornhill (1984) found that female scorpionflies preferentially mated with males that provided adequate nuptial gifts, in the form of prey items that were ingested by the female during copulation. In a sense, then, females were assessing the foraging success of males in terms of mate choice. The link between foraging theories and acquisition of nuptial prey is obvious and need not be belabored here. For species where nuptial gifts are important, it should be possible to compare statistically the foraging behavior of males (relative to some theoretical prediction, such as patch utilization) and the differential number of matings (or offspring produced) by those males.

This behavior is not confined to the insects. In snail kites (Haverschmidt 1970) and squirrel cuckoos (Skutch 1966), males have been observed to offer females a prey item and then copulate with the female while she eats. Male northern shrikes impale prey in conspicuous sites; these caches are used to attract females, and males with larger caches mate earlier and sire more offspring than those with smaller caches (Yosef and Pinshow 1989).

Males of some species of insects provide other types of nutrition to females during courtship, such as bodily fluids or spermatophores (Thornhill and Alcock, 1983). The nutrient quality of these types of nuptial gifts is thought to influence the quantity or quality of eggs produced by the females. Thus, if the fitness assumption of foraging theories is correct, the differential foraging abilities of males (relative to some prediction) should be manifest, on average, in reproductive success, via differential nutritional attributes of the "gifts".

Red-backed salamanders illustrate a different type of sexual selection directly related to foraging. Walls et al. (1989) found that female red-backed salamanders examine the dietary composition of male fecal pellets, and prefer to associate (during courtship periods) with males that have high quality diets. Thus, a male that chooses a diet one day will produce a fecal pellet in about four days that will influence his attractiveness to females. This ability for females to assess male fecal pellets may not be unique to salamanders and may provide a rapid means by which to test the fitness assumption of foraging theories. In the case of Walls et al., female salamanders preferred males that ingested high-quality prey (i.e., termites) over those that ingested lower-quality prey (i.e., ants). Once again, it should be possible to compare statistically the differential foraging choices of males with their abilities to attract courting females.

(2) Reproductive Effort. Under some circumstances, feeding rates are directly correlated with reproductive effort. For example, a series of excellent papers by Morse and Fritz (e.g., Morse 1988; Fritz & Morse 1985) showed that patch choice decisions by crab spiders have a significant effect on reproductive effort. These crab spiders catch insects that come to milkweed umbels. Several umbels are found on a single plant, and these umbels develop at different rates. In addition, newly opened umbels tend to be more productive than older umbels and therefore attract more insects. Those crab spiders choosing more productive umbels grow significantly faster than those choosing older umbels. In turn, large spiders lay bigger egg masses and their egg masses have a higher survivorship. This study is important in that it shows that animals vary in the decisions that they make, and this variation has a direct impact on reproductive success.

The crab spiders studied by Morse and Fritz lay a single clutch in their lifetime. This simplifies the analysis because there are fewer options available to a female. Things become more complicated when multiple egg masses are produced. For example, female lace bugs can invest more time in foraging, and so increase egg production rate, but this reduces egg survivorship through a reduction in time spent tending the clutch (Tallamy and Denno 1982).

Individual herring gulls specialize on different food types (e.g. garbage, mussels, or petrels), and reproductive success is strongly influenced by these feeding strategies (Pierotti and Annett 1987). Unlike the crab spiders, those specialists that forage on the energetically least profitable prey (mussels) actually have the highest reproductive success. Pierotti and Annett (1987) attributed this to the availability of nutrients in the different diets.

Courtship feeding is another example where foraging decisions can potentially affect fitness directly. For example, male common terns feed females at the nest, and feeding rates are a determinant of both the number and size of eggs that the female makes (Nisbet 1977).

(3) Parental Care. An easy way to measure payoffs to some foraging decisions is to measure the resultant growth rates. Unfortunately, this measure is complicated by reproductive tradeoffs in mature individuals and by reduced utility of growth rates in animals with determinate growth. Both of these problems are not relevant for systems where parents provide food for their young. Here the utility of foraging decisions by the adults can be evaluated in terms of growth rate of young, and growth rate in young will often affect some aspects of fitness, such as overwinter survivorship (e.g., Davies 1986).

The theoretical literature on foraging decisions by parents is fairly large (e.g., Kacelnik and Houston 1984; Houston 1987; Houston and McNamara 1985; Hussell 1988), but there have been few tests that attempt to relate directly offspring survivorship, or even growth rates, to foraging decisions. However, these few studies provide a nearly complete analysis of the fitness aspects of foraging decisions, and so provide excellent examples of how foraging decisions can be associated with direct measures of fitness.

Krebs and Avery (1985) have shown that European bee-eaters tend to bring large prey to their nestlings, and a central-place foraging model that assumes energy maximization predicts this general pattern. But the parents bring some small prey (specifically bees) which would be energetically more profitable to ignore. However, an analysis of growth rates under different diets shows that nestlings fed mixed diets have higher mean growth efficiencies than chicks fed on pure diets (Krebs and Avery 1984). Thus fitness is increased through the delivery of a "non-optimal" (vis-a-vis E/T) diet.

Two studies provide an excellent example of how life-history theory can be applied to foraging problems concerning parental care. Both assume that those foraging decisions which maximize population growth rate (r) will evolve.

Nur (1987) described a cost/benefit model used to study the adaptive significance of provisioning rates by blue tits. The model incorporated an estimate of fitness derived by Charnov and Krebs (1974) ($r=\ln(m+p)$:

m=number of offspring surviving to reproductive age, p=probability the adult survives to the following year). Using this estimate, provisioning rates can be treated as a tradeoff between adult mortality and juvenile survivorship, because both are functions of provisioning (Nur 1987). The model predicts that per-nestling feeding rates should asymptote at high nest sizes and that clutch size in broods laid late should be less than clutch size of early broods. Both predictions are fairly well supported by the data.

Sibly and McCleery (1985) studied egg brooding behavior of herring gulls. The model assumes that fitness is affected by both adult mortality (due to starvation when energy reserves are exhausted) and egg mortality (due to predation and exposure when eggs are not brooded). Several parameters were incorporated in the model, including threshold energy reserves which should cause a bird to feed when its mate is on the territory, and a lower threshold which would cause the bird to abandon the nest even if the mate is absent. The birds exploited three types of resources: mussels, garbage, and starfish. The model could also be used to evaluate which types of habitat choice would generate maximal fitness values.

OVERVIEW

The relationship between foraging decisions and fitness is a critical component of the modern approach to behavioral ecology. Despite the growth and major advances shown in this field (as illustrated in the chapters of this book), we still have relatively little empirical evidence for the fitness consequences of behavioral decisions. Our short review is intended to show that there are certain types of behaviors for which foraging decisions can be directly related to some components of fitness, and these systems allow a more direct assessment of fitness than typically found in studies of foraging behavior.

The review also points out some other, rather obvious, aspects of systems that make them particularly amenable to this type of analysis. One important property that will facilitate the evaluation of fitness effects is a limited number of alternative behaviors. The 'courseness' of the decision is also important, both at the individual level and at the population level. The gull studies in particular illustrate that some individuals can be surprisingly inflexible in their choice of food patches. If there is variation among individuals, this inflexibility greatly facilitates the estimation of reproductive success associated with behavioral alternatives. Bookstaber and Langsam (1985) provided a theoretical treatment of the conditions under which 'coarse behavior rules' are expected to evolve, and these ideas would be worth pursuing. In this respect, it would be interesting to know how often differences in diet choice between individuals is genetically related (e.g., see Arnold 1981).

The study of adaptation in diet choice has been an iterative process. The field started with simple models with simple (and restrictive) assumptions (Pyke et al. 1977). These models were useful in illustrating that diet choice could be studied with this technique. We are now re-evaluating these assumptions. Part of this process is a study of the mechanisms of diet choice decisions; these include the mechanisms used to gather information about resources and the mechanics of digestive processes. Another part of the process is to understand the selective pressures driving the evolution of foraging decisions. We now have a better understanding of the range of potential optimization criteria that might influence the evolution of behavioral traits. Studies of foraging decisions that directly influence reproductive success will provide new insights. But this will still be only a step in the iterative process. Fitness will always be difficult to measure; even empirical measures of life-time reproductive success have their problems (Grafen 1988). None-the-less, the closer we can come to estimating the impact of diet choice on some components of fitness, the more we will understand the diversity of foraging decisions.

ACKNOWLEDGMENTS

Many of the ideas in this paper, and our interest in integrating foraging studies with tests of fitness, were derived from group discussions at the NATO Advanced Study Institute on Behavioral Mechanisms of Food Selection. We thank the many participants who fed our curiosity with their own ideas, and especially Leslie Real for his stimulating approach to the problems of determining "fitness" of individuals.

REFERENCES

Arnold SJ (1981) The microevolution of feeding behavior. In: Kamil AC, Sargent TD (eds) Foraging Behavior: Ecological, Ethological, and Psychological Approaches. Garland,New York,p 409-453

Boake CRB (1986) A method for testing adaptive hypotheses of mate choice. Am Nat 127:654-666

Bookstaber R, Langsam J (1985) On the optimality of coarse behavior rules. J theor Biol 116:161-193
Charnov EL, Krebs JR (1974) On clutch size and fitness. Ibis 116:217-219
Fritz RS, Morse DH (1985) Reproductive success and foraging of the crab spider Misumena vitia. Oecologia 65:194-200
Davies NB (1986) Reproductive success of dunnocks, Prunella modularis, in a variable mating system. I. Factors influencing provisioning rate, nestling weight and fledging success. J Anim Ecol 55:123-138
Gilliam JF and Fraser DF (1987) Habitat selection under predation hazard: test of a model with foraging minnows. Ecology 68:1856-1862
Grafen A (1988) On the uses of data on lifetime reproductive success. In: Clutton-Brock TH (ed) Reproductive Success: Studies of Individual Variation in Contrasting Breeding Systems. Univ Chicago, Chicago,p454-471
Hamilton WD, Zuk M (1982) Heritable true fitness and bright birds: a role for parasites? Science 218:384-387
Haverschmidt F (1970) Notes on the snail kite in Surinam. Auk 87:580-584
Houston A (1987) Optimal foraging by parent birds feeding dependent young. J theor Biol 124:251-274
Houston A, McNamara JM (1985) A general theory of central place foraging for single-prey loaders. Theor Pop Biol 28:233-262
Houston A, Schmid-Hempel P, Kacelnik A (1988) Foraging strategy, worker mortality, and the growth of the colony in social insects. Am Nat 131:107-114
Hussell DJT (1988) Supply and demand in tree swallow broods: a model of parent-offspring food-provisioning interactions in birds. Am Nat 131:175-202
Kacelnik A, Houston AI (1984) Some effects of energy costs on foraging strategies. Anim Behav 32:609-614
Kessel EL (1955) The mating activities of balloon flies. Syst Zool 4:97-104
Krebs JR, Avery MI (1984) Chick growth and prey quality in the European bee-eater (Merops apiaster). Oecologia 64:363-368
Krebs JR, Avery MI (1985) Central place foraging by the European bee-eater, Merops apiaster. J Anim Ecol 54:459-472
Lucas JR (this volume) Time scale and diet choice decisions.
McNamara JM, Houston AI (1982) Short-term behaviour and lifetime fitness. In: McFarland DJ (ed) Functional Ontongeny. Pitman,London,p 60-87
Morse, DH (1988) Relationship between crab spider Misumena vatia nesting success and earlier patch-choice decisions. Ecology 69:1970-1973

Nisbet ICT (1977) Courtship feeding and clutch size in common terns Sterna hirundo. In: Stonehouse B, Perrins P (eds) Evolutionary Ecology. University Park, Baltimore,p101-109

Nur N (1987) Parents, nestlings and feeding frequencies: a model of optimal parental investment and implications for avian reproductive strategies. In: Kamil AC, Pulliam HR, Krebs JR (eds). Foraging Behavior. Plenum,NY,p 457-475

Pierotti R, Annett C (1987) Reproductive consequences of dietary specialization and switching in an ecological generalist. In: Kamil AC, Pulliam HR, Krebs JR (eds). Foraging Behavior. Plenum, New York, p 417-442

Pulliam HR (1975) Diet optimization with nutrient constraints. Am Nat 109:765-768

Pyke GH, Pulliam HR, Charnov EL (1977) Optimal foraging: a selective review of theory and tests. Q Rev Biol 52:137-154

Schmid-Hempel P, Wolf T (1988) Foraging effort and life span of workers in a social insect. J Anim Ecol 57:509-521

Schmid-Hempel P, Kacelnik A, Houston AI (1985) Honeybees maximize efficiency by not filling their crop. Behav Ecol Sociobiol 17:61-66

Sibly R, McCleery R (1985) Optimal decision rules for herring gulls. Anim Behav 33:449-465

Skutch A (1966) Life history notes on three tropical American cuckoos. Wilson Bull 78:139-149

Stephens DW, Krebs JR (1986) Foraging theory. Princeton University Press, Princeton, New Jersey

Tallamy DW, Denno RF (1982) Life history trade-offs in Gargaphia solani (Hemiptera: Tingidae): the cost of reproduction. Ecology 63:616-620

Thornhill R (1981) Panorpa (Mecoptera: Panorpidae) scorpionflies: systems for understanding resource-defense polygyny and alternative male reproductive efforts. Ann Rev Ecol Syst 12:355-386

Thornhill R (1984) Alternative female choice tactics in the scorpionfly Hylobittacus apicalis (Mecoptera) and their implications. Am Zool 24:367-383.

Thornhill R, Alcock J (1983) The evolution of insect mating systems. Harvard University Press, Cambridge, Massachusetts

Walls SC, Mathis A, Jaeger RG, Gergits WF (1989) Male salamanders with high-quality diets have faeces attractive to females. Anim Behav: to be published

Werner EE, Gilliam JF (1984) The ontogenetic niche and species interactions in size-structured populations. Ann Rev Ecol Syst 15:393-426

Werner EE, Gilliam JF, Hall DJ, Mittelbach GG (1983) An experimental test of the effects of predation risk on habitat use in fish. Ecology 64:1540-1548

Yosef R, Pinshow B (1989) The effects of cache size on mate choice and reproductive success in Northern Shrikes (Lanius excubitor). Amer Ornithol Union Abstracts:176

ACTIVE DIET SELECTION OR PASSIVE REFLECTION OF CHANGING FOOD AVAILABILITY: THE UNDERWATER FORAGING BEHAVIOUR OF CANVASBACK DUCKS

John P. Ball
Delta Waterfowl and Wetlands Research Station
R.R. 1 Portage la Prairie
Manitoba, R1N 3A1 Canada
(and Dept. of Zoology, University of Guelph, Guelph, Ontario, N1G 2W1).

INTRODUCTION

An omnivore is an animal with morphological and physiological adaptations that enable it to exist on plant and animal foods (Westoby 1978), although at any given time the diet of an individual may not be mixed. Canvasback ducks (Aythya valisineria) are omnivores that dive and search for prey by using tactile cues. Few investigators have studied tactile predators (see Elner and Hughes 1978, Hughes 1979); fewer still have studied the foraging strategies of omnivores.

The decision rules of omnivores may not be simply intermediate between those of carnivores and herbivores. For carnivores, prey are usually nutritionally equal (Westoby 1978) and nutritionally dense, whereas for herbivores, foods are rarely nutritionally equal and are usually bulky for a given intake of nutrients (Westoby 1978, Belovsky 1984). The diets of carnivores thus may be relatively more constrained by search and handling times, whereas the diets of herbivores seem to be more often shaped by the constraints of digestion rate and nutrition (see reviews in Belovsky 1984, Pyke 1984). Examination of an omnivore's "decision" at what trophic level to feed may highlight the limitations of existing models of food selection.

NATO ASI Series, Vol. G 20
Behavioural Mechanisms of Food Selection
Edited by R. N. Hughes

"Classical" optimal foraging models (Krebs and McCleery 1984) assume that predators attempt to maximize net intake rate and that a predator cannot search and exploit simultaneously, giving rise to the "principle of lost opportunity" (Stephens and Krebs 1986:11). Predators should assess potential gains from exploiting a given prey against the potential loss of the opportunity to do better -- in economic terms, this loss is the "opportunity cost" of that particular prey. No opportunity is lost by eating the most favourable prey because the best possible outcome from rejecting it is immediately to find another of the same type. Conversely, the time spent consuming an inferior item may be invested more profitably in finding and consuming a superior food item.

"Classical" optimal foraging models also assume that recognition times for prey are negligible, and predict that the inclusion of a prey item in the diet does not depend how often the predator encounters it, but only on the characteristics of higher-ranked prey (Krebs and McCleery 1984, Pyke 1984, Stephens and Krebs 1986). These models can, however, be modified to incorporate recognition time. Non-negligible recognition times can lead to the inclusion of low-ranked prey in the diet (Elner and Hughes 1978, Hughes 1979, and Houston et al. 1980). In terms of opportunity cost, if the encounter rate with unprofitable prey is sufficiently high, the predator should eat every one encountered because it forgoes only the opportunity to find and eat another prey, probably of the same unfavourable type.

All animals must take some time to recognize food (Pyke 1984), but it is usually negligible for visual predators. Elner and Hughes (1978) considered a case where a tactile predator had a non-negligible recognition time. Shore crabs (Carcinus maenas) had to lift each prey before deciding whether to consume it, and these authors (and subsequently Hughes 1979) showed how this situation would lead to a broadening of the diet. Canvasbacks search for food by tactile cues (pers. obs.) as does the closely-related Pochard (Aythya ferina) (Suter 1982 in Draulans 1987). Furthermore, prey items are recognized only as they are grasped in the bill. However, in a series of preliminary trials, I was unable to detect any recognition time for Canvasbacks so it appears that they can recognize food by touch at least as quickly as a visual predator can by sight. Canvasbacks took the same time to reject a food item as they did to swallow it. Under these conditions, predators suffer no lost opportunity in terms of time by eating a poor-quality food item. Because diving birds must return to the surface for oxygen, the time available to find a second prey of any type is limited and they should exhibit even less selectivity as dive duration increases (Houston and McNamara 1985).

Here, I report the results of experiments designed to evaluate how Canvasbacks, as tactile omnivores, feed. Specifically, 3 competing hypotheses about the behavioural mechanisms of food selection by Canvasbacks are tested:

(1) they forage randomly in the environment and consume foods as encountered (effectively <u>no</u> selection or "passive acceptance"),

(2) they select patches, but within these, eat every prey encountered ("patch selection"), or

(3) they select foods at the level of individual items by actively rejecting non-preferred prey ("individual prey selection").

METHODS

The underwater feeding behaviour of Canvasbacks was observed through windows in the sides of a large (2 m wide X 2 m deep X 5 m long) concrete aquarium. This depth is comparable to that observed at which Canvasbacks feed in the wild (Takekawa 1987). The data reported here are from 12 adult Canvasbacks (7 females) reared from eggs collected from the wild. All birds were allowed several hundred dives for the same foods under the same conditions as would later be used in the experiments to obviate the confounding influence of learning on diet selection (Hughes 1979). Canvasbacks feed primarily on benthic foods (Bartonek and Hickey 1969, Jarvis and Noyes 1986, Takekawa 1987), so only benthic prey were provided. Four foods differing in digestibility and nutritional composition were used in the experiments: sago pondweed tubers (Potamogeton pectinatus) which are often a major component of the diet of wild Canvasbacks (Anderson 1985), beetle larvae (Tenebrio spp.) because immature insects can also be a major food (Bartonek and Hickey 1969), and two seeds differing in size and thickness of seed coat (wheat and corn) because Canvasbacks eat a variety of seeds (Cottam 1939, Korschgen 1955, Bartonek and Hickey 1969).

The bottom of the aquarium was completely covered with 20 removable trays (1 m X 0.5 m) filled with 10 cm of sand to simulate the benthos of a water body. Before each trial, 4 randomly-selected trays were removed from the aquarium and the top 3 cm of sand was removed. Food items were counted into these trays, the top layer of sand replaced, and the trays lowered back into the aquarium. Thus, in each trial, only 4 of 20 trays contained food, but these were visually indistinguishable from those without food.

Birds were allowed to forage in the tank for 1 hour, 2 or 3 times per day. This schedule was chosen to mimic the duration and frequency of foraging bouts by an incubating female duck (Afton 1978). The same schedule was applied to both sexes. Animals may use hunger as a measure of recent prey availability (Charnov 1976), so birds were not allowed access to food between trials to avoid influencing their estimates of food abundances. On 4 or 5 consecutive days, each bird had to dive to obtain all its food from the tank. Between trials, birds were removed from the tank while trays were replenished. After a trial, trays were removed, uneaten foods were recovered by sieving, and food consumption was calculated by subtraction of food recovered from food provided.

In Experiment 1, each of 4 randomly-selected trays held different prey, providing 4 patches of prey at the same density (15 per tray). If Canvasbacks are non-selective feeders, we would expect equal numbers of all 4 food items to be eaten (in the simplest case where encounter rates with different prey are equal). Conversely, if Canvasbacks select their diets (regardless of whether they do this by selecting patches or selecting individual prey), differing numbers of the 4 foods would be eaten.

A second experiment was designed to further assess if Canvasbacks passively accept foods or actively select them. In this experiment, birds were offered 2 combinations of prey abundance. In both of these combinations, prey densities were arbitrarily set much higher than in the previous experiment to ensure that the birds did encounter even the smallest foods repeatedly. For the first combination (hereafter, the "baseline"), prey per tray were: 60 Sago tubers, 60 insect larvae, 60 kernels of corn or 200 grains of wheat. The second combination of prey abundance was similar to the baseline except that the abundances of wheat and corn were increased fourfold. If Canvasbacks passively accept foods, 4 times more wheat and corn would be eaten in the second combination than in the baseline. A significant difference from this fourfold increase would indicate that the birds were actively selecting prey (either by revisiting patches where preferred prey were found or by rejecting non-preferred prey already grasped by the bill).

A third experiment determined whether Canvasbacks select their diet at the level of patches of food, or at the level of individual food items (if they do select). I performed two independent tests in this experiment, but they shared a common design and differ only in the numbers of prey provided. The first test employed the same total numbers of prey as in the baseline of Experiment 2 above, but in some of these trials, one-quarter of each prey type was placed in each of the 4 food trays. Thus, 60 sago tubers, 60 insect larvae, 60 kernels of corn and 200 grains of wheat were available in the bottom of the aquarium, but in some trials each food had its own tray, whereas in the other trials the 4 kinds of prey were mixed together in each of 4 trays. In the second test, the foods were provided in the same "mixed vs separate" scheme, but the total numbers of prey provided were 60 sago tubers, 60 insect larvae, 240 kernels of corn and 800 grains of wheat. Only trials where all 4 "separate" food trays were encountered by the bird were considered in the analysis to permit direct comparisons with the "mixed" trials in which all 4 foods were encountered in a single tray. If diet selection occurred only at the level of patches, "mixed" and "separate" trials would differ; if selection occurred at the level of individual food items, they would be the same. Chi-squared tests (Sokal and Rohlf 1973) were used to test for differences in the frequencies of prey items eaten in each experiment.

RESULTS

In Experiment 1, Canvasbacks ate 7.6 Sago tubers, 5.1 insect larvae, 0.4 grains of corn and 2.3 grains of wheat ($n = 28$ trials). Thus, consumption of foods was not equal ($Chi^2 = 8.0$, $p < 0.05$); sago tubers were eaten more often than expected, and corn was avoided.

In Experiment 2 (if no selection occurred), I predicted that 4 times more wheat and corn would be eaten in Combination 2 than in the baseline. The prey eaten in Combination 2 differed from this predicted diet ($Chi^2 = 42.1$, $p < 0.005$), indicating that Canvasbacks did not consume prey in proportion to their altered availability (Table 1).

Table 1. Prey available and prey eaten for 2 combinations of food densities offered to foraging Canvasbacks (Experiment 2).

	BASELINE (n = 67)		COMBINATION 2 (n = 25)	
	Prey Available	Prey Eaten	Prey Available	Prey Eaten
Sago	1 tray of 60	21.1	1 tray of 60	14.5
Larvae	1 tray of 60	29.3	1 tray of 60	31.8
Corn	1 tray of 60	0.2	1 tray of 240	2.8
Wheat	1 tray of 200	40.4	1 tray of 800	63.1

In Experiment 3, if the diets consumed within each test do not differ between "mixed" and "separate," Canvasbacks select diets at the level of individual food items. In Test 1 of Experiment 3, diets did not differ between the 2 combinations of foods offered ($Chi^2 = 3.7$, $p > 0.10$, Table 2). Similarly, in Test 2, diets did not differ ($Chi^2 = 2.9$, $p > 0.10$, Table 2).

Table 2. Prey available and prey eaten for different combinations of prey abundance offered to foraging canvasbacks in either separate trays or mixed together in the same food tray (Experiment 3).

TEST 1

	Combination 1 (MIXED) (n = 148)		Combination 2 (SEPARATE) (n = 67)	
	Prey Available	Prey Eaten	Prey Available	Prey Eaten
Sago	4 trays of 15	18.3	1 tray of 60	21.1
Larvae	4 trays of 15	21.2	1 tray of 60	29.3
Corn	4 trays of 15	1.5	1 tray of 60	0.2
Wheat	4 trays of 50	35.5	1 tray of 200	40.4

TEST 2

	Combination 1 (MIXED) (n = 58)		Combination 2 (SEPARATE) (n = 25)	
	Prey Available	Prey Eaten	Prey Available	Prey Eaten
Sago	4 trays of 15	18.9	1 tray of 60	14.5
Larvae	4 trays of 15	24.4	1 tray of 60	31.8
Corn	4 trays of 60	3.9	1 tray of 240	2.8
Wheat	4 trays of 200	65.4	1 tray of 800	63.1

DISCUSSION

For ducks, diving is very energetically expensive (Woakes and Butler 1983, 1986, Takekawa 1987), so strong selection for efficient feeding might be expected to have occurred. In the wild, Canvasback diets sometimes exhibit dramatic changes, but to what degree this represents a change in food item selection, a change in the use of areas where different foods are found, or simply a change in the foods available, is difficult to determine because too many factors change concurrently and because the underwater feeding of diving ducks cannot be observed. For example, in the Carolinas, Perry and Uhler (1982) reported changes in Canvasback diets from 99.5% plant matter to 86% invertebrates by volume. The birds may have simply been tracking changes in food availability (Munro and Perry 1981, Perry et al. 1981), but without underwater observations on prey encounter rates and on visitation rates to patches holding different benthic prey, few firm conclusions can be reached from existing field studies.

Experiment 1 indicates that, when offered equal numbers of foods, Canvasbacks differentially selected certain food items. However, it is possible that encounter rates with different prey were not equal even though all were offered at the same density (30 prey per m^2) because a duck's probing bill may encounter larger food items more often than smaller ones. The problem is compounded by the fact that Canvasbacks encounter their benthic prey beneath the surface of the substrate where encounters with individual prey cannot be directly observed. Furthermore, 30 prey per m^2 of the smallest food items (wheat and corn) may be so low that it would be hard for any predator to detect. Morse (1980) commented on the difficulty in determining whether a prey item is ignored because it is of low rank or because it is not encountered often enough to be perceived as food. However, because wheat (the smallest prey) was eaten considerably more frequently than corn, it does not seem likely that Experiment 1 was confounded by encounter rates varying with prey size.

Experiment 2 (Table 1) also supports the hypothesis that Canvasbacks select their diets: the birds did not simply track a fourfold change in the abundance of wheat and corn. The results from Experiment 1 may or may not be influenced by differential encounter rates with prey of different sizes, but in Experiment 2, a tactile predator must have encountered 4 times more prey when the density of that prey increased fourfold.

Both tests of Experiment 3 support the hypothesis that the diet selection indicated by Experiments 1 and 2 in fact occurs at the level of selection of individual food items rather than selection of patches where different foods predominate. If Canvasbacks selected their diet only by selecting patches (and eating all foods encountered within that patch), the diets from "mixed" and "separate" trials would have differed. However, the observed diets in mixed and separate trials were the same, indicating that the spatial distribution of prey was not relevant to diet selection. Canvasbacks thus selected prey at the level of individual food items.

Clearly, these tactile omnivores do alter their diets and they do so at the level of individual prey (by spitting out non-preferred prey), not just by selecting patches of food.

Why spit out non-preferred (but nevertheless quite edible) prey already grasped in the bill? There would seem to be no "lost opportunity" because these omnivores save no time by spitting out non-preferred prey -- ejecting and ingesting take the same time. The principle of "lost opportunity" may be the key to understanding the behaviour of this omnivore (and perhaps other omnivores), but it is the loss of another kind of opportunity. An omnivore "decides" on whether it will feed as a herbivore or a carnivore, and this affects what constrains its diet. If it feeds as a strict carnivore, it is likely to be constrained by search and handling time; if it feeds as a strict herbivore it is likely constrained by the need to obtain a nutritionally balanced diet and by its digestive constraints. Perhaps the simplest hypothesis to explain the ejection by Canvasbacks of non-preferred prey is a digestive constraint (e.g. Belovsky 1984). For Canvasbacks, eating a low-quality food item seems to incur an "opportunity cost" -- the inability to ingest a high-quality food item if one is found because the digestive tract is full. In general then, an "optimally-behaving" omnivore (with equivalent rejection and ingestion times like these Canvasbacks) should eat all foods encountered unless (as the data presented here indicate) eating "food 2" prevents it from eating "food 1" by contributing to a bottleneck in digestion.

"Classical" optimal foraging theories emphasize the importance of search and handling time but typically ignore differences between prey other than their yield of energy per unit time (Pyke 1984). Linear programming approaches (e.g. Belovsky 1984) emphasize the nutritional differences between prey as well as the digestive constraints on the predator. It is probably no coincidence that the former approach has had considerable success predicting the diet of carnivores (see Pyke 1984), whereas the latter has proven highly successful with herbivores (e.g. Belovsky 1984). The study of omnivores thus seems particularly appropriate for testing between these two approaches to understanding the diets of animals. Furthermore, experiments could manipulate the abundances of plant and animal prey in order to test optimality models of an omnivore's decision to feed as a carnivore, herbivore, or something intermediate (as Canvasbacks did in this study). Such an approach may shed light on the limits of our current models of decision making by animals.

SUMMARY

Canvasback ducks are omnivores that find and recognize their benthic foods by tactile cues. The observation of their underwater foraging provides an interesting test of behavioural decision-making vs the passive consumption of foods encountered. I observed adult Canvasbacks diving in a large concrete and glass aquarium as they consumed 4 foods differing in size, digestibility and nutrient quality.

The results of 3 experiments indicate that Canvasbacks actively select their diet and that they must do so by differentially ejecting non-preferred foods already grasped in the bill. This rejection of captured prey suggests that Canvasbacks are constrained by their digestive capacity. Thus, even though Canvasbacks consumed both plant and animal matter, they were subject to a constraint more often associated with herbivores than with carnivores.

ACKNOWLEDGEMENTS

Funding was provided by the North American Wildlife Foundation through the Delta Waterfowl and Wetlands Research Station, a Natural Science and Engineering Research Council grant A7757 through T.D. Nudds, an Ontario Graduate Scholarship, the Canadian Wildlife Service, and the University of Guelph. M. Evans, P. Hockey, J. Leafloor, C. Neill, T. Nudds and A. Wyss assisted or commented on the manuscript.

LITERATURE CITED

Afton AD (1978) Incubation rhythms and egg temperatures of an American Green-winged teal and a renesting Pintail. Prairie Naturalist 10:115-119

Anderson MG (1985) Social behavior of breeding canvasbacks (Aythya valisineria): male and female strategies of reproduction. Phd Thesis University of Minnesota 281p.

Bartonek JC and JJ Hickey (1969) Food habits of canvasbacks, redheads, and lesser scaup in Manitoba. Condor 71:280-290

Belovsky GE (1984) Herbivore optimal foraging: a comparative test of three models. Am Nat 124:97-115.

Charnov EL (1976) Optimal foraging: attack strategy of a mantid. Am Nat 110:141-151

Cottam C (1939) Food habits of North American diving ducks. U.S. Dept of Agric Tech Bull 643. 140p.

Draulans D (1987) Do tufted duck and pochard select between differently sized mussels in a similar way? Wildfowl 37:49-54.

Elner RW and RN Hughes (1978) Energy maximization in the diet of the Shore Crab, Carcinus maenas (L.) J Anim Ecol 47:103-116

Houston AI and JM McNamara (1985) A general theory of central place foraging for single-prey loaders. Theor Pop Biol 28: 233-262

Houston AI, JR Krebs, and JT Erichsen (1980) Optimal prey choice and discrimination time in the great tit (Parus major L.) Behav Ecol Sociobiol 6:169-175

Hughes RN (1979) Optimal diets under the energy maximization premise: The effects of recognition time and learning. Am Nat 113:209-221

Jarvis RL and JH Noyes (1986) Foods of canvasbacks and redheads in Nevada: paired males and ducklings. J Wildl Manage 50:199-203

Korschgen LJ (1955) Fall foods of waterfowl in Missouri. MO Dept Conserv P-R Ser 14. 41p.

Krebs JR and RH McCleery (1984) Optimization in Behavioural Ecology. pp. 91-121 In Behavioural Ecology An Evolutionary Approach, 2nd Ed. Blackwell Scientific Publications, Boston, MA

Morse, DH (1980) Behavioral mechanisms in ecology. Harvard University Press, Cambridge MA. 383p.

Munro RE and MC Perry (1981) Distribution and abundance of waterfowl and submerged aquatic vegetation in Chesapeake Bay. US Fish and Wildl Serv FWS/OBS78/D-X391. 180p.
Perry MC and FM Uhler (1982) Food habits of diving ducks in the Carolinas. Proc Annu Conf Southeast Fish and Wildl Agencies 36:492-504.
Perry MC, RE Munro, and GM Haramis (1981) Twenty-five year trends in diving duck populations in Chesapeake Bay. Trans North Am Wildl and Natur Resour Conf 46:299-310.
Pyke GH (1984) Optimal foraging theory: A critical review. Ann Rev Ecol Syst 15:523-575
Sokal RR and FJ Rohlf (1973) Introduction to Biostatistics. WH Freeman and Company, San Francisco CA
Stephens DW and JR Krebs (1986) Foraging Theory. Princeton University Press, Princeton, NJ
Takekawa JY (1987) Energetics of canvasbacks staging on an Upper Mississippi River pool during fall migration Phd Thesis Iowa State University, Ames Iowa 189p.
Westoby M (1978) What are the biological bases of varied diets? Am Nat 112:627-631
Woakes AJ and PJ Butler (1983) Swimming and diving in tufted ducks Aythya fuligula, with particular reference to heart rate and gas exchange. J Exper Biol 107:311-329
Woakes AJ and PJ Butler (1986) Respiratory, circulatory and metabolic adjustments during swimming in the tufted duck, Aythya fuligula. J Exper Biol 120:215-231

ORAL DISCUSSION – J.P. Ball

Q. (Zimmer–Faust): What is the role of taste in prey discrimination ?
A. I can't tell whether birds use taste or tactile cues. However, some previous anatomical work has indicated that ducks have few taste receptors, but many tactile receptors, so touch would seem the most likely. I hope to determine which sensory cues are used in a subsequent experiment using artificial food pellets.

Q. (Sih): There was no effect of patchiness – were the birds unable to learn the position of food patches ?
A. There were some effects of patchiness, but not on the diet. The birds did seem to have good spatial memory; over the course of a trial, the birds learned the location of the food patches, as well as which food they contained. Patchiness may have altered feeding efficiency, but my results indicate the patchiness did not affect the diet.

Q. (Sih): Did they change patch choice ?
A. Yes, At the start of a trial, the birds had no idea of the location of the buried foods, so the first dives were distributed at random. However, the birds soon learned which trays held food and revisited them more often than empty patches. Furthermore, in those trials in which the foods were in separate patches, the birds also seemed to learn which ones held preferred foods and returned to them more often than to patches holding non–preferred prey.

Q. (Belovsky): It is true that there has been little emphasis on onmivores, even through they face a wider range of decisions than herbivores or carnivores. However, in reality it is not a dichotomous decision between foraging as a 'herbivore' or 'carnivore'. Instead there is a continuum between these two end points and the omnivore must choose some point along this continuum.
A. I agree. In fact, my data clearly show that Canvasbacks did not choose the diet of a strict herbivore or carnivore, but an intermediate diet. I probably overstated the carnivore–herbivore distinction in my introduction, but I do think that omnivores are ideal for testing between foraging models, precisely because thay face a wider range of decisions and have a correspondingly wider range of solutions available to them.

Q. (Hobbs): How much do your plant and animal foods differ in quality ?
A. They differ more in some respects than others. Rate of digestion differs; in order of increasing gut transit time, the foods are sago pondweed tubers, insect larvae, wheat, and lastly corn. Energy content per gram is fairly similar for the three plant foods but higher for the insect larvae. The larvae have much more protein per gram than the seeds or tubers. This last point has

some interesting implications because Canvasbacks lose considerable heat when diving. The 'waste' heat from digestion of protein–rich foods may be useful in offsetting this loss, and if so, plant and animal foods may differ less in energy per gram than simpler analyses might indicate. In my experiments, foods also differed in size, so tubers and larvae would probably be the most favourable in terms of net energetic profitability. I am currently evaluating foraging models to determine which gives us the best insight into the diet choice of this omnivore.

Q. (Kacelnik): An analysis of pauses at the surface could be used to exclude some of the possible models. For example, if pauses are caused by a digestive constraint we expect to see a pattern in the pauses.
A. Although neither experiments nor statistical analysis are completed, it is very much my impression that this pattern does occur. Consistent with the digestive bottleneck hypothesis the length of the feeding bouts (i.e. the time to fill up) seems to be inversely related to food density, whereas the pause between foraging bouts (the time to empty the upper digestive tract) seems constant for a given food.

Q. (Kacelnik): Was behavior learned ? Differences between trials could be caused by a lag with previous experience if the birds didn't immediately recognize that food availability had changed from the previous trial.
A. Trials were performed in a random order, so all possible sequences of trial were conducted. The food densities in preceeding trials did not seem to affect diet choice. Furthermore, within a trial, the revisitation rates to patches having different prey leads me to believe that the birds could evaluate the densities (and locations) of the four foods within the first 5 – 10 minutes of the hour–long tiral. Overall, I don't think that a lag in learning was a major influence.

Q. (Okamura): Do the ducks usually take invertebrates with tubers ?
A. In the wild it is difficult to tell, but invertebrates and tubers are often found in different locations. In my experiments, the birds would immediately head for the surface to handle a tuber, but would sometimes continue to dig after finding and eating the much smaller larvae. Therefore, they would only take a larva on the same dive as a tuber if they found the animal prey first. The larvae are large enough to make incidental ingestion with tubers unlikely.

TERRITORIAL SALAMANDERS EVALUATE SIZE AND CHITINOUS CONTENT OF ARTHROPOD PREY

Robert G. Jaeger
Department of Biology
University of Southwestern Louisiana
Lafayette, Louisiana 70504, USA

Feeding territories provide an interesting area in which to study the behavioral mechanisms of food selection, because theories of territorial economics (Brown, 1964) and of foraging economics (Stephens and Krebs, 1986) become closely allied. Foraging decisions by the territorial defender may be influenced not only by changing densities of various prey types, but also by other environmental fluctuations, by intrusions from food competitors and predators, and by opportunities for courtship. Yet if defending a particular area is crucial to a territorial resident's future fitness, then the ability of the resident to move to another area for foraging is limited. The red-backed salamander (Plethodon cinereus) is a territorial species confronted with several conflicting demands on its foraging abilities.

Plethodon cinereus (family Plethodontidae, the lungless salamanders) is widely distributed in forests of eastern North America. It is strictly terrestrial, lacking any aquatic life-stage, and it can reach high densities (mean of 2-3 individuals per square meter) in the leaf litter or under rocks and logs on forest floors (Jaeger, 1979). Being amphibians and lungless, red-backed salamanders must maintain a moist skin, and so their foraging behavior is greatly affected by alternating periods of rainy and rainless weather. During and shortly following a rainfall, salamanders forage in the leaf litter and their stomachs are usually packed with prey (Jaeger, 1972, 1980). During rainless periods, though, the leaf litter dries and salamanders are forced into moist patches under rocks and logs (for individuals that have established feeding territories) or into fossorial crevices (for those that have not established such territories) (Jaeger, 1979; Mathis, to be published). Territorial individuals apparently are able to crop a small but renewable number of prey

NATO ASI Series, Vol. G 20
Behavioural Mechanisms of Food Selection
Edited by R. N. Hughes

during dry periods (Jaeger, 1980) under and around cover objects while those without territories, in underground refugia, essentially are restricted from foraging until the next rain (Fraser, 1976). Periodic rainfall, then, establishes a scenario whereby prey remain abundant in the leaf litter but vary temporally in availability to the salamanders. Territorial behavior, by males and females separately, appears to be primarily directed at protecting prey from competing salamanders, and it involves pheromonal advertisement (Jaeger, 1986; Jaeger et al., 1986; Horne and Jaeger, 1988), visual agonistic displays (Jaeger, 1984; Horne, 1988), biting of intruders (Jaeger, 1981), and consequent exclusion of intruders (Jaeger et al., 1982b).

In an attempt to elucidate the interaction of territorial defense and foraging tactics of red-backed salamanders, a number of experiments were performed in the laboratory. Briefly, these experiments revealed the following information. (1) Given two species of dipterans at varying densities, the salamanders elected an indiscriminate diet at low prey densities but specialized on the larger prey species at higher densities; specialization occurred by "ignoring" the smaller flies when they were encountered. The salamanders switched from pursuit to ambush tactics with increasing prey densities (Jaeger and Barnard, 1981). (2) Salamanders used encounter rates in assessing densities of the dipteran prey (Jaeger et al., 1982a). (3) Salamanders learned through foraging experience to assess profitabilities of the prey types (Jaeger and Rubin, 1982). (4) Salamanders sacrificed initial caloric yield until they had established marked (with pheromones) territories and then altered to a higher sustained caloric yield (Jaeger et al., 1981). (5) When salamanders in established territories were exposed to increasing threats of intrusion (by pheromones of conspecifics or by conspecifics themselves), the residents increasingly switched from a specialized diet (as above) at high prey densities to an indiscriminate diet at the same prey densities. Time subtracted from foraging was added to time devoted to territorial defense (Jaeger et al., 1983).

Red-backed salamanders feed on a broad variety of invertebrates inhabiting the detritus of forest floors (Jaeger, 1972).

Consequently, their diets include termites (Isoptera) and springtails (Collembola), which are lightly armored with chitinous exoskeletons, and ants (Hymenoptera: Formicidae) and beetles (Coleoptera), which are heavily armored with exoskeleton. Digesting the chitin of ingested prey is difficult for many predators, and so the amount of chitin encasing a prey item may strongly influence the profitability of that item from the viewpoint of the predator. For Plethodon cinereus, the chitinous content of a prey item apparently does not affect the salamander's assimilation efficiency (A.E.) for that item; Bobka et al. (1981) reported that both chitinous ants and less chitinous dipterans (Drosophila) were assimilated equally at a given temperature (about 80% A.E. for both prey at 20 C rising to 90% A.E. for both at 10 C). However, Jaeger and Barnard (1981) noted that passage time through the digestive tract of P. cinereus is much faster for the dipterans (approximately 70 h) than for the ants (approximately 112 h) at 15 C. Passage time logically determines the rate at which new prey can be ingested. For red-backed salamanders, which obtain most of their prey during often short periods of rainfall, maximizing ingestion rate (and passage time) during those periods may be critically important in maximizing net energy yield from foraging.

The above concepts led to a hypothesis by Jaeger and Barnard (1981) concerning diet selection by P. cinereus. When prey are abundantly available, during rainfalls, salamanders should specialize, thus minimizing chitin in their diets. (1) If two prey items differ only in size (such as two species of flies), salamanders should choose the larger one because of its greater gross energy content and relatively less chitin (from surface to volume ratios). (2) For prey types differing in armor (such as termites and ants) but equal in size, the less armored one should be chosen. (3) Given a choice of a larger, heavily armored prey and a smaller prey with reduced chitin, the salamander should choose the prey type with the higher net energy profit per unit of digestion time. These three points assume equal assimilation efficiencies and energy expended per ingested prey. During times when prey are low in availability, such as rainless periods, salamanders should be less selective, and heavily armored prey should increase proportionally in the diet. The goal of the research reported here was to provide a first test of Jaeger and Barnard's (1981) hypothesis, utilizing P. cinereus in a natural forest habitat.

DIET DIVERSITY

In order to identify the major and minor taxa of invertebrates found in the diet of Plethodon cinereus, I utilized data previously collected in 1970 for other research purposes (Jaeger, 1980). Quadrates 15 x 100 m were established on the forest floor of Black Rock Mountain, Shenandoah National Park, Virginia. Fifteen quadrates were randomly sampled, one on each of 15 consecutive days from 25 July through 8 August, but no quadrate was sampled more than once. In each sample, 30-35 adult red-backed salamanders were collected, were immediately killed by anesthesia, and were preserved (including stomachs) in 10% formalin. The contents of their stomachs were later removed, and the prey contained therein were identified to taxa and were measured to determine the volume of each prey item (Table 1). More details concerning collecting and preserving the salamanders and processing the ingested prey were given in Jaeger (1980).

Table 1 shows that red-backed salamanders have a large diversity of prey types in their diet. The insects were even more impressive in diversity when partitioned by families: COLEOPTERA (Anobiidae, Cantharidae, Carabidae, Chrysomelidae, Cicindelidae, Coccinellidae, Cucujidae, Curculionidae, Elatidae, Nitidulidae, Scarabaeidae, Staphylinidae, Tenebrionidae); COLLEMBOLA (Entomobryidae, Sminthuridae); HEMIPTERA (Miridae, Nabidae); HOMOPTERA (Aphididae, Cicadellidae, Delphacidae, Membracidae); HYMENOPTERA (Braconidae, Ceraphronidae, Cynipidae, Eucharitidae, Eulophidae, Figitidae, Formicidae, Ichnuemonidae, Mymaridae, Platygasteridae, Pompilidae, Pteromalidae, Scelionidae, Trichogrammatidae); ISOPTERA (Rhinotermitidae); LEPIDOPTERA (larvae not identified to taxa); ORTHOPTERA (Blattidae, Gryllidae); PSOCOPTERA (Eupsocidae). The DIPTERA (Brachycera, Cyclorrhapha, Nematocera) could not be identified to family level.

Table 1.--Percentages (by number and volume) of invertebrate taxa found in the stomachs of 519 salamanders collected over 15 consecutive days. Total number of prey identified was 4752.

TAXA	% BY NUMBER	% BY VOLUME
Nematoda	0.02	0.002
Gastropoda	2.02	6.048
Oligochaeta	0.34	17.652
Acarina	50.02	9.126
Araneae	2.67	7.482
Isopoda	0.02	0.005
Chilopoda	0.97	8.451
Diplopoda	0.34	1.260
Insecta		
Coleoptera	2.36	19.452
Collembola	13.09	3.428
Diptera	20.85	7.474
Hemiptera	0.21	1.298
Homoptera	0.35	2.011
Hymenoptera	5.49	9.425
Isoptera	0.15	0.085
Lepidoptera (larvae)	0.04	5.402
Orthoptera	0.08	0.863
Psocoptera	0.80	0.536

The data in Table 1 demonstrate that some prey (such as Acarina) comprise a large proportion of the diet by number but not by volume while other prey (such as Coleoptera) comprise a large proportion of the diet by volume but not by number. The data in Table 1 will later be used to formulate analyses of experimental results concerning choice of diet by red-backed salamanders.

DIET SELECTION

The experimental test of Jaeger and Barnard's (1981) hypothesis of diet selection was performed during 1982 in the southern Blue Ridge Mountains of Virginia. I compared the size distribution of prey ingested by salamanders and of prey available in the environment under conditions of both wet and dry forest litter.

Methods

Two 15 x 100 m quadrates were established on the forest floor adjacent to each other. On 26 July, 12 h of heavy rain fell, and I randomly chose one quadrate for study. Forty Tanglefoot traps were placed on the forest floor (in randomly selected positions within the quadrate), where red-backed salamanders were known to forage during wet periods (Jaeger, 1972, 1979). Each trap consisted of a 15 cm diameter circle of filter paper coated with Tanglefoot spray, a sticky material commercially developed for entrapping invertebrates. Each trap was placed on the leaf litter with an inverted petri dish serving as an "umbrella", and litter was placed over the dish. Twenty-four hours later, 31 traps were recovered (nine had been destroyed) and were stored separately in 70% ethanol. Thirty-one red-backed salamanders were then collected from the litter, each salamander being at 0.5-2.0 m distance from a trap (n = 31 salamanders and 31 traps). Salamanders were killed immediately and were preserved as in the 1970 analysis (see above). Because red-backed salamanders require about 24 h to pass prey through their stomachs at 17 C (mean soil temperature upon capture), both stomach contents and contents of the traps should reflect 24 h of salamander foraging and "trap foraging".

No rain fell from 27 July until 2 August. On 31 July, the leaf litter appeared to be dry, so 40 additional Tanglefoot traps were placed in the second quadrate. These traps were positioned next to randomly selected rocks; because red-backed salamanders apparently forage under and near cover objects (which act as patches of moisture) during dry periods (Jaeger, 1972, 1979), these traps were intended to sample prey available to salamanders confined to a small distance around their cover objects. After 24 h, 30 traps were recovered (10 had been destroyed) and preserved (as above), and 30 salamanders were collected from under rocks within 0.5-2.0 m of the traps and were immediately killed and preserved (n = 30 salamanders and 30 traps). Thus, the sampling protocol for both wet and dry days was the same (0.5-2.0 m between traps and salamanders).

To determine whether my visual estimates of "wet" and "dry"

days were correct, I used a Soilmoisture Probe model 2900 (tensiometer) to record ambient moisture at 40 randomly selected positions within each quadrate while salamanders were being collected. Each preserved salamander was measured for snout-vent length (SVL) and head width, and its stomach was removed and opened. The length and mid-body width of each intact prey item was measured using a dissecting scope fitted with an ocular micrometer, and the volume of each item was estimated by converting to the volume of a cylinder or a sphere, as appropriate. Prey items removed from traps were measured similarly. The number of salamanders collected each day was deemed sufficient for analysis of stomach contents, because Fraser (1976) showed that approximately 21 stomachs of *Plethodon cinereus* were necessary to estimate prey volume diversity. Trapped prey that were too wide to be ingested by salamanders (i.e., greater than the salamanders' head widths) were deleted from the samples, as were trapped prey that were smaller than the smallest item ingested by any salamander.

These data allowed me to compare the volume-distribution and taxa of prey ingested under two conditions of mobility by salamanders (wet versus dry days) with the volume-distribution and taxa of prey available to salamanders in the leaf litter.

Sizes of Prey

As anticipated, the forest floor was significantly wetter on the rainy sampling day than on the rainless sampling day (Table 2). The salamanders collected on the dry and wet days did not differ significantly either in snout-vent length or in head width (Table 2); this indicates that individuals from the two sampling days were capable of eating the same total volumes and individual sizes of prey.

Salamanders on the dry day had significantly fewer prey per stomach than those on the wet day and significantly less total volume of prey per stomach (Table 2). Therefore, as found in an earlier study (Jaeger, 1972), the dry-day salamanders had less success at foraging than did the wet-day conspecifics. The number of prey "captured" by the traps was significantly less on the dry day (218/30 traps, $\bar{x}$ = 7.3/trap) than on the wet day (450/31

Table 2.--Means (and sample sizes) for soil moisture, snout-vent length (SVL), and head width of salamanders, number and total volume of prey ingested, and volume of individual prey items found on traps and in stomachs for dry and wet days.

TAXA	$\bar{x}$/DAY DRY	$\bar{x}$/DAY WET		n/DAY DRY	n/DAY WET	P	
Soil moisture	18	0	cb	40	40	0.0002	*
SVL	31	32	mm	30	31	0.243	
Head width	4.1	4.0	mm	30	31	0.372	
No. prey/stomach	1.5	16.8	prey	30	31	0.0002	*
Total vol. prey/ stomach	8.1	44.1	mm^3	30	31	0.0002	*
Vol. each potential prey (on traps)	1.8	1.5	mm^3	218	450	0.941	**
Vol. each prey (in stomachs)	5.3	2.6	mm^3	46	522	0.0008	*

* Significant difference, Mann-Whitney U test, two-sided, alpha = 0.05. ** Variances were homoscedastic between days.

traps, $\bar{x}$ = 14.5/trap): Mann-Whitney U test, P = 0.002 two-sided.

The distribution of volumes of potential prey items found on the traps did not differ significantly between the 218 dry-day invertebrates and the 450 wet-day invertebrates (Table 2); thus, the sizes of potential prey available to salamanders on dry and wet days apparently were the same. However, the mean volume of each prey item found in the stomachs of salamanders was significantly greater on the dry day than on the wet day (Table 2); therefore, wet-day salamanders ingested, on average, smaller prey items than did dry-day salamanders.

In summary, salamanders ingested more prey, greater total volumes of prey, but smaller individual prey items, during wet-day foraging than during dry-day foraging.

Because the volume-distributions of potential prey on traps were not statistically different between the two days (Table 2), the prey volumes from traps are pooled into a single set in Fig. 1 for ease of visual inspection. However, Table 3 utilizes the dry-day and wet-day traps separately for statistical comparisons between stomachs and traps on those two days.

The dry-day salamanders ingested significantly larger prey items than those found on the dry-day traps (Fig. 1, Table 3), indicating that "hungry" salamanders did specialize on larger

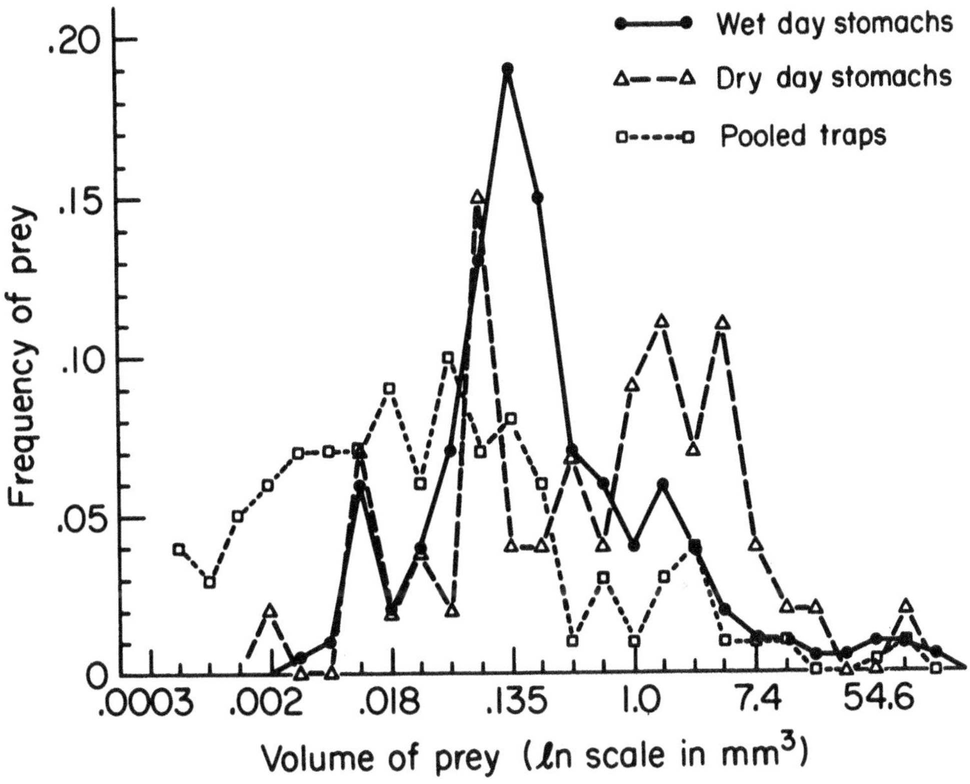

Figure 1.--Frequency of prey as a function of size-class (volume) of prey from (a) stomachs of salamanders collected on a "wet" day (superior foraging time), (b) stomachs of salamanders collected on a "dry" day (poor foraging time), and (c) traps pooled for the two days.

prey, rather than resorting to a strictly generalist diet. This occurred because the salamanders virtually excluded from their diet prey less than 0.002 mm^3 while concentrating on prey of two volumes: 0.08 mm^3 and from 1.0-4.5 mm^3 (Fig. 1).

The wet-day salamanders also ingested significantly larger prey than those found on the traps (Fig. 1, Table 3). These salamanders failed to ingest prey of less than 0.004 mm^3 and were essentially unimodal in choice of prey-volume (0.135 mm^3), biased toward the larger of the most abundant prey-volumes on the traps (Fig. 1).

Therefore, both dry- and wet-day salamanders took propor-

Table 3.--Comparisons of volumes of ingested prey and trapped potential prey on dry and wet sampling days; n = number of prey in the sample.

DAY	$\bar{x}$ VOL. (mm^3) EACH ITEM INGESTED	n	$\bar{x}$ VOL. (mm^3) EACH ITEM ON TRAPS	n	P
Dry	5.3	46	1.8	218	0.0001 *
Wet	2.6	522	1.5	450	0.002 *

* Significant difference, Mann-Whitney U test, two-sided, alpha = 0.05.

tionally larger prey than were available in the forest litter (Table 3), but the wet-day salamanders chose prey of smaller size than their dry-day conspecifics (Table 2).

Taxa of Prey

So far, I have demonstrated that red-backed salamanders shifted from ingesting smaller prey items on a superior foraging wet day to larger prey items during a poor foraging dry day. Jaeger and Barnard's (1981) hypothesis would predict that this change in diet represented a shift in prey taxa eaten: from smaller, lightly armored prey on the wet day to larger, heavily armored prey items on the dry day. I now test this hypothesis with taxonomic data from those two days.

Based on the large sample of stomach contents in Table 1, I chose four taxa for comparisons between the wet and dry days: Coleoptera and Hymenoptera (both with relatively hard exoskeletons and relatively large in volume per individual) and Collembola and Diptera (both with relatively soft exoskeletons and relatively small in volume per individual). A discussion of arthropod cuticle and chitin may be found in Miller (1980). Individuals of the four taxa combined from wet- and dry-day traps were of the following sizes (volumes): Coleoptera -- range = 0.95-92.47 mm^3 ($\bar{x}$ = 17.92 mm^3); Hymenoptera -- range = 0.08-69.91 mm^3 ($\bar{x}$ = 3.75 mm^3); Collembola -- range = 0.04-1.81 mm^3 ($\bar{x}$ = 0.43 mm^3); Diptera -- range = 0.13-12.57 mm^3 ($\bar{x}$ = 2.06 mm^3).

I computed the percentages (by volume) of Coleoptera, Hymenoptera, Collembola, and Diptera found in the stomachs and on the

Table 4.--The percentages (of volumes) of four taxa of insects found in the stomachs of salamanders and on the traps from the wet day and the dry day. Although only four taxa of prey are included here, the percentages are based on all taxa found in the samples.

TAXA	WET DAY % PREY IN:		DRY DAY % PREY IN:	
	STOMACHS	TRAPS	STOMACHS	TRAPS
Coleoptera	2.9	18.9	8.5	12.1
Hymenoptera	1.5	13.3	10.0	11.7
Collembola	7.5	7.1	3.9	4.1
Diptera	9.9	8.1	3.3	2.7
All other	78.2	52.6	74.3	69.4

traps during the wet day and the dry day. Statistical comparisons of the results were not performed because of the small number of prey (n = 46) found in the stomachs (n = 30) on the dry day. However, Table 4 provides information for useful inferences.

On the dry day when salamander foraging was relatively poor, the stomachs and traps had similar proportions of the various taxa (Table 4). Thus, the salamanders apparently foraged indiscriminately among these four taxa of prey.

On the wet day when foraging was superior, far less proportional volumes of Coleoptera and Hymenoptera showed up in the stomachs of salamanders than were on the traps (Table 4). However, Collembola and Diptera were approximately proportionally equal for stomachs and traps on that day.

DISCUSSION

Food Selection and Territoriality

These data support the hypothesis that red-backed salamanders tend to eliminate prey items with hard exoskeletons from their diets during superior (wet) foraging periods, even though those prey items may be among the largest (in body size) available to the salamanders. The data cannot, of course, determine whether such a shift in diet maximizes net energy (or nutritional) yield for the salamanders. A more highly controlled experiment is required to do that, where the behavior of a foraging salamander can be closely observed and its energetic costs and benefits during foraging can be derived. Such experiments are extremely difficult to perform in natural leaf litter habitats and so must be confined to the laboratory, as in Jaeger and Barnard (1981). Still, red-backed salamanders do employ mechanisms of food selection that are likely related to the amount of time necessary to pass food items through the digestive tract.

The rate of food passage is important to these salamanders because of the constraint of "pulse feeding". When the leaf litter on the forest floor is wet, red-backed salamanders move into it for foraging, but this opportunity may last for only a few days following a rain, depending on evaporation (which in turn is related to ambient temperature, relative humidity, and wind speed). Thus, salamanders are confronted with rich foraging periods that may be separated by variable periods of poor foraging, when territorial individuals are confined to the vicinity of cover objects and nonterritorial individuals apparently cease foraging underground. During a rich foraging period, a salamander that eats lightly armored prey, such as dipterans, can pass a greater volume of prey through its digestive tract in a given period of time than an individual that eats heavily armored prey, such as ants, although equal volumes of flies and ants yield the same energetic benefits to the salamander (Bobka et al., 1981). Thus, salamanders seem to avoid having hymenopterans and coleopterans clogging the digestive tract when softer prey are readily available. During drier times when prey are scarcely available, any type of prey item seems to suffice, and armored prey are ingested.

One intriguing problem is how salamanders differentiate among taxa of potential prey that vary in thickness (and chitin content) of their exoskeletons. Salamanders may use some innate recognition template (visual or olfactory), being genetically programmed to differentiate between, say, coleopterans and collembolans. Although *Plethodon cinereus* ingests a broad diversity of prey types (Table 1), it is possible that individuals employ some general guidelines to differentiate among major groups. If so, olfactory cues are likely to be important, because Roth (1987) reported that "visual prey schemata can be rather simple" for salamanders. Indeed, P. cinereus can locate prey purely on the basis of olfactory cues (David and Jaeger, 1981), although visual cues are also important (Jaeger et al., 1982a; Roth, 1987).

Jaeger and Rubin (1982) showed that red-backed salamanders can learn to judge the profitabilities of dipteran prey through foraging experience, and several experiments have documented the importance of early learning and a "sensitive period" during the ontogeny of feeding behavior (see review in Roth, 1987). However, a more finely tuned understanding of the behavioral mechanisms of food selection by terrestrial salamanders must await more research on sensory modes utilized by salamanders in choosing prey from the vast diversity in the environment.

Although knowledge of the territorial behavior of red-backed salamanders is still fragmentary, general concepts are presently falling into place. Males and females establish separate territories under, and perhaps around, rocks and logs on the forest floor, but many smaller sized (presumably younger) individuals lack territories (Mathis, to be published). During dry periods, salamanders with territories feed indiscriminately (by taxa) on the scanty prey therein, although indiscriminate foraging may not be true in particularly rich territories, such as under logs containing soft bodied termites. When rain or fog occurs, salamanders move into the leaf litter and have more flexibility in seeking rich patches of prey, such as earthworms, dipteran larvae, spiders, springtails, termites, and lepidopteran larvae. Yet territorial individuals are probably prevented from wandering far from their territories because of the risk of territorial takeovers from nonterritorial "floaters" (Nunes and Jaeger, 1989). Thus, terrestrial, forest floor salamanders are ultimately constrained in foraging mobility by the risk of desiccation and may be proximally constrained, even during the best of foraging times, by the need for territorial defense. However, red-backed salamanders reap a large reward from their fluctuating environment; population biomass of P. cinereus in a forest can exceed that of all species of birds combined during peak avian density of the breeding season (Burton and Likens, 1975). The low metabolic costs, high assimilation efficiency, and foraging skills of the salamanders contribute to their high densities and wide distribution in eastern North America.

Food Selection and Mate Selection

Fecal material is a consequence of foraging, and females of *Plethodon cinereus* seem to take an inordinate interest in fecal pellets of males during the courtship season. Walls et al. (1989) found, in a laboratory experiment, that females used their snouts (where chemosensory structures occur) to "squash" male fecal pellets. Females then chose to reside near pellets that were a residue of a high quality diet (soft exoskeleton termites) compared to pellets from a low quality diet (hard exoskeleton ants). Walls et al. (1989) also found that in a natural forest habitat, males that had fed on termites had proportionally greater access to gravid females than did males that had fed on ants. Apparently females make some estimate of male quality per se or of male territorial quality in a simple way: by assaying male fecal material. These results suggest that a male's choice of diet may affect his fitness not only indirectly through his own health, but also directly through mate choice by females.

ACKNOWLEDGMENTS

I thank Charmaine Bordelon and Bryant Buchanan for reading an earlier draft of this chapter and the personnel at Shenandoah National Park for granting permission to conduct research in the park during 1970. Part of the analyses of data was performed while supported by grant BSR-8815181 from the National Science Foundation.

REFERENCES

Bobka MS, Jaeger RG, McNaught DC (1981) Temperature dependent assimilation efficiencies of two species of terrestrial salamanders. Copeia 1981:417-421

Brown JL (1964) The evolution of diversity in avian territorial systems. Wilson Bull 76:160-169

Burton TM, Likens GE (1975) Salamander populations and biomass in the Hubbard Brook Experimental Forest, New Hampshire.

Copeia 1975:541-546
David RS, Jaeger RG (1981) Prey location through chemical cues by a terrestrial salamander. Copeia 1981:435-440
Fraser DF (1976) Empirical evaluation of the hypothesis of food competition in salamanders of the genus Plethodon. Ecology 57:459-471
Horne EA (1988) Aggressive behavior of female red-backed salamanders. Herpetologica 44:203-209
Horne EA, Jaeger RG (1988) Territorial pheromones of female red-backed salamanders. Ethology 78:143-152
Jaeger RG (1972) Food as a limited resource in competition between two species of terrestrial salamanders. Ecology 53: 535-546
Jaeger RG (1979) Seasonal spatial distributions of the terrestrial salamander Plethodon cinereus. Herpetologica 35:90-93
Jaeger RG (1980) Fluctuations in prey availability and food limitation for a terrestrial salamander. Oecologia 44:335-341
Jaeger RG (1981) Dear enemy recognition and the costs of aggression between salamanders. Am Nat 117:962-974
Jaeger RG (1984) Agonistic behavior of the red-backed salamander. Copeia 1984:309-314
Jaeger RG (1986) Pheromonal markers as territorial advertisement by salamanders. In: Duvall D, Muller-Schwarze D, Silverstein RM (eds) Chemical signals in vertebrates, vol IV. Plenum, New York, pp 191-203
Jaeger RG, Barnard DE (1981) Foraging tactics of a terrestrial salamander: choice of diet in structurally simple environments. Am Nat 117:639-664
Jaeger RG, Barnard DE, Joseph RG (1982a) Foraging tactics of a terrestrial salamander: assessing prey density. Am Nat 119:885-890
Jaeger RG, Goy JM, Tarver M, Marquez CE (1986) Salamander territoriality: pheromonal markers as advertisement by males. Anim Behav 34:860-864
Jaeger RG, Joseph RG, Barnard DE (1981) Foraging tactics of a terrestrial salamander: sustained yield in territories. Anim Behav 29:1100-1105
Jaeger RG, Kalvarsky D, Shimizu N (1982b) Territorial behaviour of the red-backed salamander: expulsion of intruders. Anim Behav 30:490-496
Jaeger RG, Nishikawa KC, Barnard DE (1983) Foraging tactics of a terrestrial salamander: costs of territorial defence. Anim Behav 31:191-198
Jaeger RG, Rubin AM (1982) Foraging tactics of a terrestrial salamander: judging prey profitability. J Anim Ecol 51: 167-176
Mathis A (to be published) Territoriality in a terrestrial salamander: the influence of resource quality and body size. Behaviour
Miller TA (ed) (1980) Cuticle techniques in arthropods. Springer, Berlin Heidelberg New York
Nunes V, Jaeger RG (1989) Salamander aggressiveness increases with length of territorial ownership. Copeia 1989:712-718
Roth G (1987) Visual behavior in salamanders. Springer, Berlin Heidelberg New York

Stephens DW, Krebs JR (1986) Foraging theory. Princeton University Press, Princeton
Walls SC, Mathis A, Jaeger RG, Gergits WF (1989) Male salamanders with high-quality diets have faeces attractive to females. Anim Behav:to be published

ORAL DISCUSSION – R. Jaeger

Alicia Mathis has found that territory quality is strongly related to male size, in that larger males obtain better quality territories. Termites, which have little exoskeleton but are rich in B vitamins and fats, make up a small proportion of the diet but are critical to male reproductive success. A male with a termite log in the forest apparently has it made. The digestive bottleneck caused by eating chitinous prey might be obviated if salamanders were able to voluntarily defaecate faster, as Mathis found that males do in response to females. Although the ability to accelerate defaecation has been tested only for a termite diet, a similar mechanism might operate for a diet consisting of alternate soft– and hard–bodied prey (i.e. defaecating early ingested hard–bodied prey quickly to make room for later ingested soft–bodied prey items). There were no clear systematic differences between the individuals specializing on soft–or on hard–bodied prey on dry days, when non–territorial salamanders emerged from deeper in the soil. Gut–passage times might be compared for wet and dry days, but it will be difficult to test an optimality model, because the costs of capture are not yet quantified in natural habitats of salamanders.

ARE RULES OF THUMB SUFFICIENT FOR THE STARLING'S CHOICE OF FOOD ACCORDING TO GAIN IN ENERGY?

Werner Schuler
Zoological Intitute II
University of Göttingen
Berliner Strasse 28
D-3400 Göttingen
Federal Republik of Germany

If we want to understand the ecological significance of predator-prey relationships, e. g. the impact of birds on insect populations, we must know the population dynamics of the prey species and the foraging and prey choice behaviour of their predators. With birds eating insects direct observations in the field are rather difficult, therefore we have to rely to a large extent on indirect evidence. Analyses of stomach contents were a first approach. Descriptions of food brought to the young by parent birds and comparisons of that to the prey available in the habitat revealed far more details (Tinbergen 1981). Another line of evidence starts from the investigation of the prey's defensive means. Our knowledge of defensive chemicals of arthropods has increased enormously in the last two decades (Pasteels et al. 1983). We can assume that some of these are effective against birds, even though the evidence for this is still rather incomplete (Pasteels et al. 1983, Dettner 1987).

If we compare the different lines of evidence we can on the one side state that insects such as the monarch butterfly, which contains cardiac glycosides and other defensive chemicals (Rothschild 1985), are protected against birds. On the other side we find in the literature despriptions of a starling (Sturnus vulgaris) catching and eating this insect (Brooks 1952). Similarly the starling is enumerated as one of the birds preying upon wasps (Fry 1984), which appear to be well protected by their stinging apparatus and venoms.

Thus the different lines of evidence seemed to contradict each other. However, only as long as prey choice was conceived as a process governed by static rules: "If the predator is able

NATO ASI Series, Vol. G 20
Behavioural Mechanisms of Food Selection
Edited by R. N. Hughes

to break through the prey's defence, then this defence cannot protect against it". The possibilty remains that the same defence is effective against other species of predators. However, this view is not very satisfying. Here the introduction of gain in energy and time costs as criteria for the foraging predator (Emlen 1966, MacArthur and Pianka 1966), i. e. the development of optimal foraging theory, was a decisive improvement. Prey choice is now considered as a dynamic process and therefore a selective advantage against a given predator is conceivable, even though this is able to overcome a means of defence. This was demonstrated by Davies (1977), who has shown that the spotted flycatcher (*Ficedula hypoleuca*) can eat bumble bees after removing their stinging apparatus, but does only prey upon this insect at low ambient temperatures, when other ones which require less handling time and energy are not flying.

The concept of optimizing gain in energy per unit of time helps us to understand that a bird can accept or reject a prey species depending on the availability of alternative prey. Moreover this concept is well suited for the quantitative description of foraging decisions in the field, if the prey can be treated in first approximation as belonging to one type only, as Joost Tinbergen (1981) has shown for starlings. By further experiments and refinement of theory Kacelnik (1984) was able to decide, whether the starling is maximizing its own gain in energy, minimizing its efforts, or optimizing the gain for himself or for its family.

But how does a bird like the starling accomplish such precise solutions for foraging problems? The bird or generally the predator is thought to arrive at optimal or nearly optimal solutions by administering rules of thumb (Krebs and McCleery 1984).

Rules of thumb are mechanisms by which animals solve foraging problems. Stephens and Krebs (1986: 172ff) consider in their review only simple behavioural rules such as "take the largest" or "leave after g seconds of unsuccessful search". Similarly the design of laboratory experiments indicates that experimentors think of rules which rely upon purely mechanical measures such as the frequency, size or weight, and distribution of prey types which are all of the same quality (Schuler 1987). In the following I will describe some experiments which indicate that the starling is able to measure energy gain more directly. Most experiments deal with the response to sugar solutions, which at first sight appears to be governed only by the sense of taste. However, I will try to demonstrate that responses to sugars in the starling have to do with its ability to choose food according to gain in energy.

Preferences for sugars in the starling

My starting point was the statement that the starling does not avidly select the common sugars in solution (Kare 1971). This statement did not seem to agree to the fact that this species can consume considerable amounts of sweet fruits such as cherries and grapes. Therefore I began testing starlings with sugar solutions (Schuler 1983).

The usual method for such experiments is the two-bottle preference test, for which a test solution and water as a control are offered simultaneously. The effects of position and sequence of solutions have to be balanced by an appropriate design. Percent preference, i. e. test solution as percentage of total fluid intake, can be used as a measure of response. This score is significantly greater than 50 %, if the solution is preferred over water, it is significantly less than 50 %, if the solution is rejected, and about 50 %, if the response is indifferent.

Figure 1 shows that starlings preferred 0.5 M and 1.0 M solutions of glucose over water and responded indifferently to the lower concentrations tested. Furthermore it shows that they responded to fructose solutions in the same way and that they rejected 0.5 M and 1.0 M solutions of sucrose, whereas they were also indifferent to the lower concentrations of this sugar. The rejection of sucrose can be attributed to the mechanism of food aversion learning (Schuler 1983; see also Martinez del Rio and Stevens 1989) and is not important for our present problem.

The preferences for glucose and fructose are significant. However, they occur only at higher concentrations and are not very pronounced.

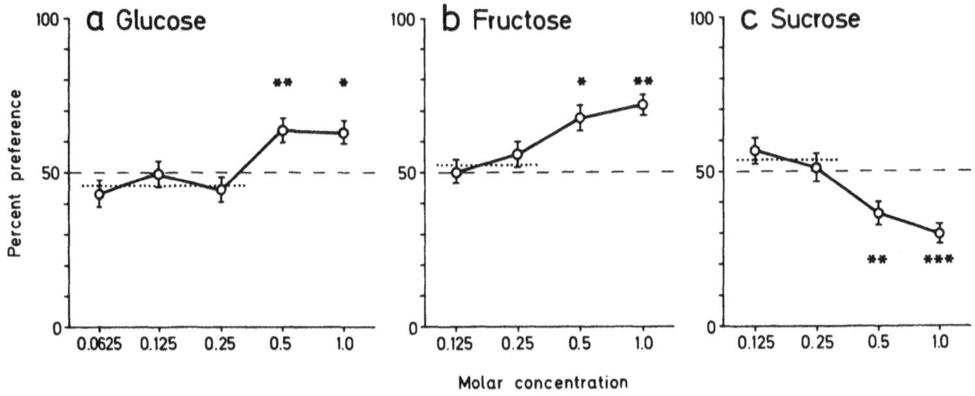

Fig. 1. Results of two-choice preference tests with starlings. Mean ± SE for n = 40 one hour tests per concentration (four with each of ten birds). .... Average of the concentrations with mean values near 50 %, indicating an indifferent response. * $P < 0.05$; ** $P < 0.01$; *** $P < 0.001$ for the distance to this average; Scheffé test subsequent to analysis of variance. (From Schuler 1983)

Since some researchers found pronounced preferences if they offered test solution and water in two differently coloured bottles, this was tested with the starlings. Colour preferences were measured in control tests before and after the sugar tests. It turned out that they changed systematically in the course of such experiments: the drinking fountain containing glucose or fructose solution during the tests was preferred thereafter. Thus a learning process had taken place in which sugar solutions acted as a reward.

To analyse this process more thoroughly, learning experiments were planned. Here each starling had a green and a blue drinking fountain the position of which was exchanged regularly. In two control sessions (No. 01, 02) water was offered in both fountains to determine individual colour preferences. Thereafter each bird was offered 0.5 M glucose for six sessions (No. 1 to 6) in the fountain with the less preferred colour and thereafter for six more sessions (No. 1' to 6') in the other fountain. The same experiment was carried out with 0.5 M fructose. In the sucrose experiment the 0.5 M solution was offered in the preferred fountain at first and afterwards in the less preferred one. (For further details see Schuler 1983).

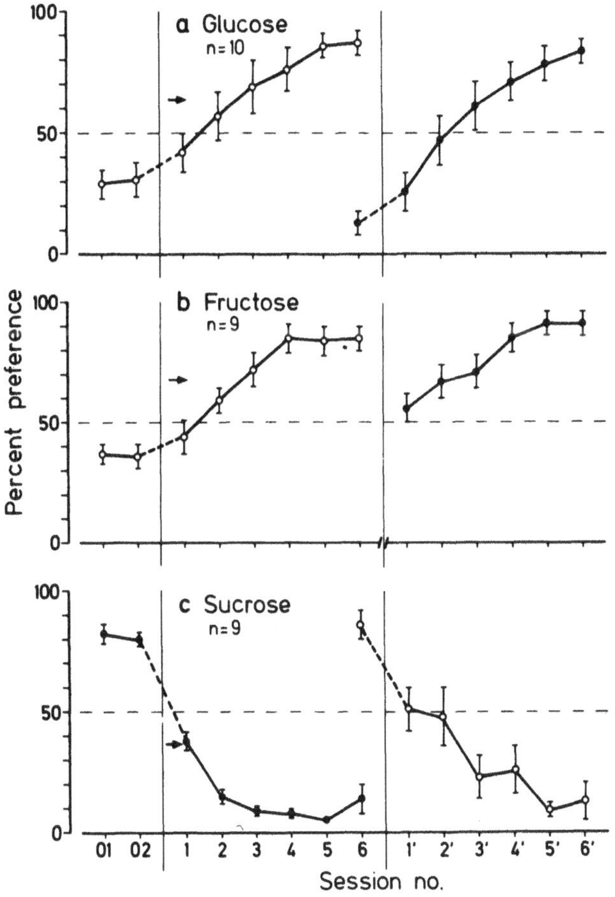

Fig. 2. Preference values (mean ± SE) for sugar solutions in learning experiments in which the starlings were offered two differently coloured drinking fountains. White dots: Fountains with the colour rejected by the individual starling during the control sessions (01, 02) with water in both fountains. Black dots: fountains with the preferred colour. Arrows: Preference values for the same concentration from the preference tests shown in figure 1. For further details see text. (From Schuler 1983)

Figure 2 shows that in the experiment with glucose the starlings gradually learned to prefer the fountain rejected during the control sessions when it contained the solution and that they gradually relearned to prefer the other fountain when the solution was offered in that one. With fructose they similarly learned to prefer the fountain, initially rejected during control sessions, and relearned to prefer the other fountain when the solution was transferred to that. In the sucrose experiment they learned to reject the fountain containing the solution in both parts of the experiment.

The preferences became very pronounced. However, these are learned and not immediate responses. Moreover they refer primarily to colour and not to the taste of the solution, since in the glucose experiment the starlings chose according to colour but not according to taste, when in session no. 1' the solution was offered for the first time in the other fountain. After this they only gradually relearned to prefer the solution.

So we have here learned preferences for colours which are caused by a rewarding effect of sugar solutions. But this effect did not seem to be present immediately after tasting.

Since taste responses can be expected to follow immediately after fluid uptake, two-choice drinking tests of only 5 min duration were carried out (see Schuler 1983). In these tests no significant preference for 0.5 M fructose was found in contrast to comparable tests with quinine, which was rejected immediately. This was further evidence for the notion that the marked preferences shown in the learning experiment were not exclusively taste responses.

Therefore glucose and fructose must act as rewarding stimuli in a different way. One has to think of positive feedback from the digestive system or from metabolism, e. g. of an increase of the level of blood glucose. Such postingestional stimuli play an important role in the control of feeding (e. g. Morgan 1965). Furthermore they can act as delayed unconditioned stimuli (Rozin and Rozin 1981).

These experiments indicate that starlings choose sugar solutions because of postingestional effects rather than just because of sweet taste. If we consider that by drinking sugar solutions they took up considerable amounts of energy, we can interpret their behaviour in these tests as a choice according to gain in energy. Gain apparently could not be measured by simple behavioural rules referring to mechanical measures such as size, weight or volume, since these would not have been adequate for this choice problem.

Thus these experiments which started from the investigation of the starling's response to sweet taste led to the hypothesis that this species can measure gain in energy more directly than by simple behavioural rules.

From this hypothesis three predictions can be derived: If the starling's response to sugar solutions is a choice according to gain in energy, ..

.. this preference should depend on the alternative food available during preference tests;

.. it should be able to prefer solid food containing additional energy in form of sugar;

.. it should be also able to prefer food containing additional energy in other form.

These prediction were tested in the following experiments (Schuler 1987).

The preference for sugar solution depends on the available food.

In a learning experiment of the type shown in Figure 2, I tested whether starlings develop a preference for a fountain containing 0.5 M fructose solution, when they were given simultaneously an easily digestible and energy rich food. For this an experimental group of 5 starlings were given 15 g mealworms each per session in addition to their usual food, which was turkey starter pellets. This was the only food given to the 5 starlings of the control group.

Two starlings of the control group chose always according to colour. The remaining three learned to prefer the fountain containing the solution in both parts of the experiment. In contrast to this, the experimental group did not develop a preference neither when the solution was offered in the fountain rejected during control sessions, nor when it was offered in the preferred one.

However, the experimental birds consumed only very small quantities of both fluids, probably because mealworms contain about 60 % water. Therefore the reason for the lack of preference could have been that they tasted the solution only about two times per 1,5 h test.

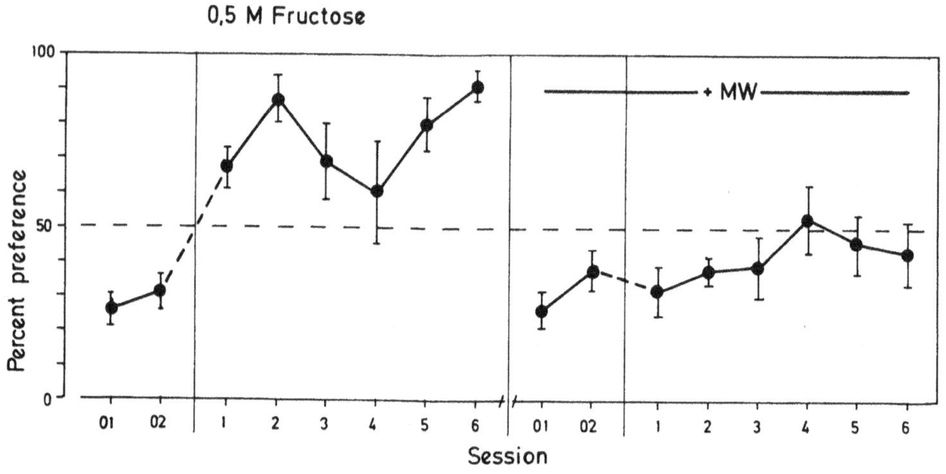

Fig. 3. Influence of the food offered on sugar preference. Mean ± SE of 8 starlings. **+MW:** Dried mealworms were offered as an additional food during these sessions. (From Schuler 1987)

To exclude this possiblity in a second experiment the mealworms given were first killed and dried. Eight starlings were offered after two control sessions 0.5 M fructose in the fountain with the less preferred colour for six sessions, during which they were given as food turkey starter pellets only. After this control experiment they were given additionally dried mealworms in a corresponding series of sessions.

Figure 3 shows that the starlings developed a marked preference for the fountain containing the solution in the course of the control experiment and that they responded indifferently to the solution when they were given dried mealworms during the experimental sessions. Here, too, they consumed less fluids than in the control sessions. As, however, the mean amount consumed corresponds to drinking five times from the solution per 1,5 h test, it can be stated that they did not show a sugar preference here, even though they tasted the solution repeatedly.

Preference for food containing sugar.

Here I tested whether starlings prefer a food enriched with sugar. Fluid food colours were used to prepare a green and a red test food from turkey starter pellets. To one of the test foods fructose and glucose was added. Nine starlings were offered a choice between these two food types twice a day for two hours. For five of them the green food contained the sugars in the first and the red food in the second part of the experiment. The remaining four birds were offered red food containing the sugars at first and enriched green food thereafter.

In the first part of the experiment, which lasted for five days, seven of the nine starlings learned to prefer the food containing sugars irrespective of its colour. Similarly in the second part, in which they were tested for four days, eight of them succeeded. The preference was significant in both cases ($p < 0.05$ and $p < 0.005$, respectively; Wilcoxon tests). So these starlings showed that they were able to prefer food enriched with sugar.

Preference for food containing fat.

If the starlings prefer food containing sugar because of its energy content, they should be able also to prefer a food type which contains digestible energy in other form. To test this prediction nine starlings were offered a choice between green and yellow test food prepared from turkey starter pellets, one of which was enriched with salad oil.

Figure 4 shows that the starlings learned to prefer the green food when oil was added to it, and chose the yellow alternative when that type was enriched with oil. After that paraffin oil, which has a similar viscosity as salad oil but no digestible energy, was added to the green food. Here the starlings did not develop a preference. When finally the tests with salad oil were repeated they again preferred the green or yellow food depending on which contained the fatty oil.

Thus the starlings demonstrated that they are able to prefer food rich in energy, even if this energy is not contained in sugar.

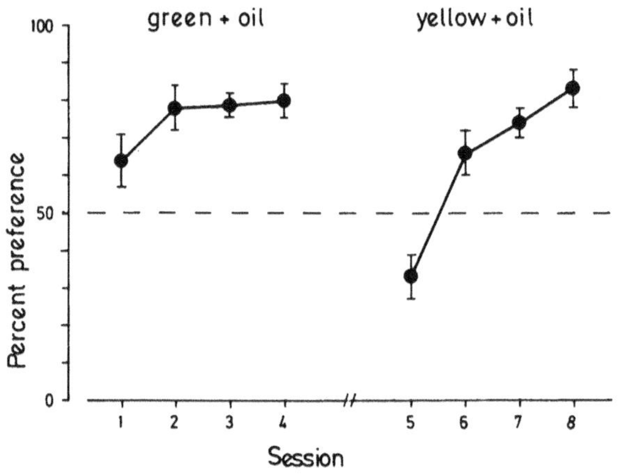

Fig. 4: Responses of 9 starlings in a two-choice experiment with green and yellow food one of which was, as indicated, enriched with oil. (After Schuler 1987)

The fact that the starlings responded in these experiments as was predicted, is evidence in favour of the hypothesis that they are able to choose according to gain in energy. In my opinion the simplest explanation for their behaviour is that they are able to measure gain in energy by postingestional effects such as by a rise in the level of blood glucose or blood fat. The experiments do not exclude the possibility that the starlings used their sense of taste for the preferences shown, preferring food containing sugar because of its sweet taste and food containing oil because it is "tasty" in some other way.

However, the finding that sugar preferences were not shown by the starlings when they could gain energy by eating insects, cannot be explained by a simple taste response. In any case in these experiments gain in energy could not be measured by simple behavioural rules referring to secondary cues such as colour, size or weight.

Trying to find an immediate response to sugar solutions

In the two-bottle drinking tests described the starlings showed preferences caused by a rewarding effect of sugar which did not seem to be its sweet taste. Therefore Dietmar Zinner and I tried by another method to demonstrate immediate responses to sugars, which would be the best evidence for a reponse to the taste quality sweet (Zinner 1985, Schuler 1987). Here in each test the starlings were given one drinking fountain for only five minutes each, which contained a sugar solution in the experimental and water during the alternating control tests. All behavioural events of drinking, eating, etc. were registered by means of an event recorder. Fluid consumption during experimental tests as percentage of fluid intake during experimental plus corresponding control tests was calculated as a preference score. Since fluid uptake appeared to influence uptake of food, the relative duration of eating, i. e. duration of eating during experimental tests as percentage of the duration of eating during experimental plus corresponding control tests was calculated accordingly.

Figure 5 shows in its upper half that the starlings showed a preference for some concentrations of fructose and glucose in these experiments (it was significant for 0.25 M and 0.5 M fructose, as well as for 0.125 M, 0.25 M and 0.5 M glucose). The time course of drinking acts showed that in the best case, i. e. with 0.5 M glucose, a significant difference between experimental and control tests could be found only ten seconds after the start of the test. This is evidence for an immediate response to the solution and therefore indicates that the starling is able to respond to the sweet taste of sugars.

However, the preferences shown here are again not very pronounced and refer only to part of the concentrations tested.

The lower half of figure 5 shows that the uptake of the sugar solutions had a more marked effect on the relative duration of eating. This effect was greatest with 1.0 M solutions for which the starlings did not show a drinking preference. Since the relative duration of eating as a function of sugar concentration was qualitatively different from fluid preference as a function of sugar concentration, the influence of the sugars on eating cannot be a direct taste reponse.

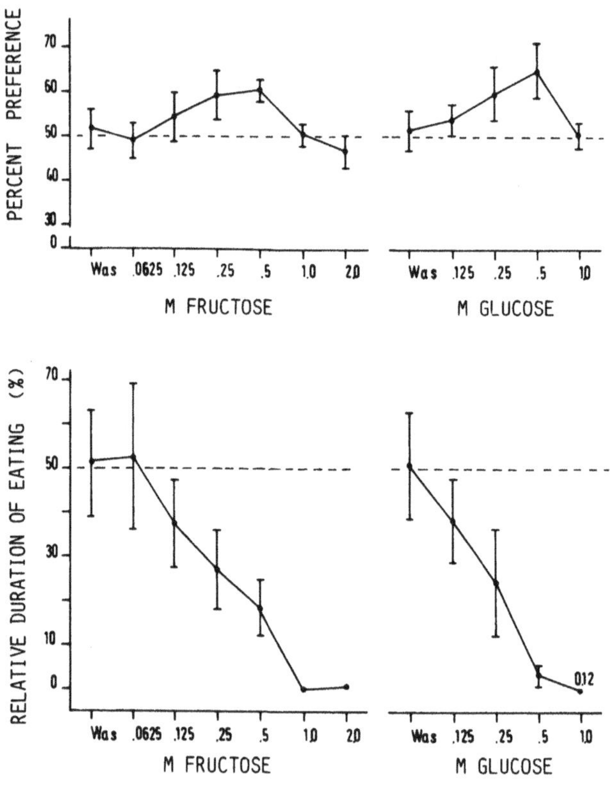

Fig. 5: Reponses of starlings to sugar solutions in one-choice tests of 5 min duration as compared with alternating control tests with water. Was: Water was offered in experimental and control tests. (After Zinner 1985)

These experiments show that uptake of higher concentrations of glucose and fructose, i. e. uptake of fluids rich in energy, influences the uptake of food considerably. I think this is further evidence for the notion that this species is able to measure gain in energy not only by secondary cues.

Discussion

The experiments described give evidence for the notion that the starling has the ability to measure energy gain by postingestional effects. Nevertheless, rules referring to the frequency, weight or volume, and distribution of prey are useful for many foraging decisions. For some problems, however, they must be supplemented by a more direct evaluation of energy gain: Prey types differing in quality, e. g. in digestible energy per unit of weight or volume, can only be differentiated by such a measure. Furthermore all novel prey types have to be evaluated in such a way before their usefulness can be compared to those already known to the predator.

Physiologists know effects of the uptake of nutrients measured by a rise in the level of blood glucose or blood fat on the regulation of feeding (Schmidt 1980). Therefore the possibility that effects of energy uptake measured in such a way may influence foraging, food choice, and other forms of behaviour should be considered by behavioural ecologists and ethologists. I think that an influence of nectar supply on territorial behaviour in sunbirds (Gill and Wolf 1975) or an influence of sugar concentration on their singing behaviour (Pflumm et al. 1984) can be best explained in this way. Moreover, if foraging decisions do not only depend on time costs and the amount of prey gained (which could be both measured by rules of thumb), but also on the energy expenditure during foraging (e. g. Cowie 1977, Kacelnik and Houston 1984), the decisive net gain appears to be evaluated best by a common biochemical measure of both energy gain and expenditure.

Summary

An analysis of preferences for sugar solutions in the starling led to the hypothesis that this species can choose according to gain in energy measured by postingestional biochemical effects. As predicted from this hypothesis starlings (1) did not develop a preference for a sugar solution if food rich in energy was available; (2) were able to choose food enriched in energy in the form of sugar; (3) were also able to choose food enriched with energy in the form of oil.

Measuring energy gain more directly than by rules of thumb appears to be an important supplement to these rules for some foraging decisions, e. g. the evaluation of food types differing in quality.

Acknowledgements. I am very grateful to Alex Kacelnik and Bob Vadas for helpful comments on the first version of this paper.

References

Brooks M (1952) Starlings, Sturnus vulgaris, eating monarch butterflies. Auk 69: 89
Cowie RJ (1977) Optimal foraging in great tits (Parus major). Nature, Lond. 268: 137-139
Davies NB (1977) Prey selection and search strategy of the spotted flycatcher (Muscicapa striata). A field study of optimal foraging. Anim Behav 25: 1016-1033
Dettner K (1987) Chemosystematics and the evolution of beetle chemical defenses. Ann Rev Entomol 32: 17-48
Emlen JM (1966) The role of time and energy in food preferences. Am Nat 100: 611-617
Fry CH (1984) The bee-eaters. Calton
Gill FB, Wolf LL (1975) Economics of feeding territoriality in the golden-winged sunbird. Ecology 56: 333-345
Kacelnik A (1984) Central place foraging in starlings. I. Patch resident time. J Anim Ecol 53: 283-299
Kacelnik A, Houston AI (1984) Some effects of energy costs on foraging strategies. Anim Behav 32: 609-614
Kare MR (1971) Comparative study of taste. In: Beidler LM (ed) Handbook of sensory physiology. Vol IV/2. Berlin, pp 278-292

Krebs JR, McCleery RH (1984) Optimization in behavioural ecology. In Krebs JR, Davies NB (eds) Behavioural ecology. 2nd ed. Oxford, pp 91-121
MacArthur RH, Pianka ER (1966) On optimal use of a patchy environment. Am Nat 100: 603-609
Martinez del Rio C, Stevens BR (1989) Physiological constraints on feeding behavior: Intestinal membrane disaccharidases of the starling. Science 243: 794-796
Morgan PM (1965) Physiological psychology. 3rd ed. New York
Pasteels JM, Grégoire J-C, Rowell-Rahier M (1983) The chemical ecology of defense in arthropods. Ann Rev Entomol 28: 263-289
Pflumm W, Comtesse H, Wilhelm K (1984) Sugar concentration and the structure of the sunbird's song. Behav Ecol Sociobiol 15: 257-261
Rothschild M (1985) British aposematic lepidoptera. In: The moths and butterflies of Great Britain and Ireland. Vol 2, pp 1-62
Rozin E Rozin P (1981) Food selection. In: McFarland D (ed) The Oxford companion to animal behaviour. Oxford, pp 209-214
Schmidt FR (1980) Durst und Hunger: Allgemeinempfindungen. In: Schmidt FR, Thews G (eds) Physiologie des Menschen. Berlin, pp 337-345
Schuler W (1983) Responses to sugars and their behavioural mechanisms in the starling (Sturnus vulgaris L.). Behav Ecol Sociobiol 13: 243-251
Schuler W (1987) Untersuchungen zur Bedeutung der unmittelbaren Erfahrung und des Lernens im Funktionskreis Nahrung beim Star. Habilitationsschrift. Göttingen
Stephens DW, Krebs JR (1986) Foraging theory. Princeton
Tinbergen JM (1981) Foraging decisions in starlings (Sturnus vulgaris L.) Ardea 69: 1-67
Zinner D (1985) Untersuchungen zur Frage der Ontogenese von Geschmacksreaktionen und zum Einfluß frühkindlicher Erfahrung auf die Nahrungswahl des Stars (Sturnus vulgaris L.). Diplomarbeit. Göttingen

ORAL DISCUSSION – W.Schuler.

Q. (DeMott) Did you try using artificial sweeteners?
A. We tried them and found no effect; in general birds don't respond to artificial sweeteners. We tried glycogen, which is not sweet, and got some effect, but not as strong as with sugars.

Q. (Kacelnik) I disagree with your interpretation in terms of rules of thumb (ROT). If the animal was using glycogen levels in the liver, or the turnover of ATP, I would still call that a ROT. The only thing I would not call a ROT is the currency that natural selection is maximizing. i.e. fitness. All others are ROT cues that correlate to fitness.
A. I agree, if you call *all* mechanisms by which animals solve foraging problems ROT. However, usually ROT are called simple behavioural rules (e.g. 'take the largest'). My point is that starlings can assess energy gain more directly. Moreover, birds can assess not just energy or time but also effort; we should look for the point where effort and gain converge and this could be the level of blood glucose.

Q. (Jørgensen) Weren't your sugar concentrations higher than those seen in nature? With unnaturally high sugar concentrations your animals may have become addicted to the sugar and lost their appetite.
A. A related species (*Sturnus sturninus*) feeds on nectar; and starlings eat fruit, so they do encounter high concentration of sugar in nature.

Q. (Ball) Have you given them a choice between oil and sugar to see if they can assess energy?
A. I have not tried it. Oil and water would make an emulsion that would not work if offered as a fluid. However, it would be interesting to compare foods containing sugar and oil, respectively.

PREY SELECTION AND PROCESSING IN A STOMATOPOD CRUSTACEAN

Roy L. Caldwell and Michael J. Childress
Department of Integrative Biology
University of California at Berkeley
Berkeley, CA 94720
USA

I. INTRODUCTION

Stomatopod crustacea constitute one of the most highly specialized groups of marine predators. Their greatly enlarged and highly modified second thoracopods provide them with the weaponry to subdue and process even the most active, armored, and/or dangerous prey (Caldwell and Dingle, 1976). Their sophisticated visual (Cronin and Marshall, 1989) and chemosensory (Caldwell and Lamp, 1981; Caldwell, 1982, 1985) systems permit the detection and identification of potential prey. And their learning ability allows them to benefit from previous experience gained feeding on different animals (Reaka, 1980; Full, et al., 1989). Here, I will: 1) briefly review the biology of stomatopods relevant to their foraging behavior; 2) discuss field studies on members of one genus, Gonodactylus, showing that stomatopods can have a considerable influence on the population structure, distribution and abundance of prey species; and 3) present laboratory data that demonstrate the ability of stomatopods to adjust their behavior in response to the costs and benefits associated with selecting different sized individuals of one prey species.

II. BIOLOGY OF STOMATOPODS

Stomatopods, or mantis shrimp as they are commonly known, are the only living representatives of an old lineage that diverged from malacostracan stock approximately 400 million years ago. The group is entirely marine. Most of the estimated 500 to 800 species live in tropical to mid-temperate shallow habitats. Stomatopods are typically secretive, occupying self-excavated burrows or pre-existing cavities in coral or rock, and are rarely encountered by casual observers (Reaka and Manning, 1981). However, mantis shrimp can occur in considerable numbers (Dingle et al., 1977; Steger, 1987) and may exercise a sufficient predatory force to shape local communities (Caldwell et al., 1989).

NATO ASI Series, Vol. G 20
Behavioural Mechanisms of Food Selection
Edited by R. N. Hughes

All species of stomatopods are dioecious and most exhibit little sexual dimorphism. Females typically tend their eggs until hatching, and in some groups, early larval stages may spend several more days in the female's cavity. The free-swimming, predatory larvae spend from a few days to several months in the plankton before they settle and molt to postlarvae. Although stomatopods exhibit indeterminate growth, maximum adult size varies considerably across species, ranging from small burrowing nannosquillids such as Pullosquilla that achieve a maximum length scarcely over 1 cm to the largest described species, Lysiosquilla maculata, that may reach 39 cm in length. The majority of species, however, falls into the 3 to 10 cm size range.

A. Raptorial Appendages

The most significant development in the evolution of stomatopods, and one that appears to dictate much of their biology, is the enlargement and modification of the second pair of thoracopods. In modern stomatopods, they form a lethal pair of raptorial appendages used in prey capture and processing, construction, and intra- and interspecific agonistic interactions.

Caldwell and Dingle (1975,1976) divided stomatopods into two functional groups, spearers and smashers, based on the morphology and use of the raptorial appendages. Spearing appears to be the primitive condition and representatives are found in all four superfamilies. Spearing second thoracopods tend to be slender, have a flat, blade-like propodus, and terminate in dactyls armed with from 2 to 17 barbs. Spearers typically specialize on soft-bodied prey such as fish or shrimp. These are seized by or impaled on the open dactyl spines with an explosive extension of the raptorial appendage. The leading edge of the propodus frequently bears sharp, stiff spines so that the barbed dactyl can be folded back against it, securely trapping the struggling prey. The spearing raptorial appendages are designed for speed, slicing through the water in 3 to 4 msec to capture even the most adroit prey that come within their range.

In contrast, the raptorial appendages of smashers are designed to deliver crushing blows to break through the armored defenses of prey such as crabs, bivalves, and gastropods. In these animals, the merus of the second thoracopod is larger, containing a more powerful muscle mass to power the strike. The stout propodus is better able to withstand the forces generated on impact. The dactyl is usually unarmed, terminating in a sharp tip, and has a highly calcified, inflated heel which bears the brunt of strike. The strike is usually delivered with the dactyl folded tightly against the propodus, the heel making contact with the target. However, when encountering less armored quarry, smashers can strike with the dactyl unfolded, impaling the victim on the needle sharp tip with a rapier-like thrust. Smashing most likely evolved from spearing and appears in several different lineages, although it is most common in the Gonodactyloidea and Lysiosquilloidea.

B. Sensory Systems

In conjunction with the development of potent raptorial appendages that allowed stomatopods to become active predators, they have evolved acute visual and chemosensory senses. The visual system is one of the most spectacular in the animal kingdom. The stalked, independently moving compound eyes have a number of morphological and physiological adaptations permitting the localization of prey. The structural arrangement of the ommatidia provides an unusual range-finding system that permits binocular or even trinocular vision within a single eye (Manning et al., 1984; Marshall, 1988). In sit-and-wait predators such as many spearers, this monocular stereopsis is probably used in directing and triggering the strike. In species such as smashing gonodactylids that actively stalk and pursue their prey, it may be used at greater distances. Perhaps even more unusual is the ability of some stomatopods to process spectral information. Cronin and Marshall (1989) have recently found that the eyes of some mantis shrimp have up to 10 spectral classes of photoreceptor. These, coupled with the presence of colored filters that occur within some ommatidia (Marshall, 1988), should permit a sophisticated form of color vision. This system even includes receptors for polarized as well as ultraviolet light. While psychophysical tests have yet to be performed to determine just how well and what stomatopods see, gonodactylids are able to distinguish visually among individual mantis shrimp that, to a human eye, vary only slightly with respect to color and pattern (Caldwell, in prep.).

It is not unreasonable to assume that these same visual capacities are used in the detection and selection of prey. For example, G. chiragra appear to be able to distinguish between snails and hermit crabs in the same snail shells based on the pattern of movement. In laboratory studies, we were able to show that G. chiragra preferred snails over hermit crabs in the same size and species of gastropod shell (Caldwell and Dingle, 1976). To determine if the stomatopods were using the form of motion to distinguish between the two types of prey, threads were attached to two identical empty snail shells and pulled them at the same over-all speed in a line 20 cm in front of a stomatopod burrow. One shell was pulled smoothly, mimicking the gliding movement of a snail. The other was pulled with a series of short jerks, similar to the motion of a walking hermit crab. Of the 10 animals tested, 8 attacked the snail mimic (Binomial Test, $p < .1$), even though the motion of the hermit crab mimic appeared more noticeable. Simple experiments such as this, combined with many observations of diurnal stomatopods hunting in the field, suggest that these animals rely heavily on visual information when foraging and have the capacity to detect prey at distances of at least several body lengths.

The chemosensory capabilities of stomatopods are no less remarkable. The paired antennules and antennae are the primary chemosensory organs used by mantis shrimp. The triramous antennules are used to sample the water column and to examine objects. When odors are detected in the water, animals increase their rate of sampling by performing frequent short vertical flicks of the flagella. Objects are investigated by an extension of the antennules to make contact, followed by a series of rapid sweeping movements.

There appears to be a fair degree of coordination between the eyes and the antennules. One eye will quickly orient towards an object that moves into its visual field, followed immediately by the ipsilateral antennule taking up the same orientation. Conversely, odor released within the scanning range of an antennule will elicit an orientation response, followed by the eye moving to the same alignment. This coupling of the eye and antennule is, at least in part, mediated by four or five long hairs that extend vertically from the proximal trunk of the antennule and which make contact with the eye when it is aligned with the antennule. Removing these hairs disrupts the coordination of the eye and antennule (Caldwell, unpublished).

The antennules play several different roles once the processing of prey commences. Immediately prior to a strike, they are lowered, brought together and are used to tap the prey to determine if it is sufficiently stable to warrant striking. This appears to be a mechanical, not a chemosensory function. However, once the mantis shrimp strikes, it uses the antennules to examine the prey's surface to detect damage and/or the presence of edible tissue. Animals also occasionally sweep the immediate vicinity with the antennules to locate pieces that may have been dropped or to find and recover the prey if it was knocked a considerable distance by the force of the strike. The antennae seem to be employed primarily during prey capture and processing to detect pieces of food that are knocked loose by a strike. They are ideally positioned for this purpose, hanging down along the side of the head with the tips making contact with the substrate.

Stomatopods, including species such as gonodactylids that hunt primarily using visual cues, rely heavily on chemosensory and tactile inputs to process prey. This is not surprising given that most prey are consumed inside a dark cavity or burrow. However, even highly visual species, when feeding in the open, still switch to using chemosensory and tactile cues even though visual inputs are available.

There is little information available concerning what chemical cues stomatopods use to detect prey. In the laboratory, gonodactylids respond by actively searching when water that contained live snails or crabs is introduced into their aquaria. There is even some indication that they can follow the trail laid down by a snail moving across the substrate (Childress, unpublished). Also, Caldwell and Lamp (1981) demonstrated that G. festae could recognize and avoid the odor of an octopus emitted from a cavity. However, perhaps the best indication of their chemosensory ability comes from work on intra- and interspecific recognition of other stomatopods. Several Gonodactylus can identify the odor of other individual stomatopods with which they have had previous experience and use this information to mediate future aggressive interactions (Caldwell, 1979, 1982, 1985). Gonodactylus can distinguish among individuals of their own species as well as among individuals of other species within the genus. Gonodactylus can also determine the sex of an individual using odor cues, but apparently cannot establish reproductive or molt condition (Caldwell, 1986). Recent work (Caldwell, in prep.) demonstrates that members of a mated pair remember one another and will not attempt to evict their previous partner from a cavity for up to 2 weeks after the union has been dissolved. These experiments reveal a chemosensory ability unusual for crustacea and suggest that the capacity for similar distinctions among potential prey might also be possible.

C. Learning

There have been few studies on the learning ability of stomatopods. Reaka (1980) did show that gonodactylids could quickly learn and remember characteristics of a home cavity, and the work cited above on individual recognition reveals the ability to rapidly learn the behavioral characteristics of an opponent. Recent studies in our laboratory have shown that stomatopods can learn to associate color or texture with the presence or absence of a food reward within an artificial snail (LaMon and Caldwell, in prep.). Finally, we have evidence that adult Odontodactylus scyllarus, a large Indo-Pacific smasher, can learn to more efficiently opening a species of snail with which they have had no experience. When we gave O. scyllarus large Tegula funebralis, a local California snail, the stomatopods were very inefficient breaking them open. However, after several days, the number of strikes taken to process the prey was markedly reduced. Observations showed the improvement resulted from the animals learning where to strike the shell to break off the entire outer whirl (Revis and Caldwell, unpublished).

III. PREY CAPTURE AND PROCESSING

Few studies have dealt specifically with the feeding behavior of stomatopods (but see Reaka, 1985, 1987; Hamano and Matsuura, 1986a, 1986b; Caldwell et al., 1989; Basch, in press;). As discussed above, most spearing species specialize on soft-bodied prey and many are sit-and-wait predators, ambushing passing fish or crustacea from their burrow entrance. For example, members of the genus Lysiosquilla feed on fish. Individuals construct large burrow systems in sandy substrates and conceal the entrance with a sand-mucus cap which contains a small hole in the center. During the day, the stomatopod sits just below the entrance with only its eyes and antennules, both camouflaged with sand-like coloration, protruding through the entrance. When a fish passes within striking distance, the animal strikes through the cap, impaling the victim, and drags it down the burrow where it is consumed. At night, animals remove the mucus-sand plug and hunt from the open burrow entrance, using their sand-colored antennal scales to fill the opening. Presumably, the strike is more effective not having to explode through the cap; in low light, the need for camouflage is not as great. Burrows are usually occupied by mated pairs. Males have larger raptorial appendages and eyes, do most of the hunting, and provision the females. Females spend most of their time deep inside the burrow, brooding eggs and typically do not hunt unless they have lost their mate or hunting success has been poor. In the later case, the male and female will occupy separate entrances to improve their success rate.

A few spearers hunt away from the burrow. Pseudosquilla ciliata is perhaps the best known example. Individuals of this species reach a length of 10 cm, are diurnal, cryptically colored, have excellent vision, and inhabit shallow back-reefs and sandy grass flats. P. ciliata can frequently be seen foraging up to several meters from their burrow and are particularly active on the incoming tide when water depths are so shallow that large fish predators have been forced off the reef flat. This common predator feeds primarily on soft-bodied prey such as shrimp, small fish, and annelids, but it also will take small crabs and even stomatopods, especially if these animals have been trapped and/or disabled by prolonged exposures during low tide (Hatziolos, 1979; Caldwell, unpublished).

The large, temperate-water stomatopod, Hemisquilla ensigera also may forage far from its burrow. This species does not fit into the smasher-spearer dichotomy and appears to be a feeding generalist. It lacks both the multi-spined dactyl of spearers and the heavily calcified dactyl heel of smashers, possessing a more blade-like dactyl. It digs burrows in muddy substrates at depths from 5 to more than 30 m and feeds on a broad range of prey including annelids, mollusks, crustaceans, fish, and even echinoderms (Basch, in press; Caldwell, unpublished). Its undifferentiated raptorial appendages apparently enable it to capture and process a greater array of items. Its large size (up to 25 cm total length) provides greater protection from predators. It is not uncommon during the day to find large animals foraging up to 20 m or more away from their burrows. While some small, soft-bodied prey may be consumed where they are found, larger items, and those that require considerable processing, are returned to the burrow where they are broken apart and consumed.

Smashers, such as members of the genus Gonodactylus, generally take armored prey such as crabs, snails, and hermit crabs, although many will take softer bodied carrion if it is available. Also, for a few days following a moult when the animals cannot strike effectively, or if both raptorial appendages have been damaged, most gonodactylids will actively search for more easily processed food such as worms, weakly armored crustacea, and thin-shelled clams. Interestingly, Gonodactylus missing only one raptorial appendage can process armored prey and do not appear to alter their diet selection (Childress and Caldwell, in prep.).

A few gonodactylids display highly specialized modes of foraging. For example, G. festae is found on the Pacific coast of Central America where fish predation is intense and visibility is frequently poor. In one population studied by the senior author at Naos Island, Panama, individuals rarely left their cavities to feed. While they quickly took snails and crabs that passed near their cavity, they would not pursue animals that were more than a couple of body lengths from the entrance. A common mode of foraging was grazing on barnacles. In this habitat, barnacles recruited rapidly and covered the rocks in which G. festae lived. However, there was usually a barren ring immediately surrounding the cavity entrance. Observations showed that the stomatopods extended out of the cavity and smashed in the tops of the barnacles, picking out the flesh. However, they rarely fed in this way more than one or two body-lengths from the cavity, presumably due to heavy fish predation (Caldwell, unpublished).

More typically, gonodactylids display a mix of sit-and-wait and central-place foraging. They either wait in the cavity entrance for suitable prey to approach within grasp, or leave the cavity and search the immediate vicinity for prey that are captured and returned to the cavity for processing. The variables involved in switching between these two types of foraging have not been formally investigated, but hunger, stage in the moult cycle, and potential risk of predation appear important in making this decision. Animals that have recently fed are reluctant to leave their cavity to take prey, but will readily accept food placed near the cavity entrance (Caldwell, unpublished). The same is true of recently moulted gonodactylids which are vulnerable to predation and to losing their cavities to competitors (Steger and Caldwell, 1983). Additionally, many gonodactylids restrict their foraging away from the safety of the cavity to times when water depths are sufficiently shallow to preclude large fish from hunting on the reef flat (Caldwell et al., 1989). Once a prey item is selected, it is unusual for a gonodactylid to process and consume it in the open. Typically, the food is returned to the home cavity, or sometimes to an unoccupied shelter, where it is eaten. Occasionally, large items that will not fit through the entrance must be broken apart before they can be pulled into the cavity. But even then, animals will often strike the prey from the safety of the burrow.

Gonodactylus attack and process various types of prey differently. A crab is usually stalked, then subjected to a lunging attack coupled with several strikes in quick succession, both raptorial appendages firing simultaneously. Often the strikes are aimed at the major chelapeds, knocking them off to prevent retaliation. Once the crab is subdued, it is returned to the cavity where the stomatopod breaks it into pieces and cuts away the flesh using its sharp, serrated mandibles.

Snails and hermit crabs are treated differently. Once an animal is selected, it is simply picked up and carried to the safety of a cavity. The stomatopod then manipulates the prey using its maxillipeds until an orientation suitable for striking is achieved. The shell is rocked back and forth until a stable position is found, a final check for stability is made with the antennules, and the stomatopod strikes. However, in contrast to attacks on crabs, the gonodactylid usually strikes snails and hermit crabs using one appendage at a time and works at a slow, relatively constant rate, rarely striking more than 3 or 4 times per minute. The raptorial appendages are fueled by high-energy arginine phosphate stores that are depleted by several strikes in quick succession. The paced striking while processing prey, using one appendage at a time, is adapted for the retention of striking reserves that might be needed to fend off competitors (Full et al., 1989).

IV. EVIDENCE THAT STOMATOPODS CAN AFFECT PREY POPULATIONS

There have been few field studies on the consequences of stomatopod predation. Steger and Bemis (in prep.) observed colonies of Lysiosquilla sulcata in the Society Islands and noted a decrease in the number of prey fish in the immediate vicinity of stomatopod aggregations. Reaka (1987) reported that adult Gonodactylus eat and suppress the recruitment of juvenile stomatopods where adult densities are high. To date, the only manipulative study to examine the effects of stomatopod predation was on Gonodactylus bredini in Thalassia beds off Galeta Point on the Atlantic coast of Panama (Caldwell et al., 1989; Roderick et al., in prep.). G. bredini are extremely abundant in this habitat, living in pieces of calcareous algae and coral rubble and occurring at densities up to 20 m^2. Preliminary observations of stomatopods captured in the act of eating showed that the most common prey were several species of snails, most in the genus Cerithium, and hermit crabs occupying Cerithium shells. Not surprisingly, larger stomatopods tended to consume larger prey.

To investigate what effect the stomatopods might be having on prey populations, we selected six, 7 by 7 m plots and removed all pieces of rubble that might contain stomatopods. Then, to each plot, we added 36 artificial concrete cavities that were designed to attract 35-45 mm gonodactylids (see Steger, 1985). Three of the plots were allowed to recruit stomatopods into the cavities. Three served as controls, the entrances to the cavities plugged with corks. Every 10 days, prey populations were sampled on each plot. At the end of 40 days, the control and experimental plots were switched by exchanging the plugged and unplugged cavities and sampling continued for another 30 days.

The artificial cavities rapidly recruited 35-45 mm G. bredini so that the experimental plots maintained an average density of 0.5 stomatopods per m^2 in this size class. Pre-manipulation densities of stomatopods in this size range on these plots were 0.2 stomatopods per m^2. However, even though we more than doubled the density of large G. bredini in the plots, this is still considerably less than the 1.3 stomatopods per m^2 in this size class recorded by Steger (1985) for this area. The difference is due to the fact that we deliberately selected plots with little rubble and hence few G. bredini. There were no other large stomatopods living in the experimental or the control plots due to the lack of suitable cavities. It is unlikely that gonodactylids passing through the area fed in these areas due to the lack of shelter.

Throughout the entire study, the presence or absence of stomatopods had no effect on the abundance of large (>16mm) snails and hermit crabs. However changes were noted in the abundances of small (2-8mm) and medium (9-16mm) prey on the experimental compared to the control plots. During the first 40 days of the study, there was a massive recruitment of small Cerithium into all plots. We estimate that each day each stomatopod ate an average of 12.1 small and 1.5 medium-sized gastropods and 5.3 small hermit crabs. During the second part of the study, there were fewer small snails available. The stomatopods responded by taking a significantly greater number of medium-sized snails (8.1 small and 3.8 medium gastropods and 5.4 small hermit crabs per stomatopod per day). These rates of consumption are similar to those estimated for similar-sized G. bredini using underwater video and laboratory data (Caldwell et al., 1989). These data show that G. bredini are selecting their prey, and that even at relatively low densities, this stomatopod can have a significant impact on the size and structure of a prey population.

V. EVIDENCE THAT STOMATOPODS SELECT PREY

There have been very few studies on prey preference in stomatopods. Hamano and Matsuura (1986a) examined the distribution and abundance of potential prey and the stomach contents of the spearer, Oratosquilla oratoria in Hakata Bay, Japan. They found that the major food in the diet was Macrura, although reproductive females also took a fair proportion of bivalves Within the Macrura, penaeids, alpheids, and Crangon affinis were identified in the stomachs and a calculated electivity index suggested that these prey were taken roughly in proportion to their density in the bay.

However, in the laboratory, Hamano and Matsuura (1986b) demonstrated that individual O. oratoria preferentially attacked and consumed penaeid and alphaeid shrimp with a specific size relationship to the stomatopod. Using morphological data on the second thoracopod and equations modified from those developed by Holling et al. (1976) for raptorial predators, they calculated the minimum and maximum prey size (cross-sectional diameter) that a given size O. oratoria could capture. The optimum prey size was then assumed to be the mean between the minimum and maximum. When stomatopods were given an array of live shrimp to feed on, they consumed animals close to the predicted optimum. When presented with stationary models, they approached a fairly broad size range of models, but attacked only those close to the predicted optimal size.

This study nicely demonstrates that O. oratoria selectively attack those prey it is most likely to be able to capture and hold. Shrimp that are too small can slip between the dactyl spines and shrimp that are too large cannot be grasped between the dactyl and propodus. The major costs associated with feeding in an ambush predator such as O. oratoria are probably associated with the actual attack. While striking is not energetically expensive, the animal risks detection by both potential prey and predators when it strikes. A miss reduces the chances of future catches and also increases the danger of predation. We therefore might expect the stomatopod to be fairly selective in the prey it attacks. Processing costs are probably minimal for spearers taking soft-bodied prey such as shrimp and it is less likely that diet selection, at least for this type of prey, would be based on such considerations.

However, in smashers such as Gonodactylus that specialize on armored prey, the time and effort taken to consume these items can be considerable. We might expect these animals to base prey selection, at least in part, on processing costs. In an early study on Gonodactylus chiragra, Caldwell and Dingle (1976) found that the number of strikes required to open a snail or hermit crab increased with shell size, the amount of edible tissue (wet weight) available to a stomatopod increased with shell size, and that for 80mm G. chiragra, the maximum amount of tissue per strike was gained when the stomatopod selected 10-12mm snails. Hermit crabs had the same peak, but yielded less tissue than did snails at any given shell size. When G. chiragra were allowed to select snails or hermit crabs from a range of shell sizes, they took the size that maximized tissue per strike. Also given prey in the same size shells, they preferred snails to hermit crabs. These results suggested that a simple prey selection model that maximizes fitness by maximizing net rate of energy (Pyke, 1984) might be appropriate. However, no measurement of the energetic cost of prey processing was attempted.

Full et al. (1989) measured prey selection and the energetic cost of processing different sized prey in G. bredini. We used large (50mm) stomatopods captured in the field and allowed them to feed on snails (Tegula funebralis) for at least 3 weeks prior to any experiments to ensure familiarity with the prey. For all experiments, the snails were divided into three size classes, small (5-6mm), medium (8-9mm) and large (11-12mm).

Prey selection trials were conducted by presenting the stomatopod simultaneously with one snail from each size class. We then scored which snail was first handled and which, if any, was consumed. These stomatopods handled the medium and large Tegula with equal frequency and were less likely to approach the small ones. However, they consumed significantly more medium sized snails. Presenting stomatopods with individual snails of different sizes, we measured the tissue consumed, handling time, the number of strikes required to process the shell, and net oxygen consumed. The later was recorded using an oxygen electrode inserted into a closed metabolic chamber. As expected, all 4 variables increased with snail size. We then calculated 3 different prey selection indices as a function of snail size: 1) energetic return (dry mass of prey tissue / unit handling time); 2) energy gain / energy used (caloric value of prey tissue / energetic cost of handling); and 3) net energy gain (caloric value of prey tissue minus caloric value of energy expended in prey processing) / handling time. All three indices have frequently been used to assess prey selectivity (Pyke, 1984).

No significant difference was observed in any of the 3 energetic indices over the range of snail sizes used. Therefore, the observed selection of prey by G. bredini cannot be explained simply by the maximization of any of these cost/benefit measures. Energy return/handling time is the simplest model used. Elner and Hughes (1978) found that it successfully predicted prey selection for crabs eating mussels. However, in G. bredini, tissue consumed and handling time increased proportionally with size of prey. Therefore their ratio did not vary with snail size and stomatopods should not be able to distinguish among prey solely on this index within the range of sizes tested (5-12 mm). Energy gained/energy expended includes the actual costs of handling prey. However, the number of strikes required to open snails increased with size and the energetic cost of each strike remained constant. Again, the ratio did not vary across this size range of snails. Net energy gain/handling time includes both the energetic benefit and the cost of processing along with handling time. This index is generally accepted as the most appropriate currency of fitness and has been reasonably successful at predicting prey selection in a variety of animals (Mittelbach, 1981). However, given that the costs and benefits that go into the index (energy gained, energy expended, handling time) all increased proportionally, the index did not vary across the range of shell sizes used. It should be pointed out that when the stomatopods ate smaller snails, there was a slight decrease in processing time and energy expenditure. However, tissue gained declines rapidly with size so all 3 indices decline. Not surprisingly, G. bredini do not take smaller shells unless larger alternatives are unavailable. Similarly, G. bredini of this size have considerable difficulty breaking open Tegula larger than 16mm and both handling time and energy expenditure increase more rapidly than does tissue gained. Again, the stomatopods avoid larger snails, if possible (Caldwell, unpublished).

Given that the G. bredini described above were more selective in taking prey than the indices predicted, we might ask what other factors influence prey choice. In this experiment, we did not consider the costs associated with searching. Animals were given prey just outside their burrow. Field observations indicate that G. bredini may make several trips a day, travelling up to several meters to acquire food (Caldwell et al., 1989). Typically, they can carry only one snail at a time. If animals must travel a considerable distance to find any prey, and if prey are clumped, as is often the case with hermit crabs and snails, they might be expected to take larger items to reduce the number of trips required. If we couple to this the risks of predation while out in the open and the chances that a competitor might take over a stomatopod's cavity while it is away searching, there might be even more pressure to increase prey size and reduce time away from the cavity. On the other hand, transport costs associated with returning very large snails to the home cavity could become so great (more energy and/or time required to move the object) that there might be a ceiling on the size of prey carried back to the cavity. Another possible factor could be wear and tear on the raptorial appendage. It is conceivable that processing larger shells might increase the risk of appendage failure or wear. While the ratio of strikes per tissue gained is constant over a fairly broad range of snail sizes, the forces generated hitting stronger shells might produce more erosion or actual breakage. A gonodactylid that loses a raptorial appendage suffers a serious competitive disadvantage in defending its cavity (Berzins and Caldwell, 1983). The loss of both appendages not only further decreases competitive ability, but eliminates readily available armored prey from the diet. Such an animal must molt at least twice before its raptorial appendages are functional and has a poor chance of surviving in the field during the 3 months this will take. We have few data on raptorial appendage failure as a function of prey size. However, gonodactylids forced to feed exclusively on large snails do appear to have more erosion of the dactyl heels than do animals given smaller snails or less armored prey (Caldwell, unpublished).

VI. PREY SELECTION AND THE EFFECT OF DISTANCE TRAVELLED

Considering that distance travelled might be one of the variables affecting prey selection, the following laboratory experiment was performed. Gonodactylus were collected in the field and returned to Berkeley where they were housed individually until needed. Thirty-two individuals, representing 3 species were used, 24 G. bredini and 5 G. oerstedii from Panama and 3 G. chiragra (sp.?) from Lizard Island, Australia. Animals ranged from 44-68 mm total length, although only 2 G. chiragra were greater than 56 mm. All 3 species are found in the low intertidal and live in cavities. All feed primarily on gastropods, crabs, and hermit crabs. Since no differences in feeding behavior were observed in the laboratory, the data for the 3 species are pooled in this analysis.

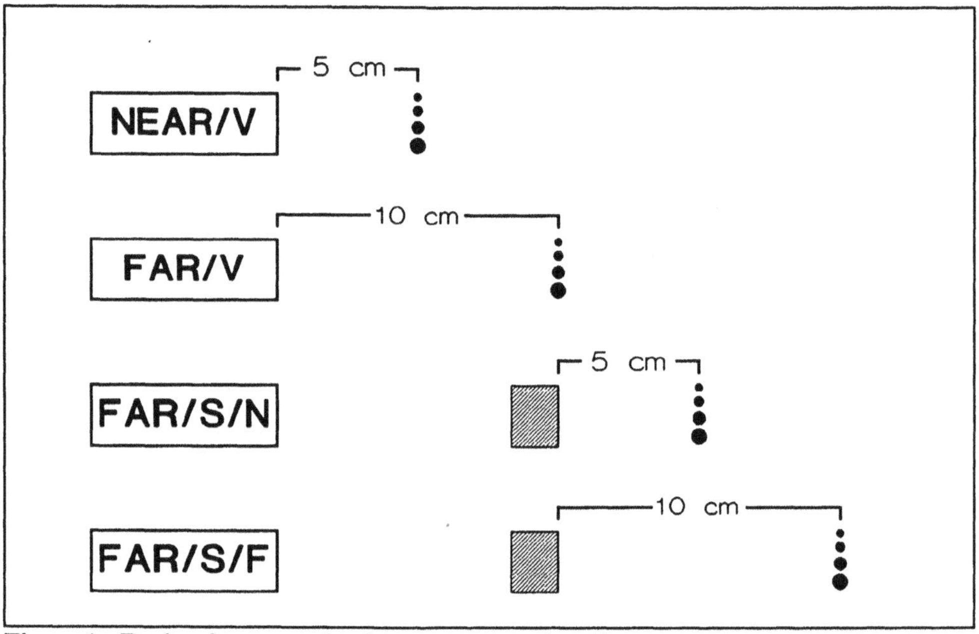

Figure 1. Design for prey selection tests at varying distances.

Prey used in this experiment consisted of 8, 12, 16, and 20 mm Tegula funebralis collected in California. The snails were killed to preclude any movement effects. Prior to the experiment, each animal was fed all sizes of snails for a minimum of one week so that they had experience opening them. To begin the experiment, stomatopods were placed individually into 50 l aquaria. The tanks were filled to a depth of 20 cm and the bottoms covered with a layer of fine coral sand. Each animal was provided with a 10 cm long, 1.5 cm diameter opaque plastic tube that served as a shelter. The stomatopods spent most of their time in these tubes and all prey were processed in them.

After a 24 h acclimation period, the stomatopod was presented with 4 snails, one from each size class, placed in a line in front of the cavity. Position was controlled. Four different placements were used. In the first, the snails were positioned in plain view approximately one body length (5 cm) directly in front of the cavity (NEAR/V). In the second, the snails were placed 10 cm in front of the cavity, again in plain sight of the

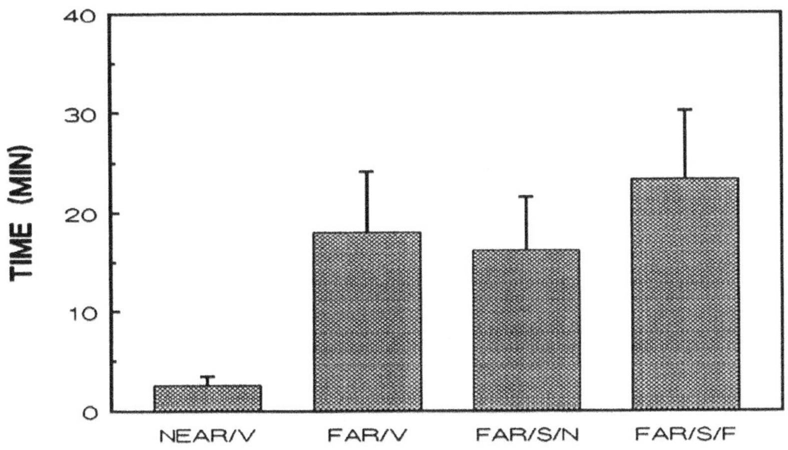

Figure 2. Time to leave cavity and make first contact with prey.

stomatopod when it was sitting in the cavity entrance (FAR/V). In the third condition, we attempted to control for changes in the apparent size of the prey when viewed from a greater distance. We placed a visual barrier 10 cm from the entrance. The snails were positioned 5 cm beyond the barrier so as the stomatopods rounded the corner, the prey were visible from the same distance as in NEAR/V. This condition was designated FAR/S/N. The final treatment was similar to FAR/S/N, but the 4 snails were 10 cm beyond the visual barrier (FAR/S/F) (See Fig. 1). Animals were presented one of the 4 treatments each morning and each afternoon. If they failed to select a snail within 1 h, the treatment was repeated again in a later session. The order of presentation was balanced. No attempt was made to control hunger, although animals that selected a snail were allowed to consume it. We recorded time to first contact with a snail, which snail was initially examined, and which snail was selected and eaten.

The stomatopods came out of their cavities and examined snails more quickly when the prey were only 1 body length away than they did for any of the FAR conditions (Friedman 2-way Analysis of Variance, $X^2 = 28.6$, df = 3, $p < 0.001$). Times to approach the snails were not significantly different among the 3 FAR treatments (Fig. 2).

The stomatopods contacted significantly smaller snails when the prey were only one body length outside the cavity (Friedman 2-way Analysis of Variance, $X^2 = 16.8$, df =

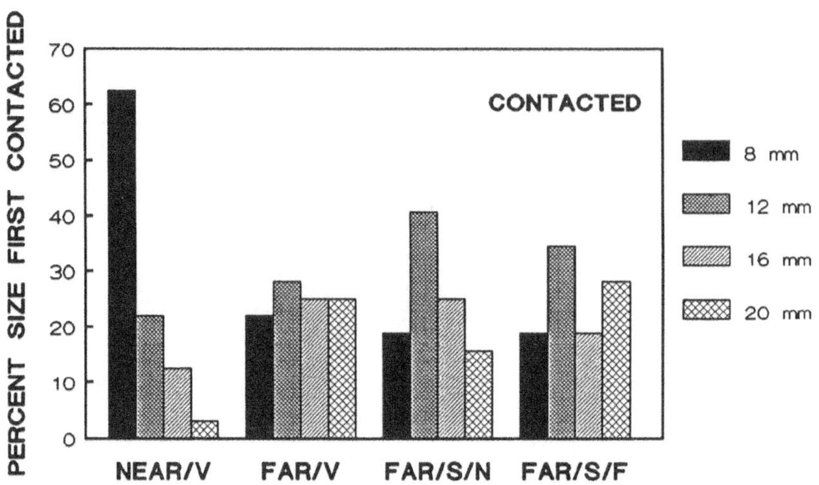

Figure 3. Percent different sized prey initially contacted. (N = 32)

3, $p < 0.001$). More 8 mm snails were contacted in NEAR/V ($X^2 = 26.2$, df = 3, $p < .001$). In all FAR conditions, more intermediate sized snails were initially examined, but the differences were not significant (Fig. 3).

The stomatopods appeared to be making their first assessment of the snails visually. After looking at the line of prey, often for several minutes, they usually moved rapidly and directly to one snail. Once prey were contacted, the speed with which the stomatopods selected and returned to the cavity with a snail depended on the proximity of cover. In both NEAR/V and FAR/S/C, 10 of 32 (31.3%) of the animals examined more than one snail before returning. In both conditions, the Gonodactylus could maintain contact with cover (tube or visual shield) while examining the snails. When the animals had to travel away from cover, they made their selections more quickly and fewer individuals examined more than one snail (FAR/V, 7 of 32, 21.9%; FAR/S/F, 4 or 32, 12.5%).

The stomatopods were more precise in selecting prey after handling (Fig. 4). Again, significantly more small snails were consumed in the NEAR/V condition (Friedman, $X^2 = 22.6$, df = 3, $P < 0.001$). A greater number of 8 mm snails were taken in NEAR/V ($X^2 = 33.8$, df = 3, $p < 0.001$), but more 12 mm snails were eaten in the FAR

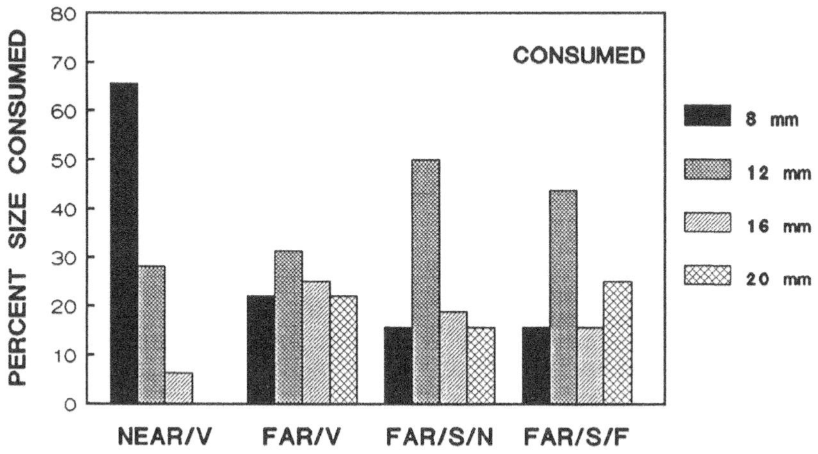

Figure 4. Percent different sized prey consumed. (N = 32)

conditions. This difference was significant only in FAR/S/N (X^2 = 10.8, df = 3, P < 0.02).

Because different sized Gonodactylus have different absolute preferences for the size of prey taken, the most appropriate analysis is to examine the choices by each individual across treatments, noting whether the animal took smaller, larger, or the same size snails as conditions changed (Table 1). Significance was tested using a 2-tailed Sign Test with Bonferroni's correction for multiple comparisons. Stomatopods contacted and consumed significantly larger snails whenever they travelled more than one body length from the cavity. Apparent size of the prey was not important under these conditions since there was no difference between the sizes of snails selected at FAR/S/N and FAR/S/F, but there was a significant difference between NEAR/V and FAR/S/N. It appears that when a stomatopod must physically leave its cavity and travel to retrieve prey, it selects a larger item than it does if it can simply reach out of the cavity and seize a snail.

The reason Gonodactylus take larger prey when they must leave the cavity is unclear. Perhaps by taking larger items, stomatopods reduced the number of foraging trips required to meet energy requirements. As discussed above, a stomatopod outside its cavity is at considerable risk from predation. Also, gonodactylids enjoy a positional advantage inside a cavity and it is difficult for another stomatopod to evict them. Should a competitor gain entry into an animal's cavity while it is away foraging, it is unlikely that the original resident could evict the intruder when it returns (Caldwell, 1987).

Table I. Changes in individual preference for snail size. N = 32 in all cases.

	SMALLER	LARGER	SAME	P
FIRST SNAIL CONTACTED				
NEAR/V vs FAR/V	4	21	7	.01
NEAR/V vs FAR/S/N	6	22	4	.02
NEAR/V vs FAR/S/F	4	21	7	.01
FAR/S/N vs FAR/S/F	7	12	13	NS
FIRST SNAIL CONSUMED				
NEAR/V vs FAR/V	3	21	8	.001
NEAR/V vs FAR/S/N	2	23	7	.001
NEAR/V vs FAR/S/F	1	22	9	.001
FAR/S/N vs FAR/S/F	7	9	16	NS

It should be noted, however, that Gonodactylus, even when foraging away from their home cavity, do not take snails larger than those they can efficiently open. The range of snails that can be consumed while maintaining the maximum energetic cost/benefit ratio provides some latitude, within which Gonodactylus can respond to other factors affecting fitness related to foraging.

This experiment only examined prey selection as a function of leaving the home cavity over relatively short distances. In the field, animals often travel several meters to find food (Caldwell et al., 1989). Transport costs may become more important as distance travelled increases. We are currently conducting experiments to examine this question. Also, we have preliminary evidence that stomatopods alter their foraging behavior in the presence of potential competitors for cavities. They are more reluctant to leave their cavity if another stomatopod is detected in the area (Childress and Caldwell, unpublished). We do not know, however, if stomatopods actually alter their selection of prey in response to this competitive threat.

Many studies have shown that central place foragers alter their foraging behavior when at risk to predation (Sih, 1982; Gilliam and Fraser, 1987; Abrahams and Dill, 1989. In most cases, these responses involve when and where individuals search for food. However, animals also may alter the size or type of prey taken (Freed, 1981; Martindale, 1982). In these cases, risk of predation constrains search/and or processing time and appears to make the foragers less selective. This results in predators returning with less profitable prey that more reflect availability.

The situation appears different in Gonodactylus. In our study, individuals select larger snails when they are forced to venture further from a safe haven even though availability was constant across conditions and search costs (time and distance travelled) did not vary with respect to the different items. If anything, transport costs (visibility, time, and energy) should have been greater for larger snails. However, we do not have a currency with which we can factor these varibles into calculating the net profitability associated with eating a particular size of prey. Risks associated with actually subduing different sized snails are negligible. Also, since processing takes place in the cavity, risks associated with processing different sized snails are also minimal. Probably most important is time away from the cavity which increases both the risk of being attacked by a predator and as well as losing the home cavity to a competitor. Given that Gonodactylus often find their prey clumped in dense aggregations, it should take little additional time to select a larger snail. As long as the shell is not sufficiently large to impede transport or processing, the number of required foraging trips, and thus risk, will be reduced while profitability remains high.

Finally, these laboratory experiments were conducted in the absence of predators and competitors. The stomatopods had encountered neither for several weeks. Yet they still altered their foraging behavior when required to leave a cavity to feed, and acted as if they perceived, and were compensating for, risks associated with venturing from the cavity. Whether this change in behavior is rigidly programmed, or the stomatopods are responsive to different levels of predation and competition, remains to be determined.

VII. CONCLUSION

The evolution of the second pair of thoracopods into powerful raptorial feeding appendages has had a profound influence on every aspect of a stomatopod's biology. Foraging is no exception. Stomatopods provide an excellent opportunity to relate functional morphology, physiology, behavior, and ecology to feeding. Studies such as those by Caldwell and Dingle (1975) and Hamano and Matsuura (1986a,b) demonstrate the relationship between the form and function of the raptorial appendage and prey choice. Research by Full et al. (1989) shows that it is possible to measure energetic costs and benefits associated with feeding, and to relate these to prey selection. Studies by Reaka (1987) and Caldwell et al. (1989) point the way to understanding the consequences of feeding behavior on community structure. Our own work described here reveals the ease with which stomatopods can be studied in the laboratory to gain a better understanding of their behavioral plasticity and what variables are important affecting prey choice. We need now to integrate these diverse studies of stomatopod foraging into one comprehensive picture of how foraging behavior affects fitness, and to elucidate through what channels evolution proceeds to find 'optimal' solutions.

ACKNOWLEDGEMENTS

We thank Dr. Gloria Caldwell for commenting on the manuscript. The Smithsonian Tropical Research Institution and the Lizard Island Research Station provided logistical assistance. Funding was provided by NSF Grant BNS -85-17573 to RLC.

REFERENCES

Abrahams MV, Dill LM (1989) A determination of the energetic equivalence of the risk of predation. Ecology 70:999-1007

Basch LV (1989) Aspects of the ecology and behavior of the stomatopod Hemisquilla ensigera californiensis (Gonodactyloidea, Hemisquillidae). In: Biology of Stomatopods E A. Ferrero (ed) Selected Symposia and Monographs UZI 3 Mucchi, Modena (Italy) (in press)

Berzins IK, Caldwell RL (1983) The effect of injury on the agonistic behavior of the stomatopod, Gonodactylus bredini (Manning). Mar Behav Physiol 10:83-96

Caldwell RL (1979) Cavity occupation and defensive behaviour in the stomatopod Gonodactylus festae: Evidence for chemically mediated individual recognition. Anim Behav 27:194-201

Caldwell RL (1982) Interspecific chemically mediated recognition in two competing stomatopods. Mar Behav Physiol 8:189-197

Caldwell RL (1985) A test of individual recognition in the stomatopod Gonodactylus festae. Anim Behav 33:101-106

Caldwell R L (1986) The deceptive use of reputation by stomatopods. In: Deception: Perspectives on Human and Non-Human Deceit. Mitchell RW, Thompson NS (eds) State University of New York Press, pp. 129-145

Caldwell RL (1987) Assessment strategies in stomatopods. Bull Mar Sci 41:135-150

Caldwell RL, Dingle H (1975) Ecology and evolution of agonistic behavior in stomatopods. Naturwissenshaften 62:214-22

Caldwell RL, Dingle H (1976) Stomatopods. Scientific American, January, 1976, pp. 80-89.

Caldwell RL, Lamp K (1981) Chemically mediated recognition by the stomatopod Gonodactylus bredini of its competitor, the octopus joubini. Mar Behav Physiol 8:35-41

Caldwell RL, Roderick G, Shuster S (1989) Studies of predation by Gonodactylus bredini. In: Biology of Stomatopods. Ferrero EA (ed) Selected Symposia and Monographs UZI 3 Mucchi, Modena pp. 117-131

Cronin TW, Marshall NJ (1989) A retina with at least ten spectral types of photoreceptors in a mantis shrimp. Nature 339:137-140

Dingle H, Caldwell RL, Manning RB (1977) Stomatopods of Phuket Island, Thailand. Phuket Marine Biological Center Research Bulletin 20:1-20

Elner RW, Hughes RN (1978) Energy maximization in the diet of the shore crab, Carcinus maenas. J Anim Ecol 47:103-116

Freed LA (1981) Breeding biology of House Wrens: new views of avian life history phenomena. Ph.D. dissertation, Univ. of Iowa

Full RJ, Caldwell RL, Chow SW (1989) Smashing energetics: prey selection and feeding efficiency of the stomatopod, Gonodactylus bredini. Ethology 81:134-147

Gilliam JF, Fraser DF (1987) Habitat selection under predation hazard: test of a model with foraging minnows. Ecology 68:1856-1862

Hamano T, Matsuura S (1986a) Food habits of the Japanese mantis shrimp in the benthic community of Hakata Bay. Bull Japan Soc Sci Fish 52:787-794

Hamano T, Matsuura S (1986b) Optimal prey size for the Japanese mantis shrimp from structure of the raptorial claw. Bull Japan Soc Sci Fish 52:1-10

Hatziolos ME (1979) Ecological correlates of aggression and courtship in the stomatopod Pseudosquilla ciliata. Ph.D. dissertation, Univ. of California at Berkeley, 252 pp

Holling CS, Dunbrack RL, Dill LM (1976) Predator size and prey size: presumed relationship in the mantid Hierodula coarctata Saussure. Canad J Zool 54:1760-1764

Manning RB, Schiff H, Abbott BC (1984) Cornea shape and surface structure in some stomatopod Crustacea. J Crustacean Biol 4:502-513

Marshall NJ (1988) A unique colour and polarization vision system in mantis shrimps. Nature 333:557-560

Martindale S (1982) Nest defense and central place foraging: a model and experiment. Behav Ecol Sociobiol 10:85-89

Mittelbach GG (1981) Foraging efficiency and body size: a study of optimal diet and habitat use by bluegills. Ecology 62:1370-1386

Pyke GH (1984) Optimal foraging theory: a critical review. Ann Rev Ecol Syst 15:523-575

Reaka ML (1980) On learning and living in holes by mantis shrimp. Anim Behav 28:111-115

Reaka ML (1985) Interactions between fishes and motile benthic invertebrates on reefs: the significance of motility vs. defensive adaptations. Proc Fifth Int Coral Reef Congr, Tahiti 5:439-444

Reaka ML (1987) Adult-juvenile interactions in benthic reef crustaceans. Bull Mar Sci 4:108-134

Reaka ML, Manning RB (1981) The behavior of stomatopod Crustacea, and its relationship to rates of evolution. J Crust Biol 1:309-327

Sih A (1982) Foraging strategies and the avoidance of predation by an aquatic insect. Ecology 63:786-796

Steger R (1985) The behavioral ecology of a Panamanian population of the stomatopod, Gonodactylus bredini (Manning). Ph.D. dissertation, Univ. of California at Berkeley, 114 pp

Steger R (1987) Effects of refuges and recruitment on gonodactylid stomatopods, a guild of mobile prey. Ecology 68:152-1533
Steger R, Caldwell RL (1983) Intraspecific deception by bluffing: A defense strategy of newly molted stomatopods (Arthropoda: Crustacea). Science 221:558-60

ORAL DISCUSSION – R.L.Caldwell.

Q.(Bazely) Cavities are limiting. When you added more burrows, did individuals immigrate from outside the area ?
A. Yes, in another experiment, colonists were continually removed every other day for 6 months. We removed over 2000 individuals with no apparent change in colonisation rate or size distribution.

Q. (Jaeger) Concerning takeovers: when salamanders leave their territory they deposit a lot of pheromone to dissuade intruders. Do stomatopods do something similar ?
A. Yes, they leave odours, and neighbours then avoid intruding. The odours only last for a few minutes but feeding trips away from the burrow also last for only a few minutes.

Q. (Sih) What about floaters? Do they avoid resident odours ?
A. Floaters might stick their heads in and see if they get hit, but they generally won't take the risk. It depends on their condition. Floaters are often recently moulted or wounded individuals that have reduced fighting ability.

Q. (Godin) Do foragers evaluate prey–encounter rates in determining whether to come out to feed?
A. We haven't looked at that yet, but there are clear hunger effects. Our experimental animals were hungry. Well–fed animals would not emerge from their burrows to forage.

TIME SCALE AND DIET CHOICE DECISIONS

Jeffrey R. Lucas
Department of Biological Sciences
Purdue University
West Lafayette IN 47907
U.S.A.

INTRODUCTION

The development of optimal foraging theory has in large part involved an evaluation of the time scales over which animal decision-making processes work. Early models primarily considered long-term averages of short-term static processes. As I discuss below, recent modelers have proposed that animal decisions consider longer and more complicated time scales. The new models are more specific and more realistic in their representation of foraging behavior, but the predictions from these models require that foragers either anticipate future events and regulate their behavior accordingly, or that they use rules which perform well under expected future conditions. The addition of time scale to foraging models raises the empirical question of how well animals forecast events that should affect current decisions. The question is particularly important because some evidence suggests that animals are incapable of planning future rewards that will arrive after a period of 1 hr (Timberlake 1984; Timberlake et al. 1987). In this review I will consider time scales implied by several foraging models and review studies that address the problem of the use of time scales by foragers. I will focus primarily on the time scale of diet choice decisions. Krebs & Kacelnik (1984; also see Kacelnik & Krebs 1985) considered this topic primarily for the problem of patch utilization and Staddon (1984) discussed time horizons and memory.

A decision can be thought of as the commitment of a certain period of time to a particular behavior pattern. In an evolutionary sense, the decision should reflect the costs and benefits of the alternative behaviors available to the forager. The relative value of alternatives may change dynamically with changes in state (e.g. energy stores or water balance) and environmental conditions (e.g. prey encounter rates, predator risk or time constraints); the optimal decision should track these changes (Mangel & Clark 1988). The time-scale of a decision can be considered to be the length of time into the future that a forager weighs potential events in its evaluation of alternative

NATO ASI Series, Vol. G 20
Behavioural Mechanisms of Food Selection
Edited by R. N. Hughes

behaviors.

The development of the classic diet choice model focussed attention on three predictions. (1) The addition of a prey type into the diet depends on the encounter rate of only higher ranking prey. (2) As prey abundance decreases, the diet should become more catholic. (3) The forager should show no partial preferences (Pyke et al. 1977). Another important prediction from the model is that the pattern of time allocation to foraging has no effect on diet choice; time minimizers (which minimize the time spent gathering a given fixed amount of energy) and energy maximizers (which maximize the amount of energy they gather in some constrained amount of time) should not differ in their choice of prey (Schoener 1971). Even under time constraints, mean and variance in the maximum available foraging time have no effect on the predicted diet under the assumptions of the model (Lucas & Schmid-Hempel 1988). If these predictions are correct, time-sharing between foraging and alternative non-foraging behaviors will have no effect on diet choice decisions.

The validity of the model rests on three major assumptions. (1) Lost opportunities (i.e. loss of preferred prey caused by the pursuit or handling of less preferred prey) are independent of foraging time. (2) Net rate of energy intake is an appropriate optimization criterion. (3) Behavior is steady state; sampling effects, acquisition of behavior, satiation, and other changes in state are unimportant.

The predictions of the classic diet choice model, while not always supported quantitatively, have been surprisingly accurate (Krebs et al. 1983). One prediction which has received little support, the lack of partial preferences, could be seen as the result of stochasticity in local conditions or behavioral rules, and as such partial preference could be seen as consistent with the model (Stephens 1985). However, subsequent work has shown that the assumptions are often violated. More importantly, the study of diet choice has focussed on details of foraging decisions which make time scale a critical component of the foraging decision.

In this review, I will emphasize the effects of potential future events on diet choice decisions. The paper will be organized around a discussion of the three assumptions listed above and how violations of these assumptions should affect the time-scale of diet choice decisions.

(1) LOST OPPORTUNITIES

The standard diet choice model can be rearranged to show that the choice of a prey item depends on the relative value of an encountered prey compared to the relative value of prey items of higher quality which could be missed if the encountered item were taken (Lucas 1983). In the model, the value of these lost opportunities is considered to be independent of foraging time. But if the forager is foraging under time constraints, the value of lost opportunities may change as foraging time decreases because fewer prey will be missed as time left runs out (Lucas 1985). If lost opportunities are taken into account by the forager, then it should be more likely to take anything that comes along as foraging time decreases. At least three studies have looked for this effect by imposing constraints on foraging time and measuring the resultant effect on diet choice. In addition to testing for effects of time constraints, there are two additional advantages to these studies. First, the time scale of the decisions is unambiguous because it is set by the researcher. Also, if the animal changes its diet within the bout, it does so in anticipation of some future event, assuming that no other complications influence the expression of choice.

Great tits were trained on a conveyor-belt apparatus to expect access to food for 30 or 60 s, and were given a choice of small or large fly pupae. As predicted, the tits were more likely to accept a small prey as the time left declined (Lucas 1987). Blue jays were trained in an operant chamber to search projected images for two kinds of moths on tree backgrounds. A peck at the appropriate key when either type of moth was projected resulted in the delivery of a mealworm. Prey profitability was manipulated as a difference in handling time. Given 10 or 20 min of foraging time, the blue jays were more likely to accept either moth type at the end of the bout than in the middle of a bout (Yoerg & Kamil 1988). Shrews were offered large and small mealworm pieces in a grid of covered pots and allowed to forage on this grid for 3, 6 or 9 min. The shrews were more selective overall during the 9 min bouts than during the 3 or 6 min bouts. They also showed a decrease in selectivity at the end of the 9 min bout, although the effect was only significant if data from all shrews were combined (Barnard & Hurst 1987).

All three studies found the same trends, but there were several interesting complications. However, I will argue that all the trends were real effects of time constraints. None-the-less, three factors other than time constraints appear to have affected these results: satiation effects, the expression of alternative behaviors, and sampling.

As with most foraging studies, the birds and shrews were food-deprived before the experiment and all ate enough food during one day's sessions to show satiation effects. In general, starved animals tend to be less selective than sated animals (e.g. Bence & Murdoch 1986; Charnov 1976; Perry 1987; Rechten et al. 1983; Snyderman 1983). If the foraging bout (here defined as the total amount of uninterrupted time available for foraging) is long enough, satiation should cause the animal to be more selective as the bout progresses. Thus satiation effects should counter the predicted end-of-bout effect caused by a change in lost opportunities. In addition, all three experiments used several bouts in succession, and so satiation effects should cause the foragers to be more selective over the entire experimental session. Within-bout effects consistent with the satiation hypothesis were exhibited by both blue jays and shrews; both were less selective when presumably hungriest, at the beginning of the bout, compared to the middle of the bout. In contrast, the tits were more selective at the beginning of the bout, but they did show a between-bouts effect, ignoring more small prey in later bouts. However, within-bout effects of satiation are unlikely in the tit experiments because they took so few prey during a bout.

Barnard & Hurst (1987) suggested that satiation could have caused the end-of-bout effect shown by the shrews. They divided each bout into 3 min intervals and, using discriminant function analysis, tested for a decrease in selectivity within each interval. No decrease in selectivity was detected in any interval and therefore they concluded that the end-of-bout effect must be caused by some other factor. Rechten et al. (1983) showed that the cost of discrimination errors (rejecting high quality prey that are judged to be lower quality) is high for hungry animals, and so hungry foragers should be less selective. Barnard & Hurst (1987) suggested that satiated animals should also make more discrimination errors. This is because sated animals should show lower priority to feeding and because there is little cost to a sated animal in taking the wrong prey. If this hypothesis is correct, the observed unimodal pattern of selectivity could be caused solely by satiation effects, with the animals less selective at the beginning of a long bout due to hunger, and less selective at the end of the bout due to satiation. While a decrease in selectivity with satiation has been demonstrated for two insects, damselfly naiads (Akre & Johnson 1979) and an intertidal beetle (Richards 1983), there is no evidence that the same should hold true for satiated vertebrates. On the contrary, considerable evidence shows the opposite trend (see above).

In addition, the end-of-bout effect may turn out to be significant if the analysis were expanded to include intervals greater than 3 min (this is not reported in Barnard & Hurst 1987). Indeed, if constraints on the measurement of time intervals (e.g. Gibbon 1977) are incorporated into the model, longer foraging bouts are expected to cause the end-of-bout effect to be less pronounced and of longer duration (Lucas 1985). Diet choice by the great tits showed this trend (Lucas 1987).

Most published data indicate that sated foragers are more selective, and therefore are *less* likely to make discrimination errors (as defined above). This still leaves open the possibility that a reduction in the priority of foraging affects diet choice. In fact, the change in feeding priority proposed by Barnard & Hurst (1987) appears to be correct; both the shrews and great tits spent less time foraging as they ate more food, a common response to satiation (McCleery 1977). This decrease in foraging time correlated with an increase in selectivity, but the change in selectivity was not due to the animals spending more time performing other, non-foraging, behaviors. This can be illustrated by comparing diet choice of individual great tits.

All birds left the belt empty-billed before the end of any given bout, and thus treated the experimental bouts of 30 or 60 s duration as a series of shorter foraging bouts. There were individual differences in the mean duration of these short foraging bouts and these differences correlated with selectivity. Hungry birds that spent less time at the belt were less selective than birds spending more time at the belt. This effect is similar to that of time constraints: selectivity decreased with a decrease in time allocated to foraging (Lucas 1987). These data are important for two reasons. First, the expression of alternative behaviors has a marked effect on diet choice decisions; thus, time minimizers and energy maximizers should not necessarily exhibit the same diet choice decisions. Second, satiation appears to affect diet choice decisions independently of the priority of foraging *per se*.

The last factor that could complicate diet choice decisions in these studies is sampling. Given two patches of unknown quality and known constraints on foraging time, theory predicts that the amount of time the forager samples the patches should correlate with the amount of available foraging time (Krebs et al. 1978; see Plowright & Plowright 1987 for a correction of the model). If the birds were sampling prey types before selecting any one type, they should have been less selective at the beginning of the longer bout because sampling is more valuable then. Depending on the variance in subjective time-keeping of the foragers, the opposite is predicted based solely on reasons of lost opportunity (Lucas 1985). Alternatively, satiation effects should be the same at the beginning of any bout, irrespective of its duration. In fact, two opposite trends were found. Blue jays were less selective at the start of 20 min bouts compared to 10 min bouts (Yoerg & Kamil 1988), while great tits were less selective at the start of 30 s bouts compared to 60 s bouts (Lucas 1987). However, the tits usually took only a single prey in a bout and rarely took more than two, so there was simply no time to sample prey. So these results do not refute the sampling hypothesis. Unfortunately, results have not been unequivocal in previous studies that specifically tested for sampling. Data reported by Houston et al. (1982) and Dow & Lea (1987) showed the predicted effect, but the design of both experiments has recently been questioned and a similar experiment on pigeon foraging has failed to find any effect of time horizon (Shettleworth & Plowright 1989).

Despite the presence of confounding effects, the results of the three studies show that when short time horizons are imposed on these foragers, they regulate diet choice decisions appropriately. Unfortunately, no experiments of this type have been reported where the time horizon was greater than 30 min, although field data on diet choice of intertidal snails (*Acanthina punctulata*) show that snails will change their diet over a 7 hr low-tide cycle (they only forage at low tide; Menge 1974).

Timberlake (1984; Timberlake et al. 1987) used a different experimental design to test for the time scale of patch choice. Rats were offered a food patch that delivered food at a diminishing rate. A second patch was available which delivered food continuously but only a fixed interval after the experiment started. If the rat values the alternative patch, then it should leave the depleting patch sooner when the alternative patch turns on sooner. Two sets of experiments showed that rats are not sensitive to future rewards 30 min or more into the future. Most of the current models assume that animals look much further into the future than just 30 min, yet none of the experimental evidence presented so far contradicts Timberlake's conclusions.

(2) NET ENERGY INTAKE AS THE OPTIMIZATION CRITERION

The classic diet choice model can be rearranged to show that the choice of a prey item depends on the relative value of an encountered prey item compared to the relative value of prey items of higher quality that would be missed if the currently encountered item were taken (Lucas 1983). Thus the model predicts that the decision should be made on the time scale of the handling time of alternative prey, assuming that net energy intake rate is an appropriate optimization criterion and that the foraging conditions are stable.

More recent work has shown that net energy intake rate is not an appropriate optimization criterion under many circumstances. Houston et al. (1988) show for social insects that maximization of net intake rate does not maximize net fitness. The tradeoff between worker longevity and foraging performance reduces the utility of maximizing intake rate; workers can contribute more to the colony over their lifetime if they maximize the efficiency of their foraging (energetic gains/energetic expenditures) than if they expend energy at a faster rate to maximize net intake rates. Foraging decisions of honeybees support this model; the amount of sugar solution collected by honeybees is explained better by maximization of energy efficiency than by maximization of net energy intake rate (Schmid-Hempel 1987; Schmid-Hempel et al 1985; Schmid-Hempel & Schmid-Hempel 1987). Unfortunately, this model has not been tested with diet choice decisions.

Caraco (1980, also see Real 1980) suggested that energy intake has different effects on fitness depending on the state of the animal; a joule to a starving animal is more of a jewel than a joule to a satiated animal. The result is that variance in reward associated with a given prey type will influence the inclusion of that prey type into the diet. This is not predicted by the classic diet choice model. In non-reproductive seasons especially, the use of survival as an optimization criterion is a better approximation of fitness because it incorporates differences in prey value caused by differences in energy reserves (McNamara & Houston 1982; Pulliam & Millikan 1982). More importantly, this increases the time scale of the problem, from the duration of handling time to the time over which survival is affected by foraging decisions. The time considered by most theoretical studies is 24 hr (e.g. Caraco 1980; Stephens 1981).

There is very good evidence that foragers are sensitive to variance in reward offered by different types of prey (or patches). For example, shrews, juncos and white-crowned sparrows tend to be risk-prone when the expected 24 hr energy budget is negative and risk-averse when the 24 hr energy budget is positive (Caraco et al. 1980; Caraco 1981, 1983; Barnard & Brown 1985a; but see Wunderle & Cotto-Navarro 1988 for a counter-example). Thus these foragers are using time scales longer than that implied in the classic diet choice model. However because all these experiments ran substantially less than 24 hr, these data give no indication as to the time-scale of the foraging decisions. This is an important point because the predictions of the models are sensitive to time scale.

Evidence that animals use foraging decisions that actually anticipate future requirements is provided by the regulation of body size. Chicks, house sparrows and white-crowned sparrows eat more food during the day when their night-time energy budget increases, rather than tolerating a greater depletion of fat reserves the following morning (Kendeigh et al. 1969; Hayne et al. 1986). Hayne et al. (1986) found that chicks reared under cold night-time temperatures (15°C) had the same body mass at dawn as chicks reared under constant warm conditions (35°C). In addition, variance in weight loss at night did not correlate with a chick's feeding rate the subsequent morning. Thus decisions about weight gain clearly anticipate energy expenditures in the subsequent 24 hr period.

(3) STEADY STATE CONDITIONS

One of the properties of the classic diet choice model, and also of the original models of risk-sensitivity (e.g. Caraco 1980, Stephens 1981), is a fixed foraging decision; once a decision is made the animal does not alter it. An alternative possibility is that animals update their decision policies throughout the day (or whatever time period).

There are several reasons why the rate of change of behavioral decisions should be of interest. From a mechanistic point of view, the rapidity with which foraging decisions are made could give some clues as to the underlying mechanisms that regulate foraging behavior. It should also give some clues as to the validity of possible rules of thumb that animals use to make decisions. But the rate of change in behavior is also important theoretically. A dynamic change in foraging decisions potentially influences a number of predictions of foraging models. Rapid changes in behavior may cause preferences to appear to be more variable than they actually are (e.g. what appear to be partial preferences can result from deterministic foraging rules that change rapidly; Lucas 1985, McNamara & Houston 1987). For risk sensitive foraging decisions, dynamic decision policies could alter the predicted level of energy reserves at which the forager should switch from risk-averse to risk-prone behavior (Houston & McNamara 1982). Thus the time-scale implied by the risk-sensitivity of foragers described above may be different than that suggested by fixed-decision models.

The effect of multiple time scales on diet choice is evident in learned responses to prey encounter rate. For example, great tits trained for several days on fixed mean encounter rates reject more small prey at higher encounter rates of large prey; this effect can last a period of days (Krebs et al. 1977). Great tits trained for several weeks on different ratios of prey sizes respond the same way, showing a higher degree of selectivity when the size difference between prey types is larger (Lucas 1987). This unsurprising fact has been documented for a number of organisms (Krebs et al. 1983). While these effects occur over periods of days, there are also demonstrable effects on the time scale of seconds. Given Poisson-distributed prey with a mean encounter rate of 0.049/s for large prey, great tits are significantly more likely to take a small prey with an increase in the time since they last took a large prey (Lucas 1987; Elner & Hughes 1978 report similar data for the shore crab, *Carcinus maenas*). Even after weeks of training at the same encounter rate, great tits alter selectivity based on variation in single events that occur with a mean (and standard deviation) of 20 s apart; there is no asymptotic diet choice at this level of analysis. Under somewhat different conditions, grey plovers show changes in diet selectivity within a period of 8 s (Pienkowski 1983), again due to perceived changes in prey density.

These observations are entirely consistent with current models of learning (e.g. the linear operator; Kacelnik et al. 1987; McNamara & Houston 1987b), and in this respect, tell us nothing new. The results are important in that they illustrate that diet choice decisions have a component that is updated over extraordinarily short time scales. Krebs & Kacelnik (1984) suggested that under some circumstances prey choice in tits is not affected by short runs of bad luck. If this were true, it would be appropriate to model diet choice problems as a Poisson process and consider only changes in the mean encounter rate. This feature is incorporated in virtually all models of diet choice. Unfortunately, even when tits are offered prey that are Poisson-distributed with a fixed mean encounter rate, they respond as if prey arrival were clumped and vary their diet accordingly (Lucas 1987). If we are interested in quantitative predictions, fine-scale features of this sort must be taken into consideration.

The response of birds to fixed or variable prey encounter rates shows that behavior is clearly affected by events that occur in the past over several time scales. Theory predicts the same for future events, but the data are not conclusive.

Houston & McNamara (1985; McNamara & Houston 1986) explicitly discussed risk-sensitivity for foragers that are capable of altering diet choice decisions throughout the day. Assuming that the forager should make decisions that maximize survival rate over a 24 hr period, the forager has two competing requirements that must be met: it must survive starvation in the near-term by avoiding some lower lethal limit of energy reserves and it must also store enough energy by dusk to survive the night (Stephens & Krebs 1986 call this the 'lazy L' model). This model is fundamentally different than the previous models because the animal is assumed to base its decision on the probability of events that could occur over a range of time scales, from minutes to days or more. The probability of starvation over the different time scales will determine how strongly the animal weighs them, with priority given to immediate starvation threat.

There is some circumstantial evidence that diet choice decisions are this complicated. For example, it is well documented that short-term changes in diet choice can arise from satiation effects; starved animals tend to be less selective than sated animals. This phenomenon has been described as non-optimal, impulsive foraging by starved animals (e.g. Snyderman 1983a) and as non-optimal foraging by sated animals that are not maximizing net energy intake (e.g. Rechten et al. 1983). If animals regulate foraging decisions over longer periods of time and if they use optimization criteria that are different than maximization of net energy intake, then neither of these contradictory statements may be correct. Houston & McNamara (1985; also see Kagel et al. 1986) propose that this shift in diet choice with satiation is a change in risk-sensitivity with a change in energy reserves. If so, these data suggest that diet choice decisions for risk sensitive foragers can be regulated on the time scale of minutes. Unfortunately, I have found no experiments on risk-sensitive diet choice that have specifically tested for a dynamic shift in diet choice decisions over a period of 24 hr or longer.

However, Wunderle et al. (1987) published some data that indicate that dynamic shifts in risk occur. They offered bananaquits either variable or constant nectar rewards and tested the preference for either over the day. The bananaquits were never risk-prone, but both birds shifted to *significant* risk-aversion at dusk. Wunderle et al. (1987) suggested that these data are not in accord with the predictions of the fixed-decision risk models because the birds lost weight from one morning to the next. But these data do suggest that risk-sensitivity changes over the day, and so a dynamic model may be more relevant. If so, the predicted patch choice would depend not only on current energy budget but also on current body size and time of day. No continuous measures of body mass were reported.

The evidence I have presented so far for the time-scales of risk sensitive decisions has been entirely circumstantial. For example, animals have been shown to be sensitive to risk, suggesting that they are probably not maximizing immediate payoff. But the time-scale associated with this decision, while theoretically important, has not been empirically tested. Indeed, there are proximate reasons why animals should be risk sensitive even if the time scale is no different than that suggested by the classic diet choice model (e.g. handling constraints, Harder & Real 1987, or constraints on the measurement of time, Gibbon et al. 1988). This is underscored by operant studies that suggest that animals tend to choose alternatives that maximize near-term profit (Staddon 1983; Shettleworth 1988).

One way to test the ability of a forager to regulate foraging decisions in anticipation of future events in foraging decisions is to provide an animal with a cue that correlates with a change in conditions, then fail to provide this change. Under these conditions a change in behavior is clearly anticipatory. Barnard & Brown (1985b) developed an ingenious experimental design of this type to test for the perceived effects of competition on risk-sensitivity. Shrews were given a choice between two patches that differed in reward variance under conditions where the expected daily energy budget was positive. A potential competitor was placed behind a Plexiglass wall, in sight of the forager but unable to affect the distribution of food in the available food patches. Under control conditions (no competitor present), the shrews chose the lower variance reward (Barnard & Brown 1985a). In contrast, the perceived threat of competition caused the shrews to be indifferent in their choice of patches and also caused them to feed at higher rates. Therefore, the shrews changed their foraging pattern in anticipation of a decrease in prey abundance and increase in the variation in food reward provided by the patches. The time scale for this shift in behavior is unknown but it is likely to be fairly short.

Two pieces of evidence suggest that time-scales for risk-sensitive foragers can be quite long. The first comes from regulation of body size and nutrient composition of the diet. Birds and mammals have been shown to store more body fat in winter in general, and especially during winter cold spells (e.g. King & Farner 1966; Dawson et al. 1983). In addition, birds that forage on the ground tend to store more fat than tree-feeding species (Rogers 1987). Thus, birds with higher energy demands and those with (presumably) higher variance in food availability tend to eat more food and so store more fat. In addition, the rate of weight gain over the course of the day varies between species. Chicks, house sparrows and white-crowned sparrows gain weight rapidly in the morning then maintain that weight or even lose some weight before dusk (Hayne et al. 1986; Kendeigh et al. 1969). Chicks also choose a diet high in carbohydrate in the morning and shift to a diet higher in protein by afternoon (Hayne et al. 1986). In contrast, hummingbirds gain weight at nearly a constant rate throughout the day (Wolf & Hainesworth 1977). If these results are due to a higher variance in resource levels available to chicks and sparrows than to hummingbirds, then hummingbirds should show a narrower diurnal range of risk-sensitivity than chicks and sparrows. This would reconcile the difference in risk-sensitivity between nectar-feeding bananaquits, which are risk-averse even when losing weight (Wunderle et al. 1987; Wunderle & Cotto-Navarro 1988), and seed-eating juncos and white-crowned sparrows, which are risk-prone when losing weight and risk averse when gaining weight (Caraco 1981, 1983, Caraco et al. 1980; but see Staddon 1987).

The second set of results comes from a related problem, caching. A number of species of animals cache food in winter, presumably as a hedge against periods of scarce resources (Sherry 1985). A model of caching behavior, similar in several respects to the model used by Houston & McNamara (1985), predicts the optimal sequence of caching decisions over the course of the day (Lucas, in prep). The specific caching decision I will discuss is whether to cache a food item taken from a resource patch or to eat the item at the patch. I assume that the forager's decisions are based on the criterion of maximizing long-term survival rate, where long-term is greater than several days (under most conditions the solution stabilizes after 4 or 5 days). The model predicts that caching behavior should change dynamically with changes in state (here body fat) and time of day. In particular, caching rate should decline at the end of the day and should also be low at dawn if the bird is sufficiently light weight. Results from Carolina Chickadees exhibit this pattern (Fig 1). The study of caching behavior has the advantage that the animal is clearly planning for future food requirements when it caches a seed. Thus the data are consistent with a shift in time horizon, from near-term at dawn when the forager is low-weight, to far-term at midday when the forager caches seeds for future use, to near-term at dusk when energy must be stored to survive the night and when cached seeds are less valuable to the bird due to the risk of over-night pilferage.

A similar conclusion can be drawn from data on the expected proportion of seeds cached per day by birds of different weights. Under energetic stress, birds are predicted to cache little at low weights due to the immediate threat of starvation, and they should cache most at high levels of energy reserve because starvation threat is always a threat in the future. The opposite correlation is expected in more favorable environments. Here immediate starvation is not a threat, so low weight birds should cache to reduce longer-term starvation threat and fat birds are expected to reduce caching rates.

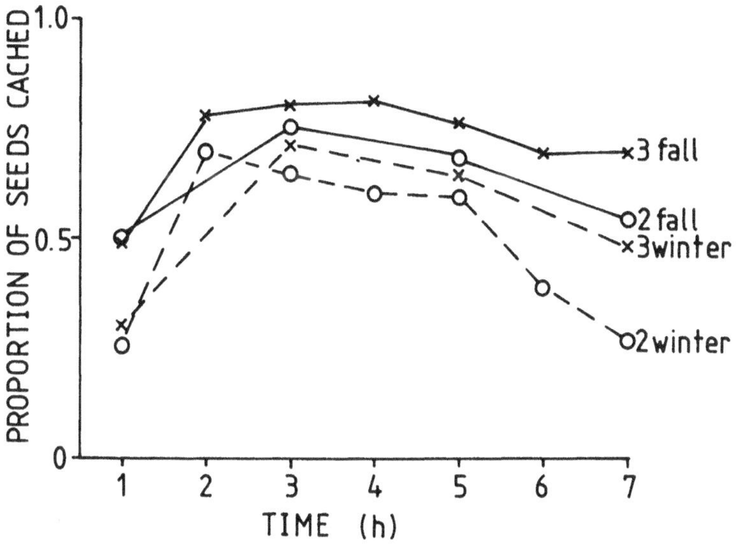

Fig 1. Proportion of seeds taken from the feeder which are cached by Carolina Chickadees over the course of a day. 'Time' starts when the bird is put in the aviary. Two birds (2 and 3) were tested in two seasons. Data from Walter & Lucas (in prep).

Two factors should facilitate the reduction in caching rates of heavy birds. First, cached seeds are available for only a few days before they are pilfered or forgotten. In favorable environments, birds with sufficient fat stores may not need a cached seed within this time period. But more importantly, starvation threat for a heavy bird is minimal, so other effects on survival are expected to dominate the solution. For example, caching increases the handling time of a seed, and therefore may place the bird at risk of predation for longer. Since starvation threat is minimal for a heavy bird, it should eat seeds instead of caching them and thereby reduce predation risk.

In the laboratory, chickadees tend to cache less at heavier weights (Fig 2). Only one significant correlation in the opposite direction was observed. The bird showed this trend in only one season (Fall), was the smallest bird tested (number 2 in Fig 3), had an unusually high variance in body weight, and showed the typical negative correlation when she was heavier (Winter in Fig 3). These data support the notion that animals can update caching decisions based on the expectation of future conditions (e.g. foraging time left in the day) and that differences in energy reserves can qualitatively change foraging decisions.

In addition to the long-term trends in caching, we also have evidence of quite short-term effects on caching and diet choice decisions. Lynda Peterson and I are now conducting an experiment on diet choice of tufted titmice and our preliminary data point out some interesting trends. The titmice were offered small and large pieces of mealworms on a conveyor belt in an experiment that lasted 15 min each day. The encounter rate of large prey decreased linearly with time, from 1/30 sec at the beginning of the bout to 0/30 sec at the end. The question is whether the birds will switch from specialists to generalists in response to the reduced encounter rate (see Lucas & Schmid-Hempel 1988).

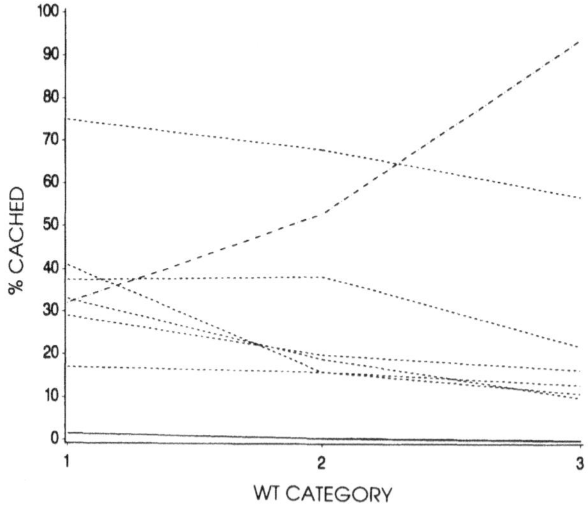

Fig 2. Percent of seeds taken from the feeder which are cached by Carolina Chickadees. 'Wt Category' represents the lower (1), middle (2) and upper (3) third of body weights measured for each of 5 birds tested (3 were tested on different feeder-access schedules giving 8 sets of data). Dotted lines represent a significant negative relationship between % cached and weight (logistic regression, $P<0.05$). Dot-dash pattern represents significant positive relationship. Solid line represents no significant trend. Data from Walter & Lucas (in prep).

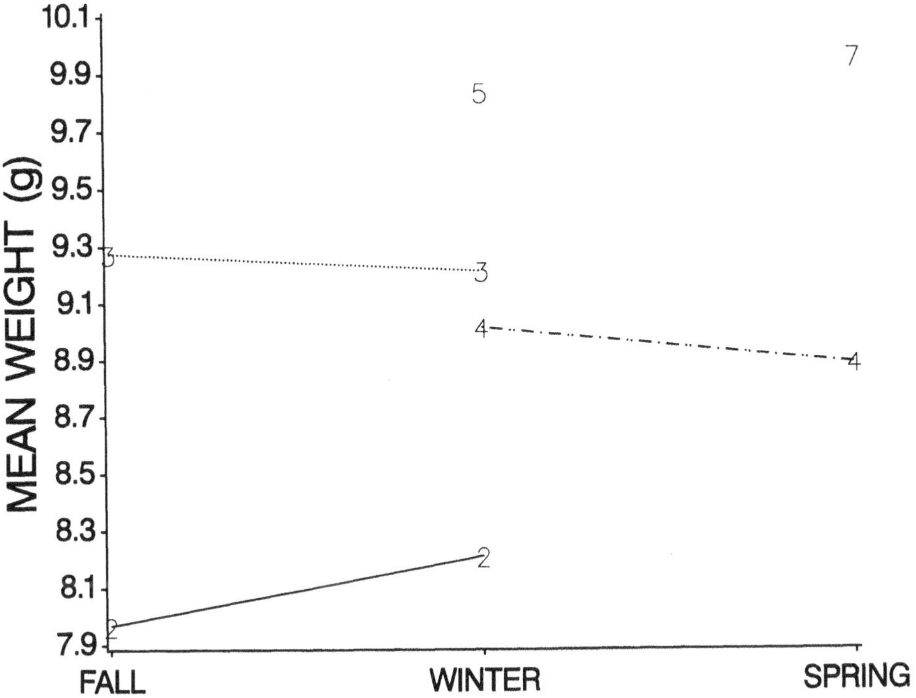

Fig 3. Mean body weight of Carolina Chickadees tested in the caching experiment. Numbers represent the identification number of each bird. Lines connect mean weights of the same bird run in different experiments. Data from Walter & Lucas (in prep).

The answer is that they do quite the opposite. The two birds we have tested so far became *more* selective in the bout (Fig 4A); this is the same effect of satiation on diet choice discussed above. For one of the birds, initial weight also affected diet choice. When the bird was fat, it was overall more choosy than when the bird was thin (Fig 4A, 4B). Thus two levels of hunger (fat stores and gut content) act in the same way: selectivity decreased with hunger level.

The titmice also cached in the aviary; this gives us the ability to ask what short-term factors influence caching under these conditions. Both birds never cached when fat (Fig 4C); a result similar to that exhibited by the chickadees in experiments lasting a full day. When light-weight, the birds only cached later in the bout. This is analogous to the diurnal caching pattern in the chickadees, in which caching rate increased only after at least several minutes of access to food. The titmice also tended to cache more when large prey came in close succession (Fig 4D). So with caching behavior, the birds respond to every time scale we have analyzed.

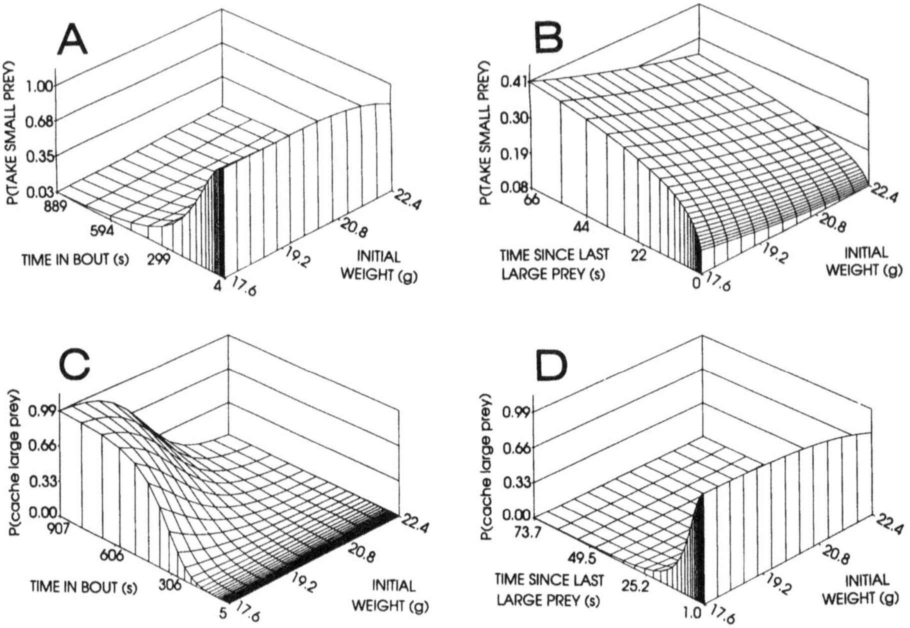

Fig 4. Best fit surfaces from maximum likelihood, logistic regression models of the probability of a titmouse taking a small prey from the conveyor belt (A, B) or the probability of caching a prey taken from the belt (C, D). 'Time in bout'=time since the experiment started, 'initial weight'=weight at the start of the experiment. Results are from a single individual. Data from a second bird are similar except for the effects of initial weight on the probability of taking small prey (A; see text).

SUMMARY

Optimal diet theory has changed substantially since the development of the classic diet choice model. One of the fundamental properties of several types of models, including models of risk and dynamic programming, is that the time scale associated with diet decisions is fairly long. This is especially true of models using survival as an optimization criterion. In addition, dynamic programming models assume that (1) the animal can respond rapidly to environmental changes and changes in state, and (2) that there is no single time scale to a decision, instead there are a range of time scales which are weighted depending on the probability of given events (e.g. starvation) occurring within these time periods.

Despite the growing understanding that time scale is important, there are few empirical studies that actually show that animals can regulate diet choice in anticipation of future events. All of these show that animals clearly respond to changes in prey density, time constraints and satiation within periods of seconds to minutes. For time scales of longer duration, data show that animals can regulate levels of body fat based on the expected energy expenditure in the forthcoming 24 hr period. Diurnal patterns in caching behavior also illustrate the same effect. Acquisition of behavior can be shown to have quite long lasting effects and it can also be shown to have effects over different time scales. But no studies of diet choice *per se* can unequivocally show effects for long time periods into the future. Indeed, several authors suggest that animals are unable to respond to time horizons as short as an hour in operant chambers (e.g. Timberlake et al. 1987; Shettleworth 1988).

Are the time scales of diet choice decisions really this short? Operant studies and studies of diet choice under time constraints require that the animal remember both the duration of the experiment and the length of time that has passed since the experiment has started. Results from studies of psychophysics suggest that the longer the duration of the interval, the less likely the animal is to predict either measure accurately (Gibbon 1977). This problem in the measurement of time will complicate any of the operant studies (see Gibbon et al. 1988). In the field, animals may be provided with external cues that reduce the problem of measuring time intervals. For example, the level of the tide correlates with the expected time left to forage for intertidal foragers; of course the position of the sun is an excellent cue for the amount of time remaining in the day. Thus the negative results from operant studies do not imply that time horizons will necessarily be short. Some circumstantial evidence suggests that animals can use fairly long time scales, but there are few good studies of diet choice that take detailed observations for long periods of time. Indeed, the results of the time constraint studies done to date suggest that the task will be daunting. Relevant state variables (e.g. body mass and satiation) must be systematically monitored continuously for the duration of the experiment. Even within a period of 30 s, diet choice decisions are affected by a host of factors, including perceived variance in encounter rate and the expression of alternative behaviors. The data analysis has to be detailed enough to consider the effects of these variables. None-the-less, as time scales become more ingrained in the way we think about foraging decisions, long-term monitoring is likely to become more important to test some fundamental unsupported assumptions about time scale and diet choice.

ACKNOWLEDGEMENTS

I would like to thank Rick Howard, Lynda Peterson, Patricia Rabenold, and Howard Whiteman for their comments on this manuscript. Lynn Walter collaborated on the chickadee work and Lynda Peterson collaborated on the titmice work.

LITERATURE CITED

Akre BG, Johnson DM (1979) Switching and sigmoid functional response curves by damselfly naiads with alternative prey available. J Anim Ecol 48:703-720

Barnard CJ, Brown CAJ (1985a) Risk-sensitive foraging in common shrews (*Sorex araneus* L). Behav Ecol Sociobiol 16:161-164

Barnard CJ, Brown CAJ (1985b) Competition affects risk-sensitivity in foraging shrews. Behav Ecol Sociobiol 16:379-382

Barnard CJ, Hurst JL (1987) Time constraints and prey selection in common shrews *Sorex araneus* L. Anim Behav 35:1827-1837

Bence JR, Murdoch WW (1986) Prey size selection by the mosquitofish: relation to optimal diet theory. Ecology 67:324-336

Caraco T (1980) On foraging time allocation in a stochastic environment. Ecology 61:119-128

Caraco T (1981) Risk-sensitivity and foraging groups. Ecology 62:527-531

Caraco T (1983) White-crowned sparrows (*Zonotrichia leucophrys*): foraging preferences in a risky environment. Behav Ecol Sociobiol 12:63-69

Caraco T, Martindale S, Whittam TS (1980) An empirical demonstration of risk-sensitive foraging preferences. Anim Behav 28:820-830

Charnov EL (1976) Optimal foraging: attack strategy of a mantid. Am Nat 110:141-151

Dawson WR, Marsh RL, Yacoe ME (1983) Metabolic adjustments of small passerine birds for migration and cold. Am J Physiol 245:R755-R767

Dow SM, Lea SEG (1987) Sampling of schedule parameters by pigeons: tests of optimizing theory. Anim Behav 35:102-114

Elner RW, Hughes RN (1978) Energy maximization in the diet of the shore crab, *Carcinus maenus*. J Anim Ecol 47:103-118

Gibbon J (1977) Scalar expectancy theory and Weber's law in animal timing. Psychol Rev 84:279-325

Gibbon J, Church RM, Fairhurst S, Kacelnik A (1988) Scalar expectancy theory and choice between delayed rewards. Psychol Review 95:102-114

Hayne H, Rovee-Collier C, Gargano D (1986) Ambient temperature effects on energetic relations in growing chicks. Physiol Behav 37:203-212

Harder LD, Real LA (1987) Why are bumble bees risk averse? Ecology 68:1104-1108

Houston AI, Kacelnik A, McNamara JH (1982) Some learning rules for acquiring information. In: McFarland DJ (ed) Functional Ontogeny. Pitman,London,p 140-191

Houston AI, McNamara JM (1982) A sequential approach to risk-taking. Anim Behav 30:1260-1261.

Houston AI, McNamara JM (1985) The choice of two prey types that minimizes the probability of starvation. Behav Ecol Sociobiol 17:135-141

Houston A, Schmid-Hempel P, Kacelnik A (1988) Foraging strategy, worker mortality, and the growth of the colony in social insects. Am Nat 131:107-114
Kacelnik A, Krebs JR (1985) Learning to exploit patchily distributed food. In: Sibly RM, Smith RH (eds) Behavioural Ecology: ecological consequences of adaptive behaviour. Blackwell,Oxford,p 189-205
Kacelnik A, Krebs JR, Ens B (1987) Foraging in a changing environment: an experiment with starlings (*Sturnus vulgaris*). In: Commons ML, Kacelnik A, Shettleworth SJ. Quantitative Analyses of Behavior, vol VI: Foraging. Erlbaum,Hillsdale NJ,p 63-87
Kagel JH, Green L, Caraco T (1986) When foragers discount the future: constraint or adaptation? Anim Behav 34:271-283
Kendeigh SC, Kontogiannis JE, Mazac A, Roth RR (1969) Environmental regulation of food intake by birds. Comp Biochem Physiol 31:941-957
King JR, Farner DS (1966) The adaptive role of winter fattening in the white-crowned sparrow with comments on its regulation. Am Nat 100:403-418
Krebs JR, Kacelnik A (1984) Time horizons of foraging animals. Ann NY Acad Sci 423:278-291
Krebs JR, Kacelnik A, Taylor P (1978) Test of optimal sampling by foraging great tits. Nature 275:27-31
Krebs JR, Stephens DW, Sutherland WJ (1983) Perspectives in optimal foraging. In: Brush AH, Clark GA (eds) Perspectives in Ornithology. Cambridge,New York,p 165-216
Lucas, JR (1983) The role of foraging time constraints and variable prey encounter in optimal diet choice. Am Nat 122:191-209.
Lucas JR (1985) Time constraints and diet choice: different predictions from different constraints. Am Nat 126:680-705
Lucas JR (1987) The influence of time constraints on diet choice of the great tit, *Parus major*. Anim Behav 35:1538-1548
Lucas JR, Schmid-Hempel P (1988) Diet choice in patches: time-constraint and state-space solutions. J theor Biol 131:307-332
Mangel M, Clark C (1988) Dynamic modeling in behavioral ecology. Princeton, Princeton
McCleery, RH (1977) On satiation curves. Anim Behav 25:1005-1015
McNamara JM, Houston AI (1982) Short-term behaviour and lifetime fitness. In: McFarland DJ (ed) Functional Ontongeny. Pitman,London,p 60-87
McNamara JM, Houston AI (1986) The common currency for behavioral decisions. Am Nat 127:358-378
McNamara JM, Houston AI (1987a) Partial prey preferences. Anim Behav 35:1084-1099
McNamara, JM, Houston AI (1987b) Memory and the efficient use of information. J Theor Biol 125:385-395
Menge, JL (1974) Prey selection and foraging period of the predaceous rocky intertidal snail, *Acanthina punctulata*. Oecologia 17:293-316
Perry DM (1987) Optimal diet theory: behavior of a starved predatory snail. Oecologia 72:360-365
Pienkowski MW (1983) Changes in the foraging pattern of plovers in relation to environmental factors. Anim Behav 31:244-264
Plowright CMS, Plowright RC (1987) Oversampling by great tits? A critique of Krebs, Kacelnik and Taylor's (1978) "Test of optimal sampling by great tits". Can J Zool 65:1282-1283
Pulliam HR, Millikan GC (1982) Social organization in the nonreproductive season. Avian Biol 6:169-197
Pyke GH, Pulliam HR, Charnov EL (1977) Optimal foraging: a selective review of theory and tests. Q Rev Biol 52:137-154
Real LA (1980) Fitness, uncertainty, and the role of diversification in evolution and behavior. Am Nat 115:623-638

Rechten C, Avery MI, Stevens TA (1983) Optimal prey selection: why do great tits show partial preferences? Anim Behav 31:576-584
Richards LJ (1983) Hunger and optimal diet. Am Nat 122:326-334
Rogers CM (1987) Predation risk and fasting capacity: do wintering birds maintain optimal body mass? Ecology 68:1051-1061
Schmid-Hempel P (1987) Efficient nectar collecting by honeybees: I. Economic models. J Anim Ecol 56:209-218
Schmid-Hempel P, Kacelnik A, Houston AI (1985) Honeybees maximize efficiency by not filling their crop. Behav Ecol Sociobiol 17:61-66
Schmid-Hempel P, Schmid-Hempel R (1987) Efficient nectar collecting by honeybees: II. Response to factors determining nectar availability. J Anim Ecol 56:219-227
Schoener TW (1971) Theory of feeding strategies. Ann Rev Ecol Syst 2:369-404
Sherry DF (1985) Food storage by birds and mammals. Adv Study Behavior 15:153-188
Shettleworth SJ (1988) Foraging as operant behavior and operant behavior as foraging: what have we learned? Psychol Learn Motiv: Adv Res Theor 22:1-49
Shettleworth SJ, CMS Plowright (1989) Time horizons of pigeons on a two-armed bandit. Anim Behav 37:610-623
Snyderman M (1983) Optimal prey selection: the effects of food deprivation. Behav Anal Letters 3:359-369
Staddon, JER (1983) Adaptive behavior and learning. Cambridge, Cambridge
Staddon, JER (1984) Time and memory. Ann NY Acad Sci 423:322-334
Staddon JER, Reid AK (1987) Adaptation to reward. In: Kamil AC, Pulliam HR, Krebs JR (eds). Foraging Behavior. Plenum,NY,p 497-523
Stephens DW (1981) The logic of risk-sensitive foraging preferences. Anim Behav 29:628-629
Stephens DW (1985) How important are partial preferences? Anim Behav 33:667-669
Stephens DW, Krebs JR (1986) Foraging theory. Princeton, Princeton
Timberlake W (1984) A temporal limit on the effect of future food on current performance in an analogue of foraging and welfare. J Exp Anal Behav 41:117-124
Timberlake W, Gawley DJ, Lucas GA (1987) Time horizons in rats foraging for food in temporally separated patches. J Exp Psychol: Anim Behav Proc 13:302-309
Wolf LL, Hainesworth FR (1977) Temporal patterning of feeding by hummingbirds. Anim Behav 25:976-989
Wunderle JM, Castro MS, Fetcher N (1987) Risk-averse foraging by bananaquits on negative energy budgets. Behav Ecol Sociobiol 21:249-255
Wunderle JM, Cotto-Navarro Z (1988) Constant vs. variable risk-aversion in foraging bananaquits. Ecol 69:1434-1438
Yoerg SI, Kamil AC (1988) Diet choices of blue jays (*Cyanocitta cristata*) as a function of time spent foraging. J Comp Psychology 102:230-235

ORAL DISCUSSION - J. R. Lucas

Q. (Kacelnik): In your last graphs, are you showing between–individual or within–individual variation in size (between the 17 and 22 grams)?
A. These are single, individual titmice.

Q. (Kaclenik); That is enough variation to actually kill a great tit. If you begin with a 19 gram bird and reduce its weight by 20%, say to 16 grams, this would be enough to kill the bird.
A. When the titmice are around 17 grams, they will eat almost anything, but when they are around 22 grams, their diet choice changes considerably. None–the–less, they can survive this range of body weights. It would be nice to have some great tits in the States since we have no tits that don't cache.

Q. (Hughes): Your birds are able to change tactics over a few seconds. How would you relate this to some natural situation where time is running out ?
A. There is a big effect of social environment on the amount of time you have available to forage. Lots of birds, subordinates for example, have only short times available for foraging. Birds are also engaged in other behaviours. Birds engaged in other behaviours time–share between foraging and these other behaviours, and this reduces foraging bout length. The important thing is that the more birds time–share, the less selective they become.

FOOD REQUIREMENT AND RISK-SENSITIVE FORAGING IN SHORTFALL MINIMIZERS

C.J. BARNARD,

Animal Behaviour Research Group,
Department of Zoology,
University of Nottingham,
University Park,
Nottingham NG7 2RD

NATO ASI Series, Vol. G 20
Behavioural Mechanisms of Food Selection
Edited by R. N. Hughes

FOOD REQUIREMENT AND RISK-SENSITIVE FORAGING IN SHORTFALL MINIMIZERS

Risk-sensitive foraging, generally characterized as the response of predators to variance in food reward (see Stephens and Krebs 1986) in choosing where and what to eat, is now well-established both in theory and empirically (see Stephens and Krebs 1986, Real and Caraco 1986 for recent reviews). While foraging models incorporating risk-sensitivity have been applied to a number of foraging situations (Real and Caraco 1986), the commonest scenario involves a high energy demand predator which risks starvation if there is a shortfall in its food supply. If such a predator makes foraging decisions so as to minimize the probability of a shortfall, reward variance is likely to be an important criterion on which decisions are based (e.g. Caraco 1981, Stephens 1981, McNamara and Houston 1982, Stephens and Paton 1984). In this paper, I look at some of the circumstances in which shortfalls and risk-sensitive choices might be expected, and at the evidence that such choices are made by predators. I also look at the effect of risk-sensitivity in foraging on decisions in other life history contexts.

Food requirement and risk-sensitivity

For any given mean reward from a feeding sites/prey types, risk-sensitive foraging theory for shortfall minimizers predicts a switch from risk-aversion (avoiding high variance) to risk-proneness (choosing high variance) as the ratio of the predator´s food requirements to its expectation of reward (expected mean) increases. A now classic series of experiments by Caraco and co-workers (e.g. Caraco et al 1980, Caraco 1981, 1983) with small passerine birds provided early evidence for the predicted switch with changing expectation relative to requirement. However, other predators such as rats (Rattus norvegicus) and bumble bees (Bombus spp.), which are less likely to starve in the event of a shortfall, have tended to be consistently risk-averse (see Stephens and Paton 1984, Stephens and Krebs 1986).

Although the apparent difference between predators in risk-sensitivity makes sense in terms of presumed differences in the risk of starvation, it may be that normally risk-averse predators become risk-prone if their demand is increased sufficiently. Differences in risk-sensitivity between predators may thus be due partly to differences across studies in the degree of food demand imposed. In a series of experiments, Barnard and co-workers have tested some of the predictions of shortfall-minimizing risk-sensitive models using two predators which differ widely in metabolic rate and thus presumed food demand: a small mammal, the common shrew (Sorex araneus) and a small freshwater fish, the bitterling (Rhodeus sericus).

Intake relative to requirement (IRR) and risk-aversion/risk-proneness

In both species, animals were given a choice between two feeding stations, one yielding food items at a constant rate (number of items per visit) and the other yielding food items at a variable rate, though sometimes with the same mean reward rate as the constant station. Full details of apparatus and techniques can be found in Barnard and Brown (1985a,b) (common shrews) and Young et al (in press) (bitterlings). In both cases, constant and variable stations were situated in different parts of a tank or choice chamber and were separated by one or more partitions so that animals had to make unequivocal choices from a separate compartment.

IRR and choice of feeding station (equal mean rewards) The first series of experiments tested the prediction that shortfall minimizers should switch their preference for constant or variable feeding stations according to their food requirement relative to expectation at the stations. Requirement was manipulated by allowing animals to meet or exceed their rate of food intake under ad libitum feeding conditions or constraining them to fail to meet it. Following the period of ad libitum feeding, animals were trained to expect constant or variable reward rates with the same mean at two sequentially available feeding stations. In some cases animals were trained while running at/above their hourly requirement as estimated from ad libitum feeding, in others they were trained while running below. Following training, animals were deprived of food and then tested by being given a series of simultaneous choices between constant and variable stations. Preferences were calculated as the proportion of visits to each of the two stations during the test period (see Barnard and Brown 1985a, Young et al in press). The relative positions of constant and variable stations was always randomized between tests and individuals.

Barnard and Brown (1985a) estimated the food requirement of seven common shrews in terms of the number of mealworm (*Tenebrio molitor*) segments consumed per hour over two days of *ad libitum* feeding. Shrews were then provided with a choice between two food pots which yielded the same mean reward of one mealworm segment per visit but differed in variance in reward (0 at the constant station, 2 at the variable station) and were trained as above with a food deprivation priod of 90min between training and

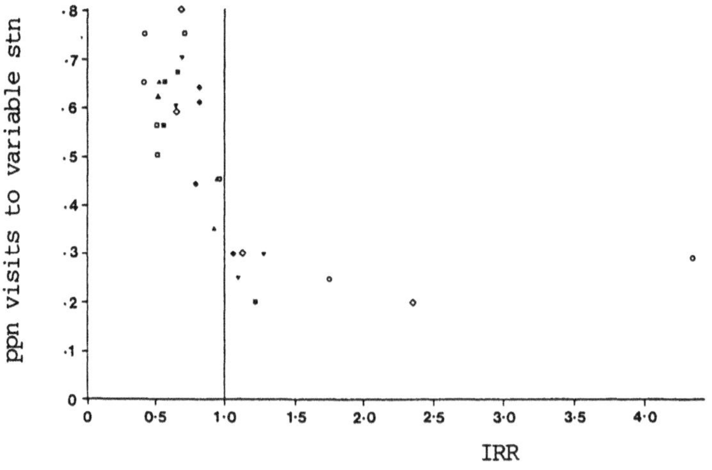

Figure 1 - The relationship between IRR (see text) during training and the proportion of visits to the variable station during tests by common shrews (F=35.60, p<0.01). Symbols represent different individuals. From Barnard and Brown (1985a).

testing. The results of the test procedure (Figure 1) showed that, as predicted for shortfall minimizers, shrews were more likely to choose the variable station (tended to risk-proneness) when they had experienced an intake rate below their estimated requirement during training. Conversely, when they met or exceeded their estimated requirement, shrews were more likely to visit the constant station (tended to risk-aversion). However, when data from all shrews were taken into account (Figure 1), the switch from risk-aversion to risk-proneness was not a step-function. Instead, the degree of risk-aversion/proneness varied continuously with IRR, though the results suggested that risk-aversion when IRR>1 was stronger than risk-proneness when IRR<1.

Young et al (in press) used a similar requirement estimation, training and testing procedure for bitterlings except that food was provided as commercial trout pellets in two surface ring patches (feeding stations), one at either end of an experimental tank. Following estimation of hourly food requirement, fish were

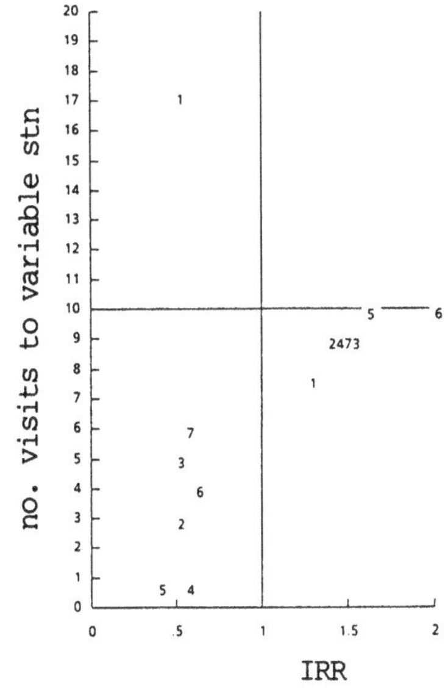

Figure 2 - The relationship between IRR and the number of visits to the variable station by bitterlings during tests. Numbers refer to different individual fish. From Young et al (in press).

trained and tested under different conditions of both food deprivation and reward rate (so that they met or failed to meet their estimated requirement) at the two feeding stations. As with the shrews, one station yielded a constant reward, the other a variable reward. Figure 2 shows the number of visits to the variable station by each fish during tests plotted against IRR . during training. Out of the seven fish tested, six (nos. 2-7) visited the variable station significantly ($t=5.75$, $p<0.01$) less often when running below requirement (48h deprivation followed by training below requirement) than when running above (3h deprivation followed by training above requirement), while the remaining fish (1) showed the opposite tendency. Despite the clustering of points close to the indifference line of 10 visits when IRR>1, fish showed a significant departure towards fewer visits to the variable station both when running above ($t=3.35$, $P<0.05$) and when running below ($t=6.80$, $p<0.002$) requirement.

The increase in risk-aversion with decreasing IRR among most of the fish is the opposite response to that predicted for shortfall minimizers and found in common shrews (Figure 1) and small passerines (e.g. Caraco 1981, 1983, Caraco et al 1980). There are at least two possible explanations. Firstly, since the metabolic rate of bitterlings is considerably lower than that of shrews, it may be that the cost of being forced to run below requirement at the level of deficit experienced during the experiment was small. Indeed, the estimate of requirement may well be too high since fish under ad libitum feeding conditions may be taking more than they need simply to survive (the important criterion in shortfall minimizing models) in order to fuel growth. If deficit cost is small, it may pay to avoid the risk of increasing it further by getting nothing on the next visit; risk-aversion thus increases. Alternatively, increased risk-aversion may reflect a changing trade-off between sampling and exploitation with different degrees of food deficit (Thomas et al 1985). When deficits are relatively low (fish are meeting or exceeding their requirement), fish can afford to check out other feeding stations while spending most of their time at the current best station (in this case the constant station). As deficit increases, however, it becomes more important to maximize immediate intake rate, so fish tend to stay at the constant station and the number of visits to the variable station drops. If food deficit is extremely high, the priority may be to reduce it as rapidly as possible (see McCleery 1978). The best option may then be to go to whichever station is likely to yield the greatest reward. Increasing the fishes´ food deficit further might thus be expected to to result in a switch to risk-proneness. This prediction was tested on one of the fish (7) which showed increased risk-aversion below requirement and on a further six fish (A-F) which had not been involved in the first experiment (Young et al in press).

In the case of Fish 7, deficit was increased by increasing the deprivation time prior to training and testing below requirement from 48h to 168h. Requirement estimation, training and testing procedures for Fish A-F were similar to those for Fish 7 except

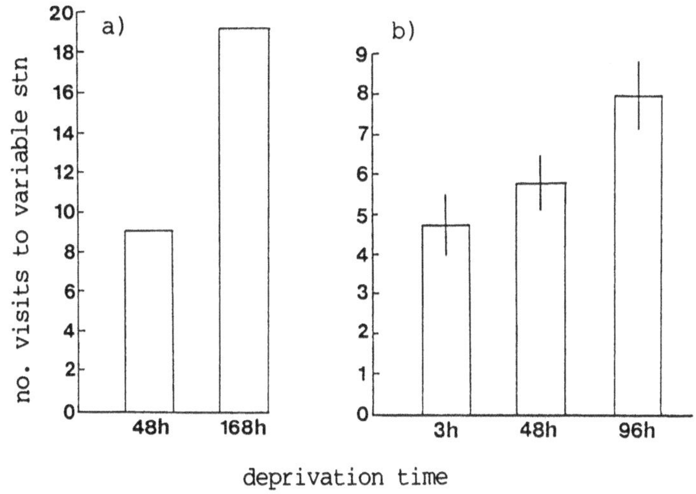

Figure 3 - a) The number of visits to the variable station by Fish 7 (see text) following 48h and 168h food deprivation while running below estimated requirement. b) as a) but for Fish A-F after 3h, 48h and 96h deprivation. b) from Young et al (in press).

that fish were subjected to deprivation periods of 3h, 48h and 96h prior to training and testing below requirement. Figure 3 shows the number of visits to the variable station by Fish 7

(Figure 3a) and Fish A-F (Figure 3b) following different periods of deprivation. In both cases, the tendency to visit the variable station increased with deprivation time, supporting the prediction of a switch to risk-proneness when deficits were high.

IRR and choice of feeding station (mean and variance variable)

In the experiments discussed so far, predators have faced a choice between two feeding stations yielding the same mean reward rate. In the real world, of course, both mean and variance in reward will vary within and between feeding sites and the relationship between them will determine the optimal policy. The interaction between mean and variance in determining choice, however, depends on whether a predator is risk-averse or risk-prone.

Risk-averse predators prefer the mean reward with certainty to the probability distribution of rewards (with the opposite being the case for risk-prone predators) (Stephens and Krebs 1986). However, there may be a trade-off between mean and variance such that variance in reward is tolerated if it is compensated by an increase in the mean. Such trade-offs in risk-averse predators have been demonstrated in bumble bees (Bombus edwardsii) (Waddington et al 1981) and juncos (Junco phaeonotus) (Caraco and Lima 1985) and formalized in variance discounting models of risk-sensitive foraging (see e.g. Stephens and Paton 1984). Clayton (1989) tested the prediction of a trade-off between mean and variance in risk-averse bitterlings. Using the same requirement estimation, training and testing procedures as before, Clayton presented six fish running above their requirement with a choice between two feeding stations, constant and variable. In a series of tests, the mean reward at the variable station was varied so that it was equal to or greater than the mean at the constant station but with, as nearly as possible, a constant associated variance (range 2.0-2.9). Two-way analysis of variance revealed significant effects of both mean reward ($F=3.68$, $p<0.05$) and individual fish ($F=7.34$, $p<0.01$) on the tendency to visit the variable station and there was a significant, positive relationship for each fish between the tendency to visit the variable station and the mean reward at the station (Figure 4).

When running above their estimated requirement, therefore, fish were more likely to visit the variable station if variance in reward was offset by a higher mean.

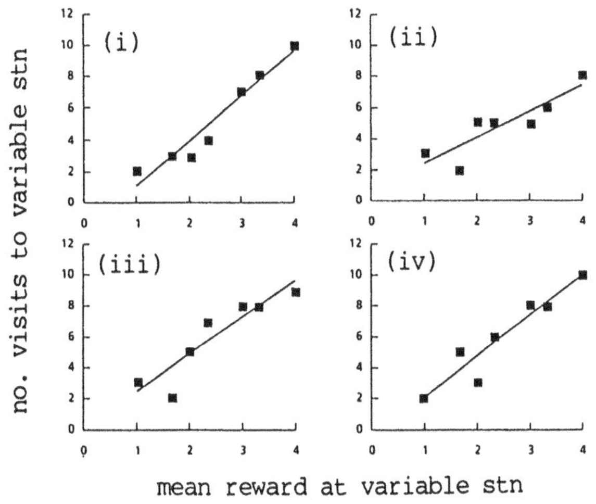

Figure 4 - The relationship between the number of visits to the variable station by four bitterlings and the mean reward at the station. F=96.51 for Fish (i); 19.56 for Fish (ii); 24.77 for Fish (iii) and 45.86 for Fish (iv); $p<0.01$ in all cases. After Clayton (1989).

The predictions for risk-prone predators are more complicated. Barnard et al (1985) developed a model which predicted risk-sensitive foraging decisions in high energy demand predators faced with interruptions to their foraging behaviour that increased their chances of starvation. In the model, the mean reward is usually more important than the variance in reward so that the higher of a pair of means should almost always be preferred. When energy reserves are high (and the risk of starvation in a shortfall therefore low), the predator should be risk-averse. When reserves are low it may be optimal to be risk-prone, though the tendency to risk-proneness should decrease as the mean reward increases because the mean alone then allows reserves to be built up to survive a shortfall.

Barnard et al (1985) tested the model using common shrews

running below their estimated requirement. Shrews were again presented with a choice between a constant and a variable feeding station but this time both the mean and the variance in reward at the variable station were varied so that the mean was equal to or smaller or greater than the mean at the constant station (Table 1). Preferences for the variable station relative to the constant station were then measured. The predictions were as follows:

1) When the mean reward at the variable station is greater than at the constant station, shrews should prefer the variable station. However, preference for the variable station should decrease with increasing variance as the risk of a low reward (and the chance that the lower mean at the constant station would be a better bet) increases.

2) When the mean reward at the variable station is lower than at the constant station, shrews should prefer the constant station, but preference should decrease with increasing variance as the chance of a large reward increases.

Reward schedule	Mean reward per visit	Variance in reward per visit
0,1,1	0.67	0.22
0,0,2	0.67	0.89
0,1,2	1.00	0.67
0,0,3	1.00	2.00
0,2,2	1.33	0.89
0,1,3	1.33	1.60
0,0,4	1.33	3.60

Table 1 - Reward schedules presented at the variable station during training and testing. Figures in the first column are the numbers of mealworm segments received on each of three successive visits (order randomized between sets of three). From Barnard et al (1985).

The results (Figure 5) broadly supported the predictions. Given a choice of two feeding stations, shrews tended to choose the station yielding the higher mean reward (as predicted by both Barnard et al's (1985) model and classical mean maximizing models (Stephens and Krebs 1986)). However, their choice was influenced by the variance in reward rate at the variable station. When preference was plotted against the variance experienced during training (and thus the shrews' expectation of variance during tests) (Figure 5a), there was a decrease with increasing variance when the mean reward at the variable station equalled or exceeded that at the constant station, with shrews showing a significant preference only when variance was low. By contrast, when the mean at the variable station was lower than that at the constant station, shrews were significantly more likely to visit the

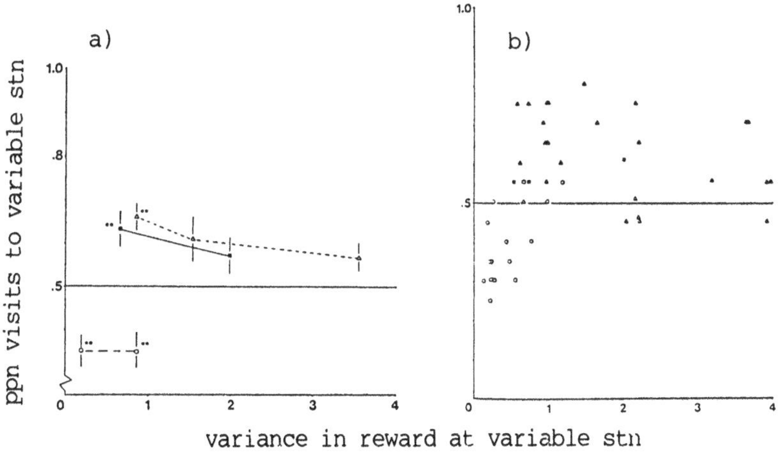

Figure 5 - a) The mean proportion of visits to the variable station by common shrews as a function of the variance in reward at the variable station during training. Open triangles, mean reward greater at the variable station; solid squares, mean reward the same at both constant and variable stations; open circles, mean reward greater at the constant station. Bars represent standard errors. **, p<0.01 (t-test comparing deviation from 0.5). b) The proportion of visits to the variable station as a function of variance in reward experienced during tests. Symbols as in a). After Barnard et al (1985).

constant station regardless of variance. As in the experiments with equal mean rewards, however, the assumption in Figure 5a is that, during tests shrews chose feeding stations on the basis of the reward probability distribution experienced during training. With stations available simultaneously (tests) rather than sequentially (training), the reward probability distribution experienced during tests was likely to differ from that during training and was also likely to differ between individuals for any given treatment. Figure 5b shows a similar analysis to that in Figure 5a except that the abscissa is now the variance in reward experienced at the variable station during tests. The figure shows a strongly significant ($F=13.27$, $p<0.01$) positive effect of variance on the tendency to visit the variable station when reward rate there was lower than that at the constant station and a significant ($F=2.24$, $p<0.05$) negative effect of variance when the mean was greater. There were only three cases in which the mean reward equalled that at the constant station. The results of the second analysis thus support Barnard et al´s (1985) predictions about the responses of shrews to the mean and variance in reward at the variable station, but suggest that animals based their choice on their immediate rather than earlier experience of reward probability distribution.

Effects of skewness in reward probability distribution

The experiment by Barnard et al (1985) suggested that preference for the variable station decreased with increasing variance when the mean reward at the station was higher than that at the constant station to avoid the risk of a low reward. This suggests that shrews may be sensitive not only to variance in reward but also to reward skewness (of which an increased chance of obtaining a very low or very high reward is one consequence). Caraco and Chasin (1984) developed a model which predicted response to skewness independently of the variance in risk-sensitive predators and found experimental evidence of

Mean	Variance at Station 1	Variance at Station 2
0.67	0.22 (0,1,1)	0.89 (0,0,2)
1.00	0.67 (0,1,2)	2.00 (0,0,3)
1.33	0.89 (0,2,2)	3.60 (0,0,4)

Table 2 - The mean and variance in reward (items per visit) at two variable stations. Figures in parentheses are the numbers of mealworm segments received on three successive visits to each station (order randomized between sets of three).

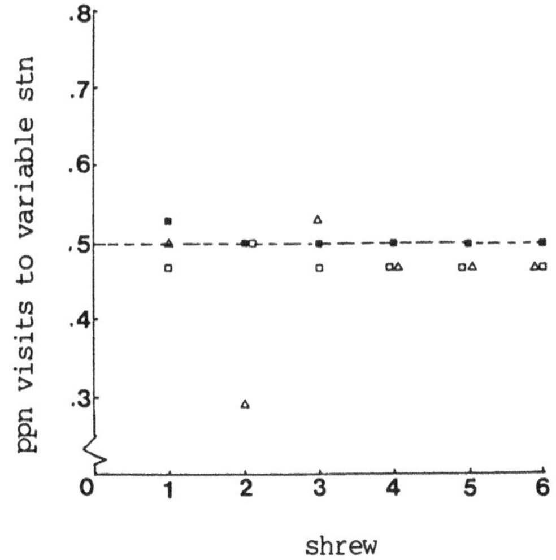

Figure 6 - The proportion of visits to the station with the higher variance by shrews when the mean reward at two stations was 0.67 (open triangles), 1.00 (solid squares) and 1.33 (open squares) items per visit.

sensitivity to skewness in white-crowned sparrows (<u>Zonotrichia leucophrys</u>). Barnard and Brown (unpubl. data) repeated Barnard et al´s (1985) experiment but this time provided shrews running below their estimated requirement with a choice between two variable stations which were similar in mean reward but differed in variance and skewness (Table 2). In each case, the higher variance was associated with the greater skewness (probability of a zero reward).

If skewness had no effect on preference then, given that shrews were running below requirement, the results of the previous experiments would lead us to predict preference for higher variance. In fact, Figure 6 shows that, for all three levels of mean reward, shrews were indifferent in their choice of station. While there may be other explanations for the apparent indifference (e.g. failure to discriminate between variances or a range of low reward values), it may thus be that the higher variance was traded-off against the risk of getting nothing, suggesting that risk-prone shrews are sensitive to reward skewness.

IRR, reward variance and patch residence time

Most predators exploit prey which are patchily distributed. In patchy environments, a predator's net rate of food intake over a foraging bout is a function of its allocation of time and energy to searching within and travelling between patches and its gain rate while in patches (e.g. Charnov 1976, Krebs and McCleery 1984). Where gain functions are deterministic, the amount of food a predator obtains will be specified by the amount of time it spends in each patch. Variation in intake over a foraging bout can thus be introduced by changes in residence time per patch and/or travel time between patches. Under these conditions, deterministic optimal foraging models such as the initial form of the marginal value theorem (Charnov 1976) predict patch residence times which maximize net intake rate during a foraging bout. However, if it is important to minimize the risk of starvation rather than maximize net intake rate, optimal residence times may depend on expected IRR and the consequences of changes in patch residence time for variance in intake rate (Stephens and Charnov 1982). In other words, patch residence time may be determined in a risk-sensitive fashion analogous to the choice of feeding station in the previous experiments.

Stephens and Charnov (1982) have modelled risk-sensitive patch residence times for situations where patch quality is constant but travel time between patches varies randomly so that variance in travel time and expected IRR determines optimal patch residence times. In Stephens and Charnov´s model, the gain function within patches is deterministic so that when residence time is constant variance in intake rate is zero. In the real world, this is unlikely to be the case. Even where patch quality is uniform, intake rate over any given time in a patch is likely to vary as a result of e.g. temporal variation in searching efficiency or runs of bad luck. Under these conditions, variance in intake rate per patch for any given residence time is less predictable and is likely to vary in a complex way with changes in residence time. Barnard and Brown (1987) examined the relationship between patch residence time and variance in intake rate in common shrews foraging in a uniform patchy enviroment in which travel time between patches was negligible and effectively constant. They predicted that, if residence time affects variance in intake rate, shrews failing to meet requirement would choose residence times that tended to maximize variance in intake rate.

Barnard and Brown (1987) trained shrews to forage on a grid of peat-filled wells divided into patches of four by Perspex walls. Individual food requirement was estimated over two days by allowing animals free access to single mealworm segments laid on the surface of the peat in each well. Once requirement had been estimated, shrews were trained to expect a mealworm segment in two randomly chosen wells within each patch while running either at/above requirement (free access to patches) or below requirement (access to patches restricted by temporarily closing the door to the subject´s home tank). Following 90min food deprivation, shrews were then tested by being allowed free access to the grid with the same distribution of mealworm segments. All behaviour on the grid during tests was video-recorded and recordings later analyzed for residence times and intake rates.

To test the assumption that patch residence time influences variance in intake rate, Barnard and Brown (1987) carried out a stepwise partial regression analysis with variance in intake rate per patch as the dependent variable and mean and variance in patch residence time, mean intake rate per patch and IRR as independent variables. Shrews were entered as dummy variables (1-5) to control for individual variation. Table 3a(i) shows that, when the strong, positive effects of mean intake rate per patch and variance in residence time were taken into account, mean residence time had a significant negative effect on variance in intake rate. IRR had no significant effect on intake rate when patch residence time was taken into account. On the basis of these results, therefore, Barnard and Brown (1987) predicted

a) **Variance in intake**	Mean intake rate	Mean res. time	Variance in res. time	df
(i) without competition	60.78**	-12.08**	7.42**	3,16
(ii) with competition	6.19**	-	-	1,18

b) **Patch residence time**	No. wells searched	No. prey eaten	No. patch in seq.	IRR	df
(i) without competition	73.48**	114.24**	5.73**	3.99**	5,394
(ii) with competition	46.36**	110.58**	-	-	3,387

Table 3 - a) F-ratios from stepwise partial regression analysis for the relationship between variance in intake rate per test and the independent variables shown. IRR was also included in the analysis but did not enter the equation. Data for 20 tests with and without competition. Minus sign indicates a negative partial regression coefficient, dash means variable included in analysis but not entered into equation. **, $p<0.01$. b) as a) but for the relationship between patch residence time and the independent variables shown. Data for 400 (i) and 391 (ii) patch visits.

shorter residence times when IRR was low. (It is important to point out, however, that the negative relationship between residence time and variance in intake rate here is purely empirical and applies only to the situation in Barnard and Brown´s experiment. There is no a priori reason to suppose that residence time and variance in intake rate will generally be negatively related). A second partial regression analysis, controlling for potentially confounding effects of search intensity, handling time, satiation and individual variation (Table 4b(i)) showed that, as predicted, shrews spent less time per patch when IRR was low.

Spending less time per patch when IRR was low may have been a means of increasing the chance of a high return. Little time is spent in ´bad´ patches which yield little or nothing and shrews move on rapidly after a quick find in a ´good´ patch (where ´bad´ and ´good´ are determined by whether the shrew happens to search the right place in the right wells first) and increase the chance of finding another food patch in the immediate future. Since shrews also achieved a higher mean rate of intake by spending less time per patch (mean and variance were positively correlated, Table 3), the question arises as to why they did not always search at this speed. Barnard and Brown (1987) suggest the answer may be the cost of increased energy expenditure (see Barnard et al 1983). When reserves are low, it may be worth paying the cost of a high searching rate to reduce a dangerous deficit as quickly as possible. When reserves are high, however, the deficit, if any, is small and it is less worthwhile spending energy to reduce it (see McCleery 1978). The positive correlation between mean and variance in intake may also mean that, when reserves are high and shrews risk-averse, the benefits of an increased mean are offset by an increase in the associated variance.

Competition and risk-sensitivity

The experiments discussed so far, and earlier work by other authors (see above), have focused on the relationship between risk-sensitivity and energy balance. Risk-sensitive foraging decisions made in these cases are assumed to be based on the predator's current ability to met its requirements from potential feeding sites or prey types. However, it is reasonable to suppose that high energy demand predators, at least, which run a serious risk of starvation in the event of a shortfall, might also take into account factors that are likely to influence their ability to meet demand in the future. In other words, we might also expect risk-sensitivity in relation to a predator's expected ability to meet its requirement.

Barnard and Brown (1985b) argued that one obvious factor that might be expected to influence future ability to meet demand is competition. Most predators do not have exclusive access to limited food supplies, but compete for them with others of the same or different species. Competition might be expected to influence risk-sensitivity in at least two ways. Firstly, except where food is rapidly renewing, competition increases the rate of food depletion and therefore the probability that, at any given time, a predator will be unable to meet its energy requirement. In this sense, it is simply one factor affecting current energy balance and thus, potentially, risk-sensitivity in relation to current balance. Secondly, and more interestingly, depletion by competitors is likely to influence the reward probability distribution across feeding sites, so reducing both the mean reward and the predictability of reward in the future. If depletion through competition is sufficient to make it likely that a predator will fail to meet its energy requirement in the near future, it may pay predators, on detecting competition, to increase their tendency towards risk-proneness for, lower levels of current requirement.

Competition and choice between feeding stations Barnard and Brown (1985b) tested the second idea by providing common shrews with a choice between constant and variable feeding stations. They used the same reward schedule as Barnard and Brown (1985a) and trained shrews to expect to meet or fail to meet their estimated food requirement. In this experiment, however, shrews were trained and tested in the presence or absence of apparent competitors (other shrews which could be perceived by subject animals as they visited stations, but which did not have access to the stations - in this case because they were restrained in perforated Perspex enclosures). By using apparent competitors, Barnard and Brown (1985b) were able to distinguish between responses to the expectation of resource depletion (second effect of competition) from responses to actual depletion (first effect). On the basis of the above argument and Barnard and Brown's (1985a) results, shrews might be expected to be risk-prone both with and without competitors when IRR<1 but also risk-prone when IRR>1 if competitors were present. Figure 7 shows the

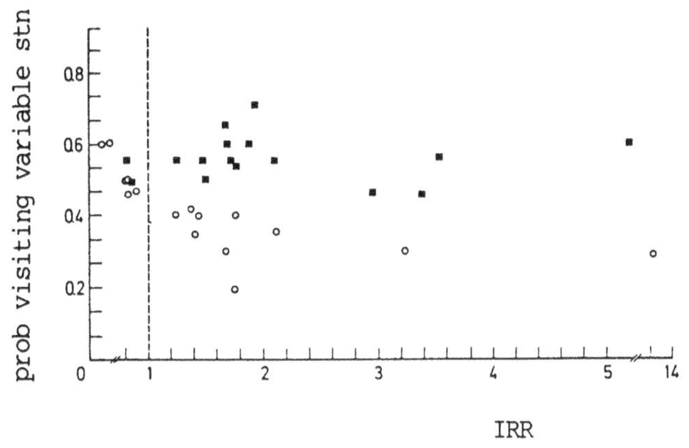

Figure 7 - The relationship between IRR during training and the proportion of visits to the variable station by common shrews with (solid symbols) and without (open symbols) apparent competitors. From Barnard and Brown (1985b).

relationship between IRR during training and the proportion of visits to the variable station during tests when competitors were present and when they were absent. In the absence of competitors, the tendency to visit the variable station declined with increasing IRR. This confirmed the results of Barnard and Brown (1985a) except that, while shrews were significantly risk-averse when IRR>1, they were not significantly risk-prone when IRR<1 (the proportion of visits to the variable station did not differ significantly from 50%). When IRR<1, therefore, shrews appeared to be risk-indifferent. When competitors were present, however, shrews were risk-indifferent irrespective of IRR and were significantly ($t=4.85$, $p<0.001$) more likely to visit the variable station when IRR>1 than in the absence of competitors (Figure 7). The prediction that competition would increase risk-proneness for any given level of expectation relative to requirement was thus borne out.

Competition and residence time in patches Barnard and Brown (1987) (see Table 3 (without competition)) presented evidence of risk-sensitivity in choice of patch residence time by shrews foraging in a uniform patchy environment with constant (negligible) travel time between patches. Following Barnard and Brown (1985b), it is thus reasonable to suppose that competition would affect choice of patch residence time in an analogous way to choice of feeding station. In the situation investigated by Barnard and Brown (1987), competition would be expected to <u>reduce</u> patch residence time for higher levels of IRR because reduced residence time correlated with an increase in variance in intake rate (Table 3). To test this, Barnard and Brown´s (1987) procedure was repeated, but this time an apparent competitor was introduced into a perforated Perspex tunnel running the length of the grid of patches. The perforations meant that subjects and apparent competitors could make visual, auditory, olfactory and limited tactile contact, but the competitor, which was fed only in a home tank attached to the grid, could not deplete the food supply on the grid. The relationship between IRR and patch residence time was analyzed using similar stepwise partial regression techniques to those used by Barnard and Brown (1987).

Table 3 (with competition) shows that, while the number of wells searched and prey taken per patch still had a trivial positive effect on residence time, there was no effect on residence time of the position of a patch in the sequence of visits or IRR. As in Barnard and Brown (1985b), therefore, IRR was not a good predictor of foraging behaviour under conditions of apparent competition. However, shrews did not, as it was predicted they would, spend less time in patches when a competitor was present. In fact, there was a significant ($t=2.43$, $p<0.02$) tendency for mean residence time to be longer. This was not due to subjects simply spending more time in activities other than foraging (e.g. vigilance, interacting with competitors) when competitors were present (see also Barnard et al 1983). Instead residence time appeared to increase because shrews took more prey per patch ($t=2.85$, $p<0.01$). This is in line with the results of a study of time budgeting by common shrews under condition of competition (Barnard et al 1983) which showed that the biggest change in behaviour when a competitor was present was an increase in the number of prey ingested during a foraging bout. The greater number of prey taken per patch appeared to be attributable to more intensive searching of individual wells (there were no differences between tests with (this experiment) and without (Barnard and Brown 1987) in the number wells searched per patch or the number of patches searched). The effect of competition on residence time, however, was complicated by the fact that, when a competitor was present, there was no longer a relationship between patch residence time and variance in intake rate (Table 3). The presence of a competitor thus appeared to remove any interrelationship between variance in intake rate, IRR and patch residence time.

Risk-sensitive foraging and life-history trade-offs

While variance in food reward may influence foraging decisions, it may also impose a constraint on decisions made in other contexts. Thus, for instance, high food reward variance at a given site may reduce the quality of a breeding territory at that site and therefore the tendency to maintain and defend it. There may thus be a trade-off in site preference between food reward variance and other measures of site quality. Young et al (in press) tested this possibility with bitterlings by offering a male different combinations of food reward variance and mussel (Anodonta sp.) shell size. Female bitterlings lay their eggs in mussel shells which are defended by males prior to fertilization (Marshall 1965). Eggs are more likely to be found in larger shells (Maitland 1977), suggesting that the quality of a potential breeding site to a male can be measured, at least approximately, in terms of mussel shell size.

The subject male, in breeding colouration, was chosen blind from among the fish (2-7) showing increased risk-aversion when IRR<1 in Figure 2. The male was introduced into a test tank as before but this time a female was placed behind a clear partition at each feeding station. The selected females were of equal size and similar appearance. They were visible to the male through

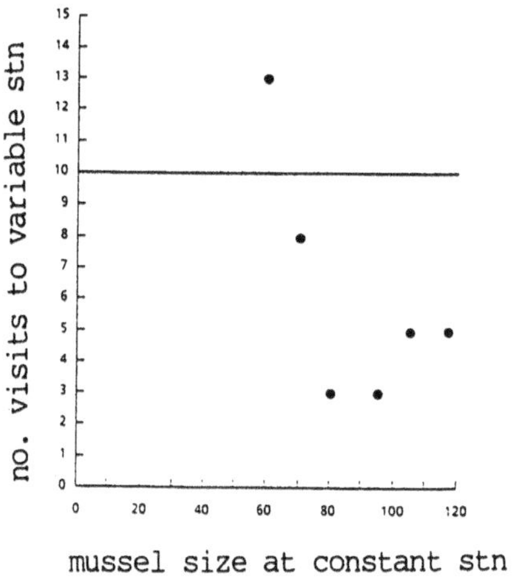

Figure 8 - The relationship for a male bitterling between the number of visits to the variable station during tests and the size of mussel in the constant station compartment (see text). From Young et al (in press).

the partitions and the male regularly oriented towards, approached and interacted with each female across the partition when in the compartments housing the feeding stations. Requirement estimation, training and testing procedures were the same as for Figure 2, except that the male was trained to run below requirement only (when increased risk-aversion was expected). The training and testing procedure was repeated six times, each with a different combination of mussel shell sizes available in the constant and variable station compartments. The same sized mussel (117mm in length) was always available in the compartment containing the variable station, but a shell ranging in size from 117-60mm was available in the compartment containing the constant station. Again, which compartment contained the constant station and which the variable was randomized between training and testing sessions.

Figure 8 shows the relationship between the number of visits to the variable station by the male during tests and the size of mussel available at the constant station. When the same sized mussels were available in the constant and variable compartments, the male made only 20% of its visits to the variable station. Bias towards the constant station was maintained as mussel size in the constant station compartment was reduced to 80mm. Further reduction, however, resulted in an increased tendency to visit the variable station so that when mussel size at the constant station was only 60mm, 65% of visits were to the variable station. Although the experiment was performed on only one fish, the results thus suggest a trade-off between the potential feeding and breeding qualities of a site and that variance in food reward can be an important factor in the trade-off.

Summary

The experiments discussed above provide broad support for shortfall-minimizing models of risk-sensitive foraging but show that the relationship between food requirement and response to variance in reward may sometimes be more complex than the straight switch between risk-aversion and risk-proneness predicted by classical shortfall models. The predicted switch occurred in common shrews which, like Caraco´s buntings, are high energy demand endotherms, but the response of low energy demand, ectothermic bitterlings was characterized by an initial increase in risk-aversion with decreasing intake relative to requirement (IRR). However, more extreme deprivation resulted in bitterlings becoming risk-prone. The results of the bitterling experiments suggest that IRR and food deprivation interact in determining response to variance in reward, presumably through their combined affect on current energy balance and systematic experiments to quantify the interaction are planned.

There is evidence of the trade-offs between mean and variance in reward predicted by shortfall-minimizing models for both risk-averse and risk-prone predators. Risk-averse bitterlings were more likely to visit a variable station as the mean reward there increased relative to the mean at a constant station. Risk-prone shrews tended to visits whichever of two (constant or variable) stations yielded the higher mean when variance in reward at the variable station was low, but preferences decreased when variance was high, the interpretation being that shrews were responding to the increased chances of a very high or very low reward depending on the direction of their preference with low variance. However, the evidence for sensitivity to reward skewness among shrews is inconclusive.

Current energy balance is one factor which may lead to risk-sensitivity in foraging decisions. In predators which are hard-pressed to find enough food to maintain themselves, and which thus face a substantial risk of starving if there is a shortfall in supply, factors affecting the chances of finding enough food in the immediate future (and thus the risk of being unable to survive a shortfall should it arise) may be another important influence on risk-sensitivity. One such factor is competition, since competitors may reduce the predictability of food supplies in the future and thus the value of the predator´s expectation based on previous experience of food availability. In two contexts, choosing between constant and variable feeding stations and choosing residence times in food patches, common shrews showed changes in risk-sensitivity when apparent competitors were present. When choosing between feeding stations, shrews tended to make more visits to the variable station when IRR>1 if competitors were present than if they were absent, though the change was to risk-indifference rather than risk-proneness. The effect of apparent competition on patch residence time was to increase time per patch, a change which was the opposite of that predicted from the relationship between residence time and reward variance in the absence of competition. However, this was complicated by the fact that, when competitors were present, there was no longer a relationship between residence time and variance and food reward variance. In keeping with the effects of apparent competition on foraging behaviour found in earlier experiments with common shrews, the increase in patch residence time appeared to be due to increased foraging intensity when competitors were present.

Finally, an experiment with bitterlings during their breeding season suggested that risk-sensitive foraging decisions may be important in other contexts, in this case choice of breeding site by males. A risk-averse male appeared to tolerate a lower quality breeding site to maintain preference for a constant reward feeding station. Only when breeding site quality at the constant station dropped to just over half that at the variable station did the fish switch its preference.

ACKNOWLEDGMENTS

I thank Jane Hurst for helpful comments on the manuscript.

REFERENCES

Barnard C J, Brown C A J (1985a) Risk-sensitivity in foraging common shrews (Sorex araneus L.). Behav Ecol Sociobiol 16, p. 161-164.

Barnard C J, Brown C A J (1985b) Competition affects risk-sensitivity in foraging shrews. Behav Ecol Sociobiol 16, p. 379-382.

Barnard C J, Brown C A J (1987) Risk-sensitive foraging and patch stay time in common shrews, Sorex araneus L. Anim Behav 35, p. 1255-1257.

Barnard C J, Brown C A J, Gray-Wallis J (1983) Time and energy budgets and competition in the common shrew (Sorex araneus L.). Behav Ecol Sociobiol 13, p. 13-18.

Barnard C J, Brown C A J, Houston, A I, McNamara, J M (1985) Risk-sensitive foraging in common shrews: an interruption model and the effects of mean and variance in reward rate. Behav Ecol Sociobiol 18, p. 139-146.

Caraco T (1981) Energy budgets, risk and foraging preferences in dark-eyed juncos (Junco hyemalis). Behav Ecol Sociobiol 8, p. 213-217.

Caraco T (1983) White-crowned sparrows (Zonotrichia leucophrys): foraging preferences in a risky environment. Behav Ecol Sociobiol 12, p. 63-69.

Caraco T, Chasin M (1984) Foraging preferences: response to reward skew. Anim Behav 32, 76-85.

Caraco T, Lima S L (1985) Foraging juncos: interaction of reward mean and variability. Anim Behav 33, p. 216-224.

Caraco T, Martindale, S, Whittam, T S (1980) An empirical demonstration of risk-sensitive foraging preferences. Anim Behav 28, p. 820-830.

Clayton H (1989) Risk-sensitive foraging in bitterlings. Unpubl BSc Thesis, Univ Nottingham.

Maitland P S (1977) Freshwater Fishes of Britain and Europe. Hamlyn, London.

Marshall N B (1965) The Life of Fishes. Weidenfeld, London. and Nicholson.

McCleery R H (1978) Optimal behaviour sequences and decision-making. In: (J.R. Krebs & N.B. Davies eds) Behavioural Ecology: An Evolutionary Approach, p. 377-410. Blackwell, Oxford.

McNamara J M, Houston, A I (1982) Short-term behaviour and life-time fitness. In: (D.J. McFarland ed) Functional Ontogeny, p. 60-87. Pitman, London.

Real L A, Caraco T (1986) Risk and foraging in stochastic environments. Ann Rev Ecol Syst 17, p. 371-390.

Stephens D W, Charnov E L (1982) Optimal foraging: some simple stochastic models. Behav. Ecol. Sociobiol 10, p. 251-263.

Stephens D W, Krebs J R (1986) Foraging Theory. Princeton Univ Press, Princeton.

Stephens D W, Paton S R (1984) How constant is the constant of risk-aversion? Anim. Behav. 34, p. 1659-1667.

Thomas G, Kacelnik A, van der Meulen J (1985) The three-spined stickleback and the two-armed bandit. Behaviour 93, p. 227-240

Waddington K D, Allen T, Heinrich B (1981) Floral preferences of bumblebees (Bombus edwardsii) in relation to intermittent versus continuous rewards. Anim Behav 29, p. 779-784.

Young R J, Clayton H, Barnard C J (in press) Risk-sensitive foraging in bitterlings (Rhodeus sericus): effects of food requirement, and breeding site quality. Anim Behav.

THE COMBINED EFFECTS OF LEARNING AND HUNGER IN THE FEEDING BEHAVIOUR OF THE FIFTEEN-SPINED STICKLEBACK (*SPINACHIA SPINACHIA* L.)

Marion I. Croy & Roger N. Hughes
School of Biological Sciences
University College of North Wales
Bangor
Gwynedd
U.K.
LL57 2UW.

INTRODUCTION

Learned changes in foraging behaviour may accentuate density-dependent predation of alternative prey. For example, if learned handling ability is reinforced by prey encounters, but forgotten in their absence, foragers may become more efficient at dealing with specific prey when these are abundant (Cunningham and Hughes 1984). As a result, prey profitabilities could be transposed in rank, perhaps causing a switch of diet (Hughes 1979).

Hunger is another factor potentially capable of influencing foraging efficiency (e.g. Werner and Hall 1974; Kislalioglu and Gibson 1976a): it is known to increase feeding rate (e.g. Brett 1979; Godin 1981), activity level (e.g. Beukema 1968; Calow 1974), reaction distance (e.g. Holling 1966; Ware 1972) and the size range of eaten prey (e.g. Ivlev 1961; Werner and Hall 1974; Kislalioglu and Gibson 1976b). Being a motivational mechanism, hunger is an abstraction not quantifiable per se, but an index of hunger can be derived from some measurable behavioural or physiological state, such as readiness to feed or gut fullness.

In the present study, the fifteen-spined stickleback,

NATO ASI Series, Vol. G 20
Behavioural Mechanisms of Food Selection
Edited by R. N. Hughes

Spinachia spinachia, was used to investigate changes in foraging behaviour resulting from experience with specific prey and the extent to which such changes are modified by hunger state. Data were entered into simple models to predict whether or not learning, in the presence of fluctuating hunger level, may cause profitabilities of different prey to become transposed in rank, so creating a mechanism for switching.

METHODS

Spinachia spinachia (103-124 mm total body length) were presented with contrasting prey: the more robust, predominantly benthic amphipod *Gammarus locusta* and the more delicate, pelagic brine shrimp, *Artemia salina*. To standardise the fish length / prey length ratio, 101-110 mm fish received prey of length 12.0-13.5 mm; 111-120 mm fish, prey of 13.5-15.0 mm; and 121-124 mm fish, prey of 15.0-17.5 mm. Any effects of prior experience with prey resembling *Gammarus* or *Artemia* were eradicated by maintaining fish on a diet of mussel flesh for two months before experimentation.

Aquaria (600 x 250 x 250 mm), furnished with gravel and artificial weed, were filled with seawater at 10-12°C. Overhead lighting was provided by 40 W fluorescent tubes on a 12-h light 12-h dark regime and aquaria were screened on three sides, leaving one side clear for observation. Experimental trials began at a consistent time of day for each fish, which was housed in a separate aquarium. Foraging behaviour was categorised into (1) attack and handling efficiency and (2) behavioural components of attack and handling, monitored by direct observation, using a stop-watch, and by video recording respectively. Separate experiments examined the effects on foraging behaviour of experience, hunger and the acquisition of learned skills at different initial levels of hunger.

To investigate the effect of learning on attack and handling efficiency, each of eight replicate fish was subjected to eight daily trials of 30 min duration. Timed observations began as soon as the fish orientated to the first prey item presented. Frequencies of captures, misses, manipulations and duration of prey handling were measured. Attack efficiency was expressed as the proportion of attacks leading to capture and handling efficiency as the proportion of grasps followed by swallowing, without the prey being spat out and reorientated. Eaten prey were replaced immediately until fish, now satiated, ceased to orientate and fixate (Brett 1971). Fish were not fed between trials. Behavioural changes associated with the first prey item eaten in successive trials, were used as measures of learning independent of changes in hunger level. To investigate the effect of hunger, each trial was divided into consecutive 10 min periods and the difference in attack and handling efficiency between the first and third periods noted. Separate trials were alloted to the diets *Gammarus*, *Artemia* and *Gammarus* alternating with *Artemia*.

<u>Basic foraging sequence</u> Components outside basic sequence

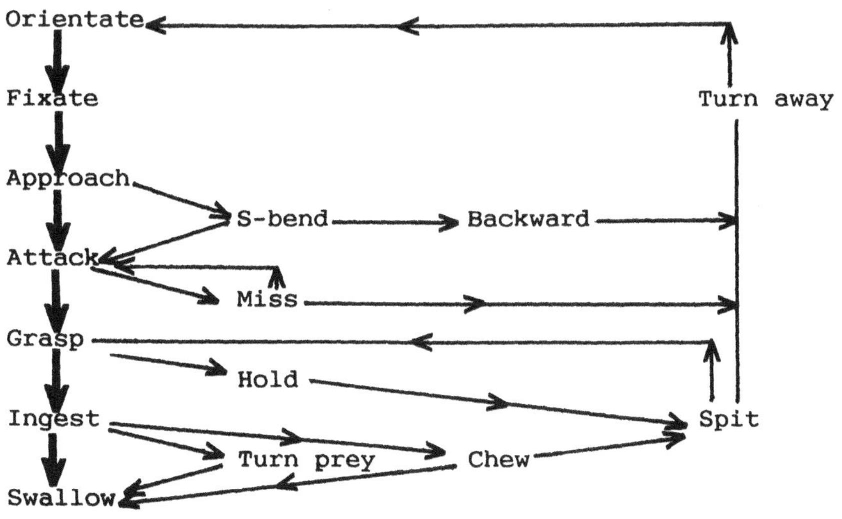

Figure 1.
Flow diagram of components involved in the foraging sequence of sticklebacks.

The same experimental protocol, but with video recording, was used to investigate the effect of learning on behavioural components of attack and handling. Fifteen behavioural components were recognised, seven belonging to a basic sequence (Kislalioglu 1975) involved in the successful attack and ingestion of a prey item and others being associated with less efficient foraging behaviour (Fig. 1).

Definition of foraging components:
Orientate: movement of the body towards the prey; all sequences begin with this activity.
Fixate: focusing with both eyes on the prey; fish is usually stationary.
Approach: swimming towards the prey; may be rapid with dorsal and ventral fins held close to the body.
S-bend: adoption of S-shaped posture before striking at the prey; commonly preceded by stopping.
Attack: a strike at the prey.
Grasp: an attack that succeeds in contacting the prey; may be subdivided according to the region contacted (head, middle or tail).
Miss: an attack that fails to contact the prey.
Hold: prey is held in the fish's jaws and may be crushed.
Backwards: fish swims backwards while still facing the prey; prey may be held in the jaws.
Spit: prey is ejected from the jaws and is commonly turned; the distance ejected from the fish is variable.
Ingest: prey is sucked or manipulated into the mouth.
Turn prey: prey is turned around inside the mouth before swallowing; occurred only in fish experienced with a familiar prey, but was too rare to be included in the analysis.
Chew: prey is chewed inside the jaws before swallowing; movement suggests breaking prey on the gill rakers, the teeth being small.
Swallow: prey is swallowed.
Turn away: movement away from the prey with a half-turn of the body.

To investigate the effect of initial hunger level on

learning, the amount of flesh required to satiate each individual was first determined. In a series of seven daily trials, fish were fed known quantities of mysid flesh until they ceased to respond to food. The mean wet mass of flesh required to satiate each fish was used as a standard to calculate the mass of flesh required to achieve 50% and 95% satiation. Prior to experimentation, fish were starved for 24 hours, then either fed appropriate rations of mysid flesh to reach 50% and 95% satiation, or kept starved. Twelve replicate fish at each of the three initial hunger levels were then presented either with *Gammarus* or *Artemia* in daily trials, between which they were satiated with mysid flesh. Frequencies of captures, misses, manipulations and handling times were measured, together with estimations of attack and handling efficiency, for each group of fish. Rate of learning was measured as the number of trials (days) required for handling time to decrease to an asymptote. To obtain sufficient data for statistical analysis, the entire procedure was repeated with the same fish after three months, and the results pooled. The experimental protocol was repeated using video recording to investigate the effect of initial hunger level on behavioural components of attack and handling, as defined above.

The combined effects of learning and hunger on the profitability of prey during feeding sequences was investigated by incorporating the experimental data into two simple models. A version of Hughes' (1979) model was used for the learning component. Handling time for the first prey item taken in a trial was expressed as a negative exponential function of the cumulative number of feeding trials, while omitting any effect of forgetting between encounters (M.I. Croy, unpubl. data). Appropriate parameters for each initial hunger level were also included (Table I). Thus:

$$T_h = T_{min} + (T_{max} - T_{min})e^{-kN} \qquad (1)$$

where T_h = handling time; T_{min} = minimum value of handling time attainable by learning; T_{max} = maximum value of handling time unmodified by learning; k = instantaneous rate of decrease of

T_h per trial; N = number of consecutive trials.

The effects of satiation were modelled as a function of the cumulative number of prey encountered during the final trial (by which time learning had reached an asymptote):

$$T_h = T_{max}[1 - ((T_{max} - T_{min}) / T_{max})e^{-kN}] \qquad (2)$$

Where N = number of prey; k = instantaneous rate of decrease of T_h. Hunger was resumed at the start of each trial (M.I.Croy, unpubl. data).

Since calorific values per unit mass for *Gammarus* and *Artemia* differ only by some 9% (Paffenhofer 1967; Cummins and Wuycheck 1971), profitability was calculated as the ratio of dry felsh yield / handling time (Charnov 1976). Dry mass of presented items was estimated from logarithmic regressions of prey mass on length (Croy and Hughes submitted, a).

RESULTS

Effect of Experience on Attack and Handling Efficiency.

Full presentations and statistical analyses of data described in this section and the next are given in Croy and Hughes (submitted, a). Attack frequency decreased among successive trials whereas attack efficiency for the first prey eaten per trial increased (Fig. 2), improving by 51% for *Gammarus*, 17% for *Artemia* and 8% for *Gammarus* presented alternately with *Artemia*.

Handling time decreased among successive trials (Fig. 3), the mean handling time for the first item eaten per trial decreasing by 68% (7th trial) for *Gammarus*, 85% (5th trial) for *Artemia* and 59% (6th trial) for *Gammarus* presented alternately with *Artemia*. Handling time was significantly longer for *Gammarus* presented in an alternating diet than for *Gammarus* or *Artemia* alone and the decrease in mean handling time for *Gammarus* was significantly less in the alternating than in the pure diet.

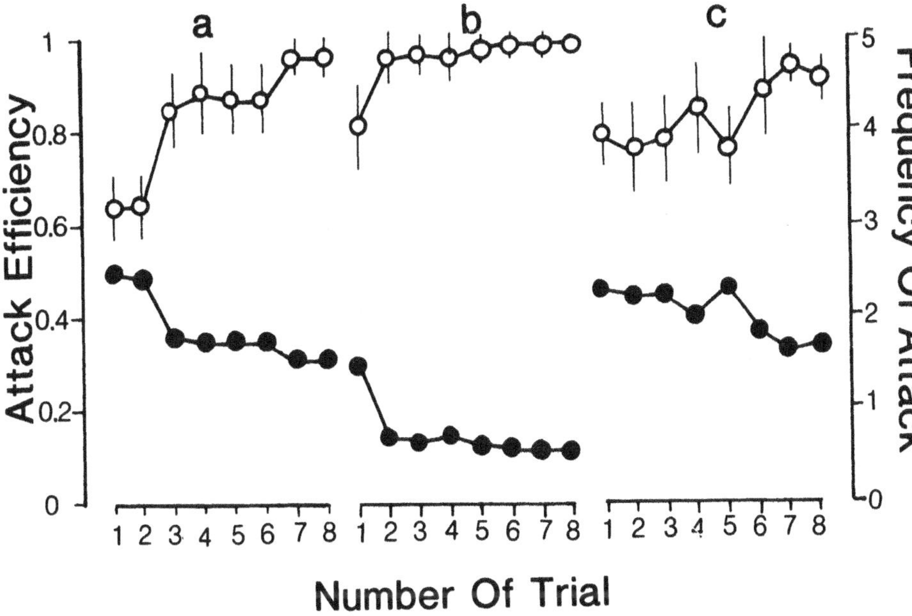

Figure 2.
Mean attack efficiency (○) and mean frequency of attacks (●) for the first prey item taken in successive feeding trials; error bars are for 95% confidence intervals. (a) *Gammarus*, (b) *Artemia*, (c) *Gammarus* presented alternately with *Artemia*.

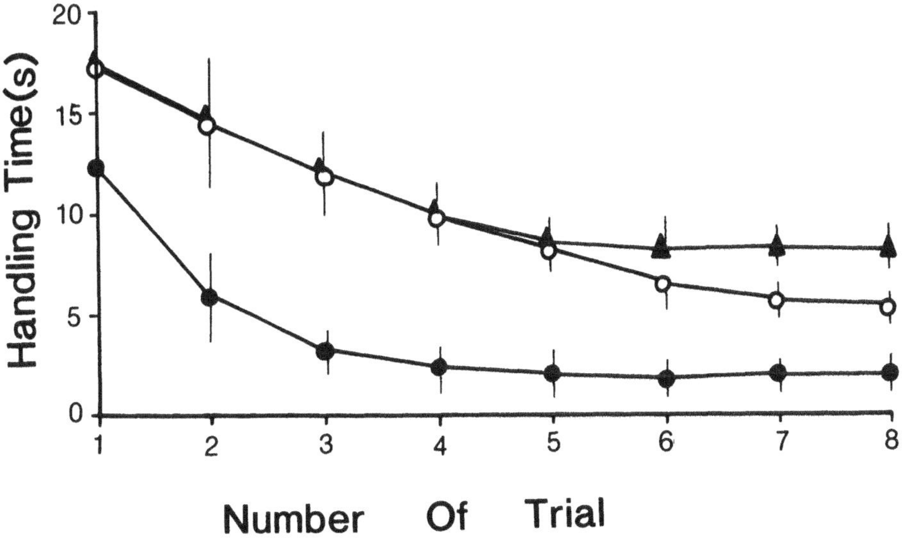

Figure 3.
Mean handling time for the first prey item taken in successive feeding trials; error bars are for 95% confidence intervals. *Gammarus* (○), *Artemia* (●), *Gammarus* presented alternately with *Artemia* (▲).

Handling efficiency (Fig. 4) increased by 116% (8th trial) for *Gammarus* and 83% (8th trial) for *Gammarus* alternating with *Artemia*, although the difference between these treatments was not statistically significant. *Artemia* was ingested in any orientation, with a corresponding handling efficiency close to unity.

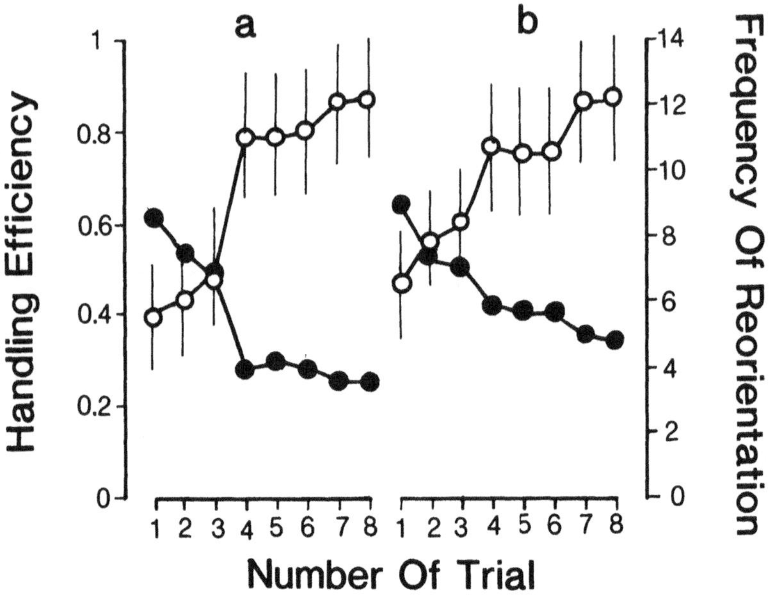

Figure 4.
Mean handling efficiency (○) and mean frequency of reorientations (●) for the first prey item taken in successive feeding trials; error bars are for 95% confidence limits. (a) *Gammarus*, (b) *Gammarus* presented alternately with *Artemia*.

Effect of Learning on Behavioural Components of Attack and Handling.

Naive fish attacked *Gammarus* and *Artemia* in any orientation, but the frequency of head-on attacks increased significantly in later trials (Fig. 5). 80% of attacks (Trial 8) were directed to the head for *Gammarus* and 69% for *Artemia*. Prey were more often ingested immediately when grasped head-on and therefore were more likely to be swallowed without reorientation. The frequency of components Miss, S-bend, Chew, Hold and Spit decreased in later trials as fish became more skillful in the attack (Fig. 5).

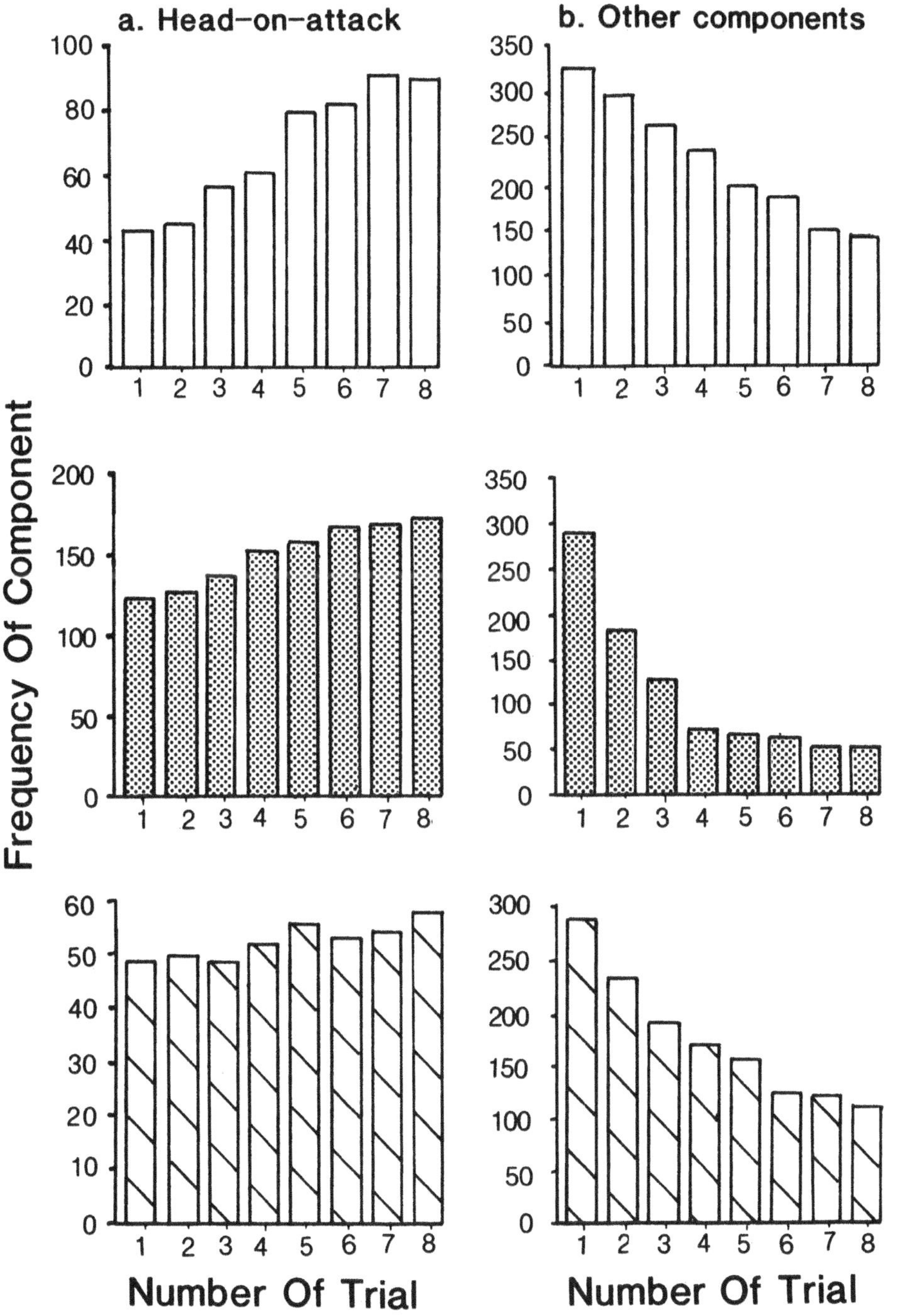

Figure 5.
(a) Frequencies of Head-on attacks for successive feeding trials and (b) frequencies of other components in addition to the basic foraging sequence. *Gammarus* (□), *Artemia* (▩), *Gammarus* presented alternately with *Artemia* (◪).

Effect of Hunger on Attack and Handling Efficiency.

Attack efficiency for *Gammarus* decreased within most trials presenting *Gammarus* alone or *Gammarus* alternately with *Artemia* (Mann-Whitney U-test, N=256 (pooled for *Gammarus* and the alternating diet), Z=7.74, $P<0.001$). Attack efficiency for *Artemia* did not change significantly (Mann-Whitney U-test, N=128, Z=0.76, $P>0.4$).

Handling time for all prey types increased significantly within trials (Mann-Whitney U-test, N=768 (pooled for all diets), Z=9.38, $P<0.001$, Fig. 6).

Handling efficiency decreased within trials for *Gammarus*, either presented alone or alternately with *Artemia* (Mann-Whitney U-test, N=256 (pooled for *Gammarus* and the alternating diet), Z=9.28, $P<0.001$).

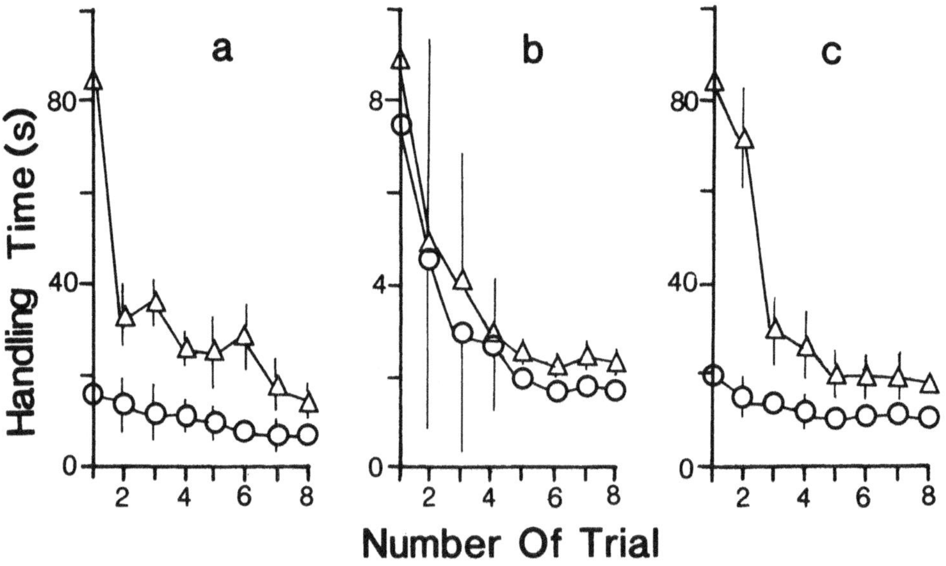

Figure 6.
Mean handling time for successive feeding trials, each subdivided into equal thirds; error bars are for 95% confidence limits. (a) *Gammarus*, (b) *Artemia*, (c) *Gammarus* presented alternately with *Artemia*. First third (○), final third (△).

Effect of Learning, at Different Initial Levels of Hunger, on Attack and Handling Efficiency.

Full presentations and statistical analyses of data

described in this and the next section are given in Croy and Hughes (submitted, b). Attack efficiency increased as fish gained experience with the prey. In the treatment *Gammarus*, attack efficiency was modified by initial hunger level: starved, inexperienced fish attacking *Gammarus* with a success rate of 59% and 95% satiated fish with a success rate of 42%. Moreover, the subsequent increase in attack efficiency among trials was modified by initial hunger level, improving by 69% (7th Trial) for starved fish, 100% (8th Trial) for fish fed to 50% satiation and 104% (8th Trial) for fish fed to 95% satiation. Attack efficiency was independent of initial hunger level when feeding on *Artemia*.

Handling time decreased as fish gained experience with the prey (Fig. 7). In the treatment *Gammarus*, the mean handling time decreased by 70% (7th trial) for starved fish, 67% (7th Trial) for fish 50% satiated and 64% (8th Trial) for 95% satiated fish. In the treatment *Artemia*, handling time decreased by 82% (5th Trial), 70% (5th Trial) and 67% (6th Trial) respectively. Handling time for starved fish was consistently less than for fish initially fed to 95% satiation.

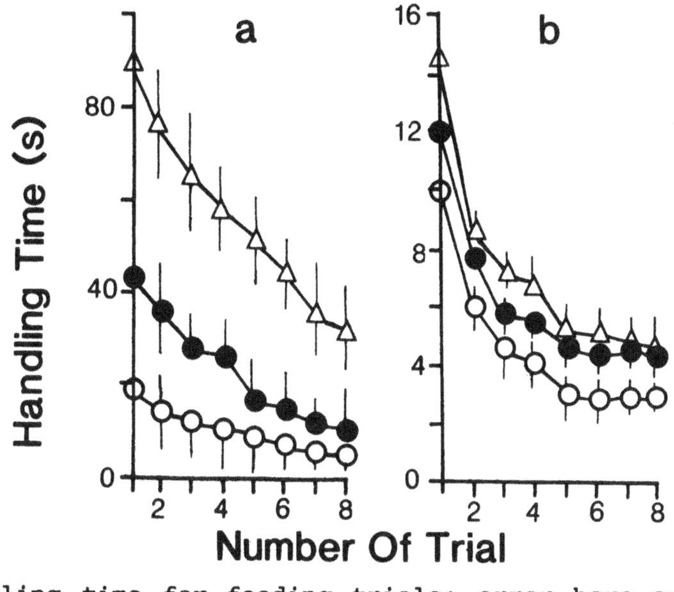

Figure 7.
Mean handling time for feeding trials; error bars are for 95% confidence limits. (a) *Gammarus* (b) *Artemia* at one of three initial hunger levels: starved (○), 50% satiated (●), 95% satiated (△).

Handling efficiency increased significantly as fish gained experience with *Gammarus*, improving by 73% (starved fish), 72% (50 % satiated) and 37% (95% satiated); it was also modified by initial hunger level. Handling efficiency for *Artemia* was close to unity throughout.

Effect of Learning, at Different Initial Levels of Hunger, on Behavioural Components of Attack and Handling.

In all treatments, head-on attacks became more frequent and attacks to the middle or tail less frequent as fish became more skillful. There was a concurrent decrease in the frequency with which *Gammarus* were reorientated before being swallowed, involving Hold and Spit and in the frequencies of Miss, S-bend and Chew (components additional to the basic foraging sequence). In the treatment *Gammarus*, the frequencies of Miss, S-bend, Chew, Hold and Spit were significantly less for starved fish than those fed to 50% or 95% satiation, but in the treatment *Artemia*, only Miss, S-bend and Chew were significantly less for starved fish.

Combined Effects of Learning and Hunger on Prey Profitability.

Constants and parameters for the learning and hunger equations (see Methods) are given in Table I.

Table I.
Constants and parameters for the learning and hunger equations.

<u>Learning sub-model</u>

Prey presented	Hunger level	T_{min}(s)	T_{max}(s)	k
Gammarus	starved	5.51	20.02	0.18
Gammarus	50% satiated	14.00	40.27	0.14
Gammarus	95% satiated	32.92	86.29	0.13
Artemia	starved	1.83	9.69	0.23
Artemia	50% satiated	3.65	10.20	0.14
Artemia	95% satiated	4.71	12.76	0.14

<u>Hunger sub-model</u>

Prey presented	Hunger level	T_{min}(s)	T_{max}(s)	k
Gammarus	starved	5.51	25.54	0.10
Artemia	starved	1.83	11.96	0.04

Amalgamation of the learning and hunger sub-models predicted a cyclical pattern of handling time and hence profitability, for both *Gammarus* and *Artemia* (Fig. 8). Increased profitabilities, resulting from learning, stabilised by Trial 5 for *Artemia* and Trial 7 for *Gammarus*. Once maximum handling efficiency had been attained, profitability became simply a function of current hunger level. Thus, in earlier foraging trials, the initial increase in profitability can be attributed to learning and its subsequent decrease to satiation.

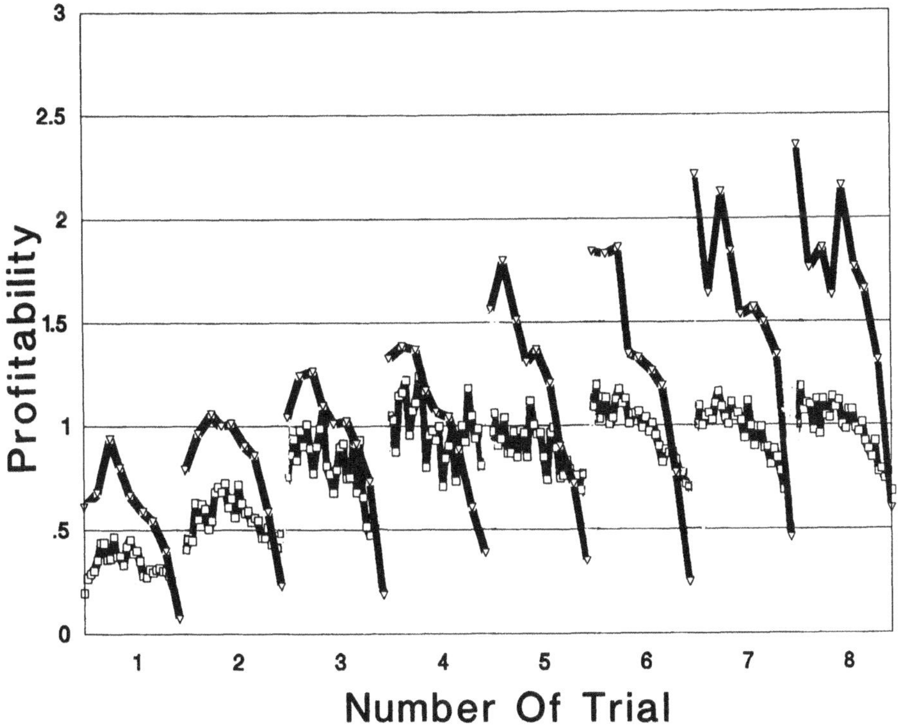

Figure 8.
Mean profitability for successive presentations of prey for eight consecutive feeding trials. *Gammarus* ▬▽▬ , *Artemia* ▬□▬ .

Profitability was also dependent on the forager's initial hunger level: lower levels resulted in proportionately longer handling times for *Gammarus*, bringing the profitabilities of *Gammarus* and *Artemia* closer together (Fig. 9). Thus, although starved fish improved their foraging efficiency faster and after fewer prey encounters than fish fed to 95% satiation, disparate initial profitabilities meant that transposition of prey could occur only after at least five trials for starved fish. Transposition occurred after only three trials for fish fed to 50% satiation and two trials for fish fed to 95% satiation, when initial profitabilities were more comparable.

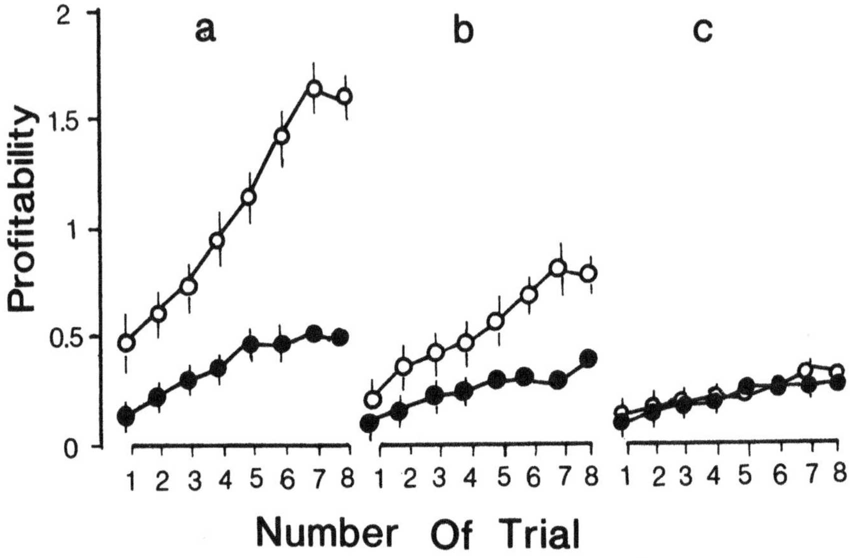

Figure 9.
Mean profitability for prey taken in trials at one of three initial hunger levels: (a) starved, (b) 50% satiated or (c) 95% satiated. *Gammarus* (O), *Artemia* (●).

DISCUSSION

Behavioural Aspects of Learning

In the present study, inexperienced *Spinachia* feeding on *Gammarus* performed long and complex foraging sequences, frequently attacking and reorientating prey several times before swallowing them. *Artemia*, however, being more easily handled, were never reorientated before swallowing and typical foraging sequences were short, containing fewer components than with *Gammarus* or an alternating diet. As *Spinachia* became more skillful, foraging sequences became simpler and shorter, involving fewer of the components Miss, S-bend, Spit, Hold and Chew, outside the basic foraging sequence. Consequently, attack and handling efficiency increased and handling times were reduced.

Rates of improvement may be correlated with the degree of skill required to catch or handle specific prey. *Artemia*, for example, are slow-moving, soft-bodied prey, caught and eaten with minimal skill. *Gammarus*, in contrast, are fast-moving and possess a tough cuticle, requiring considerable skill to be apprehended and eaten. *Spinachia* increased their skills faster when foraging on *Artemia* than on *Gammarus* (Croy and Hughes submitted, a). The learning process thus was less efficient when foragers were fed *Gammarus* alternately with *Artemia*.

Learning, Hunger, OFT and Switching

When prey are encountered in rapid succession during a foraging bout, the effect of learning, which increases prey profitability, would be opposed by the effect of satiation. Although satiation has been reported to have little effect on handling time in certain cases (Holling 1966; Beukema 1968), it is clearly associated with increased handling time in a variety of species, including the fifteen-spined stickleback (Kislalioglu and Gibson 1976; Croy and Hughes submitted, b; present study), bluegill and green sunfish (Werner 1974) and rainbow trout (Ware 1972). Thus transposition is likely to depend critically on the intrinsic rates of increase or decrease of profitability caused by learning or satiation and on the initial difference in profitability of the two prey, itself related to initial hunger level. As a result, opportunities for switching created by learned improvements in prey-handling behaviour could be small if predators frequently begin foraging bouts when very hungry (Hughes 1979). This may especially be so since the alternate handling of different prey reduces learning efficiency (Cunningham and Hughes 1984; present study), so that large numbers of prey might need to be encountered before profitabilities could become transposed.

In many natural situations, foragers experience a wide range of hunger states while searching for patchily distributed prey (Parsons et al 1970; Bailey et al 1975; Sibert 1979; Godin 1981). In the case of *Spinachia*, mean stomach fullness ranged from 25% to 75% for fish caught and examined at low tide (Kislalioglu 1975). Only 0.5-9% of fish, however, had nearly empty stomachs on any sampling occasion. Seasonal differences accounted for much of the variation, a greater proportion of nearly empty stomachs being recorded during winter. Switching in *Spinachia*, therefore, would seem not to be ruled out by frequent states of intense hunger.

In a series of experiments where prey were offered at different relative abundances, *Spinachia* did indeed switch between *Gammarus* and *Artemia*, with individual fish feeding on the most common type of prey (M.I.Croy, unpublished data). In addition to changes in handling efficiency, behavioural mechanisms promoting area-restricted searching also increased the frequency of encounter with specific prey (*Artemia* swim in the water column, whereas *Gammarus* remain near the aquarium substratum). Hence area-restricted searching, together with learned handling skills developed over successive daily trials, probably facilitated switching in these *Spinachia*, even when initial hunger levels were high.

Clearly, this mechanism of switching would not operate where the profitabilities are too different to become transposed. Under such circumstances, other mechanisms of switching such as the formation of 'searching images', where the forager selectively attends to certain characteristics or stimuli inherent to specific prey (Lawrence and Allen 1983), could still be important.

SUMMARY

As a result of experience, fifteen-spined sticklebacks

began to attack specific prey more efficiently. This was caused by changes in success of apprehending prey, in handling behaviour and in handling time; the rate of improvement of foraging efficiency was different for individuals feeding on different types of prey and was modified by initial hunger level. Starved fish acquired attack and handling skills faster than those fed to 50% and 95% satiation. With *Gammarus* as prey, mean handling time decreased by 70% (7th trial) for starved fish and 64% (8th trial) for fish fed to 95% satiation; for *Artemia*, handling time decreased by 82% (5th trial) and 67% (6th trial), respectively. Increased foraging efficiency included several mechanisms. Behavioural components additional to the basic foraging sequence were performed less frequently, correlating with reduced handling time. Learning and hunger influenced foraging efficiency independently, experience increasing and declining hunger level reducing profitability. Transposition of prey profitabilities thus depends critically on the intrinsic rates of increase and decrease of profitability associated with the independent, opposing effects of learning and hunger, and on the forager's initial level of hunger. This could limit the circumstances for dietary switching.

ACKNOWLEDGMENTS

We thank J.-G. J. Godin, M.Milinski and M.J. Dickie for comments on an earlier draft of the manuscript. The work was supported by an S.E.R.C. Research Training Award to M.I.C.

REFERENCES

Bailey JE, Wing BL, Mattson CR (1975) Zooplankton abundance and feeding habits of pink salmon, (*Oncorhynchus gorbuscha*), and chum salmon,(*Oncorhynchus keta*), in Traitors Cove, Alaska, with speculations on the carrying capacity of the area. U.S. Fish Wildl Serv Fish Bull 73:846-861

Beukema J (1968) Predation of the three-spined stickleback, (*Gasterosteus aculeatus*): The influence of hunger and experience. Behaviour 31:1-126

Brett JR (1971) Satiation time, appetite and maximum food intake of the sockeye salmon, (*Oncorhynchus nerka*). J Fish Res Bd Can 28:409-415

Brett JR (1979) Environmental factors and growth. In:Hoar WS, Randall DJ, Brett JR (eds) Fish physiology, Vol VIII. Academic Press, New York, p599

Calow P (1974) Some observations on locomotory strategies and their metabolic effects in two species of freshwater gastropods, *Ancylus fluviatilis* and *Planorbis contortus*. Oecologia 16:149-161

Charnov EL (1976) Optimal foraging: attack strategy of a mantid. Am Nat 110:141-151

Croy MI, Hughes RN (submitted, a) The role of learning and memory in the feeding behaviour of the fifteen-spined stickleback (*Spinachia spinachia* L.). Anim Behav

Croy MI, Hughes RN (submitted, b) The influence of hunger on feeding behaviour and on the acquisition of learned foraging skills in the fifteen-spined stickleback (*Spinachia spinachia* L.). Anim Behav

Cummins KW, Wuycheck JC (1971) Calorific equivalents for investigations in ecological energetics, 18. Mitt Int Ver Limnol, Stuttgart

Cunningham PN, Hughes RN (1984) Learning of predatory skills by shorecrabs, (*Carcinus maenas*), feeding on mussels and dogwhelks. Mar Ecol Prog Ser 16:21-26

Godin J-GJ (1981) Effect of hunger on the daily pattern of feeding rates in juvenile pink salmon, *Oncorhynchus gorbuscha*. J Fish Biol 19, 63-71

Holling CC (1966) The functional response of invertebrate predators to prey density. Mem Entomol Soc Can 48:1-86

Hughes RN (1979) Optimal diets under the energy maximisation premise: the effects of recognition time and learning. Am Nat 113:209-221

Ivlev VS (1961) Experimental ecology of the feeding of fishes. Yale University Press, New Haven, Connecticut

Kislalioglu M (1975) The feeding ecology and behaviour of inshore fishes. PhD thesis, University of Stirling

Kislalioglu M, Gibson RN (1976a) Prey 'handling time' and its importance in food selection by the fifteen-spined stickleback, *Spinachia spinachia*. J Exp Mar Biol Ecol 25:151-158

Kislalioglu M, Gibson RN (1976b) Some factors governing prey selection by the fifteen-spined stickleback, *Spinachia spinachia*. J Exp Mar Biol Ecol 25:159-169

Lawrence ES, Allen JA (1983) On the term 'search image'. Oikos 40:313-314

Paffenhofer G-A (1967) Calorific content of the larvae of the brine shrimp, *Artemia salina*. Helgolander wissenschaftliche Meeresuntersuchungen 16:130-135

Parsons TR, LeBrasseur RJ, Barraclough WE (1970) Levels of production in the pelagic environment of the Strait of Georgia, British Columbia: a review. J Fish Res Bd Can 27:1251-1264

Sibert JR (1979) Detritus and juvenile salmon production in the Nanaimo estuary: II Meiofauna available as food to juvenile chum salmon, *Oncorhynchus keta*. J Fish Res Bd Can 36:497-503

Ware DM (1972) Predation by rainbow trout (*Salmo gairdneri*): The influence of hunger, prey density and prey size. J Fish Res Bd Can 29:1193-1201

Werner EE (1974) The fish size, prey size, handling time relation in several sunfishes and some implications. J Fish Res Bd Can 31:1531-1536

Werner EE, Hall DJ (1974) Optimal foraging and the size selection of prey by the bluegill sunfish, (*Lepomis macrochirus*). Ecology 55:1042-1052

ORAL DISCUSSION – Croy M.I.

Q. (Kacelnik): The animals with alternating experience learned more slowly than the animals with only one prey type. I can think of 3 factors which influence this.

1) For a given trial in alternating feeding they have fewer encounters for any particular prey type

2) The encounters in alternating feeding are more spaced in time and so learning may be slower per encounter.

3) Interference from learning about a different prey type in the alternating feeding trial.

A. The relationship between handling time and the number of encounters is of course not a simple function of the rate of learning, but includes the effect of satiation. Handling times were significantly longer for *Gammarus* alone than for *Gammarus* presented in an alternating diet for a given encounter number. Thus, both the effects of interference and interval length between encounters may influence learning.

Q. (Bazely): Would your fish be more likely to choose the lower profitability prey if they had never encountered it before ?

A. Fish inexperienced with *Gammarus* and *Artemia* will attack the more profitable prey (*Gammarus*) more frequently when given the choice.

Q. (Galis): There is a decrease in profitability with satiation in your experiments. Was this due to predators adopting less costly foraging tactics as they became satiated ?

A. Although foraging tactics did not change as *Spinachia* became satiated, the frequency of components additional to the basic sequence increased. Consequently, handling times increased and profitability decreased. In laboratory experiments investigating the influence of hunger on feeding behaviour, fish approaching satiation fed less efficiently than when hungry.

Q.(DeMott): How much carry over is there in learning with two fairly similar prey ? Both prey were ingested head first; do the animals which learned to attack the *Gammarus* head first also attack the *Artemia* head first?

A. The experiment was designed as a randomised complete blocks ANOVA. Individuals were fed on one of the three diets, and so it is not possible to compare learning in individuals experienced with *Gammarus* alone and *Artemia* alone.

THE ROLE OF BEHAVIOUR AND MORPHOLOGY IN THE SELECTION OF PREY BY PIKE

Paul JB Hart+ and Stellan F Hamrin*

+Department of Zoology
University of Leicester
University Road
Leicester LE1 7RH UK

INTRODUCTION

Explaining and predicting prey selection has been one of the major tasks of optimal foraging models (Hart, 1986; Stephens & Krebs, 1986). Most models assume that optimal choice is brought about by some behavioural process. For example the basic prey model as used by Werner and Hall (1974), assumes that animals are able to rank prey by their profitabilities and to make decisions about which prey type to include in the diet. There is no implication from the models that an animal must be able to calculate complex optimisation relations but in so far as the forager uses a rule of thumb to make a foraging decision, behaviour is the main process involved (Stephens & Krebs, 1986).

---

*Department of Ecology, Limnology, University of Lund,
Box 65, S-221 00, Lund, Sweden.

NATO ASI Series, Vol. G 20
Behavioural Mechanisms of Food Selection
Edited by R. N. Hughes

In such models, morphological features of foragers are usually treated, if at all, as constraints; the animal should choose its diet according to the prey algorithm so long as a prey type falls within the maximum jaw size of the predator, although jaw size does play a part in determining profitability. Differences in morphology can have important consequences for the optimal diet as shown for sunfishes by Werner (1977). Bluegill sunfish (*Lepomis macrochirus*) have a body form which makes them more efficient at exploiting small pelagic prey such as *Daphnia*, whilst the Largemouth Bass *(Micropterus salmoides)*, with its fusiform body is more suited to capturing fish and other fast moving prey. In each species, the morphology constrains the optimal diet but behaviour is still the process by which the optimal diet is selected.

In this paper we propose to examine a more important role for morphology in prey selection. Recent work with the northern pike *(Esox lucius)* will be used to show that the observed pattern of prey selection could be determined by the fish's swimming and prey catching performance rather than by its judgements about prey profitabilities. The fish might still be maximising it's net energy gain from the feeding process, but not in relation to handling time which is often assumed to be a major cost of feeding (Stephens & Krebs, 1986).

EVIDENCE FOR PREY SIZE SELECTION BY PIKE.

Pike have a popular reputation for gluttony and voraciousness (Hart, 1988). There is evidence from field and laboratory studies that the reputation is misleading. In what follows we will concentrate exclusively on prey size selection. Evidence for selection of prey species is reviewed by Raat (1988).

Field evidence for prey size selection. Several studies have shown that prey size increases with pike size (Raat, 1988). In the River Frome, Dorset, UK, the size of dace (*Leuciscus leuciscus*) eaten by pike increased linearly (Mann, 1982). The relationship was well defined explaining 71% of the variance. The absence of points below and to the right of the regression line is circumstantial evidence for prey selection, although not conclusive, as differences in habitat use by large and small pike could influence the availability to them of different sized prey.

An analysis of pike food intake for the full range of pike sizes (Diana, 1979) showed that over 60% of pike stomachs contained only one prey item. As pike captured a prey only every 2-4 days in summer, it is unlikely that prey size was a very small proportion of pike weight. Taken together with the relationship between pike and prey size, the occurence of only one item per stomach adds further support to the conclusion that pike select the range of prey sizes they consume.

The sizes of prey taken by wild pike will be influenced by the species available. As a consequence, the mean size of prey taken by a given sized pike over a year will change in response to the availability of different sized prey. In the River Frome minnows *(Phoxinus phoxinus)* were eaten all year round, but sea trout (*Salmo trutta*) and pike, which offer a larger range of sizes, were taken mainly in the late summer and autumn (Mann, 1982). As a result a sample of pike stomachs taken in early spring might show a mean prey size rather less than that observed in late summer.

Laboratory studies of prey size selection by pike . Prey size makes a difference to a pike's energy budget (Hart & Connellan, 1984). In a 2.5m^3

tank, pike were allowed to capture minnows freely from schools of ten individuals closely matched for size. Over a period of weeks in the aquarium, each of ten pike were offered a range of prey sizes between 0.5 and 5.0g. Each pike consumed a unique average size of prey over its aquarium life. The results showed that pike growth rate was a function of weight of food eaten and the number of seconds it took the fish to capture each gramme of prey consumed. When prey were large fewer were needed over a 24h period, so reducing the time spent catching the required daily intake. Pursuit time was the same for all sizes of prey but there was a significant increase in handling time with prey size. Larger prey still yielded a greater weight caught per unit time pursuing and handling than did the smallest prey.

No of each size S	L	Prey size	Cover absent	Cover present	Total
2	8	S	8	6	14
		L	1	3	4
5	5	S	8	10	18
		L	1	0	1
8	2	S	18	11	29
		L	1	1	2
Total		S	34	27	61
		L	3	4	7

Table 1. The numbers of small and large rudd eaten by pike under six different treatment combinations of cover and prey size availability. Each figure is the total for four replicates.

A second experiment was designed to determine the influence of availability of prey size and cover on the pike's selection (Hart & Hamrin, 1988). Four pike were housed individually in seperate halves of two 3m diameter children's paddling pools. The pools were outdoors and subject to ambient temperature and light. Each pike was treated to six treatment combinations

(Table 1), each exposure lasting four days. Three different combinations of ten rudd (*Scardinius erythrophthalmus*) were offered, with and without cover; two small (mean weight=6.3g) and eight large (mean weight=13.3g), five of each size and eight small with two large. Under all conditions, the pike ate significantly more small than large rudd and the numbers eaten were far greater than expected if the pike had eaten prey in relation to their abundance ($G=69.14$, $df=5$, $p<0.001$).

EXPLAINING SIZE SELECTIVITY.

The first attempt at explaining the selectivity observed was through an optimal diet model. The time taken to catch the daily diet had a significant effect on pike growth which implied that different prey sizes had distinguishable effects on a pike's energy budget. To follow up the earlier experiments two more were done, each one designed to test the utility of optimal diet models in explaining prey size selection by pike. The first set of new trials were done in the laboratory to determine the time costs of catching different sized rudd. We were working on the assumption that pike selected small rudd in preference to large because the small were more profitable, in the sense that they yielded more energy per unit manipulation or handling time. At this stage we assumed that the time taken to manipulate prey before swallowing was the appropriate measure of cost.

Measuring capture costs . Each of twenty one pike was offered a rudd which varied between 0.80 and 19.90% of its body weight. Mean weight of pike was 88.4g±29.6 (SD) and each pike was used only once. The experiments were done in an indoor 200l glass aquarium at 18° C under artificial illumination

(12:12 L:D). For each trial the time spent pursuing the prey before attack and the time taken to manipulate the prey before swallowing were recorded. As pursuit did not vary with prey size for pike catching minnows (Hart & Connellan, 1984), we analysed only differences in manipulation times for pike catching rudd. Manipulation time increased with rudd size (Figure 1) and the increase could be most simply described by the linear regression MT = -54.29 + 40.12PS ($r^2 = 0.81$, b different from 0, $p < 0.05$).

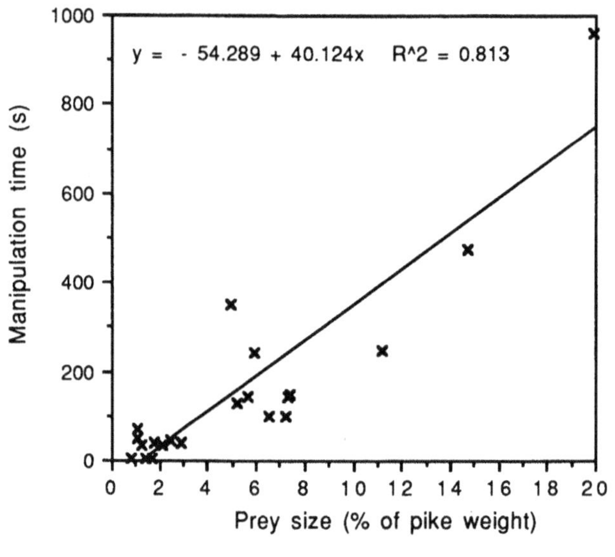

Figure 1. Manipulation time in seconds plotted against prey size expressed as a percentage of pike weight. The straight line is the regression of manipulation time on prey size.See text for the equation.

The energy content of rudd was measured and increases with size, in a relationship which also can be explained by a linear regression, E = -6.19 + 6.08PS ($r^2 = 0.84$, b different from 0, $p < 0.05$) (Figure 2).

The energy gained per unit of manipulation time (profitability or E/h) decreased slightly with increasing rudd size but a regression fitted to the data produced a slope that was not different from zero.

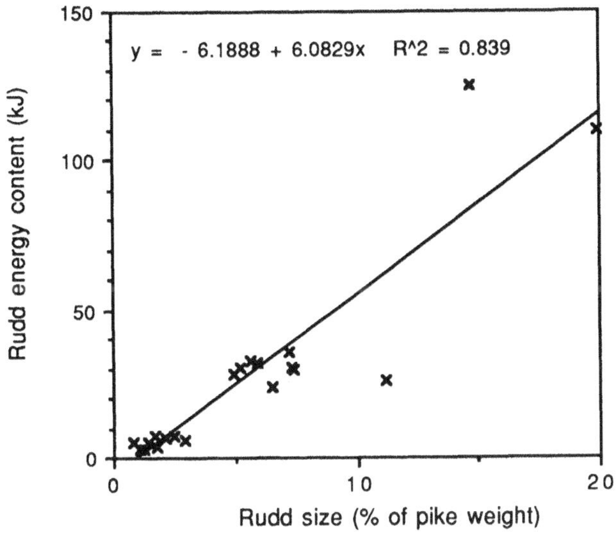

Figure 2. Energy content of rudd in kJ plotted against the size of prey expressed as a percentage of pike weight. The straight line is the regression of energy on prey size. See text for equation.

The theoretical context of further experiments. The simplest optimal diet model, called the Basic Prey Model or BPM by Stephens and Krebs (1986), only applies to a foraging environment in which prey are encountered sequentially. Pike in our pools had both sizes of prey in their sight all the time. A modification of theory to account for simultaneous occurence of prey was developed by Stephens *et al* (1986). The alternative model also assumed that the predator had only one chance to make a choice from a flock or school containing two types of prey. This makes the model highly appropriate to pike catching prey from a school with two prey types, as an attack will usually break the school up and cause the prey individuals to hide.

The principles of the model are shown in Figure 3. It is assumed that there are two prey types, p1 & p2 with corresponding energy content per attack of e1 & e2, such that e1 < e2. Also associated with each prey type is a handling time, h1 and h2. The profitability of p1 is greater than that for p2, or e1/h1 > e2/h2. An energy maximising forager should choose the prey type giving greatest e/h+s, where s is the search time for the prey school. At high encounter rates (short search times) prey type 1 yields the greatest e/h but beyond some critical encounter rate, λc at which e1/h1 = e2/h2, the forager should switch to the prey type giving the greatest energy reward .

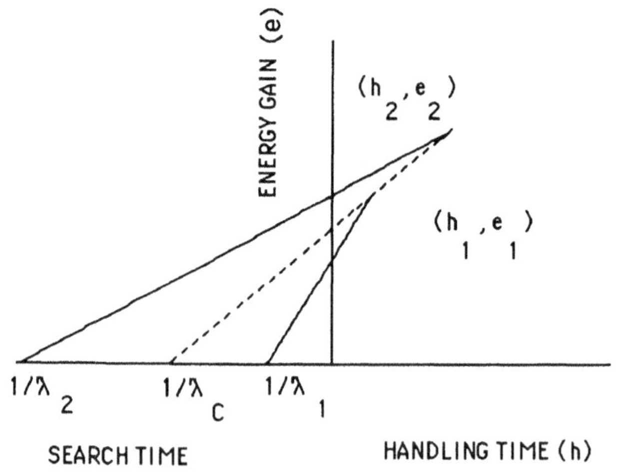

Figure 3. The way in which prey profitability and search time changes the optimal prey choice according to the model of Stephens et al (1986). The location of each prey in e, h space is shown together with the slope which defines e/h values for a given search time. The dotted line shows the search time or encounter rate which makes both prey types equally profitable. Further details are explained in the text.

Testing the model with pike. 1. Methods . To test the model of Stephens *et al* (1986) we used the same experimental system used earlier (Hart & Hamrin, 1988) but slightly altered the treatments to which the fish were subjected. Four pike weighing an average of 95.8g±8.0 (SD) were installed in the four pool halves. The two treatments used were the presence or absence of cover and the encounter frequency with the prey school. Apart from the encounter frequency and the area of the artificial weed bed, all other details were the same as for our earlier experiment. At all times the school of ten rudd contained 5 small (mean weight 7.54±0.96) and five large (mean weight 13.73g±2.15) individuals. The weed bed was enlarged to 1.5 x $1m^2$ so that it covered 42% of the bottom area as opposed to 28% in our first experiment. Encounter with prey was arranged to be either continuous or for four 20 minute periods each 24h. To prevent the pike from seeing or having access to prey for the rest of the time, they were enclosed in an opaque tube 50cm in diameter and 50cm deep which was lowered over the pike and weighted down to prevent it being overturned. The end of the tube out of the water was covered with fine-meshed plastic netting to prevent the fish from jumping out.

The frequency of prey releases was determined by applying the theory depicted in Figure 3. Using the mean rudd and pike weights for the experiment, the handling times and energy gains for small and large rudd were estimated from the relationships shown in Figures 1 & 2. Projecting the line of equal E/h to the ordinate gave the critical encounter rate of 65s (Figure 4). We chose a time interval much larger than the critical value to make it more likely that a switch would be detected if it was to occur.

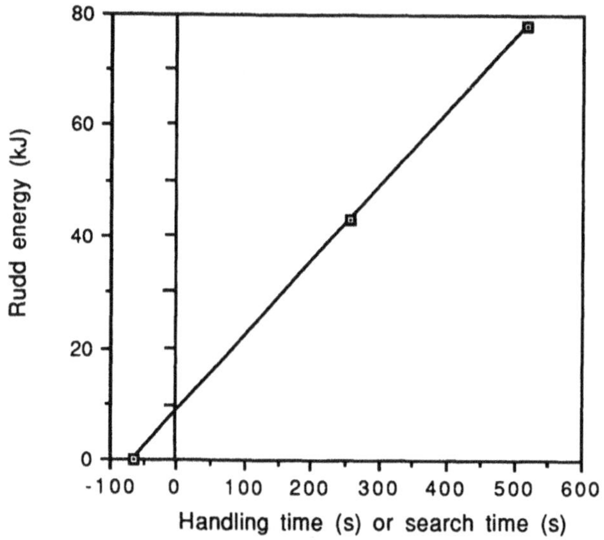

Figure 4. Estimating the critical encounter rate with prey which determines at what point the pike should switch from small to large prey.

Enclosed pike were released on a schedule which was designed to spread the encounters with prey evenly over the daylight hours but to include an element of unpredictability. The hours of daylight were divided into four periods: dawn (about 0530) to 0900, 0900 to 1300, 1300 to 1700 and 1700 to nightfall (about 2100). One release was arranged for each period, but within the period, the time of release was chosen at random.

The experiment took place over a four week period between 1st and 26th August 1988 in Lund, Sweden. The four treatment combinations (Table 2) were spread evenly over pool halves and the four weeks. Each treatment was replicated four times. In this way changes in uncontrolled variables such as daylength, temperature and pool were controlled for. See Hart and Hamrin (1988) for further details of how data were collected and prey set out.

2. Results . The experimental outcome is shown in Table 2. The number of small prey eaten was significantly more than the number of large prey ($H = 13.04$, $p = 0.0006$, $df = 1$) but there was no difference across treatments in the number of small prey consumed ($H = 0.84$, $p > 0.1$, $df = 3$). (H is a measure of variation among sample rank means and is distributed as chi -square, Meddis, 1984). These results echo those found in the earlier experiment (Hart & Hamrin, 1988).

Encounter rate with prey	Prey size	Cover absent	Cover present	Total
Continuous	S	7	7	14
	L	2	2	4
4 x a day	S	9	5	14
	L	0	2	4
Total	S	16	12	28
	L	2	4	6

Table 2. The numbers of small and large rudd eaten by pike under four different treatment combinations of cover and prey encounter rate. Each figure is the total for four replicates.

3. Discussion. Pike did not switch to larger prey when the frequency of encounter with rudd schools decreased below the critical value. The latter will be sensitive to variability in handling time and rudd energy content, and to the correct identification of manipulation time as the best measure of cost.

Our experimental design was conservative in that the frequency we chose for the intermittent encounter with prey was much larger than the estimated critical encounter rate. For example if the difference in energy content between small and large rudd is decreased by 50% the critical encounter rate changes to about 400s. This is still well within our average encounter frequency of 13,000s. If the variation in large rudd energy was really 50% greater than we estimated it to be, then large prey would have always been more profitable and pike should never have chosen small fish.

A more important problem arises if manipulation time is not the best estimate of handling time. Pike take hours to digest a single large fish, during which time they do not respond to prey. Digestion time may be a better estimate of handling time.

An additional problem arises from the fact that we did not devalue e1 and e2 by the number of attacks per capture when calculating λ c from Figure 3. When pike attacked minnows varying from 0.5 to 5.0g, attacks per capture were constant at 2.7 (Hart & Connellan, 1984). This is not to say that the same constancy held for attacks on different sized rudd, but as we have no evidence to the contrary, we assumed it did. The fallacy of this assumption will become evident later.

We simulated search time for prey by shutting the pike in a tube. Pigeons do not count time absent from a Skinner box as part of the experiment. In the same way pike could have regarded time in the tube as having nothing to do with searching for prey (Houston, pers comm). We regard this as unlikely. In the natural environment, pike will be out of sight of prey at times, even when searching, and it would be wrong to argue that when out of sight of prey, search mode was switched off.

A MECHANISM FOR SELECTIVITY.

Seeing that prey selection is not fully accounted for by the model, we would like to explore a possible mechanism for the selection observed. Pike are mostly ambush predators, creeping up on their quarry and then leaping upon them in a sudden rush (Webb & Skadsen, 1980; Hart & Connellan, 1984). The ability to accelerate rapidly and out-sprint the prey is one of the features of the hunt (Webb, 1976, 1978). The success of the predator's final lunge will also be affected by the abilities of the prey to detect the advancing pike and accelerate away (Webb, 1976).

Pike are very reactive to any fish-shaped object that shows a side-to-side movement of the tail (Clarke, 1986). Pike do not show changes in response behaviour to different sized prey (Hart and Colgan, unpublished data). Consequently we assume that pike attacked small and large rudd with equal probability in our experiments. Pike took more small prey because they were easier to hunt down and capture, an assertion that needs substantiating.

The fast-start of fish has been divided into three phases (Webb, 1978). The first two are the critical ones in getting the fish to full speed. Phase three is much more variable in form and duration. A study of seven teleosts, including the tiger musky (*Esox spp*), showed that distance travelled (S) during the first two phases of fast-start was a function of fish length such that;

$$S1 = 0.18L^{0.94}\ \mathrm{cm}$$

$$S2 = 0.38L^{1.01}\ \mathrm{cm}$$

Knowing the average size of small and large rudd and of the pike we have calculated the distances the three groups of fish would swim during fast-start. Small rudd had a mean length of 7.6cm, large rudd of 9.6cm and pike 26.1cm. Applying the power function showed that the three groups would travel 4.1, 5.2 and 14.1cm respectively during the first two phases of fast -start.

The consequences of these travel distances can be examined to estimate the vulnerability of the two sizes of prey to capture. We start with the simplest case in which the pike attacks the rudd from behind and the rudd accelerates away along the same axis of locomotion as the attacking pike. We assume that pike start their final strike when they are about 12cm from their prey (Webb and Skadsen, 1980) . Prey take a time to respond to the advancing predator and we assume that rudd would start their escape some 20ms after the pike had begun to approach. The lag would mean that the pike would have travelled about 2cm by the time the rudd responded. Given these assumptions, the pike would be able to overtake the small rudd as $10+4.1 \approx 14.1$cm. The same would not be true for the large rudd which would have swum out of reach of the pike by the end of the period of fast-start locomotion.

Webb and Skadsen (1980) found that the mean strike angle during a tiger musky attack was 80 ± 8^0 and the aiming point on the prey was 30-50% along the prey's length from the head. This point is the centre of mass of the fish's body which hardly moves during the first phase of fast-start (Webb, 1978, 1983, 1986), usually brought about by the body being first contracted into a C or J shape. This leads to large lateral movements of the head and tail but leaves the centre of mass almost stationary. Assuming an attack from the side, it is still likely that small rudd would be more vulnerable than large. The reason for the difference depends on the relationship between the timing of the pike attack and the distance the prey is able to cover.

The time taken by fast-start phases 1 and 2 are linear functions of length (Webb, 1978). To travel 12cm our 26cm pike would have taken 210ms which is close to the mean attack duration of 200ms observed by Webb and Skadsen (1980) for tiger muskies averaging 22.6cm in length. For the last 80ms of the strike, tiger muskies were not able to change the direction of their attack. During the first 20ms of the pike's advance the rudd would not respond and during the next 50ms it would not move forward. The fast-start of small rudd would take 110ms but for only 60ms of that would the fish be moving forward. The pike would then have about (210-20-50)-80=60ms to alter course in response to the rudd's escape direction. For small rudd the distance they could move would be only 4.1cm which would barely take it beyond the path swept by the wide mouth of the pike. The width of the pike's jaws in the Hart and Hamrin (1988) study ranged between 1.64 and 2.24cm. The escape response of large rudd would take 120ms but because the fish could expect to travel 5.2cm it would be carried further from the path vulnerable to the pike's mouth.

The interaction between pike morphology and fast-start performance and rudd fast-start capacities defines zones of interception, a field with probability of capture decreasing from the centre outwards (Figure 5). The probability of capture is a decreasing function of prey size and the distance covered during fast-start.

The width of the pike's jaw could also be important when it comes to handling the prey. Once captured a rudd is held sideways in the pike's mouth. To swallow the prey easily, the height, or dorsal to ventral dimension, of the prey would be most critical. Pike jaw width varied between 1.64 and 2.24cm whilst the height of small prey varied between 1.56 and 2.34cm. Large prey had heights ranging between 2.43 and 3.14cm. These dimensions could also be critical during capture as the esocids are known to vary their jaw movements with the type of attack they make (Rand and Lauder, 1981) .

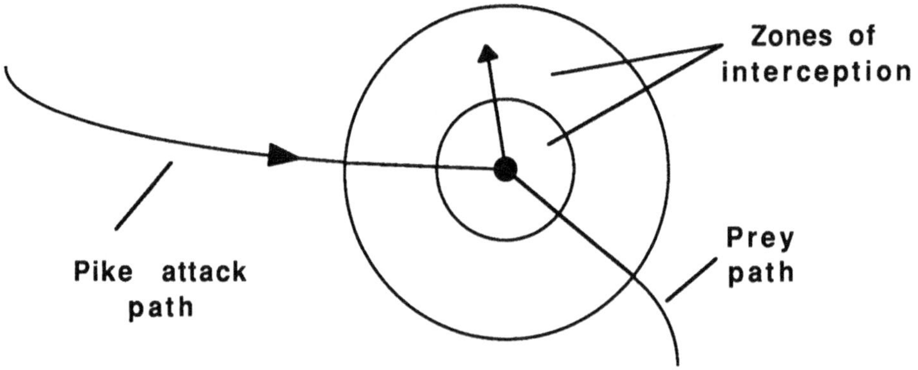

Figure 5. A representation of the zones of interception defined by the relative swimming capacities of pike and rudd. The inner circle is the area within which a small rudd would be found during its fast-start. The larger circle provides the same for large rudd. The pike's jaw width would cover most of the inner circle, so making the probability of prey capture high.

GENERAL DISCUSSION.

Based on our consideration of fast-start performance by pike and rudd, we conclude that it is unlikely that pike are as successful at catching large rudd as they are at catching small ones. As a result es (energy gained per attack) will be much greater than e_1. This alone could change λ_c in Figure 4 so rendering our test of Stephens *et al* (1986), less convincing.

Our estimate of λ_c would be further changed if digestion time rather than manipulation time, is the true handling time. In the first prey selection experiment (Hart and Hamrin, 1988), the mean time until the next meal, given a small prey had just been eaten, was 1.7 days (n=10). The equivalent gap for large prey was 2.75 days (n=4). These extended "handling" times would change the time scale of Figure 4 from seconds to days.

Handling and search time may be the most costly aspects of prey capture for fish searching for and handling abundant and easily caught prey. Pike, like other ambush predators, catch prey that are highly mobile and hard to catch. The pressure on the pike is to get to the prey before the prey gets to cover or out of reach (Webb, 1986). This has put pressure on pike, through evolutionary time, to develop its ability to accelerate fast with a great deal of energy being used. Other predatory fish, such as rainbow trout, smallmouthed bass (*Micropterus dolomieui*) and rock bass (*Ambloplites rupestris*) do not attack prey at the maximum speeds they can attain whilst tiger musky do (Webb, 1984). The three non-esocids also have less muscle mass, expressed as a percentage of body weight, than the tiger musky. These differences can be related directly to diet, the musky being the only fish which lives entirely on fast moving prey.

The tendency might be to conclude that the concept of energy maximisation is not applicable to pike. We have shown that the principle is still valid for the species but costs are not those most often assumed for other predators. Our study indicates that costs must always be defined in the context of the biology of the species under study.

SUMMARY.

Most models of prey choice are based on behavioural processes. Morphological features enter behavioural models in the form of constraints within which the behaviour must be performed. Data from field studies and experiments are presented on prey size selection by pike. We then describe an experiment testing with pike a behaviourally based optimal diet model. The model does not account for the fish's behaviour. We propose that the prey size selection observed can best be explained in terms of the relative fast-start swimming performances of predator and prey. The smaller size of prey are eaten most frequently because they are easier to catch. The pike may still be maximising net energy but in terms of energy gained per unit digestion time rather than per unit handling time.

Acknowledgements. The 1988 experiment on pike choice was funded by grants to PH from the University of Leicester, the Royal Society and the Swedish Fisheries Board. The determinations of manipulation time were made by Perola Johanesson and Ulf Bergendorf. We would like to thank Peter Shelton and Andrew Sih for commenting on the manuscript. Sebastian Diehl must be thanked for the loan of his 'A-nyckel'.

REFERENCES

Clarke I (1986) The interaction of factors influencing the predatory behaviour of the pike Esox lucius L. PhD Thesis, University of Leicester.
Diana JS (1979) The feeding pattern and daily ration of a top carnivore, the northern pike (Esox lucius) . Can J Zool 57:2121-2127
Hart PJB (1986) Foraging in teleost fishes. In: Pitcher TJ (ed) The behaviour of teleost fishes. Croom Helm, London, p553
Hart PJB (1988) The murderous patriarch. New Scientist 119:53-57
Hart PJB, Connellan B (1984) The cost of prey cature, growth rate and ration

size in pike (Esox lucius) as functions of prey weight. J Fish Biol 25:279-291
Hart PJB, Hamrin SF (1988) Pike as a selective predator. Effects of prey size, availability, cover and pike jaw dimensions. Oikos 51:220-226
Mann RHK (1982) The annual food consumption and prey preference of pike (Esox lucius) in the River Frome, Dorset. J Anim Ecol 51:81-95
Meddis R (1984) Statistics using ranks: A unified approach. Basil Blackwell, Oxford, p449
Raat AJP (1988) Synopsis of biological data on the northern pike, Esox lucius Linnaeus, 1758. FAO Fish Synop 30 Rev 2:178p
Rand DM, Lauder GV (1981) Prey capture in the chain pickerel, Esox niger : correlations between feeding and locomotor behaviour. Can J Zool 59:1072-1078
Stephens DW, Krebs JR (1986) Foraging Theory. Princeton University Press, Princeton p247
Stephens DW, Lynch JF, Sorenson AE, Gordon C (1986) Preference and profitability: theory and experiment. Amer Nat 127:533-553
Webb PW (1976) The effect of size on the fast-start performance of rainbow trout Salmo gairdneri, and a consideration of piscivorous predator-prey interactions. J exp Biol 65:157-177
Webb PW (1978) Fast-start performance and body form in seven species of teleost fish. J exp Biol 74:211-226
Webb PW (1983) Speed, acceleration and manoeuvrability of two teleost fishes. J exp Biol 102:115-122
Webb PW (1984) Body and fin form and strike tactics of four teleost predators attacking fathead minnow (Pimephales promelas) prey. Can J Fish Aquat Sci 41:157-165
Webb PW (1986) Locomotion and predator-prey relationships. In: Feder ME, Lauder GV (eds) Predator-prey relationships. Perspectives and approaches from the study of lower vertebrates. Chicago University Press, Chicago, p198
Webb PW, Skadsen JM (1980) Strike tactics of Esox. Can J Zool 58:1462 -1469
Werner EE (1977) Species packing and niche complementarity in three sunfishes. Amer Nat 111:553-578
Werner EE, Hall DJ (1974) Optimal foraging and the size selection of prey by the bluegill sunfish (Lepomis macrochirus) . Ecology 55:1042-1052

ORAL DISCUSSION – P.J.B.Hart.

Questions were raised about the presentation of prey, the apparent unselective feeding by the pike and the weighting of prey profitability by attack success rate.

The prey in the orginal pike experiment were adjusted to constant availability. The aim in the main experiment was to distinquish the mechanism by which the pike selects its prey, notwithstanding that some theoreticians consider that what an animal actually ends up eating may have nothing to do with the decisions which led to the choice. The lack of choice seen in the experiment may imply that there was no active behavioural decision by the pike. Capture success might be incorporated in the test of Stephen's model by replacing energy per unit handling time by energy per encounter, or by looking at the conditional probabilities of each stage of prey capture recorded on film.

HOW IMPORTANT ARE NUTRIENT CONSTRAINTS IN OPTIMAL FORAGING MODELS OR ARE SPATIAL/TEMPORAL FACTORS MORE IMPORTANT?

Gary E. Belovsky
Department of Biology and School of Natural Resources
University of Michigan
Ann Arbor, Michigan 48109-1115
USA

Optimal foraging theory (henceforth called OFT) perhaps has achieved some of the greatest successes in quantitatively predicting observed patterns (*i.e.* diets) in nature, than any other ecological theory (Schoener 1986, Stephens and Krebs 1986). This is not to say that OFT has been accepted without reservation, nor that the predictive successes of the theory are unambiguous (Gray 1986, Pierce and Ollason 1987). Two major complaints about OFT studies have been espoused. First, predicted diets may not be as close to the observed values as implied by investigators. Second, investigators sometimes invoke *post hoc* explanations when predicted and observed values do not closely correspond.

When investigators have not been satisfied with the congruence between OFT predictions and observed diets and *post hoc* explanations have been sought, potentially ignored nutritional constraints have been the most common candidate for a *post hoc* explanation. For example, if energy is the principal currency sought by a forager, it can be argued that categorizing foods by their energy value may not be sufficient to describe diet choices. In addition, other nutrients may be required along with energy or the efficient use of energy may depend upon the mix of energy sources (Krebs and Avery 1984, Hill 1988). The number of OFT studies, however, that *a priori* examine nutritional requirements other than as the principal currency hypothesized to be sought by the forager (eg. energy) are few.

Only ten species of mammals (*Odocoileus hemionus*, *Alces alces*, *Lepus americanus*, *Macropus rufus*, *Macropus robustus*, *Oryctolagus cunniculus*, *Ovis aries*, *Bos taurus*, *Ammospermophilus leuceurus* and *Homo sapiens*) have had OFT applied to their diet choices with explicit *a priori* nutritional constraints employed in addition to the major currency hypothesized (eg. energy, protein, water, etc.) (Belovsky 1978, 1984a, 1987a, Belovsky et al. in prep, Hill 1988, Karasov 1985, Spalinger 1980). There are several reasons for this paucity of OFT studies employing nutritional constraints.

NATO ASI Series, Vol. G 20
Behavioural Mechanisms of Food Selection
Edited by R. N. Hughes

First, while numerous nutritional requirements can be listed for any animal based on its nutritional physiology (eg. Robbins 1983), these requirements do not necessarily operate as constraints on the animal's diet choices. Many, if not most nutritional requirements, may be satisfied incidental to the acquisition of the currency sought by the forager. For example, in the course of maximizing energy intake, human hunter-gatherers exceed their daily protein requirements, making the inclusion of a protein constraint unnecessary (Belovsky 1987a). Therefore, in many cases nutritional constraints may not have to be considered.

Second, when nutritional constraints must be included in an OFT model, questions arise as to how the constraint must be incorporated in the model. Do specific constraint equations have to be developed for the nutritional requirement or can existing equations in the OFT model be modified to include the need to satisfy the additional nutritional requirements?

Third, when the predictions from simple OFT models are not in accord with observed diets, the most parsimonious explanation for the model's failure may not require the invocation of additional nutritional constraints. OFT models are often applied in "cookbook fashion" without inquiring whether or not the assumptions used to construct the OFT model (Stephens and Krebs 1986) are appropriate for the forager in its environment. In particular, the feeding time constraint, a universal constraint to all OFT models, can take many forms depending upon the spatial and temporal distributions of foods in the environment (Belovsky et al. 1989). However, many investigators fail to realize this and employ an OFT model that is not appropriate for the forager and its environment. Therefore, poor agreement between predicted and observed diets may be due to the use of an inappropriate OFT model, rather than the need to consider other nutritional requirements or the failure of OFT.

The first two concerns listed above will be examined in this paper using the existing OFT studies in the literature. The third concern will be examined using the foraging of cowbirds (*Molothrus ater*) on insects and seeds in the field.

How Common Are Nutritional Constraints?

From OFT studies conducted by my students and me (Belovsky 1978, 1981, 1984a, b, 1986, 1987a, b, c, Belovsky and Ritchie in press, Belovsky and Slade 1987, Belovsky et al. in prep, Ritchie 1988, Schmitz in prep), only eight cases out of 108 studies (24 species in a variety of habitats, seasons and years) had to consider nutritional constraints other than energy. This is less than 8% of our studies. These studies employ the optimization technique of linear programming (henceforth called LP), where three constraints are always employed: feeding time, digestive capacity and energy requirements (see Belovsky 1984c for a review and Belovsky et al. 1989 for a discussion of how LP models relate to other OFT models).

LP is a simple optimization method for dealing with constraints (Westoby 1974, Pulliam 1975, Altman and Wagner 1977, Belovsky 1978). The only assumption needed in constructing an LP model is that the constraint equations are linear. This means that the foods are either substitutable (interchangeable in satisfying the constraint) or essential (only one satisfies the constraint) (Tilman 1983). If the foods are complimentary (the foods synergistically interact in satisfying the constraint), then the constraint equations become non-linear which necessitates the use of non-linear programming, a slightly more difficult optimization technique, or the non-linear problem can be reformulated into a series of linear problems defined for restricted sets of parameter values (Intriligator 1971).

By defining a set of linear constraint equations arising from assumptions about what physiological (eg. digestive capacity, thermal budget, etc.) and behavioral processes (eg. ability to locate and recognize foods, etc.) restrict a forager's food choices, a diet usually can be predicted that maximizes or minimizes some goal "sought" by the forager. The two goals most commonly employed in these models are energy maximization and feeding time minimization (Schoener 1971). Energy maximization is the goal if the forager's fitness increases with greater intake of some nutrient (i.e. food limitation). Feeding time minimization is the goal if the forager's fitness increases by satisfying its nutritional requirements in less time, leaving more time to hide from predators, look for mates, care for young, etc. With the constraints and goal stipulated, a mathematical algorithm can be used to find the diet which optimizes the goal. In two dimensions the optimum solution occurs at the intersection of two constraint lines or at the intersection of a constraint line and an axis. Several LP model examples are presented in Fig. 1 and will be more fully reviewed below.

The 100 studies that do not consider nutritional requirements other than energy, have the predicted and observed diets that are based on food categories, rather than species (eg. grass vs. forb, meat vs. vegetable, etc.), highly correlated ($r^2 = 0.96$). In all but two of these studies, the foragers appear to be nutrient maximizers (maximum daily energy intake is sought); in the other two cases (males during the mating season) feeding time minimization appears to be selected by the forager.

In the eight studies which include nutritional constraints in addition to energy, the predicted and observed diets are not highly correlated if the additional nutritional constraint is ignored ($r^2 = 0.31$) and the predicted and observed diets deviate by an average of 15.6 $\pm$ 6.4%. When the additional nutritional constraint is included, the

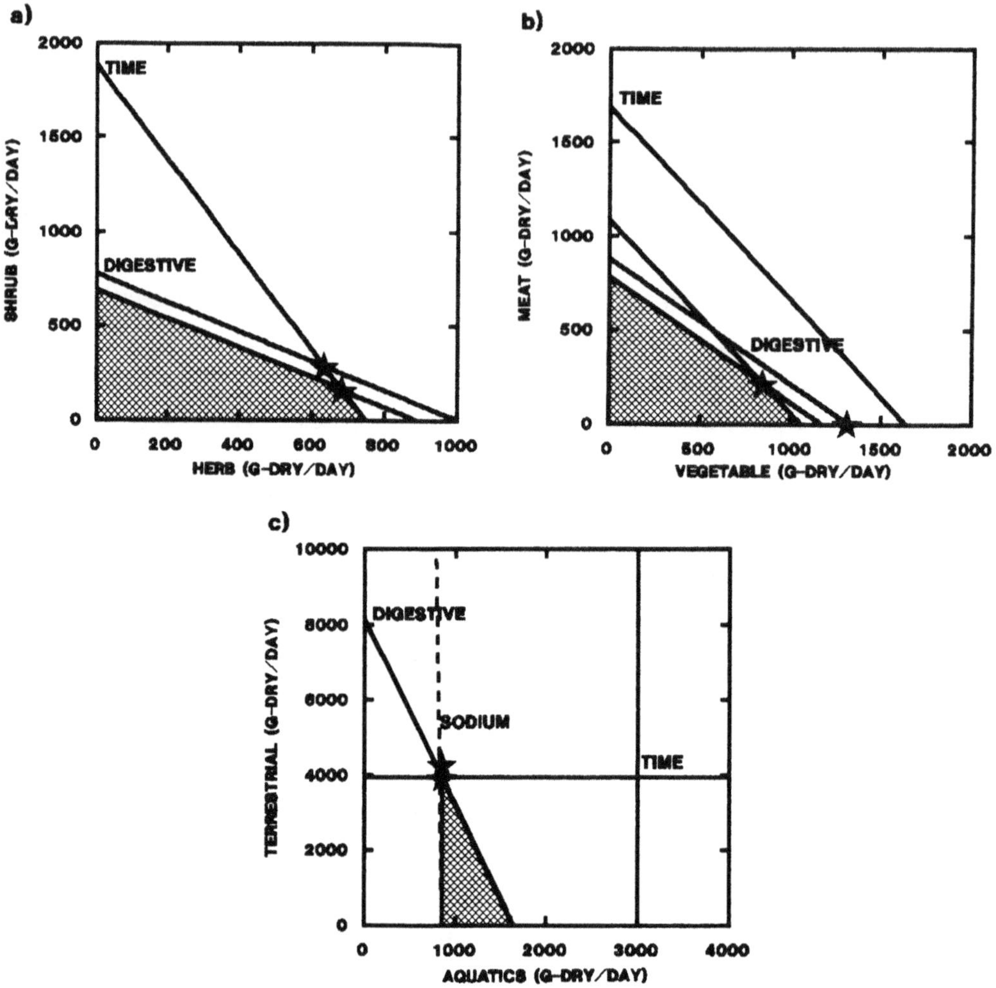

Figure 1. Examples of three LP models that include additional nutrient constraints are presented (see text). Each constraint is labeled. Parallel sets of lines demonstrate how the additional nutrient constraint changes other constraints (the constraints affected by the additional nutrient constraint are closest to the origin). When the additional nutrient constraint results in an additional constraint line, rather than modifying an existing constraint, this constraint is shown as a dashed line. The possible diet combinations that satisfy all the forager's constraints are depicted by the cross-hatched region. The energy constraints are not shown since the solutions with and without the additional nutrient constraint (stars) are for an energy-maximizing goal, which makes the energy constraint inoperative. The models are for red kangaroos (a), Kade San, human hunter-gatherers (b), and moose (c).

correlation is much improved (r^2 = 0.91) and the average deviation between the predicted and observed diets is only 2.8 $\pm$ 2.4%. This 560% better prediction of the observed diets by including the additional nutritional constraint only reduces the forager's maximum energy intake by 4.7 $\pm$ 3.6%.

Such a small reduction in energy intake in response to satisfying additional nutritional requirements can be very important to a forager whose fitness is based upon surplus energy for reproduction and survival. Ritchie (1988, in press) found that Columbian ground squirrels (*Spermophilus columbianus*) have an approximate margin of 8-10% between their maximum energy intake and failing to reproduce, and Belovsky (1978) found a less than 2% margin for moose (*Alces alces*).

From the eight studies employing additional nutritional constraints, only two nutrients emerge as constraining. Water intake had to be considered in arid environments for: two species of kangaroos, cattle, sheep and rabbits in the Australian Outback, and the Kade San, human hunter-gatherers, in the Kalahari desert of Africa. Sodium intake had to be considered in a region of highly glaciated basaltic shield: moose and snowshoe hare in the boreal forest of North America (Isle Royale). LP studies reported by Karasov (1985) for antelope ground squirrels (*Ammospermophilus leucerus*) and Spalinger (1980) for mule deer (*Odocoileus hemionus*) indicate that water and protein may be constraining in arid environments.

Are nutrient constraints more important at certain times of the year? This is difficult to assess since studies over an animal's annual feeding cycle have been conducted for only three species (*Alces alces, Lepus americanus* and *Spermophilus columbianus*), all temperate zone animals. One might expect the summer foods (growing plants) of the temperate zone to be more nutritious than the non-growing winter plants, and that storage of some nutrients during the summer might be required. *Alces alces* and *Lepus americanus* had to acquire adequate sodium in the summer by consuming particular foods to supply their winter needs; this required an additional constraint in the LP model. The LP model for *Spermophilus columbianus* did not require any additional nutrient constraints to explain its diet choice, even though this animal hibernates for up to eight months and must store adequate nutrients. Also, the majority of studies were conducted on summer feeding for temperate species and additional nutrient constraints did not have to be considered to explain diet choice.

The above review does not argue that other nutrients are not constraining elsewhere, but serves to illustrate the obvious nature of the types of environments, seasons and types of nutrients where the additional nutritional constraints might have to be considered. Furthermore, because these LP models do not consider the species composition of the diet, it might be possible that certain species of food are consumed to satisfy particular nutrient requirements or avoided to reduce exposure to anti-nutrients (toxins and digestion-inhibiting substances). This, however, is glossed over in the LP model's gross representation of food choice which is based upon average conditions.

How Are The Additional Nutritional Constraints Incorporated Into OFT Models?

A misrepresentation of LP diet models is that they are constructed under the assumption that only one nutrient can be considered in the currency to be optimized (eg. energy) and additional nutrients must be included as minimum amounts that only need to be satisfied (Hill 1988). While this is the way that most LP models have been constructed, this does not have to be the case. There is no reason that the currency to be optimized cannot be the maximum ingestion of several nutrients that must be consumed in a ratio or the sum of several nutrients composing each food, where the consumption of each nutrient can be converted into an appropriate fitness unit.

LP - OFT models to date have included additional nutrients as minimum requirements that must be attained. How these models are constructed depends upon how the nutrient is obtained by the forager in its food choices and the amount of information available to the investigator.

1. To satisfy the minimum nutrient requirement, the forager must select food items within each food category that contain more of the nutrient. Red kangaroos (Fig. 1a) select the green plants of plant types (herbs and shrubs) rather than dried plants in order to obtain their water requirements. In this case the nutrient constraint can be incorporated (but not explicitly included) by using the characteristics (eg. cropping rates, passage through the digestive tract, digestive fill, energy value, digestibility, etc.) of green plants in the model. In Fig. 1a, these changes lead to reductions in the model's digestive capacity constraint (constraint line moved towards the origin), since the green plants are bulkier (i.e. contain more water), utilizing more of the digestive capacity per unit dry matter of food ingested. This is the simplest way to include nutrient constraints, and does not require a specific constraint for the nutrient requirement (water) that defines the mechanisms determining nutrient needs and acquisition. LP models for red and hill kangaroos, sheep, cattle and rabbits in the Australian Outback have been constructed using this method (Belovsky et al. in prep).

2. A specific food is consumed to obtain the nutrient requirement and this food is the only source of the nutrient. Kade San, human hunter-gatherers, must extract water from tubers to satisfy their daily water requirement (Tanaka 1976). While the

water that is squeezed from the tubers and drunk will be absorbed quickly from the digestive tract since it is free-water and therefore, will not affect the tract's capacity to hold food, the tubers are also consumed to get the moisture that could not be extracted. The consumption of the processed tubers will reduce digestive capacity for other foods; the time needed to find, harvest and process these tubers will reduce the daily time available to find, harvest, process and ingest other foods; and the tubers do provide some energy reducing the amount of energy required from other foods (Belovsky 1987a). These changes in the LP model's constraints are depicted in Fig. 1b. This model construction provides more detail of how the nutrient constraint is satisfied, but the constraint still is not explicitly included in the model.

3. One or several foods might supply a nutrient requirement as well as other nutrients. In this case the consumption of each food must be "traded off" to find the best combination of foods that satisfy the nutrient requirement and maximize the intake of another nutrient. Under these conditions, the nutrient requirement must be explicitly included in the model as an additional constraint. Fig. 1c presents the addition of a nutrient constraint, sodium, for moose (Belovsky 1978). This approach has also been employed for snowshoe hare sodium requirements (Belovsky 1984a) and mule deer protein requirements (Spalinger 1980). While providing detail on the mechanisms that determine nutrient constraints and how they are satisfied, this modelling approach also requires the most knowledge about the forager's nutrient demands and food characteristics which often are not readily available.

Each of the above modelling approaches depends upon the manner in which the nutrient acts as a constraint and the amount of detailed information available to the modeler. The first two methods may not be very satisfying to many nutritional physiologists since they do not explicitly include the constraint in the model; rather the model is simply modified to include the consumption of a food needed to satisfy the nutritional requirement. Even the third method may not satisfy many nutritional physiologists since the additional nutritional constraint is taken as a fixed amount independent of the forager's diet choices and activity. However, to most ecologists interested in understanding the dietary selections of free-ranging animals and how diet choice might influence population dynamics, habitat selection, etc., the additional detail accorded by including the nutritional kinetics in the model may not be necessary.

The addition of all the detail sought by nutritional physiologists might make OFT models unwieldy due to the inclusion of nested sets of equations representing the detailed kinetics of forager nutrition. This might leave us with the "Gordian Knot" of most system's models, unable to discern the major driving forces. Even more troublesome would be the capricious predictions of these megalithic models since each equation for nutritional kinetics would be based upon estimates, and these estimates will lead to the propagation of error throughout the model. As Stephens and Krebs (1986) argue, what might be a single value or constraint for constructing an OFT model will constitute the entire research program for another biologist. Therefore, invoking parsimony in investigating OFT and employing OFT models may be necessary.

The Problem With Invoking *Post Hoc* Nutrient Constraints

Oftentimes investigators have invoked additional nutrient constraints in their studies of OFT, when predicted and observed diets do not match as well as anticipated (see Krebs and Avery 1984, Hill 1988). The philosophical problems with using *post hoc* explanations to save unsubstantiated hypotheses has been well discussed (Stephens and Krebs 1986), but I wish to address here whether or not potentially ignored nutrient constraints are the most likely explanation for the failure of OFT when OFT might indeed be the correct hypothesis.

While an ignored nutrient constraint could easily lead to OFT predicted diets failing to match observed diets (see Fig. 1), an incorrect assumption concerning any other constraint could also be the cause. My students and I have employed four categories of constraints in our LP models: digestive capacity, feeding time, energy requirements, and additional nutrient requirements. In most of our 108 studies, the feeding time and digestive capacity constraints are the most important constraints for diet predictions. Furthermore, we have come to believe that the feeding time constraint may be the most critical constraint for OFT models.

The functional form of the feeding time constraint depends upon how foods are encountered by the forager in space and time, which depends upon the distribution of the foods in the environment and the forager's search behavior (Belovsky et al. 1989). Several simple possibilities will be reviewed below (Fig. 2).

The classical OFT model employed in most studies (Stephens and Krebs 1986) was developed under the assumption that the foods are encountered randomly or uniformly. If the foods are distributed in the environment randomly or uniformly, a randomly searching forager can search for all foods at the same time (i.e. simultaneous search *sensu* Schoener 1971). The functional form of this feeding time constraint is depicted in Fig. 2a and the two line segments composing the constraint intersect where the quantities of the two foods are in the proportion encountered by the forager.

In most of the studies by my students and me, we have found that foods are not distributed uniformly or randomly in space or time, so a randomly searching forager cannot search for all food types at the same time (i.e. non-simultaneous search *sensu* Belovsky 1978).

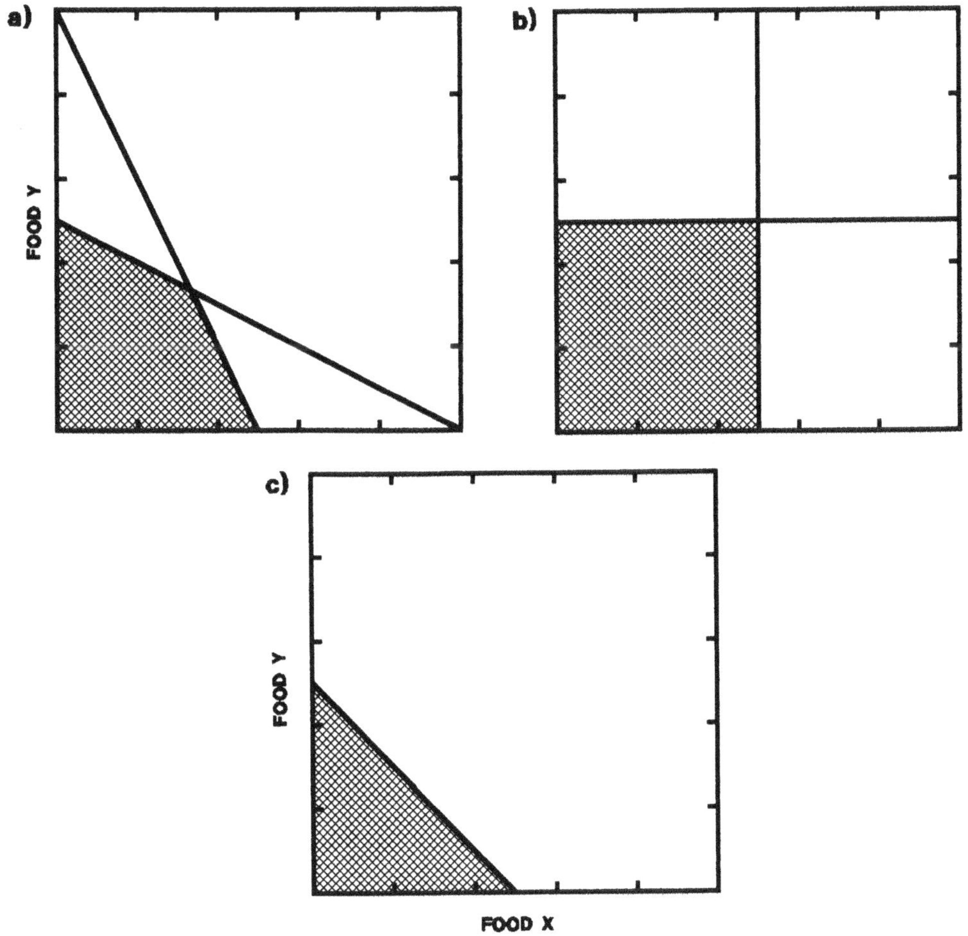

Figure 2. The different functional forms for the simple feeding time constraints discussed in the text are presented for the case of two foods that the forager can select: simultaneous search (a), temporal non-simultaneous search (b), and spatial non-simultaneous search (c). The cross-hatched regions contain all diet combinations that do not violate the feeding time constraint. These functional forms and explicit equations for each can be found in Belovsky et al. (1989).

The functional form for a feeding time constraint based upon temporal non-simultaneous search (the foods do not co-occur at the same time) is depicted in Fig. 2b. Because the foods are not encountered at the same time, feeding on one does not reduce feeding on the other. Examples of how temporal non-simultaneous search might arise for a forager are numerous (eg. different insects may be encountered by bats at dawn and dusk vs. at night, moose cannot feed on aquatic and terrestrial plants at the same time because each food type is encountered in a different habitat which provides a different thermal regime that limits the moose's activity to different times of day, etc.).

The functional form for a feeding time constraint based upon spatial non-simultaneous search (the foods do not co-ocur in space) is depicted in Fig. 2c. It is easy to envision this type of feeding time constraint when each food type tends to occur only in patches with other items of the same type (eg. even if moose could feed on aquatic and terrestrial plants at the same times of day, they would experience spatial non-simultaneous search since aquatic and terrestrial plants do not occur in the same location). In this case, feeding on one food type reduces the time for feeding on the other because the forager cannot be in two places at once. Another way that spatial non-simultaneous search can emerge is if the forager has to change its search or handling behavior for each food (eg. to hunt for deer a mountain lion must wait in ambush, but to hunt rabbits it must actively seek and pursue). An excellent example of this is for human hunter-gatherers who require different tools to harvest vegetable foods and to hunt, and both tool sets cannot be carried and used by the same individual at the same instant in time.

The functional forms for each of the search modes (simultaneous search, spatial non-simultaneous search, temporal non-simultaneous search) are different and can be combined to express various foraging environments (Belovsky et al. 1989). The different feeding time constraints lead to different OFT diet predictions. Therefore, the correct choice of a feeding time constraint is critical for testing OFT models.

An example of this problem and the pitfall of invoking a *post hoc* additional nutrient constraint can be addressed using data on cowbird foraging at the National Bison Range, Montana, USA, to construct OFT models. During the period June - August, cowbirds consume insects and seeds (Bent 1958). In the western United States, the insects primarily consumed by cowbirds are grasshoppers (Acrididae) (Beal 1900).

Methods of parameter estimation: While the data used to construct the OFT model will be discussed more fully elsewhere (Belovsky and Slade in prep), a brief discussion of methodology follows. Two methods were employed to measure model parameters (Table 1, Fig. 3 - 4): field observations and laboratory experiments.

1. Field observations of birds were conducted in two ways: observation of captive cowbirds foraging in an aviary (0.32 m^2 area: 45 cm X 45 cm X 70 cm) and observations of free-roaming cowbirds.

The small aviaries had a large mesh floor which could be placed over the natural vegetation without disturbing it and the density of grasshoppers could be experimentally manipulated (1-10 in aviary). This small contained system appeared to be adequate to study cowbird hunting behavior. First, the foraging behavior of the ground-hunting cowbirds did not appear to be different in the aviaries and the field in terms of time spent searching and distance travelled during the duration of an experiment (less than 5 minutes). Second, the grasshoppers were able to escape from the birds and remain hidden.

Because the birds were very tame, observations on birds in the aviary could be made at a distance of 2m, using a stopwatch to compute the time between encounters (attacks) for grasshoppers at different densities and body masses (functional responses: Fig. 3a-c), and different sexes (Fig. 4a). Birds were deprived of food for 3h prior to being placed in the aviary and no experiments were conducted after the bird consumed 4 or 5 grasshoppers to avoid observations when the bird was satiated. The field density of grasshoppers of different sizes and species was known from field samples (18.6/m^2 of vegetated ground, Belovsky et al. in press). Finally, these experiments were also used to compute the cowbirds' capture success for grasshoppers of different body masses (captures/attack: Fig. 4b).

The aviaries were also used to measure the encounter rate with seeds (millet) at different experimental densities. Because seeds were difficult to eliminate from the natural vegetation under the aviaries and difficult to recover at the end of the experiment, we used a plastic grass mat in the aviary that was "sown" with known densities (mass) of millet. At the end of the experiment, the mat could be removed and the remaining millet could be weighed. The functional response for seeds could then be computed (Fig. 3d). The natural density (mass: 14.4 g/m^2) of seeds was estimated by removing the top 2 cm of soil from 0.01 m^2 plots and then applying a flotation technique to recover the seeds (Pulliam 1981). In these aviary experiments the birds were allowed to forage for only 5 min so they would not deplete the available seeds; no more than 7% and usually less than 2% of the seed biomass was removed.

Observations of free-roaming cowbirds were conducted by an observer with a spotting scope at a prairie site in August. The observer sampled all hourly periods equally from dawn to sunset recording the number of cowbirds observed and the number actively foraging (Fig. 4c). These observations were used to compute the time spent feeding in the field using methods reported by Belovsky and Slade (1986). Cowbirds were observed to be foraging from 0700 - 1900 hours with very restricted

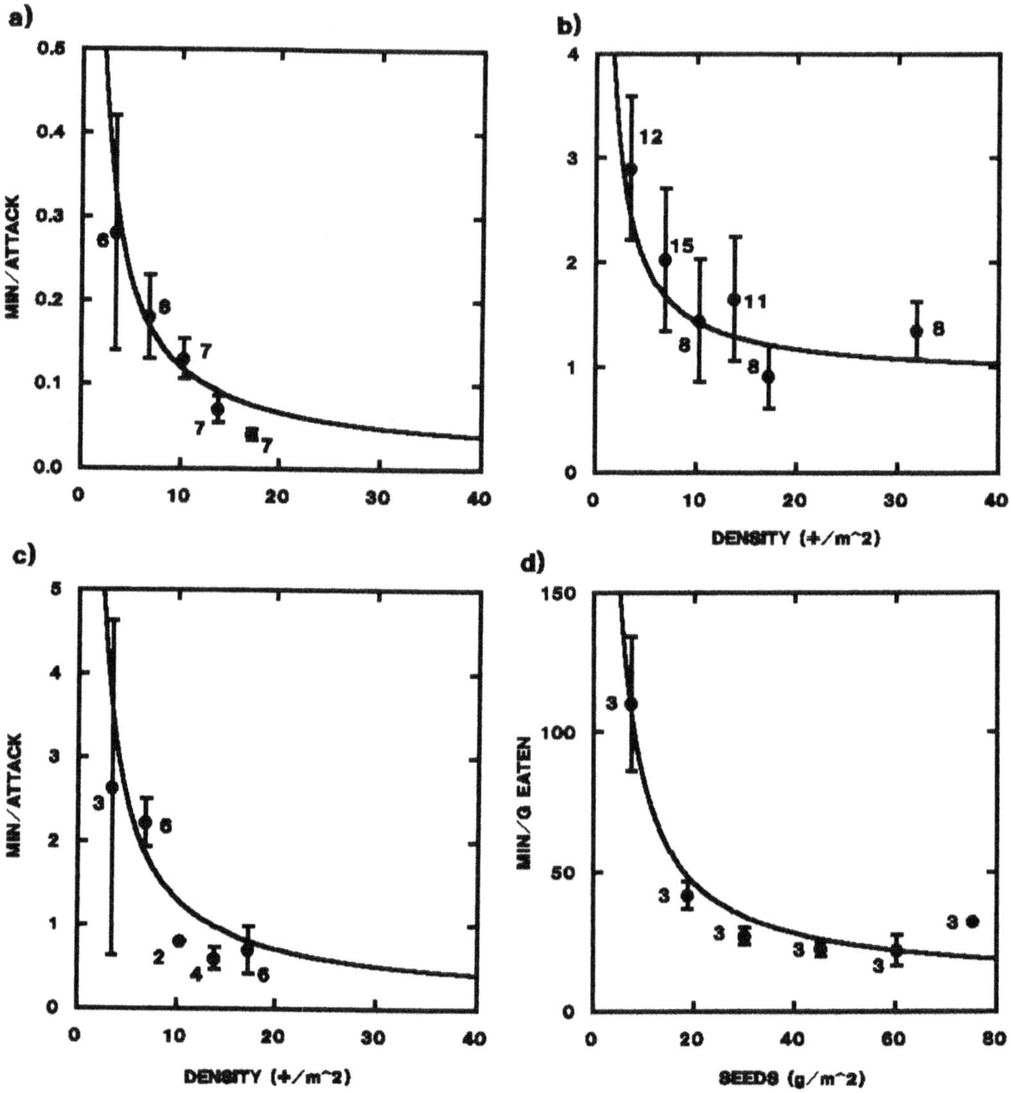

Figure 3. The functional response curves for cowbirds feeding on grasshoppers of different sizes are presented [(a) small: < 100 mg, (b) intermediate: 100 - 400 mg, (c) large: > 400 mg)] as the time to find and attack a grasshopper (± sd and sample size) at different densities. Because an attack may be unsuccessful, these functional responses represent the bird's encounter rate with grasshoppers. These results indicate that encounters with small grasshoppers are greater than encounters with large grasshoppers which are greater than encounter with intermediate grasshoppers (encounters increase as time between attacks decreases). The functional response of cowbirds feeding on millet seeds is presented (d) as the time required to consume a gram of seed (± sd and sample size). Seed density is measured as g/m^2.

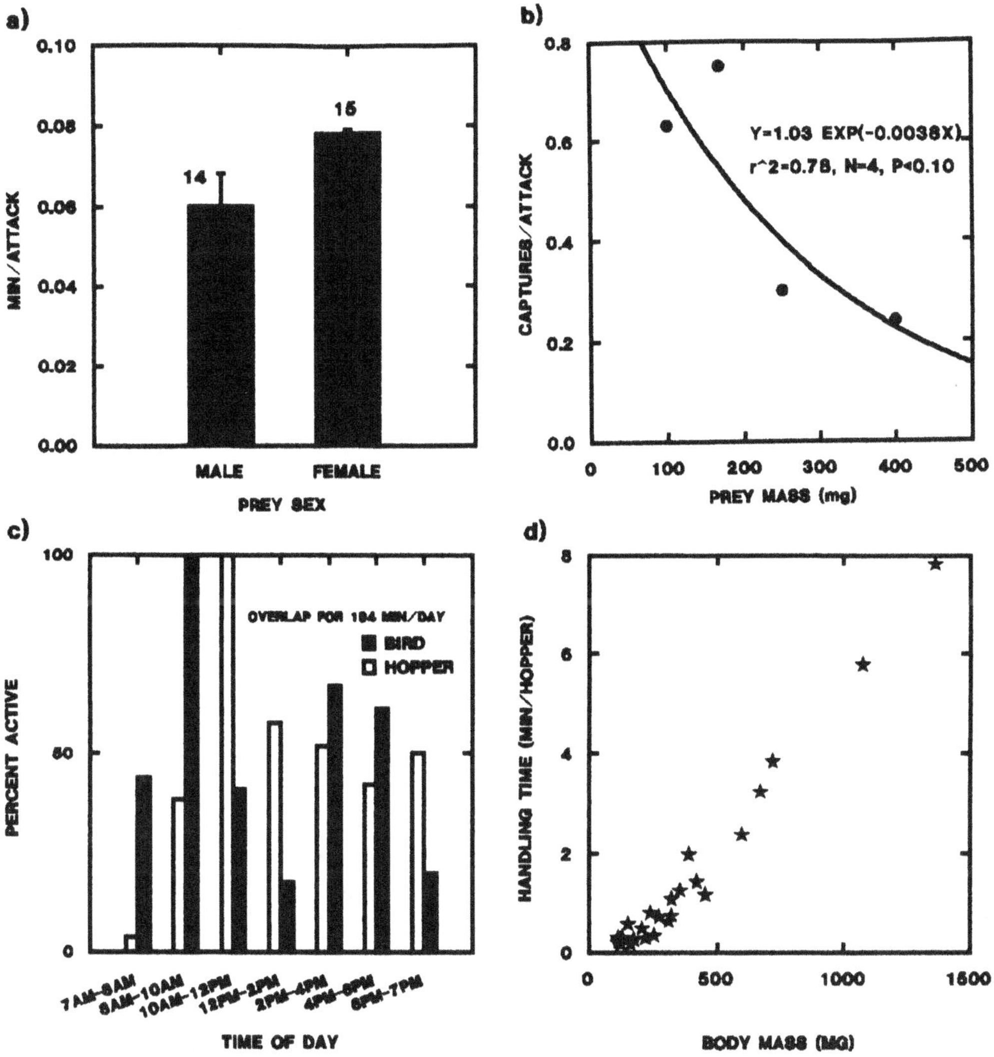

Figure 4. (a) The time for a cowbird to find and attack a grasshopper (± sd and sample size) at a constant density (5 grasshoppers/aviary) is compared for male and female grasshoppers, indicating that males are encountered 30% more frequently (less time between attacks). (b) The capture success of cowbirds hunting grasshoppers is presented as a function of the grasshopper's body mass. Capture success is the number of captures divided by the observed number of attacks. These results indicate that small grasshoppers are easier to capture. (c) The daily activity patterns (percent of all individuals active over the day) are presented for cowbirds in August (146 bird-hours) and grasshoppers (179 grasshopper-hours) in July - August. (d) The time cowbirds require to handle (kill and ingest) grasshopper species-sexes of different body masses is presented, indicating that handling time increases with grasshopper size.

Table 1. The parameters (+ 1 s.d.) used to construct the OFT models for cowbirds.

Digestive Parameters:		
Crop capacity :	0.94 g wet ± 0.07	N=5
Retention time:		
seeds-	4.4 hours ± 0.4	N=5
hoppers-	0.7 hours ± 0.1	N=5
Bulk:		
seeds-	1.03 g wet/g dry ± 0.01	N=6
hoppers-	2.70 g wet/g dry ± 0.01	N=6
Feeding Time Parameters:		
Feeding time:		
with seeds available-	420 min/day	(no estimate of sd)
with hoppers available -	194 min/day	(no estimate of sd)
Cropping rate (search + handling):		
seeds-	41.9 min/ g dry ± 10.9	N=15
hoppers-	60.27 min/g dry ± 49.57	N=62
Energy Requirement Parameters:		
Metabolism:	12.6 kcal/day	Lustick 1970
Gross energy:		
seeds-	5.0 kcal/g dry	Golley 1961
hoppers-	6.3 kcal/ g dry	Bailey and Mukerji 1977
Digestibility:		
seeds-	60.70% ± 8.4	N=6
hoppers-	77.80% ± 7.6	N=6

activity from 1100 - 1500 hours and again at 1800 - 1900 hours. The reduced activity between 1800 - 1900 was probably due to roosting preparation for the night, and the 1100 - 1500 reduction in activity was probably due to abiotic conditions causing excessive heat gain. The cowbirds in the aviaries were observed to stop feeding and begin panting when the air temperature exceeded 30° C. The air temperature in the field usually exceeded this value from 1100 - 1500 during our study. Similar findings for birds have been reported (Clark 1987).

2. Laboratory studies with cowbirds were used to obtain most of the OFT model parameters. In small cages (16 cm X 16 cm X 38 cm), individual birds

were fed different grasshopper species and seeds (millet and sunflower). By providing grasshoppers or seeds *ad libitum*, total daily food consumption could be measured. Comparing the dry mass of food consumed with the dry mass of cloacal output, digestibility could be computed; the product of this value with the energy content of each food reported in the literature provided the net energy for the birds.

In some studies birds were deprived of food for 3 h and then provided with food *ad libitum*. The maximum amount of food consumed before the bird quit feeding due to satiation was assumed to be the capacity of its crop. The foods (freshly killed grasshoppers and seeds) were also dyed with concentrated beet juice by allowing the foods to sit in the juice for 30 min. The time from first ingestion until purple-stained cloacal output was observed was used as a measure of digestive passage time. Since purple-stained cloacal output was only observed at one time and not in subsequent eliminations, mean and maximum passage times were assumed to be equal.

The number of times the bird can empty its crop each day can be computed as the time it spends active over the day divided by the measured passage time plus one additional emptying during the inactive nocturnal period. This estimate multiplied by the estimated crop capacity is the digestive capacity constraint. Interestingly, this value is very similar to the measured daily food consumption (Fig. 6a), illustrating that it may be unnecessary to understand the details of digestive physiology kinetics to construct an adequate OFT model constraint, contrary to Hobbs' claims (this volume). Digestive capacity constraints for bird foraging also have been reported by others (Gartshore et al. 1979, Sorensen 1984).

The captive birds were very tame and they would take individual live grasshoppers and seeds from the investigator and consume them. This permitted very accurate measures of handling times for different species and sexes of grasshoppers and for seeds (millet) that were individually weighed before giving them to the birds (Fig. 4d). Handling time was defined as the time it took the individual bird to ingest the food item after taking it from the observer. This included the time to kill the grasshopper, dismember the hind legs and wings of intermediate and large species, and decapitate and eviscerate the large grasshoppers. For seeds, handling time included the time to remove the seed hull. While seed handling could not be observed in the field, the handling of grasshoppers by field birds appeared to follow the same process.

Results: The data presented in Table 1 and Fig. 3 - 4 can be used to solve OFT (Stephens and Krebs 1986) models with different feeding time constraints and to compare their predictions with the cowbirds' observed diet.

Since grasshoppers and seeds usually co-occur in the native prairie, a logical assumption is that these two foods can be searched for simultaneously. This would make the classical OFT model the most appropriate one to use (Stephens and Krebs 1986). The graphical solution to the classical model is presented in Fig. 5a. By ranking the potential grasshopper prey and seeds from "best" to "worst" based upon their ratio of net energy value to handling time ("best" = highest ratio), the average rate of energy intake for the cowbird can be computed for a diet containing a given food item and all higher ranked items when encountered. This reflects the classical OFT rules that the "best" foods will always be consumed upon encounter and foods are either consumed in proportion to their encounter or never consumed (no partial preferences). The greatest average rate of energy intake is the optimum diet. In this case, cowbirds should consume seeds and all but the very largest grasshoppers.

Cowbirds, however, are observed to eat the very largest grasshoppers. Many investigators might think this to be a minor aberation and the OFT model has successfully predicted the cowbird's diet. But the largest grasshoppers are some of the most preferred by birds in the field (Belovsky et al. in press).

Beal (1900) measured diets of hundreds of cowbirds collected throughout the western U.S. prairies based upon the volume of the crop and stomach composed of different foods. Because the volume of food in the crop and stomach composed of each food is a function of how much has been consumed and how quickly it passes from the digestive organs (foods retained for longer periods will be overestimated in

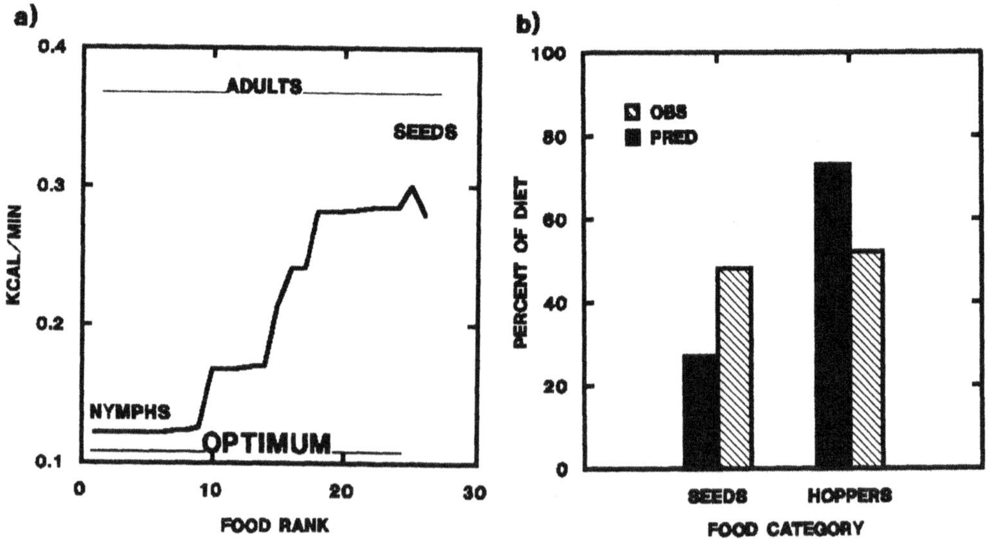

Figure 5. The graphical solution to the classical OFT model for cowbirds is presented (a) along with the predicted and observed cowbird diets (b).

a) **DIGESTIVE CONSTRAINT:**

PHYSIOLOGICAL MEASURES: **OBSERVED:**

SEEDS: $$\left(\frac{17\ \text{hours/day}}{4.4\ \text{hours/fill}} + 1\ \text{fill/day}\right) \times .94\ \text{g/fill} = 4.57\ \text{g/day}$$ $4.57 + .07$ g/day (N=6)

HOPPERS: $$\left(\frac{17\ \text{hours/day}}{.7\ \text{hours/fill}} + 1\ \text{fill/day}\right) \times .94\ \text{g/fill} = 23.77\ \text{g/day}$$ $24.80 + 4.13$ g/day (N=6)

(BASED ON ACTIVITY OVER A 12 HOUR PERIOD EACH DAY IN THE FIELD)

$$\frac{1.03\ \text{g-wet/g-dry}}{3.71\ \text{fills/day}} \times \text{SEEDS g-dry} + \frac{2.70\ \text{g-wet/g-dry}}{17.44\ \text{fills/day}} \times \text{HOPPERS g-dry} < .94\ \text{g/fill}$$

b) **FEEDING TIME CONSTRAINT:**

DETERMINANTS OF DAILY FEEDING TIME IN THE FIELD:

A. OBSERVED FEEDING FROM 0700 - 1900 HOURS
B. NO FEEDING WHEN AIR TEMPERATURE > 30 °C
C. ACTUAL FEEDING FOR ONLY 7 HOURS/DAY
D. GRASSHOPPERS ONLY AVAILABLE (ACTIVE) FOR 3.23 HOURS/7 HOURS

$$41.90\ \text{min/g-dry} \times \text{SEEDS g-dry} + 60.27\ \text{min/g-dry} \times \text{HOPPERS g-dry} < 420\ \text{min/day}$$

$$60.27\ \text{min/g-dry} \times \text{HOPPERS g-dry} < 194\ \text{min/day}$$

c) **ENERGY CONSTRAINT**

$$5\ \text{kcal/g-dry} \times .61\ \text{digestibility} \times \text{SEEDS g-dry} +$$
$$6.3\ \text{kcal/g-dry} \times .78\ \text{digestibility} \times \text{HOPPERS g-dry} > 12.6\ \text{kcal/day}$$

Figure 6. (a) The digestive capacity constraint is developed from the parameters in Table 1. First, the parameters are used to compute the amount of grasshoppers and seeds that the cowbirds can consume in captivity based upon the digestive capacity parameters (see text) and these are compared with the observed daily consumption, showing no significant difference (t-test). Second, the equation for the digestive capacity constraint is presented for cowbirds in the field (see text). (b) The feeding time constraint is developed from the parameters in Table 1, Figure 4c and the text. First, the pattern of feeding activity is presented. Second, the equations that define the feeding time constraint are presented (the first for temporal non-simultaneous search for grasshoppers and the second for spatial non-simultaneous search for grasshoppers and seeds). (c) The energy requirement constraint for cowbirds is presented based upon the parameters in Table 1.

the diet study), the OFT diet predictions must be corrected for differential passage (seed intake X digestive fill of seeds/digestive fill of grasshoppers or 0.28/0.15: see Table 1 and Figure 6). When the OFT predictions are compared with the measured diet (Fig. 5b), seeds are found to be underestimated by 180%. One might question the comparison of the diet measured in 1900 with a model constructed in 1988 at a single site, but the OFT model's predictions are very insensitive to realistic ranges of grasshopper abundances (see below).

At this junction, there is a high probability that an investigator who had only considered simultaneous search and the classic model of OFT might invoke some unknown *post hoc* additional nutrient requirement that can be satisfied only by the consumption of seeds in order to salvage the OFT model. Another possibility is that the investigator rejects the notion of OFT for cowbirds. However, the assumption of simultaneous search may not be appropriate.

While seeds and grasshoppers do co-occur, they cannot be searched for simultaneously. First, when birds look for seeds, they employ a different pattern of search than when they look for grasshoppers. When the birds search for seeds they bend over and carefully scan the ground; when they search for grasshoppers they scan at a distance in front for grasshoppers to jump or fly and then pursue then to the landing site. These two search patterns, I believe, are mutually exclusive (spatial non-simultaneous search). Second, there are periods through the day that the bird's foraging activity and grasshopper activity (Belovsky and Slade 1986) do not coincide (Fig. 4c) (temporal non-simultaneous search). This is very important since it appears that the cowbirds must see the grasshoppers move in order to recognize them as prey and when inactive the grasshoppers tend to bury themselves in the litter at the base of grass clumps, refusing to move unless "rooted" out. Therefore, the appropriate OFT model for this system is not the classical model, but a model which has a time constraint that is a composite of spatial and temporal non-simultaneous search.

Using the data in Table 1 and Fig. 3 - 4, an LP model of OFT can be constructed with three constraints: digestive capacity (Fig. 6a), feeding time (Fig. 6b) and energy requirements (Fig. 6c). The graphical presentation of the LP model appears in Fig. 7. The possible diet combinations of seeds and grasshoppers that satisfy all three constraints is cross-hatched. The predicted energy-maximizing and time-minimizing diets are also presented along with the observed diet (Beal 1900).

Using a Monte Carlo simulation (100 iterations) that varies the model parameters simultaneously within their 95% confidence intervals (Table 1), the energy-maximizing diet is found to have a standard deviation of $\pm$ 6.7%. Furthermore, decreasing the grasshopper density by 50% only leads to a 10% decrease for grasshoppers in the diet and a 50% increase in grasshopper density only leads to an

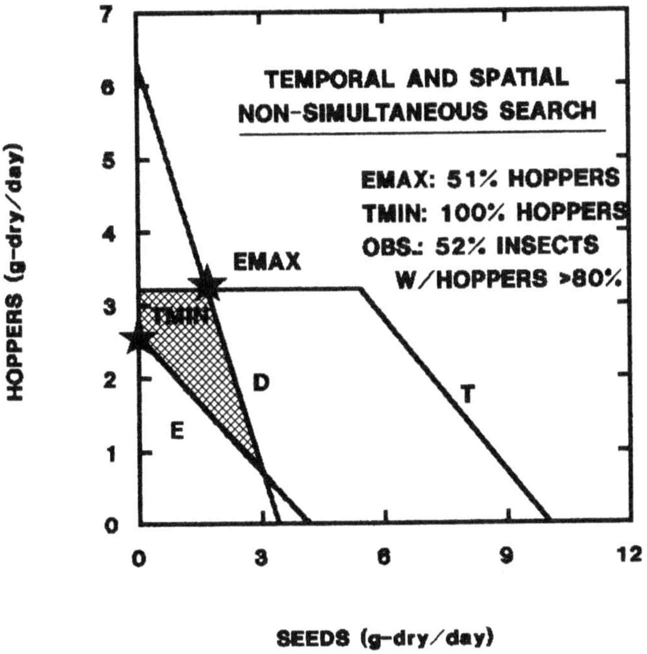

Figure 7. A graphical solution to the LP model of cowbird foraging is presented. Each constraint equation is labeled: D = digestive capacity (Fig. 6a), T = feeding time (Fig. 6b), and E = energy requirement (Fig. 6c). The cross-hatched region contains all diet combinations of seeds and grasshoppers that satisfy the constraints. The stars represent the solutions to the energy-maximizing (EMAX) and time-minimizing (TMIN) goals. Also presented are the predicted and observed diets (Beal 1900), indicating that cowbirds appear to forage as energy maximizers.

11% increase in grasshoppers in the diet. These results indicate that the OFT model's predictions are rather insensitive to parameter changes within the confidence intervals.

We might conclude that cowbirds are energy maximizers in their diet choice, given the close agreement between the energy maximizing diet and the observed diet. This illustrates the importance of applying the correct assumptions in constructing an OFT model, especially for the feeding time constraint, before attempting to test OFT.

The classical OFT model, however, is appropriate for examining cowbird foraging if we ask a different question: when cowbirds choose to hunt for grasshoppers, which prey items should be attacked? In this case, non-simultaneous search no longer applies and simultaneous search is the appropriate assumption. The

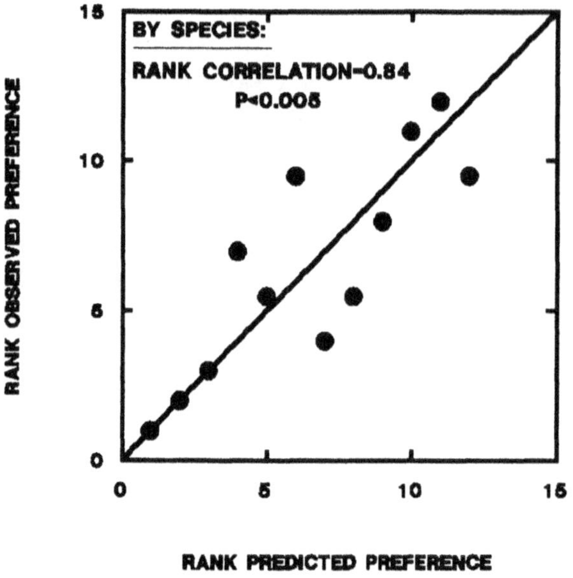

Figure 8. The classical OFT model is used to predict the cowbird's grasshopper diet (see text). The predicted and observed preferences (see text) are compared using Spearman Rank Correlation.

classical OFT model under these circumstances predicts that all grasshoppers should be attacked upon encounter, as observed (Belovsky et al. in press). The rank of the cowbirds' preference (proportion of diet composed of a grasshopper species/relative abundance of the grasshopper species) can be compared with the rank of intensity of predation on a grasshopper species (Belovsky et al. in press). A close agreement between the model's predictions and the observed values is obtained (Fig. 8).

Conclusion

I have attempted to discuss in this paper several statements of caution regarding the use of OFT models and the invocation of nutrient constraints in addition to the nutrient that serves as the currency to be optimized.

First, the number of cases where additional nutrient constraints need to be included in OFT models is very small. Eight cases out of 108 studies conducted by my students and me needed to include additional nutrient constraints. This implies that most nutrient demands for a forager are satisfied incidently with the forager's optimization of some other nutrient currency. While these additional nutritional requirements might be of interest, it might not be necessary to include them in constructing a model of diet choice, in particular OFT models.

Second, if additional nutrient constraints need to be included in an OFT model, there are simple ways to accomplish this. While these simple methods do not include the detailed mechanisms studied by nutritional physiologists, they have the advantage of requiring less information, an important consideration for field studies of OFT. More important, they do not require sets of nested equations defining the nutritional kinetics which are based upon numerous estimates. In constructing models, we must remember that as more detail is added to the model, there are more ways that the model's predictions can be in error. This is particularly a problem since small errors in parameter estimation will be carried to the next calculation, and the next; eventually propagating error throughout the entire OFT model and magnifying the original error. This is an important consideration for the verification of a model.

Third, when OFT models do not meet expectations of precision for diet predictions, I believe that there are other alternatives to invoking *post hoc* additional nutrient constraints or claiming that OFT is not applicable. Rather, many investigators have not critically examined whether the assumptions used to construct the OFT model they employ are appropriate for the environment and forager that they are studying. This is especially true for the feeding time constraint which depends upon the forager's search behavior and the distribution of foods in the environment (Belovsky et al. 1989). Consequently, many OFT studies may be misapplying the model chosen by the investigator; it is my opinion that this is particularly the case for the use of the classical OFT model which is based upon the assumption that foods are encountered randomly or uniformly.

Fourth, OFT is a collection of models appropriate for different environments and foragers. To test OFT properly, the appropriate model must be employed and the model must be simple enough to see whether or not it portrays the essential elements of the forager's dietary choices. The inclusion of complex nutritional and physiological kinetics may improve the model's precision. More likely, the addition of numerous parameters with their associated measurement errors will make the model's predictions erratic, obscure the major factors controlling diet choice by a forager, and at worst lead to predictions that are so poor that OFT might be incorrectly rejected.

OFT shows great promise in helping ecologists understand feeding behavior. It also helps to link behaviorial and physiological studies with population ecology. But there is no single prescription for OFT models, rather there exists a plurality of models, each appropriate for different environments and foragers. As ecologists testing the concepts of OFT, we must be aware of this, but still keep our models as simple as possible so that we can search for generalities.

ACKNOWLEDGEMENTS:

I wish to thank J. B. Slade for aiding in the cowbird research and commenting on the manuscript. Comments on the manuscript by Fred Provenza and Ian Gordon were very helpful. The work was supported by NSF, USDA, and Rackham School, University of Michigan.

LITERATURE CITED:

Altman SA, Wagner SS (1978) A general model of optimal diet. In Chivers DJ, Herbert J (ed) Recent Advances in Primatology (Vol 1). Academic Press, London, p 407-414

Bailey CG, Mukerji MK (1977) Energy dynamics of *Melanoplus bivittatus* and *M. femurrubrum* in a grassland ecosystem. Can Ent 109: 605 - 614

Beal FEL (1900) Food of the bobolink, blackbirds, and grackles. Biol Surv Bull USDA #13

Belovsky GE (1978) Diet optimization in a generalist herbivore: the moose. Theor Pop Biol 14: 103-134

___________ (1981) Food plant selection by a generalist herbivore: the moose. Ecology 62: 1020-1030

___________ (1984a) Snowshoe hare optimal foraging and its implications for population dynamics. Theor Pop Biol 25: 235 - 264

___________ (1984b) Summer diet optimization by beaver. Amer Midl Natur 111: 209 - 222

___________ (1984c) Herbivore optimal foraging: a comparative test of three models. Amer Natur 124: 97 - 115

___________ (1986) Optimal foraging and community structure: implications for a guild of generalist grassland herbivores. Oecologia 70: 35 - 52

__________ (1987a) Hunter-gatherer foraging: a linear programming approach. J Anthro Arch 6: 29 - 76

__________ (1987b) Foraging and optimal body size: an overview, new data and a test of alternative models. J Theor Biol 129: 275 - 287

__________ (1987c) An optimal foraging model for wild herbivores. In Rose M (ed) Herbivore Nutrition Research: Second International Symposium on the Nutrition of Herbivores. Occasional Publication of the Australian Society of Animal Production, p 227

__________, Slade JB (1986) Time budgets of grassland herbivores: body size similarities. Oecologia 70: 53 - 62

__________, Slade JB (1987) The role of plant distributions on herbivore diet choice: a comparison of wild and domestic herbivores. In Rose M (ed) Herbivore Nutrition Research: Second International Symposium on the Nutrition of Herbivores. Occasional Publication of the Australian Society of Animal Production, p 87

__________ Ritchie ME (in press) Geographical and seasonal variation in ground squirrel foraging. Oecologia

__________, Slade JB, Stockhoff BA (in press) Susceptibility to predation for different grasshoppers: an experimental study. Ecology

__________, Ritchie ME, Moorehead J (1989) Foraging in complex environments: when prey availability varies over time and space. Theor Pop Biol

Bent AC (1958) Life histories of North American blackbirds, orioles, tanagers, and allies. Smithsonian Inst US Nat Mus Bull #211

Clark L (1987) Thermal constraints on foraging in adult european starlings. Oecologia 71: 233 - 238

Gartshore RG, Brooks RJ, Somers JD, Gilbert FF (1979) Temporal changes in gullet food passage in penned red-winged blackbirds: significance for research in feeding ecology. Can J Zool 57: 1592 - 1596

Golley F (1961) Energy values of ecological materials. Ecology 42: 581 - 584

Gray RD (1986) Faith and foraging: a critique of the "paradigm argument from design". In Kamil AC, Krebs JR, Pulliam HR (ed) Foraging behavior. Plenum Press, New York, p 69

Hill K (1988) Macronutrient modifications of optimal foraging theory: an approach using indifference curves applied to some modern foragers. Human Ecology 16: 157-197

Intriligator, MD (1971) Mathematical optimization and economic theory. Columbia Press, New York

Karasov WH (1985) Nutrient constraints in the feeding ecology of an omnivore in a seasonal environment. Oecologia 66: 280 - 290

Krebs JR, Avery MI (1984) Chick growth and prey quality in the european bee-eater. Oecologia 64: 363 - 368

Lustick S (1970) Energetics and water regulation in the cowbird. Physiol. Zool. 43: 270-287

Pierce GJ, Ollason JG (1987) Eight reasons why optimal foraging theory is a complete waste of time. Oikos 49: 111-118

Pulliam HR (1975) Diet optimization with nutrient constraints. Amer Natur 109: 765 - 768

_________ (1980) Do chipping sparrows forage optimally? Ardea 68: 75 - 82

Ritchie ME (1988) Individual variation in the ability of Columbian ground squirrels to select an optimal diet. Evol Ecol 2: 232 - 253

_________ (in press) Optimal foraging and fitness in Columbian ground squirrels. Oecologia

Robbins CT (1983) Wildlife nutrition. Academic Press, London

Schoener TW (1971) Theory of feeding strategies. Ann Rev Ecol Syst 2: 369 - 404

___________ (1986) A brief history of optimal foraging theory. In Kamil AC, Krebs JR, Pulliam HR (ed) Foraging behavior. Plenum Press, New York, p

Schmitz O (submitted) Optimal diet selection by white-tailed deer: responses to environmental change. J Wildl Mgmt

Sorensen AE (1984) Nutrition, energy and passage time: experiments with fruit preference in european blackbirds. J Anim Ecol 53: 545 - 557

Spalinger DE (1980) Mule deer habitat evaluation based upon nutritional modelling. MA Thesis, University of Nevada, Reno

Stephens DW, Krebs JR (1986) Foraging theory. Princeton Univ Press, Princeton

Tanaka J (1976) Subsistence ecology of Central Kalahari San. In Lee RB, DeVore I (ed) Kalahari hunter-gatherers. Harvard University Press, Cambridge, p 98

Tilman D (1982) Resource competition and community structure. Princeton University Press, Princeton

Westoby M (1974) An analysis of diet selection by large generalist herbivores. Amer Natur 112: 627- 631

ORAL DISCUSSION - G.E. Belovsky

Q. (Sih): The energy maximisation model has done well in predicting the proportion of intake from broad categories, say forbs versus grass, but it hasn't done very well in predicting among different forbs. Are nutrient constraints likely to be important within a category like forbs ?
A. When we examine selection of foods within a category, other nutrient constraints might be important in determining selection of specific plant species. However, the herbivore's search for food items within a category also becomes simultaneous, which might account for the selection of individual food items. The question of nutrient constraints controlling the selection of food items has not been studied, while the change to simultaneous search has been shown to be important.

Q. (Kacelnik): I'm impressed by the precision of observed and predicted (from LP model) results in the 108 species studied. I have studied only one species and yet I get enormous confidence intervals in predictions and have managed to explain only a very small proportion of variances in foraging behaviour.

A. I also obtain very large confidence intervals based upon the standard errors of the parameter values. These confidence intervals can be as large as 30 – 40% of the mean value predicted by the model. However, the close agreement that you are referring to reflects the means of all studies compared with the observed diets. This is the difference between the confidence that we can place on a single experiment versus repeated independent experiments.

Q. (Peterson) In marine systems plant chemistry is seen as being very important in dictating choice of herbivores. Are marine systems different from terrestrial systems.
A. Secondary compounds may be important, but I have not included them in a linear programming model. I have not found it necessary to consider them in the cases that I have studied.

Q. (Bazely): The terrestrial–insect people have recently started to move away from the classic arguments about secondary compounds towards a more purely nutritionally–based explanation of diet choice.
A. I agree and this may require a re–evaluation of the role that secondary compounds play in diet choice. However, the mammalian ecologists are currently paying more attention to secondary compounds.

Q. (Jumars) Many secondary compounds are digestive inhibitors and reduce absorption rate.
A. Yes, if the secondary compounds simply reduce digestibility, they do not have to be considered since they simply change the nutritional value of the forage and do not impose any additional constraints on the forager.

Q. (Hobbs): The Linear Programming technique assumes that there are no associative effects among the items ingested. If you have specific substances which reduce overall digestibility then LP is violated.
A. Associative effects would lead to synergisms. If this does occur, LP would be violated because the constraint functions would become non–linear rather than linear, but at present we would have to know how strong they are to determine whether or not the non–linearities will change the LP model's predictions. However, if you are correct and there are strong associative effects, the non–linear model can be solved using non–linear programming, a more general optimization technique.

ECOLOGICAL AND MORPHOLOGICAL ASPECTS OF CHANGES IN FOOD UPTAKE THROUGH THE ONTOGENY OF *HAPLOCHROMIS PICEATUS*

Frietson Galis
University of Leiden
Division of Ecological Morphology and Department of Population Biology
P.O. Box 9516 2300
RA Leiden
The Netherlands

The importance of integrating ecology and functional morphology has been recognized for a long time (Lack 1948; van der Klaauw 1951; Hespenheide 1973). Ecological studies provide the basis for the determination of the function of animal structures (van der Klaauw 1951; Bock and Von Wahlert 1965; Dullemeijer 1974). If it can be quantitatively shown how anatomical changes affect fitness, then this would be an important contribution towards the understanding of animal structures (Alexander 1988). Functional morphological studies, moreover, contribute to ecological understanding if they can reveal constraints on the optimisation of behaviour (Werner 1974; Gould and Lewontin 1979; Grant and Grant 1982; Zweers 1982; Barel 1983; Hoogerhoud 1987). The problem here is that researchers are usually specialists in either ecology or functional morphology. This tends to result in a simplification of the research in one of the disciplines (see for examples Motta 1988).

Studying the ontogeny of an organism offers the possibility of comparing a gradually transforming series of stages. During this transformation, the organism must continue to live. It is an intriguing question whether the level of performance of the animal at any stage is hampered by this demand of continued performance during transformation, as has been suggested, for example, De Jongh (1968) and Gould and Lewontin (1979). In the present study different ontogenetic stages of *Haplochromis piceatus* Greenwood and Gee are compared, both in an ecological way and in a functional morphological way, with the focus on feeding.

NATO ASI Series, Vol. G 20
Behavioural Mechanisms of Food Selection
Edited by R. N. Hughes

In the first part,stomach contents were analysed to determine which food types were eaten by fishes of different size. In the second part, foraging experiments were designed, based on the stomach content analyses. The aim of these experiments was to investigate whether the observed differences in diet are the result of an optimal prey choice. Experimental results were compared both with stomach content analyses and with predictions based on Charnov's (1976) optimal foraging model. The third part contains the first results of a functional morphological analysis of the pharyngeal jaw apparatus. This is to provide an explanation, on the basis of morphological factors, of constraints on the efficiency of food uptake.

MATERIALS AND TECHNIQUES

Specimens for dissection.

H. piceatus specimens were caught at different dates between 1–2–1979 and 29–7–1980, on 2–9–1984 and 28–2–1984. All specimens were preserved in 70% ethanol after being fixed in 4% formalin.

Stomach content analyses.

The stomach was taken separately and the intestine was divided into four sections of equal length. Food items (insects and crustaceans) were counted and classified to the lowest possible taxonomical level. The life stage of the insects was determined according to Verbeke (1957) and Luttik & Hoogerhoud (pers. comm.).

The length of the ingested copepods and cladocerans in 36 fish was measured using a Wild M5 stereomicroscope with a micrometer eye piece. If the number of copepods or cladocerans greatly exceeded 64, only 64 randomly selected specimens were measured, since in 15 sections of the digestive tracts of 5 fishes, measurements on such random samples presented no significant differences from measurements on the total number (U test; $p>0.05$).

In the calculation of the size of the eaten copepods and cladocerans, values were pooled for the entire digestive tract since there was no significant difference between the size of cladocerans and copepods in 84.3% ($n = 121$) of the sections (U test; $p>0.05$). In the remaining 15.7% of cases, items at the beginning of the digestive tract were slightly larger than those at the end. It seems likely that these differences are due to the disintegration effect of the digestion process. It is, however, possible that they reflect temporal fluctuations in food availability.

No significant differences were found in the size of ingested copepods and cladocerans caught on different dates (Friedman two–way analysis of variance; $p>0.05$), and data were pooled for subsequent statistical analysis.

Foraging experiments.

The data of this part were taken from Galis and de Jong (1988).

The predator.

Parental *H. piceatus* specimens were caught as juveniles in January 1984 in the Mwanza Bay of Lake Victoria (Tanzania), and were shipped by air to Leiden (For a taxonomic analysis of the Mwanza Gulf population see Witte 1987). F1–, F2 and F3–generation of fish were used in the experiments, as well as one specimen that was caught in the wild. Results can be directly compared, since no significant differences were found between similarly obtained data of fishes from different generations. (F–test and Mann–Whitney U test; $p>0.05$). Every week the standard length of the experimental fishes was measured and used to calculate fish volume through the following regression equation: $Y=0.282X+0.035$ ($r=0.099$; $p<0.05$) in which Y is the cube root of fish volume (ml) and X is the standard length (cm). This regression is based on the standard length and volume of 31 fish of different sizes (2.1–6.0 cm SL). The volume was determined by submersion. The experimental fish were fed zooplankton supplemented with Tetramin when necessary. Care was taken that both prey items used in the experiments, *Daphnia magna* and *Chaoborus* (see below) were regularly and alternatingly offered to ensure experience, and to avoid both a training bias on one prey item only (Dill 1983) and long term learning (Werner et al, 1981). We found that a carefully balanced diet is particularly important because if only *Chaoborus* larvae are offered for some time, large fish lose any interest in *D. magna*, even when no other food is available.

The prey.

We chose two prey types that differ considerably in size: *D. magna* does not occur in Lake Victoria and there *H. piceatus* feeds on *Bosmina longirostris*, *Chydorus* sp., *Moina micrura* and *Diaphanosoma* (see results: stomach content analyses). However, for many particulate feeders, the capture success for different cladoceran species is approximately equal, with the exception of *Diaphanosoma* which is more difficult to capture (but not as difficult as copepods) (Confer and Blades 1975; Kettle and O'Brien 1978; Drenner et al. 1978; Bohl 1982; Schmidt and O'Brien 1982; Wright and O'Brien 1984). Also, the chemical composition of these prey types does not vary much (Vijverberg and Frank 1976). Therefore we feel that *D. magna* is a reliable substitute for the cladocerans eaten in the lake. The length (taken as the distance from the tip of the head to the base of the caudal spine) of 25 specimens was measured under a Nikon profile projector V–16A. Two length classes were used in the experiments, 0.9–1.2 mm and 1.2–1.6 mm. The mean dry weight of the smallest size class was determined on a Mettler H54 balance after 10 days drying at 80 C. Mean dry weight was 11.19 g. The caloric value of *D. magna* is 5640+60 Cal/g dry weight (Cummins and Wuycheck, 1971) therefore one *D. magna* of this size class has an average energy content of 0.26 J. Because the natural prey species, *Chaoborus anomalus* and *C. edulis*, do not occur in the Netherlands, a mixture of fourth instar larvae of *C. cristallinus*, *C. pallidus*, *C. flavicans* and *C. obscuripes* was used as insect prey. Thus we ensured a sufficient supply of prey items throughout the year. Mixtures of the above mentioned species of *Chaoborus* larvae can also be easily purchased in aquarium shops. Because they are difficult to distinguish, individual identification of the larvae used in the experiments was not feasible. Since the experiments could be repeated without significantly different results (Wilcoxon matched pairs signed–ranks test, $p>0.05$), the use of a mixture of species is not problematical. Mean dry weight of a larva, determined as described for *D. magna*, is 0.64 mg (n=200). The caloric value of *Chaoborus* is 4936 Cal/g (Cummins & Wuycheck 1971), therefore one *Chaoborus* larva has an average energy content of 13.37 J.

The model.

Charnov's (1976) model was used to predict prey selection (for more detials see Galis and de Jong 1988.). This model predicts that when two prey types are available, the less profitable prey type should not be eaten as long as the abundance of the most profitable prey type is sufficient to meet the dietary needs of the predator. A similar conclusion was reached by Schoener (1971), Pulliam (1974) and Werner and Hall (1974). Instead of using the net rate of energy intake, we calculated the amount of energy ingested and neglected energy costs of the food uptake. We think that this had little influence on the results since Stein et al. (1984) found that for pumpkinseed sunfish (*Lepomis microlophus*, size: 130–220 mm), the energy costs of handling prey, which involved crushing molluscs, were only 0.25% to 2.6% of the prey's energetic content. In another case, Drost et al. (1987) found that for larval carp (*Cyprinus carpio*), the energy costs of the actual attack are only a fraction of a percent of the energetic content of the attacked prey.

Experimental set up:

Three series of experiments were carried out. Series 1 with *Chaoborus* larvae as prey, Series 2 with *D. magna* as prey and Series 3 with both species as prey. Six, five and eight *H. piceatus* individuals were respectively used in Series 1, 2 and 3.

In the Mzanza Gulf, densitites of fourth instar *Chaoborus* larvae were found to vary from 0.003 to 0.6 larvae per litre (Goldschmidt et al. 1988, Goldschmidt and Luttik unpublished data). Galis and de Jong (1988) found that densities of *Chaoborus* larvae varying from 0.35 to 4 larvae per litre did not influence the uptake efficiency of *H. piceatus* in light, as long as the absolute number of larvae was sufficient to avoid depletion; A density of 2 larvae per litre was offered in Series 1.

Akiyama (1977) and Goldschmidt and Hoogenboezem (unpublished data) found that densities of cladocerans in the Mwanza Gulf vary from 0.21 to 30 individuals per litre. Total densities of cladocerans and copepods varied from 4 to 150 individuals per litre. Because densities varying from 2.5 to 30 individuals per litre did not influence the uptake efficiency of *H. piceatus* in light (Galis and de Jong 1988) a density of 30 *D. magna* per litre was offered in Series 2. In Series 3 30 *D. magna* and 4 *Chaoborus* larvae per litre were offered because these densities do not limit prey uptake rate.

An experimental time of 45 minutes plus the time until capture of the next prey item was chosen, because no effect of satiation was measured within three hours of feeding for fishes of these sizes (Galis and de Jong 1988). Because light intensity (except for total darkness) does not influence prey uptake rate (Galis and de Jong 1988) a high illumination level was chosen to facilitate good observation. The experiments were performed in 20 and 40 litre aquaria. The size of the aquaria did not significantly influence the results (Wilcoxon matched pairs signed-rank test, $p > 0.05$). Behavioural observations were made with the help of a behaviour recorder or videorecorder (Sony Umatic).

The following behavioural parameters were measured: 1) Experiment time (see above), which can be divided into search time and handling time. In our experiments search time can be neglected because searching occurred during processing time and therefore did not require extra time (Galis and de Jong 1988). 2) Attack time, the time from the moment the fish approaches a prey until it has taken the prey into its mouth. 3) Processing time (not including attack time) the time the prey spends in the buccal cavity, was indirectly measured as the total processing time divided by the number of prey items eaten excluding the last one whose moment of capture determined the end of the experiment. This indirect estimation was necessary, because *H. piceatus*, with the exception of very small individuals, simultaneously processes several prey items in the buccal cavity. The space of the buccal cavity seems to limit the number of prey items that can be ingested within a short period of time (sometimes fishes try to suck additional prey into an already full buccal cavity; the prey item is thereupon released, or, in the case of *Chaoborus*, is sometimes held out of the mouth and is slowly taken into the buccal cavity). 4) Total number of attacks. 5) Percentage of successful attacks. 6) Duration of a successful attack. 7) Total attack time spent per successful attack.

Pharyngeal jaw apparatus.

Because at higher prey densities food uptake rate is almost exclusively limited by processing time (Galis and de Jong 1988 and this study) and because the storage capacity of the stomach is not limiting within the timespan of the experiments (Galis, unpublished data), food uptake is limited by the speed at which the food is processed and passed from the buccal cavity into the oesophagus. Because the pharyngeal jaw apparatus plays a crucial role in the food processing and transport to the oesophagus (Liem 1973), this apparatus was selected for a functional morphological analysis.

For the reconstruction of the muscle attachments, serial sections were made of heads in a non-expanded situation, i.e. a closed mouth and a non-abducted suspensorium. The heads were embedded in plastic following the technique described by Anker (1974). Sections were stained with Luxol Fast Blue and Picro-fuchsine. Reconstructions have been made from the serial sections at a magnification of 25x. Worklines of the muscles were taken to be between the estimated centres of the attachments of the muscles. The centres of attachment were estimated by taking the centre of the middle vertical cross section of the attachment. Coordinates of these centres were determined using a Calcomp 9100 x,y tablet in combination with the 3D Reconstruction Package Mac Reco (Otten, 1988).

The ventral side of the parasphenoid bar (part of the neurocranium, see Barel, Witte and van Oijen 1976) was taken as a reference line for the position of the muscles. At two places the most ventral position of the parasphenoid bar was determined: rostrally at the place of attachment of the palatine-vomerine notch ligament on the skull (R_1 in Fig.9, see Anker 1986), and dorsally at the place of the opening in the neurocranium through which the arteria carotica passes (R_2 in Fig.8).

Average length of muscle filaments was estimated using a Wild M5 stereomicroscope with micrometer eye piece. Dry weight of the muscles was determined after 48 hrs at 80°C using a Mettler UM3 balance.

Allometric relations

Allometric relations were computed between fish volume (Y) and muscle filament length (X_1), muscle dry weight (X_2) and dry weight/filament length ($X3_)$ (as an indication of muscle force) using the expression $Y=aX_i$. If X values are of the same dimension as Y values (volume) b=1 indicates isometry, b<1 indicates negative allometry and b>1 indicates positive allometry. Symmetrical 95% confidence limits were computed.

RESULTS

Stomach content analyses

H.piceatus feeds predominantly on cladocerans and copepods. Of the fish that had food in their digestive tract (86% n = 156) 94% contained cladocerans and copepods (cladocerans: mainly Diaphanosoma excisum, Moina micrura, Bosmina longirostris and Chydorus spec. and to a lesser extent Ceriodaphniae cornuta and Daphnia spec.; copepoda (cyclopoida and calanoida), 39% insects (mainly Chaoborus anomalus or edulis and to a lesser extent chironomids), 20% algae (mainly Melosira) and 18% detritus.

14% of the fish had an empty digestive tract, all adult females except for a single male. H.piceatus is a mouth brooder and brooding females never have food in their stomach. Because most specimens with an empty digestive tract were caught in the preferential brooding time (Witte, in prep.) and all had relatively small gonads (as expected for brooding females) we think that they had been brooding, but lost their brood during capture. In captivity, disturbance sometimes causes brooding fish to expel their brood.

Not surprisingly, large fish eat more than small fish (fig.1).

Changes in diet through the ontogeny

Cladocerans and copepods. The average size of the eaten cladocerans and copepods increased significantly with fish volume (Fig. 1). This increase in food particle size is caused by a greater preponderance of large species in the diet of large fishes. The proportion of calanoids among copepods increases with fish size (as volume: $y = x^{0.006} + 1$; $r = 0.41$, $n = 36$, $p<0.05$) and so does the proportion of Diaphanosoma excisum and Moina micrura among cladocerans ($r = 0.24$, $n = 36$, $p<0.05$, ($y = 0.08x^{0.30}$) and with standard length: $r = 0.45$, $n = 36$, $p<0.05$, $y = (-0.34x + 28.13)^{-1}$. There was no trend in the ratio of the overall number of cladocerans to the number of copepods with fish volume ($r = 0.10$ $n = 36$, $p>0.05$ and with standard length, $r = 0.05$, $n = 36$, $p>0.05$). In all cases, far more copepods than cladocerans had been eaten. On average the ratio of cladocerans to copepods was 0.14 (s.d.=0.11).

Insects. As fish grow, they tend to eat insects more often (Fig.2). Again there is a shift in the frequency of eaten species. Large fishes eat more Chaoborus (3rd and 4th instar larvae and pupae) than small fishes. Small fish eat more

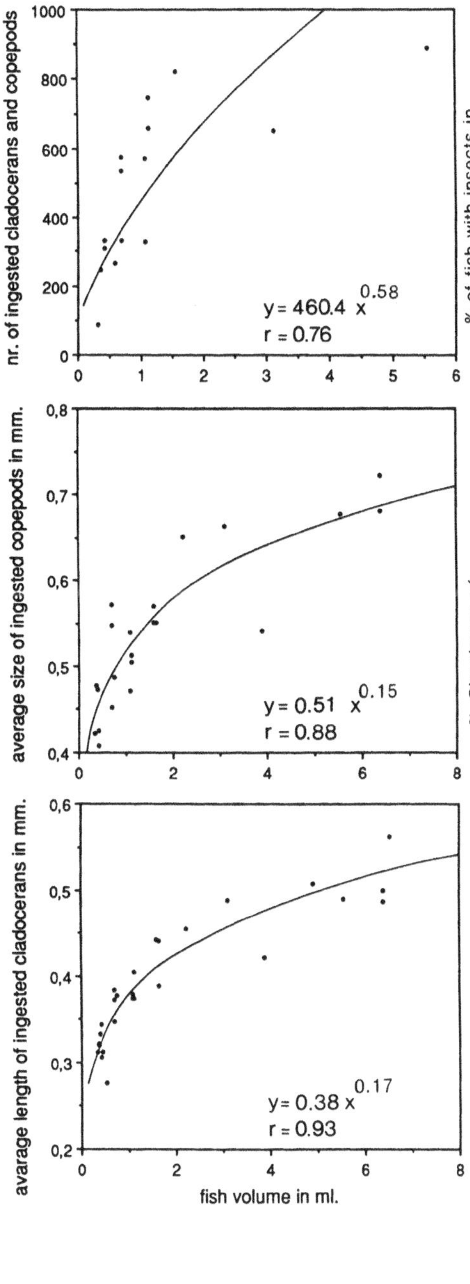

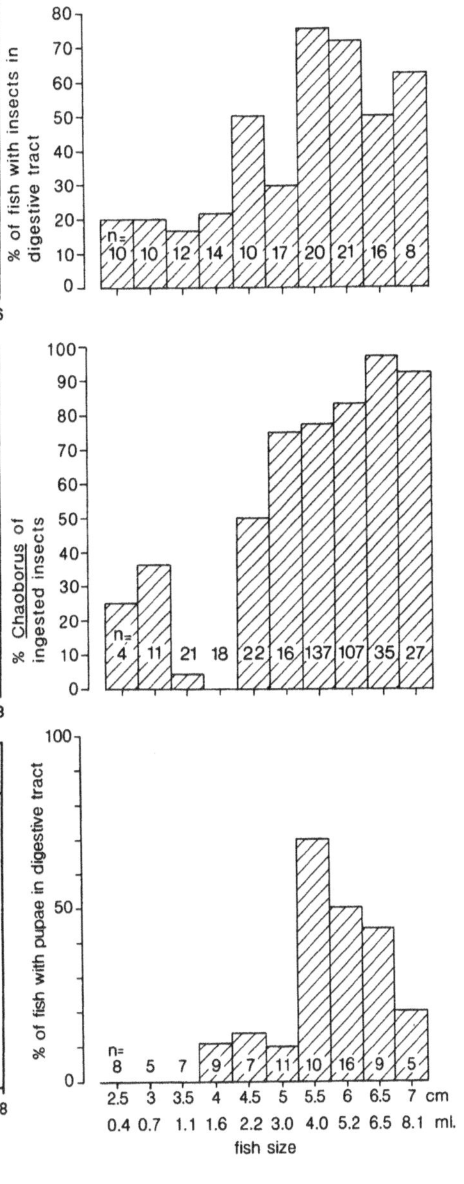

Fig.1. Relationship between fish volume and (a) number of ingested cladocerans and copepods in stomachs that were filled for at least 75% with cladocerans and copepods; (b) mean size of ingested copepods and (c) mean size of ingested cladocerans ($p<0.05$).

Fig.2. Frequency distributions of stomach contents.

chironomid larvae (mainly 2nd and 3rd instar). Fish smaller than 1.64 ml volume (4.05 cm standard length) never ate Chaoborus pupae (Fig. 2).

Foraging experiments.

Chaoborus larvae (Series 1). Uptake of Chaoborus per hour increased with the volume of the fish ($Y=10.17X+0.71$, $r=0.996$, $p<0.05$) and thus energy uptake rate increased linearly with volume (Fig.3). Interestingly, the energy uptake/min.gr body weight is about the same for all ontogenetic stages and is not significantly correlated with body weight (Table 1). During development, H. piceatus' success of capture increases: The duration of an attack decreases (Fig.4) and the chance of success increases ($\ln y=0.16 \ln x+4.40$, $r=0.41$, $p<0.05$). Success rate is variable and low for the young fish and invariably high for larger fishes. The increase in the food uptake with body size of Chaoborus is, for a minor part, the result of a decrease in average attack time per larva eaten, but is mostly the result of a decrease in the mean processing time (Fig.8) which constitutes on average 99.4% (sd=0.07) of the handling time per larva.

D. magna (Series 2). The results resemble those of Chaoborus. The uptake of D. magna increases linearly with volume of the fish ($Y=4.119X+1.353$, $r=0.97, p<0.05$) and thus the energy uptake rate per minute (Fig.3). However, the uptake of energy/min.gr body weight does not remain constant as is the

Table 1. Energy uptake rate. Chaoborus

Volume of fish(ml)	J/min.gr(x)
0.17	2.89
0.36	2.15
1.26	2.43
1.47	2.35
1.62	2.56
1.69	2.48
1.75	2.18
1.82	2.61
1.95	2.13
5.17	2.28

$r=-0.31$, $P>0.05$;
$x=2.41$ (sd=0.6)

Table 2. Energy uptake rate Daphnia magna

Volume of fish(ml)	J/min.gr(x)
0.06	2.62
0.13	1.87
0.20	2.83
0.22	2.14
0.36	1.80
0.56	1.98
0.56	2.30
0.56	1.93
0.90	1.76
1.26	0.96
1.41	1.87
2.45	0.97
3.79	1.27
3.79	1.35
4.68	1.02

$r=-0.74$, $P<0.05$
($Y=-28X+2.16$)

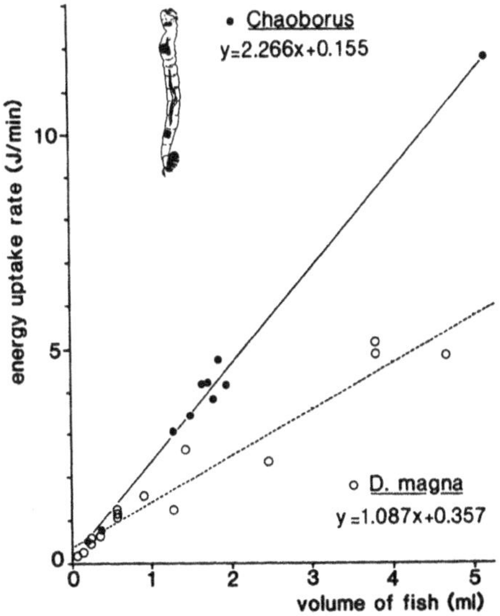

Fig.3. Relationship between fish volume and rate of energy intake from Chaoborus ($r=0.99$, $P<0.05$) and D. magna ($r=0.97$, $P<0.05$).

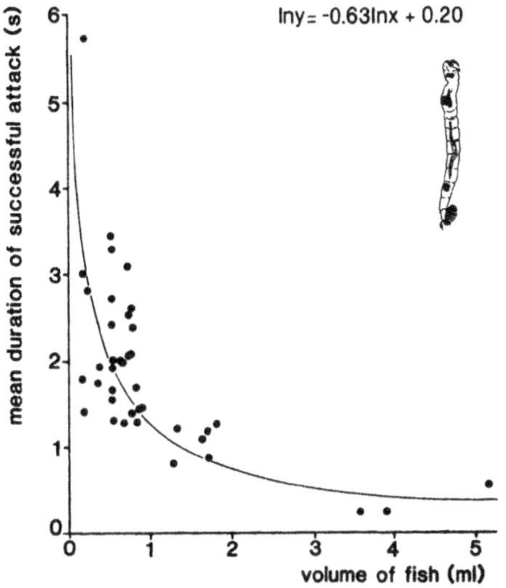

Fig.4. Relationship between fish volume and mean duration of successful attack on Chaoborus ($r=-0.71$, $P<0.05$).

Fig.5. Relationship between fish volume and percentage of successful attacks on D. magna ($r=0.87$, $P<0.05$). Black circles represent experiments with D. magna of 0.9-1.2 mm. Open circles represent experiments with D. magna of 1.2-1.65 mm (not included in regression analysis).

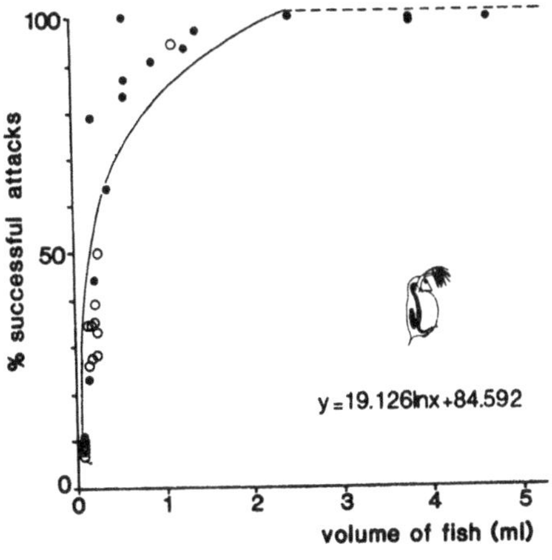

Fig.6. Relationship between fish volume and processing time for D.magna ($r=-0.80$, $P<0.05$).

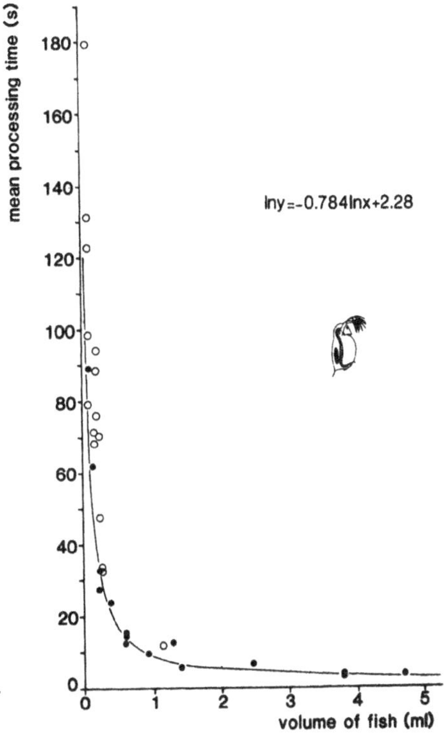

case with Chaoborus but decreases (Table 2). Again, the duration of an attack decreases with fish size ($\ln y=-0.28\ln x-1.64$, $r=-0.90$, $p<0.05$) and the percentage of successful attacks increases (Fig.5). The variation for young fishes is much lower than in experiments with Chaoborus. Though processing time (Fig.6) again constitutes most of the handling time, attack time is relatively more important, especially for small fish. From a fish volume of 0.06 to 1.26 ml the percentage of attack time decreases from 10.5% to 1.2%. From 1.3 to 4.7 ml the percentage attack time increases to 5.2% but this is probably a measuring error, since the attacks are too short (less than 0.2s) for a reliable measurement with our behaviour recorder.

Prediction for the mixed-prey experiment (Series 3). With increasing fish size, the regression lines for energy yield per time unit of D. magna and Chaoborus are different (Fig 3). Up to 0.49 ml there is no significant difference in energy uptake for both prey (F-test, $p>0.05$). Above 0.49 ml the difference is significant and increases with fish size. According to Charnov's model (1976) fish larger than 0.49 ml should specialize entirely on Chaoborus. Fish smaller than 0.49 ml should generalize and take D. magna and Chaoborus.

Results of the mixed-prey experiments (Series 3): Fig.7 shows that fish smaller than 0.49 ml ate both Chaoborus and D. magna as predicted: the percentage of Chaoborus in the prey eaten varies from 7.7 to 60%, with 60% for a fish of 0.46 ml and 87.5% for a fish of 0.49 ml. Because a Chaoborus larva contains 50 times more energy than a single D. magna, even when Chaoborus makes up only 8% of the items eaten it still provides 81% of the total energy intake. Fish larger than 0.49 ml ate mainly Chaoborus as predicted and virtually all ingested energy is derived from this prey. The energy uptake rate in these experiments did not differ from those in which only Chaoborus larvae were eaten (F test, $p>0.05$). If we assume similar handling times for D. magna in mixed- and single-prey experiments, the handling times of Chaoborus larvae and the number of attacks on Chaoborus larvae per time unit do not significantly differ from those with only Chaoborus as prey (F test, $p>0.05$).

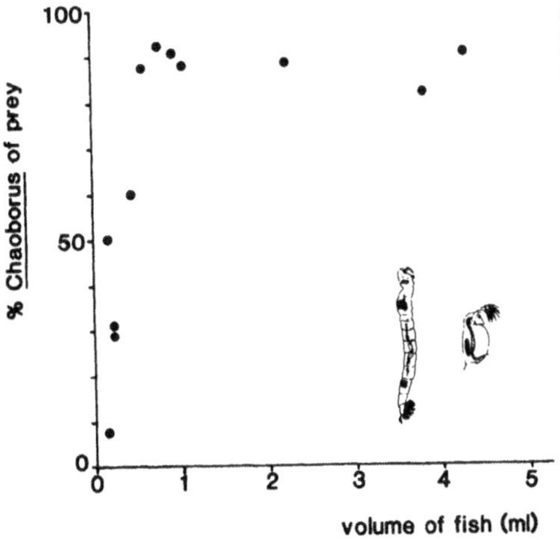

Fig.7. Relationship between fish volume and percentage Chaoborus of ingested prey in mixed-prey experiments.

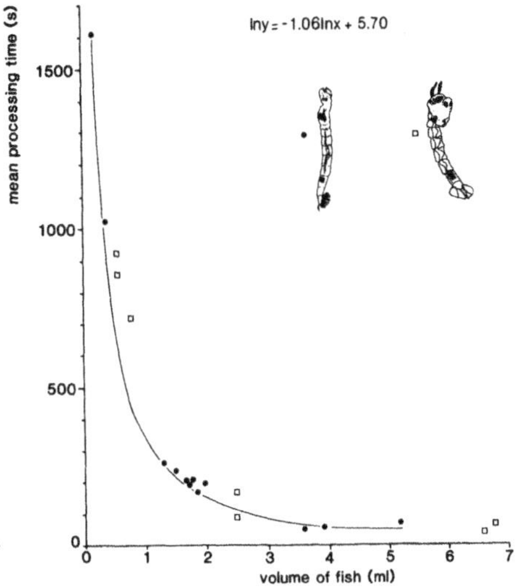

Fig.8. Relationship between fish volume and mean processing time for Chaoborus larvae (r=-0.996, p<0.05) and Chaoborus pupae (not included in regression analysis).

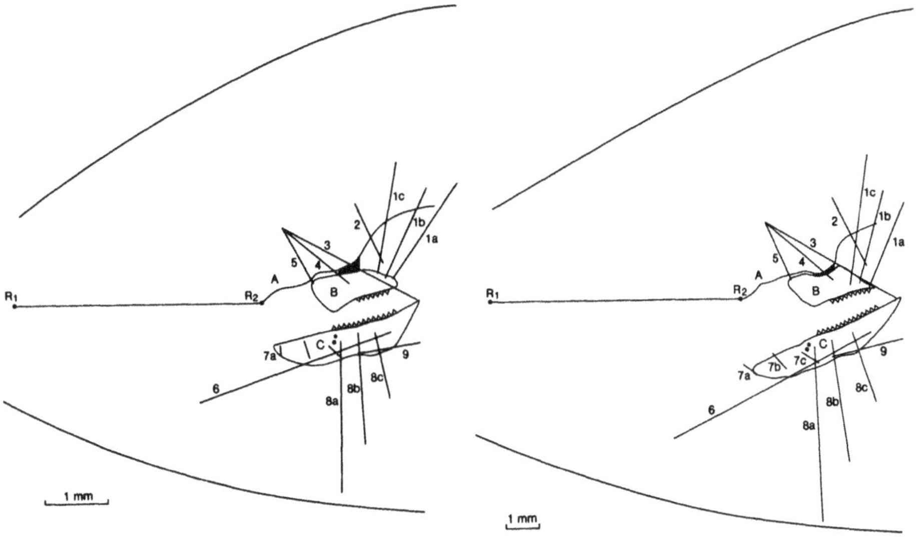

Fig.9. Projection in the median plane of pharyngeal jaws and worklines of muscles of, to the left a specimen of 0.4 ml and to the right of 3.0 ml. For names of the muscles, see corresponding numbers in Table 3. Numbers with (a), (b) and (c) refer to (a) most ventral border, (b) workline and (c) most dorsal border of muscle.

Table 3. Allometric relation between fish volume and:

	Muscle dry weight			Fiber length			Dry weight/ Fiber length	
1	0.92	i	n=13	0.29	i	n=14	0.63	i
2	0.92	i	n=13	0.41	+	n=15	0.51	-
3	1.03	i	n=14	0.35	i	n=12	0.67	i
4	0.98	i	n=12	0.37	+	n=14	0.61	i
5	1.05	i	n=14	0.39	+	n=17	0.66	i
6	0.84	i	n=6	0.41	i	n=10	0.85	+
7		-			-			-
8	1.09	+	n=11	0.34	i	n=13	0.75	
9	1.16	+	n=11	0.31	i	n=15	0.85	+
10	0.96	i	n=13	0.41	i	n=13	0.55	-

1.m.retractor dorsalis, 2.m.levator posterior, 3.m.levator externus 4.m.levator internus lateralis 5.m.levator internus medialis 6.m.pharyngohyoideus 7.m.transversus ventralis 8.m.pharyngocleithralis externus 9.m.pharyngocleithralis internus 10.m.geniogyoideus (for terminology, Anker 1978). i indicates isometry, + indicates significant positive allometry and - indicates significant negative allometry.

Pharyngeal jaw apparatus

Table 3 shows that most parameters increase isometrically or slightly allometrically. Fig.9 shows the projection in the median plane of the pharyngeal jaws and the worklines of nine muscles involved in the movement of the jaws. Table 4 lists the angles between the worklines and the reference line. The

Table 4. Angle between workline of muscles and reference line.

Muscles	Specimen of 0.41 ml	Specimen of 3.0 ml
m.retr.dl.	63°	75°
m.lev.post.	64°	61°
m.lev.ext.4	27.5°	30°
m.lev.int.med.	60°	64°
m.lev.int.lat.	38°	39°
m.ph.hy.	18.5°	26°
m.transv.vl.	84°	45.5°
m.ph.cleit.ex.	86.5°	81°
m.ph.cleit.int.	9°	13°

main differences between the small and large specimens are found for the musculus retractor dorsalis, the m. pharyngohyoideus and the m. transversus ventralis. However, because the last muscle runs in transverse direction, slight changes in the position of either the lower pharyngeal jaw or the fourth gill arch (to which the muscle is attached) have a large impact on the angle measured in the median plane. The relative position of the articulation tissue of the skull and the upper pharyngeal jaw indicates a similar position in both fish. The length of the lower pharyngeal jaw is relatively longer in the larger specimen (positive allometry).

DISCUSSION

Food choice in nature and foraging experiments

In nature, large H.piceatus specimens tend to eat larger prey than small specimens and to eat more prey that are difficult to catch such as Diaphanosoma excisum, Moina micrura, calanoids and Chaoborus pupae (for capture ability see Confer and Blades 1975; Kettle and O'Brien 1978; Drenner et al. 1978; Bohl 1982; Schmidt and O;Brien 1982; Wright and O'Brien 1984). These differences in diet are likely to be due to differential food selection, because the data are based on ontogenetic series taken from each catch. The only absolute difference in diet of small and large fish is that fish smaller than 1.64 ml never eat Chaoborus pupae.

The increasing preference (sensu Sih this volume) through the ontogeny of H.piceatus for insects agrees well with the experimental results which show that during growth the selection of Chaoborus instead of D.magna becomes increasingly more profitable.

Is food choice optimal?

The results of the mixed-prey experiments are in agreement with the predictions of Charnov's model based on the results of the single prey experiments, except that fish larger than 0.5 ml never totally exclude D.magna. Similar partial preferences that are not predicted by the model are reported from many studies (e.g. Werner and Hall 1974; Krebs et al. 1977; Elner and Hughes 1978). Krebs and McCleery (1984) and Stephens and Krebs (1986) discuss seven different explanations that have been proposed in the literature for this partial preference. Except for, possibly, a nutritional constraint, or an avoidance of toxins (see Pulliam 1975 and Tinbergen 1980), none of the explanations is likely to have caused the discrepancy with the model (see Galis and de Jong 1988).

Apart from this small deviation, Charnov's model predicts the behaviour of H.piceatus remarkably well. The model, however, did not agree with Persson's (1985) findings. Persson found that Perca fluviatilis feeds more efficiently on Chaoborus than on D. magna when offered separately, as is the case for H. piceatus specimens larger than 0.5 g. However, P.

fluviatilis generalizes when both prey items are simultaneously offered. According to Persson this is due to the inability of *P.fluviatilis* to cope simultaneously with prey species exhibiting different anti-predation behaviour, which results in different handling times and encounter rates in the mixed-prey experiments. Behavioural observations show that *H.piceatus* employs the same feeding technique for *Chaoborus* and *D. magna*. This may explain the difference with *P. fluviatilis*. Like *H. piceatus* several molluscivorous cichlids show an optimal prey choice (Hoogerhoud 1987; Slootweg 1987). A high selection pressure towards an optimal prey choice through the ontogeny is likely because a high growth rate lowers the age of reproductive maturity and it shortens the time spent in the vulnerable young stages, thereby increasing survival up to the age of reproductive maturity.

Pharyngeal jaw apparatus and food uptake

Considering the more than sevenfold increase in volume, the similarities in shape between the two stages are striking (Fig.9 , Table 4). The isometrical or slightly allometric growth of most muscles, also indicates similarity in shape. These results coincide with the gradual nature of the increase in uptake efficiency of both *Chaoborus* larvae and *D. magna*.
The shift to a more vertical position of the m. retractor dorsalis through ontogeny interestingly compares with a difference in position of this muscle between mollusc crushing cichlids and piscivores. Preliminary data (Witte and Hoogerhoud, unpublished material) show that the position is more vertical in mollusc crushers. In the insectivorous *H.burtoni* this muscle is active during mastication (food preparation by the pharyngeal jaws) and swallowing (food transport by the pharyngeal jaws) and not during abduction-protraction (the widening and forward movement of the jaws) (Liem 1973). It seems, therefore, likely that the more vertical position in adults indicates a larger ability for crushing during mastication. In the field, *H. piceatus* specimens smaller than 1.64 ml never eat pupae (Fig.2). Preliminary foraging experiments with *Chaoborus* pupae show that handling time for pupae is longer than for larvae in small fish (Fig.8; note that the difference should be read in vertical direction) but not for larger fish. The lack of pupae

in the diet of small *H. piceatus* in the field is thus likely to be the result of a morphological constraint on the force that the pharyngeal jaws can exert in vertical direction.

Does ontogeny pose constraints on prey choice?

The good agreement between the optimal prey choice model and the results of the mixed-prey experiments for all sizes of *H.piceatus* does not indicate a developmental constraint on prey choice within the size range of specimens used in this study. The slight disagreement caused by the partial preference is unlikely to be caused by a developmental constraint, since partial preference is such a common phenomenon in adult animals.

Does ontogeny pose constraints on uptake efficiency?

When feeding on *Chaoborus*, the efficiency of food uptake by *H. piceatus*, measured as energy per unit time per gram body weight, is equal for individuals differing as much as thirty times in body volume (0.17-5.17 ml, see table 1) . This means that for all fish sizes the same energetic reward per gram body weight exists. These unexpected results relativize in this case, the enormous importance attributed to the constraints posed by ontogeny on the optimal performance of animals by *e.g.* Gould and Lewontin (1979).

When *H.piceatus* feeds on *D.magna*, the energetic reward per unit body volume decreases with increasing body volume because the processing time per ingested Joule does not decrease as much as the body weight increases ($\ln Y=-0.77X+3.6$, $r=0.98$, $P<0.05$; Y=total processing time/total ingested energy; X=body volume). However, compared to metabolic rate change, the decrease in feeding efficiency is less: Winberg (1956) and Paloheimo and Dickie (1966) found that resting metabolic rate scales with body weight (volume) to the power of approximately 0.8 (for juvenile and adult fishes), whereas Brett and Groves (1979) found, for active metabolism, a mass exponent near unity. When *H. piceatus* feeds on *Chaoborus*, the mass exponent of the energy rate is near unity (0.96), and on *D. magna* it is 0.79. This means that through ontogeny,a feeding bout on *D.magna* becomes somewhat less productive and one on *Chaoborus* somewhat more productive, if we assume a metabolic mass exponent between 0.8 and 1.

The diminished reward per ml body volume when a fish is feeding on D. magna can be due to size constraints *per se*. A possibity is that transporting particles into the oesophagus with the pharyngeal jaw teeth is more efficient for large particles than for small ones. Because D. magna become relatively smaller as a fish grows, this is possibly an example of a constraint due to the relative size of predator and prey. Such a size constraint is not a developmental one because it is independent of the transformation itself. Another possibility is that the demand of continued performance during transformation is constraining the reward rate at some stage (a true developmental constraint). If the latter is true, adaptive peaks would be expected interspersed with less adaptive stretches (less adaptive only if considered in isolated context, not given the necessity of transformation, see Real this volume). It seems more likely that the diminished reward is due to a size constraint *per se* since the uptake rate of D. magna linearly increases with the volume of the fish without observable irregularities. A quantitative model of the forces involved based on the (almost isometrically scaling) morphological parameters measured in this study should give more information on this and on other possible morphological constraints.

SUMMARY

Food selection was studied through the ontogeny of Haplochromis piceatus both by analysing stomach contents and by carrying out feeding experiments. Stomach contents revealed that larger fish eat, on average, larger prey items that are more difficult to catch than smaller fish. These differences in diet agree well with the experimental results, which show that during growth the selection of Chaoborus instead of Daphnia magna becomes increasingly more profitable.

The results of the prey choice experiments are in agreement with the predictions of Charnov's optimal prey choice model, except for a slight difference caused by partial preferences.

Food intake per unit body weight and per unit time of different sizes of H.piceatus was equal when feeding on Chaoborus but decreased with increasing fish size when feeding on D. magna. The latter decrease is likely to be due to size constraints and not to developmental constraints.

Most morphological parameters of the pharyngeal jaw apparatus increased isometrically or slightly allometrically. The results indicate that the lack of insect pupae in the diet of small fish may be the result of a morphological constraint on crushing ability.

ACKNOWLEDGEMENTS

I thank Jacques van Alphen, Gerrit Anker, Kees Bakker, Kees Barel, Jim Gilliam, Simon Goss, Peter de Jong and Jaap de Visser for their comments on the manuscript. I thank John Hughes for correcting the english, Gerrit Anker and Frans Witte for helpful discussions, Frans Witte and Jaco Voorham for the use of their unpublished data.

REFERENCES

Akiyama T, Kajumulo AA, Olsen S (1977) Seasonal variations of plankton and physicochemical conditions in Mwanza Gulf, Lake Victoria. Bull Freshw Fish Res Lab 27:49-61

Alexander R McN (1988) Functional and ecological morphology. Neth J Zool 38:3-22

Anker GCh (1974) Morphology and kinetics of the head of the stickleback, _Gasterosteus aculeatus_. Trans Zool Soc Lond 32:311-416

Anker GCh (1978) The morphology of the head-muscles of a generalized Haplochromis species: _H. elegans_ Trewavas 1933 (Pisces, Cichlidae). _Neth J Zool_ 28:234-271

Anker GCh (1986) The morphology of foints and ligaments in the head of a generalized _Haplochromis_ species: _H. elegans_ Trewavas 1933 (Teleostei, Cichlidae) III. The hyoid and the branchiostegal apparatus, the branchial apparatus and the shoulder girdle apparatus. Neth J Zool 37:394-427

Barel CDN (1983) Towards a constructional morphology of cichlid fishes (Teleostei, Perciformes). Neth J Zool 33:357-424

Barel CDN, Witte F, Oyen MJP van (1976) The shape of the skeletal elements in the head of a generalized _Haplochromis_ species: _H.elegans_ _Trewavas_ 1933 (Pisces, Cichlidae). Neth J Zool 26:163-265

Bock WJ, Wahlert G von (1965) Adaptation and the form-function complex. Evolution 19:269-290.

Bohl E (1982) Food supply and prey selection in planktivorous Cyprinidae. Oecologia (Berlin)53:134-138

Brett JR, Groves TDD (1979) Physiological energetics. _In_:"Fish Physiology " (Hoar WS, Randall DJ, Brett JR, eds) Vol 8, pp:279-352. Academic Press, New York

Charnov EL (1976) Optimal foraging: attack strategy of a mantid. Am Nat 110:141-151

Confer JL, Blades PI (1975) Omnivorous zooplankton and planktivorous fish. Limnol Oceanogr 20:571-579

Cummins KW, Wuycheck JC (1971) Caloric equivalent for investigations in ecological energetics. Mitt Int Ver Limnol 18
Dill LM (1983) Adaptive flexibility in the foraging behavior of fishes. Can J Fish Aquatic Sci 40:398-408
Drenner RW, Strickler JR, O'Brien WJ (1978) Capture probability: the role of zooplankter escape in the selective feeding of planktivorous fish. J Fish Res Board Can 35:1370-1373
Drost MR Bogaart JGM van den (1987) The energetics of feeding strikes in larval carp (Cyprinus carpio). J Fish Biol
Dullemeijer P (1974) Concepts and approaches in Animal Morphology. Van Gorcum Assen
Elner RW, Hughes RN (1978) Energy maximization in the diet of the shore crab, Carcinus maenas. J Anim Ecol 47:103-116
Galis F, Jong PW de (1988) Optimal foraging and ontogeny; food selection by Haplochromis piceatus. Oecologia 75:175-184
Goldschmidt T, Witte F, Visser J de (1988) Ecological segregation of zooplanktivorous haplochromines (Pisces, Cichlidae) from Lake Victoria. Thesis, University of Leiden
Gould SJ, Lewontin RC (1979) The spandrels of San Marco and the Panglossian paradigm: a critique of the adaptationist programme. Proc R Soc Lond B 205:581-598
Grant BR, Grant PR (1982) Niche shifts and competition in Darwin's finches: Geospiza conirostris and congeners. Evolution:637-657
Hespenheide HA (1973) Ecological inferences from morphological data. Ann Rev Ecol Syst 4:213-229
Hoogerhoud RJC (1987) The adverse effects of shell ingestion for molluscivorous cichlids, a constructional morphological approach. Neth J Zool 37:277-300
Hoogerhoud RJC (1987) Ecological morphology of some cichlid fishes. Thesis, University of Leiden, The Netherlands
Jongh HJ de (1968) Functional Morphology of the jaw apparatus of larval and metamorphosing Rana Temporaria L.. Neth J Zool 18:1-103
Kettle D, O'Brien WJ (1978) Vulnerability of arctic zooplankton species to predation by small lake trout (Salvelinus namaycush). J fish Res Board Can 35:1495-1500
Krebs JR, Erichsen JT, Webber MI, Charnow EL (1977) Optimal prey selection in the great tit (Parus major). Anim Behav 25:30-38
Krebs JR, McCleery RH (1984) Optimization in behavioural ecology. In: Krebs JR, Davies NB (eds) Behavioural ecology. An evolutionary approach. Second edition. Blackwell Scientific Publications, Oxford
Klaauw CJ van der (1951) Bau, Lebensweise, und Milieu der Tieren. Die Grundlagen einer Oekologischer Morphologie. Acta Soc pro Fauna et Flora Fennica 67(2)
Lack D (1948) Darwin's finches. An essay on the general biological theory of evolution. Harper and Brothers, New York
Liem KF (1973) Evolutionary strategies and morphological innovations: cichlid pharyngeal jaws. Syst Zool

22:425-441
Motta PJ (1988) Functional morphology of the feeding apparatus of ten species of Pacific butterflyfishes (Perciformes, Chaetodontidae): an ecomorphological approach. Env Biol Fish 22:39-67
Otten E (1988) A manual for the 3D reconstruction Package MacReco written for Macintosh TM computers. Bloemsingel 10 NL-9712KZ Groningen. The Netherlands
Paloheimo JE, DIcikie LM (1966) Food and growth of fishes. III Relations among food, body size and growth efficiency. J Fish Res Board Can 23:1209-1248
Persson L (1985) Optimal foraging: the difficulty of exploiting different feeding strategies simultaneously. Oecologia (Berlin) 67:338-341
Pulliam HR (1974) On the theory of optimal diets. Am Nat 108:59-74
Pulliam HR (1975) Diet optimization with nutrient constraints. Am Nat 109:765-768
Real LA (this volume) Predator swithching and the interpretation of animal choice behavior: the case for constrained optimization.
Schoener TW (1971) The theory of feeding strategies. Ann Rev Ecol Syst 2:369-404
Schmidt D & O'Brien WJ (1982) Planktivorous feeding ecology of arctic grayling (Thymallus articus). Can J Fish Aquat Sci 39:475-482
Slootweg R (1987) Prey selection by molluscivorous cichlids foraging on a schistosomiasis vector snail, Biomphalaria glabrator. Oecologia 74:193-202
Sih A (this volume) Interacting effects of predator and prey behaviour in determinig diets.
Stein RA, Goodman CG, Marschall EA (1984) Using time and energy measures of cost in estimating prey value for fish predators. Ecology 65:702-715
Stephens DW, Krebs JR (1986) Foraging theory. Princeton University press. Princeton, New Jersey
Tinbergen JM (1980) Foraging decisions in starlings (Sturnus vulgaris L). Ardea 69:1-67
Verbeke J (1957) Chaoborida (Diptera, Nematocera). Stades immatures et adultes. Résult Sci Explor Hydrobiol Lacs Kivu, Edouard, Albert (1952-1954)3:183-201
Vijverberg J, Frank Th H (1976) The chemical composition and energy contents of copepods and cladocerans in relation to their size. Freshwat Biol 6:333-345
Werner EE (1974) The fish size, prey size, handling time relation in several sunfishes and some implications. J Fish Res Bd Can 31:1531-1536
Werner EE, Hall DJ (1974) Optimal foraging and the size selection of prey by the bluegill sunfish (Lepomis macrochirus). Ecology 55:1042-1052
Werner EE, Mittelbach GG, Hall DJ (1981) The role of foraging profitability and experience in habitat use by the bluegill sunfish. Ecoloogy 62:116-125
Winberg GG (1956) Rate of Metabolism and Food Requirements of Fishes. Nauch Tr Belorussk Gos Univ Imeni V I Lenina, Minsk (Fish.Res.Board Can, Transl Ser No 194, 1960)
Witte F (1987) From Form to Fishery. An Ecological and Taxonomical Contribution to Morphology and Fishery of Lake Victoria Cichlids. Thesis, University of Leiden
Wright DI, O'Brien WJ (1984) The development and field test of

a tactical model of the planktivorous feeding of white crappie (Pomoxis annularis). Ecol Monogr 54:65-98
Zweers GA (1982) Pecking of the pigeon (Columba Livia L.). Behav 81:173-230

ORAL DISCUSSION – F. Galis

Galis explained that as in temperate lakes, the Lake Victoria *Chaoborus* were nocturnally active and were buried in the bottom during the day, but that the fish could visually delect their prey under low light conditions. Only in the absence of moonlight was visual detection on *Chaoborus* prevented by darkness, although the fish is still able to consume *Chaoborus*.A short discussion of Lewontin and Gould's ideas of phylogenetic and ontogenetic constraints took place, with Galis reporting that she detected no developmental constraints, although some may be present. In response to a question about developmental plasticity, Galis reported that some cichlid species exhibited a radically different mouth structure when allowed to develop in an aquarium rather than in natural field conditions. She had not studied whether oxygen uptake varied ontogenetically but intends to study the gill area. She expressed the opinion that enhanced visual acuity was not a factor in explaining high predation rates of larger fish in these experiments.

PREDICTING ONTOGENETIC NICHE SHIFTS IN THE FIELD: WHAT CAN BE GAINED BY FORAGING THEORY?

Lennart Persson
Department of Animal Ecology
University of Umeå
S-901 87 Umeå
Sweden

Foraging theory has been developed greatly since the pioneer papers on this topic (MacArthur and Pianka 1966; Emlen 1966). Even though one of the original aims was to advance an economic theory of diet and habitat choice of animals in the field (MacArthur 1972), surprisingly few field studies have included quantitative predictions of diet choice in the field. For example of the 35 field studies testing the classical diet model listed by Stephens and Krebs (1986), only 9 included quantitative predictions. Considering only experimental studies reduces this figure to 3. One reason for this scarcity of field tests of optimal foraging theory is that foraging studies during the last decade have focused more on the purely behavioural aspects of animal foraging than community organization. For this purpose, laboratory experiments have been regarded as more powerful and appropriate than field studies (Stephens and Krebs 1986, but see Schoener 1987 for artificial laboratory experiments). This behavioural research has led to the questioning on experimental and theoretical grounds of the many assumptions in the classical diet and patch models, such as sequential encounter, complete information and constancy of encounter rate and handling time. Recent foraging theory has also to some extent rejected static optimization in favour of dynamic optimization (Gilliam 1982; Houston and McNamara 1986; Mangel and Clark 1986). However, from the population ecologist´s point of view it can be argued that simplifications have to be made, and that the incorporation of even the classical foraging models into population models may involve a significant advancement compared to previous population models (see Comins and Hassell 1979 for the patch model; Gleeson and Wilson 1986 for the diet model). Such an incorporation also provides a mechanistic basis for population and community processes, a basis often lacking in traditional approaches (Wilson 1976; Schoener 1986; Tilman 1986; Osenberg and Mittelbach 1990).

Most field studies testing optimal foraging models have been qualitative and are therefore of limited value as several models may predict the same qualitative pattern. Furthermore, to make predictions based on the classical diet model, it is necessary that encounter rates and

NATO ASI Series, Vol. G 20
Behavioural Mechanisms of Food Selection
Edited by R. N. Hughes

handling times have been estimated independently for each prey type, as independence is assumed in the model. This is an approach mainly restricted to the work of Werner and coworkers (Mittelbach 1981; Werner and Mittelbach 1981; Werner et al. 1983; Osenberg and Mittelbach 1990; but see Persson and Greenberg 1990b). The results of this group have generally been viewed as a strong support for the utility of optimal foraging models (i.e. the classical diet model) in predicting (and understanding) prey and habitat use of animals in the field (Werner and Mittelbach 1981; Werner 1984). The influential success of the approach has been related to the accuracy of prediction of optimal diets (despite the observed inclusion of non profitable prey) and habitat shifts (even though time delays between predicted and observed shifts have been common). A third factor used as a support for the usefulness of optimal foraging models for understanding population and community processes is the positive correlation between predicted net energy intake based on the diet model and individual growth performance (Mittelbach 1983; Mittelbach 1988, Mittelbach et al. 1988; Osenberg and Mittelbach 1990). One basic difference between approaches using foraging models as a tool for a mechanistic understanding of population processes and classical competition studies is that encounter rates (efficiencies) of different predators for different prey are considered instead of simple prey biomass/density estimates (see Werner and Hall 1977; 1979). In fact it may to some extent be suggested that it is this explicit consideration of species and size specific encounter efficiencies that at the field experimental level have been the most fruitful part of the interaction between behavioural, population and community ecology rather than optimal foraging per se.

An area where a mechanistic approach based on foraging theory is potentially rewarding is the prediction of niche shifts in organisms such as fish which undergo undergo substantial shifts in resource use through ontogeny (Mittelbach 1981; Gilliam 1982; Werner 1988; Persson and Greenberg 1990a, b). Such predictions may also include the effect of habitat specific predation risk (Gilliam 1982; Werner and Gilliam 1984; Werner and Hall 1988). The purpose of this paper is to discuss the degree to which optimal foraging models (i.e. the classical diet model) will increase our understanding of ontogenetic niche shifts compared to a mechanistic approach based on encounter efficiencies only (the term mechanistic is here and in the following text used in relation to population/community processes). The evaluation is based on an experimental study of ontogenetic niche shifts in perch (*Perca fluviatilis*). The experiment was carried out in a system lacking predators, hence the effects of predators on resource use by foragers were not considered. I will discuss the application of optimal foraging theory with respect to 3 aspects: diet breadth, habitat shifts and the net return rate-growth performance relationship.

Ontogenetic niche shifts in perch - a short summary

Perch potentially undergo substantial niche shifts during ontogeny. These shifts involve an increase in prey size related to an increase in body size. Young perch feed on pelagic microcrustaceans (zooplankton), then benthic macroinvertebrates, and finally fish if they become large enough (Johansson and Persson 1986; Persson 1988). Because these different prey types inhabit different habitats, the ontogenetic niche shifts may be relatively discrete (Persson and Greenberg 1990a). It has been experimentally demonstrated that roach (*Rutilus rutilus*), an efficient zooplanktivore, will affect the age and size at which perch shift from pelagic to benthic habitat by depressing pelagic prey availability to low levels. This roach density-dependent diet and habitat shift will in turn result in reduced growth rate of perch (Fig. 1) (Persson 1986; Persson and Greenberg 1990a).

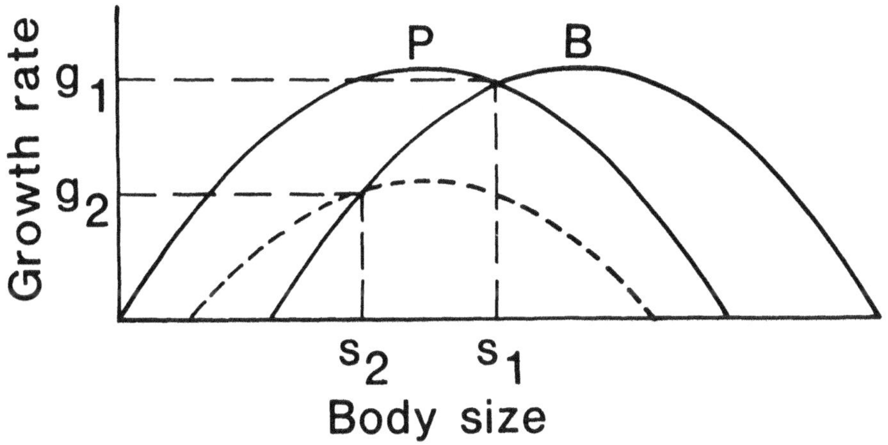

Fig. 1. Effect of planktivorous roach on ontogenetic niche shifts in perch. Solid curves represent growth rates in relation to body size for perch utilizing pelagic (P, zooplankton) and benthic (B, macroinvertebrates) habitats in the absence of the competitor (s_1 is the optimal size for shifting habitat with corresponding growth rate g_1). Dotted curve represents the effect of roach on the profitability of pelagic habitat (s_2 is the optimal size for shifting habitat with corresponding growth rate g_2).

Experimental design

An extensive description of the field experiment is given in Persson and Greenberg (1990a, b), hence only a very short background related to the principle experimental design is given here. The experiment was conducted in a shallow (0.8 m) pond (area 2000 m^2) that was divided into 9 enclosures, each with an area of 156 m^2. To each enclosure, we added young-of-the-year (0+) perch and one-year-old (1+) perch in the same densities in all enclosures. While the densities of perch were constant between enclosures, roach densities were varied between five different densities, each except the highest roach density replicated once. The addition of both 0+ and 1+ perch to the enclosures was done to study the effects of increasing roach density on size class interactions in perch, an issue which is outside the scope of this chapter (see Persson and Greenberg 1990a for this). The densities of the two size classes of perch and the highest density of roach used resembled that normally found in nearby highly productive lakes (Persson 1986a). Samples for diet and growth analyses of 0+ perch were taken twice, once in the middle and once at the end of the experiment. At the same time samples were taken of the two main prey types, pelagic (zooplankton) and benthic (macroinvertebrate) prey.

Roach is a very efficient forager on pelagic prey, but a poor forager on benthic macroinvertebrates. It was therefore possible to selectively remove zooplankton at different rates by varying the density of roach. The success of this selective reduction of pelagic prey was verified by significant roach density effects on pelagic prey biomass, while the biomass of benthic prey was similar in all enclosures (Persson and Greenberg 1990a). As a consequence of the roach-induced reduction in pelagic prey, we expected that perch in the enclosures with the highest roach densities would feed mainly on benthic macroinvertebrates, and perch in the enclosures with lowest roach densities would feed mainly on pelagic cladocerans. In addition, we expected that the roach density at which perch shifted to macroinvertebrates would decrease during the experiment due to the continuous depletion of pelagic prey by roach. In agreement with these predictions, 0+ perch in the enclosures with the two highest roach densities had shifted from feeding mainly on pelagic prey to feeding mainly on benthic prey by the middle of the experiment. By the end of the experiment, we observed the expected temporal effect as 0+ perch at the next lower roach density had also shifted to benthic prey. The frequency distributions of the proportions of pelagic prey in the diet in the different enclosures were bimodal on both sampling dates, suggesting that the shift from feeding mainly on pelagic to benthic prey was indeed discrete (Persson and Greenberg 1990a, b).

The foraging model

The optimal foraging model that was used to estimate net energy intake (J/s) of perch from the two habitats was presented by Mittelbach (1981). This foraging model is essentially the same as those developed by Charnov (1976) and several others (i.e. a basic prey model, Stephen and Krebs 1986), with costs for handling and searching explicitly stated in the currency of energy per unit time:

$$E_j/t = \frac{\sum_{i=1}^{n} \lambda_{ij} e_{ij} - C_{sj}}{1 + \sum_{i=1}^{n} \lambda_{ij} h_{ij}} \qquad (1)$$

E_j is energy intake in habitat j (J), t time (s) λ_{ij} is the encounter rate with prey size i in habitat j, e_{ij} is the net energetic gain from consuming prey size i in habitat j, C_{sj} is energetic cost of searching (J/s) in habitat j and h_{ij} is the time spent handling prey size i in habitat j. e_{ij} equals $Ac_{ij} - C_h h_{ij}$ where A is the assimable fraction minus proportional cost for Specific

Dynamic Action (SDA), c_{ij} is the energetic content of prey size i in habitat j and C_h is the energetic cost of handling prey i (J/s). Prey sizes and types were added to the model starting with the prey size with the highest relative profitability until E/t was maximized. The assimilable fraction was assumed to be 70% and the costs for SDA 14% of the energy ingested (Elliott 1979; Rice et al. 1983). The energetic cost of handling and searching were derived from the equations given in Rice et al. (1983) and Persson (1987).

The above foraging model was used to quantitatively predict (1) the optimal diet breadth for perch feeding in the pelagic and benthic habitat and (2) the pelagic prey density at which 0+ perch were expected to shift from pelagic to benthic feeding (Mittelbach 1981, Werner et al. 1983).

Predation cycle and laboratory experiments

To generate quantitative predictions of optimal diet breadths and habitat shifts, encounter rates and handling times for the most common prey types in the two habitats were estimated in laboratory experiments. The basic prey models assume that encounter rate and handling time for different prey types are independent, but independent estimates may be difficult or even impossible to obtain in the field. An advantage with using laboratory estimated encounter rates for each prey type and size separately is that this procedure will provide the required independent estimates of encounter rates and handling times. A disadvantage is that it is only practicable to carry out experiments with a limited number of prey taxa and sizes and hence extrapolations of these laboratory-estimated encounter rates and handling times to the field may be crude.

Prey use by a predator is essentially a function of 3 important components: (1) the probability of encounter between predator and prey, (2) the probability that the predator attacks an encountered prey, and (3) the probability that an attack results in the consumption of the prey (O'Brien 1979; Greene 1983; 1986; Osenberg and Mittelbach 1990). The first component sets the background against which the animal's selection of prey will take place. This component is also in many cases the most difficult to measure empirically (see Osenberg and Mittelbach 1990 for an extensive discussion of the problem). In the present study, laboratory estimated attack rates were used to estimate encounter rates. Because attack rate is a function of both the probability of encountering a prey item and the probability that an encountered prey item is attacked, several precautions were taken to make it almost certain that an encountered prey was attacked. First, the laboratory experiments were run for each prey type and prey size separately to minimize the confounding influence of decision making by the forager (i.e. all prey in a trial had the same relative profitability). Second, fish were deprived of food for 24 hrs before each trial to increase the motivation to capture an encountered prey. Each trial was also only run for a short period to minimize satiation effects. Third, the fish had been trained on each prey type and size for a week to maximize encounter efficiency (see Mittelbach 1981; Persson 1987).

Laboratory experiments were run for both zooplankton and benthic macroinvertebates. Two prey types from each habitat were used: for pelagic prey the cladoceran (*Daphnia*) and the copepod (*Cyclops*) and for benthic prey *Sialis* and chironomids. Cyclopoid copepods differ from cladoceran zooplankton in being more capable of evading a predator attack and, as a consequence, the handling time is longer and attack coefficient lower for this prey type compared to cladocerans (Persson 1987). Although they both are more cryptic than pelagic prey, *Sialis* and chironomids differ in that chironomids burrow deeper into the sediment. Other prey in the benthic habitat were classified as more like one of the above benthic groups depending on their habitat microdistribution (see Persson and Greenberg 1990b).

Table 1. (a) An example for perch of the relation between prey encounter rate (λ_{ij}) and prey density (D), prey size (l) and fish length (L). The multiple regression for encounter rate had the following form: $\ln \lambda_{ij} = a+b*\ln D+c*\ln l+d*\ln L$. (b) The relation between handling time and the ratio prey length to predator length. The equation was: $h = a \exp[b(l/L)10^3)]$. (Data from Persson and Greenberg 1990b).

		Overall significance		
(a) Encounter rate (*Sialis*)		F	P	n
a ± 1SE	-19.3±1.95	145.3	<0.0001	113
b ± 1SE	1.088±0.053			
c ± 1SE	2.189±0.547			
d ± 1SE	1.045±0.249			
(b) Handling time (*Sialis*)		F	P	n
ln a ± 1SE	-0.345±0.159	139.9	<0.0001	215
b ± 1SE	11.01±0.93			

For pelagic prey, a Holling type II functional response curve was fitted to the data:

$$C=\frac{ax}{1+ah} \tag{2}$$

where C is capture rate, a the attack coefficient, x prey density and h handling time. The encounter rate (λ) equals ax. Handling times, which were too short to be measurable were estimated as the reciprocal of the maximum feeding rate (Ware 1972; Mittelbach 1981; Persson 1987). For benthic prey, empirical fits of encounter rate as a function of prey density were derived from the relationship between capture rate and prey density (Persson 1987). Encounter rates were operationally defined as the number of prey captured divided by total trial time minus time spent handling. For benthic prey, handling times were measured directly.

By varying prey density, prey size and predator size for each prey type, encounter rates could be obtained as a multiple function of these variables (for an example, see Table 1). The multiple regressions for the different prey types were based on more than 430 experiments (Persson 1987, Persson and Greenberg 1990b). When applied to predict diet breadths and net return rates from the two habitats, encounter rates and handling times were adjusted for temperature conditions in the field on the sampling dates (see Persson and Greenberg 1990b).

Growth performance in relation to resource bases in the two habitats

One argument for the usefulness of the diet model has been that a substantial degree of the variation in the forager´s growth performance has been explained by their predicted foraging net return rates (Mittelbach 1983, 1988; Mittelbach et al. 1988; Osenberg and Mittelbach 1990). However, a large part of the variance in the prey use by the predator may be explained by considering size-specific encounter rates only (Osenberg and Mittelbach 1990; Sih this Vol.). Considering this issue, I compared the predictive power of a regression of perch growth rate on resource base for these different methods of calculating: (1) biomass of prey, (2) net energy intake assuming that perch captured every prey encountered (encounter return rate) and , (3) net energy intake assuming that perch optimized its energy intake (optimal foraging return rate).

The situations in the middle and at the end of the experiment differed qualitatively from each other with respect to habitat-dependent growth performance. In the middle of the experiment a significant amount of the variation in growth rate could be related to pelagic prey resource levels, while no effect of benthic resource levels was present. Conversely, a significant amount of the variation in perch growth rate during the second half of the experiment was explained by benthic prey resources, while this was not the case for pelagic prey (Persson and Greenberg 1990b). The shift from pelagic to benthic prey resources being the most important factor for perch growth performance was partly related to a very low return rate from pelagic prey in most enclosures on the latter sampling date. To examine the influence of pelagic resource availability on perch growth rate, the comparison between prey biomass, encounter return rate and optimal foraging return rate as predictors of perch growth rate was restricted to the first half of the experiment. Analogously, the comparison regarding benthic prey was restricted to the second half of the experiment.

For pelagic prey, the regression of growth on net return rates based on optimal foraging considerations gave a slightly better fit (r^2=0.52, F=7.4, P=0.03) than a fit based on encounter return rate (r^2=0.51, F=7.3, P=0.03), which in turn gave a slightly better fit than a regression with prey biomass as the independent variable (r^2=0.49, F=6.7, P=0.04) ((Fig. 2a). However, as is apparent from the r^2 values, the improvements were minimal and far from significant.

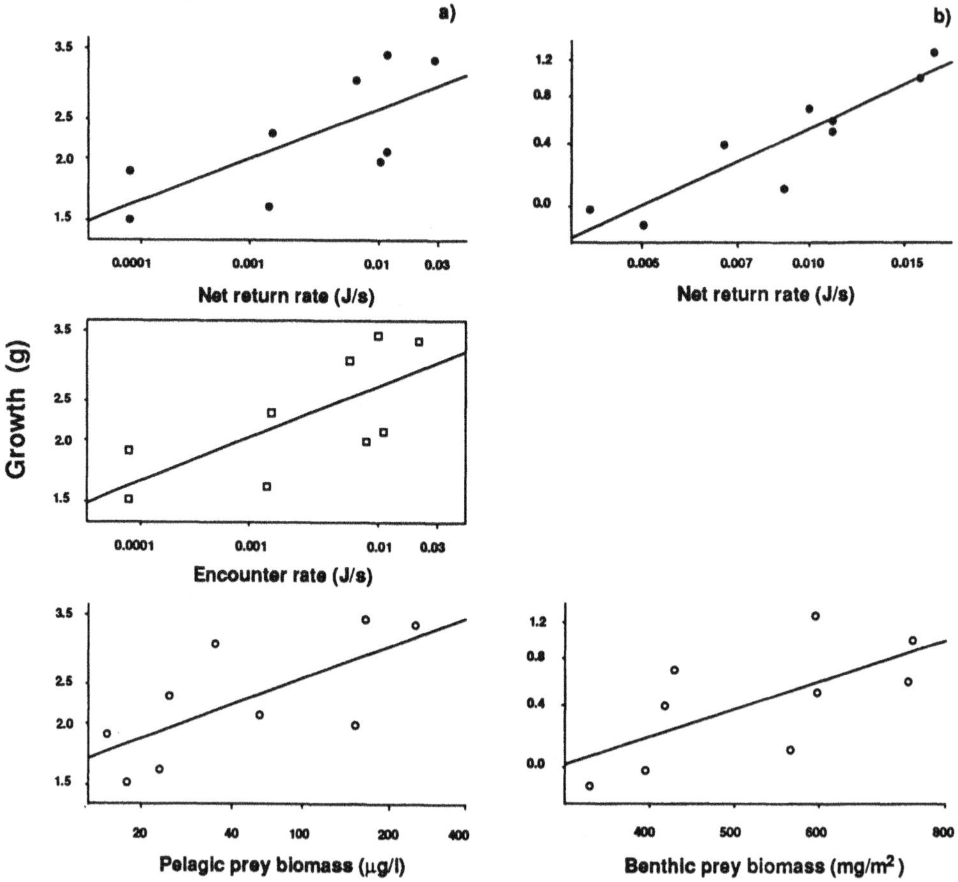

Fig. 2. a) Growth rate of perch during the first half of the experiment as a function of optimal foraging return rate (above), encounter return rate (middle) and biomass of pelagic prey (bottom). **b)**. Growth rate of perch during the second half of the experiment as a function of optimal foraging return rate (above) and biomass of benthic prey (bottom).

Perch were in most cases expected to include even the smallest size class of benthic prey sampled (see next section and Persson and Greenberg 1990b). For benthic prey, estimates of optimal foraging return rates will thus be the same as estimates of encounter return rates. Consequently, the comparison was in this case restricted to a comparison of net return rates and resource availability expressed as biomass. In contrast to the situation for pelagic prey, a significantly higher amount of the variation in perch growth rate was explained when net return rate was used as the independent variable ($F=16.8$, $P<0.001$) (Fig. 2b).

In conclusion, the data lend support to the idea that an inclusion of predator encounter rate for prey of different types and sizes will improve predictions of forager growth performance compared to predictions based on simple prey biomasses. The existence of this improvement is, at least according to the data presented here, dependent on prey crypticity. For prey with relatively low crypticity (i.e. pelagic prey) the improvement in prediction is only slight, while for the more cryptic benthic prey, the improvement can be quite substantial.

Table 2. Mean sizes (mm) of pelagic prey (cladocerans) in the environment, as encountered and in the diet of perch in enclosures where perch were feeding mainly on pelagic prey on the two sampling dates. For calculations of mean sizes of prey as encountered and as ambient in the environment, size distributions have been truncated at a lower size limit of 0.3 mm (the smallest size the perch could handle). Encountered size distributions have been calculated based on laboratory experiments in which different sizes of cladocerans were used (see Persson and Greenberg 1990b).

Enclosure	Date					
(roach density g/m^2)	September			October		
	Env	Enc	Diet	Env	Enc	Diet
0	0.39	0.47	0.65	0.40	0.45	0.70
0	0.41	0.49	0.62	0.45	0.48	0.59
5	0.46	0.52	0.62	0.40	0.52	0.61
5	0.36	0.40	0.57	0.39	0.45	0.64
10	0.41	0.49	0.72			
10	0.43	0.47	0.49			

Passive consumption or active selection

Although a few studies have gone as far as to discuss the growth consequences of foraging behaviour in relation to the diet model, the overwhelmingly majority of foraging studies have restricted their attention to diet and habitat use. The first question to ask with respect to diet use is whether an active prey selection by 0+ perch occurred or not. We have previously shown that the mean size of cladocerans in the diet of perch decreased with increasing roach density (Persson and Greenberg 1990b). Although indicative, these data still do not prove that perch actively selected larger pelagic prey when their availability was high, because perch taking larger pelagic prey at low roach densities than at high could simply reflect a changed resource size spectrum along the density gradient of roach. However, comparing the mean size of cladocerans in the diet in enclosures where perch fed mainly on pelagic prey shows that the mean size in the diet on both sample dates was significantly larger than both the mean size in the environment and the mean size encountered (Table 2). Furthermore, the ratio mean size in diet/mean size encountered decreased with increasing roach density (September F=5.9, P=0.046; October F=7.3, P=0.03).

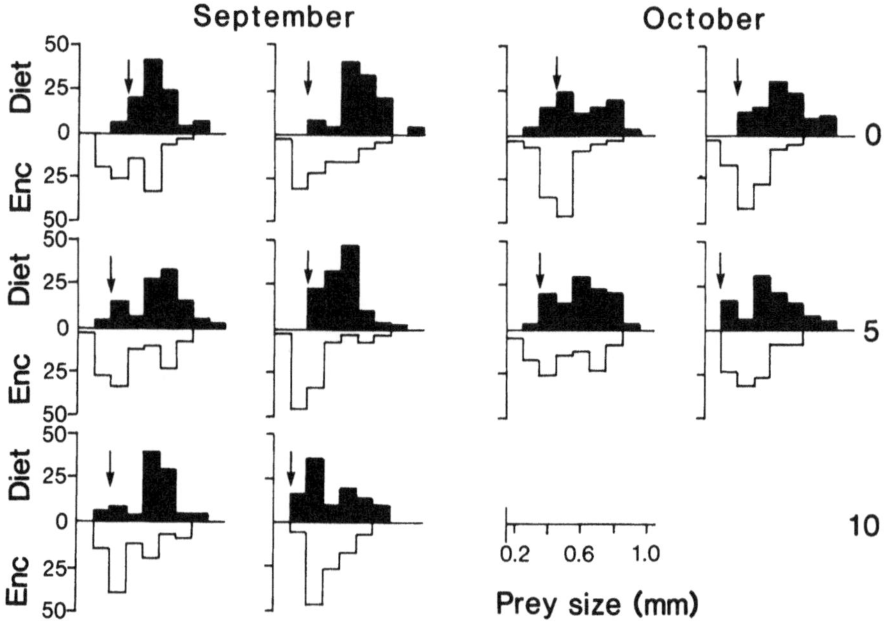

Fig. 3. Size distributions of pelagic cladocerans in the diet of perch and as encountered in the environment in enclosures where perch fed mainly on pelagic prey (0, 5, 10 g roach/m^2 in September; 0, 5 g roach/m^2 in October). Arrows denote predicted smallest prey size to be included in the diet. For method of calculating encountered size distributions, see Table 2.

The more complete test of food selection is to compare the observed smallest prey size eaten by perch with that predicted from eqn 1. Such comparisons were made for pelagic prey in the enclosures where perch were feeding mainly on pelagic prey. Similarily, comparisons were made for benthic prey in enclosures where perch were feeding mainly on benthic prey. The comparison with respect to benthic prey turned out to be of little utility, because the smallest prey size predicted to be included in the diet was usually the smallest prey size sampled in the environment (Persson and Greenberg 1990b). The comparison of the smallest size class of cladocerans predicted to be ingested and that observed, generated a close agreement between prediction and observation (Fig. 3). (An inclusion of unprofitable cladoceran sizes up to 5% in the total diet of cladocerans was not viewed as an example of partial preferences, see Persson and Greenberg 1990b). In only 1 out of the 8 cases where perch had ingested a larger smallest size class than that constrained by their physical ability (i.e. larger than 0.3 mm), did perch include a significant amount of size classes not predicted to be included in the diet (Fig. 3). Considering pelagic prey use alone, our data thus can be interpreted as a strong support for the usefulness of the basic diet model. However, the picture becomes more complex when profitabilities of the two habitats are included in the analyses.

The basic diet model and prediction of habitat use

Besides the relation between growth rate and net return rate and diet patterns, accurate predictions of habitat shifts have been advanced as a support for the usefulness of applying foraging models to the field situation (Mittelbach 1981; Werner and Mittelbach 1981; Werner et al. 1983; Benkman 1987). In general these studies on habitat shifts are impressive. However, they have not clearly distinguished between the improvements in prediction resulting from a consideration of encounter rate with different prey types compared with predictions derived from a consideration of just prey density or biomass changes (mechanistic approach with no optimality argument) and predictions obtained by the additional inclusion of optimal foraging (but see Osenberg and Mittelbach 1990 for prey size use). A mechanistic approach in the former sense only sets explicit constraints on the forager´s capacity on different prey types.

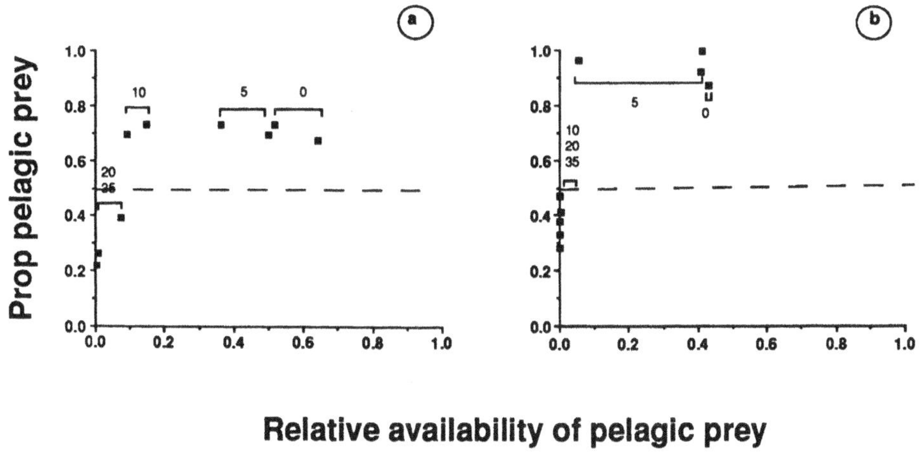

Fig. 4. Relative proportion of pelagic prey in the diet of perch versus the ratio [optimal foraging return rates from pelagic prey / (optimal foraging return rates from pelagic prey plus optimal foraging return rates from benthic prey)] in September (a) and October (b). Perch were expected to shift habitat at a relative availability of pelagic prey of 0.5. Numbers given in figure are roach densities (0, 5, 10, 20, 35 g/m^2, each density replicated once except highest roach density).

The above experiments have largely dealt with a situation where prey and habitat use have been studied in only one treatment with resource levels varying over time. In contrast, in our experiment different treatments were subjected to different levels of resource depletion (of pelagic prey) yielding, besides a temporal effect (by sampling twice), the possibility to compare different treatments at the same date.

A forager with complete information on resource levels in different habitats is generally expected to shift habitat when the return rate in the habitat presently being utilized is reduced below the return rate for an alternative habitat (Mittelbach 1981; Werner et al. 1983; Stephens and Krebs 1986). Qualitatively, the shift in habitat use of perch from pelagic to benthic habitat took place as expected. However, perch shifted from the pelagic habitat when its profitability relative to the benthic habitat was much lower than predicted (Fig. 4). Furthermore, the accuracy in prediction of habitat shifts based on encounter return rates only was not less than that based on optimal foraging return rates. Finally, besides the observed time lag between predicted and observed habitat shift the habitat shift was in addition never total, with partial preferences observed (Fig. 4).

Foraging tactics and optimal diet and habitat use

Given the restricting assumptions of basic prey models (see Stephens and Krebs 1986), the close agreement between predicted and observed smallest size class of cladocerans in the diet is noticeable. Several hypotheses can be advanced to explain this. First, predictions may turn out to be not that sensitive to violations of certain assumptions. Second, animals foraging in the field may have more complete information about actual resource levels than animals in the laboratory (see Schoener 1987). Finally, the good correspondence between predictions and observations could result from the predicted lower size class being close to the smallest prey size that perch of this size could handle (see Persson and Greenberg 1990b). A caveat is that the agreement between predicted and observed diet use is based on the assumption that (1) perch encountered pelagic prey with maximum efficiency (since the laboratory experiments were carried out under this condition) and (2) that perch resource use while foraging on pelagic prey was not influenced by the ingestion of prey from the benthic habitat (potentially profitable prey as e/h was generally higher for benthic prey).

The second assumption relates to the fact that perch never restricted their resource use completely to either habitat (Fig. 4). This means that if the optimal foraging tactic differed for the prey of the two habitats, the tactics may have interfered with each other. One aspect in which pelagic prey differ from benthic prey is that the latter are sediment-living and more cryptic to the predator. Complementing this, it has been shown that perch swimming speed while foraging on pelagic prey (*Daphnia*) is higher than that when feeding on benthic prey (Persson 1987). The swimming speed of perch in the pond experiment was quantified and it was found that swimming speed increased with return rates from pelagic prey but decrease at the highest levels (Fig. 5). The majority of theoretical models on search rates have focused on the costs and gains of different search rates where prey are viewed as energy packages all with the same energy content (Ware 1975, Norberg 1977, Sih 1984, Speakman 1986). There are differences in predictions of different models with respect to whether the search rate should increase monotonically or decrease at high resource levels (as we found for perch). However, there is a general agreement that search rate should decrease at low resource levels. As pelagic prey are less cryptic than benthic prey, the observed swimming speed pattern of perch could also be explained by different prey crypticity of pelagic and benthic prey (Andersson 1980, Gendron and Staddon 1983). Irrespective of which explanation is most likely, the observed variation in search rate is likely to affect the probability of encountering different prey types. A forager searching at a swimming speed that is intermediate between the optimum for different prey types (and/or different habitats) may reduce to below the maximum the probability of encountering a prey of either prey type. On the other hand, prey with high relative profitabilities may become available from the other prey type which could increase the total net return rate. Partial habitat use has been commonly assumed to be a sampling program by which the forager obtains information about resource levels in alternative habitats while time delays are commonly related to learning phenomena. It has been suggested elsewhere that search rate dependent encounter probabilities affect both the interpretation of partial habitat uses and delays between predicted and observed habitat shifts (Persson and Greenberg 1990b). Partial preferences may occur because they lead to a higher return rate than if the forager had allocated its time exclusively to either prey type (Persson and Greenberg 1990b). The important point in this context is that net return rates calculated on the assumption of exclusive habitat uses may be in error. Furthermore, where perch searched at a suboptimal (intermediate) swimming speed for each prey type, the close agreement that was observed between predicted and observed diet of pelagic cladocerans cannot be used as a unequivocal support for the usefulness of the basic diet model, as the predictions were based on exclusive habitat use. This is especially the case if the perch restricted their diet breadth in each habitat as a consequence of using both habitats.

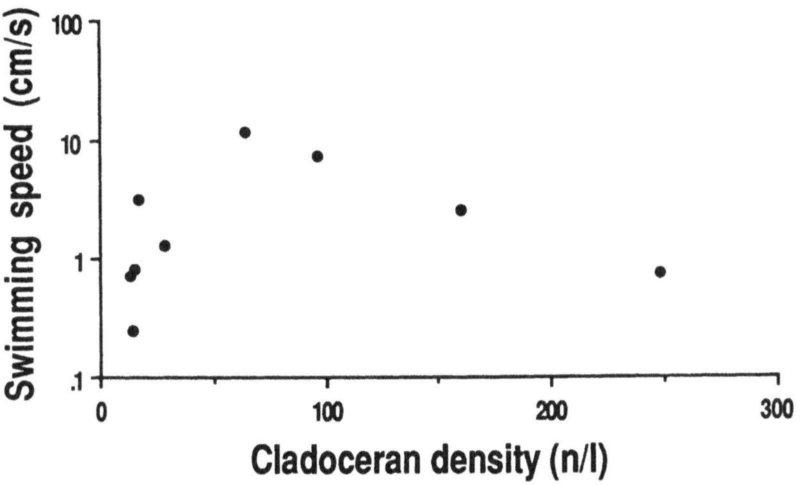

Fig. 5. Swimming speed of perch as a function of pelagic prey density.

General discussion

In population and community ecology there has been an increasing effort to introduce mechanistic models - based on concepts from behavioural ecology, physiological ecology and /or ecological morphology - when analyzing population and community processes (Moermond 1986; Schoener 1986; Tilman 1986). In some respects, the early efforts of MacArthur (MacArthur and Pianka 1966; MacArthur 1972) can be viewed as a first attempt to merge behavioural and population/community ecology (see also Werner 1984; 1986; Osenberg and Mittelbach 1990) The current focus on strictly behavioural issues has also lead to the introduction of more and more variables thought to affect animal foraging. From a population and community point of view it can be questioned if it is feasible, at least at present, to incorporate into population models all the aspects of animal foraging considered by the behavioural ecologist. The situation can be compared to the prevalent use of the logistic growth equation in population ecology: there are a number of unrealistic assumptions in the equation and there are more realistic (and more complex) alternatives, still the logistic equation is generally used in predator prey models because of its convenient mathematical properties. Similarily, in the study of coevolution of character displacement, explicit quantitative genetic models are considered preferable because they include the underlying genetic mechanisms. Still, due to the intensive numerical computations that quantitative genetic models require, there are severe practical limitations (even with today´s computer facilities) to using these quantitative genetic models (Tapper 1987).

The question regarding the relationship between behavioural and population/community ecology can be stated at a more general and epistemological level: To what extent can entities and processes at the population and community levels be reduced to behavioural entities and processes? To answer this question, it is useful to distinguish three different kinds of reductionism: (1) ontological which proposes that behavioural entities and processes underlay all population processes, (2) epistemological in which population models may be built on behavioural ecological concepts or variables and (3) methodological reductionism which postulates that the most productive way to study population and community processes is always to investigate the underlying behavioural processes (for a detailed treatment of these different kinds of reductionism, see Dobzhansky et al. 1977). Those who have argued that a reduction of population/community ecology to behavioural ecology is possible have focused on the ontological and epistemological aspects of reductionisms (i.e. Schoener 1986). However, even though ontological and to some extent epistemological reduction may be possible, methodological reductionism may not be.

While a behavioural ecologist may argue that it is a waste of time to pursue the basic prey models in the light of what we know about behaviour, this may not be the case for a population ecologist. If we return to the predation cycle, an explicit inclusion of species and size specific encounter rates for different prey types will make explicit the constraints on foraging. For example, compared to the nonmechanistic Lotka-Volterra equations, inclusion of these constraints will lead to a higher realism and accuracy in predicting habitat partitioning and specialization, including ontogenetic niche shifts. Perhaps even more important is that an understanding of the processes leading to different patterns will be provided (Schoener 1986). If resource distributions and levels are known, it will also be possible to predict abundance relationships between species and among size groups within species in different habitats and systems (see Werner and Hall 1977, 1979; Mittelbach 1984; Persson 1987, 1988; Werner and Hall 1988 for examples). In this respect the interaction between foraging theory and population/community ecology has thus already proven useful in understanding population/ community statics.

If optimal foraging tactics for different resource types do not interfere my diet data also provides partial support for the accuracy of the classical diet model to predict diet breadths of foragers in the field, although the agreement between predicted and observed habitat shift was only qualitative. From a population/community point of view, these demonstrated potentials and limitations in the model should be evaluated in relation to previous population models incorporating diet selection behaviour in the predator functional response. Previous models have shown that switching behaviour in the predator will have a stabilizing effect on predator-prey interactions, the extent to which depends on other parameters (Murdoch and Oaten 1975; Oaten and Murdoch 1975; Armstrong 1976; Taylor 1984). Incorporating the diet model into a one predator - two prey population model suggests that optimally foraging predators can have a stabilizing effect too and tend to dampen oscillations in intrinsically unstable predator-prey systems (Gleeson and Wilson 1986). However, the conditions for coexistence are more complex than in predator-prey models with traditional switching behaviour in the predator. The behaviours of switching models and the diet model differ in two major respects: (1) in the diet model the competitively superior prey must also be the most profitable to the predator and (2) the absolute abundance (of the profitable prey) is the critical variable compared to the relative abundance in switching models. Another difference is that with an optimal foraging predator, the one predator-two prey system may sometimes function as a one predator-one prey system. Switching and optimal foraging predator-prey models thus have both similarities and dissimilarities in their dynamic behaviours. A more complete evaluation of whether the diet model means a qualitative step forward in modelling predator-prey interactions must be based on experimental studies to compare which of the two mechanisms, switching or optimal foraging, give the most accurate and realistic predictions. Such experimental studies are almost lacking, but form a necessary link in future interactions between behavioural and population ecology. Such studies should also consider whether other population models including optimal foraging predators (cf. Abrams 1982, 1984) may prove more fruitful than the diet model.

Conclusions

The experiment presented here lend support to the accuracy of the diet model in predicting pelagic prey diet breadths of perch. However, predictions of habitat shifts were only qualitatively correct, with a time lag occurring between predicted and observed habitat shift. Different optimal search rates for prey of different crypticity may explain the observed patterns in addition to the commonly invoked explanations of learning and sampling. Although a deviation between quantitative predictions of habitat shift and the observed was present, the usefulness of the basic diet model should be related to its alternatives in traditional population ecology. Predictions derived from population models based on the diet model differ in some respects from previous models. Experimental studies are necessary to evaluate whether incorporating the diet model into population models means a qualitative step forward in our understanding of population and community processes.

Acknowledgements. I thank John B. Ball, Frietson Galis and Paul J. B. Hart for comments on a previous draft of the manuscript. The research has been sponsored by the Swedish Natural Science Research Council. The Swedish Natural Science Research Council also provided the travelling grant that made possible my presence at the workshop on Behavioural Mechanisms of Food Selection.

References

Abrams PA (1982) Functional responses of optimal foragers. Am Nat 120:382-390

Abrams PA (1984) Foraging time optimization and interactions in food webs. Am Nat 124:80-96

Andersson M (1980) On optimal predator search. Theor Pop Biol 19:58-86

Armstrong RA (1976) The effects of predator functional response and prey productivity on predator-prey stability: a graphical approach. Ecology 57:609-612

Benkman CW (1987) Food profitability and the foraging ecology of crossbills. Ecol Monogr 57:251-267

Charnov E L (1976) Optimal foraging: attack strategy of a mantid. Am Nat 110:141-151

Comins HN, Hassel MP (1979) The dynamics of optimally foraging predators and parasitoids. J Anim Ecol 48:335-351

Dobzhansky T, Ayala FJ, Stebbins GL, Valentine JW (1977) Evolution. WH Freeman and Company, San Fransisco

Elliott JM (1979) Energetics of freshwater teleosts. In: Miller PJ (ed) Fish phenology: anabolic adaptiveness in teleosts. Symposia of the Zoological Society of London no 44, pp 29-61

Emlen JM (1966) The role of time and energy in food preference. Am Nat 100:611-617

Gendron RP, Staddon JER (1983) Searching for cryptic prey: the effect of search rate. Am Nat 121:172-186

Gilliam JF (1982) Habitat use and competitive bottlenecks in size-structured populations. Thesis, Michigan State Univ, East Lansing

Gleason SK, Wilson DS (1986) Equilibrium diet: optimal foraging and prey coexistence. Oikos 46:139-144

Greene CH (1983) Selective predation in freshwater zooplankton communities. Int Rev gesamten Hydrobiol 68:297-315

Greene CH (1986) Patterns of food selection: implications of predator foraging tactics. Am Nat 128:824-839

Johansson L, Persson L (1986) The fish community of temperate eutrophic lakes. In: Riemann B, Søndergaard M (eds) Carbon dynamics of eutrophic, temperate lakes: The structure and functions of the pelagic environment. Elsevier, Amsterdam, pp 237-266

Krebs JR, McCleery RH (1984) Optimization in behavioural ecology. In. Krebs JR, Davies NB (eds) Behavioural Ecology: An Evolutionary Approach. Blackwell Scientific Publications, Oxford, pp 91-121

MacArthur RH, Pianka ER (1966) On optimal use of a patchy environment. Am Nat 100:603-609

MacArthur RH (1972) Geographical Ecology: Patterns and Distributions of Species. Harper & Row, Publishers, Inc., New York

Mangel M, Clark CW (1986) Towards a unified foraging theory. Ecology 67:1127-1138

McNamara JM, Houston AI (1986) The common currency for behavioral decisions. Am Nat 127:358-378

Mittelbach GG (1981) Foraging efficiency and body size: a study of optimal diet and habitat use in bluegill sunfish. Ecology 62:1370-1386

Mittelbach GG (1983) Optimal foraging and growth in bluegills. Oecologia (Berlin) 59:157-162

Mittelbach GG (1984) Predation and resource partitioning in two sunfishes (Centrarchidae). Ecology 65:499-513

Mittelbach GG (1988) Competition among refuging sunfishes and effects of fish density on littoral zone invertebrates. Ecology 69:614-623

Mittelbach GG, Osenberg CW, Leibold MA (1988) Trophic relations and ontogenetic niche shifts in aquatic ecosystems. In: Ebenman B, Persson L (eds) Size Structured Populations - Ecology and Evolution. Springer Verlag, Berlin, pp 219-235

Moermond TC (1986) A mechanistic approach to the structure of animal communities: *Anolis* lizards and birds. Am Zool 26:23-37

Murdoch WW, Oaten A (1975) Predation and population stability. Adv Ecol Res 9:1-131

Norberg RÅ (1977) An ecological theory on foraging time and energetics and choice of optimal food-searching method. J Anim Ecol 46:511-529

Oaten A, Murdoch WW (1975) Functional response and stability in predator-prey systems. Am Nat 109:289-298

O'Brien WJ (1979) The predator-prey interaction of planktivorous fish and zooplankton. Am Sci 67:572-581

Osenberg CW, Mittelbach GG (1990) The effects of body size on predator-prey interaction between pumpkinseed sunfish and gastropods. Ecol Monogr (in press)

Persson L (1986) Effects of reduced interspecific competition on resource utilization of perch (*Perca fluviatilis*). Ecology 67:355-364

Persson L (1987) The effects of resource availability and distribution on size class interactions in perch *Perca fluviatilis*. Oikos 48:148-160

Persson L (1988) Asymmetries in predatory and competitive interactions in fish populations. In: Ebenman B, Persson L (eds) Size Structured Populations - Ecology and Evolution. Springer Verlag, Berlin, pp 203-218

Persson L, Greenberg LA (1990a) Juvenile competitive bottlenecks: the perch (*Perca fluviatilis*)-roach (*Rutilus rutilus*) interaction. Ecology (in press)

Persson L, Greenberg LA (1990b) Optimal foraging and habitat shifts of perch (*Perca fluviatilis*) in an experimentally-induced resource gradient. Ecology (submitted)

Rice JA, Breck JE, Bartell SM, Kitchell JF (1983) Evaluating the constraints of temperature, activity and consumption on growth of largemouth bass. Env Biol Fish 9:263-275

Schoener TW (1986) Mechanistic approaches to community ecology: a new reductionism? Am Zool 26:23-37

Schoener TW (1987). A brief history of optimal foraging theory. In: Kamil, AC, Krebs, JR, Pulliam HR (eds), Foraging Behavior. Plenum Press, New York and London, pp 5-67

Sih A (1984) Optimal behavior and density-dependent predation. Am Nat 123:314-326

Speakman JR (1986) The optimum search speed of terrestrial predators when feeding on sedentary prey: a predictive model. J theor Biol 122:401-407

Stephens DW, Krebs JR (1986) Foraging Theory. Princeton University Press, Princeton, New Jersey

Tapper ML (1987) The coevolution of resource competition: Appropriate and inappropriate models of character displacement. Bull Soc Popul Ecol No 44:45-53

Taylor, RJ (1984) Predation. Chapman & Hall, New York London

Tilman D (1986) A consumer-resource approach to community structure. Am Zool 26:23-37

Ware DM (1972) Predation by rainbow trout (*Salmo gairdneri*): the influence of hunger, prey density, and prey size. J Fish Res Board Can 29:1193-1201

Ware, DM (1975) Growth, metabolism, and optimal swimming speed of a pelagic fish. J Fish Res Board Can 32:33-41

Werner EE (1984) The mechanisms of species interactions and community organization in fish. In: Strong DR, Simberloff D, Abele LG, Thistle AB (eds), Ecological Communities Conceptual Issues and the Evidence. Princeton University Press, Princeton, pp 360-382

Werner EE (1988) Size, scaling, and the evolution of complex life cycles. In: Ebenman B, Persson L (eds) Size Structured Populations - Ecology and Evolution. Springer Verlag, Berlin, pp 60-81

Werner EE, Gilliam JF (1984) The ontogenetic niche and species interactions in size-structured populations. Ann Rev Ecol Syst 15:393-425

Werner EE, Hall DJ (1977) Competition and habitat shift in two sunfishes (Centrarchidae). Ecology 58:869-876

Werner EE, Hall DJ (1979) Foraging efficiency and habitat switching in competing sunfishes. Ecology 60:256-264

Werner EE, Hall DJ (1988) Ontogenetic habitat shifts in the bluegill sunfish (*Lepomis macrochirus*): The foraging rate-predation risk tradeoff. Ecology 69:1352-1366

Werner EE, Mittelbach GG (1981) Optimal foraging: field tests of diet choice and habitat switching. Am Zool 21:813-829

Werner EE, Mittelbach GG, Hall DJ, Gilliam JF (1983) Experimental tests of optimal habitat use in fish: the role of relative habitat profitability. Ecology 64:1525-1539
Wilson DS (1986) Deducing the energy available in the environment: an application of optimal foraging theory. Biotropica 8:96-103

ORAL DISCUSSION – L. Persson

Persson explained that the ontogenetic shifts in fish diet can occur either gradually or precipitously. Secondly, he explained that his experiments did reveal competitive interaction between size and age classes of perch. He also reported that in situations where perch switched later than otherwise from zooplankton to benthic invertebrate diets, larger zooplankton were usually available.

THE IMPACT OF DIFFERENT GROWTH PATTERN ON THE UTILIZATION OF TREE SPECIES BY A GENERALIST HERBIVORE, THE MOOSE *ALCES ALCES*: IMPLICATIONS OF OPTIMAL FORAGING THEORY

BERNT–ERIK SAETHER
Norwegian Institute for Nature Research, N–7004 TRONDHEIM, Norway.

INTRODUCTION

Plants cannot flee from their predators. Instead they gain protection from attack by making themselves inedible. Many plants are covered with a spiny or tough surface. Others avoid herbivores by producing toxic chemical compounds which are distasteful. Although many of these characters may serve a variety of functions, their importance in deterring herbivores is generally obvious.

It is, however, difficult for the plant to achieve complete protection against herbivores through chemical or mechanical protectiv agents. For instance, in northern boreal forests mammalian herbivores are able to feed regularly on a large proportion of the available tree species (e.g. Ahlen 1975, Cederlund et al. 1980, Hagen 1983). In this paper I will suggest that the growth form of the tree species will strongly influence the impact of browsing by large mammalian herbivores.

Great differences are found in the pattern of growth of woody plants. In environments with low availability of resources the plants grow slowly but are long–lived (Chaplin 1980, Bryant et al. 1983). Their requirements for light and nutrients are low. These plants are therefore able to successfully grow and reproduce even at high densities. In evergreens, the leaves constitute a major storage organ (Chaplin et al. 1980). In contrast, species growing on fertile soils have been selected for more rapid growth, in order to successfully compete for the available light and minerals (Bryant et al. 1983). These plants are more limited by the rate of allocation of carbon to growth than by nutrients and may soon grow beyond the range of browsing mammals.

From a single tree's point of view, two aspects of the herbivore's foraging behaviour are important. Firstly, what is the probability of being detected and browsed ? Secondly, if browsed, how much of the tree's biomass will be removed? I hypothesize, based on qualitative predictions from an optimal foraging model (Vivas et al. in prep.), that the differences in the growth form of the tree species will influence the foraging behaviour of the herbivore and thereby

NATO ASI Series, Vol. G 20
Behavioural Mechanisms of Food Selection
Edited by R. N. Hughes

the impact of herbivory. Consequently, because of the differences in growth pattern, the impact of browsing in relation to height and density will differ between tree species. Such relationships will significantly influence the interactions between plants and herbivores (Coley et al. 1985, Fagerstrom 1989).

In this chapter I will examine these questions by comparing the browsing pattern of individual moose *Alces alces* on three browse species: the coniferous species *Pinus sylvestris* and the two deciduous species *Betula pubescens* and *Sorbus aucuparia*. The latter represents a fast–growing species, adapted to fertile soils, whereas the former is an evergreen which in my study area grows on poor dry soils. *B. pubescens* is assumed to represent a species with growth characteristics intermediate between these two species.

THE SYSTEM

During the summer, moose feed on large herbs (e.g. *Chamaenerion angustifolium*), grasses and leaves, especially of *Betula* spp. (Cederlund et al. 1980, Bergstrom and Hjeljord 1987). During winter the moose eat twigs. The most important browse species are *P. sylvestris*, *B. pubescens* and *Salix* spec. (Cederlund et al. 1980, Hagen 1983, Bergstrom and Hjeljord 1987).

Recently, the Fennoscandian moose has dramatically increased in number (Cederlund and Markgren 1987). In many areas, moose have become so numerous that the availability of winter food is limiting. For instance, variations in snow conditions between two years were sufficient to markedly reduce the condition of calves (Saether and Gravem 1988).

This study was conducted during the years 1986–1988 in a southern (Asnes, about 60° 30'N 12° E) and a northern study area (Troms, about 69° N 18° 30'E). In both areas the moose population increased rapidly in the beginning of the 1980s. Consequently, the winter food supply declined, resulting in extensive damage to the commercial forest in the Asnes study area.

The species composition of the browse species differed greatly between the study areas. In the Troms study area the main species were *Salix* sp., *B. pubescens*, *Prunus padus*, *Alnus incana* and *P. sylvestris* (Saether and Andersen in prep. a, b). A large proportion of the available biomass was removed by the moose by the end of the winter (Saether and Andersen in prep. a). The Asnes study area is completely dominated by large forests of *P. sylvestris*, which is intensively managed for timber harvest. In this area the winter diet of the moose is less diverse than in Troms, with *P. sylvestris* and *B. pubescens* as the predominant species. In addition some *Salix* sp. are browsed.

In both study areas, foraging behaviour was studied by following the tracks of radio–collared moose. At random intervals, a browsed tree was selected as the center of a browsing site, defined as a circle with diameter 4.0 m around the selected tree. At each browsing site the number of twigs removed and their diameter were measured on all browsed trees. In addition, the number of trees in the different height intervals was counted for each species present at the browsing site. For each moose, 20–30 browsing sites were sampled. Twigs with a diameter of 6 mm were randomly selected, clipped and were later weighed in the laboratory after being dried for 24 h at 70°C. Finally, the number of 6 mm twigs was counted in each height interval (0.5 m intervals to 3.0 m above snow level) for all species on randomly selected trees along the track.

These data can be used to estimate (1) the amount of browse removed and (2) the browse availability of the different tree species in different height classes (see Saether and Andersen (in prep. b) for computational details).

CHARACTERISTICS OF THE TREE SPECIES

On most Fennoscandian winter grounds, *B. pubescens* and *P. sylvestris* constitute the major food resource for moose. In most areas lower biomass of *S. aucuparia* is found. However, this species is among the most preferred by the moose (Bergstrom and Hjeljord 1987, Hjeljord and Gronvold 1988, Saether and Andersen in prep. b).

These trees differ in morphology. The long–shoots of *S. aucuparia* are of greater length and have larger base–diameter than the long–shoots of *B. pubescens* and *P. sylvestris* (Table 1). Thus, twigs larger than 3 mm diameter of *S. aucuparia* were heavier than corresponding (> 3mm) twigs of *B. pubescens* (Fig. 1, but see Bergstrom and Danell 1987a for an opposite trend between the two deciduous species). The larger current, annual long–shoots of *S. aucuparia* bore twigs that were more digestible than those of the other two species. Twigs of *P. sylvestris* were also of high digestibility. Fewer twigs were present in a given height class in *S. aucuparia* than in *B. pubescens* and *P. pubescens* (Saether and Andersen in prep. b).

BROWSING BEHAVIOUR AND OPTIMAL FORAGING THEORY

Several thousand times each day during the winter, a moose is faced with the question: which twig size to select? A correct choice is profoundly important in influencing the gains from foraging, because several characteristics of the twig depend on its diameter. For instance, the weight of the twig increases curvilinearly (Fig. 1), whereas digestibility decreases (Hjeljord 1987) with diameter. Thus, if a too large twig is clipped the gross energetic gain will be high,

Table 1. Summary of important morphological characteristics for moose browsing of the tree species.

Characteristics	Sorbus aucuparia	Pinus sylvestris	Betula pubescens	Locality	Source
Maximum length of current shoot (mm)	329.1[1)]		211.7[1)]	Grimsø (54°40'N)	Bergstrøm and Danell 1987a
Mean length of annual shoots (mm)		227[2)]		Umeå (63°49'N)	Niemelä and Danell 1988
	59	235	39	Sunnäs (61°N40'N)	Gröning 1980
Base diameter of annual shoot (mm)	4.9[1)]		2.4[1)]	Grimsø (59°40'N)	Bergström and Danell 1987a
		5.2[2)]		Umeå (63°N40'N)	Niemelä and Danell 1988
In vitro digestibility	29.[3)]		22[3)]	Grimsø (59°40N)	Bergström and Danell 1987a
	43.3[3,4)]		38.8[3,4)]	Troms (69°N)	Sæther and Andersen in prep.a
		46.4[3,4)]	28.7[3,4)]	Åsnes (60°30'N)	Sæther and Andersen unpubl.
	41[5)]		27[6)]	Umeå (63°46'N)	Lundberg et al. in press

1) Measured on long-shoots in the tree height of 2.5-3.0 m
2) Mean of shoots/twig on five uppermost whorls in the high productive type
3) Twigs of diameter 4 mm
4) Corrected values, estimated from the relationship with in vivo dry matter digestibility
5) Mean twig diameter 5.8 ± 0.3 mm
6) Mean twig diameter 4.2 ± 0.5 mm

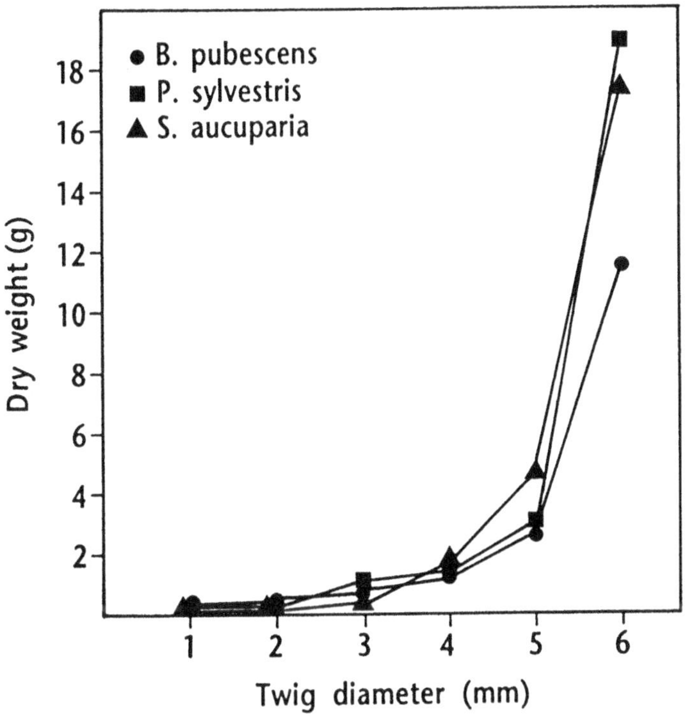

FIGURE 1 The relationship between the dry weight and twig size of the species.

but the time needed to extract the digestible parts will be long. Conversely, a small twig will be quickly digested, but only a small amount of energy will be available.

Vivas et al (in prep.) developed a model based on the theory of Engen and Stenseth (1984) to predict twig–size selection of a moose foraging optimally. This model was based on a restricted set of assumptions which severely limits its general applicability. However, in one particular situation the model was able to predict twig–size selection well, which suggests that the model includes the parameters essential in describing moose foraging behaviour. Simulations of the consequences of changing different parameters therefore help to identify the characteristics of the food plants having the greatest influence on the choice of twig–size. These sensitivity analyses show that the optimal twig–size is closely related to the clipping–rate, which in the studied environment probably depends on the density of twigs (e.g. Spalinger et al. 1988). When resource availability is increased, a moose foraging optimally should be more selective and remove a smaller fraction of the available browse. Furthermore, this relationship also depends on *how* the energetic gain is related to twig size, which is specific for the different browse–species (Table 1). In species with large long–shoots (e.g. *S. aucuparia*) the net energetic gain decreases more slowly with diameter than in species with shorter long–shoots (e.g. *B. pubescens*). This is because the moose is better able to exploit such twigs, due to their higher digestibility (Hjeljord et al. 1982, Hjeljord 1987). An optimal foraging moose should remove a larger amount of the available browse in such species with large shoots. Extension of this simple model to more realistic situations has confirmed these initial conclusions (Vivas in prep.).

Three qualitative predictions emerged from these simulations:

(1) The impact of browsing on all species should decline with browse availability because optimal foraging moose become more selective.

(2) Larger twigs of *S. aucuparia* should be selected than of *B. pubescens* and *P. sylvestris*, given the same availability of browse.

(3) Because of the different shape, a larger proportion of the available biomass should be removed in *S. aucuparia* than in *B. pubescens* and *P. sylvestris*.

(4) Since the influence of browsing decreases with the available biomass (see prediction 1), the height specific impact of browsing should differ among the tree species, due to their different patterns of growth.

RESOURCE UTILIZATION IN RELATION TO BROWSE AVAILABILITY

In order to assess the impact of moose foraging behaviour on the browse species, I have grouped the browsing sites according to their availability of browse. I wanted to examine whether the proportion of plants browsed and the utilization pattern of a browsed tree differ in relation to resource availability.

The patterns of resource utilization in relation to the available biomass of browse differed between the species (Fig. 2a). A larger amount of browse of *P. sylvestris* was removed at a browsing site than of the other two species. However, a much larger fraction of the biomass was removed when the moose browsed on *S. aucuparia* than on the other two species ($F=38.05$, d.f.=2,249, $p<0.001$)

The response of moose to changes in the amount of browse biomass available at a browsing site differed among the species (Fig. 2a). A significant linear increase in the mean amount of biomass removed at a browsing site occurred with increasing browse availability in both *B. pubescens* and *P. sylvestris* ($F=25.00$, d.f.= 1,3, $p<0.05$ and $F=19.30$, d.f.= 1,3, $p<0.05$, respectively), whereas no such relationship was found in *S. aucuparia* ($p>0.1$).

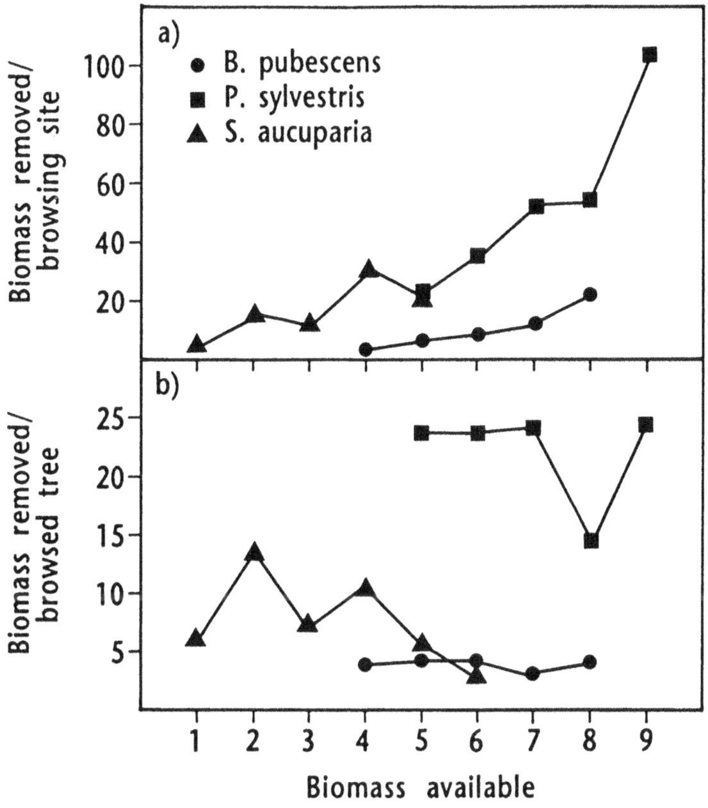

FIGURE 2 (a) The relationship between the amount of biomass removed by individual moose per browsing site and (b) the amount removed per individual moose per browse (g dry weight) in relation to the available biomass of different tree species. The availability of browse (g/50 m^2) was grouped on a logarithmic scale as follows: 1: 2.7–7.3, 2: 7.4–20.0, 3: 20.1–54.5, 4:54.6–148.4, 5: 148.5–403.4, 6: 403.5–1096, 7: 1097–2980, 8: 2981–8103 and 9: >8103.

Comparing only sites with similar amounts of browse available (Fig. 2a), smaller amounts of *B. pubescens* were removed than of *P. sylvestris* and *S. aucuparia* (F= 8.55, d.f.= 2,39, p<0.001).This result was qualiatively predicted from simulations of the optimal foraging model.

For all species, resource availability at a browsing site was closely correlated with the number of trees present. The moose browsed a larger number of trees at browsing sites with a larger biomass of browse (Fig. 3, F= 6.19, d.f.= 4,87, p<0.001; F= 4.91, d.f.= 4,87, p<0.01 and F= 4.84, d.f.= 5,62, p<0.001 for *B. pubescens*, *P. sylvestris* and *S. aucuparia*, respectively).

However, the proportion of trees that remained unbrowsed showed the reverse relationship (Fig. 3). That is, only a small fraction of the trees was browsed at the browsing sites with highest browse availability ($p<0.001$ for all three species).his relationship was closely related to tree density. The slope of the regression of the number of browsed on the number of unbrowsed trees at the browsing site was significantly less than 1 for all species ($p<0.001$). Accordingly,

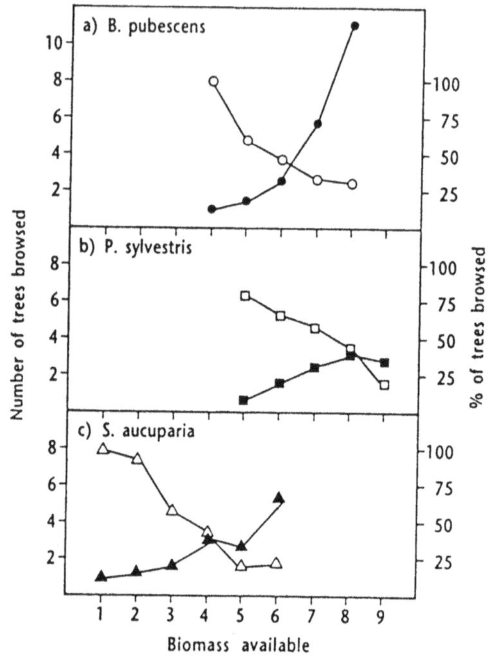

FIGURE 3 The relationship between number (filled symbols) and the proportion (open symbols) of the available trees browsed in relation to the biomass at the browsing site for (a) *Betula pubescens*, (b) *Pinus sylvestris* and (c) *Sorbus aucuparia*. The intervals of the resource availabilities are given in Fig. 2.

the proportion of trees browsed decreased with density of trees. Thus, greater density of trees and consequently greater availability of browse, increased the probability that a tree would not be browsed by an individual moose.

What happens with a tree if a moose decides to browse on it? Does the impact of the moose depend on the biomass of browse around the tree? Such an effect may occur in two ways: either by a decrease in the number of twigs browsed, or by a reduction of the size of the twigs selected. Moose clipped larger twigs of *S. aucuparia* than of the two other species (Saether in prep. a,b). Thus, a larger biomass was removed per clip in this species. However, no significant decrease in the mean clip–diameter in relation to an increase in the available biomass occurred in any of the three species (p>0.1).

The number of twigs clipped at a browsing site increased with the available biomass in *B. pubescens* and *P. sylvestris* (Fig. 4a, F=5.53, d.f.=4,87, p<0.001*)* *S. aucuparia* (Fig. 4a, F=1.34, d.f.=5,62, p>0.1). However, when the number of clips per tree was computed, no significant variation in relation to browse availability was apparent in any of the species (Fig. 4b). Accordingly, the amount of browse removed per browsed tree did not

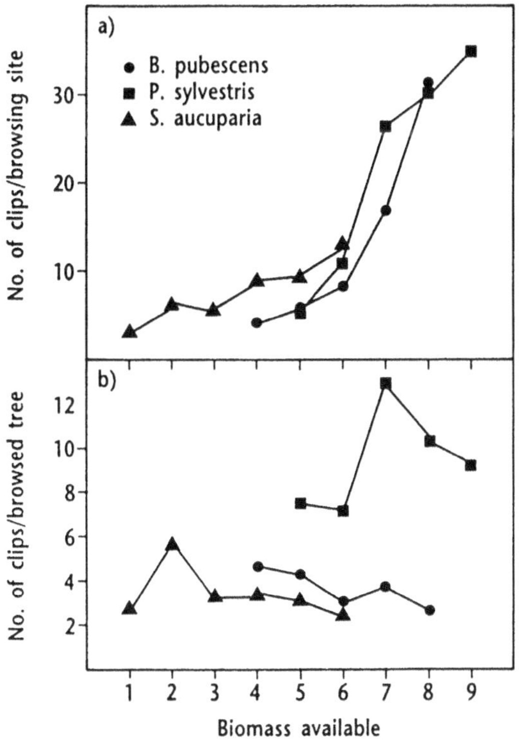

FIGURE 4 The number of clips per browsing site (a) and the number of clips per browsed tree (b) in relation to the biomass at the browsing site for the three tree species. The intervals of the resource availabilities are given in Fig. 2.

differ significantly in relation to the biomass of browse available at the browsing site (Fig. 2b). Thus, the consequences of being browsed did not differ in relation to resource availability.

The impact of being browsed differed strongly between the tree species. Most of the available biomass (70.6+8.3 (+ 1SE) %) was removed from a browsed tree of *S. aucuparia*. In contrast, the moose removed very little from *P. sylvestris* (5.3+0.6 %) and *B. pubescens* (3.5+0.7 %; difference significant, $F=6.82$, d.f.=1,181, $p<0.01$).

IMPACT OF MOOSE BROWSING IN RELATION TO TREE HEIGHT

Tree size and number determine the browse availability in the different height classes. Fewer *S. aucuparia* were found at the browsing sites containing the largest trees (Fig. 5a, $F=3.40$, d.f.=4,66, $p<0.01$). When these density estimates were combined with

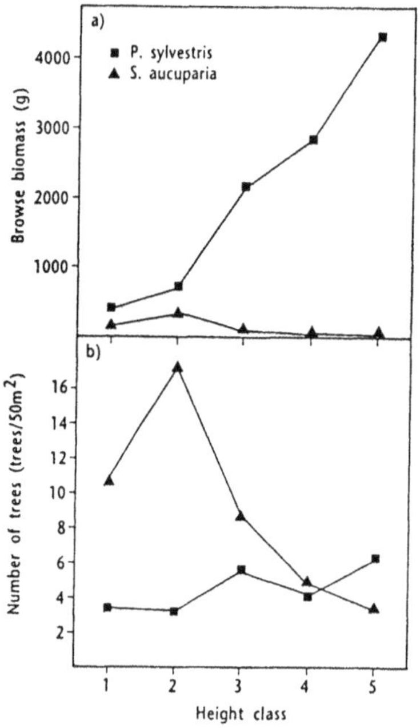

FIGURE 5 The available biomass (g dry weight/50 m^2) (a) and the number of trees (b) in relation to the average height of the trees at the browsing site for *Pinus sylvestris* and *Sorbus aucuparia*. The height intervals (m) are: 1: 0.0.49, 2: 0.50–0.99, 3: 1.00–1.99, 4: 2.00–2.99 and 5: >2.99

the data for single trees (Saether and Andersen in prep. b), the available biomass relative to tree height differed strongly between the tree species (Fig. 5b). A significant linear increase occurred in the mean biomass with height in *P. sylvestris* (F=18.64, d.f.=1,3, p<0.001). Incontrast *S. aucuparia* (F=5.50, d.f.=1,3, p

We should expect from these great interspecific differences that the pattern of resource utilization by the moose in relation to tree height should also differ between the species. Because the variance in the height of *B. pubescens* within a browsing site was large, this species was excluded from the comparisons presented below.

In *S. aucuparia* the mean number of trees browsed at a browsing site did not differ significantly in relation to the average tree height (F=0.62, d.f.=4,63, p>0.1). However, the proportion of trees browsed increased, with the exception that only a small fraction of the lowest trees were browsed (Fig. 6a). The mean clip diameter did not

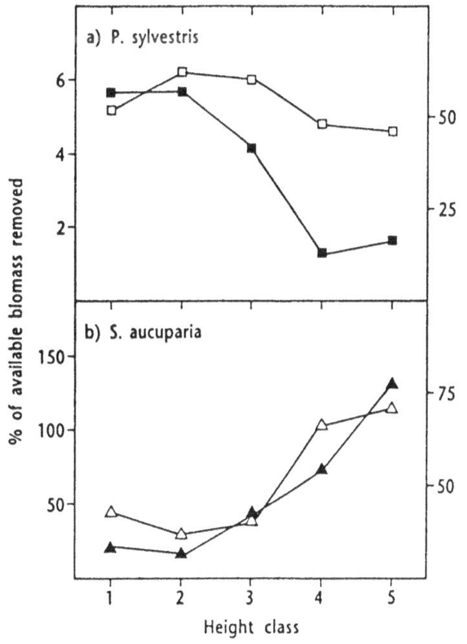

FIGURE 6 The proportion of the available biomass removed (filled symbols) and the proportion (open symbols) of the trees browsed in relation to the average height of (a) *Sorbus aucuparia* (b) *Pinus sylvestris* at the browsing site. The height intervals are given given in Fig. 5.

differ significantly in relation to tree height (Fig. 7a, F= 1.96, d.f.=4,87, p>0.01). However, a larger number of twigs were removed from taller, compared with shorter, trees (Fig. 8, F= 3.38, d.f.=4,87, p>0.05). Thus *S. aucuparia* increased with tree height (Fig. 9, F=6.65, d.f.= 4,63, p<0.001).

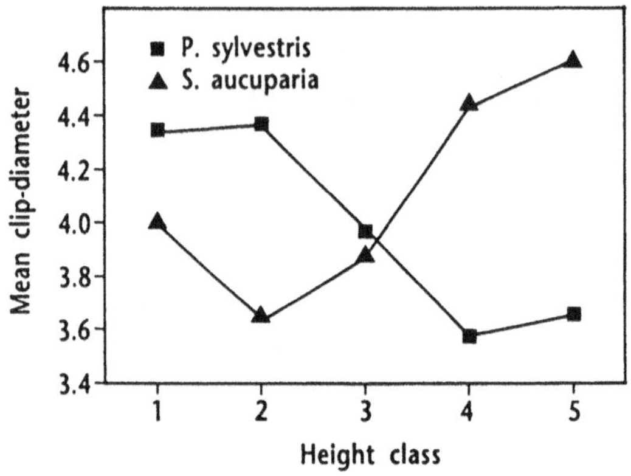

FIGURE 7 The mean clip diameter (mm) at the browsing sites in relation to the average tree height for *Sorbus aucuparia* and *Pinus sylvestris*. The height intervals are given in Fig. 5.

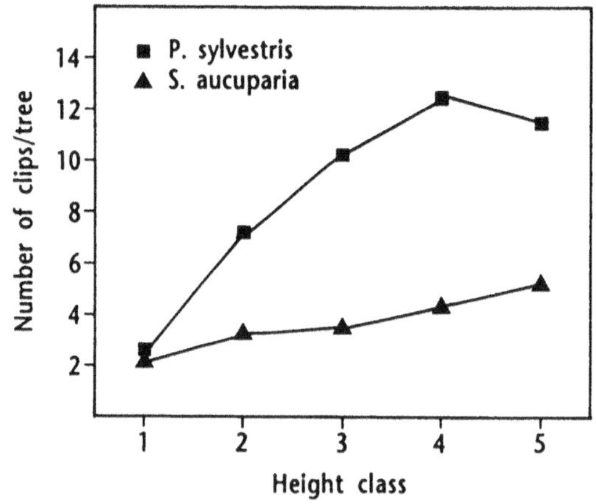

FIGURE 8 The mean number of clips per tree in relation to the average height of *Sorbus aucuparia* and *Pinus sylvestris* at the browsing site. The height intervals are given in Fig. 5.

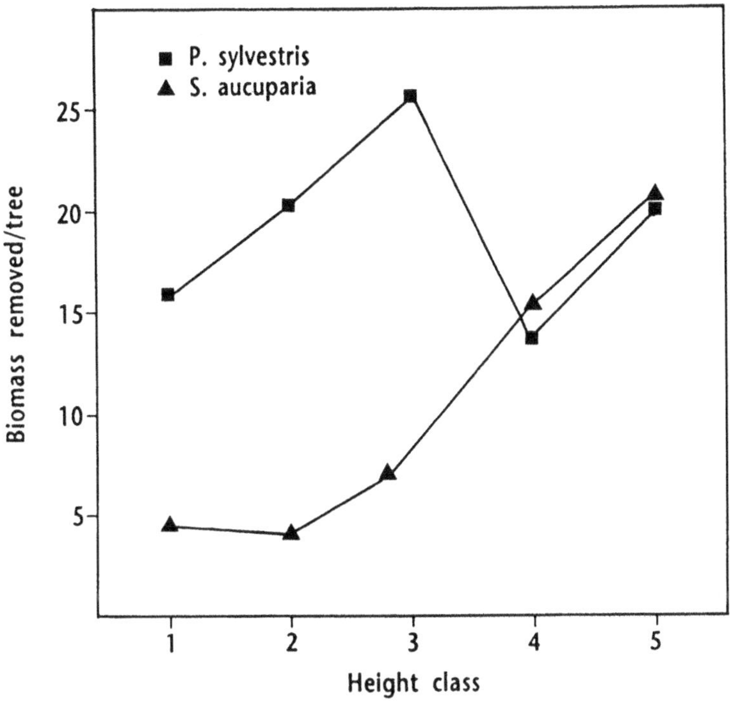

FIGURE 9 The biomass removed (g dry weight) per tree in relation to the average height of *Sorbus aucuparia* and *Pinus sylvestris* at the browsing site. The height intervals are given in Fig. 5.

The patterns of resource utilization relative to tree height in *P. sylvestris* differed from the patterns in *S. aucuparia*. The number of trees browsed did not increase significantly with tree height (F= 1.67, d.f.= 4.87, p>0.01). Hence, the proportion of the available trees browsed did not differ with the average height of the trees at the browsing site (Fig. 7, F=1.00, d.f.= 4,87, p>0.1). The mean clip–diameter decreased with tree height (Fig. 9, F= 3.81, d.f.=4.87, p<0.01).No significant differnce was found in the number of twigs removed per tree (Fig. 8, F= 1.52, d.f.= 4,87, p>0.1), although a smaller number of twigs were clipped in low trees. Thus, the amount of biomass removed from a browsed tree did not differ significantly with tree height in *P. sylvestris* (Fig. 9, F=0.76, d.f.= 4,87, p>0.1).

These differences in resource utilization imply that the consequences of being browsed by a moose differed between *S. aucuparia* and *P. sylvestris*. In *S. aucuparia* the moose removed a larger proportion of the available biomass at browsing sites with tall trees than at sites with lower trees (F= 4.42, d.f.= 4,63, p<0.01).The reverse trend is found in *P. sylvestris*, where a lower impact was found at the browsing sites with the largest average height ($F= 4.46$, d.f.= 4,87, $p<0.01$).

DISCUSSION

As has been experimentally verified on several occasions (Vivas and Saether 1987, Niemela and Danell 1988, Lundberg et al. in press), this study has demonstrated that moose alter their foraging behaviour as a response to variation in the amount of browse present. In general, the impact of moose browsing on different tree species followed the pattern expected from optimal foraging theory, i.e. increased selectivity occurred when resource availability was high. A similar success in predicting the foraging behaviour of herbivores from optimal foraging theory has previously been achieved by Belovsky (1986, but see Hobbs, this volume).

One of the most pronounced effects of increased food availability was that the proportion of trees browsed decreased (Fig. 3). A similar effect has also been demonstrated in a field experiment on moose foraging behaviour in artificial plots with different availability of *B. pubescens* (Vivas and Saether 1987). Thus, the probability of being browsed by a mammalian herbivore depends on the density of tree *per se*. This is in accordance with predictions from simulations based on models of optimal foraging moose (Vivas in prep., Vivas et al. in prep): a lesser proportion of the biomass should be removed when the resource availability increases. Thus, plant density and resource availability play an important role in the interaction between plants and herbivores (see also Lundberg 1988 for a similar argument).

The pattern of resource utilization in relation to the availability of browse differed among tree species. This was not unexpected, considering the large differences in their characteristics (Table 1, Figs. 1, 5, Saether and Andersen in prep. b). The largest consumption per tree occurred in the coniferous species *P. sylvestris*. However, the largest proportion of the available biomass was removed from *S. aucuparia*, a species highly preferred by the moose (Saether and Andersen in prep. b). Thus, in absolute terms more browse was removed from the species with large biomass, but *relative* loss of biomass was greater in the smaller species, *S. aucuparia*. Also, the impact of moose browsing increased with height in *S.aucuparia*, whereas the reversed trend was found in *P. sylvestris*. The only exception was that very small bushes of *S. aucuparia* were browsed to a lesser degree by the moose, perhaps due to the presence of compounds that prevent browsing of this highly vulnerable juvenile stage (cf. Bryant 1981). Thus, for the rapidly growing *S. aucuparia* the consequences for an individual tree being browsed by a mammalian herbivore are much greater than for *P. sylvestris* and *B. pubescens*. These patterns are consistent with the evolutionary responses to browsing predicted by Bryant et al. (1983). However, this pattern is unlikely to be related to differences in the fertility of the growth site, as predicted by Coley et al. (1985), since *P. sylvestris* is an evergreen, which in the present study area grows on dry, poor soil. Accordingly, *P.sylvestris* should be well protected against browsing by a chemical defense system. Thus, the interspecific differences in the impact of browsing can, as argued by Myers (1987), be explained as a consequence of physiological adaptations to different growth strategies, rather than as a result of an adaptive anti–herbivore defense system as such.

A consequence of this argument is that *S. aucuparia* should be more vulnerable to browsing then *B. pubescens* and *P. sylvestris*. In a study in southern Norway, Hjeljord and Gronvold (1988) found that moose browsing retarded the growth of *S. aucuparia* more than it did on *B. pubescens*. However, in plots protected from browsing, greater height increments were found in *S. aucuparia* than in *B. pubescens*.

Plants often change growth form as a response to herbivory. Such changes are often influenced by the nature and amount of branching. When the terminal bud of a branch is removed, the hormones that suppress growth at lower nodes disappears (Crawley 1983). Studies of *B. pubescens* have shown that an increase in the frequency of branched shoots is a common response to browsing (Bergstrom and Danell 1987). I suggest that one consequence of this change in structure is that it reduces the effects of browsing. The larger the numbers of branches, the higher the clipping rate will be. According to the predictions from optimal foraging models (Vivas et al. in prep), this will make the moose more selective, leaving a larger part of the plant unbrowsed. The increase in the size of the shoots following browsing (Dannel et al. 1985, Danell and Bergstrom 1989) may also influence foraging behaviour in a similar way. In fact, some data supporting this suggestion were obtained by Bergstrom and Danell (1987a): twig size selected by penned moose was positively related to the diameter of the twig.

The theory of the interaction between plants and herbivores was in the beginning stimulating, but was able to give only qualitative predictions (e.g. Rhoades 1979), which were difficult to test. Recently more complex quantitative models have been made (Coley et al. 1985, Fagerstrom et al. 1987, Fagerstrom

1989). Predictions from these models depend on the relationship between the cost of an investment (e.g. a chemical defense system) and the gain in terms of reduced impact of herbivory. For instance, the assumption (Coley et al. 1985) of a constant relative consumption rate of browse, independent of resource availability will not apply to a mammalian herbivore such as the moose. The present study has demonstrated that a general understanding of such relationship can be obtained from optimal foraging theory.

CONCLUSIONS

This study has demonstrated that the impact on browse species of foraging by individual moose depends on resource availability and the pattern of growth of the trees. Morphological differences related to such differences in growth form are likely to strongly influence the foraging behaviour of mammalian herbivores (e.g. Bergstrom and Danell 1987a). The consequences of browsing were most severe on *S. aucuparia* than on *P. sylvestris* and *B. pubescens*.

A strong relationship existed between the growth form and the height–specific impact of browsing. In the coniferous species *P. sylvestris*, which has a large biomass of twigs, the effect of browsing decreased with tree height. The reversed trend was found in *S. aucuparia*.

It is suggested that optimal foraging theory can be used to evaluate the effects of changes in the characteristics of plants on the foraging behaviour of herbivores. Such knowledge will be important for deriving more precise, quantitative predictions for the pattern of interaction between plants and herbivores.

ACKNOWLEDGEMENTS

This study was financed through a grant from the Directorate for Nature Management in Norway. I am grateful to J. E. Andersen, R. Andersen, J. Bekken, A. Gravem, M. Heim and T. Rodbergshagen for help in collecting the data. K. Danell and P. Lundberg generously provided valuable information. I appreciate helpful comments from D. Bazely, N.T. Hobbs, K. Danell, S. Albon, T. Skjogland, R. Langvatn, P. Lundberg and R. Andersen, which greatly improved the presentation.

LITERATURE REFERENCES

Ahlén I (1975) Winter habitats of moose and deer in relation to land use in Scandinavia. Swedish Wildl Res 9: 45-192.
Belovsky GE (1986) Optimal foraging and community structure: implications for a guild of generalist grassland herbivores. Oecologia (Berl) 70: 35-52.
Bergström R, Danell K (1987a) Moose feeding in relation to morphology and chemistry of six tree species. Alces 22: 91-112.
Bergström R, Danell K (1987b) Effects of simulated winter browsing by moose on morphology and biomass of two birch species. J Ecol 75: 533-544.
Bergström R, Hjeljord O (1987) Moose and vegetation interactions in Northwestern Europe and Poland. Swedish Wildl Res Suppl 1: 213-228.
Bryant JP (1981) Phytochemical deterrence of snowshoe hare browsing by adventitious shoots of four Alaskan trees. Science 213:889-890.
Bryant JP, Chapin FS III, Klein DR (1983) Carbon/nutrition balance of boreal plants in relation to vertebrate herbivory. Oikos 40: 357-368.
Cederlund G, Markgren G (1987) The development of the Swedish moose population, 1970-1983. Swedish Wildl Res Suppl 1: 55-62.
Cederlund G, Ljungqvist H, Markgren G, Stålfelt F (1980) Foods of moose and roe deer at Grimsö in Central Sweden. Swedish Wildl Res 11: 169-247.
Chapin FS III (1980) The mineral nutrition of wild plants Ann Rev Ecol Syst 11: 233-260.
Chapin FS III, Johnson DA, McKendrick JD (1980) Seasonal movement of nutrients in plants of differing growth form in an Alaskian tundra ecosystem: implications for herbivory. J Ecol 68: 189-209.
Coley PD, Bryant JP, Chapin FS III (1985) Resource availability and plant antiherbivore defense. Science 230: 895-899.
Crawley MJ (1983) Herbivory. The dynamics of animal-plant interactions. Blackwell Scientific Publ, Oxford.
Danell K Bergström R (1989) Winter browsing by moose on two birch species: impact on food resources. Oikos 54: 11-18.
Danell K, Huss-Danell K, Bergström R (1985) Interactions between browsing moose and two species of brich in Sweden. Ecology 66: 1867-1878.
Engen S, Stenseth NC (1984) A general version of optimal foraging theory: the effect of simultaneous encounters. Theor Pop Biol 26: 192-204.
Fagerström T (1989) Anti-herbivore chemical defense in plants: a note on the concept of cost. Am Nat 133: 281-287.
Fagerström T, Larsson S, Tenow O (1987) On optimal defense in plants. Func Ecol 1: 73-81.
Gröning P (1980) Simulerat bete på vårtbjørk samt morfologi och fenologi hos några viltbetesarter. Medd Växtbiol Inst Uppsala 1980 (4): 1-23. (In Swedish).
Hagen Y (1983) Elgens vinterbeiting i Norge. Viltrapport 26: 1-111. (In Norwegian).

Hjeljord O (1987) Research on moose nutrition in the Nordic countries. Swedish Wildl Res Suppl 1: 389-397.
Hjeljord O, Grønvold S (1988) Glyphosate application in forest-ecological aspects VI. Browsing by moose (Alces alces) in relation to chemical and mechanical brush control. Scand J For Res 3: 115-121.
Hjeljord O, Sundstøl F, Haagenrud H (1982) The nutritional value of browse to moose. J Wildl Mgmt 46: 333-343.
Lundberg P (1988) Functional response of a small mammalian herbivore: the disc equation revisited. J Anim Ecol 57:
Lundberg P, Åström M, Danell K (to be published) An experimental study of frequency dependent food selection: winter browsing by moose. Hol Ecol
Myers JH (1987) Nutrient availability and the deployment of mechanical defenses in grazed plants: a new experimental approach to the optimal defense theory. Oikos 49: 350-351.
Niemelä P, Danell K (1988) Comparison of moose browsing on Scots pine (Pinus sylvestris) and lodgepole pine (P. contorta). J Appl Ecol 25: 761-775.
Rhoades DF (1979) Evolution of plant chemical defense against herbivores. In: Rosenthal GA, Janzen DH (eds) Herbivores. Their interaction with secondary plant metabolites. Academic Press, New York.
Spalinger DE, Hanley TA, Robbins CT (1988). Analysis of the functional response in the Sitka black-tailed deer. Ecology 69: 1166-1175.
Sæther B-E (1987) Patterns and processes in the populatin dynamics of the Scandinavian moose (Alces alces): some suggestions. Swedish Wildl Res Suppl 1: 525-537.
Sæther B-E, Andersen (in prep a) Resource limitation in a generalist herbivore, the moose Alces alces: ecological constraints on behavioural decisions. Can J Zool (submitted)
Sæther B-E, Gravem A (1988) Annual variation in winter body condition of Norwegian moose calves. J Wildl Mgmt 52: 333-336.
Vivås HJ, Sæther, B-E (1987) Interaction between a generalist herbivore, the moose Alces alces, and its food resources: an experimental study of winter foraging behaviour in relation to browse availability. J Anim Ecol 56: 509-520.
Vivås HJ, Sæther B-E, Andersen R (in prep) Optimal twig-size selection of a generalist herbivore, the moose Alces alces: stochastic variationin prey-choice models.

ORAL DISCUSSION – B.–E. Saether

Q. (Vadas): Do you suggest that moose are prudent predators ?
A. No, only that when there's a lot of snow, the moose can't move about easily, and then the reason they don't remove all the biomass from one place is that their stomachs are absolutely full of poor quality twigs they can't digest.

Q. (DeMott): Can you comment on effects of grazing on plant species composition in the long term ?
A. Rowan grows very fast and is very competitive with birch, but can easily be killed by grazing and this has a very marked effect on species composition.

Q. (Bazely): The apparent effects that herbivore have on plants seem more like a generalised stress reaction by plants. The whole area of secondary compounds is being reviewed and may not have anything to do with herbivore defences.
A. Some plants are definitely poisonous, so maybe it's all–or–none; either the plant is eaten or not.
Bazely: It's not clear that secondary compounds are a specific adaptation to protect them from herbivory, you need to know why plants increase their secondary compounds in response to stress.

Q. (Illius to Bazely): Are you saying that in the generalised stress response the fact that the plant produces substances which are obnoxius to herbivores is just a coincidence ?
Bazely: We don't know whether it's a specific adaptation or of benefit elsewhere.
Zimmer–Faust: Maybe the plants generalize their response not knowing what's the cause of the stress. As Dan Janzen says: 'fire is a generalist herbivore'.
Real: Are there many insect herbivores in this forest ?
A. I don't know about the impact of the insects, which are abundant in the forest.

RULES AND CUES USED BY SHEEP FORAGING IN MONOCULTURES

D. R. Bazely
Edward Grey Institute of Field Ornithology
Department of Zoology
South Parks Road
Oxford OX1 3PS
Great Britain

INTRODUCTION

It has long been pointed out that the problems of forage selection faced by herbivores are complex (e.g. Tribe 1950, Arnold and Dudzinski 1978, Illius and Gordon this volume). For example, the nutritional status of potential forage species is constantly fluctuating and herbivores must somehow track these changes. In addition, unlike carnivores which meet nearly all their nutritional requirements in one meal, herbivores must select diets to meet an array of nutritional requirements e.g. to obtain essential nutrients such as sodium and phosphorus (Westoby 1974, Belovsky 1978, Stephens and Krebs 1986, Bazely 1989). Functional hypotheses which have been proposed to explain observed diet selection are (a) the need to select the most nutritious vegetation (Arnold and Dudzinski 1978) and (b) the need to avoid toxic secondary plant compounds (Bryant and Kuropat 1980, Crawley 1983).

Although the aim of optimal foraging theory (OFT) has been to investigate general rules which animals use to forage (Krebs *et al* 1983), few investigations involving vertebrate herbivores have adopted an optimality approach (those that have are: Westoby 1974, 1978, Belovsky 1978, 1984a, b, Owen-Smith and Novellie 1982). Stephens and Krebs (1986) have pointed out that patch use models may be more useful and appropriate for herbivores than prey models. Patch models assume that food is distributed in clumps, and they often predict that foraging animals will leave patches with declining intake rates before consuming all available food - a characteristic of herbivores. Also, food items encountered and eaten by herbivores are not discrete in the sense envisaged by the prey models (Stephens and Krebs 1986). Patch use models have apparently been used in only one herbivory study (Parker 1984), which found that grasshoppers

NATO ASI Series, Vol. G 20
Behavioural Mechanisms of Food Selection
Edited by R. N. Hughes

stayed in patches for longer when their inter-patch travel time was increased.

Prior to carrying out research into whether a specific patch use model, the marginal value theorem (MVT) (Charnov 1976, Krebs and McCleery 1984), could predict the foraging behaviour of sheep (*Ovis aries* L.) in monocultures of perennial ryegrass (*Lolium perenne* L.) (Bazely 1988), I investigated the nature of patchiness in these monocultures. The MVT predicts the optimal patch leaving time for a forager moving between patches of food, subject to the constraints of the gain curve for the patch and the travel time between patches. Foragers should stay in patches with the same gain function for longer when the travel time between these patches is greater in order to maximize intake rate of the currency e.g. energy for which the gain function is measured. Since the MVT deals with discrete patches of food, I had to define what constituted patchiness for a sheep in order to create an appropriate experimental situation. Since real life is so complex, it seemed sensible, in adopting an optimality approach to herbivory, to investigate diet selection in a simple foraging environment before moving on to more realistic systems. This also had the advantage of making potential rules and cues used by sheep in diet selection easier to identify: it is generally agreed that animals use various "rules of thumb" to obtain solutions to optimality problems (Houston *et al* 1982, Krebs and McCleery 1984, Stephens and Krebs 1986).

In this paper I describe experiments aimed at identifying some of the variables to which sheep may respond in making foraging decisions in monocultures. I discuss cues, possible rules, and senses (behavioural constraints) that sheep might use in selecting diets from monocultures to maximize intake rates of dry matter (biomass) and nitrogen. The latter are foraging currencies that may be of relevance to herbivores: the intake of biomass of specific digestibilities is related to energy intake (Freer 1981) while nitrogen is essential for protein synthesis (Mattson 1980, Ørskov 1982).

DIET PREFERENCES OF SHEEP GRAZING IN MONOCULTURES

I investigated the grazing behaviour and preferences of sheep presented with swards of ryegrass which differed in their visual appearance (vegetation height and brightness) and in some cases their nutritional content. The term "luminance" refers to the total amount of incoming radiation reflected from an object (Jacobs 1981) and "brightness" is taken to have the same meaning. Brightness is unrelated to "hue" or "colour", which refers to the specific wavelength characteristics of an object. Sheep are apparently monochromats i.e. they have black and white vision (Tribe and Gordon 1949, Bazely and Ensor 1989).

The experiments were carried out at the University Field Laboratory, Wytham, Oxford from 1986-1987.

Response of sheep to variation in height and brightness in a perennial ryegrass sward created by *in situ* sward manipulations

Methods

Three 25 m x 25 m paddocks sown with perennial ryegrass (var. Melle) in April 1985 were used in these experiments. Apart from an initial application of fertiliser at the time of sowing, the paddocks were unfertilised.

Variation in sward height was created by allowing a heavily grazed and then mowed 25 m x 25 m paddock to grow for 4-6 weeks. Ryegrass in the vicinity of sites where grazing sheep had urinated or defaecated grew up after mowing to be much taller than the background sward which was nitrogen poor. These tall patches were also darker green compared with the lighter green background sward. One paddock was treated in this way in July 1986 and two paddocks in June 1987.

The number of identifiable tall patches in each paddock was counted prior to the experiment in each paddock. In 1986 five of these patches were selected at random. The sward height of each patch was measured as well as the height of shorter grass 50 cm from the edge of the patch and that of five additional sites selected at random throughout the paddock in the short sward ($n = 5$ sward heights taken at each site or patch). The use of sward heights to characterize ryegrass pastures is a well established technique (Hodgson *et al* 1986). I compared the height of shorter swards near tall patches with that of shorter swards further away from these patches in order to demonstrate the highly localized effect of nutrient input from faeces and urine on height; these data demonstrated whether the the nutrient input declined abruptly beyond a certain point or had diminishing effect. All patches and sites in the short sward were marked with bamboo poles.

Three ewes (Border Leicester X Swaledale) were placed in the paddock and observed for a period of one hour for as many uninterrupted 15 minute grazing periods as possible - sheep sometimes ceased feeding in order to ruminate and a 15 minute observation period had to be terminated. In each 15 minute period each sheep was scored at 15 second intervals as to whether it was grazing in a tall patch of ryegrass. The experiment was terminated after one hour, and the heights of the marked sites were remeasured. In addition, the number of identifiable tall patches that had been grazed was scored.

In 1987 the same procedure was followed, but ten tall patches in each paddock were marked and ten points were selected at random for measurement in the short background sward. The sample size was increased because there were more tall patches in paddocks and so perhaps a decrease in the likelihood of all patches being visited by sheep.

Different groups of three sheep were observed in each paddock over two consecutive days. During 15 minute observation bouts each sheep was scored at 15 second intervals for its behaviour in the following three categories: (1) grazing in a tall patch, (2) grazing in the shorter background sward, (3) other activities, such as walking or standing. After the initial hour of observation the sheep remained in the paddock for another hour after which the experiment was terminated and the sheep removed.

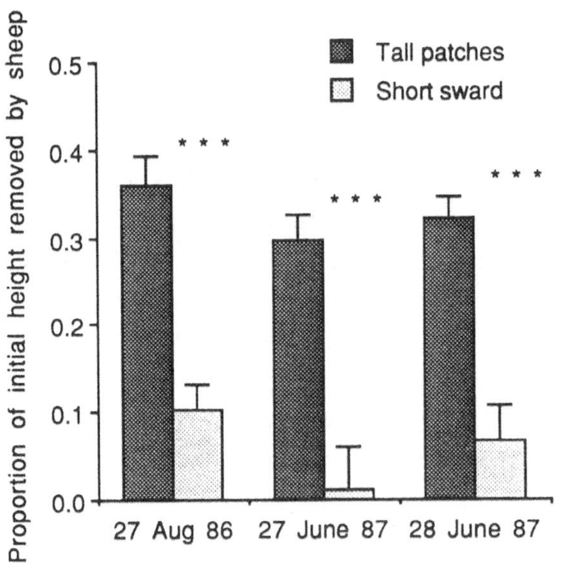

Figure 1. Proportionate decreases in the initial sward heights of tall patches and areas in the shorter background sward of perennial ryegrass paddocks on different experimental days. (Proportions are means; bars are standard errors; n = 5 in 1986; n = 10 in 1987; *** denotes $p < 0.001$).

Results

The mean heights of the tall ryegrass patches created by nutrient inputs from the sheep were 9-13 cm greater than the background sward 50 cm away and elsewhere in the paddocks (Mean sward height (cm) in each paddock: tall patches - 21.5 (1986), 15.1 (1987 pad 1), 20.9 (1987 pad 2); short swards adjacent to tall patches - 11.8, 6.5, 8.7; short sward selected at random - 10.2, 6.6, 7.6).

In all paddocks, all marked tall patches were intensively grazed by sheep and their heights decreased by 30-40% (5 - 8 cm) compared with 1-10% (0.2 - 1.0 cm) in the short sward 50 cm away (Figure 1). Proportionate decreases in height were significant in tall patches compared with short swards (one-way ANOVA: 27 Aug 1986 $F_{1,9} = 26.5^{***}$, 27 June 1987 $F_{1,19} = 26.5^{***}$, 28 June 1987 $F_{1,19} = 28.4^{***}$). Proportion data were transformed with the angular transformation, arcsin $\sqrt{(\text{proportion})}$, where required (Rohlf and Sokal 1981, Table 5).

In 1986, sheep spent 33-48% of a total of 30 minutes of observation time (2 x 15 min. periods) grazing in tall patches (Figure 2a). In 1987, the amount of grazing time was separated from time spent in other activities, and results are presented as the percentage of grazing time (15 second spot observations) spent feeding in tall or short swards (Figures 2b,c). Sheep spent from 50 - >90% of grazing time feeding in tall patches. When time spent grazing in tall patches was calculated as a proportion of total observation time, it varied from 27-70% in paddock 1 (27 June) and 60-85% in paddock 2 (28 June) (n = 30 and 45 minutes total observation time respectively).

Tall patches comprised 4 - 7% of the total paddock areas. An expected number of observation points for sheep grazing in tall patches and shorter background sward was calculated for each paddock based on these percentages. I corrected for possible interdependence of consecutive 15 second interval data points by dropping 3 out of 4 consecutive points. The observed and expected values of where sheep were/should be grazing were compared using Fisher's Exact tests. In both 1986 and 1987, all sheep spent significantly more time (observation points sampled at one minute intervals) grazing in tall patches than would have been expected if grazing time was dependent only on the area of the paddocks occupied by these patches (sheep 220: $p = 0.011$; sheep 212: $p = 0.003$; other seven sheep $p < 0.001$). After 1-2 hours, 70-80% of identifiable tall patches (there were 31 to 227 patches per paddock) were grazed, which was another indication that sheep were particularly attracted to select them.

In summary, sheep in all experiments preferred tall, dark green patches, and

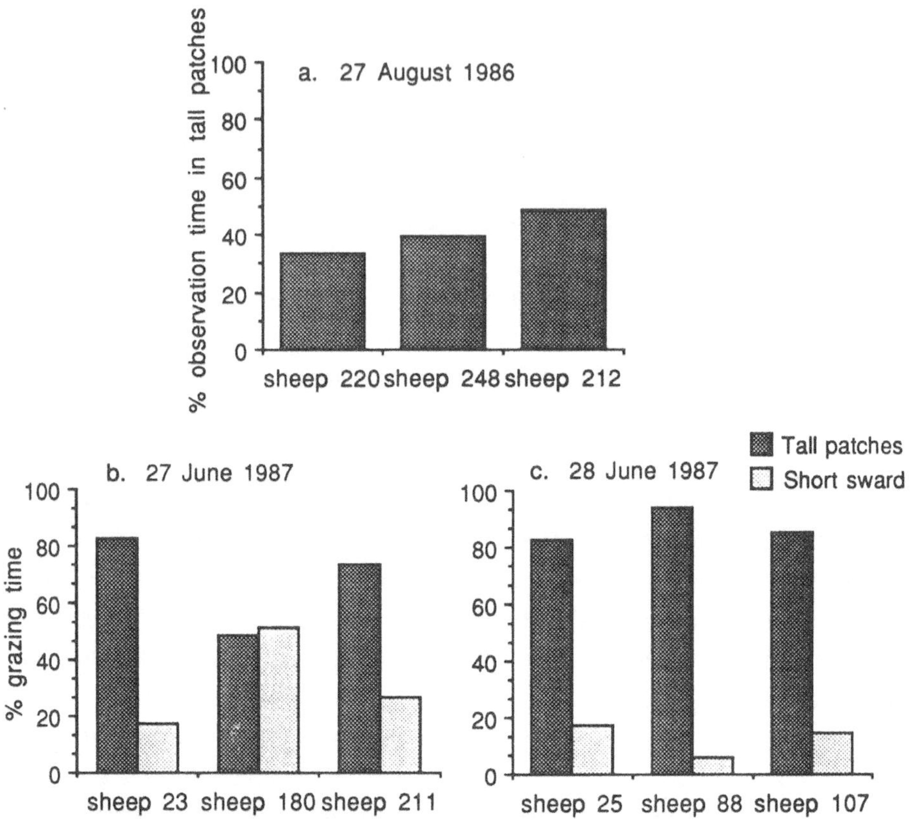

Figure 2. Percentage of time spent grazing by sheep in tall patches (sward) and shorter background sward of perennial ryegrass on different experimental days. (a) 1986 (30 minute total observation period). (b) paddock 1 1987 (30 minute total observation period). (c) paddock 2 1987 (45 minute total observation period). (Sheep were sampled at 15 second intervals.)

spent significantly more time than expected grazing in them. This was reflected in the greater proportionate decrease in height of these patches compared with adjacent short, light green swards. Thus the sheep were obviously discriminating between sward types within this monoculture which were clearly different to my eye. In further experiments I investigated whether height (Bazely 1988) and brightness by themselves were preferred, since they were clearly correlated in this experiment.

The relationship between brightness of perennial ryegrass and its biochemical characteristics

Methods

I investigated a possible basis of the consistent preference of sheep for darker green areas of ryegrass in the experiment described below. It aimed at determining whether there was a relationship between brightness and nutritional quality of ryegrass. Trays of ryegrass received different levels of fertiliser. Their brightnesses were quantified and related to biochemical analyses of the vegetation.

In June 1987, 35 trays of perennial ryegrass (var. Parcour) were sown at 2 g seed per tray. Seven trays were assigned to one of five fertiliser treatment levels:

(1) Control - received no fertiliser

(2) Low treatment - 0.5 g ammonium nitrate (NH_4NO_3) salts, 0.5 g potassium phosphate (K_2HPO_4) salts

(3) Medium 1 treatment - 1 g of each of NH_4NO_3 and K_2HPO_4 salts

(4) Medium 2 treatment - 2 g of each of NH_4NO_3 and K_2HPO_4 salts

(5) High treatment - 5 g of each of NH_4NO_3 and K_2HPO_4 salts.

Groups of trays received the above amounts of fertilising salts on 4-5 occasions, every 4-12 days. All trays were trimmed to 5 cm heightat four and six weeks after sowing.

Brightness, or the total amount of light reflected from trays of ryegrass receiving different amounts of fertiliser, was characterized by two methods:

(1) Analysing photographic transparencies of ryegrass with a Macbeth TD 504 Densitometer.

(2) Measuring the reflectance spectra of ryegrass using an ISCO model SR Spectroradiometer.

The densitometry method was more versatile, since slides could be taken and analysed later. A Kodak grey card of known reflectance (18%) was be photographed in each frame to allow between-frame calibration and to control for varying light conditions. Use of spectroradiometry was more limited since readings had to be taken on live vegetation, although it allowed characterization of hue as well as brightness.

The densitometer measured the brightness (total amount of light reflected) of different trays of ryegrass. A transmission densitometer passes a beam of light through the transparency and characterizes the "blackness" of images with a density reading (Wakefield 1970).

$$\text{Density} = \log_{10}(\text{Opacity}) \qquad \text{Opacity} = \text{inverse of transparency} = \frac{L_i \text{ (incident light)}}{L_t \text{ (transmitted light)}}$$

The higher the density, the darker or more opaque the image on the transparency and the lower the total amount of light reflected from that image.

Seven weeks after sowing, trays from different treatment groups were photographed under natural light conditions with a Hasselblad twin lens reflex camera. Agfachrome colour slide film was used. In order to extend the range of hues and brightnesses of ryegrass, cores were taken from "natural" dark green ryegrass patches from the ryegrass paddocks and also photographed.

The live biomass from three replicate trays in each treatment group was harvested, washed and dried at 80°C in a drying oven and analysed for total nitrogen, fibre and water soluble carbohydrate content (see Bazely 1988 for procedures). The water content of vegetation was also calculated.

In September 1987 the reflectance spectra of the ryegrass from remaining trays in each group, and from ryegrass cores sampled from dark green and light green areas of the paddocks, were characterized with the spectroradiometer equipped with a 3 foot long fibre optics extension probe. The fertiliser treatments had been maintained at 15-20 day intervals. Spectral intensity readings ($\mu W\ cm^{-2}\ nm^{-1}$) of the amount of light relected in 12.5 nm wavelength bands from 400 - 750 nm were collected from ryegrass samples. Correction factors used were calculated by ISCO, Nebraska, U.S.A. in June 1986. Measurements were made under controlled light conditions. After each ryegrass sample was measured with the spectroradiometer, it was washed, dried and chemically analysed.

Results

The relationships between fertiliser treatment and chemical characteristics of ryegrass are shown in Table 1. Trays receiving least amounts of fertiliser had the lowest nitrogen content and the highest water soluble carbohydrate contents. Water content of ryegrass from trays measured in July 1987, and subjected to similar growth conditions, also varied according to fertiliser treatment.

Densitometer readings varied significantly among trays from different fertiliser treatment groups and ryegrass cores from the paddocks ($F_{4,22} = 31.1^{***}$). The darkest green trays were from the high level fertiliser group and had the highest density readings (Table 1a). Regressions of densitometer readings on total nitrogen and water soluble carbohydrate content of ryegrass were highly significant

Table 1 a. Brightness and chemical characteristics of perennial ryegrass grown under different fertiliser treatment regimes, measured in July 1987. (chemical analyses: n = 3 for fertiliser treatments and n = 5 for ryegrass samples from paddocks; densitometer readings: n = 3 - 6 depending upon size of grass tray in slide).

Fertiliser treatment/ sample source	Densitometer reading: colour slide film		Water content (% fresh wt)		Total nitrogen (% dry wt)		WSC† (mg/g dry wt)		% MADF††	
	mean	SE	mean	SE	mean	SE	mean	SE	mean	SE
No fertiliser	2.06	0.02	85.7	0.2	1.7	0.1	229.4	6.2	20.6	0.8
Low level	2.37	0.05	89.6	0.3	2.5	0.1	73.0	19.5	25.3	2.2
Medium 1 level	2.56	0.05	89.2	0.7	3.0	0.3	48.3	15.3	23.9	0.7
Medium 2 level	-	-	90.7	0.4	4.4	0.2	32.9	12.5	23.1	0.9
High level	2.66	0.04	90.9	0.9	5.1	0.1	15.1	2.6	21.9	0.3
		$F_{4,10}$:	13.8***		71.8***		45.9***		2.4 n.s.	
		Fisher's PLSD:	1.8		0.52		40.3		3.7	
Paddock ryegrass										
Dark green core	2.70	0.11	83.6	0.7	3.9	0.2	15.0	6.0	26.8	1.1

Table 1 b. Brightness and chemical characteristics of perennial ryegrass grown under different fertiliser treatment regimes, measured in September 1987 (n = 1 for each chemical analysis and n = 3 or 1 for peak reflectance values).

Fertiliser treatment/ sample source	Reflectance at 575 nm ($\mu W\ cm^{-2}\ nm^{-1}$)		Total nitrogen (% dry wt)	WSC† (mg/g dry wt)	% MADF††
	mean	SE			
No fertiliser	.248	.000	1.58	152.0	25.7
No fertiliser	.202	.011	1.50	210.4	22.9
Low	.228	.020	2.02	160.2	22.2
Medium 1	.155	.007	2.20	165.8	23.9
Medium 2	.157	.005	3.50	21.2	23.9
High	.161	.006	4.21	55.1	22.5
Paddock ryegrass					
Light green	.177	.-	1.90	201.6	19.9
Dark green	.106	.-	5.61	39.2	17.2
Dark green	.101	.003	5.28	46.6	17.0

†Water Soluble Carbohydrate content
†† Modified acid-detergent fibre

(nitrogen: regression $F_{1,11} = 29.7^{***}$, $r = 0.85^{**}$ df = 11; water soluble carbohydrate: regression $F_{1,11} = 100.4^{***}$, $r = 0.95^{**}$ df = 11). Regressions of density on water content and modified acid-detergent fibre (MADF) were not significant (water content: regression $F_{1,11} = 3.1$ n.s., r = 0.47 n.s. df = 11; MADF: regression $F_{1,11} = 1.7$ n.s., r = 0.37 n.s. df = 11).

The shape of reflectance spectra curves determined with the spectroradiometer was very similar for all ryegrass samples measured and the peak value of spectral intensity (μW cm^{-2} nm^{-1}) occurred at 575 nm in all cases (Figure 3). Thus, the curve for each sample measured has been characterized in Table 1b by its peak reflectance value at 575 nm. The main visual difference among samples was not in the hue or colour, but brightness. Samples from trays receiving no fertiliser and from light green areas of the paddock were the brightest, in that they reflected most incoming radiation from the controlled lighting and they had the highest peak reflectance values. The regressions of peak reflectance on

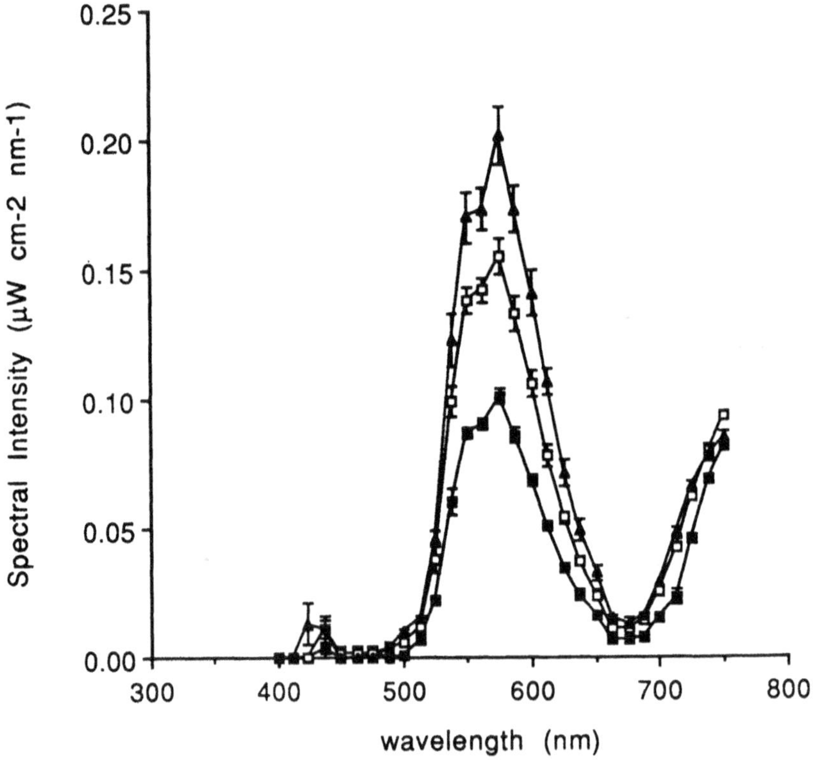

Figure 3. Spectral intensity (μW cm^{-2} nm^{-1}) showing amount of incoming radiation reflected from dark green ryegrass from paddocks (■), ryegrass in the medium 1 fertiliser group (□), and ryegrass in the control group receiving no fertiliser (▲). (mean ± SE, n = 3).

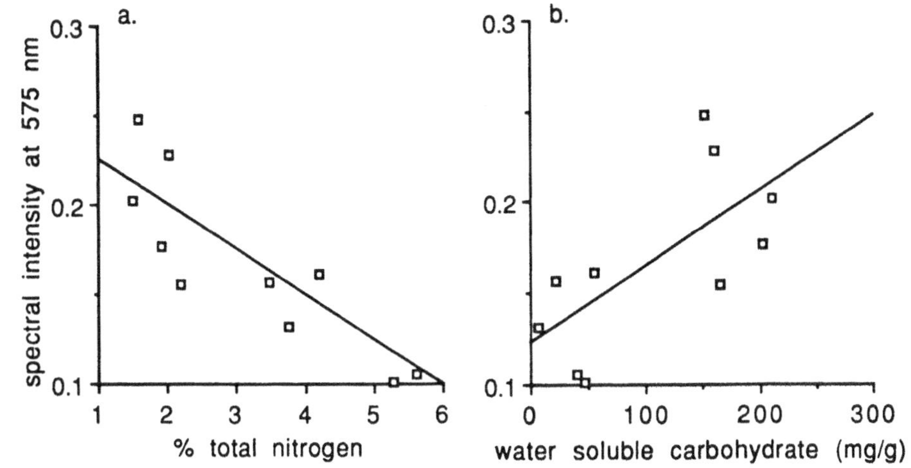

Figure 4. Relationship between spectral intensity (amount of light reflected - $\mu W\ cm^{-2}\ nm^{-1}$) and (a) total nitrogen content (% dry weight), (b) water soluble carbohydrate content (mg/g dry weight) of perennial ryegrass. (Lines are from simple regressions (a) $y = 0.253 - 0.027x$, (b) $y = 0.123 + 0.0004x$).

nitrogen content and water soluble carbohydrate content were significant (nitrogen: regression $F_{1,8} = 23.2^{**}$ $r = 0.86^{**}$ df = 8; water soluble carbohydrate: regression $F_{1,8} = 6.9^{*}$ $r = 0.68^{*}$ df = 8) (Figure 4) while that on fibre (MADF) was not ($F_{1,8} = 2.1$ n.s. $r = 0.46$ n.s. df = 8).

In groups of ryegrass grown under the same watering regime, increased brightness was negatively correlated with water content ($F_{1,10} = 40.5^{***}$, $r = 0.90^{**}$ df = 10; dark green paddock core excluded from regression). The darkest trays, receiving the greatest amounts of fertiliser, had the highest water content (Table 1a).

Responses of sheep in indoor choice trials between dark green and light green ryegrass

Methods

In view of the preceding results it was possible that in the paddock experiments, sheep were responding to variation in brightness of ryegrass because it was a cue to nutritional quality. I was interested in determining how fixed this preference for dark green ryegrass was. The next experiment investigated what decision a sheep would make when given a straight choice between ryegrass of two different brightness. I predicted that they would always prefer darker green trays if this was a rule used to select nutritionally (i.e. higher nitrogen) patches of ryegrass.

In the experimental set-up a sheep was allowed access to an inner pen from an outer holding pen when a gate was raised. Two trays of ryegrass that differed in brightness were placed at the far end of the inner pen. When the gate was raised, the sheep entered the pen and was allowed 30 seconds to graze from either or both trays. The amount of time spent grazing in different trays was recorded as was the number of times that the sheep switched back and forth. At the end of the 30 second period, the sheep was removed from the paddock. The two trays were then removed from the inner pen and replaced with two more trays. After a three minute interval, the sheep was again allowed to feed from the trays for 30 seconds. Two further trials of 30 seconds at three minute intervals were carried out. Fifteen sheep were tested in this way. They were deprived of food for 3-5 hours prior to testing.

Trays of ryegrass were 20 cm x 35 cm in size and were planted four months beforehand at 0.8g seed (var. Parcour) per tray. Trays of light green ryegrass had not been fertilised and were covered with black polythene sheeting three to four days prior to the choice trials. This resulted in etiolation and lightening of leaf colour. Trays of dark green ryegrass were fertilised with either Maxicrop liquid fertiliser or with 2 g of each of NH_4NO_3 and KH_2PO_4 salts one week before the trials began. Before being fed to sheep, all trays were trimmed to 10 cm height.

The relative brightness of the grass trays was assessed from photographic slides with a densitometer. The biomass from replicate trays was analysed for its nutritional content, as well as strength of leaves, characterized as the amount of force in grammes needed to break a leaf that registered on a pesola scale.

Results

During the first trial significantly more sheep selected the dark green ryegrass first (sign test: $p = 0.018$, $n = 15$) (Figure 5a). However, in further trials, there was no significant difference between the number of sheep selecting a light green or dark green ryegrass first (trial 2: $p = 0.85$; trials 3 & 4: $p = 0.50$) (Figure 5a). There was a significant decrease over the four trials in the number of sheep feeding from dark green trays first was significant ($t = 2.38^*$ df = 14) (see Bazely 1988 for details of these and following tests).

The amount of time spent grazing from trays of dark green ryegrass in each 30

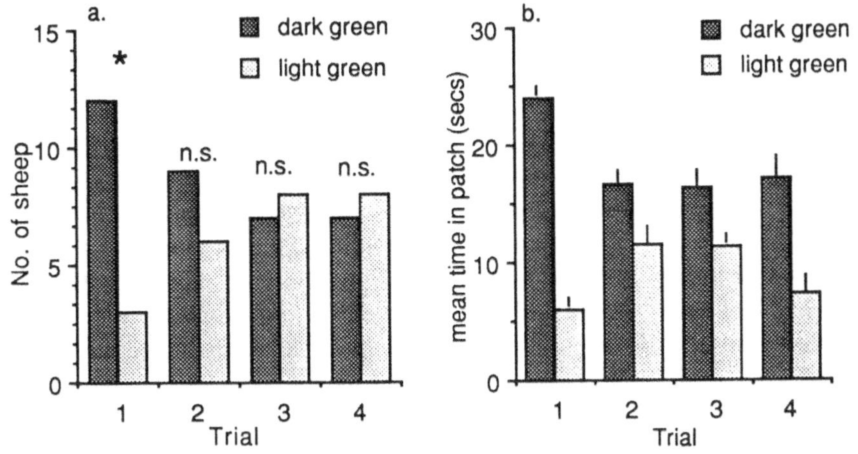

Figure 5. (a) Number of sheep choosing to graze first from dark green or light green perennial ryegrass and (b) time spent grazing by sheep (seconds) from dark green or light green perennial ryegrass during 30 second trials, in which individual sheep had free access to both trays. Each sheep was tested over 4 consecutive trials. (Times spent grazing are means; bars are standard errors; n = 15).

second period was compared with an expected allocation of 50% of time feeding in each tray if the sheep had no preference (Figure 5b). In trials 1 and 4 the sheep spent significantly greater time feeding from dark green ryegrass ($t_{trial\ 1} = 5.2^{***}$ df = 14; $t_{trial\ 4} = 2.96^{*}$ df = 14). In trials 2 and 3, sheep did not spend more time than expected feeding in either dark or light green trays ($t_{trial\ 2} = 1.33$ n.s. df = 14; $t_{trial\ 3} = 1.79$ df = 14).

Densitometer readings were highest for dark green trays (Table 2). The chemical content of dark green and light green ryegrass varied enormously (Table 2). There were in effect two categories of "dark" green ryegrass on offer to sheep with widely different nitrogen contents. These categories were related to whether trays of ryegrass were kept outside or inside the greenhouse in the autumn prior to use. While the fibre content of the light green grass was significantly higher than that of the two groups of dark green grass, the nitrogen content was in-between. Also, the breaking strength or "toughness" of dark green grass was greater. The correlation between brightness (densitometer reading) of ryegrass and its nitrogen and water content did not hold in these trials. As no record was kept of which trays of dark green ryegrass were offered to which sheep in particular trials, it was not possible to relate switches in preferences for dark and light green trays to the sheep feeding from the different groups of dark green trays.

Table 2. Physical and chemical characteristics of perennial ryegrass turves used in indoor colour choice trials, January 1986.

	Dark green (1)†			Dark green (2)††			Light green			one-way
	$\bar{x}$	SE	n	$\bar{x}$	SE	n	$\bar{x}$	SE	n	ANOVA
Grass strength (grammes required to break grass)	172.0	12.9	20	-		-	132.5	10.5	20	$F_{1,38}$ = 5.6*
Densitometer reading	1.18*	0.03	3	1.31*	0.01	3	0.81*	0.01	3	$F_{2,6}$ = 216.3***
Water content (% fresh weight)	76.8*	0.2	10	83.2*	0.8	6	86.9*	0.3	5	$F_{2,18}$ = 124.7***
Total nitrogen (% dry weight)	1.5*	0.003	10	4.2	0.7	6	3.1	0.1	5	$F_{2,18}$ = 42.8***
Fibre content (% Modified Acid Detergent Fibre)	24.8	1.1	9	22.3	0.5	6	34.8*	0.6	5	$F_{2,17}$ = 32.1***

†Turves left outdoors until January 1986 and fertilised with liquid fertiliser
††Turves brought into greenhouse in November 1985
*Significantly different from other groups at $p < 0.05$ based on Mean Significant Differences for unequal sample sizes (T-method, Sokal and Rohlf 1981).

DISCUSSION

These experiments were designed to give an understanding of how sheep perceive patchiness in monocultures. The results give some indication of possible rules and cues that sheep might use in selecting diets and whether they may actually be using rules at all.

The role of sward height in diet selection

Sheep may be expected to respond to variation in sward height because height is positively correlated with bite size (Black and Kenney 1984). The selection of taller swards would be a means of increasing intake rate. This would result in a decrease in foraging time if nutritional requirements were met within a shorter period. In field trials, total daily grazing time was observed to be less for sheep grazing taller swards than for animals on shorter swards (Allden and Whittaker 1970). Kenney and Black (1984) found that sheep preferred the shorter lengths of chopped, dried forage e.g. hay, which gave them increased intake rates.

The hypothesis that sheep should select taller areas of ryegrass in which to graze (Black and Kenney 1984) and the observation in the first experiment, that sheep preferred the tall patches of ryegrass created by nutrient inputs from sheep urine and faeces, runs counter to other observations of grazing behaviour, and a general consensus that sheep and cattle prefer to graze in shorter vegetation (Arnold 1964, 1966b; Van Soest 1982). It has been suggested that cattle may avoid taller patches of vegetation because they have been contaminated through urination and defaecation and because the ratio of nitrogen and phosphorus to other nutrients is greater and lower respectively than that to which cattle are accustomed (Plice 1951). However, when Plice's (1951) nutrient ratio hypothesis was tested by creating control patches with fertiliser, cattle preferred artificially created patches compared with those created by urination/defaecation (Marten and Donker 1964). Thus, some additional cue or rule other than height and brightness was being used by cattle to discriminate between different types of tall patch. Cattle may increase the risk of infection by internal parasites such as nematodes by feeding in contaminated patches and avoid them for this reason.

Sheep grazing in multi-species plant communities often prefer to graze in shorter areas of vegetation and avoid taller, rank areas dominated by unpalatable species e.g. the grass, *Nardus stricta* (Bakker *et al* 1983, Grant *et al* 1985. D. R. Bazely unpubl. data). In these circumstances, sheep may have used other cues to discriminate among plant species and thereafter used height to select preferred grazing sites from among acceptable patches of vegetation. It would be of interest to determine whether unpreferred species, which may have been of similar nutritional quality to preferred species, were much tougher, and whether rates of dry matter intake of sheep forced to feed from such patches were lower than for sheep feeding on preferred species.

Brightness: visual cue to nitrogen content of perennial ryegrass

The total nitrogen content of the background swards in paddocks was less than minimum levels required for maintenance of rumen microflora populations (Allden 1981, Simpson and Stobbs 1981, Van Soest 1982). Therefore, by feeding preferentially on dark green ryegrass patches, sheep increased their nitrogen intake to adequate levels. Brightness was a highly reliable cue to nitrogen and water soluble carbohydrate content of ryegrass in the field, so by using the rule "always feeded on the darkest patch" sheep would have maximized nitrogen intake. Brightness preferences have previously been demonstrated for insects. For example, Myers (1985) found that Cabbage White butterflies (*Pieris rapae*) preferred to oviposit on darker green cabbage plants (*Brassica oleracea* L.), which were higher in nitrogen and phosphorus than lighter green plants. The butterflies apparently used brightness or colour as a cue to plant quality (Myers 1985).

The importance of visual versus olfactory cues in diet selection

There is an implicit assumption in the discussion thus far that sheep were using their sense of sight to respond to variation in the visual appearance of ryegrass swards.

From his experiments investigating the role of vision in diet selection in sheep, Arnold (1966a) concluded that sight was not particularly important, and that its main function was to orient the animal to its environment since in most of his experiments, the diets selected by blindfolded animals were not significantly different in species composition or nutritional quality from those selected by control animals. Unfortunately, sight has been found to be extremely difficult to manipulate in sheep (Tribe 1950, Arnold 1966a): blindfolded sheep do not behave normally.

These experiments have been cited as demonstrating the lack of importance of vision in diet selection by sheep (e.g. Arnold and Dudzinski 1978, Arnold 1981). However, other herbivores such as the frugivorous monkey, *Callicebus torquatus*, are thought to locate food by using primarily visual cues, and to respond to both variation in colour and brightness of fruit (Snodderly 1979). Sheep have binocular vision (Whitteridge 1978) and are able to discriminate detail on a fine enough level to distinguish between the faces of familiar and unfamiliar conspecifics (Kendrick and Baldwin 1987). In addition, Arnold (1966a) described some experimental results in which blindfolded sheep grazing in monocultures did select significantly less nutritious diets.

While tactile and taste cues used in diet selection can only operate while the sheep is biting, both olfactory and visual cues may operate over distance. The relative importance of vision and olfaction in my experiments remains undetermined, but could best be investigated by impairing the sense of smell of the sheep. This may be done either by surgical removal of the olfactory lobes (Arnold 1966b) or by spraying a local anaesthetic into the nostrils (Chapple *et al* 1987). The latter method may be more viable than surgical impairment, in future experiments.

Although sheep may have used olfactory cues to discriminate between perennial ryegrass swards varying in brightness, evidence from other experimentss suggested that sheep used primarily vision (Bazely 1988). Also, in experiments testing visual discrimination in sheep, individuals were required to learn to associate a food reward with a brightness cue consisting of a dark or light card (Bazely and Ensor 1989). Olfactory cues were completely controlled and could not have influenced the high success rate of sheep performing this task.

Senses and cues used in multi-species plant communities

Brightness was a reliable cue to nitrogen and water soluble carbohydrate content within a ryegrass monoculture, but in a multi-species plant community its reliability may decline. Leaves of different plant species may overlap in brightness, yet have completely different chemical and nutritional characteristics. However, variation in brightness may still have relevance as a rule of thumb and intra-specific cue to nutritional quality if sheep are able to discriminate between plant species in some other way.

Harper (1977) suggested that herbivores may recognize the shapes of plants. Sheep apparently distinguished between different clover polymorphisms by recognizing white leaf marks (Cahn and Harper 1976). It would be of interest to determine whether sheep are able to distinguish plant species on the basis of their general appearance. It might be possible to test this by showing slides of different plant species to sheep, in experimental set-ups similar to those used by Kendrick and Baldwin (1987), or by Pietrewicz and Kamil (1981) with Blue jays (*Cyanocitta cristata*). This would control for the influence of olfactory cues.

The role of colour vision is also of interest, although it has long been assumed that sheep are monochromats (Tribe and Gordon 1949, Fontenot and Blaser 1965). Although Bazely and Ensor (1989) also failed to demonstrate colour vision behaviourally, cones, which are the photoreceptor cells associated with colour vision (Jacobs 1981) have been found in sheep retinae (C. V. Ensor and D. R. Bazely unpubl. data). Also, colour vision has been demonstrated in a variety of other ungulates such as Pygmy goats (*Capra hircus* L.) and Red deer (*Cervus elaphus* L.) which were also able to discriminate fine detail in patterns even under relatively dim light conditions (von Backhaus 1959). Thus, the question of whether sheep have and use colour vision remains unresolved.

The exact role of the senses of smell and taste in diet selection in sheep also remains to be resolved (Arnold *et al* 1980). Individuals deprived of the ability to smell and taste selected different diets compared with those in which these senses were unaffected, suggesting that smell and taste are used to discriminate between plant species (Arnold 1966b, Arnold *et al* 1980). Different plant species contain unique combinations of secondary compounds, many of which are aromatic or have distinctive tastes. Sheep have varying preferences for the the odour and taste of a variety of specific compounds present in plants, such as gramine and tannic acid (Arnold *et al* 1980). Interestingly, preferences varied according to the concentration of the compound, but this was not related to concentrations present in plants (Arnold *et al* 1980). The sense of smell does not apparently influence selection for different varieties of the same species (Tribe 1949, Arnold 1966b). No significant difference in growth was found between groups of sheep with intact senses of smell and taste and those with impaired senses, leading Arnold (1966b) to conclude that these senses were not always critically involved in total intake of nutrients.

The role of learning

Thus far, I have discussed sensory mechanisms which sheep may have used to discriminate ryegrass patches varying in brightness and height, as well as functional explanations for observed preferences. I have not considered whether decisions to select certain heights or brightnesses constituted simple, inflexible rules or whether they were based on learned relationships between cues and aspects of diet quality which were subject to modification.

It is still unclear as to just how good herbivores are at learning about the nutritional quality of diet items. In past reviews of diet selection, it was suggested that herbivores have some kind of nutritional wisdom or "euphagia" that allows them to select forage on the basis of its nutritional content and toxicity (Tribe 1950, Arnold and Dudzinski 1978, Arnold 1981). An early mechanism proposed to account for nutritional wisdom was that animals have "specific hungers" for particular nutrients which are innate and require no previous experience (Richter 1943 cited in Zahorik and Houpt 1981, Tribe 1950) i.e. animals possess specialized receptors which allow them to detect the presence of a particular nutrient or poison in food. While a specific hunger appears to exist for sodium intake in a variety of species, including sheep (Denton and Sabine 1963), it is unlikely that animals possess the great number of specialized receptors required to detect all necessary nutrients and possible toxins that may be encountered

(Zahorik and Houpt 1981).

More recently, psychologists have examined the role of learning in diet selection (Zahorik and Houpt 1981, Shettleworth 1984). Herbivores can learn to associate immediate effects of a food item with that item. (Zahorik and Houpt 1981, Olsen and Ralphs 1986). For example, cattle, sheep, ponies and goats were able to learn to avoid a novel food paired with poisoning as long as the food was presented alone, and the consequences followed immediately after ingestion (Zahorik and Houpt 1981). Zahorik and Houpt (1981) have questioned whether herbivores can relate the delayed effects, following ingestion, of a specific diet item to the item in question, since herbivores do not have discrete meals and ingest diets composed of many different plant species. However, lambs and ewes learned to associate poisoned food with gastrointestinal consequences which took effect up to two hours after ingestion (Thorhallsdottir *et al* 1987). Interestingly, the sheep never completely stopped eating the poisonous food and always included a small amount in their diet (Thorhallsdottir *et al* 1987). Consumption of some foods increased after a positive experience, demonstrating that foods which ceased to have harmful effects were also detected. The continued consumption of small amounts of food may be evidence of sampling by sheep, which Westoby (1974) suggested should be a feature of herbivore foraging behaviour: herbivores must somehow keep track of a fluctuating environment, in which the nutritional value of different plant species and parts of plants is continually changing. In the paddock experiments sheep always fed for sometime in the short, light green background sward and this may also have been sampling.

I do not know whether the preference of sheep for dark green and tall patches of ryegrass was the result of some learned association between these cues and nitrogen content or intake rate. A learned preference would allow a rule to be flexible. However, it seems unlikely that the sheep learned that the association between brightness and nitrogen was unreliable in the 30 second indoor trials with ryegrass of different brightnesses and that this caused the reduction in the preference for the dark green ryegrass. Alternative explanations exist which do not require the sheep to have changed the rule: select the darkest green ryegrass. Some other non-visual cue or rule e.g. toughness of vegetation, may have been influencing selection. Alternatively, the sheep may not have bothered to select dark green ryegrass first after trial 1 because they learned that they had access to both trays during the 30 second trial. The experimental design could have been improved by including choice trials in which sheep were denied access to the other tray once a choice was made. It is noteworthy that sheep always fed from dark green ryegrass patches for longer and that this difference was significant in two of four trials.

RELEVANCE OF THE MARGINAL VALUE THEOREM FOR THESE RESULTS

In subsequent experiments, intake curves or gain functions were determined for sheep feeding from trays of ryegrass. These were best described by a smoothly decelerating curve: as the sheep bit through the sward, bite size and therefore intake rate declined (Bazely 1988). Patches of ryegrass used in these gain function determinations were of similar size to those used in the 30 second choice trials described above. For very short travel times between such patches e.g. trays placed one metre apart in a pen, the optimal patch residence time for maximizing dry matter intake was in the order of 10-15 seconds to be spent feeding from a turf before switching to another. Thus, the MVT provides another interpretation of why sheep switched between trays of dark green and light green ryegrass. The rules used by sheep to determine when to leave a patch of ryegrass are less clear that those which they may use to select a patch initially. They may respond to changing bite size, but Illius and Gordon (this volume) found that sheep did not track rapid changes in intake rate well.

Predicted optimal patch residence times from the gain functions for various travel times between patches were similar to observed times spent by grazing sheep in patches of ryegrass transplanted into paddocks for those travel times. The MVT predicted that at inter-patch travel times of less that 30 seconds, sheep should leave the transplanted patches, which were much taller and darker green that the background sward, before fully depleting them. This type of behaviour was observed in both experiments with transplanted patches of ryegrass and in the first experiment described in this paper.

SUMMARY: AN EVALUATION OF RULES AND CUES USED BY SHEEP IN DIET SELECTION

In this paper I have suggested two possible rules that sheep may use to select forage within monocultures of perennial ryegrass (1) select the tallest ryegrass, and (2) select the darkest ryegrass. In field experiments sheep were shown to have a high preference for tall, dark green patches of ryegrass. These rules require sheep to discriminate different heights and brightnesses, which they are able to do (Bazely and Ensor 1989). Both cues (height and brightness) are directly or indirectly correlated with parameters suitable for inclusion as currencies in foraging models: brightness is correlated with nitrogen content of vegetation while height is correlated with bite size which is in turn related to dry matter intake rate and energy intake.

However, sheep were apparently not consistent in their preference for dark green ryegrass when brightness did not correlate with nitrogen content. Thus, the policy to select a certain brightness may not be a fixed rule or may be contingent on some other condition being met. Alternatively, brightness may simply not be a relevant cue for sheep.

The existence of other simple rules used by sheep in forage selection, is implicit in an observation of Arnold *et al* (1966) who pointed out the very consistent preference hierarchy of: old green leaf, green stem, dry leaf, dry stem. In addition, the observation that cattle tend to feed only in the top horizon of swards (Grant *et al* 1985) not only lends support to the "passive selection" explanation of how more nutritious forage is selected but may also be considered as another simple choice rule. Younger, more nutritious vegetation tends to occur in the upper horizons of swards, so lawn-mower like grazing (passive selection) will result in a diet of higher nutritonal content being selected than the average for the sward (Grant *et al* 1985).

Further tests of the predictions of foraging models in simplified grazing environments would be helpful in elucidating the basis of diet selection in vertebrate herbivores, as would experiments of the type described in this paper and those of Black and Kenney (1984) and Kenney and Black (1984). They have manipulated only one or two variables which may be of importance in determining how sheep make diet choices, while holding other variables constant.

ACKNOWLEDGEMENTS

I thank Robin McCleery, Alasdair Houston and Rudi Drent for helpful discussions. Beth Okamura and Gary Belovsky kindly read and commented on the manuscript. I thank Peter Ewins and Clare Ensor for all of their help with field work and Malcolm Ryder, Cyril Band, Frank Thompson and Peter Mitchell for access to various pieces of photographic equipment, densitometer and spectroradiometer. I was supported by a Royal Commission for the Exhibition of 1851 Postgraduate Scholarship and the Ernest Cook Research Fellowship in Environmental Studies (Somerville College Oxford) while carrying out this research.

REFERENCES

Allden WG (1981) Energy and protein supplements for grazing livestock. In: Morley FWH (ed) Grazing Animals: World Animal Science. B. Disciplinary Approach (1). Elsevier, Amsterdam, p 289

Allden WG, Whittaker IA (1970) The determinants of herbage intake by grazing sheep: the inter-relationship of factors influencing herbage intake and availability. Aust J Agric Res 21: 755-766

Arnold GW (1964) Plant associations affecting grazing animals. In: Crisp DJ (ed) Grazing in Terrestrial and Marine Environments. British Ecological Society Symposium no. 4. Blackwell Scientific, Oxford, p 133

Arnold GW (1966a) The special senses in grazing animals. I. Sight and dietary habits in sheep. Aust J Agric Res 17: 521-529

Arnold GW (1966b) The special senses in grazing animals. II. Smell, taste and touch and dietary habits in sheep. Aust J Agric Res 17: 531-542

Arnold GW (1981) Grazing Behaviour. In: Morley FWH (ed) Grazing Animals: World Animal Science. B. Disciplinary Approach (1). Elsevier, Amsterdam, p 79

Arnold GW, Dudzinski ML (1978) Ethology of free-ranging domestic animals. Elsevier, Amsterdam

Arnold GW, Ball J, McManus WR, Bush IG (1966) Studies on the diet of the grazing animal. I. Seasonal changes in the diet of sheep grazing on pastures of different availability and composition. Aust J Agric Res 17: 543-556.

Arnold GW, de Boer ES, Boundy CAP (1980) The influence of odour and taste on the food preferences and food intake of sheep. Aust J Agric Res 31: 571-587

Backhaus D von (1959) Visual acuity in Ungulates. Zeits Tierpsychol 16: 445-467

Bakker JP, de Leeuw J, van Wieren SE (1983) Micro-patterns in grassland vegetation created and sustained by sheep-grazing. Vegetatio 55: 153-161

Bazely DR (1988) Foraging behaviour of sheep (*Ovies aries* L.) grazing on swards of perennial ryegass (*Lolium perenne* L.). D. Phil. Thesis, Oxford University, Oxford

Bazely DR (1989) Carnivorous herbivores: mineral nutrition and the balanced diet. Trends Ecol Evol 4:155-156

Bazely DR, Ensor, CV (1989) Discrimination learning in sheep with cues varying in brightness and hue. Appl Anim Behav Sci 23: 293-299

Belovsky GE (1978) Diet Optimization in a Generalist Herbivore: The Moose. Theor Pop Biol 14: 105-134

Belovsky GE (1984a) Snowshoe Hare Optimal Foraging and its Implications for Population Dynamics. Theor Pop Biol 25: 235-264

Belovsky GE (1984b) Herbivore Optimal Foraging: a Test of Three Models. Am Nat 124: 97-115

Black JL, Kenney PA (1984) Factors affecting diet selection by sheep. II. Height and density of pasture. Aust J Agric Res 35: 565-578

Bryant JP, Kuropat PJ (1980) Selection of winter forage by subarctic browsing vertebrates: the role of plant chemistry. Ann Rev Ecol Syst 11: 261-285

Cahn MA, Harper JL (1976) The biology of the leaf mark polymorphism in Trifolium repens L. II. Evidence for the selection of leaf marks in rumen fistulated sheep. Heredity 37: 327-333

Chapple RS, Wodzicka-Tomaszewska M, Lynch JJ (1987) The learning behaviour of sheep when introduced to wheat. II. Social transmission of wheat feeding and the role of the senses. Appl Anim Behav Sci 18: 163-172

Charnov EL (1976) Optimal foraging: the marginal value theorem. Theor Pop

Biol 9: 129-136
Crawley MJ (1983) Herbivory: the dynamics of animal-plant interactions. Blackwell Scientific, Oxford
Denton DA, Sabine JR (1963) The behaviour of Na deficient sheep. Behaviour 20: 364-376
Fontenot JB, Blaser RE (1965) Symposium on factors influencing voluntary intake of herbage by ruminants: selection and intake of grazing animals. J Anim Sci 24: 1202-1208
Freer M (1981) The control of food intake by grazing animals. In: Morley FWH (ed) Grazing Animals: World Animal Science. B. Disciplinary Approach (1). Elsevier, Amsterdam, p 105-124
Grant SA, Suckling DE, Smith HK, Torvell L, Forbes TDA, Hodgson J (1985) Comparative studies of diet selection by sheep and cattle: the hill grasslands. J Ecol 73: 987-1004
Harper JL (1977) Population Biology of Plants. Academic Press, London
Hodgson J, Mackie CK, Parker JWG (1986) Sward surface heights for efficient grazing. Grass Farmer 24: 5-10
Houston A, Kacelnik A, McNamara J (1982) Some learning rules fro acquiring information. In: McFarland D (ed) Bioscience Research Reports: Research notes in animal behaviour I. Pitman Adv. Pub. Prog, Boston, p 140
Illius AW, Gordon IJ (to be published) Constraints on diet selection and foraging behaviour in mammalian herbivores. In: Hughes RN (ed) Behavoural Mechanisms of Food Selection. Springer-Verlag, Berlin
Jacobs GH (1981) Comparative Color Vision. Academic Press, New York
Kendrick KM, Baldwin BA (1987) Cells in temporal cortex of conscious sheep can respond preferentially to the sight of faces. Science 236: 448-450
Kenney PA, Black JL (1984) Factors affecting diet selection by sheep. I. Potential intake rate and acceptability of feed. Aust J Agric Res 35: 551-563
Krebs JR, McCleery RH (1984) Optimization in Behavioural Ecology. In: Krebs JR, Davies NB (eds) Behavioural Ecology: An Evolutionary Approach, 2nd edn. Blackwell Scientific, Oxford, p 91
Krebs JR, Stephens DW, Sutherland WJ (1983) Perspectives in optimal foraging. In: Brush AH, Clark GA (eds) Perspectives in Ornithology. Cambridge University Press, New York, p 165
Marten, GC, Donker JD (1964) Selective grazing induced by animal excreta. II. Investigation of a causal theory. J Dairy Sci 47: 871-874
Mattson WJ (1980) Herbivory in relation to plant nitrogen content. Ann Rev Ecol Syst 11: 119-161
Myers JH (1985) Effect of physiological condition of the host plant on the ovipositional choice of the cabbage white butterfly, *Pieris rapae*. J Anim Ecol 54: 193-204
Olsen JD, Ralphs MH (1986) Feed aversion induced by intra-ruminal infusion with larkspur extract in cattle. Am J Vet Res 47: 1829-1833
Ørskov ER (1982) Protein Nutrition in Ruminants. Academic Press, London
Owen-Smith DN, Novellie P (1982) What should a clever ungulate eat? Am Nat 119: 151178
Parker MA (1984) Local food depletion and the foraging behaviour of a specialist grasshopper, *Hesperotettix viridis*. Ecology 65: 824-835.
Pietrewicz, AT, Kamil AC (1981) Search Images and the Detection of Cryptic Prey: An Operant Approach. In: Kamil AC, Sargent TD (eds) Foraging Behaviour: Ecological, Ethological and Psychological Approaches. Garland STPM Press, New York, p 311
Plice MJ (1951) Sugar versus the intuitive choice of food by livestock. J Range Man 5: 69-74
Rohlf FJ, Sokal RR (1981) Statistical Tables, 2nd edn. W. H. Freeman, New York
Shettleworth SJ (1984) Learning and Behavioural Ecology. In: Krebs JR, Davies NB (eds) Behavioural Ecology: An Evolutionary Approach, 2nd edn.

Blackwell Scientific, Oxford, p 170
Simpson JR, Stobbs TH (1981) Nitrogen supply and animal production from pastures. In: Morley FWH (ed) Grazing Animals: World Animal Science. B. Disciplinary Approach (1). Elsevier, Amsterdam, p 261
Snodderly DM (1979) Visual discriminations encountered in food foraging by a neotropical primate: Implications for the evolution of color vision. In: Burtt EH (ed) The Behavioral Significance of Color. Garland STPM Press, New York, p 237
Sokal RR, Rohlf FJ (1981) Biometry, 2nd edn. W. H. Freeman, New York
Stephens DW, Krebs JR (1986) Foraging Theory. Princeton University Press, Princeton, New Jersey
Thorhallsdottir AG, Provenza FD, Balph DF (1987) Food aversion learning in lambs with or without a mother: discrimination, novelty and persistence. Appl Anim Behav Sci 18: 327-340
Tribe DE (1949) The importance of the sense of smell to the grazing sheep. J Agric Sci 39: 309-313
Tribe DE (1950) The behaviour of the grazing animal: a critical review of present knowledge. J Brit Grass Soc 5: 209-224
Tribe DE, Gordon JG (1949) The importance of colour vision to the grazing sheep. J Agric Sci 39: 313-314
Van Soest PJ (1982) Nutritional Ecology of the Ruminant. O & B Books, Inc., Corvallis, Oregon
Wakefield GL (1970) Practical Sensitometry. Fountain Ltd., London
Westoby M (1974) An analysis of diet selection by large generalist herbivores. Am Nat 108: 290-304
Westoby M (1978) What are the biological bases of varied diets? Am Nat 112: 627-631
Whitteridge D (1978) The development of the visual system in the sheep. Arch Ital Biol 116: 406-408
Zahorik DM, Houpt KA (1981) Species Differences in Feeding Strategies, Food Hazards and the Ability to Learn Food Aversion. In: Kamil AC, Sargent TD (eds) Foraging Behaviour: Ecological, Ethological and Psychological Approaches. Garland STPM Press, New York, p. 289

ORAL DISCUSSION - D.R.Bazely

There was concern over the negatively accelerating return–rate curves and whether the slopes had been tested. D.B. indicated that a straight fit had been used. Others thought the curves were acceptable but questioned whether significant depletion had occurred in the patches during the trials. In the short–travel–time experiments, the animals always left within the time set by the confidence limits. In the longer–travel–time experiments, animals stayed longer, as predicted by the marginal value theorum (MVT), and experienced significant depletion. This was attributed to learning by the sheep, since they "knew" there were fewer patches. The use of MVT to support the results of the renewal and depletion trials was challenged. The possible use of bulk density as a cue for leaving a patch was raised. DB responded that the sheep probably reacted to decreasing bite size. Questions were also raised about nocturnal feeding, since this occurs in many herbivores. DB thought sheep feeding at night would still maximize their food intake and be able to do so on the basis of bite size. Several questions were raised on the role of vision and whether taste might not be involved in nocturnal feeding. This was considered a likely possibility. The issue of colour vision was raised and the conclusion was that for the green and yellow cue cards used, there was no response. However sheep do have cones and so may have some colour vision. It was suggested that colour vision could be tested using negative rewards.

CONSTRAINTS ON DIET SELECTION AND FORAGING BEHAVIOUR IN MAMMALIAN HERBIVORES

A. W. Illius*
I. J. Gordon+
*Department of Agriculture
University of Edinburgh
West Mains Rd
Edinburgh EH9 3JG
Scotland

INTRODUCTION

In a recent critique of optimality theory, Gray (1986) questioned the validity and usefulness of the optimality approach to the empirical study of foraging behaviour. He called instead for an epigenetic approach integrating morphological, physiological and behavioural processes. Without denying the role of optimality theory as a quantification of the premise that natural selection leads to adaptation (see Krebs & McCleery 1984), it must be recognised that optimality is an obscure condition to the extent that constraints are poorly defined. In this paper, we attempt to clarify and quantify the morphological, physiological and behavioural constraints on foraging in mammalian herbivores.

Herbivores inhabit a world in which at first sight the food plants are more or less continuously distributed in space and time. By comparison with a carnivorous predator, whose food is patchily distributed, highly nutritious and mobile, the herbivore confronts food plants where the nutrients are more evenly distributed, at low density, and whose accessibility is further reduced by structural and chemical properties of the vegetation. Variation in these properties, such as biomass density, extent of lignification and chemical content, is important in determining the potential value of a patch of vegetation, but may not present clear visual cues to animals. Furthermore, the morphology of the incisor dentition determines the extent to which an animal can prehend and ingest plant food items from within the spatial array of vegetation. Herbivores

+Macaulay Land Use Research Institute Pentlandfield Roslin Midlothian EH25 9RF Scotland

NATO ASI Series, Vol. G 20
Behavioural Mechanisms of Food Selection
Edited by R. N. Hughes

normally practise partial consumption of food items, and the degree or extent of partial consumption is the result of mechanical interactions, variation in feeding style and behavioural responses to the perceived physical properties of the vegetation. It follows that the outcome of foraging choices may not become clear until after consumption and may even be delayed until digestion has taken place. Since the herbivore may be taking 10,000-40,000 bites from the vegetation in a day, the significance of individual foraging choices for nutrient yield is likely to be lost. So the main problem for a herbivore is how to gain and process information about its foraging choices. This is represented schematically in Fig 1, showing the channels through which information is processed and the temporal separation of prospective faculties such as sight and smell, which precede consumption, and retrospective faculties such as touch, taste, gut distension and nutrient absorption. This paper reviews how these processes impose constraints on the foraging behaviour and diet selection of ungulates.

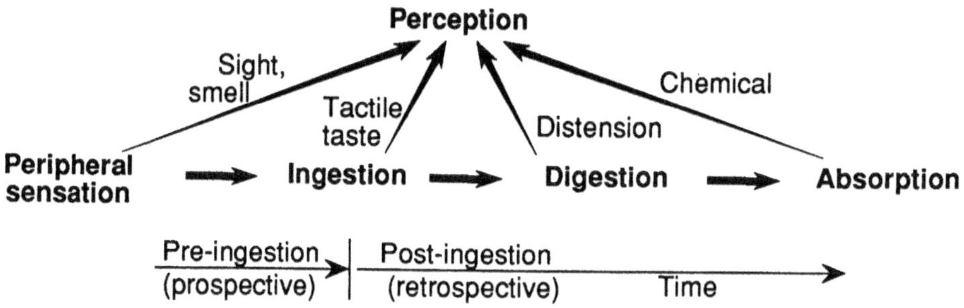

FIG 1. Flow diagram of the processing of food and information.

Since the ultimate goal of a foraging strategy is efficient nutrient assimilation, it makes sense to start by examining how food and animal characteristics determine digestion and nutrient absorption. Foraging theorists have generally ignored the influence of digestion and nutrient absorption on foraging decisions by using animal models for which it can reasonably and conveniently be assumed that digestive efficiency is high and constant for alternative food items and across individuals. The size and spatio-temporal distribution of food items have been viewed as the dominant variables influencing foraging behaviour, with some exceptions (theoretical: Sibly 1981; Penry & Jumars 1987; herbivores: Jarman 1974; Owen Smith & Novellie 1982; Belovsky 1978,1986).

The structural complexity of vegetation, with its highly variable composition and potential digestibility, necessitates the treatment of food quality as a variable equal to conventional variables in its capacity to influence choice. Semantically there may not seem to be much difference between patchily distributed nutriment (say, mealworms) and nutrients patchily distributed within a clump of vegetation. The difference is that in plants the nutrients are not separated by space but by indigestible chemical components, chiefly lignocellulose, which acts as an anti-nutrient, and this constraint pervades all herbivory.

CONSTRAINTS RELATED TO BODY SIZE

Body size and related variables have important consequences for diet selection in mammalian herbivores. An influential paper by Demment & Van Soest (1985) gave strong qualitative support to the hypothesis, originally proposed by Bell (1970) and Jarman (1974), that body size determines the ability of an animal to utilise fibrous foods, since large animals have greater gut capacity ($\propto W^1$) relative to energy requirements ($\propto W^{0.75}$). However, their model was quantitatively imperfect, lacking an explicit linkage with the plant characteristics known to influence the kinetics of digestion (Van Soest 1982). Other treatments of food quality influences on herbivore foraging behaviour also lack a specification of plant characteristics that is fully integrated with animal processes. For example, Owen-Smith & Novellie (1982) presented a model predicting the diets of kudu (a 250 kg browsing ungulate) based partly on correlations established in sheep (a 50 kg selective grazer) of nitrogen content with digestibility in foods different from those available to the kudu. Likewise turnover characteristics of foods in the rumen of kudu were estimated using sheep, and constant digestion of the cell wall fraction was assumed. These were probably the best estimates available at the time, but the same cannot always be said of other studies. Belovsky (1978,1986) recognised the possible role of animal size and digestive capacity constraints on diet choice. However, the most important constraint on forage intake in the Belovsky model, "bulkiness", is determined as the ratio of wet to dry weight; evidence from the literature (see Hanley 1980) indicates that this does not affect intake at the low water contents quoted by Belovsky (1986). In addition, his expression for energy needs is not compatible with that for food energy yield without considering fermentation and metabolic losses amounting to about 45% of digestible energy intake, the exact value depending on the food and the type of digestive system. These two inaccuracies have a big effect on the position of the constraint lines, and suggest that while the approach has been conceptually useful, the predictions are invalid in some cases (see Belovsky 1986 for discussion of the sensitivity of the model). The sordid details of the real world cannot be ignored, even by theoreticians, and far less by empiricists, since both require a sufficiently exact description of reality to distinguish between alternative hypotheses.

To achieve a better quantitative understanding of how food and animal variables influence the rate of energy yield from the gut, we developed a model of rumen kinetics to predict intake and digestion of any specified forage (Illius & Gordon, submitted). Intake in each meal was defined as the amount of food required to refill the digestive tract after the partial digestion and passage of previous meals. The clearance of food from the digestive tract is the result of the differential rates of digestion and passage of cell contents, cell wall and indigestible residues. Ruminants selectively retain large fibre particles in the rumen until digestion and physical comminution has reduced particle size to about 1 mm (Poppi, Minson & Ternouth 1981). This acts to delay passage and allow more extensive microbial digestion of a food fraction which the animal is otherwise incapable of digesting. Passage rate is the inverse of retention time, which scales with $W^{0.27}$ (Illius & Gordon, submitted), in common with the scaling of the duration of other time-related physiological variables (Taylor 1980; Peters 1983). The model determines intake and digestion for given values of cell contents, digestible cell wall and their digestion rates, and by determining passage rate from W. The resulting energy yield from digestion can then be expressed as the multiple of the maintenance energy requirements the modelled animal can obtain from an abundant food eaten at the maximum rate allowed by clearance of digesta and residues from the gut.

The model was used to determine the different foods required by an imaginary guild of four African ruminants to satisfy their maintenance requirements for energy. The animals were: Thomson's gazelle, a browser of 20 kg body mass making highly selective use of forbs; impala, feeding on browse and weighing 60 kg; wildebeest, a grazer of 180 kg; and African buffalo, a grazer of 540 kg. Characteristics for a range of qualities of browse, forbs and grass were determined from the data of Smith, Goering & Gordon (1972) and Hanley & McKendrick (1983). Browse, in the form of leaves of shrubs, is characterised by a high proportion of cell solubles but poorly digestible cell walls; grass cell solubles proportion declines to low levels with advancing maturity, but cell walls are more digestible at an equivalent phenological stage to browse; and forbs are intermediate in these respects, resembling browse more than grass. Fig 2 shows the effect of body size on the energy yield from the gut, relative to maintenance needs, when these foods are eaten at maximum daily intake. Large animals can utilise poorer quality foods, as shown by the potential digestibilities (an empirically determined standard) ranging from 0.79 to 0.7 over the body size range. This is because they can eat more and retain the cell wall fraction of food for longer, allowing more extensive digestion, in their relatively large guts. The model predicts that small animals obtain about 80% of digested energy from the cell solubles fraction, compared with about 50%

in the larger two animals. It is clear that the quality of grass which is adequate for the grazers can only provide about 75% of the requirements of the browsers, demonstrating the importance of body size for diet selection. However, more highly digestible grass, for instance fresh growth following rains, would be sufficient to sustain the smallest animals, and this is accompanied by seasonal shifts in the diets of these animals (see also Gordon & Illius 1988).

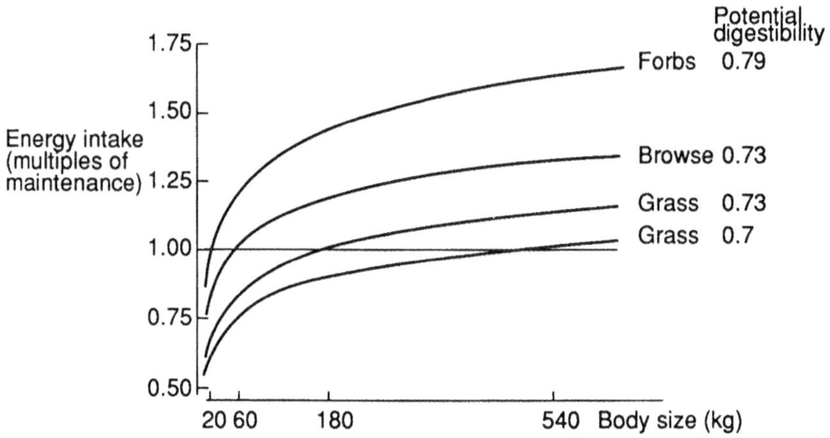

Fig 2. Relationship between body size, food quality and energy yields relative to daily maintenance energy needs.

The quality of forbs which would maintain the gazelle would provide 1.6 times maintenance if eaten by the buffalo (1.3 for the wildebeest). Why then are forbs and browse not selected by the two larger animals? The reason may be that forbs have low biomass compared with grasses, and occupy the base of the vegetation profile. Browse is widely dispersed, generally as single items (leaves, fruit) on perennial woody tissue. Grazing animals have a dental morphology which allows a broad and flattened incisor arcade to be presented to the vegetation, but prevents the accurate selection of individual plant parts, while browsers have narrower and more pointed incisor arcades, capable of greater selectivity (Gordon & Illius 1988). Even if the large grazers could select such food items, they could not contribute much to energy intake because of their small size. If it is assumed that a browser could take up to 15,000 food items in a day (Gordon 1986), then the food items of a sufficient size to meet the animals energy needs would only provide a fraction of the needs of larger animals. This is shown in Table 1, where the minimum item sizes of the forbs and browse adequate for gazelle and impala are inadequate for larger animals if all animals were to take the same number of them. This is a reasonable assumption given that large animals cannot feed much more rapidly or for longer than small animals (Gordon 1986). So the distribution of food item sizes and potential digestibility determines broadly the diets that animals of different body size must select.

Table 1. Daily energy intake, expressed in multiples of maintenance energy needs, from small items of food

	W (kg)	Forbs item size+ (mg)	Energy intake	Browse item size+ (mg)	Energy Intake
Thomson's Gazelle	20	3.6	1	8.5	0.9*
Impala	60	3.6	0.45	8.5	1
Wildebeest	180	3.6	0.21	8.5	0.46
Buffalo	540	3.6	0.09	8.5	0.2

+chosen to suit gazelle or impala
*limited by digestive capacity

What are the implications of body size for intake rate from grazed swards as compared with intake rate of browse food items of finite size? Our results quantify the disadvantages of large body size under conditions of resource depletion, where low biomass limits intake rate. If the guild of animals were to graze from short swards, maintained in a highly digestible and vegetative state by grazing pressure (McNaughton 1984; Illius, Wood-Gush & Eddison 1987) then it has been shown that the small animals will have an advantage over large animals (Illius & Gordon 1987). This was demonstrated by a model of grazing processes which describes how dental morphology and vegetation structure interact to determine the weight and composition of plant material which can be removed in a single bite. Comparing animals taking 30 000 bites in a day (close to the maximum) shows how body size and variation in vegetation characteristics such as height, biomass concentration and vertical distribution of live and dead plant parts affects the functional response (daily energy intake rate relative to requirements). The model now incorporates the relationships between animal size, digestive efficiency and physical constraints to intake examined above. The model was used to predict energy intake from a medium density sward with a majority of highly digestible components, and with 20% dead material at the base of the sward profile. The results show that short swards impose greater limitations on food intake by larger animals than smaller animals (Fig 3a).

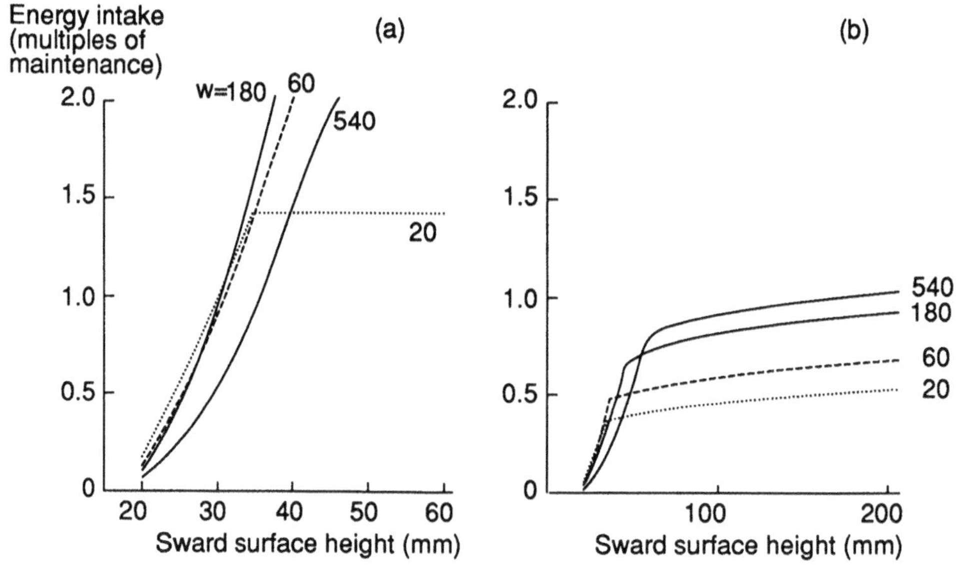

Fig 3. Sward height effects on daily energy intake of animals differing in weight on (a) vegetative and (b) senescent swards.

On swards with a surface height of 30 mm, the buffalo is severely restricted (half maintenance) while the gazelle can obtain maintenance, and the other two animals can get 0.9 of maintenance. Above this sward height, the comparison becomes complex, because differences between browsers and grazers in dental morphology begin to have an effect, and the gazelle runs into digestive capacity constraints. On sparse swards with a large proportion of dead material (Fig 3b), the combined effects of incisor morphology constraints and digestive capacity constraints cause a reversal of the scaling of intake with body size from $W^{0.36}$ (on very short swards) to $W^{0.9}$ (on taller swards). The implication of this for foraging choices is that the relative value of two alternative food patches, for example a tall mature patch of herbage and a short vegetative sward, depends not merely on the foods' properties but on the size of the animal confronted by that choice.

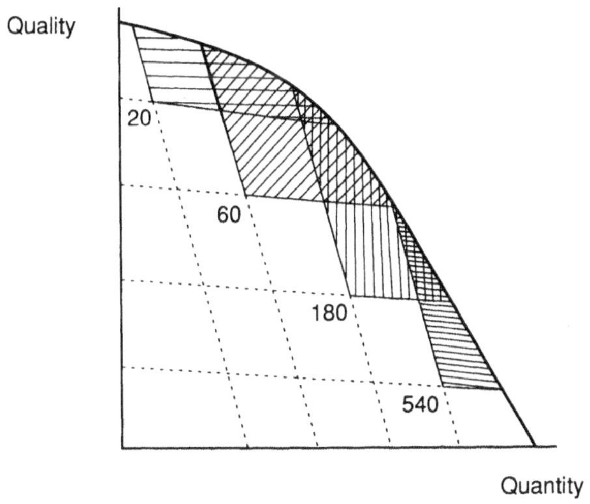

Fig 4. Feeding niches defined by food quality and quantity constraints for animals of different body size.

The combined effects of these constraints on niche selection can be generalised across the distribution of vegetation in two dimensions of quality and quantity (Fig 4). Each body size is associated with two zones: an exclusive zone in which either the quantity or quality of food is inadequate for other species; and a common zone which can also be utilised by other species. In either case, size-related constraints reinforce what must be essentially a perceptual construct of the animal - the recognised feeding niche - which is the quantity-quality space in which diet selection is least constrained.

The perceptual aspect of feeding niche selection is illustrated by dietary differences between sheep and goats inhabiting semi-natural associations of gorse and grass/clover. Goats ingest gorse, presumably recognising it as a food, but avoid clover, while sheep do the opposite, and both goats and sheep feed on grasses (Clark *et al* 1982). Other factors may also influence the perception of what things in the world are food and what are not, such as early experience (Arnold 1981; Provenza, this volume).

PATCH SELECTION

The processes which lead to diet selection from the range of plants which an animal recognises as acceptable foods may be involve, hypothetically, varying degrees of sophistication. The animal may simply recognise a broad category of potentially useful foods, and only rely on mechanical selection. This, as modelled above, would result in diet selection being determined by the blind interaction of dental morphology, feeding style and spatial distribution of plant parts. A more sophisticated animal would use information gained before, during and after ingestion to adapt its foraging tactics so as to increase its intake rate over the level achieved by mechanical selection. Decision rules which might approximate to rate maximization for mammalian herbivores could be "choose tall green grass" or "choose large browse items whose colour indicates an immature growth stage", with some judgement about the trade-offs between height and greenness or between item size and apparent immaturity. Only with perfect post-consumption information about intake rate and perfect correlation with pre-consumption cues could the animal maximise intake rate. This still falls short of maximizing nutrient intake rate since ruminants cannot recognise most nutrients on consumption (Arnold 1981) and nutrient absorption may be delayed for several hours, reducing further any association between pre-consumption cues and nutrient intake rate. This probably overstates the case against nutrient selection, because in plant foods the concentrations of different nutrients are highly correlated

(Van Soest 1980). Therefore selection of plant tissues with a high content of cell solubles would avoid tissues where the cell wall fraction dilutes all nutrients almost equally. Thus "juiciness" and lack of tough structural components may be used as indicators of nutrient density. The key difference between these levels of sophistication in diet selection is in the use of information, and we now wish to determine where on this spectrum mammalian herbivores lie.

Intake rate in grazing animals is largely influenced by sward height and density, and the depth to which an animal bites (Hodgson 1985). This determines bite weight and the bite rate or handling time during which the herbage is bitten off, chewed into a bolus and swallowed.

To determine the effect of sward height on patch selection, Clark *et al* (unpublished) offered a range of contrasts in sward height to cattle, sheep and goats. Plots of perennial ryegrass were divided into two and each half (22.5 x 4 m) mown to a predetermined height. This gave a range of nine paired height comparisons: 2 v 4, 2 v 8, 4 v 8, 4 v 16, 8 v 16, 8 v 32, 16 v 32, 2 v 2 and 8 v 8 cm. Sub-groups of twelve female cattle (400 kg), sheep (60 kg) and goats (20 kg) were observed while grazing separate plots for 30 minutes, and the grazing position of each animal (on the long or short alternative subplot) was recorded every minute. The size of the plots was adjusted beforehand to ensure that no significant patch depletion occurred over the experimental period. Bite rate and bite weight by each animal species was estimated on each of the five sward heights used in the experiment. With increasing sward height, bite rate declined by about 50%, and bite weight increased 3-4 fold, and fitted values from regressions of these variables on pre-grazing sward height were used to calculate intake rate for each species. Intake rates increased by a factor of 2 over the range of sward heights.

The taller swards were generally strongly preferred by all species. Proportions of time spent on each of the null height comparisons (2 v 2, 8 v 8 cm) were not significantly different from 0.5. The remaining results are plotted in Fig 5a, showing data and regressions of the proportion of time grazing the taller patch in relation to the ratio of the intake rates on tall and short patches. As this ratio increased, each species spent a greater proportion of time grazing the taller sward.

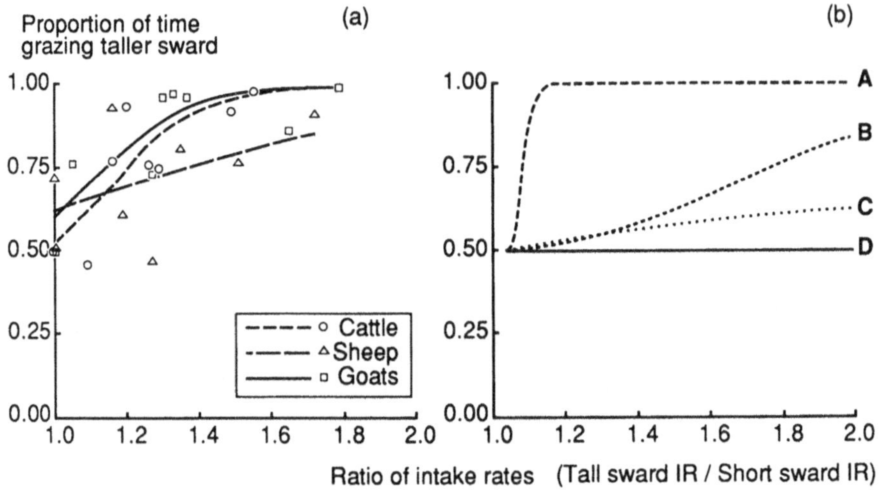

Fig 5. Proportion of time grazing the taller of two patches in relation to the ratio of the intake rates on tall and short patches. (a) experimental data (b) hypothetical patterns.

Sheep showed the lowest sensitivity to height or intake rate differences, with 0.90 selection of the taller sward being predicted from the regression to occur on a sward offering 2 times the intake rate of the alternative. Cattle were more sensitive, showing 0.90 selection at an intake rate ratio of 1.35. Increased sensitivity of larger animals is predicted by the model of grazing mechanisms (above) since compared with small animals, they experience relatively greater restrictions on short swards than long swards. However, in this experiment the smallest animals (goats) showed greatest sensitivity, spending 0.90 of the time grazing longer swards offering 1.3 times the intake rate of an alternative. The goats were found to have a different grazing style, taking shallower bites from the sward surface than the other species, especially on the taller swards, perhaps betraying their inheritance as a browsing species with a preference for selecting small bites from high biomass material, and hence greater sensitivity to sward height.

To evaluate the foraging decisions made by these three species, comparison can be made with the hypothetical patterns of response shown in Fig 5b. The four lines show the responses of (A) a highly sensitive animal having very good information about the relative values of the alternatives; (D) the other extreme, an animal indifferent to the alternatives or incapable of detecting the difference between them; (B) an animal not indifferent to the alternatives but with very poor perception of their value; and (C) an animal showing perfect matching, for reasons about which it will not serve the present discussion to speculate (see Stephens & Krebs 1986). Against this background, we can see that the experimental animals were fairly sensitive to height or intake rate differences, implying that they could perceive the alternatives reasonably well and selected the superior sward. During grazing the animals' movements about the plot often (5-10 times) took them back onto the shorter sward, but they never stayed there for long. Bearing in mind that the swards were not significantly depleted during the experiment, the fact that the animals left the long sward, and did so repeatedly, suggests that they needed repeatedly to sample the short sward by consumption, rather than relying on sight. Further work would be needed to clarify whether these results are due to the partial failure of animals in a complex short-term learning task, resulting in a poor perception or memory of what lay beyond their immediate proximity. Similar observations were made by Jarman (pers. comm.) on feeding preferences in impala. Jarman offered impala a choice of two species of *Acacia* (a genus of browse plants varying widely in thorniness and content of toxic secondary compounds), one of which was distinctly preferred. The impala ate a small amount of each before rejecting one and feeding on the other, but every few minutes they returned to the previously-rejected species and took a couple of bites before returning to the preferred species. Given the complexity of the feeding process, and variability of outcome of successive bites, it would not be surprising if information uptake and retention was imperfect. Therefore, one possible explanation of the impalas' switching behaviour is that the memory of the briefly-encountered rejected species (both its ingestive properties and location) was short-lived, and could not be refreshed merely by sight of the sample, but only by consumption. The evidence for the role of memory is circumstantial, but the involvement of memory seems to us to be a more likely explanation than alternatives.

Both this and the results of Clark's experiment suggest that herbivores may not remember much detail about one food while eating another, cannot map any recollection of briefly-encountered foods onto some spatial memory, and need to sample food by consumption rather than relying on sight. Illius *et al* (1987) have already shown that cattle graze heterogeneous swards in a manner which is consistent with sampling: that is, the extent of grazing from 3 patch types, for which there were different preferences, was partly a function of the frequency with which they were encountered. The need for consumption-sampling, rather than being able to rely on visual sampling, may impose a constraint on herbivore foraging tactics which, in a highly heterogeneous patch, would limit intake rate to the patch average.

SHORT TERM DECISION-MAKING

Clark's experiment estimated molar responses to vegetation monocultures intended to differ only in height. The following experiment (Illius *et al* in prep.) was designed to test how sheep responded to a choice between pairs of swards differing in height and grass/clover content. Short-term responses were observed more closely than is possible in the field by using swards grown with pre-determined characteristics in seed trays. Mixtures of perennial ryegrass and white clover were sown into seed trays giving a sown area of 0.13 m^2. By varying the seed ratio, different mixtures of grass and clover were obtained with nominal clover contents ranging from 0-100% in 10% intervals. The intention was to present sheep with a choice between sward trays covering the full range of contrasts in clover content, and to do this, 25 of the 65 possible combinations were chosen (Fig 6). Treatments located on a given diagonal all present the same absolute difference in clover content between swards, but at different levels of average clover content. These clover contrasts were either pre-trimmed to produce swards of nearly equal surface height (25 experiments per sheep) or differing by 4 cm in height (15 experiments per sheep). Four sheep were allowed to graze separately from pairs of swards until: (1) at least 100 bites in total had been taken; (2) both trays had been grazed from; and (3) the animal had terminated the current biting bout by turning to the alternate sward after satisfying conditions (1) and (2).

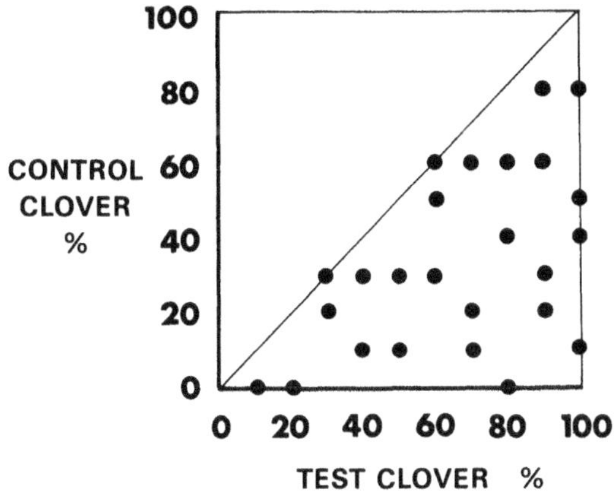

Fig 6. Contrasts between clover contents of pairs of swards offered together in the clover choice experiment.

The time at which each bite was taken was recorded to give, for each sward, the total number of bites taken, grazing time, rate of biting, and the number and length of biting bouts. A bout was defined as a run of uninterrupted bites on a sward, the animal commencing the next bout on turning to the alternate sward. Bite weight was determined by dividing the weight loss of the swards during grazing by bite number. Intake rate was calculated as the product of bite rate and bite weight of fresh material. Measurements of sward surface height before grazing and of grazed stubble height were used to determine the depth of the grazed horizon. Sward composition in the grazed horizon was assessed by inclined point quadrat. Multiple regression was used to analyse choices between swards, using the proportion of intake from one or other of the trays as the dependent variate, and fitting terms for the clover contents of each sward and the differences between swards in height or realised intake rate. Angular transformation of the dependent variates gave qualitatively very similar results to regression on untransformed variates. Differences between swards in pre-grazing height and intake rate were transformed by expressing differences as a proportion of the mean of the two swards.

The sheep grazed readily from the swards, switching between alternate swards after bouts of 20-30 bites. A subset of the results describing the combined effects of differences between swards in height and clover content is shown in Fig 7. The curves show the proportion of intake taken from swards varying in clover content when the alternate sward offered had 60% clover and was either shorter, of equal height or taller than the control sward. Considering the top curve, the proportion of intake from long swards with 60% clover was 0.67 when the alternative was short (and also had 60% clover), but intake from long swards with only 20% or 90% clover was about 0.5. The bottom curve shows that only 0.11 or 0.15 of intake was taken from short trays with 20% or 90% clover. So sheep clearly preferred longer swards with intermediate levels of clover and avoided short swards, especially if they contained low or high levels of clover.

Intake rate, which was measured on each of the pair of swards between which the sheep were switching, was affected by clover content and increased slightly with sward height (Fig 8a,b), with the added complication that the intake rate at a given clover content was influenced by the clover content of the sward alternate to that being grazed. Intake rate was maximal at 50, 65 and 80% clover when the alternate sward content was 0, 50 or 100% respectively. In switching between swards, sheep apparently carried over some of the grazing style they had developed on the sward from which they had switched.

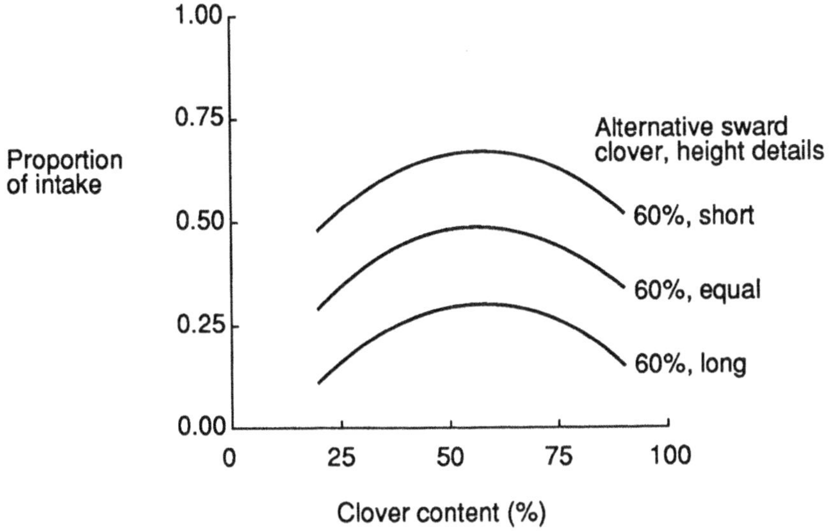

Fig 7. Proportion of intake from swards varying in clover content (x-axis) when offered with swards differing in height and with 60% clover.

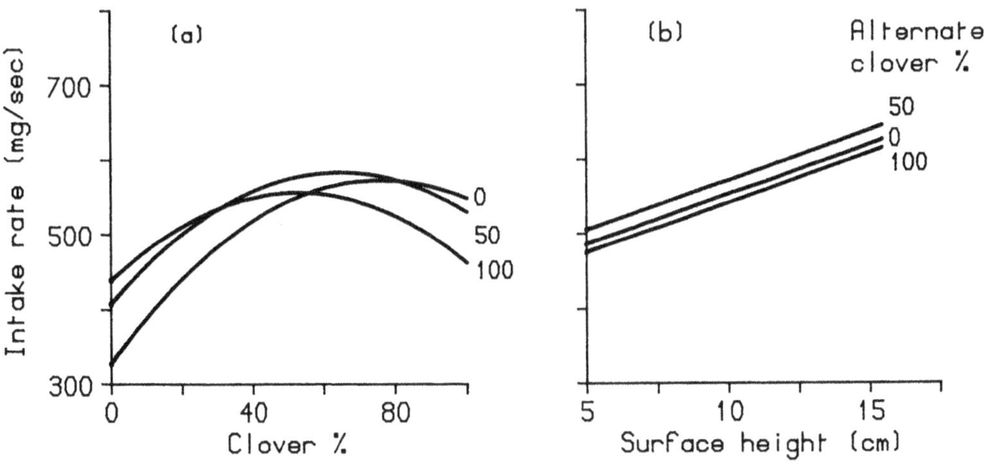

Fig 8. Intake rate in relation to (a) clover content and (b) sward height. In each, the 3 curves show intake rate when the alternate swards contained 3 different levels of clover.

The coincidence of the intake rate response and the selection response to changing clover content is striking, but superficial. Neither the proportion of time spent grazing a sward nor the proportion of intake from the sward was closely related to the *actual* differences in intake rate between the alternatives (as distinct from the *overall* effect of sward height on intake rate shown in Fig 8b). Intake rate differences only accounted for about 12% of the residual sums of squares accounted for by height differences after fitting terms for clover content. In other words, height was a good predictor of selection, but intake rate differences were not. Insensitivity to intake rate differences could be because the period of biting on each sward was too short (typically 10-40 bites per bout) for the sheep to register these differences, even though the differences ranged up to two-fold. Further evidence suggesting the poor assimilation of information came from the analysis of surface selection.

Local selection of clover from the sward surface was tested by scoring clover content over a 5 x 10 grid applied before and after grazing, to estimate the difference between pre-grazing clover contents in grid squares subsequently grazed and those ungrazed. Since selection can only take place from heterogenous swards, the variance in surface clover content was calculated, showing as expected that the highest variance occurred at intermediate mean surface clover contents. However, there was no evidence of non-random distribution of deviations in grazed clover content from the mean content. Therefore the sheep showed no sign of directing successive bites to areas of the swards which were consistently higher or lower in clover content than the sward mean. Sheep showed a preference for the average clover content of one or other of the swards during short bouts of biting from each sward, but within such a bout sheep did not take the opportunity to exploit heterogeneity in surface clover content. This suggests that the individual bite does not provide sufficient information to influence selection. On turning to a sward, the sheep grazed rapidly (c. 1 bite/sec) from a confined area before switching to another location in the same sward or in the alternate sward. It therefore seems likely that information gained from each bite is accumulated over successive bites, and that switching decisions are based on short periods of grazing rather than on individual bites. From this it would be expected that termination of a biting bout would be delayed on swards offering high rewards. This is supported by the finding that bout length was longer on taller swards ($p<0.01$) and at intermediate clover levels ($p<0.01$) although these variables only accounted for 13% of the variation in bout length (range 4-82, mean 26.8, residual SD 19.5).

The implications of these results for short term perception and decision-making can now be summarised. Given choices which required a fairly high level of discrimination to detect differences in height, clover content and intake rate, sheep did not respond to clover content and intake rate in the short term, i.e. within short bouts of biting. Height differences were more reliably detected and clover content was detected to the point that allowed selection to be exercised over a succession of bouts, but sheep did not maximize overall intake rate. Although average intake rate, as measured here, appears to provide limited information, it is nevertheless possible that instantaneous intake rate precipitated switching. This would be more plausible if the sheep had shown any sensitivity to average intake rate.

DISCUSSION OF PATCH SELECTION EXPERIMENTS

How can the results of Clark's experiment and the clover experiment be reconciled? Of the two sets of experimental conditions, those in Clark's are the more artificial, in the sense that highly homogeneous patches are not frequently encountered in nature. Natural vegetation has greater local variation in sward characteristics and can be viewed as a continuous heterogeneous patch, much like that which is addressed in the model of Arditi & Dacorogna (1988). Conditions in the clover experiment, with small, fairly similar patches, and a trade-off between height and clover content, presented realistically complicated choices. Clark's experiment provided clear visual cues in the form of large homogeneous patches, the taller of which was immediately chosen. The marked selection of the taller patches over 30-minute grazing periods does not necessarily imply the use of intake rate as a decision variable or reinforcer. It seems likely that height and other variables associated with height, which indeed are correlated with intake rate, are used as decision variables. The clover experiment examined selection over 1-2 minute periods, with frequent switching between fairly similar alternatives, and showed that sheep did not discriminate between patches offering different intake rates. Here, the visual cue of height differences were not a reliable predictor of which sward was grazed first, but did influence overall selection, so positive reinforcement from taller swards during a biting bout must be primarily tactile. We speculate that tactile cues are: ease of handling, effort of handling and bite weight. Such cues as these appeared to override any effect of intake rate. One obvious reason why intake rate is of limited use in discrimination during short term switching is that it is not exclusively a property of the patch being grazed, but is also influenced by the carry-over of the grazing style developed on the previously-grazed patch. Given longer runs of bites on homogeneous patches, as in Clark's experiment, this carry-over effect would be minimized and thus intake rate would provide a purer signal of patch profitability. An analogy to this is the signal detection analysis of Getty & Krebs (1985) who found that great tits had lagged responses to changes in prey frequency, showing the carry-over of habituation to previously-experienced prey frequencies. Grazing animals face the added complexity of partial consumption varying with sward characteristics, and it is this which may be modified by recent experience.

Signal detection theory and its terminology (McNicol 1972) can be applied, by analogy at least, to examine the selection processes used by sheep to discriminate between two swards of different height and intake rate. Had sheep used height as the sole evidence variable and criterion for selecting the superior alternative, then in the run of experiments designed to contain height differences they would have obtained 79% correct choices (higher intake rate) and 21% false alarms (lower intake rate). These are the proportions of the frequency distributions of high and low intake rate swards which were tall, and therefore would have been selected on the basis of an objective height criterion. This does not explain the animals' partial preferences, since sampling by consumption reduced the selection of taller swards to 67%. The inadequacy of height as an evidence variable, and the necessity of sampling swards by consumption (rather than visual sampling) does not result in a biased selection criterion if both alternatives are sampled equally. But consumption sampling, as a means of gaining access to the evidence variable, does reduce discrimination to the extent that some 'incorrect' choices are the result of the need for information, in addition to the false alarms subsequently arising from the use of that information. The effect is to reduce the sensitivity measure, d', the standard normal deviate separating the means of the two alternatives' distributions. The paradox of sampling is that where the difference between alternatives (signal and noise) are small, the need for consumption sampling reduces the apparent difference further. This will only occur to the extent that repeat sampling is necessary. If animals are highly perceptive, or remember and use the differences perceived during sampling to exploit a superior homogeneous patch for a prolonged period then the cost of sampling will be diluted. Frequent sampling of heterogeneous patches will prevent this attenuation of sampling costs relative to benefits. This may place severe constraints on the ability of herbivores to maximize intake rate.

This brings us to the final question, concerning the costs and benefits of sampling. Given differences in intake rate ranging up to two-fold, Clark's sheep had overall intake rates which averaged about 6% less than the maximum possible without partial preferences, and about 13% higher than unselective grazing. In the clover experiment, partial preferences resulted in intake rates being about 12% lower than the maximum possible, and 3% higher than from grazing each patch equally. The difference between the two experiments is due to the poorer short-term discrimination in the clover experiment, because the costs and benefits of sampling arise directly from the relative patch profitabilities. Given that these differences in patch profitability are of a similar order of magnitude to those encountered in natural vegetation, at least within the recognized feeding niche, the cost of partial preference arising from short-term selection tactics is not very large. This low downside risk of making ill-informed choices could explain why sheep are not better at measuring intake rate.

CONCLUSIONS

Mammalian herbivores are constrained in the diets they can select by variables associated with their body size and perceptual faculties. Body size determines the efficiency with which a given food can be consumed and utilised by setting quantity and quality constraints, and these together define the limits of the feeding niche. Within these limits, patch and food item selection operate. We speculate that constraints on short term perception limit the acquisition and use of information about intake rate itself. Perception constraints would increase the severity of the consumption-sampling constraint on foraging tactics. It appears to us that the mammalian herbivore sees a world which is divided into two categories: food and non-food; they are sensitive to differences between homogeneous patches of food; but within a patch show only limited ability in the short term to evaluate alternatives either prior to or following consumption. Without further understanding of these processes it would be premature to attempt a definition of optimal strategies for mammalian herbivores, and for this reason we prefer, at present, to develop a procedural theory of foraging.

ACKNOWLEDGEMENTS

Thanks to Dave Clark for permission to use his data, to Dawn Bazely and Fred Provenza for comments and to Ian Pitkethly, Gordon Finnie and Mary Brown for the graphics.

REFERENCES

Arditi R, Dacorogna B (1988) Optimal foraging on arbitrary food distributions and the definition of habitat patches. Am Nat 131:837-846

Arnold GE (1981) Grazing behaviour. In: Morely F (ed) Grazing animals. Elsevier, Amsterdam.

Bell RHV (1970) The use of the herb layer by grazing ungulates in the Serengeti. In: A. Watson (ed) Animal populations in relation to their food resources. Blackwell Scientific Publications. New York pp 111-123

Belovsky GE (1978) Diet optimization in a generalist herbivore: the moose. Theor Pop Biol 14:105-134

Belovsky GE (1986) Optimal foraging and community structure: implications for a guild of generalist grassland herbivores. Oecologia(Berlin) 70:35-52

Clark DA, Lambert MG, Rolston MP, Dymock N (1982) Diet selection by goats and sheep on hill country. Proc NZ Soc Anim Prod 42:155-157

Demment MW, Van Soest PJ (1985) A nutritional explanation for body-size patterns of ruminant and nonruminant herbivores. Am Nat 125:641-672

Getty T, Krebs JR (1985) Lagging partial preferences for cryptic prey: a signal detection analysis of great tit foraging. Am Nat 125:39-60

Gordon IJ (1986) The feeding strategies of ungulates on a Scottish moorland. PhD Thesis, University of Cambridge.

Gordon IJ, Illius AW (1988) Incisor arcade structure and diet selection in ruminants. Functional Ecol 2:15-22

Gray R (1986) Faith and foraging. In: Kamil AC, Krebs JR, Pulliam HR (eds) Foraging Behavior. Plenum Press, New York

Hanley TA (1980) Nutritional constraints on food and habitat selection by sympatric ungulates. PhD Thesis, University of Washington.

Hanley TA, McKendrick JD (1983) Seasonal changes in chemical composition and nutritive value of native forages in Spruce-Hemlock forest, Southeastern Alaska. US Forest Service, Portland, Res Paper PNW-312

Hodgson J (1985) The control of herbage intake in the grazing ruminant. Proc Nutr Soc 44:339-346

Illius AW, Gordon IJ (1987) The allometry of food intake in grazing ruminants. J Anim Ecol 56:989-999

Illius AW, Wood-Gush DGM, Eddison JC(1987) A study of the foraging behaviour of cattle grazing patchy swards. Biol Behav 12:33-44

Jarman PJ (1974) The social organization of antelope in relation to their ecology. Behav 48:215-266

Krebs JR, McCleery RH (1984) Optimization in behavioural ecology. In: Krebs JR, Davies NB (eds) Behavioural Ecology: An Evolutionary Approach (2nd edn) Blackwell Scientific Publications, Oxford

McNaughton SJ (1984) Grazing lawns: animals in herds, plant form and coevolution. Am Nat 124:863-886

McNicol D (1972) A primer of signal detection theory. Allen & Unwin, Sydney

Owen-Smith N, Novellie P (1982) What should a clever ungulate eat? Am Nat 119:151-178

Penry DL, Jumars PA (1987) Modeling animal guts as chemical reactors. Am Nat 129:69-96

Peters RH (1980) The ecological implications of body size. Cambridge University Press, Cambridge

Poppi DP, Minson DJ, Ternouth JH (1981) Studies of cattle and sheep eating leaf and stem fractions of grasses. III. The retention time in the rumen of large feed particles. Aust J Agic Res 32:123-137

Sibly RM (1981) Strategies of digestion and defecation. In: Townsend CR, Calow P (eds) Physiological ecology. Blackwell Scientific Publications, Oxford

Stephens DW, Krebs JR (1986) Foraging theory. Princeton University Press, New Jersey

Taylor StCS (1980) Genetic size scaling rules in animal growth. Anim Prod 30:161-165

Van Soest PJ (1982) Nutritional Ecology of the Ruminant. O & B Books Inc., Corvallis, Oregon

Smith LW, Goering HK, Gordon CH (1972) Relationships of forage compositions with rates of cell wall digestion and indigestibility of cell walls. J Dairy Sci 55:1140-1147

ORAL DISCUSSION – A.W. Illius

Q.(Kacelnik): Your procedural approach is really just highly constrained optimality.
A. I agree, but I am asking for a reorientation in viewpoint. A "strategy" implies that there is an overall leadership or some campaign, but what the animal has got for natural selection to work on are things like morphology, physiology and behaviour. I adopt the procedural approach from empirical understanding of the organism interacting with its environment. This is a valid alternative to having theory structure your empirical tests. I think the optimality approach comes later; you have to do an apprenticeship in empiricism before you are competent to couch your questions in a theoretical framework.

Q. (Saether): You have models which could be used to look at resource partitioning and community structure. Have you tried to do that yet ?
A. As yet we haven't assumed anything about the relative abundance of food–quality classes, or considered what implications there are for rates of nutrient flow in different circumstances. We get complex induced changes even within homogeneous swards, and cannot control for this. The sward can look homogeneous, but at a detailed level we get huge local variation in the sward profile.

Q. (Hobbs, Bazely): Can you justify your conclusion that memory is involved in switching between patches ? There are other possible reasons, like sampling.
A. When an animal went to the alternative patch, it did so apparently forgetting that it had been there several times before, and had rejected that patch after only a couple of bites. The fact that the animal returns, even for only a very short time, does imply that it has lost information gained previously.

DIET SELECTION BY GENERALIST HERBIVORES: A TEST OF THE LINEAR PROGRAMMING MODEL

N. Thompson Hobbs
Research Center
Colorado Division of Wildlife
317 West Prospect Street
Fort Collins CO 80526
USA

In science, as in art, we often try to create simple abstractions that faithfully represent the workings of the natural world (Hofstadter 1980; Prignogine and Stengers 1984). Ecological models built this way have many virtues, not the least of which is the clarity that emerges when excessive detail is removed from our vision of reality (Starfield and Bleloch 1986). However, although simplicity is a worthwhile objective in building models, it is less important than fidelity -- the processes we model must remain fundamentally loyal to processes in nature. Otherwise, our abstractions will obscure rather than clarify the behavior of ecological systems.

Models based on linear programming have provided simple representations of the feeding behavior of herbivores in many different ecosystems. Using three or four equations, these models have enjoyed astonishing success in predicting feeding behavior of a wide range of animals varying in body size, digestive anatomy, and feeding style (Belovsky 1978, 1981a, 1981b, 1984a, 1984b, 1984c, 1986a, 1986b). However, this success is puzzling because the formulation of extant linear programming models fails to faithfully represent knowledge of important constraints on dietary decisions.

Enormous progress has been made in understanding the way the characteristics of animals interact with characteristics of plants to constrain the food choices of generalist herbivores (see reviews of Hofmann 1973; Janis 1976; Kay et al. 1980; Bryant and Kuropat 1980; Sibly 1981; Foose 1982; Hanley 1982; Owen-Smith 1982; Van Soest 1982; Crawley 1983; Clutton-Brock and Harvey 1983; Bunnell and Gillingham 1985; Demment and Van Soest 1985; Hodgson 1985; Karasov 1986; Illius and Gordon 1987; Weston and Poppi 1987; Demment and Greenwood 1988; Karasov and Diamond 1988). Despite these advances, the equations of the linear programming model have remained essentially unchanged since the model's introduction 11 years ago. If the formulation of the model diverges from our current thinking on processes mediating dietary choices, then we may wonder whether the model arrives at what seem to be the right predictions for what are clearly the wrong reasons. Such an arrival defeats the very purpose of our endeavor; we build models to understand the behavior of natural systems, not to merely forecast their behavior (Starfield and Bleloch 1986; Swartzman and Kaluzny 1987).

NATO ASI Series, Vol. G 20
Behavioural Mechanisms of Food Selection
Edited by R. N. Hughes

Here, I examine the role of digestive constraints in influencing the outcome of a linear programming model of diet choice in a guild of generalist herbivores. I reformulate the model to include equations incorporating the effect of plant structural characteristics on food intake in ruminants and small mammals. Using this new formulation, I test the ability of the linear programming model to accurately predict dietary choices and to differentiate between time minimizing and energy maximizing strategies of feeding.

Methods

The linear programming model and its parameters. I reanalyzed the model and data of Belovsky (1986a). I chose to analyze existing information because application of linear programming requires strict adherence to a set of a priori assumptions (G. E. Belovsky, written pers. commun. Feb. 1, 1989). These include criteria on the behavior of foraging herbivores relative to the distribution of their food plants and standards for field methods used to estimate model parameters. To avoid any doubt that these assumptions were satisfied, I decided to use a data set that clearly satisfied them.

Belovsky (1986a) studied diet selection by a guild of 14 generalist herbivores feeding in a midgrass prairie on the National Bison Range in Montana, northwestern U.S.A. He used a linear programming model to predict the proportion of dicotyledonous and monocotyledonous plants in animal diets and to test the hypothesis that the objective of their feeding strategy was best characterized as time-minimizing or energy maximizing.

To test this hypothesis, Belovsky (1986a) compared observed diets with those predicted by the model. The model was formulated with two alternative objective functions: minimizing time spent feeding and maximizing total energy intake. Lack of fit between model predictions and field observations was used as a basis to reject the hypothesis that animals feed according to the goal implied by the objective function. Model formulation under the energy maximizing hypothesis was to maximize

$$E = a_1 g + a_2 f$$

subject to the constraints

$$I \geq b_1 g + b_2 f$$
$$T \geq c_1 g + c_2 f$$

where

E = intake of digestible energy (kj/day),
g = intake of monocots (g-dry/day),

f	=	intake of dicots (g-dry/day),
a_1	=	digestible energy content of monocots (kj/g-dry),
a_2	=	digestible energy content of dicots (kj/g-dry),
I	=	maximum intake (g-wet/d) allowed by gut volume and ingesta turnover rate,
b_1	=	bulk of monocots (g-wet/g-dry),
b_2	=	bulk of dicots (g-wet/g-dry),
T	=	feeding time (min/day),
c_1	=	cropping rate of monocots (g/min),
c_2	=	cropping rate of dicots (g/min).

Alternatively, the formulation of the model under the time minimizing hypothesis was to minimize

$$T = c_1 g + c_2 f$$

subject to the constraints

$$MR \leq a_1 g + a_2 f$$
$$I \geq b_1 g + b_2 f$$

where

MR = energy needed for maintenance and reproduction (kj/d) and other variables are defined as above.

When Belovsky (1986a:Fig. 3) compared predictions of the energy maximizing formulation to field observations of herbivore diets, he found excellent agreement ($r^2 = 0.99$) between the predicted proportion of dietary monocots and the observed proportion, but saw substantial lack of fit in the time-minimizing predictions of diet composition. Hence, Belovsky (1986a) concluded that during the growing season, 14 species of herbivores universally sought to maximize their daily energy intake rather than to minimize the time they spent feeding. Based on this result, he offered several inferences on the structure of this animal community.

The digestive constraint. The digestive constraint ($I \geq b_1 g + b_2 f$) is particularly important in controlling the model's behavior and the inferences derived from it. This is the case because limitations on dry matter intake imposed by gut volume and turnover represent the only constraint found in both of the alternative formulations of the model. This equation has been used in almost all applications of the linear programming model, and consequently, represents a pivotal process, not only in Belovsky (1986a), but in other papers as well (Belovsky 1978, 1981a, 1981b, 1984a, 1984b, 1984c, 1986b).

The digestive constraint equation limits daily intake of dry matter in proportion to its moisture content or "bulk". Thus, if the animal eats nothing but monocots then it can consume no more than

$$g \quad = \quad I/b_1$$

and if it eats only dicots then its dry matter intake is limited to

$$f \quad = \quad I/b_2 \, .$$

Because the parameters of the model (Belovsky 1986a:Table 3) specify that

$$b_2 \quad > \quad b_1$$

then

$$g \quad > \quad f.$$

Given the values of bulk used in the model (b_1 = 1.64, b_2 = 2.67), herbivores eating to fill could consume only 61 units of dicot dry matter for each 100 units of monocot dry matter eaten (Fig. 1). The limit on intake of a mixed diet can be solved as a weighted average of b_1 and b_2 .

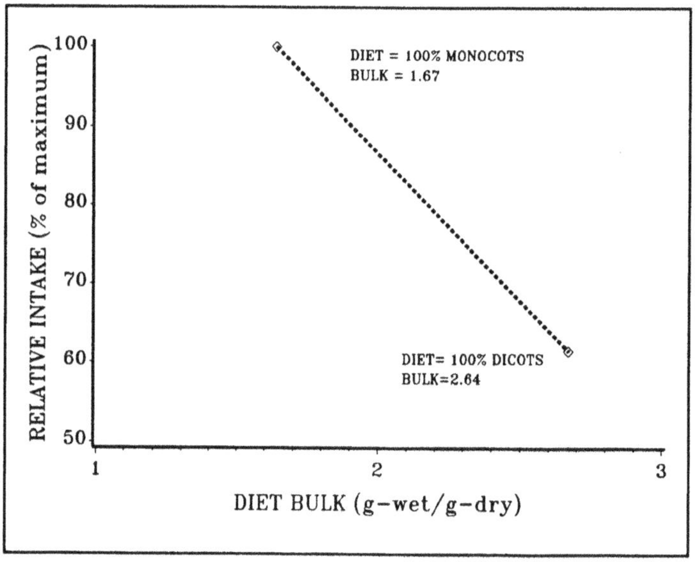

Fig. 1. The linear programming model of feeding behavior (Belovsky 1986a) is formulated such that dry matter intake is proportional to the reciprocal of the dry matter content (bulk) of the animal's diet. Because monocots have a higher dry matter content than dicots, the model specifies that animals eating to fill can consume only 61 units of dicots for each 100 units of monocots in their diets.

To illustrate the role of the digestive constraint equation in controlling model behavior, I ran the model for all 14 herbivores while varying values for bulk of monocots (b_1) and dicots (b_2). I chose 29 values of b_1 and 31 values of b_2 at evenly spaced intervals over ±1 standard deviation[1] for each coefficient and used a factorial arrangement of these values as input to the model. I used the simplex algorithm of Press et al. (1986:312-325) to solve for maxima and minima and to obtain optimum values for diet composition for each herbivore. For each combination of bulk values, I obtained 1) the mean proportion of monocots in the diets of the 14 herbivores, 2) the number of herbivores that consumed diets containing both monocots and dicots, and 3) the r^2, slope, and intercept of the regression of the observed vs. predicted diet composition. I used the SAS system for regression analysis (Freund and Littel 1986).

To examine the fidelity of the digestive constraint equation to physiological processes regulating intake, I reviewed published experiments that tested the hypothesis that forage dry matter content (g-dry/g-wet) mediates dry matter intake[2] (g-dry/kg $BW^{0.75}$) by ruminants. I also assembled correlative information on forage moisture and intake in small mammals. To construct alternative formulations to those based on bulk, I assembled extant literature on the role of forage characteristics in controlling dry matter intake[2] in ruminants and small mammals and developed simple equations representing constraints on intake based on the conclusions of these papers. I examined the behavior of the model when it was reformulated with these equations.

I chose two standards for judging the success of the linear programming model when I changed its constraint equations. First, the model must be able to predict diet composition with reasonable accuracy. This is to say there must be a significant regression between observed and predicted composition of diets that accounts for a major portion of variation in the observations. More importantly, however, the model must distinguish between solutions for time-minimizing and energy-maximizing objectives. If the model fails to do so, then it will not be able to test hypotheses on feeding strategies of herbivores.

---

[1]Choosing the appropriate range of input values included in sensitivity analysis depends on the question the analysis is intended to answer. Here, I ask the question: "How does the model respond to the temporal and spatial variation in plant moisture that we would expect to see in nature?" The appropriate input in this case depends on the probability distribution of the observations of plant moisture. Hence, I chose the standard deviation as a basis for the range of values included in my simulations. This question differs from the question "How precise are the model's estimates?", which depends on the distribution of the sample mean and suggests using the standard error as a basis for the range of input values.

[2]Synthesizing data on intake by mammals requires expressing data relative to animal body size. This can be done as ratio of intake to body mass (e.g., kg $BW^{1.0}$) or to metabolic body mass (e.g., kg $BW^{0.75}$). Published data differ in this respect, and it can be difficult to accurately adjust denominators to the same exponent when individual weights are not given. Consequently, I expressed intake data on the basis of metabolic body weight because this was the most prevalent form in the literature I reviewed. When body size was not given in papers on cattle, I assumed that a cow weighs 400 kg.

Results and Discussion

Model sensitivity to digestive constraints. The digestive constraint equation controlled model behavior. Varying the values for diet bulk over ±1 standard deviation caused the predicted average proportion of monocots in herbivore diets to vary substantially (Fig. 2A). This variation in predicted diet composition caused marked changes in the goodness of fit of predictions to observations (Fig. 2B).

When regressing observations on predictions (e.g., Belovsky 1986a:Fig. 3), three statistics are used to evaluate the extent to which the outcomes of the model fit the observations used to test those outcomes. A perfect fit is obtained when the r^2 and the slope of the regression = 1.0 and the intercept = 0.0. Out of 899 combinations of values for bulk input to the model, only one (b_1 = 1.74, b_2 = 2.57) resulted in an r^2 higher than that obtained by the original parameters (b_1 = 1.64, b_2 = 2.67) used by Belovsky 1986a (Table 1). However, the original parameters provided a slope closer to 1.0 and an intercept closer to 0.0 (Table 1). Hence, when we simultaneously consider r^2, slope, and intercept, we find that the combination of bulk values used by Belovsky (1986a) provided a better fit to the observations than any other combination included in my simulations (Fig. 2B).

Table 1. Slope, intercept, and r^2 for regression of observations of percent monocots in diets of 14 herbivores vs. model predictions of percent monocots. Values are presented for bulk values achieving the five highest r^2 values out of 899 possible combinations (Fig. 2B). Model predictions are based on the energy maximizing form of the linear programming model with bulk levels input as shown. Bulk values used by Belovsky (1986a) were dicot bulk = 2.67 and monocot bulk = 1.64.

Dicot bulk	Monocot bulk	Slope	Intercept	r^2
2.57	1.79	0.93	0.02	0.981
2.67	1.69	0.96	-0.01	0.982
2.67	1.64	0.96	0.00	0.984
2.57	1.69	0.92	0.03	0.984
2.57	1.74	0.93	0.03	0.985

Although virtually all of the variation in observed diet composition was explained by the energy maximizing formulation when bulk parameters were set at their original values, small shifts in the value of these parameters caused the r^2 to drop to <0.30 (Fig. 2B). This was the case because when the value of dicot bulk became more similar to the value of monocot bulk, monocots dropped out of the diets of herbivores (Fig. 2C). In particular, if the bulk of monocots remained constant and the bulk of dicots decreased by as little as 12%, the model predicted that 12 of the 14 herbivores would totally eliminate monocots from their diets.

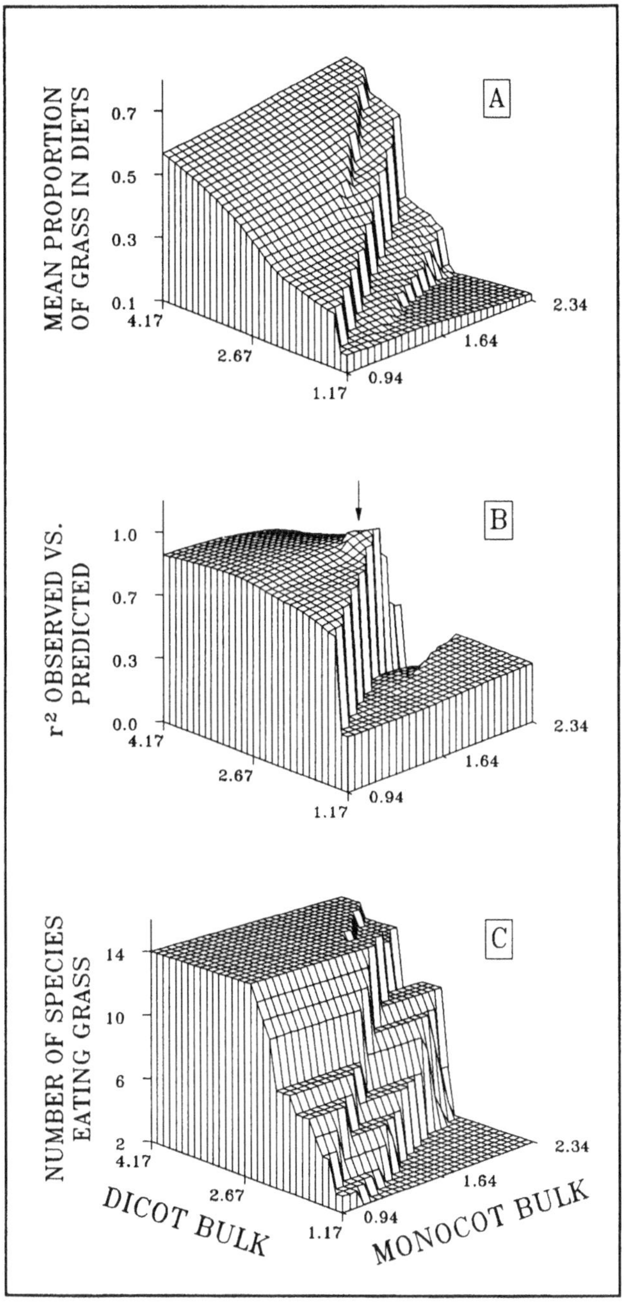

Iteratively reducing the difference between the values for bulk in monocots and dicots revealed that the model failed whenever

$$b_2 < b_1 + 0.38.$$

Outside this boundary, the time minimizing solutions became indistinguishable from the energy maximizing ones; 12 of the 14 were predicted to feed exclusively on dicots in either case. This occurred because the energy content of dicots exceeded that of monocots for all herbivores studied (Belovsky 1986a: Table 3). Obtaining a solution specifying a mixed diet required that the bulk of dicots must sufficiently exceed the bulk of monocots to provide an advantage in monocot intake compensating for its disadvantage in energy concentration (Belovsky 1986a:Table 3).

Plant bulk emerges as a sensitive parameter that must be sampled accurately if the model's predictions are to

Fig. 2. Responses of the energy maximizing formulation of the linear programming model to variation in plant bulk (g-wet/g-dry). Axis range = ± 1 SD from the mean value for bulk at the axis midpoint. The mean proportion of monocots (grass) in diets is calculated across 14 species of herbivores (A). The r^2 value is based on the regression of observed diets vs model predictions (B) for those 14 species. The r^2 obtained by Belovsky (1986a) rests on the peak (see arrow) of the surface in (B). The number of herbivores eating grass (C) is the sum of species who include >0% monocots in their diets.

accurately reflect the observed behavior of herbivores. Plant bulk is the reciprocal of dry matter content and as such is highly variable over time and space. Plant dry matter varies in a daily cycle, reaching a low point in the evening, and climbing by as much as 25% to a peak at midday (Street and Opik 1976:Fig 4.1). Sampling this variation is particularly difficult because diurnal changes in plant dry matter are superimposed on diurnal differences in periods of peak feeding for the herbivores studied (Belovsky and Slade 1986:Fig 1). In addition to this daily variation, plant dry matter changes weekly, seasonally, and annually (Larcher 1973:164-170, Fitter and May 1987:Fig 4.7). Estimating the mean bulk for monocots and dicots required summarizing variation operating at these different temporal scales during three complete growing seasons. This estimate required distilling spatial variation contributed by 9000 ha of mixed prairie varying in elevation from 820 to 1500 m. This estimate also incorporated interspecific variation resulting from dietary differences among 14 herbivores differing in body size by seven orders of magnitude and responded to phenological, physiological, and morphological differences among 55 different species of plants. However, despite this variability, a relatively small sample (4 samples/plant species, Belovsky 1986a:Table 1, 3) produced two average values that provided a nearly perfect fit of predictions to observations (Fig 2B). Given the sensitivity of the model to an inherently variable parameter (Fig. 2A,B,C), this outcome may prove difficult to reproduce.

The biological basis for the digestive constraint equation. In light of the pivotal importance of the digestive constraint equation in determining model behavior, we should be sure that it is built on a sturdy conceptual foundation. The equation is intended to represent the limitation on intake that is imposed by the filling effect of ingesta. That is, "Different foods use different amounts of digestive capacity depending on their bulk (g-wet wt/g-dry wt)" (Belovsky 1986b:55). The digestive constraint depends on the assumption that water and dry matter are additive in their effects on ingesta fill.

The basis for this assumption is infirm. In the linear programming model, maximum daily intake is limited by the total wet mass of ingesta and its turnover rate. Thus, the full form of the digestive constraint is given by:

$$F \cdot T = b_1 g + b_2 f$$

where

F = ingesta fill (g-wet),

T = turnover rate (times/day).

For most of the species studied, this formulation is mathematically incorrect. It inappropriately applies a single turnover rate to four different compartments in the gut.

Turnover rates used in Belovsky (1986a) were frequently measured with an indigestible, solid marker. Particulate markers collected in the feces allow inference to the rate of excretion of the indigestible dry matter (solids that are defecated) (Milchunas et al. 1978:Fig. 6; Mertins and Ely 1979; Warner 1981; Van Soest 1982:218-220, 228; Robbins 1983:314; France et al. 1985; Holleman and White 1989). However, ingesta fill as it is defined in the constraint equation also includes digestible dry matter (solids that are absorbed across the gut wall), as well as absorbed and excreted water. These fractions of the ingesta have their own turnover rates (Van Soest 1982: Fig. 13.1) that differ substantially from the turnover rate of the indigestible solids (Warner 1981:800; Van Soest 1982:Table 13.1) In particular, the turnover rate of water absorbed from the rumen is as much as 40 times more rapid than the turnover rates used in the digestive constraint equation (Engelhardt 1970:Fig. 1, Belovsky 1986a:Table 3)

Thus, "bulk" appears to be a misnomer. Although the digestive constraint equation is based on the notion that water and dry matter contribute equally to gut fill, proper consideration of turnover of the solid and liquid fractions suggests that water does not measurably add to the filling effects of ingesta dry matter. When herbivores chew or ruminate plant tissue, the liquid in cytoplasm and vacuoles is liberated as soon as their membranes rupture. The plant cell wall contains many large interstices (Nobel 1970:2), and as a result is quite permeable to water (Sutcliffe 1968:26, Nobel 1970:27). Any water that emerges from plant cells can be rapidly absorbed across the rumen epithelium (Hungate 1966:202, 222; Engelhardt 1970; Willes et al. 1970). Because of this rapid absorption, plant moisture content is poorly related to the filling effect of ingesta -- moist tissue is no more "bulky" than dry tissue.

Based on these concepts, we would predict that dry matter intake should be independent of bulk as it has been defined here. Empirical results support this prediction. There is no evidence that intake of dry matter decreases as bulk increases in ruminants (Fig. 3A) or small mammals (Fig. 3B). The regression of dry matter intake on forage bulk accounted for 2% of the variation in intake in ruminants (Fig. 3A) and 21% of the variation in intake by small mammals (Fig. 3B). Although the slope of the regression for ruminants failed to differ from 0, there was a significant *positive* relationship between bulk and intake in small mammals.

Dry matter intake appears to be related to forage bulk whenever forage moisture content exceeds about 80% (Butris and Phillips 1987). However, this relationship results from feedback from fermentation products of ensilage (Moore et al. 1960; Thomas et al. 1961; Waldo et al. 1965) or from effects of surface water on palatability (Butris and Phillips 1987) rather than from filling effects of plant moisture. Thus, any depressing effect of bulk on intake appears to result from confounding correlations between plant moisture and other plant characteristics. However, we may also observe positive relationships between bulk and intake of natural diets (e.g., Fig. 3C) as result of the confounding effect of plant fiber, which tends to be inversely related to plant moisture (Drozdz and Osiecki 1973).

All papers that have used the digestive constraint equation (Belovsky 1981a, 1981b, 1984a, 1984b, 1984c, 1986a, 1986b) cite Belovsky 1978 as basis for the limiting effects of bulk on intake. In an appendix to Belovsky (1978) we find two references supporting the mechanism represented in the digestive constraint equation. The first is Duckworth and Shirlaw (1958). It is difficult to understand how this paper supports the operation of the digestive constraint equation. Duckworth and Shirlaw (1958) show that dry matter intake is unrelated to forage bulk over the range of plant moisture encountered in most terrestrial

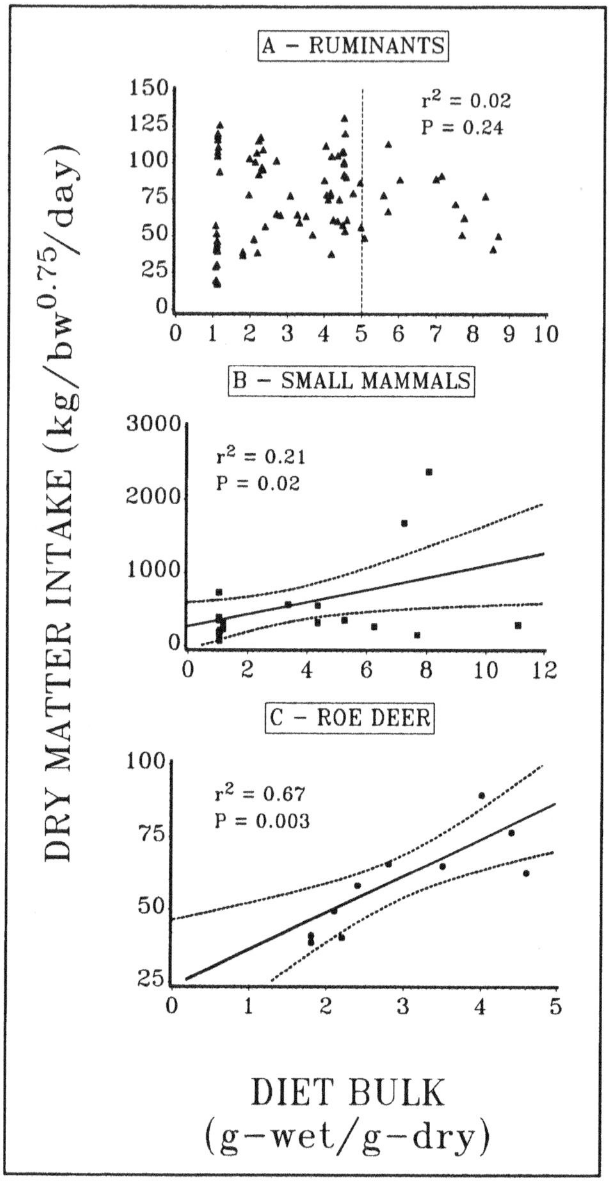

Fig. 3. Dry matter intake fails to decline as bulk increases in ruminants (A) or small mammals (B). Vertical reference line in (A) represents the approximate threshold at which intake of silage and grasses may spuriously decrease in response to surface water and ensiled fermentation products. Most terrestrial plants have bulk values to the left of this threshold. Positive correlations with bulk also occur in response to dietary fiber in natural diets of wild ruminants (C). Dashed lines are 95% confidence intervals on the mean prediction of intake (solid line) at a given value of bulk. Data for ruminants from Duckworth and Shirlaw (1958); Thomas et al. 1961; Holmes and Lang (1963); Moore et al. (1960); Waldo et al. (1965); Drozdz and Osiecki (1973); Morgan et al. (1980); Baker and Hansen 1985; Baker and Hobbs (1987); Butris and Phillips (1987); and Undersander et al. (1987). Data for small mammals from Drozdz (1968); Golley (1960); Sawicka-Kapusta et al. (1975); Gross et al. (1985); and Gross and Wunder, unpubl. Data for roe deer from Drozdz and Osiecki (1973).

plants. They show that although plant moisture may control the amount of free water an animal drinks, it has no influence on the amount of dry matter it eats (Fig. 4). The second reference cited in support of the formulation of the digestive constraint equation is Thomas et al. (1961) who concluded that moisture content of silage was spuriously correlated with intake:

> "The results of these trials suggest that the dry matter content of the forage ensiled and the resulting differences in fermentation appear to be major determinants for the amount of dry matter consumed in the form of silage rather than the dry matter content of the silage per se as it is consumed. This is true even though there is a general and positive relationship between the dry matter content of silages and their rate of consumption. This relationship is somewhat fictitious and probably more accurately relates intake to the fermentation process during silage production and the quality of the resulting dry matter." (Thomas et al. 1961:1476)

Belovsky (1978:133) acknowledged that dry matter intake will be influenced by bulk only when forage moisture is exceedingly high, but this has been overlooked in subsequent publications (Belovsky 1981a, 1981b, 1984a, 1984b, 1984c, 1986a, 1986b).

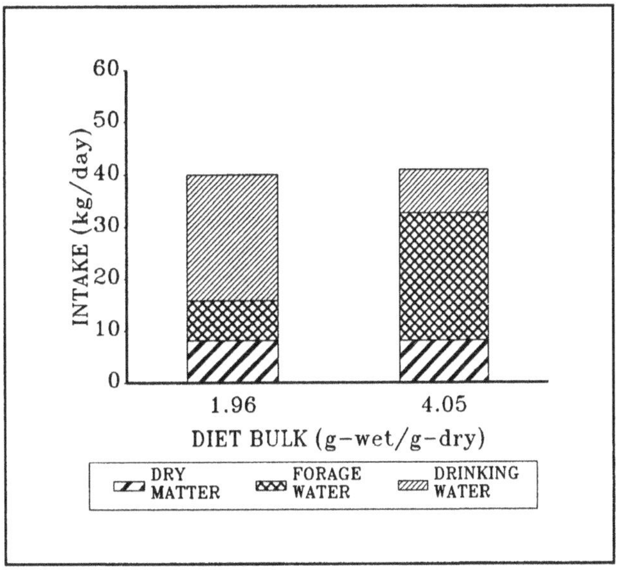

Fig. 4. Increasing bulk caused no difference in intake of dry matter of grass diets fed to cattle but did influence the form of water intake (after Duckworth and Shirlaw 1958: Table III). When animals were fed moist ("bulky") diets they drank less free water but ate the same amount of dry matter as animals fed dry diets.

Alternative constraints on dry matter intake: Implications for model behavior. The linear programming model is formulated such that dry matter intake in ruminants and small mammals is regulated by the same process. A more widely held viewpoint is that regulation of intake in animals that have anterior fermentation chambers (e.g., ruminants) differs markedly from those in which fermentation occurs in the posterior portion of the gut (e.g., small mammalian herbivores) (Janis 1976; Foose 1982; Uden and Van Soest 1982; Owen-Smith 1982; Bunnell and Gillingham 1985; Demment and Van Soest 1985; Weston and Poppi 1987).

In ruminants, unlike small mammals, forage particles are selectively retained in a forestomach fermentation chamber until they can be reduced to a size sufficiently small to be passed to the chemical stomach. Intake in ruminants consuming fibrous diets is controlled by distension of rumen caused by the filling effects of ingested dry matter (Campling and Balch 1961; Campling 1970; Thornton and Minson 1973; Baile and Forbes 1974; Grovum 1979a, 1979b). Passage of forage dry matter from the rumen is an important means for relieving fill and, hence, acts to regulate intake of fibrous diets (Mertins 1973; Mertins and Ely 1979; Welch 1982). The rate of passage of indigestible dry matter depends on the rate of reduction of forage particles (Poppi and Norton 1980; Ehle and Stern 1984; Welch 1982; Worrell et al. 1986; Pond et al. 1987; Woodford and Murphy 1988). The rate of reduction of particles, in turn, is directly related to their cell wall content because increasing fiber and lignin in cell wall render plant tissue more resistent to comminution during mastication and rumination (Balch 1971; Bae et al. 1979, 1981; Poppi et al. 1981a, 1981b; Welch 1982, 1986; Lippke 1986; Spalinger et al. 1986; Baker and Hobbs 1987). Thus, the retarding effect of cell wall on comminution of particles offers a mechanistic explanation for the inverse relationship observed empirically between dietary cell wall content and dry matter intake in wild and domestic ruminants (Fig. 5).

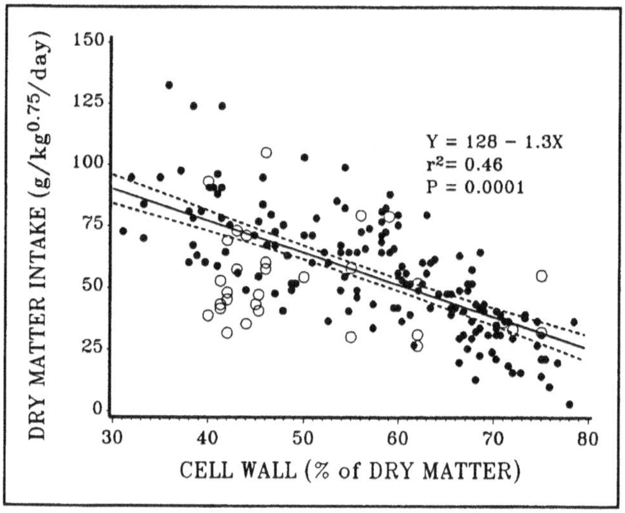

Fig. 5. Dry matter intake of forage diets in ruminants declines as dietary cell wall content increases. Dashed lines are 95% confidence intervals on the mean prediction of intake (solid line) at a given value of dietary cell wall. Diets include 128 monocots and 68 dicots. Data for domestic ruminants (closed circles) taken from review of Mertins (1973). Data for wild ruminants (open circles) from Milne et al. (1978); Foose (1982); Baker and Hansen (1985); Spalinger (1985); Baker and Hobbs (1987).

However, dicotyledonous leaves almost always contain less cell wall than do monocots (Short et al. 1974, 1975; Cook 1972; Hobbs et al. 1981; Milchunas et al. 1978; Baker and Hobbs 1982; Hanley and McKendrick 1983; Schwartz and Hobbs 1985). It follows from Figure 5 that, contrary to the predictions of the digestive constraint equation, ruminants can eat more "bulky" dicots than "compact" monocots.

Based on the relationship between dietary cell wall and dry matter intake (Fig. 5), we can derive a constraint equation for intake in ruminants that enjoys a firmer biological foundation than the one used in Belovsky (1986a) and other papers (Belovsky 1981a, 1981b, 1984a, 1984b, 1984c, 1986a, 1986b). Although it would appear that maximum dry matter intake of forage diets would occur at 0% cell wall (Fig. 5), intake becomes limited by chemostatic and thermostatic regulation when diets are low in fiber, and hence, high in digestibility (Conrad et al. 1964; Baumgardt 1970; Montgomery and Baumgardt 1965; Ammann et al. 1973; Robbins 1983:Fig 14.7). The increase in residual error in predicting intake that occurs with decreasing cell wall (Fig. 5) is due in part to the replacement of physical mechanisms by physiological mechanisms regulating dry matter intake at low cell wall levels (Mertins 1973:153). However, cell wall contents of <30% of dry matter are exceedingly rare in natural forages (Short et al. 1974, 1975; Cook 1972; Hobbs et al. 1981; Baker and Hobbs 1982; Hanley and McKendrick 1983; Schwartz and Hobbs 1985). Thus, presuming that the upper limit on dry matter intake (I, g/kg $BW^{0.75}$) occurs when cell wall contributes about 30% of dietary dry matter, we can solve for I using the equation developed in Fig. 5,

$$I = 128 - 1.3(30)$$

and we have:

$$I \geq b_1 g + b_2 f,$$

where b_1 and b_2 are unitless coefficients expressing the relative filling effect of cell wall. When the animal eats only monocots, we can solve for their filling effect (b_1) by determining the cell wall content of monocots (CWC_1) and by substituting the cell wall/intake equation (Fig. 5) for g:

$$b_1 = I/g$$
$$b_1 = I/(128 - 1.3\ CWC_1).$$

We can similarly solve for b_2 based on knowledge of the cell wall content of dicots (CWC_2):

$$b_2 = I/f$$
$$b_2 = I/(128 - 1.3\ CWC_2).$$

It follows that because monocots contain higher levels of cell wall than do dicots, then

$$CWC_1 > CWC_2$$

and hence,

$$b_2 < b_1 .$$

However, whenever

$$b_2 < b_1 + 0.38$$

the r^2 of the prediction of the energy maximizing formulation drops to <0.34 and the time minimizing and energy maximizing solutions converge. It follows from the above that the cell wall content of dicot leaves must exceed the cell wall content of monocots by several percentage points if the model is to succeed. This almost never occurs in plants of similar phenological age (Short et al. 1974, 1975; Cook 1972; Milchunas et al. 1978; Hobbs et al. 1981; Baker and Hobbs 1982; Hanley and McKendrick 1983; Schwartz and Hobbs 1985). Hence, in the case of ruminants, we are forced to conclude that when a biologically faithful abstraction of the effects of plant characteristics on intake is incorporated in the model, the model fails.

Small mammals, unlike ruminants, appear able to increase voluntary intake when dietary fiber levels increase (Sawicka-Kapusta et al. 1975; Williams et al. 1978; Gross et al. 1985), an ability common to hind-gut fermenters (Janis 1976; Demment and Van Soest 1985). Increased intake is accomplished by expanding gut fill (Gross et al. 1985; K. A. Hammond and B. A. Wunder, unpubl. data) and by an extremely rapid and variable rate of passage. Unlike ruminants, the rate of passage of the solid fraction from the fermentive chamber of small mammals exceeds that of the liquid fraction (Uden 1978). Instead of being regulated by dietary cell wall, dry matter intake by small mammals appears to be directly related to the energy density of the diet and the animal's energy requirements (Fig. 6). There appears to be an upper limit on intake that is set by the

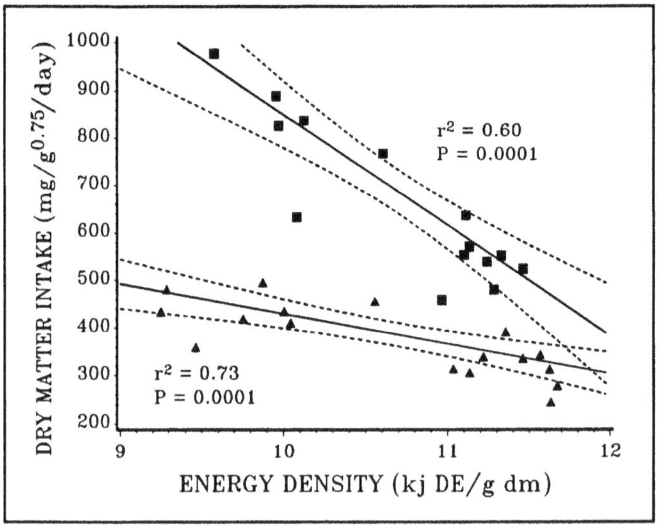

Fig. 6. Small mammals adjust dry matter intake in response to energy demands and digestibility of dietary energy. Data are for *Microtus ochrogaster* held at temperatures above and below their lower critical temperature (K. A. Hammond and B. A. Wunder, Colorado State University, unpubl.). At 5 C, animals had to expend extra energy for thermoregulation and consumed more dry matter (squares) than they did at 23 C (triangles). Animals increased intake in response decreased concentrations of digestible energy in their diets at both temperatures. Similar increases have observed for forage diets with energy densities as low as 6.7 kj DE/g dm (Batzli and Cole 1979: Fig. 1) and at temperatures of 3 vs 13 C (Batzli and Cole 1979: Table 3). Dashed lines are 95% confidence intervals on the mean prediction of intake (solid line) at a given value of energy density.

animal's energy demands (Batzli and Cole 1979:Fig. 1). However, because dry matter intake of small mammals increases as the digestible energy concentration of the diet declines (Drozdz 1968; Bienkowski and Marszalek 1974; Sawicka-Kapusta et al. 1975; Batzli and Cole 1979; Gross et al. 1985), there doesn't appear to be a digestive constraint resembling the physical limitation on intake seen in ruminants (Conrad et al. 1964; Campling 1970) until the energy density of the diet drops to about 6 kj DE/g dm (Batzli and Cole 1979:Fig. 1). Other factors, particularly plant minerals and defensive chemistry, may also constrain intake (Kendall et al. 1978; Jung and Batzli 1981; Bergeron and Jodoin 1987; Lindroth and Batzli 1984; Lindroth et al. 1986; Batzli 1986). However, as a starting point we can reasonably formulate a constraint on dry matter intake for the energy maximizing alternative such that energy intake increases with decreasing energy density up to some limit set by the animal's metabolic requirements (Batzli and Cole 1979; Gross et al. 1985; K. A. Hammond and B. A. Wunder, unpubl.):

$$MR \geq a_1 g + a_2 f,$$

where variables are as defined previously. Unfortunately, this constraint equation is the same as the objective function for the energy maximizing form of the model. Consequently, we must conclude that an energy maximizing formulation may be inappropriate for small mammals because dry matter intake is often constrained by the animal's energy needs.

Conclusions

The outcome of the linear programming model of feeding behavior in generalist herbivores depends on an equation that has no apparent biological foundation. The model fails when its equations are formulated to represent biologically realistic constraints regulating dry matter intake in herbivorous mammals. However, it can be argued that because the model offers predictions that mimic observations, then its formulation must be correct. This is to say that the model is informative because it works.

This argument is not compelling. Building revealing ecological models depends on our ability to symbolically represent the operation of processes in nature. If we formulate a model that has no biological basis and if we have the good fortune to find satisfactory agreement between the predictions of our model and our observations of nature, then we have probably learned more about serendipity than biology. To be revealing, models must be faithful to the natural world.

Because the findings of the linear programming model have been founded on equations that lack fidelity to the processes they represent, those findings may be misleading. Until the linear programming model can be shown to predict feeding behavior on the basis of a biologically sound formulation, I suggest that those who wish to make progress in understanding the feeding behavior of herbivores should look elsewhere.

Acknowledgments. I am grateful to T. A. Hanley, A. W. Illius, T. M. Pojar, T. E. Remington, C. T. Robbins, D. E. Spalinger, D. M. Swift, and C. R. Tracy for helpful discussions and for comments on early drafts. K. A. Hammond and B. A. Wunder graciously allowed me to use their unpublished data. N. W. McEwen prepared the manuscript. This work is a contribution of Federal Aid in Wildlife Restoration.

Literature Cited

Ammann AP, Cowan RL, Mothershead CL, Baumgardt BR (1973) Dry matter and energy intake in relation to digestibility in white tailed deer. J Wildl Manage 37:195-201

Bae D, Welch JG, Smith AM (1979) Forage intake and rumination by sheep. J Anim Sci 49:1292-1299

Bae DH, Welch JG, Smith AM (1981) Efficiency of mastication in relation to hay intake by cattle. J Anim Sci 52:1371-1375

Baile CA, Forbes JM (1974) Control of feed intake and regulation of energy balance in ruminants. Physiol Rev 54:160-214

Baker DL, Hansen DR (1985) Comparative digestion of grass in mule deer and elk. J Wildl Manage 49:77-79

Baker DL, Hobbs NT (1982) Composition and quality of elk summer diets in Colorado. J Wildl Manage 46:694-703

Baker DL, Hobbs NT (1987) Strategies of digestion: digestive efficiency and retention time of forage diets in montane ungulates. Can J Zool 65:1978-1984

Balch CC (1971) Proposal to use the time spent chewing as a index of the extent to which diets for ruminants possess the physical property of fibrousness characteristic of roughages. Brit J Nutr 26:383

Batzli GO (1986) Nutritional ecology of the California vole *Microtus californicus*: effects of food quality on reproduction. Ecology 67:406-412

Batzli GO, Cole FR (1979) Nutritional ecology of microtine rodents: digestibility of forage. J Mamm 60:740-750

Baumgardt BR (1970) Control of feed intake and the regulation of energy balance. In Phillipson AT (ed), Physiology of digestion and metabolism in the ruminant. Oriel Press, Newcastle

Belovsky GE (1978) Diet optimization in a generalist herbivore: the moose. Theor Pop Biol 14:76-104

Belovsky GE (1981a) Food plant selection by a generalist herbivore: the moose. Ecology 62:1020-1031

Belovsky GE (1981b) Optimal activity times and habitat choice of moose. Oecologia 48:22-30

Belovsky GE (1984a) Summer diet optimization by beaver. Amer Midl Nat 111:209-222

Belovsky GE (1984b) Summer diet optimization by beaver *Castor canadensis*. Amer Midl Nat 111:209-222

Belovsky GE (1984c) Moose and snowshoe hare competition and a mechanistic explanation from foraging theory. Oecologia 61:150-159

Belovsky GE (1986a) Optimal foraging and community structure: implications for a guild of grassland herbivores. Oecologia 70:35-52

Belovsky GE (1986b) Generalist herbivore foraging and its role in competitive interactions. Amer Zool 26:51-69

Belovsky GE, Slade JB (1986) Time budgets of grassland herbivores -- body size similarities. Oecologia 70:53-62

Bergeron JM, Jodoin L (1987) Defining "high quality" food resources of herbivores: the case for meadow voles (*Microtus pennsylvanicus*). Oecologia 71:510-517

Bienkowski P, Marszalek U (1974) Metabolism and energy budget in the snow vole. Acta Theriologica 19:55-67
Bryant JP, Kuropat PJ (1980) Selection of winter forage by subarctic browsing vertebrates: the role of plant chemistry. Ann Rev Ecol Syst 11:261-285
Bunnell FL, Gillingham MP (1985) Foraging behavior: dynamics of dining out. In Hudson RJ, White RG (eds), Bioenergetics of wild herbivores. CRC Press, Boca Raton
Butris GY, Phillips JC (1987) The effect of herbage surface water and the provision of supplementary forage on the intake and feeding behavior of cattle. Grass and Forage Sci 42:259-264
Campling RC (1970) Physical regulation of voluntary intake. In Phillipson AT (ed), Physiology of digestion and metabolism in the ruminant. Oriel Press, Newcastle upon Tyne, England
Campling RC, Balch, CC (1961) Factors affecting the voluntary intake of food by cows. 1. Preliminary observations on the effect, on the voluntary intake of hay, of changes in the amount of reticulo-ruminal contents. Brit J Nutr 15:523-530
Clutton-Brock TH, Harvey PH (1983) The functional significance of variation in body size among mammals. Spec Publ Amer Soc Mammalogists 7:632-663
Conrad HR, Pratt AD, Hibbs JW (1964) Regulation of feed intake in dairy cows. I. Change in importance of physical and physiological factors with increasing digestibility. J Dairy Sci 47:454-463
Cook CW (1972) Comparative nutritive value of forbs, grasses and shrubs. In McKell CM, Blaisdell JP, Goodin JR (eds), Wildland shrubs--their biology and utilization. USDA For Serv Gen Tech Rep Int 1, Ogden, Utah
Crawley MJ (1983) Herbivory: The Dynamics of Animal-Plant Interactions. University of Calif Press, Berkeley
Demment MW, Greenwood GB (1988) Forage ingestion: effects of sward characteristics and body size. J Anim Sci 66:2380-2392
Demment MW, Van Soest PJ (1985) A nutritional explanation for body-size patterns of ruminant and nonruminant herbivores. Amer Nat 125:641-672
Drozdz A (1968) Digestibility and assimilation of natural foods in small rodents. Acta Theriologica 13:367-389
Drozdz A, Osiecki A (1973) Intake and digestibility of natural feeds by roe deer. Acta Theriologica 18:81-91
Duckworth JE, Shirlaw DW (1958) A study of factors affecting feed intake and the eating behaviour of cattle. Animal Behaviour 6:147-154
Ehle FR, Stern MD (1984) Physical and chemical variables influencing particle passage and size reduction. In Baldwin RL, Bywater AC (eds), Modeling ruminant digestion and metabolism: proceedings of the 2nd international workshop. Univ Calif, Davis
Engelhardt Wv (1970) Movement of water across the rumen epithelium. In Phillipson AT (ed), Physiology of digestion and metabolism in the ruminant. Oriel Press, Newcastle
Fitter AH, May RKM (1987) Environmental physiology of plants. Academic Press, New York
Foose TJ (1982) Trophic strategies of ruminant versus nonruminant ungulates. PhD Diss, Univ Chicago
France J, Thornley JHM, Dhanoa MS, Siddons RC (1985) On the mathematics of digesta flow kinetics. J Theor Biol 113:743-758
Freund RJ, Littell RC (1986) SAS system for regression. SAS Institute, Cary, NC
Golley FB (1960) Energy dynamics of a food chain of an old-field community. Ecol Monogr 30:187-206
Gross JE, Wang Z, Wunder BA (1985) Effects of food quality and energy needs: changes in gut morphology and capacity of *Microtus ochrogaster*. J Mamm 66:661-667
Grovum WL (1979a) Factors affecting the voluntary intake of food by sheep. 2. The role of distention and tactile input from compartments of the stomach. Br J Nutr 42:425-436
Grovum WL (1979b) Factors affecting the voluntary intake of food by sheep. Ann Rech Vet 10:216-218
Hanley TA (1982) The nutritional basis for food selection by ungulates. J Range Manage 35:146-151
Hanley TA, McKendrick JD (1983) Seasonal changes in chemical composition and nutritive value of native forages in a spruce-hemlock forest, southeastern Alaska. US For Serv Res Pap PNW-312
Hobbs NT, Baker DL, Ellis JE, Swift DM (1981) Composition and quality of elk winter diets in Colorado. J Wildl Manage 45:156-171
Hodgson J (1985) The control of intake in the grazing ruminant. Proc Nutr Soc 44:339-346

Hofmann RR (1973) The ruminant stomach. East African Monographs in Biology, Vol. 2. East African Literature Bureau
Hofstadter DR (1980) Godel, Escher, Bach: an eternal golden braid. Vintage, New York
Holleman DF, White RG (1989) Determination of digesta fill and passage from nonabsorbed particulate phase markers using the single dosing method. Can J Zool 67:488-494
Holmes JC, Lang RW (1963) Effects of fertilizer nitrogen and herbage dry matter content on herbage intake and digestibility in bullocks. Anim Prod 5:17-26
Hungate RE (1966) The rumen and its microbes. Academic Press, New York
Illius AW, Gordon IJ (1987) The allometry of food intake in grazing ruminants. J Anim Ecol 56:989-999
Janis C (1976) The evolutionary strategy of the equidae and the origins of rumen and cecal digestion. Evolution 30:757-774
Jung HJG, Batzli GO (1981) Nutritional ecology of microtine rodents: effects of plant extracts on the growth of arctic microtines. J Mamm 62:286-292
Karasov WH (1986) Energetics, physiology, and vertebrate ecology. Trends Ecol Evol 1:101-104
Karasov WH, Diamond JM (1988) Interplay between physiology and ecology in digestion. Bioscience 38:602-611
Kay RNB, Engelhardt WV, White RG (1980) The digestive physiology of wild ruminants. In Ruckebush Y, Thivend P (eds), Digestive physiology and metabolism in ruminants. AUI, Westport, Conn
Kendall WA, Hill RR, Shenk JS (1978) Regulation of intake and utilization of carbohydrates by meadow voles. J Anim Sci 46:1641-1647
Larcher W (1973) Physiological plant ecology. Springer-Verlag, Berlin
Lindroth RL, Batzli GO (1984) Plant phenolics as chemical defenses: effects of natural phenolics on survival and growth of prairie voles *Microtus ochrkogaster*. J Chem Ecol 10:229-244
Lindroth RL, Batzli GO, Avildsen SI (1986) Lespedeza phenolics and penstemon alkaloids effects on digestion efficiencies and growth of voles. J Chem Ecol 12:713-728
Lippke H (1986) Regulation of voluntary intake of ryegrass and sorghum forages in cattle by indigestible neutral detergent fiber. J Anim Sci 63:1459-1468
Mertins DR (1973) Application of theoretical mathematical models to cell wall digestion and forage intake in ruminants. PhD Diss, Cornell Univ, Ithica, New York
Mertins DR, Ely LO (1979) A dynamic model of fiber digestion and passage in the ruminant for evaluating forage quality. J Anim Sci 49:1085-1095
Milchunas DG, Dyer MI, Wallmo OC, Johnson DE (1978) In-vivo/in-vitro relationships of Colorado mule deer forages. Colo Div Wildl Spec Rep No 43
Milne JA, MacRae JC, Spence AM, Wilson S (1978) A comparison of the voluntary intake and digestion of a range of forages at different times of the year by the sheep and the red deer (*Cervus elaphus*). Br J Nutr 40:347-357
Montgomery MJ, Baumgardt BR (1965) Regulation of intake in ruminants. 1. Pelleted rations varying in energy concentration. J Dairy Sci 48:569-574
Moore LA, Thomas JW, Sukes JF (1960) The acceptability of grass/legume silage by dairy cattle. Proc International Grassland Congress VIII:701-104
Morgan CA, Edwards RA, McDonald P (1980) Intake and metabolism studies with fresh and wilted silages. J Agric Sci 94:287-298
Nobel PS (1970) Plant cell physiology. WH Freeman, San Francisco
Owen-Smith N (1982) Factors influencing the consumption of plant products by large herbivores. In Huntley BJ, Walker BH (eds), Ecology of tropical savannas. Springer-Verlag, Berlin
Pond KR, Ellis WC, Lascano CE, Akin DE (1987) Fragmentation and flow of grazed coastal bermudagrass through the digestive tract of cattle. J Anim Sci 65:609-618
Poppi DP, Norton BW (1980) The validity of the critical size theory for particles leaving the rumen. J Agric Sci Camb 94:275-280
Poppi DP, Minson DJ, Ternouth JH (1981a) Studies of cattle and sheep eating leaf and stem fractions of grasses. II. Factors controlling the retention of feed in the reticulo-rumen. Aust J Agric Res 32:109-121
Poppi DP, Minson DJ, Ternouth JH (1981b) Studies of cattle and sheep eating leaf and stem fractions of grasses. III. The retention time in rumen of large feed particles. Aust J Agric Res 32:123-137
Press WH, Flannery BP, Teukolsky SA, Vetterling WT (1986) Numerical recipes: the art of scientific computing. Cambridge University Press, Cambridge
Prignogine I, Stengers I (1984) Order out of chaos: man's new dialogue with nature. Bantam, New York

Robbins CT (1982) Wildlife feeding and nutrition. Academic Press, New York
Sawicka-Kapusta K, Dobrolecka M, Drozdz A, Tertil R (1975) Bioenergetic parameters of experimental groups of common voles [*Microtus arvalis* (Pall. 1779)]. Ekol Pol 23:347-365
Schwartz CC, Hobbs NT (1985) Forage and range evaluation. In Hudson RJ, White RG (eds), Bioenergetics of wild herbivores. CRC Press, Boca Raton
Short HL, Blair RM, Segelquist CA (1974) Fiber composition and digestibility by small ruminants. J Wildl Manage 38:197-209
Short HL, Blair RM, Epps EA (1975) Composition and digestibility of deer browse in the southern forests. US For Serv Res Pap SO-111
Sibly RM (1981) Strategies of digestion and defecation. Pages 109-139 in Townsend CR, Calow P (eds), Physiological ecology:an evolutionary approach to resource use.
Spalinger DE (1985) The dynamics of forage digestion and passage in the rumen of mule deer and elk. PhD Diss, Washington State Univ, Pullman
Spalinger DE, Robbins CT, Hanley TA (1986) The assessment of handling time in ruminants: the effect of plant chemical and physical structure on the rate of breakdown of plant particles in the rumen of mule deer and elk. Can J Zool 64:312-321
Starfield AM, Bleloch AL (1986) Building models for conservation and wildlife management. Macmillan, New York
Street HE, Opik H (1976) The physiology of flowering plants: their growth and development. Elsevier, New York
Sutcliffe J (1968) Plants and water. Edward Arnold, London
Swartzman GL, Kaluzny SP (1987) Ecological simulation primer. Macmillan, New York
Thomas SW, Moore LA, Okamotto M, Sykes JF (1961) A study of factors affecting the rate of intake of heifers fed silage. J Dairy Sci 44:1471-1483
Thornton RF, Minson DJ (1973) The relationship between apparent retention time in the rumen, voluntary intake, and apparent digestibility of legume and grass diets in sheep. Aust J Agric Res 24:889-898
Uden P (1978) Comparative studies on rate of passage, particle size, and rate of digestion in ruminants, equines, rabbits, and man. PhD Diss, Cornell Univ, Ithica, New York
Uden P, Van Soest PJ (1982) Comparative digestion of timothy (Phleum pratense) fibre by ruminants, equines, and rabbits. Br J Nutr 47:267-272
Undersander DJ, Cole NA, Naylor CH (1987) Digestibility and rate of passage by lambs of water-stressed alfalfa. J Anim Sci 64:1813-1820
Van Soest PJ (1982) Nutritional Ecology of the Ruminant. O&B Books, Corvallis, Oregon
Waldo DR, Miller RW, Okamoto M, Moore LA (1965) Ruminant utilization of silage in relation to hay, pellets, and hay plus grain. II. Rumen content, dry matter passage, and water intake. J Dairy Sci 48:1473-1480
Warner ACI (1981) Rate of passage of digesta through the gut of mammals and birds. Nutr Abstr Rev Ser B 51:789-820
Welch JG (1982) Rumination, particle size, and passage from the rumen. J Anim Sci 54:885-984
Welch JG (1986) Physical parameters affecting rate of passage from the rumen. J Anim Sci 69:2750-2754
Weston RH, Poppi DP (1987) Comparative aspects of food intake. In Hacker JB, Ternouth JH (eds), The nutrition of herbivores. Academic Press: Sydney
Willes RF, Mendel VE, Roblee AR (1970) Water transfer from the recticulo-rumen in sheep. J Anim Sci 31:85-91
Williams PH, Shenk JS, Baker DE (1978) Cadmium accumulation by meadow voles (*Microtus pennsylvanicus*) from crops grown on sludge-treated soil. J Environ Qual 7:450-456
Woodford ST, Murphy MR (1988) Dietary alteration of particle breakdown and passage from the rumen in lactating dairy cattle. J Dairy Sci 71:687-696
Worrell MA, Clanton DC, Stroup WW, Nicholas JT (1986) Effect of harvest date on meadow hay quality. II. Particle size degradation and particulate passage from the rumen of growing cattle. J Anim Sci 63:1538-1546

ORAL DISCUSSION – N.T.Hobbs

Q. (Belovsky) Three of my papers point out that the way digestive constraint is set up does not necessarily fit the best current ideas of physiological processes. I've tried dry matter and it failed, so I stuck with something that worked even if we didn't understand why it worked. Regarding your point about testing and retesting, I found 248 tests of linear programme models, 80 by me, and only 3 didn't work, and that's because the authors didn't put in time constraints. So if 245 out of 248 worked, then maybe its telling us that something to do with bulk is working even if we don't understand precisedly what it is.

Hobbs: My feeling about building models is that we do it to understand not to predict. Your model predicts incredibly well, but you're saying the model and the digestive constraint are valid because they work, even though you don't know why.

Belovsky: No, I'm saying the model is telling us there's something there we don't understand, we should be asking more about why the model works as well as it does.

Ball: What about bulk constraints on simple carnivores ?

Hobbs: The epithelium of the mammalian gut is highly adapted to absorb water rapidly. So, except in the crops of birds which have no absorptive function, where there might be a contraint, I can't see in mammalian guts how water–related bulk can be a constraint.

A Reply To Hobbs

Gary E. Belovsky
Department of Biology and School of Natural Resources
University of MIchigan
Ann Arbor, Michigan 48109-1115
USA

Hobbs (this volume) raises two issues concerning my use of Linear Programming (LP) methods to model optimal diet choices: the models are highly sensitive to parameter values and the digestive constraint is constructed using debatable assumptions about the kinetics of digestion. He further proposes a different formulation of the digestive constraint which fails to predict diet choices using an LP model. This leads Hobbs to claim that my use of an LP model is incorrect and should be discarded, even though it has been empirically successful.

Hobbs' criticisms also have more general implications for Optimal Foraging Theory models (OFT), especially given the 'next generation' of models and the common invocation of digestive constraints found in this volume. Each of Hobbs' concerns will be addressed in turn.

Model Sensitivity

LP is a well established optimization technique (Intriligator 1971) and has been invoked to describe animal diet choices by several authors (Schoener 1971, Westoby 1974, Pulliam 1975, Covich 1976, Altman & Wagner 1978). Indeed, most OFT models can be written as LP models (Belovsky *et al.* 1989). If the sensitivity of any OFT model is examined, small parameter changes will frequently be found to change diet predictions substantially (e.g. the stepwise switch from eating only one food to eating two foods in proportion to their encounter as predicted by the classical OFT model: Stephens & Krebs 1986). Animals can change their diets dramatically between habitats, times of the day, and seasons; OFT models attempt to describe some of these changes. Therefore, by necessity OFT must be highly sensitive to subtle parameter changes that occur with changing environments. This sensitivity cannot be eliminated, since it is the nature of the problem being addressed.

NATO ASI Series, Vol. G 20
Behavioural Mechanisms of Food Selection
Edited by R. N. Hughes

When food bulk parameters were varied in my LP model (Belovsky 1986a), Hobbs claims that only 2/899 solutions provided an equal or greater explanation (r^2) of the observed diets for the 14 species and none provided a regression slope closer to 1 or intercept closer to zero. Because of these findings, he concludes that the LP results are too sensitive to be of value. Both of Hobbs' contentions, that the model is too sensitive and that only a small number of parameter values lead to good predictions, depend upon the range of bulk values used in the analysis. The figure of 2/889 (above) is based upon $\pm$ 1 standard deviation for the bulk values; $\pm$ 2 standard deviations provides a figure of 2/3600, and the observed range of bulk values observed during the study (grass: 2.32 - 1.12; dicots: 6.67 - 1.81) provides 2/1322.

What is the correct range of bulk values on which to base the sensitivity analysis? This depends upon the question being asked by the OFT model. My LP model asks the question: within each species population, can an average adult forager's average summer diet during the period of the study be explained by the average measured food characteristics that affect digestive capacity, feeding time and energy requirement constraints? Since my LP model deals with averages, the standard error is the appropriate measure of precision, whereas Hobbs' use of the standard deviation overestimates the model's sensitivity. The correct procedure gives r^2 values greater than or equal to those I report in 2/8 cases based on $\pm$ 1 standard error or 2/34 cases based on the 95% confidence interval.

Goodal (1972) argues that a Monte Carlo simulation, varying the parameter values according to 'fiducial distributions' (probability distributions), should be used to state probabilities more precisely. The LP model was solved 100 times for each of the 14 species, where each run for all species used the same bulk parameters, generated from a normal random distribution given their means and standard errors. This Monte Carlo simulation indicates that 18/100 runs provide a better or equal r^2 value than I originally reported and 46/100 values are within 0.05 of my original reported value. In 39/100 simulations the regression comparing predicted and observed values has a slope equal or better than the original and 55/100 values were within 0.05 of the original value (0.94). The distributions of r^2 and slope values generated by the Monte Carlo simulation, when plotted, show that almost all the variations in r^2 and slope estimates arise from the very sensitive solutions for just two out of the 14 species. Hobbs argues that because bulk can vary dramatically over a day, let alone seasonally, a model based upon a single average value is inappropriate. However, much of this variance will cancel out of my LP model and not influence sensitivity. This occurs because grass and forb-bulk values tend to covary, so that each constraint line can be written as a slope (ratio of food bulk values) and an intercept based on a single food. Consequently, when foods covary, the intercept value will change with variance more than the slope, whereas, Hobbs incorrectly assumes that the slope and intercept equally vary, increasing apparent sensitivity.

Formulation Of The Digestive Constraint

Hobbs disagrees with the assumptions used to construct the digestive constraint, in particular the notion that food bulk (wet mass/dry mass) determines the fill of digestive organs. What controls maximum food intake is a hotly debated topic, but distension of the digestive organs is one common explanation (Baile & Forbes 1974, for ruminants). Distension can only be relieved by the passage of food out of the organs. In this respect Hobbs and I agree. While I argue that this distension is a function of food bulk, expressed as food wet/dry mass, Hobbs' view is that of traditional rumen physiology, i.e. distension is a function of food dry mass. My assumptions have been supported by other studies (Tulloh 1966, John & Ulyatt 1987). Moreover, Butris & Philips (1987) demonstrate that increasing food wet mass decreases daily dry matter consumption. They do not attribute this to a physical limit to intake but to reduced feeding time, and they speculate that feeding time is reduced because increased water content reduces palatability. Would not a full digestive organ, however, lead to reduced feeding time (Belovsky 1981, Belovsky 1984)?

Since free-water can be rapidly absorbed from most digestive organs, I am leery of studies showing no effect when free-water is added to the food surface. For example Thomas *et al.* (1961) attempted to explain their finding that increased silage bulk reduced dry matter intake in cattle by showing that free-water added to feeds had no effect on intake. Similarly, I question Butris & Philips' (1987) study, even though it supports my assumptions, because they used free-water. When I included a water requirement in the LP model (Belovsky 1987), the ingestion of free-water was not debited against the digestive constraint, because I assumed it was rapidly absorbed. Furthermore, studies that artificially reduce food bulk by drying may not be useful for testing my assumptions, since drying can change the foods' digestive characteristics (Hungate 1966). The studies that report reduced intake as food bulk increases have used fresh-cut forages (e.g. Tolluh 1966, John & Ulyatt 1987, see below).

I now examine Hobbs' arguments (see also Hanley 1980) against my assumptions about food bulk. These arguments are that water can be removed from most digestive organs rapidly and there is no relationship between food bulk and daily food intake. Hobbs claims that water absorption is so rapid that it is 40 times faster than the passage rate of food. Based on one of his cited studies (Willes *et al.* 1970), this is true given their reported gross absorption of water in the sheep rumen (37.1 - 70.8 ml/min); however, when the maintenance of osmotic balance in the rumen by the insorption of blood-water is considered, the reported net absorption is only 5.1 - 13.4 ml/min. Sheep ingest 10 - 18 l/day of water by drinking and ingesting food (MacFarlane & Howard 1970). It would require on average 25 h for the rumen to absorb this water, while the passage rate of food out of the rumen averages about 24 h. Furthermore, this study employed free-water, not water contained in plant tissue.

The only study that I have been able to find which addresses the water uptake from the rumen, when the water is contained in fresh-cut plant tissues, is reviewed in Engelhardt (1970), one of Hobbs' main references on the rapid absorption of water. Engelhardt (1970) reports that a maximum of 2.5 ml/min is the net absorption in the first hour after feeding by sheep, but 22 h are required to absorb 50% of the plant tissue's water. This is much slower than the passage rate of food through the digestive tract. In addition, Engelhardt (1970) states that water must be continually added to the rumen (water turnover rate of 1.87 - 3.2 times/day) to keep the food hydrated for digestion, since water can pass from the rumen faster than the dry matter.

Hobbs claims, based upon data gleaned from the literature (Fig. 3), that there is no correlation between daily food intake (scaled to forager mass to the 0.75 power) and food bulk. He then claims that this disproves the formulation of my digestive constraint. Employing the LP solutions in question (Belovsky 1986a), I also find no correlation between daily dry matter intake (similarly scaled) and food bulk.

If bulk should decrease food consumption as assumed in my construction of the digestive constraint, why is this not reflected when all the LP solutions are combined? Daily food intake is not only a function of bulk, but also a function of the digestive organ capacity and the passage rate of food through the digestive organ. In the LP solution each of the 14 species has a different digestive capacity and passage rate which cannot be simply scaled to forager mass to the 0.75 power.

Hobbs, moreover, combines into a single comparison a wide variety of foods: fruits, seeds, grasses, forbs, woody twigs, hay, silage and pelleted rations. But for any animal species these foods can vary in passage rate by 300% or more. Furthermore, Hobbs combines studies that have the forager's digestive organ capacity varying seasonally. Finally, the positive relationship Hobbs finds for roe deer reflects annual diet variation since low bulk - low dry matter intake values are for winter twigs, the high bulk - high dry matter intake values are for spring/summer growing plants and the intermediate values are from autumn. The passage rates for growing plants are approximately 200% faster than for twigs, and this explains Hobbs' observed relationship, not food bulk.

Fresh-cut plants from 4 highly preferred species (2 grasses, 2 dicots) at the same phenological stage and from a common garden were fed to 5 out of the 14 species reported in Belovsky (1986a). Each plant species alone was provided *ad lib* to at least three individuals of each forager species and their daily wet and dry mass intake were measured. Comparisons cannot be made using absolute intake since it is a function of species-specific digestive organ capacity and passage rate. Therefore, the coefficient of variation in mean intake by each forager species was used to eliminate these differences. Furthermore, the coefficients of variation only reflect the differences due to food, since the variation among individuals within each forager species was eliminated (total coefficient of variation - coefficient of variation between individuals fed the same food). For all 5 forager species, the daily wet mass intake varied less than 12% among the 4 different foods and was not significantly different among 4 of the forager species (One-Way ANOVA, $P<0.05$). This was so even though the food-bulk values varied by 20-70%. Daily dry matter intake, however, varied from 18-55% and the dry matter consumption of each food was significantly different within forager species (One-Way ANOVA, $P<0.05$). Furthermore, the variance in food bulk explained 80% of the variance in dry matter intake ($r^2 = 0.80$, $N = 19$, $P<0.01$). These results indicate that at least for these 5 foragers and 4 plants, daily food wet mass intake is relatively constant and daily dry mass intake is largely explained by food bulk. John & Ulyatt (1987) report the same pattern for sheep fed fresh-cut plants. These are exactly the patterns expected from the formulation of the digestive constraint and explains why my LP model successfully predicts daily dry matter intake for the 14 species (Belovsky 1986a: Fig. 6, p. 45).

Conclusion - Comments On Serendipity And The Future Of LP

In a forthcoming review of OFT for mammals (Belovsky in press), I identify 241 tests of the LP model. Two do not work, but 239 LPs that successfully predict diet choices explain 92% of the observed dietary variation. The greatest deviation between observed and predicted diet proportions is 0.16 for average foragers in a population and 0.20 for single individuals in a population. Using Monte Carlo simulations, Z-statistics can be computed to measure the normalized deviations between predicted and observed diets. The average Z-statistic is not significantly different from zero (mean = -0.10), indicating no bias in the model to overestimate or underestimate the proportion of the most commonly consumed food in the diet. The 241 Z-statistics have a standard deviation of 1.55. A value less than 1 indicates that the diets are predicted better than expected, given the confidence intervals for the parameter values. The observed standard deviation, however, is significantly greater than 1, indicating that the model predictions are poorer than expected; i.e. there are other factors to diet choice that the model is not considering.

A model may work under one set of conditions (e.g. one forager species, environment, etc.) due to serendipity even though the model is incorrect. To avoid this pitfall, a model must be tested under a range of conditions (Smith 1970). This has been done with the LP model and it has thus far withstood the test. Rather than an outmoded theory that should be discarded as Hobbs claims, the model is 'alive and vibrant'. But although 239 successful tests of the model have been conducted, these only involve 36 species, almost all temperate (2 birds, 6 insects and 28 mammals). I do not think this constitutes a universal theory and I propose that continued testing is necessary (Belovsky 1986b: p. 63-64).

More work on the LP model lies before us. I can see a ripe opportunity for digestive physiologists and OFT practitioners to work together. Science progresses by tensions between ideas and I believe we see this here: why does the LP model appear to work with a digestive constraint constructed with debatable assumptions, and how might digestive physiology experiments be designed in light of this?

An excellent starting point for collaborative work might be the incorporation of optimal digestion models (Dade *et al.* this volume, Taghon *et al.* this volume) into the LP model. Another example is provided by Baker & Hobbs (1987) concerning winter feeding by montane ungulates on dried grasses and twigs, which presents an opportunity to construct an LP model incorporating the differential passage rates of foods (see also Spalinger *et al.* 1986). Different questions may require different models and different degrees of detail, but LP and OFT in general appear to answer some simple initial problems. Therefore, this provides a perfect setting for theoreticians to work with empiricists to expand our knowledge on diet choice.

Literature Cited

Altman SA, Wagner SS (1978) A general model of optimal diet. In Chivers DJ, Herbert J (ed), Recent Advances in Primatology (Vol 1). Academic Press, London, p 407-414

Baile CA, Forbes JM (1974) Control of feed intake and regulation of energy balance in ruminants. Physiol Rev 54: 160-214

Baker DL, Hobbs NT (1987) Strategies of digestion: digestive efficiency and retention time of forage diets in montane ungulates. Can J Zool 65: 1978-1984

Belovsky GE (1981) Optimal activity times and habitat choice of a moose. Oecologia 48: 22-30

__________ (1984) Herbivore optimal foraging: a comparative test of three models. Amer Natur 124: 97-115

__________ (1986a) Optimal foraging and community structure: implications for a guild of generalist grassland herbivores. Oecologia 70: 35-52

__________ (1986b) Generalist herbivore foraging and its role in competitive interactions. Amer Zool 26: 51-69

__________ (1987) Hunter-gatherer foraging: a linear programming model. J Arch Anthro 6: 29-76.

__________ (in press) Optimal diet choice studies with mammals. Fifth International Theriological Congress, Rome, Italy

__________, Ritchie ME, Moorehead J (1989) Foraging in complex environments: when prey availability varies over time and space. Theor Pop Biol 36

Butris GY, Philips JC (1987) The effect of herbage surface water and the provision of supplementary forage on the intake and feeding behavior of cattle. Grass and Forage Sci 42: 259-264

Covich AP (1976) Analyzing shapes of foraging areas: some ecological and economic theories. Ann Rev Ecol Syst 7: 235-257

Dade WB, Jumars PA, Penry DL (in press) Placing limits on the cost of selection and digestion in optimization modeling. This Volume

Engelhardt WV (1970) Movement of water across the rumen epithelium. In Phillipson AT (ed), Physiology of Digestion and Metabolism in the Ruminant. Oriel Press, Newcastle

Goodal DW (1972) Building and testing ecosystem models. In Riechle DE (ed), Ecological Studies Analysis and Synthesis. Springer-Verlag, Berlin

Hanley TA (1980) Nutritional constraints on food and habitat selection by sympatric ungulates. Ph.D. dissertation, Univ Washington, Seattle

Hobbs NT (in press) Diet selection by generalist herbivores: a test of the linear programming model. This Volume

Hungate RE (1966) The rumen and its microbes. Academic Press, New York

Intriligator, MD (1971) Mathematical optimization and economic theory. Columbia Press, New York

John A, Ulyatt MJ (1987) Importance of dry matter content to voluntary intake of fresh grass forages. NZ Soc Anim Prod 47: 13-16

MacFarlane WV, Howard B (1970) Water in the physiological ecology of ruminants. In Phillipson AT (ed), Physiology of Digestion and Metabolism in the Ruminant. Oriel Press, Newcastle

Pulliam HR (1975) Diet optimization with nutrient constraints. Amer Natur 109: 765-768.

Schoener TW (1971) Theory of feeding strategies. Ann Rev Ecol Syst 2: 369-404

Smith FE (1970) Analysis of ecosystems. In Riechle DE (ed), Ecological Studies Analysis and Synthesis. Springer-Verlag, Berlin

Spalinger DE, Robbins CT, Hanley TA (1986) The assessment of handling time in ruminants: the effects of plant chemical and physical structure on the rate of breakdown of plant particles in the rumen of mule deer and elk. Can J Zool 64: 312-321

Stephens DW, Krebs JR (1986) Foraging theory. Princeton Univ Press, Princeton

Taghon GL, Greene RR, Bard D (in press) Do deposit feeders vary feeding rates to maximize net rate of gain as food value changes? This Volume

Thomas SW, Moore LA, Okamotto M, Sykes JF (1961) A study of factors affecting the rate of intake of heifers fed silage. J Dairy Sci 44: 1471-1483

Tolluh NM (1966) Physical studies of the alimentary tract of grazing cattle: III. Seasonal changes in capacity of the reticulo-rumen of dairy cattle. NZ J Agric Res 9: 252-260

Westoby M (1974) An analysis of diet selection by large generalist herbivores. Amer Natur 112: 627-631

Willes RF, Mendel VE, Roblee AR (1970) Water transfer from the reticulo-rumen in sheep. J Anim Sci 31: 85-91

APPLICABILITY OF FIVE DIET–SELECTION MODELS TO VARIOUS FORAGING CHALLENGES RUMINANTS ENCOUNTER

Fred D. Provenza & David F. Balph, Utah State University

INTRODUCTION

It is common knowledge that ruminants do not forage at random, but select a diet from the plants available to them. We believe foraging environments present at least five problems or challenges to ruminants selecting dietary items: (1) variation among dietary items in kind and amount of nutritional constituents, (2) variation among potential dietary items in kind and amount of chemical defenses, (3) plant morphological defenses, (4) temporal and spatial variation in the quantity and quality of forage, and (5) exposure of ruminants to unfamiliar foraging environments. Our objective is to assess the ability of five explanations of diet selection to provide insights into the responses of ruminants to these challenges. The models are: (1) endogenously–generated hungers (euphagia), (2) immediate sensory consequences (hedyphagia), (3) body morpho–physiology and size (morphophysiology), (4) learning through foraging consequences (learning), and (5) nutritional optimization (optimal foraging). We make the assessment by first describing the diet–selection challenges and then discussing the models and their applications to the challenges.

CHALLENGES FORAGING ENVIRONMENTS PRESENT TO HERBIVORES

The first challenge the foraging environment presents is variation in the quantity of energy, protein and minerals contained in different dietary items (Van Soest 1982). Herbivores that are able to recognise nutritional deficiencies in themselves and the presence and amount of nutritional components in the foods they eat should be able to select a more nutritious diet than animals that cannot. The question is: can ruminants directly or indirectly sense nutritional components, and if so, how ?

The second challenge for ruminants is coping with plant chemical defenses. Toxins vary in kind and amount, as do nutritional constituents. Plant toxins may reduce digestibility or interfere with metabolic processes and cause death (Cheeke & Shull 1985). Some toxins may be readily sensed through taste, odor, and gastro–intestinal distress, while others may not (Provenza *et al.* 1988, 1990). The ability of herbivores to cope

NATO ASI Series, Vol. G 20
Behavioural Mechanisms of Food Selection
Edited by R. N. Hughes

with toxins is important to their nutritional health and survival. Here we ask: Can ruminants detect and avoid plant chemical defenses, and if so, through what mechanisms?

The third challenge to ruminants in the foraging environment is plant morphological defenses, such as standing dead material in some grasses (Willms et al. 1980; Norton et al. 1982), thorns in forbs (Myers 1987) and woody plants (Cooper and Owen-Smith 1986), and differences in canopy shape and architecture (McNaughton 1984; Penning 1986). Morphological defenses can inhibit foraging or decrease ingestion rate (Cooper and Owen-Smith 1986; Flores et al. 1989 a,b,c), which in turn can influence diet preference (Kenney and Black 1984). Animals that can circumvent such defenses can enhance their nutritional welfare. Thus, we ask: How effective are ruminants in coping with plant morphological defenses?

The fourth challenge is spatial and temporal variation in obtaining food efficiently. Spatial variation has been documented within and among plants of the same species and plant assemblages (e.g. Van Soest 1982; Bryant et al. 1983, 1987; Gershenzon 1984). Temporal variation occurs within a day and season for individual plant parts and within a season and over years for plant species and assemblages (e.g. Coley et al. 1985; Bailey 1988). Herbivores that position themselves in the foraging environment at the proper place and time can increase the probability of encountering nutritious food. This raises the question: How well do ruminants position themselves to increase the encounter rate of nutritious dietary items?

The fifth challenge is that of being in an unfamiliar foraging environment. Ruminants may find themselves in such a setting through abiotic or catastrophic events. Fire or rain in arid areas can rapidly and significantly alter the vegetation available. The movement of animals themselves through dispersal, migration or forced immigration, as in the case of livestock or wildlife transplants, can result in the animals foraging in unfamiliar environments. Ruminants that forage in a manner that allows them to adjust quickly to novel foraging conditions should reduce nutritional distress. Our question here is: How well do ruminants adjust to novel foraging environments?

DIET-SELECTION MODELS

The five models of diet selection examined here are not mutually exclusive, nor is the list exhaustive. They vary in the degree to which they address the five foraging challenges because of differing objectives and levels of resolution and scale (Senft et al. 1987). Our objective is not to compare the explanations per se, but to examine their relative abilities to explore and provide insights into the process of diet selection. The models cover a spectrum of approaches to the study of diet selection and have their origins in diverse fields including animal science, behavioral ecology, and psychology. Our examination includes a description of each model and how well it can be used to explain and explore diet selection in the face of each foraging challenge. Our primary interest is in ruminants, but details of their diet selection are unknown. This necessitates that we occasionally refer to non-ruminants, especially laboratory rats, for which mechanisms of diet selection are better understood. Some studies suggest that the mechanisms used by rats for selecting a diet are similar to those used by some ruminants (reviewed by Provenza and Balph 1987, 1988; Provenza et al. 1988, 1990).

Euphagia

Proponents of euphagia argue that animals have the innate ability to sense, through taste and smell, specific nutrients and toxins in plants. Through such nutritional "wisdom" animals should select nutritious foods and avoid those containing harmful plant metabolites. The euphagia model does not include any associations between the taste and odor of food and post-ingestive consequences. It developed among animal nutritionists in response to the finding that livestock (e.g. see Arnold and Dudzinski 1978) and rats (e.g. see Richter 1943) select a more nutritious diet than would be expected if the animals selected foods at random. The idea has been embraced by some researchers of plant secondary metabolites (e.g. Rhoades 1979). Euphagia clearly pertains to two of the five challenges, recognition of nutritional components and coping with plant defenses. However, it is not clear how valid is the assumption that animals can directly sense, via olfaction and gustation, specific nutrients or toxins in the foods they eat.

The only well-established, innate hunger is for sodium. During sodium deficiency, rats innately recognize the relevance of ingesting very salty materials that are normally aversive (Krieckhaus and Wolf 1968). With increasing deprivation, higher concentrations of salt are needed to make the chorda tampani nerve respond vigorously (Contreras and Frank 1979). Thus, rats perceive a particular concentration of salt as equivalent to a lower concentration prior to deprivation and prefer saltier foods. This change appears to be endogenously programmed and not dependent on previous experience with sodium deprivation. Likewise, sheep (Denton and Sabine 1963; Denton 1965) and cattle (Bell 1984) can very precisely rectify sodium deficiences, perhaps by the same mechanisms. However, with the exception of sodium, researchers have been unable to demonstrate that animals can directly sense mineral nutrients in the foods they eat (Gordon and Tribe 1951; Gordon et al. 1954; Arnold 1970; Coppock et al. 1974, 1981; Booth 1985).

Nor have researchers been able to demonstrate consistent relationships between diet selection by ruminants and compounds such as protein, soluble carbohydrates, sugars, and organic acids (reviewed by Arnold and Hill 1972). It is not likely that animals can recognize most of these compounds per se, because they do not exist in the same molecular form in the lab as in the plant (Arnold and Hill 1972). Even pure glucose in solution occurs in seven interconverting forms, each with its own specific stereochemistry (Arnold and Hill 1972). Where correlations exist between such phytochemicals and preference, they must relate to specific compounds that animals can detect or some physical property of the plant (Arnold and Hill 1972; Reichardt et al. 1987). However, even if ruminants can detect specific nutrients, it does not necessarily follow that they innately recognize the value of ingesting such compounds.

Natural selection should favor feeding behavior that minimizes the ingestion of compounds that interfere with digestive or metabolic processes (Rhoades 1979; Cooper et al. 1988). In such cases, the plant tissue could provide a signal, presumed to be the secondary metabolite, which herbivores innately recognize and avoid. For example, some argue that herbivores such as goats evolved the innate ability to recognize and avoid condensed tannins because goats are browsers, and most woody plants (about 80%; Bate-Smith and Metcalf 1957; Rhoades and Cates 1976) contain condensed tannins (Rhoades 1979). However, many plants high in condensed tannins are highly nutritious, and the diverse array of condensed tannin structures (Zucker 1983; Clausen et al. 1989) and food flavors makes it unlikely that goats are innately able to recognize and avoid plants that contain condensed tannins (Provenza et al. 1990). Indeed, goats readily accept on initial exposure current season's growth from the shrub blackbrush (Coleogyne ramosissima) which contains high levels of condensed tannins, but refuse to eat current growth on subsequent exposures (Provenza et al. 1990). One might mistakenly assume such foraging behavior is innate because goats learn so quickly to avoid current season's growth.

Thus, existing evidence suggests that euphagia generally fails to explain how herbivores either select or reject specific plant species or parts. Perhaps this is not suprising given that the properties of phytochemicals depend on their detailed structures, and general classes of phytochemicals such as proteins, carbohydrates, tannins, terpenoids, alkaloids, and resins are not likely to contain compounds that are uniformly good or bad (Arnold and Hill 1972; Freeland and Janzen 1974; Reichardt et al. 1984, 1987; Provenza et al. 1988). However, our current view of euphagia could change, because so few wild or domestic ruminants have been examined for their ability to directly sense the nutritional or toxicological properties of the foods they eat.

Hedyphagia

Some livestock nutritionists and ecologists theorize that animals obtain a nutritious diet by selecting vegetation that is immediately "pleasing" to olfactory, gustatory, and tactile senses and avoiding that which is not (Arnold and Dudzinski 1978). This is the hedyphagia model of diet selection. Proponents of hedyphagia argue that evolution progresses such that those plant compounds that are nutritious ultilmately taste good and those that are toxic ultimately taste bad, all through natural selection (Rhoades 1979).

Hedyphagia addresses how animals might obtain a nutritious diet and avoid plant chemical and morphological defenses, not directly but through association. Ruminants generally prefer plant parts that are young, green, and leafy, attributes generally associated with more nutritious vegetation, to those that are old, dead, and stemmy (Arnold and Dudzinski 1978). They often avoid plants with spines and stinging hairs that are difficult to ingest or are aversive to tactile senses. In addition, many mammals readily accept sweet and reject bitter tastes which suggests that sweet and bitter may be stimuli signifying nutrients and toxins, respectively (Bate-Smith 1972; Garçia and Hankins 1975, 1977; Grill et al. 1984; Beauchamp and Cowart 1987; but see Woods 1983). Different species of ruminants differ in their ability to discriminate and tolerate various purified compounds representative of the primary taste modalities (Arnold and Hill 1972; Church 1979; Hofmann 1989; see Schwartz and Grill 1985 for a review relative to rats), and such differences presumably affect their ability to discriminate among and tolerate different forages. Thus, hedyphagia provides a means for explaining responses of different species of ruminants to some nutrients, as well as chemical and physical defenses in cases where they have immediate positive or aversive effects on tactile senses.

Hedyphagia does not deal explicitly with fundamentally important interactions among odor, taste, and consequences of ingesting foods. As a result, the model cannot explain why herbivores prefer to ingest some plant parts that are toxic (Arnold and Dudzinski 1978; Fowler 1983; Cheeke and Shull 1985;

Provenza et al. 1988), or come to avoid others that are immediately pleasing to the senses (Garcia and Holder 1985; Provenza et al. 1990). Research on conditioned flavor aversions and preferences clearly suggests that the hedonic value (palatability) of food is relative and depends to a significant degree on post-ingestive consequences (Garcia et al. 1985; Garcia and Holder 1985; Mehiel and Bolles 1988; Tordoff and Friedman 1989a). If the food is nutritious, visceral feedback raises the hedonic value of the food. If nausea ensues, the food is avoided. While bitter and sweet are often considered feeding deterrent and attractants respectively (Garcia and Hankins 1975, 1977), preference for bitter increases when it is paired with positive post-ingestive consequences while preference for sweet decreases when it is paired with aversive post-ingestive consequences (Garcia and Holder 1985). Thus, odor and taste may be more important for enabling ruminants to identify and discriminate among subtle differences in dietary items, than for innately recognizing plants that are nutritious or toxic based on pleasing or adverse gustatory and olfactory sensations (see also Galef 1981).

Trials conducted to assess the relative palatabilities of different purified compounds or food items to ruminants have neglected the fact that foods have post-ingestive consequences (Grovum 1988). With few exceptions (Grovum and Chapman 1988; Pfister et al. 1989), researchers have not attempted to separate the effects of food flavor from those associated with post-ingestive consequences. Such studies reveal that post-ingestive consequences have direct and important influences on the palatability of specific food items. For example, the shrub bitterbrush (Purshia tridentata) contains higher concentrations of condensed tannins which are more astringent than those contained in the shrub blackbrush, yet herbivores strongly prefer bitterbrush to blackbrush (Clausen et al. 1989; Provenza et al. 1990). Bitterbrush and blackbrush differ in post-ingestive consequences. Nor have researchers attempted to determine how previous experiences with specific food flavors affect the responses of ruminants to new food flavors (Provenza and Balph 1987, 1988; see also Galef 1981). It is likely that the animals generalize or discriminate based on previous positive or negative experiences with specific food flavors (Launchbaugh and Provenza 1990).

Body Morphophysiology and Size

Ruminant nutritionists and ecologists have developed two conceptual frameworks for understanding food selection and resource partitioning based on differences in morphology. Proponents of these models argue that as a result of evolving in different environments, different ruminant species possess different morphophysiological characteristics that cause them to ingest forage that differs in physical and chemical characteristics. Although the models differ somewhat, we discuss them together in this section.

The first model is based on a classification of ruminants as concentrate selectors, intermediate selectors, and grass-roughage selectors (Hofmann 1973, 1988, 1989). Concentrate selectors (40% of ruminants) possess characteristics that enable them to ingest forages high in cell solubles such as forbs and browse, while roughage selectors (25% of ruminants) are best adapted to ingest forages high in cell wall contents such as grasses. Intermediate selectors (35% of ruminants) are morphophysiologically intermediate between the other two classes. Concentrate selectors have a narrower muzzle with a relatively deeper mouth opening than do roughage selectors. This enables concentrate selectors, not only to carefully select plant parts, but also to strip twigs of leaves. Concentrate selectors presumably rely heavily on olfaction to sample various plant parts, while roughage selectors rely heavily on taste. Concentrate selectors have larger salivary glands than do roughage selectors, presumably an adaptation for: (1) providing liquid which reduces retention time in the alimentary tract, (2) carrying cell solubles down the ventricular groove and directly into the abomasum, and (3) countering plant chemical defenses such as tannins. Concentrate selectors also have less mass and volume of reticulorumen and larger livers for detoxifying phytochemicals than do roughage selectors. As a result of these and other differences, concentrate selectors quickly pass forage through the alimentary tract, which enables them to quickly digest cell solubles and pass cell walls. Conversely, slower rates of passage enable roughage selectors to digest cell walls more efficiently. Thus, concentrate selectors eat smaller, more frequent meals than do roughage selectors. All three classes may select diets that differ in nutritional quality by selecting different plant species (e.g. Van Dyne et al. 1980; Illius and Gordon this volume) or different parts from the same plant species (Bell 1969; Janis 1976).

The second model assumes that the relationship between body size and metabolic requirements is of fundamental importance in diet selection by ruminants (Bell 1971; Jarman 1974; Geist 1974; Demment and Van Soest 1985; Demment and Longhurst 1987). Small ruminants have higher metabolic costs per unit volume of rumen than do larger ruminants. Thus, the forages they select must produce higher fermentation rates and energy yields and have higher rates of passage than those selected by larger ruminants. Higher passage rates result in lower digestion of cell walls by small ruminants. The higher the ratio of rumen volume to body weight, the more cell wall can be tolerated in the diet. While small ruminants must spend time finding and handling high quality but rare foods, the total amount of energy they require is less than that required by large ruminants and therefore they are able to spend more time obtaining such foods.

These two models assume that diet selection by ruminants is governed by morphology and physiology. The implication is that diet selection operates through sampling and that the probability of selecting the same dietary item again is governed by interactions between plant chemistry and morphology and animal morphophysiology. The models make general predictions about foraging in unfamiliar environments and foraging in environments where the abundance and nutritional quality of dietary items vary temporally and spatially (duToit and Owen-Smith 1989; see also Illius and Gordon this volume). Wider feeding tolerance among larger species leads to their use of a wider range of habitats and more even use of environmental resources as compared with smaller species, provided a diversity of habitats is available.

To the extent that classifications based on morphophysiology are able to elucidate causal relationships (see Romesburg 1981) between structure and function, they are of real value in understanding diet and habitat selection by ruminants. Broad classifications of diet and habitat selection based on characteristics of animals (e.g. duToit and Owen-Smith 1989) and plants (e.g. Cooper and Owen-Smith 1985) are appealing because they provide generality. However, it is not likely that they will ever explain diet selection in quantitative taxonomic terms (Hanley 1982) because of the complex chemical and morphological nature of plants, and because behavioral as well as morphological and physiological attributes of animals must be taken into account to understand diet selection fully (Provenza and Balph 1988; Provenza et al. 1988, 1990).

In our view, explanations for diet and habitat selection based only on body morphophysiology and size are constrained by the fact that they ignore the potential for intraspecific plasticity. Every individual within a species is different neurologically, morphologically, and physiologically (Dove 1935; Williams 1978). Part of this variation is genetic, while part results from experience and is shaped by the specific environment where an individual is reared though learning (Skinner 1981). Because individuals within a species differ, the diets and habitats they select should also be expected to vary, which is indeed the case (Arnold and Dudzinski 1978; Provenza and Balph 1987, 1988; Logue 1986). For example, while cattle and bison (Bison bison) are commonly considered grazers, often they prefer and are productive on ranges dominated by shrubs (e.g. Zimmerman 1980; Provenza et al. 1983; Waggoner and Hinkes 1986). In addition, while goats and mule deer (Odocoileus hemionus) are commonly considered browsers, often they prefer and are productive on grass diets (e.g. Willms and McLean 1978; Austin and Urness 1983; Austin et al. 1983; Malechek and Provenza 1983; Urness et al. 1983; Urness 1986). We also find that individual goats vary in ability to tolerate ingestion of blackbrush twigs high in condensed tannins (Provenza et al. 1990). Different subgroups of cattle in areas as diverse as Arizona (Martin 1979), Oregon (Roath and Krueger

1982), and Australia (Yeates and Schmidt 1974) have different and reasonably well defined home ranges year after year, which differ in plant composition. The same is true for sheep that are not herded (Hunter and Milner 1963; Key and MacIver 1980).

It would be worthwhile to study how behavioral, morphological, and physiological relationships are affected when the same species of herbivore is confronted from conception onward with foraging environments that differ in nutritional quality (Provenza and Balph 1988), for example cattle and goats reared on high concentrate versus high roughage forages. Young goats reared from 1 to 4 months of age on blackbrush, a shrub very low in nitrogen and digestibility (Provenza et al. 1983; Provenza and Malechek 1984), ingested over twice as much blackbrush during a 19-day trial as goats reared from 1 to 4 months of age on alfalfa pellets and Calfmanna, a diet high in concentrates (Distel and Provenza 1990). Striking differences existed in rumen mass and volume; the goats reared on blackbrush had 30% more rumen mass than those reared on the concentrate ration. We are currently studying the persistence of these responses.

One can envision how natural or artificial selection could lead to populations of related individuals that actually differ from each other in their ability to use different forages and habitats (Dove 1935). As a result of using different migratory routes and habitats, animals of the same species undoubtedly use forages that differ in morphological characteristics as well as concentrations of nutrients and potentially harmful metabolites (Hunter and Milner 1963; Freeland and Janzen 1974; Key and MacIver 1980; Roath and Krueger 1982; Anderson 1989). Individuals within such sub- populations may be more similar than individuals in the population at large because of being shaped genetically and through learning by the same environment. For example, the ability of sheep to metabolize copper is in part genetic (Hill 1977). Sheep that live in environments where forages contain low concentrations of copper are selected over several generations for enhanced ability to metabolize copper. However, when these sheep are placed in environments where the concentrations of copper in forages are higher, the sheep die from copper toxicity.

Learning through Consequences

The learning model is based on the assumption that the development and modification of diets is the result of the positive and negative consequences of foraging. These consequences are of two broad types: post-ingestive consequences resulting from eating specific foods and pre-ingestive experiences associated with foraging behavior. Learning provides for more flexibility in diet choice than the other models. Although it is largely mechanistic, the model is based on the premise that the inherent ability to learn about foods is the product of natural selection in changing foraging environments. The learning model is a collage of principles and findings from psychology. Only recently have attempts been made to present a cohesive explanation of diet selection based on learning and to apply the explanation to ruminants (e.g. Zahorik and Houpt 1977, 1981; Provenza and Balph 1987, 1988; Provenza et al. 1988). Because of the models multifaceted nature and relative complexity, it will require more description than any of the other models.

Learning through consequences concerns all five foraging challenges, and in some detail. The post-ingestive aspects of learning are primarily associated with variation among plants species and parts in chemical defenses and in nutritional constituents. Pre-ingestive experiences are related to plant morphological defenses, temporal and spatial variation in the forage resource, and unfamiliar foraging environments. In this section, we hope to show how various aspects of learning might be used to explore how ruminants cope with the foraging challenges.

There are hundreds of studies of animals as diverse as insects, mollusks, rats, ruminants, and primates that indicate animals can learn to associate specific foods with aversive post-ingestive consequences (e.g. see Barker et al. 1977; Braveman and Bronstein 1985; Garcia et al. 1977; Provenza and Balph 1988; Garcia 1989; Provenza et al. 1990). Conditioned food aversions occur when animals learn to avoid a particular food by relating its specific taste with nauseous feedback (Pelchat et al. 1983; Garcia and Holder 1985; Garcia et al. 1985; Garcia 1989). The odor or sight of a particular food is the signal the animal uses to avoid specific foods that elicit negative feedback (Garcia 1989). The odor and taste of the food, and the specific compound that causes aversive feedback, need not be and probably rarely are synonymous because several chemical compounds combine to produce a food's characteristic flavor.

Conditioned food aversions may be manifest in the absence of any cognitive association or memory of the feedback event, because the critical psychological process is a decrease in the hedonic value (palatability) of the odor and taste (Garcia 1989). Thus, once the flavor has been presented, delayed nauseous feedback is effective in producing a hedonic shift even if it occurs while the animal is anesthetized (Roll and Smith 1972; Bermudez-Rattoni et al. 1988), deeply tranquilized (Forthman Quick 1984), or when its electrocortical activity is depressed (Buresova and Bures 1973; Davis and Bures 1972). This shift is caused by stimulation of the emetic system of the midbrain and brainstem (Garcia 1989), which can occur through many means, including but not limited to, the cardiovascular system, cerebrospinal fluid, visceral afferent nerves, and second-order gustatory afferents (Borison 1986; Davis et al. 1986; Grahame-Smith 1986). Any physiological, emotional, or perceptual event, or any physiochemical agent that causes nausea, will establish a conditioned food aversion even at doses insufficient to produce overt signs of nausea, provided the event or agent affects the emetic system of the midbrain and brainstem (Garcia 1989).

Ruminants may have difficulty learning to avoid foods that contain toxic compounds that do not affect the emetic system (Provenza et al. 1988, 1990). For example, drugs such as nembutal, strychnine, and cyanide do not cause conditioned aversions because they only affect the nervous system and metabolic processes (Garcia et al. 1985). Likewise, gallamine, which causes neuromuscular blockage, or naloxone, which blocks the action of endogeneous opiates on pain, cause rats to form strong place avoidances but only weak food aversions (Lett 1985). Nor do allergies, bloating, and lower intestinal discomfort cause aversions (Pelchat et al. 1983; Garcia 1989). Conversely, drugs such as lithium chloride which cause nausea in mammals (Prien et al. 1971), cause rats to form strong food aversions but only weak place avoidances (Lett 1985).

While it is clear that animals readily form conditioned food aversions, some argue that the incentive to eat and drink is a result of conditioned preferences (Booth 1985). Research to determine the ability of animals to associate foods with positive post-ingestive consequences has been conducted primarily with rats. This research clearly indicates that rats form strong preferences for: (1) non-nutritive flavors paired with calories (Bolles et al. 1981; Messier and White 1984; Booth 1985; Mehiel and Bolles 1984, 1988; Simbayi et al. 1985), (2) foods paired with recovery from nutritional deficiency (Garcia et al. 1967; Zahorik et al. 1974), and (3) foods paired with recovery from post-ingestive distress (Green and Garcia 1971; Sherman et al. 1983). Ruminants may possess the same abilities, but little data exist to support or refute this contention (Provenza and Balph 1987). Ongoing studies indicate that lambs form strong preferences for non-nutritive flavors paired with calories (Burritt and Provenza, unpublished data). In addition, cattle may form conditioned preferences for supplemental protein blocks when ingesting forages low in protein (Provenza et al. 1983).

In ruminants byproducts of microbial fermentation such as volatile fatty acids and ammonia are likely candidates for producing conditioned food preferences, as are pancreatic hormones such as insulin and brain-gut hormones such as cholecystokinin. Changes in portal blood metabolites such as volatile fatty acids, beta-hydroxybutyrate, and insulin occur within 15 minutes of initiation of a meal in ruminants (Chase et al. 1976; de Jong 1986). Such changes may enable ruminants very quickly to associate specific foods with their post-ingestive consequences and to form strong preferences for foods that produce positive consequences. They may also be the primary reason that ruminants generally prefer leaf to stem, green to dead, and young to older plant tissues. The gastrointestinal tract and liver of ruminants are richly supplied with sensory receptors and afferents to the brain (see review papers in Milligan et al. 1986), and neurally-mediated signals that arise in the liver are clearly implicated in conditioned flavor preferences and the control of food intake in rats (Tordoff and Friedman 1989b). It is not likely that a single nutrient, compound, hormone, or organ will have a dominant role in producing conditioned food preferencs given the variety of nutrients needed by animals. Rather, many different mechanisms will undoubtedly be involved, as is the case with the control of food intake (de Jong 1986) and conditioned food aversions (Grahame-Smith 1986).

We categorize the pre-ingestive experiences into four types: (1) in utero, (2) mother's milk, (3) social, and (4) individual trial and error. We believe these experiences, coupled in many cases with post-ingestive consequences, shape the diets of ruminants. We begin by describing the above experiences and then show how the learned behavior might be applied by ruminants confronted with plant morphological defenses, temporal and spatial variation in nutrients and toxins in plants, and unfamiliar environments.

Animals can learn initially about the flavors of solid foods in utero and from mother's milk. Many compounds in plants quickly cross the placenta from mother to fetus (e.g. Smotherman and Robinson 1987; Keeler 1988), and their influence may have elements of both post- and pre-ingestive consequences. There is evidence that rats form preferences for (Hepper 1988) and aversions to (Stickrod et al. 1982; Smotherman 1982) food flavors based on experiences in utero, and that there may be critical periods for learning in utero (Hill and Przekop 1988). The fetal taste system of lambs is functional during the last trimester of gestation (Bradley and Mistretta 1973), and some hypothesize that fetal taste experiences affect adult taste preferences in ruminants such as goats, sheep, and cattle (Bradley and Mistretta 1973; Provenza and Balph 1988). Flavors in foods also influence the flavor of the mother's milk

(Bassette et al. 1986), which subsequently affects preferences for solid foods in rats (Galef and Sherry 1973; Capretta and Rawls 1974; Galef 1976) and ruminants (Morrill and Dayton 1978; Nolte et al. 1990). Thus, animals likely have been conditioned to prefer and avoid specific flavors that occur in foods in the environment before they ever take their first bite of solid food.

When young animals begin to forage, experienced conspecifics undoubtedly play an important role in the development of their solid food intake (e.g.; Boyd and Richerson 1985; Galef 1986). The interactions may be obvious as is the case when young ruminants initially eat and avoid foods which social models such as the mother eat and avoid (reviewed by Provenza and Balph 1987, 1988; Thorhallsdottir et al. 1987; Thorhallsdottir et al. 1989 a,b,c; Mirza and Provenza 1990). They may also be more subtle as is the case when rats learn about the foraging environment from olfactory cues of conspecifics (Galef and Heiber 1976; Galef and Wigmore 1983; Galef et al. 1984). While no data exist for ruminants on this matter, it may be that they can determine the general locations and specific plants upon which individual conspecifics have foraged. Livestock identify one another on the basis of odor, and their physiological state can be determined as well; they can smell where others have passed and how long ago others were there (Hart 1985). Thus, it is reasonable to assume that ruminants can determine through olfaction which locations and dietary items experienced conspecifics have selected while foraging.

When young ruminants begin to forage alone, it is important that they: (1) remember foods eaten in the past, and (2) sample novel foods cautiously (see Rozin 1976 for a detailed review). Young ruminants can remember specific foods that provided either aversive (Lane et al. 1989; Burritt and Provenza 1989c) or positive (Green et al. 1984; Squibb et al. 1989) consequences for at least 1 to 3 years (see Provenza and Balph 1987, 1988 for additional references and discussion). They also recognize and sample novel foods cautiously (Chapple and Lynch 1986; Chapple et al. 1987a,b; Thorhallsdottir et al. 1989a; Burritt and Provenza 1989a,b; Provenza et al. 1990). When an animal experiences aversive or pronounced positive post-ingestive consequences it should associate the consequences with novel rather than familiar foods. Again, that is exactly what sheep and goats do in conditioned food aversion experiments (Burritt and Provenza 1989a,b; Provenza et al. 1990), given as many as four familiar foods and one novel food and a delay between ingestion and consequences of up to 6 hours (Burritt and Provenza 1989b). Given the fact that ruminants spend as many as 8 to 12 hours foraging every day, and given the importance of food for survival, ruminants are probably no less able than humans to learn about and remember tens of hundreds of foods, and their locations, in the environment.

This does not negate the need to sample foods. Herbivores must sample foods, even familiar foods in familiar environments, because their nutrient content and toxicity change frequently (Freeland and Janzen 1974; Westoby 1974, 1978). In that sense, they are probably no different from humans who sample familiar fresh fruits and vegetables from a garden or the market to determine whether or not they are ripe. This undoubtedly explains why it is often difficult to maintain conditioned food aversions in ruminants (Provenza and Balph 1987; Thorhallsdottir et al. 1987; Lane et al. 1989): they possess great abilities to discriminate among subtle chemical differences within plant species based on odor and taste and quickly associate specific food flavors with post-ingestive consequences (Launchbaugh and Provenza 1990). The degree to which ruminants generalize and discriminate undoubtedly depends to a great extent on the similarity among different plant species and parts as determined by taste, odor, and post-ingestive consequences, and on the specific sensory abilities of the individual animals related to olfaction, gustation, and sight. Thus, what constitutes a novel food to a ruminant may or may not be different from our taxonomic classifications (Provenza et al. 1988).

Ruminants need not only select foods that are nutritious and avoid those that are toxic, they must also ingest forage from plants that differ greatly in canopy shape and architecture. Clearly, the relationship between animal morphophysiology and plant morphology is important, as discussed in the preceeding section. In addition, there is evidence that sheep (Flores et al. 1989 a,b,c) and cattle (Hodgson 1971) acquire the skills needed to efficiently ingest forage from plants that differ in morphology. For example, lambs with experience foraging on shrubs ingest more shrub per unit time than lambs lacking foraging experience (Flores et al. 1989a), or than lambs experienced foraging on grass (Flores et al. 1989b). Likewise, lambs with experience foraging on grass ingest more grass per unit time than lambs experienced foraging on shrubs (Flores et al. 1989b,c). Unfortunately, with the exception of these studies, little is known about the acquisition of foraging skills in ruminants. Thus, the degree to which learning affects ingestion rates by large and small ruminants foraging on a diversity of plant life forms is virtually unexplored.

In addition to learning about the consequences of ingesting specific foods, ruminants must also learn about the locations of resources in the environment. Studies of spatiotemporal memory in animals as diverse as rats (Olton 1977), cattle (Bailey et al. 1989), and monkeys (Menzel 1973, 1978) suggest that they quickly acquire cognitive maps of the environment. Thus, for example, when animals are required to search an external environment for hidden food baits, they display an objective mastery of space and time (Garcia and Holder 1975). Likewise, observations of habitat use suggest that ruminants learn and remember the locations of food, water, and shelter in many different environments used on a seasonal basis year-after-year (Hunter and Milner 1963; Zimmerman 1980; Roath and Krueger 1982). Indeed, both wild (e.g. Festa-Bianchet 1986; Cederlund et al. 1987; McNay and Doyle 1987; Andersen 1989; Gasaway et al. 1989) and domestic (e.g. Hunter and Milner 1963; Yeates and Schmidt 1974; Key and MacIver 1980; Zimmerman 1980; Roath and Krueger 1982) ruminants often form specific attachments to environments in which they know the kinds and locations of food, water, shelter, conspecifics, and predators, and these preferences are often transferred culturally from one generation to the next. Thus, it is reasonable to assume that ruminants can position themselves in the environment in a manner that increases their likelihood of encountering nutritious forages both spatially and temporally. However, research has not been conducted to elucidate how diet selection by ruminants is affected by factors that are potentially incompatible with use of preferred foraging locations, such as need for water, need to avoid predators, and desire to be with certain conspecifics. Such research may be particularly amenable to optimality analyses.

In some instances, ruminants are forced to forage in unfamiliar environments. Their performance in unfamiliar environments is undoubtedly determined, to some unknown degree, by the similarity of the new environment to the environment where they lived formerly. The greater the difference, presumably the greater the challenge because of the relatively high degree of novelty in foods and their locations in the environment. Often, both wild (Hlavachick 1970; White 1981; Hanson and Willison 1983; O'Bryan and McCullough 1985; McCall et al. 1988) and domestic (Arnold 1970; Arnold and Maller 1977; Arnold and Dudzinski 1978; Hodgson and Jamieson 1981) ruminants introduced into a new environment are less productive than conspecifics reared in the environment (see also Griffith et al. 1989). Compared with experienced animals, naive animals spend more time foraging but ingest less forage; they spend more time walking and walk for greater distances; they suffer more from predation, malnutrition, and ingestion of toxic plants (Provenza and Balph 1987, 1988; Provenza et al. 1988).

Diet selection in unfamiliar environments probably also depends on experiences early in life. As a result of conditioning early in life, animals may come to perceive certain forages and environments as desirable and others as undesirable (Tolman 1949; Garcia and Holder 1985; Garcia 1989). Thus, the value of any new food item is evaluated in terms of the animal's perception of what constitutes desirable and undesirable food (e.g. Nolte et al. 1989; Mirza and Provenza 1990). A new food may be rejected, even though it is nutritious, because the animal does not perceive its characteristics as desirable; likewise, a food of lower nutritional value may be accepted. Neurological, morphological, and physiological systems undergo a great deal of development, which is not strictly controlled genetically, from conception through the first year of life in ruminants (Provenza and Balph 1987,1988). This provides the flexibility for animals to learn during a time when they must obtain information important for survival. However, the nature of the central nervous system is such that behaviors developed during sensitive periods early in life are often difficult to change (Kandel and Schwartz 1985; Aoki and Siekevitz 1988). On this basis, young, rather than older, animals should introduce new foraging and habitat selection behaviors into a population, which is indeed the case with Japanese monkeys (Itani 1958; Kawai 1965). Likewise, young animals should adapt more readily than adults when moved from familiar to unfamiliar environments (Provenza and Balph 1988).

In our view, the potential of learning through consequences to provide insights into diet selection by ruminants is great. Research on learning clearly suggests that foods have both aversive and positive post-ingestive consequences which animals can associate with specific dietary items, and suggests mechanistic explanations for the processes. The learning hypothesis also provides a mechanistic framework in which to study the acquisition of dietary habits, foraging skills, and habitat selection patterns by ruminants. In this sense, it may ultimately help explain how and why ruminants select specific foods that are nutritious and avoid those that are not, as well as important anomalies such as voluntary intake of poisonous plants by wild (Fowler 1983) and domestic (Cheeke and Shull 1985) herbivores (Provenza et al. 1988, 1990). It may also help explain and make predictions about foraging behavior of ruminants in heretofore unknown situations (Provenza and Balph 1987, 1988). The above notwithstanding, the learning explanation does lack generality in the sense that it does not set limits to selection of specific dietary items or habitats by ruminants. Rather, it argues that individual animals are shaped through learning about the specific environment in which they are reared, within rather broad morphophysiological constraints (Provenza and Balph 1987, 1988).

Optimal Foraging Theory

Optimal foraging theory (reviewed by Stephens and Krebs 1986) is based on the premise that animals forage to maximize nutritional gain per unit cost. Specific models are generally concerned with: (1) the currency being optimized (e.g. energy, but see Huston this volume), (2) the decisions to be made (e.g. which dietary items to ingest; when to leave a particular foraging location), and (3) intrinsic (e.g. learning and memory; need to ingest a minimum amount of nutrients; inability to ingest more than a certain amount of particular secondary metabolite) and extrinsic (e.g. kinds and locations of forages in the environment) contraints that limit and define the relationship between the currency and decision variables. The theory's application typically consists of a test to determine if an animal's foraging performance matches a calculated optimum under stated conditions (Krebs et al. 1983). The model usually concerns rate maximization of energy intake and is thought to be more appropriate for carnivores than herbivores, because the nutritional quality of prey varies little relative to that of vegetation (Crawley 1983).

Optimal foraging theory differs from the other models in that it was developed deductively from several assumptions about the natural selection of behavioral traits (Dunbar 1983). Theorists, through logic, develop hypotheses about how animals should forage (behave optimally) under various conditions and constraints to increase an animals' fitness (e.g. Charnov 1976; Mangel and Clark 1986). Researchers then test aspects of the hypotheses. The majority of current tests are conducted by psychologists using operant conditioning analogs to actual foraging situations (e.g. Killeen et al. 1981), and by behavioral ecologists (see papers in this volume). The theory does not deal with causal mechanisms underlying foraging behavior.

Optimal foraging theory has been valuable in stimulating research on foraging behavior and providing a quantitative focus for that research. It may be particularly useful for studying behavior under conditions where animals are continually subject to multiple decisions concerning which several activities (e.g.

finding and selecting food, avoiding predation while looking for food, reproducing) to adopt at any time, as well as how best to undertake a chosen activity (Mangel and Clark 1986). The fact that the model has not been particularly successful in predicting diet selection by large herbivores in complex environments (Schluter 1981; Owen-Smith 1989), does not necessarily invalidate its heuristic value, provided it generates insights into foraging behavior. In that sense, we agree with Kerbs et al. (1983) that optimal foraging theory is best used as a tool to investigate foraging behavior rather than as a test of foraging optimality.

We believe the theory has some application to all of the foraging challenges. It can always be used to test a specific optimality model involving one of the challenges as a constraint (but see Stevens and Krebs 1983). Perhaps a better general application is the use of a calculated optimal performance as a benchmark in comparing the ineffeciencies of diet choice created by such challenges as positioning within the environment, foraging in unfamiliar environments, and plant chemical and mechanical defenses. It can also be applied to the challenge of temporal and spatial positioning. The calculation of the optimum time to move to a new position (Charnov 1976, but see Clark and Mangel 1984) can be used as a standard with which to compare ruminant performance in a variety of situation. There may also be value in comparing an optimal to observed rate of movement between feeding stations (see Bazley this volume), mosaics in plant assemblages, and along migratory routes. Each position could be viewed as a patch of depleting resources. May of these movements concern foraging rate maximization, but they can also involve dietary choice. When diet choice is involved, as in moving between plant types within mosaics, optimal foraging theory is applicable in the study of diet selection.

Optimal foraging theory is not well suited for determining whether or not diet choice is influenced by the nutritional composition of dietary items. If an animal maximizes or is constrained by a particular phytochemical, it may mean that the compound is: (1) sensed directly (e.g. salt), (2) sensed directly with delay (e.g. some toxins), (3) not sensed directly but associated with a cue to which the animal is attentive (e.g. high energy density associated with new growth), or (4) some spurious association. Questions regarding an animal's ability to sense nutritional components are best addressed through operant techniques (Reynolds 1968, see conclusions).

CONCLUSIONS

A comparison of the five explanations (Table 1) shows that they differ in the number of foraging challenges they address, their level of resolution, and the firmness of their foundations. This is not surprising because the explanations, with the exception of hedyphaghia, were developed independently of one another and often apply to different scales within a hierarchy (Senft et al. 1987). It is likely that most of the researchers responsible for the development of a particular explanation were unfamiliar with details of the others. However, there are relationships among the explanations that we believe are of interest.

Euphagia was developed to explain the observation that animals are able to obtain a nutritious diet. When subsequent research suggested that animals were unable to directly sense nutritional components, it was generally rejected by animal nutritionists (e.g. Arnold and Hill 1972) and psychologists (e.g. Booth 1985). However, we believe that total rejection may be premature because there could be situations where the need to recognize a specific nutrient is especially strong. Likewise, there may be a need for ruminants to innately recognize and avoid secondary metabolities that circumvent the emetic system. The fact, again, is that very few wild or domestic ruminants have been examined for their ability to directly sense nutrients or toxins in a diverse array of plants.

Hedyphagia was developed by animal nutritionists as a replacement for euphagia. It explained how animals could obtain a nutritious diet without endogenous sensing mechanisms. However, explanations based on hedyphagia are often circular in that what tastes good is defined by what is eaten. Clearly, there is need for careful analysis of the role of hedyphagia in diet selection, but such work must separate the effects of taste and odor from those of post-ingestive consequences (Grovum 1988; see also Schwartz and Grill 1985). In addition, previous dietary history must be considered as there is growing evidence that many taste-elicited responses and physiological controls over ingestion in adults have their origins early in life in both rats (Hall and Williams 1983; Schwartz and Grill 1985) and ruminants (Provenza and Balph 1987, 1988).

The morphophysiology and learning models are complementary, and in many ways inseparable, because behavior is a result of its consequences, and consequences are determined by interactions among neurological, morphological, and physiological processes. Morphophysiology provides diet selection with broad limits, propensities, and generality; while learning fine tunes diet selection within the above constraints, even down to the level of the individual animal. Together, the explanations pose some interesting questions. For example, to what extent can diet selection be manipulated by learning within the constraints of morphophysiology? Likewise, to what degree can morphophysiology be manipulated genetically to enable ruminants to select specific forages? These questions are fundamental to attempts to shape diet selection for management or conservation purposes (Provenza and Balph 1987, 1988).

The most difficult explanation for us to characterize is optimal foraging, primarily because both the currency for, and the constraints on, maximization can be so varied. These could be standard (e.g. rate of caloric gain under a predation constraint), or they could be quite atypical (e.g. rate of calcium gain constrained by mouth size or unfamiliar environment). This flexibility gives the model a great deal of breadth. The final test, though, is always the same: does the observed performance match the calculated optimum performance? The question concerns ultimate function, not proximate cause, which restricts its depth considerably. The most appropriate currency when applying the theory to diet selection by ruminants is nutrients. However, this raises the question of whether or not the animal can sense nutrients. Animals cannot maximize the gain of something they cannot sense, in any direct way.

We believe the difficulty for most of the models, and for progress on diet selection in general, is the lack of information on the capabilities of ruminants to sense plant nutritional and toxicological components, and to differentiate among dietary items. The research paradigm for sensory discrimination is well known by most experimental psychologists. A behavioral analysis laboratory for large herbivores, such as the one established by the late Ron Kilgore of New Zealand would be ideal. Through a series of discrimination trials using operant techniques (Reynolds 1968), a variety of herbivores and nutritional and toxicological phytochemicals could be tested. There is also a need for research on the capabilities of ruminants to differentiate among potential dietary items and habitats within the environment. Through well-designed and carefully-controlled experiments, researchers could gain insights into how ruminants respond to the complex foraging environments in which they live.

We believe that future research should place greater emphasis on the use of: (1) induction to establish reliable associations among sets of facts related to the process of diet selection, (2) retroduction to establish research hypotheses that are explanations for these facts, and (3) hypothetico-deduction to test the predictions, generated from the research hypotheses, using well-designed and carefully-controlled experiments (see Romesburg 1981 for a detailed review). In this way, researchers can develop theories that explain and make predictions about the process of diet selection in ruminants.

ACKNOWLEDGEMENTS

This work was supported by grants from Cooperative State Research Service (87-CRSR-2-3117) and The National Science Foundation (BSR-8614856). We thank G.W. Arnold, E.A. Burritt, J.T. du Toit, B.G. Galef Jr., I. Gordon, K.L. Launchbaugh, N. Owen-Smith and J.A. Pfister for comments on various drafts of the manuscript.

Table 1. A comparison of five explanations for diet selection by ruminants facing five foraging challenges.

Foraging Explanation	Reason Ruminants Select Dietary Items	Challenges Addressed by Explanation*	Is Explanation Mechanistic?	Does Explanation Address Individual Variation?	Are Explanation's Asssumptions Valid?
Euphagia	inherent recognition of nutrients and toxins	1, 2	yes, where applicable	no	maybe
Hedyphagia	nutrients taste good, toxins taste bad	1, 2, 3	no	no	circularity
Morphophysiology and Size	body adapted to utilize some forages better than others	1, 2, 3, 4, (5)**	yes, to a degree	no	yes
Learning	consequences	1, 2, 3, 4, 5	yes	yes	yes
Optimal Foraging	most benefit, least cost	1, 2, 3, 4, (5)	no	indirectly	not always

* Challenges: (1) nutritional variation, (2) plant chemical defenses, (3) plant morphological defenses, (4) environmental patchiness, and (5) unfamiliar environments.

** Parentheses indicate some question about how well the explanation deals explicitly with the challenges; see text for details.

REFERENCES

Andersen R (1989) Counteracting selection pressures on the migrating behavior of moose (Alces alces): the costs of culture. Behav Ecol Sociobiol in press

Aoki C, Siekevitz P (1988) Plasticity in brain development. Sci Am 256:56-64

Arnold G W (1970) Regulation of food intake in grazing ruminants. In: Phillipson AT (ed), Physiology of digestion and metabolism in the ruminant. Oriel Press, London, p 264

Arnold G W, Hill J L (1972) Chemical factors affecting selection of food plants by ruminants. In: Harborne J B (ed) Phytochemical ecology. Academic Press, London, p 71

Arnold G W, Maller R A (1977) Effects of nutritional experience in early and adult life on the performance and dietary habits of sheep. Appl Anim Ethol 3:5-26

Arnold G W, Dudzinski M L (1978) Ethology of free-ranging domestic animals Elsevier/North Holland, New York

Austin D D, Urness P J (1983) Overwinter forage selection by mule deer on seeded big sagebrush-grass range. J Wildl Manage 47:1203-1207

Austin D D, Urness P J, Fierro L C (1983) Spring livestock grazing affects crested wheatgrass regrowth and winter use by mule deer. J Range Manage 36:589-593

Bailey D W, Rittenhouse L R, Hart R H, Richards R W (1989) Characteristics of spatial memory in cattle. Appl Anim Behav Sci 23:331-340

Bailey E M Jr (1988) The importance of diagnosing poisoning from plants. In: James L F, Ralphs M H, Nielson D B (eds) The ecology and economic impact of poisonous plants on livestock production. Westview Press, Boulder, p 337

Barker L M, Best M R, Domjan M (eds) (1977) Learning mechanisms in food selection. Baylor Univ Press, Waco

Bassette R, Fung D Y C, Mantha V R (1986) Off-flavors in milk. CRC Crit Rev Food Sci Nutr 24:1-52

Bate-Smith E C (1972) Attractants and repellents in higher animals. In: Harborne J B (ed) Phytochemical ecology. Academic Press, London, p 45

Bate-Smith E C and Metcalf C R (1957) Leucoanthocyanins 3. The nature and systematic distribution of tannins in dicotyledenous plants. J Linn Soc Bot 55:669-705

Beauchamp G K, Cowart B J (1987) Development of sweet taste. In: Dobbing J (ed), Sweetness. Springer-Verlag, London, p 127

Bell F R (1984) Aspects of ingestive behavior in cattle. J Anim Sci 59:1369-1372

Bell R H V (1969) The use of the herb layer by grazing ungulates in the Serengeti. In: Watson A (ed) Animal populations in relation to their food resources. Blackwell Sci, Oxford, p 111

Bell R H V (1971) A grazing ecosystem in the Serengeti. Sci Am 225:86-93

Bermudez-Rattoni F, Forthman Quick D L, Sanchez M A, Perez J L, Garcia J (1988) Odor and taste aversions conditioned in anesthetized rats. Behav Neurosci 102:726-732

Bolles R C, Hayward L, Crandall C (1981) Conditioned taste preferences based on caloric density. J Exp Psychol: Anim Behav Proc 7:59-69
Booth, D A (1985) Food-conditioned eating preferences and aversions with introceptive elements: conditioned appetities and satieties. In: Braveman N S, Bronstein P (eds) Experimental assessments and clinical applications of conditioned food aversions. New York Acad Sci, New York, p 22
Borison H L (1986) Anatomy and physiology of the chemoreceptor trigger zone and area postrema. In: Davis C J, Lake-Bakaar G V, Grahame-Smith D G (eds) Nausea and vomiting: mechanisms and treatment. Springer-Verlag, New York, p 10
Boyd R, Richerson P J (1985) Culture and the evolutionary process. Univ Chicago Press, Chicago
Bradley R M, Mistretta C M (1973) The gustatory sense in foetal sheep during the last third of gestation. J Physiol 231:271-282
Braveman N S, Bronstein P, eds (1985) Experimental assessments and clinical applications of conditioned food aversions. New York Academy Sci, New York
Bryant J P, Chapin F S III and Klein D R (1983) Carbon/nutrient balance of boreal plants in relation to vertebrate herbivory. Oikos 40:357-368
Bryant J P, Provenza F D, Gobena A (1987) Environmental controls over woody plant chemical defenses: implications for goat management. In: Santana O P, da Silva A G, Foote W C (eds) Proc IV intnl conf on goats. Departmento de Difusao de Technologia, Brazilia, p 1005
Buresova O, Bures J (1973) Cortical and subcortical components of the conditioned saccharin aversion. Physiol Behav 11:435-439
Burritt E A, Provenza F D (1989a) Food aversion learning: ability of lambs to distinguish safe from harmful foods. J Anim Sci 67: 1732-1739
Burritt E A, Provenza F D (1989b) Ability of lambs to learn with a delay between food ingestion and consequences given meals containing novel and familiar foods. Appl Anim Behav Sci submitted
Burritt E A, Provenza F D (1989c) Food aversion learning in sheep: persistence of conditioned taste aversions to palatable shrubs (Cercocarpus montanus and Amelanchier alnifolia). J Anim Sci in press
Capretta P J, Rawls L H (1974) Establishment of a flavor preference in rats: importance of nursing and weaning experience. J Comp Physiol Psychol 88:670-673
Cederlund G, Sandegren F, Larsson K (1987) Summer movement of female moose and dispersal of their offspring. J Wildl Manage 51:342-352
Chapple R S, Lynch J J (1986) Behavioral factors modifying acceptance of supplementary foods by sheep. Res Develop Agric 3:113-120

Chapple R S, Wodzicka-Tomaszewska M, Lynch J J (1987a) The learning behavior of sheep when introduced to wheat. I. Wheat acceptance by sheep and the effect of trough familiarity. Appl Anim Behav Sci 18:157-162
Chapple R S, Wodzicka-Tomaszewska M, Lynch J J (1987b) The learning behavior of sheep when introduced to wheat. II. Social transmission of wheat feeding and the role of the senses. Appl Anim Behav Sci 18:163-172
Charnov E L (1976) Optimal foraging: the marginal value theorem. Theor Popul Biol 9:129-136
Chase L E, Wangsness P J, Kavanaugh J F, Griel L C Jr, Gahagan J H (1976) Changes in portal blood metabolities and insulin with feeding steers twice daily. J Dairy Sci 60:403-409
Cheeke P R, Shull L R (1985) Natural toxicants in feeds and poisonous plants. AVI Publ Co, Westport
Church D C (1979) Taste, appetite and regulation of energy balance and control of food intake. In: Church D C (ed) Digestive physiology and nutrition of ruminants. Volume 2 - Nutrition, O & B Books Inc, Corvallis. p 281
Clark C W, Mangel M (1984) Foraging and flock strategies: information in an uncertain environment. Am Nat 123:626-641
Clausen T P, Provenza F D, Burritt E A, Bryant J P, Reichardt P B (1989) Ecological implications of condensed tannin structure: a case study. J Chem Ecol in press
Coley P D, Bryant J P, Chapin F S III (1985) Resource availability and plant antiherbivore defense. Science 230:895-899
Contreras R J, Frank M (1979) Sodium deprivation alters neural responses to gustatory stimuli. J Gen Physiol 73:569-594
Cooper S M, Owen-Smith N (1985) Condensed tannins deter feeding by browsing ruminants in a South African savanna. Oecologia 67:142-146
Cooper S M, Owen-Smith N (1986) Effects of plant spinescence on large mammalian herbivores. Oecologia 68:446-455
Cooper S M, Owen-Smith N, Bryant J P (1988) Foliage acceptability to browsing ruminants in relation to seasonal changes in leaf chemistry of woody plants in a South African savanna. Oecologia 75:336-342
Coppock C E, Everett R W, Smith N E, Slack S T, Harner J P (1974) Variation in forage preference in dairy cattle. J Anim Sci 39:1170-1179
Coppock C E, Everett R W, Harris B (1981) From feeding to feeding systems. J Dairy Sci 64:1230-1249
Crawley M J (1983) Herbivory: the dynamics of animal-plant interactions. Univ Calif Press, Berkeley
Davis C J, Harding R K, Leslie R A, Andrews P L R (1986) The organization of vomiting as a protective reflex: a commentary on the first day's discussion. Davis C J, Lake-Bakaar G V, Grahame-Smith D G (eds) Nausea and vomiting: mechanisms and treatment. Springer-Verlag, New York, p 65
Davis J L, Bures J (1972) Disruption of saccharin-aversion learning in rats by cortical spreading depression in the CS-US interval. J Comp Physiol Psychol 80:398-402

de Jong A (1986) The role of metabolites and hormones as feedbacks in the control of food intake in ruminants. In: Milligan L P, Grovum W L, Dobson A (eds) Control of digestion and metabolism in ruminants. Prentice-Hall, Englewood Cliffs, p 459

Demment M W, Van Soest P J (1985) A nutritional explanation for body-size patterns of ruminant and nonruminant herbivores. Am Nat 125:641-672

Demment M W, Longhurst W M (1987) Browsers and grazers: constraints on feeding ecology imposed by gut morphology and body size. In: Santana O P, da Silva A G, Foote W C (ed) Proc IV intnl conf on goats. Departamento de Difusao de Tecnologia, Brazilia, p 989

Denton D A (1965) Evolutionary aspects of the emergence of aldosterone secretion and salt appetite. Physiol Rev 45:245-295

Denton D A, Sabine J R (1963) The behavior of Na deficient sheep. Behaviour 20:264-276

Distel R A, Provenza F D (1990) Early experience affects voluntary intake of blackbrush by goats. In: Abstr 43rd ann mtg soc range manage, Reno

Dove F W (1935) A study of individuality in the nutritive instincts and of the causes and effects of variations in the selection of food. Am Nat 69:469-544

Dunbar R I M (1983) Adaptation, fitness and the evolutionary tautology. In: King's College Sociology Group (eds) Current problems in sociobiology. Cambridge University Press, New York, p 9

du Toit J T, Owen-Smith N (1989) Body size, population metabolism, and habitat specialization among large African herbivores. Am Nat 133:736-740

Festa-Bianchet M (1986) Seasonal dispersion of overlapping mountain sheep ewe groups. J Wildl Manage 50:325-330

Flores E R, Provenza F D, Balph D F (1989a) Role of experience in the development of foraging skills of lambs browsing the shrub serviceberry. Appl Anim Behav Sci 23:271-278

Flores E R, Provenza F D, Balph D F (1989b) The effect of experience on the foraging skill of lambs: importance of plant form. Appl Anim Behav Sci 23:285-291

Flores E R, Provenza F D, Balph D F (1989c) Relationship between plant maturity and foraging experience of lambs grazing hycrest crested wheatgrass. Appl Anim Behav Sci 23:279-284

Forthman Quick D (1984) Reduction of crop damage by olive baboons (Papio anubis): the feasibility of conditioned taste aversion. PhD Diss Univ Calif Los Angeles

Fowler M E (1983) Plant poisoning in free-living wild animals: a review. J Wildl Dis 19:34-43

Freeland W J, Janzen D H (1974) Strategies in herbivory by mammals: the role of plant secondary compounds. Am Nat 108:269-286

Galef B G Jr (1976) Social transmission of acquired behavior: a discussion of tradition and social learning in vertebrates. In: Rosenblatt J S, Hinde R A, Shaw E, Beer C (eds) Advances in the study of behavior. Academic Press, New York, p 77

Galef, B G Jr (1981) Development of flavor preference in man and animals: the role of social and nonsocial factors. In: Aslin R N, Alberts J R, Peterson M R (eds) Development of perception: psychobiological perspectives. Academic Press, New York, p 411
Galef B G Jr (1986) Social interaction modifies learned aversions, sodium appetite, and both palatability and handling-time induced dietary preference in rats (Rattus norvegicus). J Comp Psychol 100:432-439
Galef B G Jr, Sherry D F (1973) Mother's milk: a medium for transmission of cues reflecting the flavor of mother's diet. J Comp Physiol Psychol 83:374-378
Galef B G Jr, Heiber L (1976) The role of residual olfactory cues in the determination of feeding site selection and exploration patterns of domestic rats. J Comp Physiol Psychol 90:727-739
Galef B G Jr, Wigmore S W (1983) Transfer of information concerning distant foods: a laboratory investigation of the 'information-centre' hypothesis. Anim Behav 31:748-758
Galef B G Jr, Kennett D J, Wigmore S W (1984) Transfer of information concerning distant foods in rats: a robust phenomenon. Anim Learn Behav 12:292-296
Garcia J (1989) Food for Tolman: cognition and cathexis in concert. In: Archer T, Nilsson L (eds) Aversion, avoidance and anxiety. Hillsdale, New Jersey, p 45
Garcia J, Hankins W G (1975) The evolution of bitter and the acquisition of toxiphobia. In: Denton D A, Coghlan J P (eds) Fifth internl symp olfaction and taste. Academic Press, New York, p 39
Garcia J, Hankins W G (1977) On the origin of food aversion paradigms. In: Barker L, Domjan M, Best M (eds) Learning mechanisms in food selection. Baylor Univ Press, Waco, p 3
Garcia J, Holder M D (1985) Time, space and value. Human Neurobiol 4:81-89
Garcia J, Ervin F R, Yorke C H, Koelling R A (1967) Conditioning with delayed vitamin injections. Science 155:716-718
Garcia J, Hankins W G, Coil J D (1977) Koalas, men, and other conditioned gastronomes. In: Milgram N W, Krames L, Alloway T M (eds) Food aversion learning. Plenum Press, New York, p 195
Garcia J, Lasiter P A, Bermudez-Rattoni F, Deems D A (1985) A general theory of aversion learning. In: Braveman N S, Bronstein P (eds) Experimental assessments and clinical applications of conditioned food aversions. New York Academy Sci, New York, p 8
Gasaway W C, DuBois S D, Boertje R D, Reed D J, Simpson D T (1989) Response of radio-collared moose to a large burn in central Alaska. Can J Zool 67:325-329
Geist V (1974) On the relationship of social evolution and ecology in ungulates. Am Zool 14:205-220
Gershenzon J (1984) Changes in the levels of plant secondary metabolities under water and nutrient stress. Rec Adv Phytochem 18:273-320
Gordon J G, Tribe D E (1951) The self-selection of diet by pregnant ewes. J Agric Sci 41:187-190

Gordon J G, Tribe D E, Graham T C (1954) The feeding behavior of phosphorus-deficient cattle and sheep. Brit J Anim Behav 2:727

Grahame-Smith D G (1986) The multiple causes of vomiting: is there a common mechanism? Davis C J, Lake-Bakaar G V, Grahame-Smith D G (eds) Nausea and vomiting: mechanisms and treatment. Springer-Verlag, New York, p 1

Green G C, Elwin R L, Mottershead B E, Keogh R G, Lynch J J (1984) Long-term effects of early experience to supplementary feeding in sheep. Proc Aust Soc Anim Prod 15:373-375

Green K F, Garcia J (1971) Recuperation from illness: flavor enhancement in rats. Science 173:749-751

Griffith B, Scott J M, Carpenter J W, Reed C (1989) Translocation as a species conservation tool: status and strategy. Science 245:477-480

Grill H J, Berridge K C, Ganster D J (1984) Oral glucose is the prime elicitor of preabsorptive insulin secretion. Am J Physiol 246:R88-R95

Grovum W L (1988) Appetite, palatability and control of feed intake. In: Church D C (ed) The ruminant animal: digestive physiology and nutrition. Prentice Hall, Englewook Cliffs, p 202

Grovum W L, Chapman H W (1988) Factors affecting the voluntary intake of food by sheep. 4. The effects of additives representing the primary tastes on sham intakes by oesophageal - fistulated sheep. Brit J Nutr 59:63-72

Hall W G, Williams C L (1983) Suckling isn't feeding, or is it? A search for developmental continuities. In: Rosenblatt J S, Hinde R A, Beer C, Busnel M C (eds) Advances in the study of behavior. Academic Press, New York, p 219

Hanley T A (1982) The nutritional basis for food selection by ungulates. J Range Manage 35:146-151

Hanson M T, Willison J M (1983) The 1978 relocation of Tule elk at Fort Hunter Liggett-reasons for its failure. CalNeva Wildl West Sect, The Wildl Soc, Sacramento, Calif: 43-49

Hart B L (1985) The behavior of domestic animals. W H Freeman, New York

Hepper P G (1988) Adaptive fetal learning: prenatal exposure to garlic affects postnatal preferences. Anim Behav 36:935-936

Hill D L, Przekop P R Jr (1988) Influences of dietary sodium on functional taste receptor development: a sensitive period. Science 241:1826-1828

Hill R (1977) Copper toxicity. II. Brit Vet J 133:365-373

Hlavachick B D (1970) Success and failure of antelope transplants in Kansas. Proc Antelope State Workshop 4:11-15

Hodgson J (1971) The development of solid food intake in calves. I. Effect of previous experience of solid food, and the physical form of the diet, on the development of food intake after weaning. Anim Prod 13:15-24

Hodgson J, Jamieson W S (1981) Variations in herbage mass and digestibility, and the grazing behavior and herbage intake of adult cattle and weaned calves. Grass Forage Sci 36:39

Hofmann R R (1973) The ruminant stomach: stomach structure and feeding habits of East African game ruminants. East Afr Monogr Biol, E A Lit Bureau, Nairobi

Hofmann R R (1988) Anatomy of the gastrointestinal tract. In: Church D C (ed) The ruminant animal. Prentice Hall, Englewood Cliffs, p 14
Hofmann R R (1989) Evolutionary steps of ecophysiological adaptation and diversification of ruminants - a comparative view of their digestive system. Oecologia 78:443-457
Hunter R F, Milner C (1963) The behavior of individual, related and groups of south country Cheviot hill sheep. Anim Behav 11:507-513
Itani J (1958) On the acquisition and propagation of a new food habit in a troop of Japanese monkeys at Takasaki-Yama. Primates 1:84-98
Janis C (1976) The evolutionary strategy of the equidae and the origins of rumen and cecal digestion. Evolution 30:757-774
Jarman P J (1974) The social organization of antelope in relation to their ecology. Behaviour 48:215-266
Kandel E R, Schwartz J H (1985) Principles of neural science. Elsevier, New York
Kawai M (1965) Newly-acquired pre-cultural behavior of the natural troop of Japanese monkeys on Koshima Islet. Primates 6:1-30
Keeler R F (1988) Livestock models of human birth defects, reviewed in relation to poisonous plants. J Anim Sci 66:2414-2427
Kenney P A, Black J L (1984) Factors affecting diet selection by sheep. I. Potential intake rate and acceptability of feed. Aust J Agric Res 35:551-563
Key C, MacIver R M (1980) The effects of maternal influences on sheep: breed differences in grazing, resting and courtship behavior. Appl Anim Ethol 6:33-48
Killeen P R, Smith J P, Hanson S J (1981) Central place foraging in Rattus norvegicus. Anim Behav 29:64-70
Krebs J R, Stephens D W, Sutherland W J (1983) Perspectives in optimal foraging. In: Clark G A, Brush A H (eds) Perspectives in ornithology. Cambridge University Press, New York, p 165
Krieckhaus E E, Wolf G (1968) Acquisition of sodium by rats: interaction of innate mechanisms and latent learning. J Comp Physiol Psychol 65:197-201
Lane M A, Ralphs M A, Olsen J D, Provenza F D, Pfister J A (1989) Conditioned taste aversion: potential for reducing cattle loss to larkspur. J Range Manage in press
Launchbaugh K L, Provenza F D (1990) Can plants practice mimicry as an anti-herbivore mechanism? In: Abstr 43rd ann mtg soc range manage, Reno
Lett B T (1985) The pain-like effect of gallamine and naloxone differs from sickness induced by lithium chloride. Behav Neurosci 99:145-150
Logue A W (1986) The psychology of eating and drinking. W H Freeman, New York
Malechek J C, Provenza F D (1983) Feeding behavior and nutrition of goats on rangelands. World Anim Rev 47:38-48
Mangel J, Clark C W (1986) Towards a unified foraging theory. Ecology 67:1127-1138

Martin S C (1979) Evaluating the impacts of cattle grazing on riparian habitats in National Forests of Arizona and New Mexico. In: Forum--grazing and riparian/stream ecosystems. Trout Unlimited, Denver, p 35
McCall T C, Brown R D, De Young C A (1988) Mortality of pen-raised and wild white-tailed deer bucks. Wildl Soc Bull 16:380-384
McNaughton S J (1984) Grazing lawns: animals in herds, plant form, and coevolution. Am Nat 124:863-886
McNay R S, Doyle D D (1987) Winter habitat selection by black-tailed deer on Vancouver Island: a job completion report. B C Ministry Environ Park, Victoria B C, IWIFR-34
Mehiel R, Bolles R C (1984) Learned flavor preferences based on caloric outcome. Anim Learn Behav 12:421-427
Mehiel R, Bolles R C (1988) Learned flavor preferences based on calories are independent of initial hedonic value. Anim Learn Behav 16:383-387
Menzel E W (1973) Chimpanzee spatial memory organization. Science 182:943-945
Menzel E W (1978) Cognitive mapping in chimpanzees. In: Hulse S H, Fowler H, Honig W K (eds) Cognitive processes in animal behavior. Erlbaum, Hillsdale, p 375
Messier C, White N M (1984) Contingent and non-contingent actions of sucrose and saccharine reinforcers: effects on taste preference and memory. Physiol Behav 32:195-203
Milligan L P, Grovum W L, Dobson A, eds (1986) Control of digestion and metabolism in ruminants. Prentice-Hall, Englewood Cliffs
Mirza S N, Provenza F D (1990) Interactions with the mother modify diet selection by lambs differing in age. In: Abstr 43rd ann mtg soc range manage, Reno
Morrill J L, Dayton A D (1978) Effect of feed flavor in milk and calf starter on feed consumption and growth. J Dairy Sci 61:229-232
Myers J H (1987) Nutrient availability and the deployment of mechanical defenses in grazed plants: a new experimental approach to the optimal defense theory. Oikos 49:350-351
Nolte D L, Provenza F D, Balph D F (1989) The establishment and persistence of food preferences in lambs exposed to selected foods. J Anim Sci in press
Nolte D L, Provenza F D, Balph D F (1990) Early experience with flavors affects lambs' preferences for foods with those flavors. In: Abstr 43rd ann mtg soc range manage, Reno
Norton B E, Johnson P S, Owens M K (1982) Increasing grazing efficiency on crested wheatgrass. Utah Sci 43:110-113
O'Bryan M K, McCullough D R (1985) Survival of black-tailed deer following relocation in California. J Wildl Manage 49:115-119
Olton D S (1977) Spatial memory. Sci Am 236:82-98
Owen-Smith N (1989) Optimal foraging models for ungulates: assumptions, predictions and potential developments. In press in: Symposium on optimal foraging applied to mammals. Fifth Theriological Conf, Rome, Italy

Pelchat M L, Grill H J, Rozin R, Jacobs J (1983) Quality of acquired responses to tastes by Rattus norvegicus depends on type of associated discomfort. J Comp Physiol Psychol 97:140-153
Penning P D (1986) Some effects of sward conditions on grazing behavior and intake of sheep. In: Gudmundsson O (ed) Grazing research at northern latitudes. Plenun Press, New York, p 219
Pfister J A, Provenza F D, Manners G D (1989) Ingestion of tall larkspur by cattle: separating the effects of flavor from post-ingestive consequences. J Chem Ecol in press
Prien R F, Caffey E M, Klett C J (1971) Lithium carbonate: a survey of the history and current status of lithium in treating mood disorders. Dis Nerv Syst 32:521-531
Provenza F D, Malechek J C (1984) Diet selection by domestic goats in relation to blackbrush twig chemistry. J Appl Ecol 21:831-841
Provenza F D, Balph D F (1987) Diet learning by domestic ruminants: theory, evidence and practical implications. Appl Anim Behav Sci 18:211-232
Provenza F D, Balph D F (1988) The development of dietary choice in livestock on rangelands and its implications for management. J Anim Sci 66:2356-2368
Provenza F D, Bowns J E, Urness P J, Malechek J C, Butcher J E (1983) Biological manipulation of blackbrush by goat browsing. J Range Manage 36:513-518
Provenza F D, Balph D F, Olsen J D, Dwyer D D, Ralphs M H, Pfister J A (1988) Toward understanding the behavioral responses of livestock to poisonous plants. In: James L F, Ralphs M H, Nielsen D B (eds) The ecology and economic impact of poisonous plants on livestock production. Westview Press, Boulder, p 407
Provenza F D, Burritt E A, Clausen T P, Bryant J P, Reichardt P B, Distel R A (1990) Conditioned flavor aversion: a mechanism for goats to avoid condensed tannins in blackbrush. Am Nat in press
Reichardt P B, Bryant J P, Clausen T P, Wieland G (1984) Defense of winter-dormant Alaska paper birch against snowshoe hares. Oecologia 65:58-69
Reichardt P B, Clausen T P, Bryant J P (1987) Plant secondary metabolities as feeding deterrents to vertebrate herbivores. In: Provenza F D, Flinders J T, McArthur E D (eds) Proceedings-symposium on plant herbivore Interactions. USDA For Serv Intermtn Res Sta Gen Tech Rep INT-222
Reynolds G S (1968) A primer of operant conditioning. Scott, Foresman Co, Glenview
Rhoades D F (1979) Evolution of plant chemical defense against herbivores. In: Rosenthal G A, Janzen D H, (eds) Herbivores: their interactions with secondary plant metabolites. Academic Press, New York, p 4
Rhoades D F, Cates R G (1976) Toward a general theory of plant antiherbivore chemistry. Rec Adv Phytochem 10:168-213
Richter C P (1943) Total self-regulatory functions in animals and human beings. Harvey Lecture Series 38:63-103
Roath L R, Krueger W C (1982) Cattle grazing and behavior on a forested range. J Range Manage 35:332-338

Roll D L, Smith J C (1972) Conditioned taste aversion in anesthetized rats. In: Seligman M E P, Hager J L (eds) Biological boundaries of learning. Appleton-Century-Crofts, New York, p 98
Romesburg H C (1981) Wildlife science: gaining reliable knowledge. J Wildl Manage 45:293-313
Rozin P (1976) The selection of foods by rats, humans and other animals. In: Rosenblatt J S, Hinde R A, Shaw E, Beer C (eds) Advances in the study of behavior. Academic Press, New York, p 21
Schluter D (1981) Does the theory of optimal diets apply in complex environments? Am Nat 118:139-147
Schwartz G J, Grill H J (1985) Comparing taste - elicited behavior in adult and neonatal rats. Appetite 6:373-386
Senft R L, Coughenour M B, Bailey D W, Rittenhouse L R, Sala O E, Swift D M (1987) Large herbivore foraging and ecological hierarchies. Biosci 37:789-799
Sherman J E, Hickis C F, Rice A G, Rusiniak K W, Garcia J (1983) Preferences and aversions for stimuli paired with ethanol in hungry rats. Anim Learn Behav 11:101-106
Simbayi L C, Boakes R A, Burton M J (1985) Acquired preferences for flavors mixed with nutritive and non-nutritive sweet solutions. Neurosci Letters 22(Suppl) S158
Skinner B F (1981) Selection by consequences. Science 213:501-504
Smotherman W P (1982) Odor aversion learning by the rat fetus. Physiol Behav 29:769-771
Smotherman W P, Robinson S R (1987) Prenatal influences on development: behavior is not a trivial aspect of fetal life. Develop Behav Pediat 8:171-175
Squibb R C, Provenza F D, Balph D F (1989) Effect of age of exposure on consumption of a shrub by sheep. J Anim Sci in press
Stephens D W, Krebs J R (1986) Foraging theory. Princeton Univ Press, New Jersey
Stickrod G, Kimble D P, Smotherman W P (1982) In utero taste/odor aversion conditioning in the rat. Physiol Behav 28:5-7
Thorhallsdottir A G, Provenza F D, Balph D F (1987) Food aversion learning in lambs with or without a mother: discrimination, novelty and persistence. Appl Anim Behav Sci 18:327-340
Thorhallsdottir A G, Provenza F D, Balph D F (1989a) Ability of lambs to learn about novel foods while observing or participating with social models. Appl Anim Behav Sci in press
Thorhallsdottir A G, Provenza F D, Balph D F (1989b) The role of the mother in the intake of harmful foods by lambs. Appl Anim Behav Sci in press
Thorhallsdottir A G, Provenza F D, Balph D F (1989c) Social influences on conditioned food aversions in sheep. Appl Anim Behav Sci in press
Tolman E C (1949) There is more than one kind of learning. Psychol Rev 56:144-155
Tordoff M G, Friedman M I (1989a) Drinking saccharin increases food intake and preference - III. Sensory and associative factors. Appetite 12:23-36

Tordoff M G, Friedman M I (1989b) Drinking saccharin increases food intake and preference - IV. Cephalic phase and metabolic factors. Appetite 12:37-56

Urness P J (1986) Use of crested wheatgrass for big game. In: Johnson K L (ed) Crested wheatgrass: its values, problems and myths. Utah State Univ, Logan, p 147

Urness P J, Austin D D, Fierro L C (1983) Nutritional value of crested wheatgrass for wintering mule deer. J Range Manage 36:225-226

Van Dyne G M, Brockington N R, Szocs Z, Duek J, Ribic C A (1980) Large herbivore subsystem. In: Breymeryer A I, Van Dyne G M (eds) Grasslands, systems analysis and man. Cambridge Univ Press, Cambridge, p 269

Van Soest P J (1982) Nutritional ecology of the ruminant. O & B Books Inc, Corvallis

Waggoner V, Hinkes M (1986) Summer and fall browse utilization by an Alaskan bison herd. J Wildl Manage 50:322-324

Westoby M (1974) An analysis of diet selection by large generalist herbivores. Am Nat 108:290-304

Westoby M (1978) What are the biological bases of varied diets? Am Nat 112:627-631.

White J (1981) Trouble on Angel Island. Outdoor Calif 42:9-12

Williams R J (1978) You are extraordinary. In: The art of living. Berkeley Books, New York, p 121

Willms W, McLean A (1978) Spring forage selected by tame mule deer on big sagebrush range, British Columbia. J Range Manage 31:192-199

Willms W, Bailey A W, McLean A (1980) Effect of burning Agropyron spicatum in the autumn on the spring foraging behavior of mule deer and cattle. J Appl Ecol 17:69-84

Woods S C (1983) Conditioned hypoglycemia and conditioned insulin secretion. Adv Metabol Disorder 10:485-495

Yeates N T M, Schmidt P J (1974) Beef cattle production. Butterworth, Sidney

Zahorik D M, Houpt K A (1977) The concept of nutritional wisdom: applicability of laboratory learning models to large behaviors. In: Barker L M, Best M, Domjan M (eds) Learning mechanisms in food selection. Baylor Univ Press, Waco, p 45

Zahorik D M, Houpt K A (1981) Species differences in feeding strategies, food hazards, and the ability to learn food aversions. In: Kamil A C, Sargent T D (eds) Foraging behavior. Garland, New York, p 289

Zahorik D M, Maier S F, Pies R W (1974) Preferences for tastes paired with recovery from thiamine deficiency in rats: appetitive conditioning or learned safety? J Comp Physiol Psychol 87:1083-1091

Zimmerman E A (1980) Desert ranching in central Nevada. Rangelands 2:184-186

Zucker W V (1983) Tannins: does structure determine function? An ecological perspective. Am Nat 121:335-365

ORAL DISCUSSION – F.D. Provenza

Q. (Bazely) I am impressed with your data on responses of ruminants to plant toxins.
A. From the psychological literature, it is evident that the sort of thing natural selection has favored goes across a broad range of groups. If an animal encounters hazards such as toxins in the environment, it needs mechanisms for dealing with them; it doesn't matter what kind of animal, at that level there is a great generality.

Q. (Sih) Can you trick the animals by making familiar foods the ones that are nasty ?
A. That's very difficult to do, because discrimination is based upon odour and taste, and ruminants have very great abilities in that regard. The experimenter would have to be very clever in order to trick them. Ruminants sample quickly and consequently are able to discriminate subtle changes within plants: their choice is based on that. They can be fooled occasionally, however.

Q. (Kacelnik) Theory can be important; all of us, from different disciplines, should listen to each other.
A. Theory is clearly important, but I'm wondering what heuristic value it has to generate good research questions in the context of work that's being done on ruminants at present.

THE CHARACTERISTICS OF ALGAE IN RELATION TO THEIR VULNERABILITY TO GRAZING SNAILS

TREVOR A. NORTON & NICOLA L. MANLEY, University of Liverpool

Introduction

A large number of studies have shown that many marine invertebrates feed selectively, preferring some algae and shunning others. For example, the littorinid snails that abound on British shores are highly selective in their choice of food, but different species of *Littorina* rank the same food plants differently (Table 1).

The benefit gained by an animal from its food depends on its ability to manipulate and ingest the food, and to digest, absorb and assimilate the nutrients. Each of these steps constitutes a potential obstacle to the grazer and provides scope for the development of effective anti–herbivore defences by the alga.

The aim of the present paper is to examine some of the properties of seaweeds that influence their vulnerability to grazers, and to assess the extent to which such properties determine the food preferences of certain snails of the genus *Littorina*.

Physical Defences

Thallus size and morphology

Most marine snails are relatively small and slow moving, and because of the position of the mouth, they must occupy the substratum they browse. Unfortunately, seaweeds produce slippery thalli that flail and flap violently in the surf so that many sedentary grazers are unable to maintain a secure purchase on their surface. Whilst *Patina pellucida*, *Lacuna vincta*, *Littorina obtusata* and *L. mariae* characteristically inhabit the thalli of laminarian or fucoid algae, sometimes causing devastating damage (Fralick *et al* 1974, Norton 1971, Thomas & Page 1983), closely related species such as *Patella vulgata* or *L. littorea* are rarely found on these erect seaweeds and largely confine their foraging to microalgae growing on the rock surface, or to drift plants cast up on the shore (Norton *et al* 1989, Watson & Norton 1985a). For *L. littorea* the broad thallus of *Ulva lactuca* is readily pinned down to provide a relatively stable accessible food substrate whereas a single tubular inflated frond of *Enteromorpha intestinalis* may be too narrow to carry the entire foot of the snail and the plant's strong inherent buoyancy often frustrates efforts to hold it in place.

The most vulnerable algae are often large plants with broad flat thalli or tiny plants including the early microscopic stages of larger seaweed. Algae ofintermediate size (e.g. erect forms 1–10cm tall) may be eaten less often, if they are too large to be

NATO ASI Series, Vol. G 20
Behavioural Mechanisms of Food Selection
Edited by R. N. Hughes

Introduction

A large number of studies have shown that many marine invertebrates feed selectively, preferring some algae and shunning others. For example, the littorinid snails that abound on British shores are highly selective in their choice of food, but different species of *Littorina* rank the same food plants differently (Table 1).

The benefit gained by an animal from its food depends on its ability to manipulate and ingest the food, and to digest, absorb and assimilate the nutrients. Each of these steps constitutes a potential obstacle to the grazer and provides scope for the development of effective anti-herbivore defences by the alga.

The aim of the present paper is to examine some of the properties of seaweeds that influence their vulnerability to grazers, and to assess the extent to which such properties determine the food preferences of certain snails of the genus *Littorina*.

Physical defences

Thallus size and morphology.

Most marine snails are relatively small and slow moving, and because of the position of the mouth, they must occupy the substratum they browse. Unfortunately, seaweeds produce slippery thalli that flail and flap violently in the surf so that many sedentary grazers are unable to maintain a secure purchase on their surface. Whilst *Patina pellucida, Lacuna vincta, Littorina obtusata* and *Littorina mariae* characteristically inhabit the thalli of laminarian or fucoid algae, sometimes causing devastating damage (Fralick et al 1974, Norton 1971, Thomas & Page 1983), closely related species such as *Patella vulgata* or *Littorina littorea* are rarely found on these erect seaweeds and largely confine their foraging to microalgae growing on the rock surface, or to drift plants cast up on the shore. (Norton et al 1989, Watson & Norton 1985a). For *Littorina littorea* the broad thallus of *Ulva lactuca* is readily pinned down to provide a relatively stable accessible food substrate whereas a single tubular inflated frond of *Enteromorpha intestinalis* may be too narrow to carry the entire foot of the snail and the plant's strong inherent buoyancy often frustrates efforts to hold it in place.

The most vulnerable algae are often large plants with broad flat thalli or tiny plants including the early microscopic stages of larger seaweed. Algae of intermediate size (eg erect forms 1-10cm tall) may be eaten less often, if they are too large to be

TABLE 1. The edibility (Palatability Index) of seaweeds to snails in laboratory experiments (compiled from Watson and Norton, 1985a, & Norton, unpubl.) The Palatability Index compares the consumption of the test alga with a standard food, *Ulva lactuca*. Significance of difference Wilcoxon rank test. *0.01 < *P* < 0.05: ** 0.001 < *P* < 0.01: *** *P* < 0.001

	Palatability Index (Amount consumed/amount of *Ulva lactuca* consumed)		
	L. littorea	*L.obtusata*	*L.mariae*
Cladophora rupestris	0	0	0
Entermorpha intestinalis	0.54*	-	-
Ulva lactuca	1.00	1.00	1.00
Ascophyllum nodosum	0	2.05	0.94
Fucus ceranoides	0.21***	6.20***	-
F.serratus	0	3.63*	3.10**
F.spiralis	0.17***	9.67**	5.36**
F.vesiculosus	-	3.72**	4.04
Pelvetia canaliculata	0.04***	3.51**	2.51*
Corallina officinallis	0	0	0
Laurencia pinnatifida	0.11***	-	-
Mastocarpus sellatus	0	-	-
Polysiphonia lanosa	0	0	0
Porphya umbilicalis	0.19**	-	-

readily browsed from the rock or too buoyant to be pulled down, yet too insubstantial for many herbivores to inhabit. Steneck & Watling (1982) have attempted to relate features such as the size, morphology and anatomy of seaweeds to their susceptibility to grazing.

It must be remembered however, that even the largest seaweeds begin as tiny juveniles and that they may therefore be differentially vulnerable at different stages during their development. There is evidence that small (<3cm) plants of *Fucus* are more susceptible to grazing by *Littorina littorea* than larger plants (Lubchenco 1983), and without doubt snails foraging on the rock surface can wreak havoc amongst the lawns of germlings which develop at certain times of year.

Thallus toughness.

A major factor influencing edibility is the toughness of the outer part of the thallus, for this barrier must be breached by any successful grazer. Algae differ very considerably with respect to the ease with which their thalli can be punctured or abraded and this often correlates with their susceptibility to specific herbivores (Littler et al 1983; Padilla 1985; Watson & Norton 1985).

Subjective assessments of apparent 'toughness' are misleading however, for different parameters of 'toughness' do not necessarily go hand in hand. An alga may be difficult to puncture, but is easily abraded or vice-versa, (Table 2), This is because puncture resistance and abrasion resistance measure different things. Puncture resistance is dependent on the strength of the outermost cortical layer, whereas resistance to abrasion is related to general cell size in the cortex, which influences the ratio of wall material to cytoplasm (Watson & Norton 1985b). Which parameter is of most relevance to defence against herbivores depends on the species of grazer under consideration. For *Littorina littorea*, a seaweed's edibility is closely correlated with the ease with which it is punctured, but is not related to it's abrasion resistance; for *L.mariae* the reverse is true (Watson & Norton 1985b). These differences are related to the capability and modes of operation of the radulae of the two species. *Littorina littorea* can employ its sharp denticular cusps to tear the thallus, a mode of action which relies heavily on the initial penetration of the tissue (Watson & Norton 1985b). *Littorina obtusata* on the other hand uses the cusps to abrade the thallus.

TABLE 2. Two assessments of the relative 'toughness' of the thalli of a variety of seaweeds. All figures represent a percentage of the value obtained for the toughest species, designated 100%. (for details see Watson and Norton 1985a).

Species	Puncture strength	Abrasion resistance
Porphyra umbilicalis	13	13
Ulva lactuca	25	44
Fucua spiralis	30	91
Fucus vesiculosus	30	100
Pelvetia canaliculata	48	18
Fucus serratus	53	72
Mastocarpus stellatus	63	13
Ascophyllum nodosum	100	20

Molluscan radulae vary very considerably in the structure and arrangement of their teeth and in the manner in which they are applied to the substratum. Steneck & Watling (1982) have attempted to relate the capabilities of different types of radula to the possible diet of the animals that possess them. Their survey of the diets of a variety of molluscs indicates that those with a taenioglossan radula feed largely on microalgae or delicate filaments, but hardly ever on leathery macrophytes, and none are known to eat calcareous algae (Fig. 1). Such animals have quite soft chitinous teeth which must set limits on the toughness of the food that can be tackled. By contrast, the teeth on the radulae of Docoglossan molluscs are impregnated with iron oxide and are as hard as human teeth (Fig. 1). Such species are adept at browsing directly on the surface of rock and some can consume the thalli of calcareous algae (Fig.1). Clearly the size of the animal, and particularly the size of the musclature that operates the radula, may be significant. For example, adult *Littorina obtusata* can readily excavate the thallus of fucoid algae, but juvenile specimens, or adults of the smaller species *Littorina mariae*, have much less powerful buccal musclature and consume only microalgal epiphytes, being unable to break into the underlying seaweed (Watson & Norton 1987).

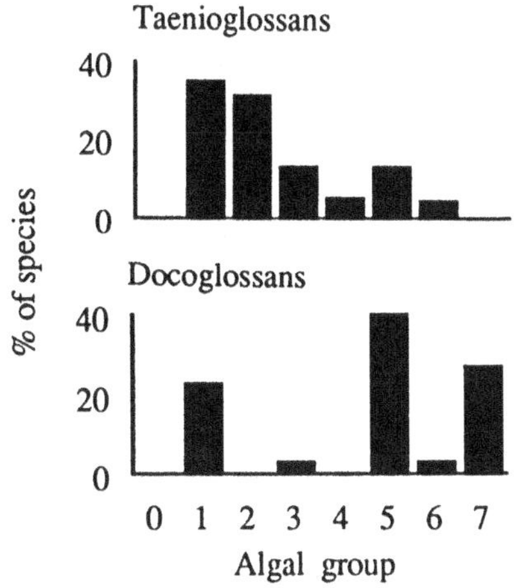

Test material	Hardness (Moh's scale)
Human fingernail	2.0
Littorina littorea	2.0 - 2.5
Littorina mariae	2.0 - 2.5
Littorina obtusata	2.0 - 2.5
Acmaea testudinalis	4.0 - 4.5
Patella aspera	4.5 - 5.0
Patella depressa	4.5 - 5.0
Patella vulgata	4.5 - 5.0
Human teeth	5.0

Figure 1. The dietary spectrum of molluscs with different types of radulae (n=21-48 species in each case, after Steneck and Watling, 1982), in relation to the hardness of the radula teeth (after Hawkins *et al.*, 1989). The algal groups are, from least to most tough: 1 microalgae, 2 filaments, 3 foliose, 4 corticate, 5 leathery, 6 articulate calcareous algae, 7 calcareous crusts (for details see Steneck and Watling, 1982).

Rasping on rock, or on calcareous algae induces severe wear on the animal's teeth. The radula is however, a conveyer belt for the continual replacement of teeth. Even non-calcareous seaweeds such as *Mastocarpus stellatus* and *Chondrus crispus* are dauntingly tough (Bertness et al 1983, Watson & Norton 1985b). Browsing on such tissue causes visible tooth wear (Fig 2), the degree of which has been used as an indication of the toughness of the alga being abraded (Bertness et al 1983). Unfortunately the degree of radula wear does not necessarily correlate with for example, the scratch resistance of the algal tissue.

It would also be misleading to stress the relative toughness of algal tissue to the exclusion of its other properties. Fleshy seaweeds are flexible and initially stretch under applied forces. Thus they require suprisingly more energy to break them than might be anticipated (Padilla 1985).

Likewise, the hardness of the snails teeth is by no means a complete description of their cutting capabilities, for the size, disposition and cutting angle are also critical (Padilla 1985). Furthermore, many molluscs have at their disposal more than one means of feeding. *Littorina littorea* for instance can bite as well as rasp (Watson & Norton 1985a) and may use either or both of these mechanisms to attack a particular seaweed.

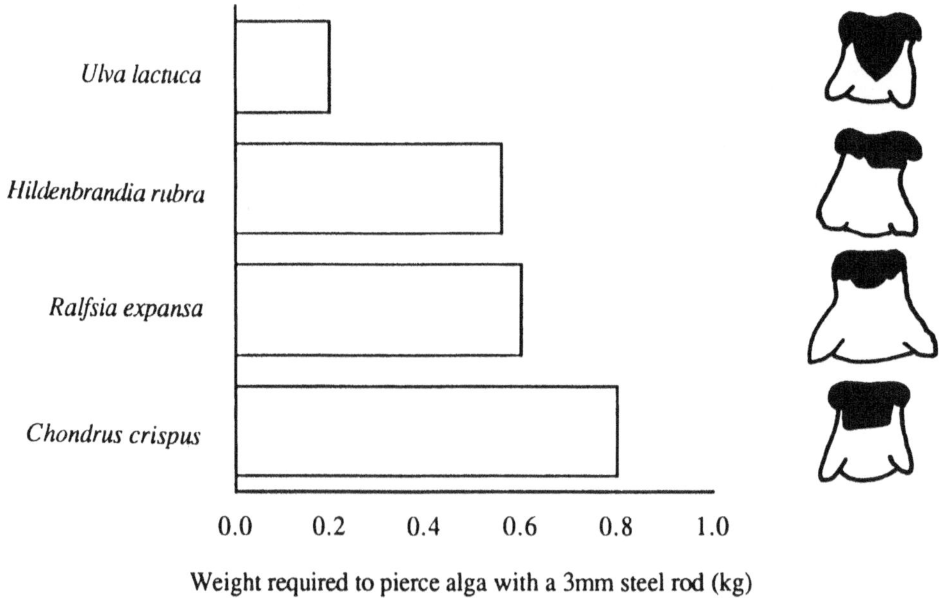

Figure 2. The amount of wear visible on the central (Rachidian) tooth of the radula of *Littorina littorea* after the snails had been fed for two weeks on algae exhibiting different degrees of "toughness" (after Bertness *et al.*, 1983)

All parts of a seaweed's thallus are not necessarily equally accessible, nor equally palatable to littorinid grazers; herbivores may attack the tougher parts of the thallus only after consuming the more tender portions. D.P. Cheney (unpublished) observed that in early spring *Littorina littorea* may consume the margins of the frond of *Fucus vesiculosus* leaving the midrib intact. Similarly, *L.littorea* grazes the thallus of *Pelvetia canaliculata* selectively, preferring the frond tips to the basal tissue. This may relate to variations in the toughness of the thallus, for the tips of plants of *Pelvetia* are more easily punctured than is the basal tissue (Watson & Norton 1985b). The vegetative and reproductive tissue of the same plant may also be differentially susceptible to grazing (Manley & Norton unpublished). It is however difficult to test whether swollen mucilaginous reproductive receptacles are more or less tough than the vegetative thallus. D.P. Cheney (unpublished) observed that although vegetative plants of *Chondrus crispus* were not consumed by *L.littorea*, reproductive plants were, but only after they had shed their spores. Perhaps the open ostioles allow ingress through the otherwise impenetrable epidermis.

Texture versus taste.

The effects of texture and 'taste' have been differentiated by comparing the rankings of intact plants and extracts of the plants impregnated into starch gel when offered to littorinids. Extracts of *Ulva lactuca* stimulate the feeding of all the snails tested, where as extracts from fucoid algae have more varied effects (Fig 3). For example, *Ascophyllum* extracts stimulate the feeding of *Littorina obtusata*, have no effect on *Littorina mariae* and depress browsing in *Littorina littorea* (Fig 3). The general correspondence of preferences obtained with intact plants and plant extracts suggests that chemical cues are often responsible for determining food preferences.

However, some algae that are inedible when intact (eg. *Cladophora rupestris* and several calcareous plants) are readily consumed as extracts (Padilla 1984, Watson 1983), which indicates that it is only their physical characteristics that protect them. Conversely, extracts from *Laurencia pinnatifida, Ascophyllum nodosum, Ralfsia verrucosa, Hildenbrandia rubra* and *Chondrus crispus* remain as unpalatable as the intact plants. Also, germlings of *Ascophyllum* too tiny to possess effective physical defences are as repellent as the adult plants (Watson & Norton 1985a). The inference that we draw is that such plants are protected largely by chemical factors.

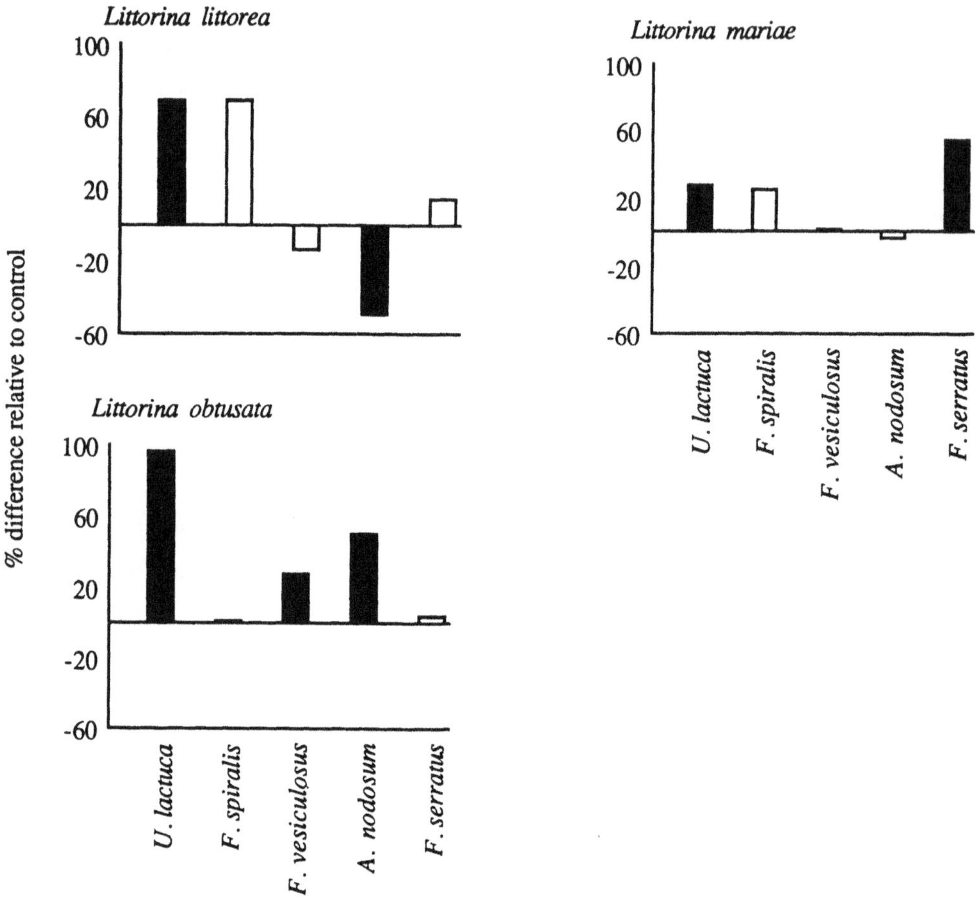

Figure 3. The number of radula scrapes on the surface of starch gel impregnated with extracts of various algae after five snails were left to browse for 48 hours at 10°C ±2°C, 16:8 photoperiod. (For each species of snail and alga the experiment was replicated 10 times using extracts from 10 different plants. The extraction was made with ethanol, dried, then resuspended in seawater. Solid bars indicate the difference from the control (gel impregnated with seawater) was significant at 95%, Wilcoxon signed rank test.)

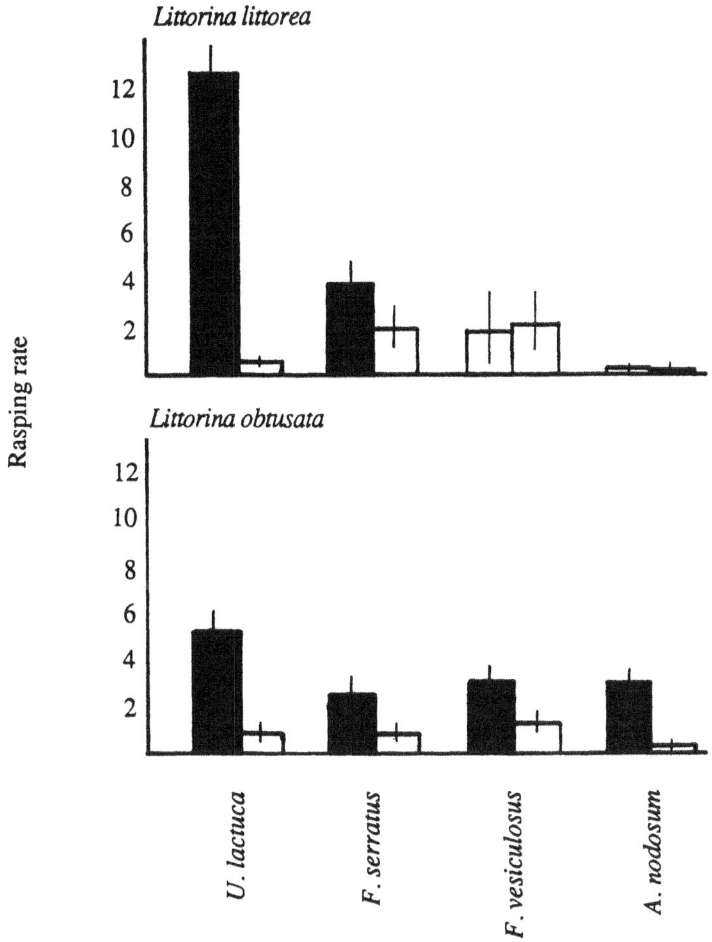

Figure 4. The number of radula rasps made during the first minute after the snails were exposed to 1ml. extract from various algae (left-hand bar) compared to the rasping of the snails exposed to 1ml. filtered seawater (right-hand bar) (n=10 in each case) at 10°C $\pm$2^{o}C. Lines indicate standard errors. Solid bars indicate significant differences from the control ($P \leq 0.05$ using Analysis of variance tests).

Chemical defences

Stimulants, attractants and repellents.

Littorinids are not only selective feeders, they also often reject algae without leaving any sign of having physically sampled them (Watson & Norton 1985a). When submerged the snails are heavily reliant on olfactory and gustatory stimulii (Croll 1983), and they are almost certainly responding to exudates from the algae which may enable them to select or reject food items from a distance, thus obviating the need for excessive foraging.

Extracts of algae applied to the head of snails often stimulates them to begin to feed or accelerates feeding behaviours (Fig 4). Arousal occurs within 20-60 seconds and is maintained for one to two minutes. Extracts of all the algae tested stimulate *Littorina obtusata* (P0.05 - >0.001, Wilcoxon signed rank test), but although the feeding of *Littorina littorea* is strongly enhanced by extracts of its preferred food *Ulva* (P<0.001), it is slightly enhanced by extracts of *Fucus serratus*, and unaffected by those of *Fucus vesiculosus*, and *Ascophyllum* .

In laboratory experiments most individuals of *Littorina littorea, L.mariae* and *L.obtusata* given a choice of filter papers impregnated with exudates of different species of algae, move onto that containing the exudate of the more edible plant (Table 3). In an apparatus to test longer distance perception (see Watson 1983), the same three species of snail respond rapidly to exudates from some marine algae by reorientating themselves and moving off in a consistent direction, sometimes at an accelerated rate of crawling. They clearly differentiate between the exudates from different seaweeds and respond by crawling towards some and away from others (Table 4). Moreover, different species of snails may respond differently to the same alga. They appear to be attracted towards their favoured food plants or to their preferred algal habitat (Norton et al 1989). Particularly striking are the contrasting responses of different species of snail to exudates of *Ascophyllum*. *Littorina littorea* is repelled as effectively by the exudates as by the plant itself, whereas *L.obtusata* is strongly attracted towards the exudates and *L.mariae* is indifferent to them.

Polyphenols and their effects

Red seaweeds contain a variety of substances including brominated phenols that

TABLE 3. The responses of snails to filter paper impregnated with exudates of various algae or with seawater. Values represent the number of snails in each of 3 sections of a 30 x 20cm test dish after 20 minutes. All were initially placed in the sector lacking filter papers and designated 'neutral' NS = not significant, *0.01 < *P* < 0.05: *** *P* < 0.001

Snails	Source of exudate	+ exudate	neutral	- exudate	Significance level
L. littorea	*Fucus serratus*	13	6	13	NS
	Ascophyllum nodosum	3	9	20	***
	Entermorpha intestinalis	21	6	5	***
	Ulva lactuca	18	9	5	*
L. mariae	*Fucus serratus*	18	9	5	*
	Ascophyllum nodosum	12	9	11	NS
	Entermorpha intestinalis	13	10	9	NS
	Ulva lactuca	21	6	5	***
L. obtusata	*Fucus serratus*	12	9	11	NS
	Ascophyllum nodosum	21	7	4	***
	Entermorpha intestinalis	15	9	8	NS
	Ulva lactuca	12	10	10	NS

are potentially distasteful to herbivores (Kain & Norton 1989) whereas ephemeral green algae such as *Ulva* seem to be unprotected. The brown fucoid seaweeds that dominate many temperate shores contain polyphenols at concentrations significantly higher than those found in other brown algae (Norton et al 1989, Ragan & Glombitza 1986). Phenolic compounds are well-known grazer repellents in higher plants (eg Levin 1976, Swain 1977), and it has been claimed that polyphenols produced by *Ascophyllum* may act in a similar manner (Geiselman & McConnell 1981).

Polyphenols owe their efficacy as repellents to their astrigency, which is assumed to result from cross-linking between them and proteins or glycoproteins in the mouth of the herbivore (Goldstein & Swain 1965). The cross-linking precipitates the grazer's salivary proteins, or immobilizes enzymes, rendering the plant tissue unpalatable (Bate-Smith 1973). We have found little correlation between the total polyphenol content of the tissue and its astringency (Norton et al 1989). This is almost certainly because it is the high molecular weight polyphenols that have the greatest protein-binding capacity and therefore the highest astringency and presumably the greatest defence against grazers.

TABLE 4. The movement of snails in darkness at a temperature of 10°C ± 2°C in shallow (5 mm), flowing (5 cm s^{-1}) water containing exudates of various algae. The water was 'charged' by culturing 15cm^3 of the test alga in 25l of filtered seawater for 60 mins at 10 - 12 °C and an irradiance of 5Wm^{-2}. Control snails were exposed to filtered seawater treated similarly but not exposed to algal exudates . NS = not significant, **0.001 < P < 0.01: *** P < 0.001.

Snails	Source of exudate	n	Towards	Away	Significance of difference (Wilcoxon-Signed Rank Test)
L. littorea	*Fucus serratus*	30	17	13	NS
	Ascophyllum nodosum	29	5	24	***
	Entermorpha intestinalis	27	23	4	***
	Ulva lactuca	28	22	6	***
	Controls	58	30	28	NS
L. mariae	*Fucus serratus*	54	48	6	***
	Ascophyllum nodosum	41	20	21	NS
	Entermorpha intestinalis	51	32	19	***
	Ulva lactuca	47	39	8	***
	Controls	100	50	50	NS
L. obtusata	*Fucus serratus*	43	34	9	***
	Ascophyllum nodosum	53	45	8	***
	Entermorpha intestinalis	54	33	21	***
	Ulva lactuca	46	26	20	**
	Controls	41	18	23	NS

Our assessments clearly indicate that the tissues of *Ascophyllum nodosum* and *Pelvetia canaliculata* are markedly more astringent than those of other intertidal fucoids (Norton et al 1989), which correlates well with the apparent repugnance exhibited by *Littorina littorea* for these plants (Watson & Norton 1985a). *Ascophyllum* is however, readily consumed by *L.obtusata.* Moreover, the palatability of *Fucus serratus* to *Littorina littorea* is reduced by the presence of extracts of other more astringent algae such as *Ascophyllum* (Fig 5), and 12 commercially available polyphenols had a similar effect at concentrations as low as 0.01 mg per ml. (P>0.001, 3 way analysis of variants). These compounds had a similar effect on the feeding of *L. obtusata,* but extracts of astringent algae did not suppress it's appetite (Fig 5).

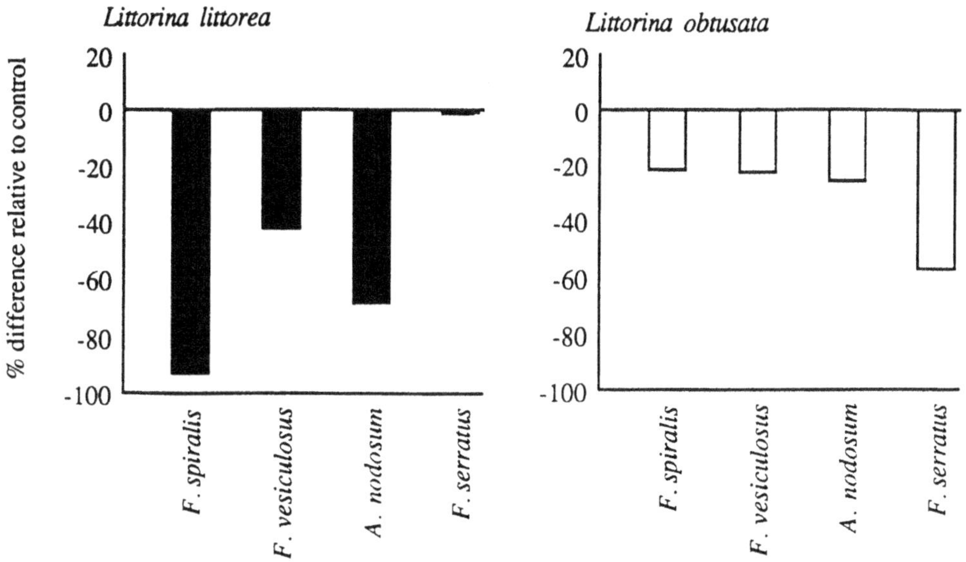

Figure 5. The consumption rate of snails feeding on 2g. *Fucus serratus* tissue whilst exposed to extracts from various algae at 10°C ±2°C (n=10 in each case). Solid bars indicate a significant difference from the control (filtered seawater in place of the extract) ($P \leq 0.05$ using Wilcoxon signed rank tests).

Discussion

Seaweeds possess a range of physical and chemical characteristics that discourage potential grazers. Such defences present a challenge to grazers to overcome them.

No defence is impenetrable; even the most redoubtable physical barrier can be breached. Calcareous seaweeds for example, would be considered very tough whatever assessment is used, and this is sufficient to defeat the efforts of most benthic grazers. Nonetheless, some acmeaed limpets, abalones and chitons consume calcareous algae almost exclusively (Barnes & Gonor 1973, Branch 1975, Clokie & Norton 1974, Padilla 1984, Steneck 1982). Such animals are clearly undaunted by the toughness of the thalli or by their low nutritive content which, it is claimed, renders them poor value for the effort required to consume them (Littler 1976, Littler & Littler 1980, Steneck 1982). Several grazers have such an affinity for a particular calcareous food plant that their larvae settle preferentially on its thallus (Barnes & Gonor 1973, Morse & Morse 1984, Rumrill & Cameron 1983, Steneck 1982).

Similarly, a particular chemical defence cannot act effectively against every type of grazer. Several molluscs habitually consume seaweeds that are known to be repellent to other grazers, and they can detoxify the active chemicals or even utilise them for their own defence (Hay & Fenical 1988, Littler et al 1986, Norris & Fenical 1981, Stallard & Faulkener 1974).

More surprising are the diverse reactions of taxonomically very closely related grazers to the same plant. This is dramatically demonstrated in the responses of littorinids to *Ascophyllum nodosum*. *Littorina mariae* is indifferent to *Ascophyllum* and its exudates, whereas *L. littorea* will starve rather than consume *Ascophyllum* (Watson & Norton 1985a), and is repelled by its exudates. In contrast ,a third species , *Littorina obtusata*, consumes *Ascophyllum* or its extracts in preference to other seaweeds, and is strongly attracted to its exudates. It would appear that *Ascophyllum* has developed effective defences against generalist herbivores such as *Littorina littorea*, but that these defences have been overcome by the more specialist grazer *Littorina obtusata,* which lives and feeds predominantly on *Ascophyllum.*

The behaviour of *Littorina obtusata* may exemplify an evolutionary 'arms race' which encourages the specialist herbivore to overcome, even to utilise, the defences of the food plant. Indeed, by grazing on *Ascophyllum* it may even stimulate the plants defences, for Van Alstyne (1988) has shown that natural grazing damage to the thallus of *Fucus distichus* leads to an increase in the polyphenolic content of the injured plant which can deter some species of *Littorina.* The relatively limited mobility of *L. obtusata* restricts its scope for widespread foraging. However, by overcoming the defences of *Ascophyllum*, which is large, abundant and very long lived, *L. obtusata* has secured a reliable source of both food and habitat.

Little is known of the mechanism by which *Littorina obtusata* detoxifies the chemical deterrents of *Ascophyllum*. Once such a substance has been ingested by a herbivore it must reach a specific site within the animal to exert toxic or deterrent effects. A specialist herbivore may counteract the chemical by preventing it from reaching the crucial site either by rapidly degrading and excreting the substance or by channelling it away for storage in specialized tissues or organelles. Alternatively, the chemical may be unable to interact with the target organ because of subtle modifications in the organ (Brattsten, 1979). In terrestrial invertebrates the detoxification mechanisms that have been most thoroughly investigated are mixed function oxidase (MFO) systems. These are known to catalyse numerous oxidative reactions and show a remarkable non-specificity as a large variety of chemicals are accepted as substrates. MFO enzymes are thought to be able to adjust rapidly to the presence of external chemicals by induction, which initiates the production of an enzyme protein. Induction is known to be triggered by plant secondary metabolites (Brattsten, 1979). Many species of marine invertebrates are known to possess an MFO enzyme system (see Brattsten 1979) and they have been implicated in the breakdown of polyphenols in insects (Miles 1969). It is possible that a similar system is functioning in *Littorina obtusata* to allow it to consume astringent fucoid tissue.

The physiological effects of polyphenols on herbivorous marine snails are unknown, but it is known that the tannins from higher plants act on insects by inhibiting feeding or causing cell damage, rather than by reducing digestibility (reviewed by Hay & Fenical 1988).

Under ideal conditions in the laboratory littorinid snails can detect algal exudates when at least a metre away from the source (Van Dongen 1956). It would however, be much more difficult to detect the concentration gradient of an exudate amidst the turbulent water motion of the shore. In culture even aerating the water may confuse the directional response of an aplysian (Frings & Frings 1965).

Some exudates from brown algae do operate effectively in the sea. It has been found that *Stypopodium zonale* releases grazer-repellent substances into the water (Gerwick & Fenical 1981) and that even highly palatable seaweeds acquire significant protection from grazers simply by growing within 10 cm of a plant of

Stypopodium (Littler et al 1986). Indeed, being close to *Stypopodium* is as protective as living next to fire coral (Littler et al 1987). Similarly, polyphenols produced by *Ascophyllum* leak into the surrounding water, especially during spring and summer , when the water above the algal beds becomes significantly enriched with phenolics. (Sieburth & Jensen 1968, Sieburth & Tootle 1981). In laboratory experiments *Littorina littorea* is strongly repelled by exudates of *Ascophyllum* (Table 4).

On the shore the multitudinous exudates emanating from various algae may confuse the herbivore. Certainly in culture the presence of exudates of more astringent algae can reduce the apparent edibility of algae normally preferred by *Littorina littorea* (Fig 5). It is perhaps significant that the algae that are most attractive to littorinids are often present on the shore in huge quantities. The magnitude of the chemical signals that emanate from large monospecific stands of such algae may be sufficient to swamp any interference from subsidary members of the flora.

Summary

Seaweeds possess a range of physical and chemical characteristics that discourage potential grazers and these play an important role in determining the varied food preferences and behaviours of different species of *Littorina.*

Seaweeds vary considerably in the toughness of their outer surface. However, as different measures of 'toughness' yield different results, it is only when the method of assessment mimics the way in which the grazer attacks the thallus that the plant's 'toughness' correlates with its susceptibility to a particular species of herbivore. Some intact plants that are rejected by snails because they are too tough, are readily consumed when offered as extracts impregnated into a soft substrate.

Many seaweeds also appear to possess chemical defences that can be perceived by snails from afar. *Littorina littorea* for example is repelled by exudates of *Ascophyllum nodosum,* and the presence of extracts from this plant lowers the attractiveness of normally favoured food. The edibility of various tissues from different species of brown algae is discussed in relation to the astringency of polyphenol constituents which act as grazer repellants.

References

Barnes J R, Gonor JJ (1973) The larval settling response of the lined chiton *Tonicella lineata*. Mar Biol 20: 259-264

Bate-Smith E C (1973) Haemanalysis of tannins: The concept of relative astringency. Phytochemistry 12: 907-912.

Bertness M D, Yund P O, Brown A F (1983) Snail grazing and the abundance of algal crusts on a sheltered New England rocky beach J Exp Mar Biol Ecol 71:147-164.

Branch G (1975) Intraspecific competition in *Patella cochlear* Born J Anim Ecol 44:263-282

Brattsten L B (1979) Biochemical defence mechanisms in herbivores against plant allelochemicals. In Rosenthal G A, Janzen D H (eds) Herbivores. Their interactions with plant secondary metabolites Academic Press: 199-270

Clokie J J P, Norton T A (1974) The effect of grazing on the algal vegetation of pebbles from the Firth of Clyde. Br Phycol J 9:216

Croll R P (1983) Gastropod chemoreception. Biol Rev 58:293-319

Fralick R A, Turgeon K W, Mathieson A C, (1974) Destruction of kelp populations by *Lacuna vincta* (Montagu). Nautilus 88: 112-114

Frings H, Frings C (1965) Chemosensory basis of food finding behaviour in *Aplysia juliana* (Mollusca Opistobranchia) Biol Bull 128:211-217

Geiselman J A, McConnell O J (1981) Polyphenols in the brown algae *Fucus vesiculosus* and *Ascophyllum nodosum*. Chemical defence against the marine herbivorous snail *Littorina littorea* J Chem Ecol 7: 1115-1133.

Gerwick W H, Fenical W (1981) Ichthyotoxic and cytotoxic metabolites from the brown alga *Stypopodium zonale* J Org Chem 46: 22-27.

Goldstein J L, Swain T(1965) The inhibition of enzymes by tannins. Phytochemistry 4: 185-192.

Hay M E, Fenical W (1988) Marine plant-herbivore interactions: the ecology of chemical defence Ann Rev Ecol Syst 19:111-145.

Hawkins S J, Watson D C, Hill A S, Harding S P, Kyriakides M A, Hutchinson S, Norton T A (1989) A comparison of feeding mechanisms in microphagous, herbivorous, intertidal prosobranchs in relation to resource partitioning. J Moll Stud 55:151-165.

Kain J M, Norton T A (1989) Marine ecology. In Cole K M, Sheath R (eds) The Biology of Red Algae. Cambridge Univ. Press.

Levin D A (1976) The chemical defences of plants to pathogens and herbivores. Ann Rev Ecol Syst 7:121-159.

Littler M M (1976) Calcification and its role among the macroalgae. Micronesica 12:27-41.

Littler M M, Littler D S (1980). The evolution of thallus form an survival strategies in benthic marine macroalgae: field and laboratory tests of functional form model. Am. Nat. 116:25-44.

Littler M M, Littler D S (1981) Intertidal macrophyte communities from Pacific Baja California: relatively constant versus environmentally fluctuating systems. Mar Ecol Prog Ser 4:145-158.

Littler M M, Littler D S, Taylor P R (1987) Animal-plant defense associations: effects on the distribution and abundance of tropical reef macrophytes. J Exp Mar Biol Ecol 105:107-121.

Littler M M, Taylor P R, Littler D S (1986) Plant defense associations in the marine environment. Coral Reefs 5:63-71.

Lubchenco J (1983) *Littorina* and *Fucus*: Effects of herbivores, substratum heterogeneity and plant escapes during succession. Ecology 64:1116-1123

Miles P W (1969) Interaction of plant phenols and salivary phenolases in the relationship between plants and *Hemiptera.* Entomol Exp App 12:736-744.

Morse A N C, Morse D E (1984) Recruitment and metamorphosis of *Haliotis* larvae induced by molecules uniquely available at the surfaces of crustose red algae. J Exp Mar Biol Ecol 75:191-215.

Norris J N, Fenical W (1986) Chemical defense in tropical marine algae. In: Rutzler K, Macintyre I G (eds) Atlantic barrier reef ecosystem at Carrie Bow Cay, Belize. Smithsonian Contribs Mar Sci 12:590-626.

Norton T A (1971) An ecological study of the fauna inhabiting the sublittoral marine alga *Saccorhiza polyschides* (Lightf) Batt. Hydrobiologia 37:215-231

Norton T A, Watson D C, Hawkins S J, Manley N L, Williams G A (1989) Scraping a living: a review of littorinid grazing Hydrobiologia.

Padilla D K (1984) The importance of form: differences in competitive ability, resistance to consumers and environmental stress in an assemblage of coralline algae. J Exp Mar Biol Ecol 79:105-127.

Padilla D K (1985) Structural resistance of algae to herbivores. Mar Biol 90:103-109.

Ragan M A, Glombitza K W (1986) Phlorotannins, brown algal polyphenols. Progr Phycol Res 4:129-241.

Rumrill S S, Cameron R A (1983) Effects of gamma-aminobutyric acid on the settlement of larvae of the black chiton *Katherina tunicata* Mar Biol 72:243-247.

Sieburth J McN, Jensen A (1968) Studies on algal substances in the sea. I Gelbstoff (humic material) in terrestrial and marine waters. J Exp Mar Biol Ecol 2:174-189.

Sieburth J McN, Tootle, J L (1981) Seasonally of microbial fouling on *Ascophyllum nodosum* L (LeJol) *Fucus vesiculosus* L *Polysiphonia lanosa* (L) Tandy and *Chondrus crispus* Stackh J Phycol 17:57-64.

Stallard M O, Faulkener D J (1974a) Chemical constituents of the digestive gland of the sea hare *Aplysia californica.* I Importance of diet. Comp Biochem Physiol 49B:25-35.

Steneck R S (1982) A limpet - coralline alga association: adaptations and defenses between a selective herbivore and its prey. Ecology 63:507-522.

Steneck R S, Watling L (1982) Feeding capabilities and limitation of herbivorous molluscs: A functional group approach. Mar Biol 68:299-319.

Swain T (1977) Secondary compounds as protective agents. Ann Rev Plant Physiol 28: 479-510.

Thomas M L H, F H Page (1983) Grazing by the gastropod *Lacuna vincta* in the lower intertidal area of Musquash Head, New Brunswick, Canada. J Mar Biol Ass U K 63:725-736.

Van Alstyne K L (1988) Herbivore grazing increases polyphenolic defenses in the intertidal brown alga *Fucus distichus* Ecology 69:655-663.

Van Dongen A (1956) The preference of *Littorina obtusata* for Fucaceae. Arch Neer Zool 11:373-386.

Watson D C (1983) Seaweed palatability and selective grazing by littoral gastropods. PhD thesis University of Glasgow.

Watson D C, Norton T A (1985a) Dietary preferences of the common periwinkle *Littorina littorea* (L) J Exp Mar Biol Ecol 88:193-211.

Watson D C, Norton T A (1985b) The physical characteristics of seaweed thalli as deterrents to littorine grazers. Bot Mar 28:383-387.

Watson D C, Norton T A (1987) The habitat and feeding preferences of *Littorina obtusata* (L) and *Littorina mariae* Sacchi et Rastelli. J Exp Mar Biol Ecol 112:61-72.

COMPARATIVE FORAGING BEHAVIOR OF TROPICAL AND BOREAL SEA URCHINS

Robert L. Vadas
Dept. of Botany and Plant Pathology
University of Maine, Orono,
Maine 04469, U.S.A.

INTRODUCTION

Sea urchins are important structuring forces in marine ecosystems. They control the abundance and distribution of algae (Jones and Kain 1967; Dayton 1975; Pearse and Hines 1979) and influence the life form and species diversity of benthic assemblages (Paine and Vadas 1969; Sammarco *et al.* 1974; Vance 1979). Sea urchins primarily occupy hard substrates, although several species regularly inhabit seagrass communities. Here, too, they can influence plant assemblages (Ogden *et al.* 1973b). They also control epifaunal invertebrate communities (Guida 1976;Sammarco 1980; Ayling 1981; Briscoe and Sebens 1988).

Urchin diets reflect the broad range of effects on benthic communities. Regular sea urchins feed mainly on plants but most are omnivores (Lawrence 1975). Even those responsible for overgrazing of algal or seagrass beds are regarded as omnivores or generalists. Most urchins occupy two or three trophic levels, from detritivores to carnivores, and some are cannibals. Larson *et al.* (1980) termed them "facultative specialists" whereas Andrew (1989) called them "forced generalists".

Sea urchins are simple animals, consisting of gonads and gut surrounded by a calcareous test. The neural system consists of only an ectoneural layer and an internal neural ring (Sloan and Campbell 1982). Despite their anatomical simplicity and omnivorous habit, their foraging strategies are enigmatic. Efficient exploitation of a wide range of foods is thought to be costly, and generalists are

NATO ASI Series, Vol. G 20
Behavioural Mechanisms of Food Selection
Edited by R. N. Hughes

less efficient than specialists at competing for specific prey (Patridge and Green 1985). In addition, most omnivores lack preferences and adaptations related to specialist feeding (McFarland 1981). Nonetheless, numerous studies indicate that regular sea urchins make complex behavioral "decisions" (Ayling 1978; Garnick 1978; Schiel 1982; Carpenter 1984; Vadas *et al.* 1986), which include foraging, predator avoidance, and maintenance. Paradoxically, some urchins not only select their prey but also exhibit "nutritional wisdom" (Vadas1977; Larson *et al.* 1980).

Here I discuss sea urchin foraging strategies and provide a framework for analysing foraging and other behavior. I use a comparative approach because urchins occupy several habitats, and it is likely that common foraging strategies have evolved in similar (parallel) habitats. Latitudinal analyses can reveal patterns in ecological processes including algal–herbivore interactions (Gaines and Lubchenco 1982), and may explain the "feeding paradox". I begin by comparing productivity, biomass and grazing in relevant habitats along latitudinal gradients. Using assumptions based on these patterns, I advance several hypotheses and graphical models from which I make generalized predictions. I then analyzse urchin movements, habitats and life styles and relate them to feeding behavior. Last, I evaluate the benefits of selective and other foraging strategies.

HABITAT COMPARISONS: BIOMASS AND GRAZING

Potential food availability is clearly different between tropical–subtropical and boreal–temperate waters (hereafter tropical and boreal, respectively) (Table 1). Benthic algal productivity is higher in tropical waters, but standing algal biomass is orders of magnitude higher in boreal regions. Hard substrates in tropical areas are dominated by small, turf–like filamentous algae (Carpenter, 1981), and large foliose thalli are rare except on exposed coral reefs (Connor and Adey 1977). In contrast, boreal regions contain abundant macroalgae, including giant forms in areas of upwelling, and intertidal zones are rich in algal biomass (Mann 1973). The biomass potential of tropical intertidal habitats is reduced, however, because of grazing, lower tidal amplitudes, and solar radiation. Patterns for seagrass communities are more easily generalized: productivity ishigher in tropical grass beds, but biomass is similar in both regions (Zieman and Wetzel 1980).

Table 1. Potential availability of benthic algae and seagrasses in different biogeographic regions

Region	Productivity $Kg.m^{-2}.yr^{-1}$	Biomass $Kg.m^{-2}$
Algae		
tropical	1–50	0.01–3.0*(0.1)
no. temp/boreal	5–15	1–32 (10)
so. temp/boreal	10–21	6–18
arctic	0.1	0.1–0.2
Seagrasses		
tropical	0.3–16	0.01–8.1 (1.5)
no. temp/boreal	2–10	0.01–2 (0.4)

1 Numbers in parentheses are typical values.
2 Based on Mann 1973, 1982; Kirkman 1984; Zieman and Wetzel 1980; Dunton and Schell 1986; Connor and Adey 1977; Carpenter 1981; Rogers and Salesky 1981.
3 '*' maximum values (extreme, wave–exposed algal pavements).

Biomass patterns (Fig. 1) are not independent of herbivore activity (Hatcher and Larkum 1983). Intensive grazing by fish and urchins is a major cause of reduced plant abundance intropical waters (Randall 1965; Ogden and Lobel 1978; Littler *et al.* 1983). Furthermore herbivory may be more intense now than at any time in the past (Steneck 1983). Urchin grazing also is a major cause of erosion of coral (Lawrence and Sammarco 1982). Conversely, boreal regions have fewer excavating herbivores, especially grazing fish (but see Choat 1982). Northern boreal waters experience periodic outbreaks of intensive grazing by urchins (Breen and Mann 1976; Foreman 1977; Hagen 1987) and some areas have remained as coralline pavements for decades (Himmelman 1980; Steneck and Vadas unpublished). However, most boreal areas are covered, at least seasonally, with lush growths of macroalgae. Algal biomass may be lower in some (Ayling 1981; Schiel 1982), but not all (e.g., Castilla and Moreno 1982; Anderson and Velimirov 1982), south temperate waters.

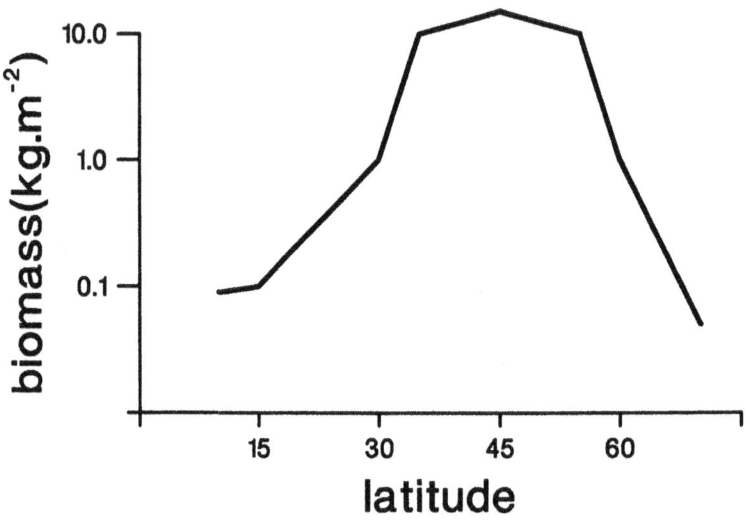

Figure 1. Latitudinal pattern of predictable seaweed biomass (based on Table 1).

Outbreaks of intensive grazing by urchins in seagrass beds have also occurred in both regions (Camp *et al.* 1973; Bak and Nojima 1980). However, intensive grazing on grasses is uncommon except at the edge of reefs (Randall 1965; Ogden *et al.* 1973b). Moore *et al.* (1963) report regular, but not intensive, grazing on seagrasses in the Florida keys. In the Miskito Cays intensive grazing was observed only in local patches (Vadas *et al.* 1982). The intensity of urchin grazing on *Posidonia oceanica* in the Mediterranean Sea is variable (Ott and Mauer 1977; Kirkman and Young 1981; Verlaque 1984).

LATITUDINAL PATTERNS and HYPOTHESES

Energy and Diet Breadth

Assuming that present biomass patterns reflect past (evolutionary) environments, then animals in biomass–rich habitats should expend less energy on search and survival and thereby make finer adjustments to their prey. Because of this, and increased metabolic costs at higher temperatures, energy expenditures should be higher in tropical waters, hence inversely related to latitude (Fig. 2). Two predictions from the "search energy" hypothesis (and biomass pattern) for the tropics are: i) less energy will be obtained per unit search time, which should result

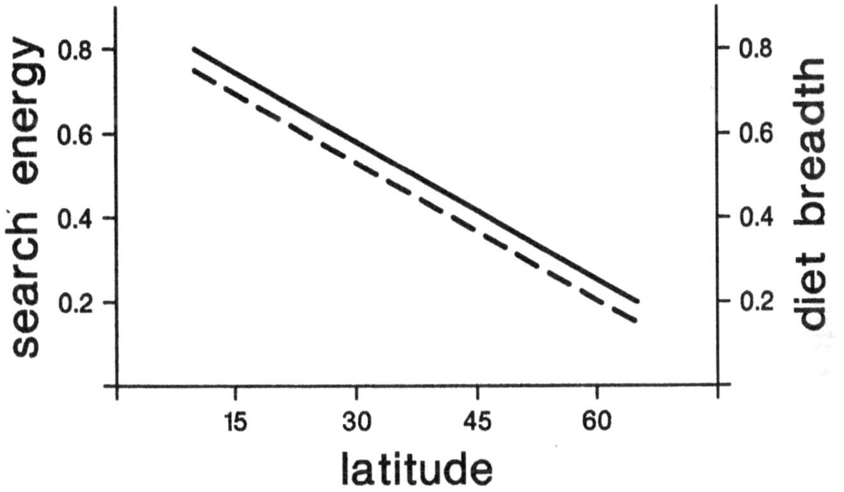

Figure 2. Proposed patterns of relative search energy costs (solid line) and predicted diet breadth (dashed line) vs latitude (based on Fig. 1 and presumed higher metabolic expenditure in tropical waters, and the widespread occurrence of intense grazing and plant defences in tropical regions, relative scale).

in limited movements (to conserve energy, see Dance 1987) and ii) diet breadths should be wider and result in generalist feeding. With the exception of disease or ecological catastrophe the patterns of seaweed biomass in boreal waters should constitute a predictable resource (sensu Menge 1972) for urchins.

In tropical waters the reduced availability of algae should intensify competition between benthic herbivores. Given this and the above patterns, diet breadths should be greater in tropical regions ("diet breadth" hypothesis, Fig. 2). Competition for a limited algal supply should lead to: i) either increased generalist feeding or extreme niche specialization in tropical urchins, and ii) the widespread occurrence of defensive responses in tropical plants (Paul and Hay 1986; Hay and Fenical 1988), and thus should further reinforce generalist feeding or niche diversification.

Predation and Sea Urchin Foraging

Predators can influence the foraging strategies of herbivores in at least three ways. They can i) limit the time herbivores spend feeding and hence the net intake of food, ii) modify behavior, including diet selection and iii) alter temporal feeding patterns and affect the quality of prey available (nocturnal vs diel leaves etc., see Hay *et al.* 1988). Sea urchins modify their behavior, movements, and diet breadth in thepresence of predators (Ogden *et al.* 1973b; Nelson and Vance 1979; Vance and Schmidt 1979; Carpenter 1984).

The intensity of predation in marine environments generally increases toward the tropics (Paine 1966; McClanahan and Muthiga 1989) (Fig.3). In Caribbean waters large numbers of fish, crustaceans, and gastropods prey on both juveniles and adults (Randall 1967; Vermeij 1978). Fewer mobile predators feedon adult urchins in boreal waters, although sea otters (Lowry and Pearse 1973; Estes *et al.* 1978) and at times fish (Cowen 1983) control urchins. Some boreal regions have never had major (sea otter) predators (Castilla and Moreno 1982). Juvenile urchins, however, may sustain high mortality (Tegner and Dayton 1981; Keats *et al.* 1987; Witman 1985) Nonetheless, predators seem less effective in controlling urchin populations in boreal waters (Himmelman and Steele 1971; Choat 1982). Despite sporadic dieoffs (Miller and Colodey 1983), populations levels are high and seemingly more variable than in the tropics. The causes of the dramatic shifts

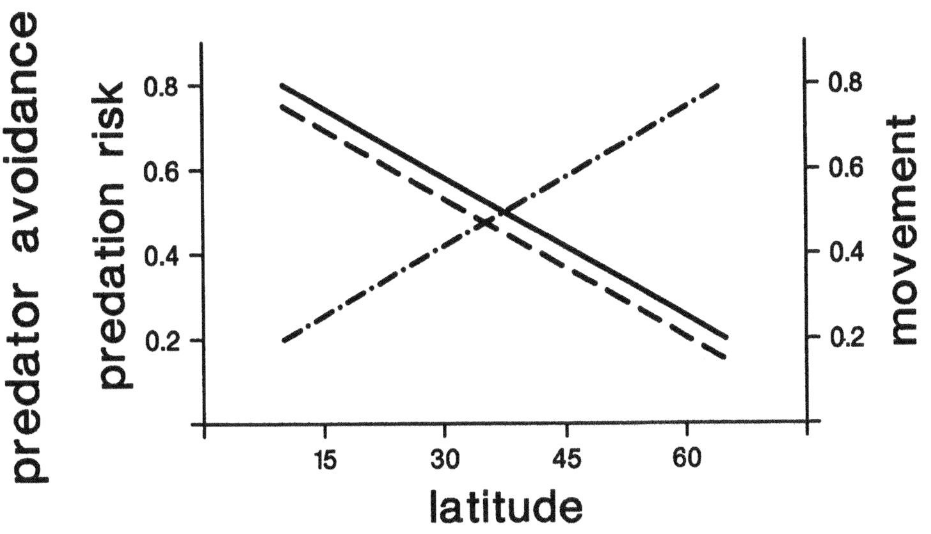

Figure 3. Predicted patterns of sea urchin movements (dot– dashed line) and predator avoidance behavior (dashed line) with latitude and the generalized threat from predators (solid line) (based on Fig. 2, literature, predator–risk theory, and the notion that predation is higher during movements and migrations).

in boreal populations (from kelp to urchin barrens) are obscure (Elner and Vadas 1990). Although predation on herbivores may well have been reduced in recent centuries by overharvesting of fishery stocks (Hay 1984; Barnard 1986), herbivores probably have not had time to evolve to the altered intensities, and therefore differences in predator avoidance behavior might be expected between the two regions.

The increased risk of predation in the tropics is predicted to result in i) restricted or greatly limited movements and increased predator avoidance behavior ("predation risk" hypothesis) and ii) a positive correlation between "innate" movements (*sensu* Tertschnig 1989) and latitude ("movement" hypothesis) (Fig. 3). Avoidance behavior is assumed to be better developed in animals regularly exposed to predators, and is consistent with predator–risk theory (Hughes 1988) and the notion that predation is higher during movements and migrations. Both hypotheses predict that diets of tropical urchins will i) be broader and ii) include more low–quality prey (Vance and Schmitt 1979). The movement hypothesis is consistent with the idea of higher search costs in the tropics. Conversely, in boreal regions foraging should be less costly and less constrained (by predators), and hence urchin diets should be narrower and include higher–quality prey.

An alternative to sedentary behavior is "innate" defensive movements (Tertschnig 1989). These are energetically costly but may have survival value. This strategy would be marginal, however, when food was scarce. Sea urchins move slowly; predator avoidance and short–term fleeing movements, effective against some predators (Vadas *et al.* 1986), would be futile against mobile predators. Alternatives for minimizing predation risk include sedentary behavior, combined with sheltering or camouflage (many species carry shell fragments, algae or debris on their aboral surface), burrowing, and temporal movements (crepuscular etc.). Burrowing occurs on softer substrates (dead coral or beach rock in the tropics and limestone or sandstone at higher latitudes) and results from feeding activity and water movement (Lawrence and Sammarco 1982). Regardless of cause or location, burrowing relegates the animal to a sit–and–wait strategy (Vadas, 1977). A corollary to sitting is lower selectivity and a diversified diet. Departures from generalized feeding might occur where territories are defended (Grunbaum *et al.* 1978; Maier and Roe 1983; Tsuchiga and Nishihera 1985).

Predictions from the predator risk and movement hypotheses were analysed by contrasting activities related to avoidance (Table 2). Five of the eight categories (nocturnal movements, agonism, territoriality, burrow dwelling and homing) were more common in tropical urchins. Surprisingly, camouflage and innate movements were evenly distributed. However, Lawrence and Sammarco (1982) have questioned the utility of concealment in urchins. Boreal species were more active

Table 2. Latitudinal comparison of activity and behaviors potentially related to predator avoidance in shallow water sea urchins (expressed as the number of reports and species involved)[a].

Activity / Behavior	Tropical		Boreal–Temperate	
	# Rep.	# Spp.	# Rep.	# Spp.
Daily (24 hrs) Movement[b]	>10	7[e]	>10	10
Nocturnal Movement[b]	>20	10	>7	5
Agonism	7	3	1	1
Territoriality	3	2	1	1[c]
Inhabit Burrows	>20	11	4	2
Homing	5	4	1	1
Camouflage	>10	4	>10	4
Innate Movements	>10	6[d]	>25	7

a) No attempt was made to proportion these by the total number of species in a region. These likely represent minima for several categories but should not greatly affect trends (an extensive review is planned elsewhere, see Jangoux and Lawrence 1982 and Lawrence 1975; 1987 for reviews). Aggregation was not considered due to uncertainty in assessing its nature (see Vadas *et al.* 1986).
b) Not reported in most papers.
c) Questionable record, may only be a feeding range?
d) All are associated with seagrasses.
e) most show pronounced nocturnal activity.

during the day, supporting (?) the notion that visual predators may be less of a threat in these waters. In summary, tropical species are largely nocturnal, whereas boreal species show less specificity for particular periods (the absence of patterns may also reflect inadequate night observations in colder waters). Agonistic behavior was reported only in burrow or crevice dwellers, which suggests that a shelter or food source was being protected. Refuges may provide protection from predators and wave action (Lissner 1980). Although limited, these data are consistent with predictions of the predator risk and movement hypotheses.

Mobility and Preferences

The above patterns (Figs. 2,3) suggest that understanding innate movements may be the key to interpreting foraging behavior. Hence, the contrasting life styles and feedingbehavior of sedentary vs nomadic forms will form the basis of comparisons within–and–between regions. Assuming that sedentary life styles lead to broader diets, then the converse, searching or nomadic behavior, is predicted to lead to increased selectivity and a narrower diet. Directed movements in predators are more efficient than random excursions (Hughes 1980), so enhanced sensory perception might exist in nomadic forms. The globuliferous pedicellaria of some sea urchins are chemosensory and respond to algae (Sloan and Campbell 1982). Some urchins respond to the the organic matter sloughed from the fronds of seaweeds (Vadas 1977; Mann *et al.* 1984) Directed movements also increase the probability of further prey encounters independent of prey preferences (Partridge and Green 1985). Hence, highly mobile urchins should have a greater probability of encountering prey.

My main hypothesis then, is that highly mobile urchins have better–developed chemosensory mechanisms and prey preferences. I regard avoidance of prey as selective behaviour. A corollary is that the diets of urchins should reflect the distances normally searched for food. I therefore focus on the degree of "apparent or innate mobility" and postulate that the degree of preference is positively correlated with the degree of mobility ("mobility preference" hypothesis, Fig. 4). I assume that the energy return to urchins seeking (and capturing) prey is greater than that expended in search, and that directed movements are more efficient than random walks or sitting (Hughes 1980).

Analyses of Mobility and Feeding Behavior

Published and unpublished data on feeding and behavior were used to analyse predictions from the mobility preference hypothesis. Analyses focused on feeding

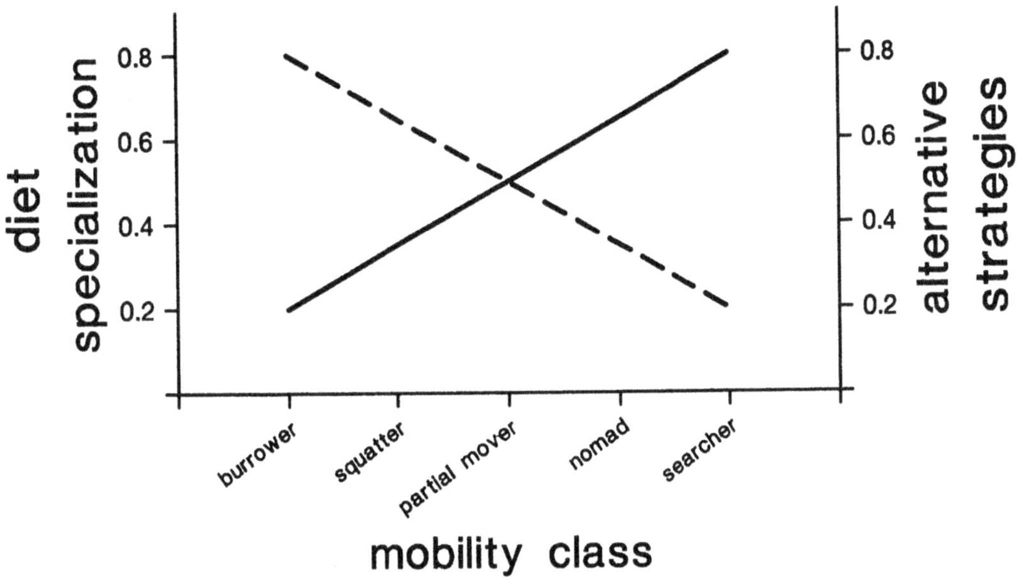

Figure 4. Predicted degree of diet specialization (solid line) and alternative survival strategies (dashed line) as a function of mobility patterns. (The five mobility and preference categories are based on published comments on behaviour, habitats, habits and personal observations: alternative nutrient acquisition and survival strategies are based on the assumption that if "all else is equal", there should be compensatory mechanisms for the limited movements, foraging capabilities and intake of quality prey by urchins in persistently overexploited environments).

studies but were augmented by natural history observations related to movements and foraging. Data on gut analyses, growth and reproduction, and ecological effects are extensive (Lawrence 1975; Jangoux and Lawrence 1982; Lawrence 1987); fewer studies have addressed foraging behavior. It was difficult to analyze the diverse and unstructured (for these purposes) data on urchin feeding. Some report preferences for specific items based on habitat alone. In other cases only a few prey were tested, often less than three. Multispecies comparisons provide the most realistic field simulations, but are problematic for parametric analyses (Peterson *et al.* 1989). Some prey are given as groups, e.g., algae, detritus and various kinds of grit, that may have been ingested passively. Data from each experiment were treated as separate observations e.g., three data points were obtained for *Parechinus angulosus* (Anderson and Velimirov 1982). Evidence that sea urchins select their prey comes primarily from laboratory studies, but is supplemented by field observations (Lawrence and Sammarco 1982; Andrew and Stocker 1986). Numerous gut analyses exist (Lawrence 1975; DeRidder and Lawrence 1982), and are useful for analyses of foraging behavior where prey availability is also reported (Vadas 1985).

To analyse the mobility preference hypothesis, sea urchins were categorized as: burrowers, squatters, partial movers, nomads, or searchers. Burrowers live in holes or crevices (in rock, coral, etc.), and rarely forage more than a few centimeters from the refuge. Squatters reside in shelters (depressions, amongst boulders, holdfasts, etc.) or less commonly in open areas and are relatively immobile. Partial movers occur in kelp or sea grasses, under coral overhangs, on large boulders, or in open areas and exhibit limited movements usually during specific periods (nocturnal, diurnal, etc.). Nomads occupy a variety of habitats, ledges and walls in seaweed covered or open areas, and make regular movements. Searchers travel extensively and occupy several habitats, or are reported as highly mobile. The distinctions in publications and in nature are not as clear as I tried to make them. Also populations in the same region may be characterized differently. For example, *Strongylocentrotus droebachiensis* was usually a nomad or searcher, but at one area in Maine (consisting of two study sites) it exhibited searching behavior at low densities ($<0.5 m^{-2}$) and sedentary behavior at high densities (ca 30–60 m^{-2} Vadas *et al.* in prep). Similarly *S. franciscanis* and *S. purpuratus* have stationary (crevice habitat) and mobile (aggressive) phases (Dayton and Tegner 1984).

Sea urchins were also categorized according to their degree of food preference. Objective assessment was difficult and relied on diet width, degree of preference, strength of rankings and comments by authors. Urchins were placed into one of five preference categories 1) very weak, 2) weak, 3) moderate,4) strong, and 5) very strong. These categories ranged from broad generalists, (very weak) showing little or no evidence of selection to specialists (very strong), exhibiting strong preferences and consistent rankings for a few prey.

Forty data points were obtained for tropical and subtropical urchins (Fig. 5). The food preference rankings were positively correlated with mobility classes (rs= 0.668; P <0.05). Approximately two–thirds of the observations ranged from burrowers to partial movers, and no species were classified as searchers. The shift toward sedentary behavior is consistent with the movement and mobility–preference hypotheses. Burrowers and squatters showed few preferences and little variation. There was, however, considerable variation among urchins exhibiting

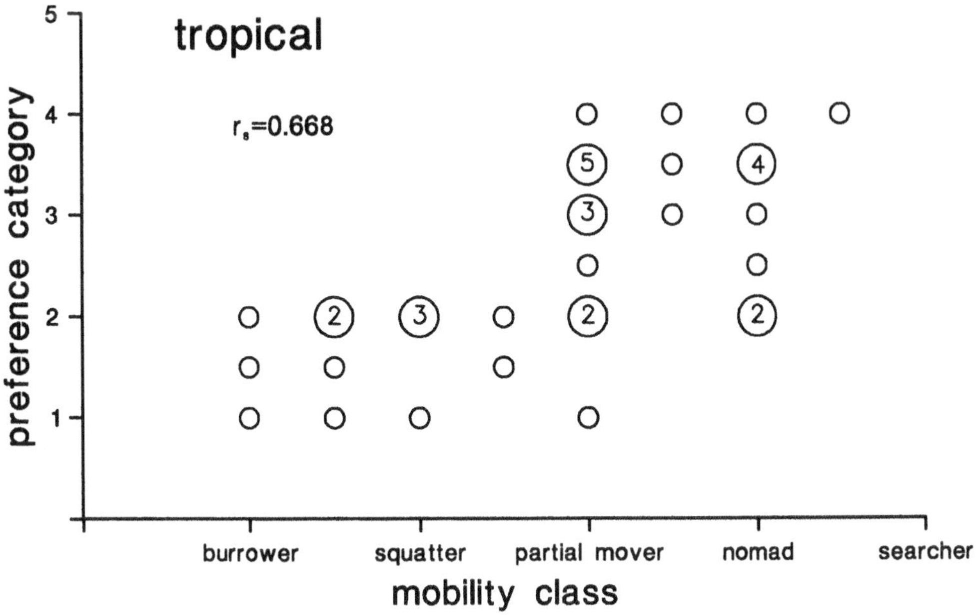

Figure 5. Relationship between mobility classes and preference categories for tropical sea urchins (see double asterisked (**) references).

movements (usually grass bed species), which do not show strong preferences in the field because of the scarcity of edible (non–defended) algae (Ogden and Lobel

1978; Vadas *et al.* 1982). One urchin, *Tripneuestes esculentus*, exhibited strong preference in the laboratory for *Turbinaria*, an uncommon alga in seagrass beds (Ogden *et al.* in prep). It is now clear that avoided species possess chemical or physical defences (Hay and Fenical 1988).

There is a positive correlation between the rank of urchin feeding preferences and mobility in boreal forms (rs=0.599; $P < 0.05$, n= 47; Fig. 6). There is a shift towards increased movement, and no urchins were listed as burrowers. However, some populations of *Strongylocentrotus purpuratus* live in crevices or burrows (Ebert 1968; Maier and Roe 1983), and individuals tested from exposed tide pools were less discriminate than species categorized as partial movers or searchers (Vadas 1977). Animals exhibiting regular movements, generally, had well–developed preferences. These data are consonant with the movement and mobility–preference hypotheses. Given the opportunity most species are selective (Sloan and Campbell 1982), which suggests that selective behavior may be deeply embedded (genetically) for omnivores. However, as these data are few and biased by numerous studies on *Strongylocentrotus* spp., further speculation may be premature.

The changing activity patterns of individual species contributed to the variation in these ranked correlations. For example, Klinger and Lawrence (1985) noted that *Lytechinus variegatus* moves more when algae are abundant, whereas Tertschnig (1989) suggested that movement in *Tripneustes esculentus* was unrelated to food availability. In *Paracentrotus lividus* diurnal movement was interpreted as predator avoidance in Ireland (Ebling 1966). However, in France the same species moved nocturnally while predators were active (Dance 1987). The relationship of movement to feeding is not always clear, as Dance noted. Lastly, my feeding and mobility categories artificially break natural continua. Nonetheless, they appear to have some meaning and heuristic value.

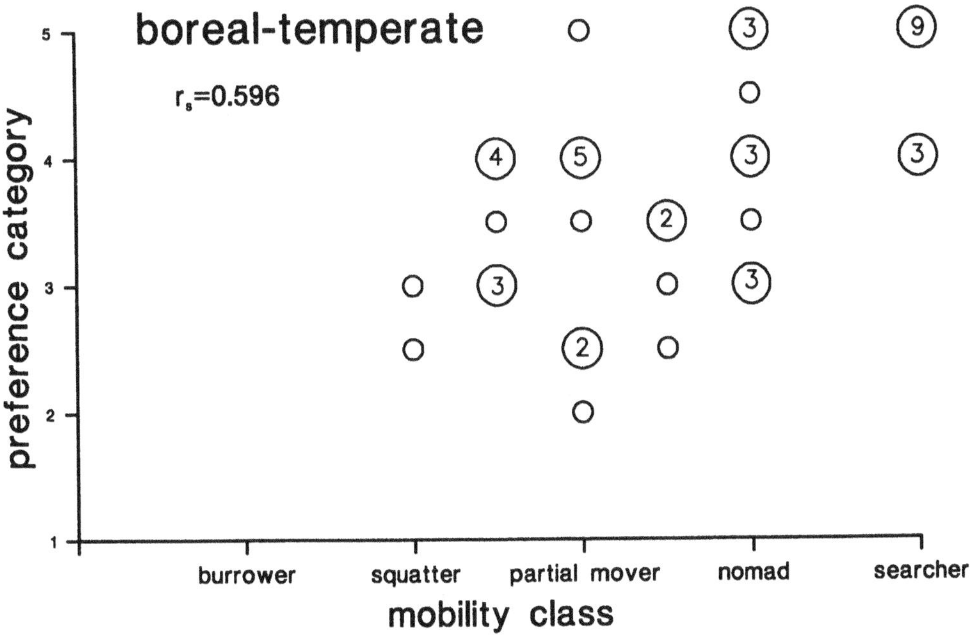

Figure 6. Relationship between mobility classes and preference categories for boreal–temperate sea urchins (see asterisked (*) references).

Alternative Strategies and Nutrient Acquisition

Assuming that urchins in overexploited habitats are at a disadvantage, and that ecological or evolutionary compensation exists at different latitudes, then alternative strategies to promote survival should exist in urchins long exposed to intensely grazed, e.g., tropical, environments. This might take the form of compensatory mechanisms to promote growth or reproduction, or conversely, reduce mortality. Formally the "alternative strategy" hypothesis states that alternative survival or nutrient acquisition systems should be better developed in sedentary and tropical forms, i.e., inversely related to mobility and hence latitude (Fig 4). This hypothesis could be generalized to include species that have evolved with low or unpredictable food supplies.

The data on alternative strategies are given by species and region in Table 3. These data should be interpreted cautiously because of the different orientations and uneven distributions of studies. There are thirteen potential compensatory mechanisms; eight physiological, four foraging and one ecological process. Almost all of the mechanisms are present in sea urchins from both regions. Only sedentary behavior showed a clear increase in the predicted direction. Greater reallocation of resources, especially in feeding mechanisms (Ebert 1980; Black *et al.* 1984) was expected in tropical urchins. Conversely, density–independent responses were more common in boreal species. No information was found on the use of dissolved organic matter (DOM) by tropical forms but it probably occurs, as in other invertebrates in impoverished waters. Also it was difficult to obtain data on faecal pellet coatings in boreal species. I have assumed that the absence of coatings represents an energy saving. However packaged faeces are thought to be an efficient means of passing wastes and controlling water volume (see DeRidder and Lawrence 1982). It is unlikely that DOM and faecal coatings form major energy paths, but if they did, they would cancel each other. Thus, these data alone do not support the predictions of the alternative strategy hypothesis.

DISCUSSION

Movements and Life Styles

Sea urchin mobility patterns and life styles are useful indicators of feeding behavior. Sedentary urchins, especially tropical ones, show little selectivity and tend to be broad generalists. Mobile forms range from generalists to specialists but exhibit stronger preferences. The degree of food selection by mobile urchins appears to be more pronounced than in nomadic terrestrial herbivores. However, in marine systems, large–scale vegetation disturbances are common (see Dayton and Tegner 1984b) and preferred or high quality biomass is frequently transported long distances into "sinks" (Mattison *et al.* 1977; Harrold and Reed 1985). Therefore, at times in boreal environments, sitting may also be a plausible alternative to movement, but for reasons different than in the tropics.

Table 3. Latitudinal Comparison of Alternative Nutrient Acquisition and Survival Strategies in Shallow Water sea urchins (expressed as the number of reports and species involved)[a]

Strategy / Mechanism	Tropical # Rep.	Tropical # Spp.	Boreal–Temperate # Rep.	Boreal–Temperate # Spp.
Foraging				
Sit & Wait/Sedentary	>25	12	>10	5
Shift Diet[b]	>10	6	>10	7
Cannibalism/Predation[c]	1	1	4	1
Reallocate Resources[d]	4	2	3	1
Physiology				
Reduce metabolism	1	1	>5	3
Reabsorb test	2	2	4	3
Reabsorb gonads	2	2	5	3
Alter rate of food passage	2	2	3	3
Uncoated fecal pellets[e]	1	1	–	?
DOM uptake	–	?	5	4
Nitrogen Fixation (gut)	1	3	7	4
Gut flora/fauna	+		+	
Ecology				
Density–independent	1	1	>10	4

a) See footnote a) in Table 2. b) Usually toward encrusting organisms, detritus and sediment. c) On other sea urchin species.d) Refers to functional (teeth) and body morphology. e) *Tripneustes ventricosus* did not produce coated pellets on some algal diets (Ogden *et al.* in prep); most regular urchins except cidaroids, produce encapsulated faeces.

Sedentary behavior may be adaptive, especially in tropical regions or where grazing is intensive. Relative growth rates of *Eucidaris*, a sedentary urchin, are significantly lower than those of *Diadema* and *Tripneustes* (McPherson 1968), categorized here as partial movers or nomads. These differences are inherent and not due to food availability. There is considerable flexibility in the range of behaviors that particular urchins can employ: e.g., *Lytechinus variegatus* was highly mobile when fed *ad libitum* but sedentary when starved (Klinger and

Lawrence 1985). *Diadema antillarum* is nocturnal but in the absence of predation pressure becomes diurnal (Carpenter 1984). The behavior and foraging patterns of *Strongylocentrotus franciscanis* and *S. purpuratus* can range from squatting to searching (Lowry and Pearse 1973; Mattison *et al.*1977; Vadas 1977; Dayton and Tegner 1984a).

Some movements may reflect both foraging and predator avoidance (Vance and Schmitt 1979; Carpenter 1984), whereas others are difficult to explain. Tertschnig (1989) interpreted extensive nocturnal excursions by *Tripneustes vetricosus* among seagrasses as innate defense against a nocturnal predator, *Cassis* spp.. These movements are energetically costly but may promote survival. Tropical urchins inhabiting sea grasses usually move extensively and eat detrital rather than living grass (Ogden1976; Tertschnig 1989). However, when given the opportunity *T.ventricosus* shows moderate to strong preferences (Ogden *et al.* in prep). Dance (1987) suggested that the behavior of *Paracentrotus* in grass beds was a compromise between foraging and conserving energy.

Niche diversification occurs in some tropical urchins: habitat segregation has been reported for sedentary species in several regions (Neil and Larkum 1965; Lewis and Storey 1984; Niell 1988), and there is little habitat overlap between sedentary and mobile forms. There is no evidence for trophic niche segregation in sedentary forms. Partial movers and nomads overlap more extensively and diet segregation may occur in some genera, e.g. *Tripneustes.* This species resides and feeds primarily in grass beds, but also ranges over beach rock and coral (Herring 1972; Ogden and Lobel 1978). *Lytechinus* also occurs in both habitats, but has a much broader diet which includes lower–quality prey (Keller 1983). Niche segregation is not limited to tropical forms (Vasquez *et al.* 1984; Contreras and Castilla 1987), but fewer reports exist for boreal species.

Foraging Strategies and Effects

The foraging strategies common to tropical and boreal urchins may be ancient or the result of convergence. Boreal species prefer certain algal groups, especially kelps and a few red algae. Some tropical urchins (e.g., *Tripneustes ventricosus*, Ogden *et al.* in prep), also prefer brown algae, but in the field this behavior is preempted by availability. Thus, foraging strategies in the tropics may be analogous to strategies employed in intensely exploited boreal environments (e.g., New Zealand, Norway, Nova Scotia), where urchins may be "forced generalists" (Andrew 1989).

Although selective feeding exists in urchins, its role in the field is uncertain (Andrew and Stocker 1986). Urchin foraging produces predictable floristic changes in several boreal regions (Vadas 1968; Mann 1973; Dayton 1975; Himmelman 1980; Himmelman *et al.* 1983). Preferential grazing was also responsible, in part, for the absence of *Saccorhiza polyschides* from Lough Ine (Norton 1978). In the San Juan Islands of Washington (Vadas 1968; unpublished) *S. droebachiensis*, at densities of 8–10 m^{-2} cleared the annually recruited population of *Nereocystis luetkeana* (a preferred alga), although foraging was based on availability (this result paralleled the episodic grazing event on *Nereocystis* in British Columbia, Foreman 1977). At lower densities (3 m^{-2}) urchin grazing eliminated the dominant, preferred kelps and produced a disclimax association of *Agarum* spp. Tropical studies also suggest that selective foraging influences community structure (Ogden 1976; Carpenter 1981; Vadas *et al.* 1982).

Selective feeding on a dominant organism can affect community structure under certain conditions (Lubchenco and Gaines 1982). In New Zealand, *Evechinus chloroticus* strongly prefers the kelp *Ecklonia radiata*, but rarely has the opportunity to eat it (Ayling 1978; Schiel 1982; Andrew and Stocker 1986). Persistently high urchin densities prevent kelp from becoming established and maintain those areas as sponge and coralline assemblages. There are also availability problems with preferred kelps that have long stipes and floating fronds (Lowryand Pearse 1973; Vadas 1977). Here urchins often harvest drift and do not directly influence algal community structure. Intermediate kelp assemblages also exist, depending on drift and urchin densities (Mattison *et al.* 1977; Dean *et al.* 1984).

Selective Feeding: Strategy or Ghost ?

The occurrence of selective foraging by sea urchins in persistently overexploited habitats is not easy to explain. First, intense grazing might be a recent phenomenon resulting from human exploitation of predatory species and release from predation pressure. Hence, there would not have been enough time for selection to have eliminated this behavior. Second, selective feeding may be deeply embedded (genetically) and thus be a relic feature, perhaps from the Mesozoic. The late Palaeozoic appears to have been dominated by large algal forests and has been termed the "Age of Algae" (Steneck 1983). However, intense grazing existed, by modern sea urchins, since the mid–Mesozoic and by fish (scarids etc.), since the early Cenozoic. These herbivores were capable of denuding fleshy macroalgae and damaging calcareous forms (Steneck 1983). It is difficult to envision the adaptive significance of selective feeding under the present, intense grazing pressure of the tropics (Hay *et al.* 1988). Third, patchy or intermittant pulses of algae, analogous to present day reversals in kelp and urchins, could have existed in the past and have periodically reinforced selective behavior. These temporal patches could result from the recent series of glacial–interglacial periods or perhaps from shorter term phenomena, e.g. recent sea urchin die–offs (Miller and Colodey 1983; Lessios *et al.* 1984). The latter may be the most plausible explanation given the Cenozoic radiation of marine herbivores.

Several advantages accrue to sea urchins that feed selectively. In boreal urchins absorption efficiences were significantly higher on preferred items (Fuji 1962; Vadas 1977). Since this is the first step in the conversion of food to tissue, most of which is gonad, there is a direct pay–off for efficient nutrient incorporation. Accordingly high feeding rates generally occur with items readily assimilated (Larson *et al.* 1980). Absorption efficiencies in tropical urchins were lower and varied with food type but were weakly correlated with preferences (Lowe and Lawrence 1976; Ogden *et al.* in prep).

Somatic growth and reproduction (gonad volume) also increase differentially on foods ranked high in the diet. For boreal species correlations between food preference and either growth or reproduction are positive (Fuji 1967; Larson *et al.* 1980; Keats *et al.* 1984; Andrew 1986). This aspect has been poorly studied in tropical urchins, but *Tripneustes ventricosus* grew and reproduced best on the preferred alga (Ogden *et al.* in prep).

Why Selective Foraging is not Universal

Various constraints may limit the ability of animals to forage optimally and maximize their energy intake in the field (Hughes 1988). Selective feeding in sea urchins can be confounded by several variables including: food availabilty (Dix 1970; Vadas 1977; Andrew and Choat 1985), density (Keats *et al.* 1984; Andrew 1989); and, in some environments, predator avoidance (Nelson and Vance 1979; Carpenter 1984). Urchins have an enormous number of options, all of which prolong survival, even in overexploited habitats. These behavioral and physiological mechanisms allow them to survive without food for up to 12 months (North W.J. pers. comm. 1965). Sea urchins can also reduce their oxygen uptake (Johansen and Vadas 1967; Webster 1975), absorb their gonads (Larson et al. 1980), and remain at relatively high densities for years (Johnson and Mann 1982; Andrew 1989). When food is scarce they can shrink their test or alter their feeding apparatus (Ebert 1968, 1980; Black *et al.* 1984; Levitan 1988) and scavenge nutrients. Urchins also utilize amino acids and dissolved organic matter (Pearse and Pearse 1973), and can adopt a "sit and wait" strategy, which, when combined with reduced oxygen uptake, can further reduce metabolic demands (Vadas 1977). Urchins in both regions can obtain nitrogen from nitrogen fixing bacteria in the gut (Guerinot and Patriguin 1981; Odintsov 1981). Physiological flexibility is also suggested as the major reason sea urchins are not limited by density–dependent processes (Andrew 1989). In addition, these strategies reduce the need for, and confound analyses of, selective feeding in the field.

These alternative, energy–saving, strategies are present in boreal and tropical regions, and apparently in all mobility groups. Their ubiquity suggests that intensely grazed environments (presumably responsible for these adaptations) have long existed, and may support Steneck's (1983) arguments concerning the antiquity of excavating herbivores. Morphological, behavioral and physiological flexibility appears to be adaptive (Johnson and Mann 1982) and consistent with the idea that blooms and barrens are temporal phenomena.

Present–day herbivorous echinoids evolved during the Mesozoic and Cenozoic (Vermeij 1978) and likely arose from cidaroids which are sedentary, and ingest detritus and sediment (McPherson 1969). They lack the reinforced teeth of herbivorous species and are incapable of feeding on tough or calcareous algae. Alternative nutrient acquisition features may be primitive or may have evolved secondarily following the demise of the Palaeozoic algal forests, which presumably coincided with the rise of excavating grazers and the radiation of calcareous encrusting algae (Steneck 1983). Regardless of evolution, their parallel occurrence in boreal and tropical species draws into question assumptions about compensatory responses and ecological equivalence across latitudes. "All else is apparently not equal", fast may be better, and urchins capable of selective foraging may still be favored under present environments.

In summary, studies on sea urchin foraging have been fragmentary and lacking in theory. The breadth and specificity of urchin diets transcend the diets of most terrestrial omnivores. Urchin search and mobility patterns may also be different from those of terrestrial forms, where nomads are considered generalists. The ability to feed selectively enhances the opportunities for optimal foraging, but constraints, such as food availability and predation, are common in the field, especially in the tropics. The ease with which urchins shift strategies further confounds our understanding of foraging. As Andrew (1989) suggests, a different perspective may be required for understanding sea urchins and their fluctuations. Similarly, a different approach to foraging behavior may help to explain the dynamics of kelp forests and urchin barrens.

Summary

Latitudinal differences in available algal biomass place constraints on urchin foraging in tropical and selected boreal environments. I argue that tropical urchins have higher search costs, especially in intensely grazed habitats, and have adopted a more sedentary life style than boreal urchins that have predictably higher biomass levels. Reduced metabolic costs and predictable food levels are thought to have encouraged the evolution of search and chemosensory behavior in boreal species, and hence, greater opportunity for selective foraging. Comparison of mobility and foraging behavior in tropical and boreal urchins support the hypothesis of enhanced movement and selective feeding in boreal forms. Conversely, sedentary behavior and broader, low–quality diets are found in tropical species. Urchins in overexploited tropical habitats do not have compensatory mechanisms or alternative strategies (different from boreal species) to deal with food limitation.

ACKNOWLEDGEMENTS

I thank Mike Burrows and Pat Vadas for assistance with the manuscript, and Bob Elner and Chuck Hollingworth for valuable criticism. I also thank John Gray and the University of Oslo, and Roger Hughes and the University College of North Wales for logistic and other support. These efforts were financed by a faculty grant, the Sea Grant Program and the Maine Agricultural Experiment Station of the University of Maine.

BIBLIOGRAPHY

Abbott DP, Ogden JC, Abbott IA (1974) Studies on the activity pattern, behavior, and food of the echinoid *Echinometra lucunter* (Linnaeus) on beachrock and algal reefs at St. Croix, US Virgin Is. West Indies Lab, Spec Publ 4: p. 111 **

Anderson RJ, Velimirov B (1982) An Experimental investigation of the palatability of kelp bed algae to the sea urchin *Parechinus angulosus* Leske. P.S.Z.N.I: Mar Ecol 4:357–373 *

Andrew NL (1986) The interaction between diet and density in influencing reproductive output in the echinoid *Evechinus chloroticus* (Val.). J Exp Mar Biol Ecol 97:63–79

Andrew NL (1989) Contrasting ecological implications of food limitation in sea urchins and herbivorous gastropods. Mar Ecol Prog Ser 51:189–193

Andrew NL, Choat JH (1985) Habitat related differences in the survivorship and growth of juvenile sea urchins. Mar Ecol Prog Ser 27:155–161

Andrew NL, Stocker LJ (1986) Dispersion and phagokinesis in the echinoid *Evechinus chloroticus* (Val.). J Exp Mar Biol Ecol 100:11–23

Ayling AL (1978) The relation of food availability and food preferences to the field diet of an echinoid *Evechinus chloroticus* (Valenciennes). J Exp Mar Biol Ecol 33:223–235 *

Ayling AM (1981) The role of biological disturbance in temperate subtidal encrusting communities. Ecology 62:830–47

Bak HP, Nojima S (1980) Immigration of a tropical sea urchin, *Astropyga radiata* (Leske), in a temperate eelgrass, *Zostera marina* L., patch: Its feeding habit and grazing effect on the patch. Amakusa Mar Biol Lab, Kyushu Univ 5 (2):153–169

Barnard M (1986) Sea, salt and sweat: a story of Nova Scotia and the vast Atlantic Fishery. Nova Scotia Dept of Fish, Halifax, p111

Black R, Codd C, Herbert D, Vink S, Burt J, (1984) The functional significance of the relative size of Aristotle's lantern in the sea urchin *Echinometra mathaei* (de Blainville) J Exp Mar Biol Ecol 77:81–97

Bonsdorf E, Vahl O (1982) Food preference of the sea urchins *Echinus acutus* and *E. esculentus*. Mar Behav Physiol 8:243–248 *

Breen PA, Mann KH. (1976) Destructive grazing of kelp by sea urchins in Eastern Canada. J Fish Res Bd Can 33:1278–1283

Briscoe CS, Sebens KP (1988) Omnivory in *Strongylocentrotus droebachiensis* (Muller) (Echinodermata: Echinoidea): predation on subtidal mussels. J Exp Mar Biol Ecol 115:1–24 *

Camp K, Cobb SP, Van Breedveld JF (1973) Overgrazing of seagrasses by the regular urchin, *Lytechinus variegatus*. Bioscience 23:37–38

Carpenter RC (1981) Grazing by *Diadema antillarum* (Philippi) and its effects on the benthic algal community. J Mar Res 39:749–765**

Carpenter RC (1984) Predator and population density control of homing behavior in the Caribbean echinoid *Diadema antillarum*. Mar Biol 82:101–108

Castilla JC, Moreno CA (1982) Sea urchins and *Macrocystis pyrifera*: experimental test of their ecological relations in southern Chile. In: Lawrence JM (ed) Echinoderms: Proc Intnl Conf Tampa Bay, Balkema, Rotterdam: 257–263

Choat JH (1982) Fish feeding and the structure of benthic communities in temperate waters. Ann Rev Ecol Syst 13:423–449

Choat JH, Schiel DR (1982) Patterns of distribution and abundance of large brown algae and invertebrate herbivores in subtidal regions of northern New Zealand. J Exp Mar Biol 60:129–162

Connor JL, Adey WH (1977) The benthic algal composition, standing crop, and productivity of a Caribbean algal ridge. Atoll Res Bull 211:1–15

Contreras S, Castilla JC (1987) Feeding behavior and morphological adaptations in two sympatric sea urchin species in central Chile. Mar Ecol Prog Ser 38: 217–224

Cowen RK (1983) The effect of sheephead (*Semicossyphus pulcher*) predation on red sea urchin (*Strongylocentrotus franciscanus*) populations: an experimental analysis. Oecologia 58:249–55

Dance C (1987) Patterns of activity of the sea urchin *Paracentrotus lividus* in the Bay of Port–Cros (Var, France, Mediterranean). Mar Ecol 8(2):131–142

Dayton PK (1975) Experimental studies of algal canopy interactions in a sea otter–dominated kelp community at Amchitka Island, Alaska. Fish Bull 73:230–7

Dayton PK, Tegner MJ (1984a) The importance of scale in community ecology: A kelp forest example with terrestrial analogs. In: Price WP, Slobodchikoff CN,G aud WS, (eds). A New Ecology: Novel Approaches to Interactive Systems. J Wiley New York

Dayton PK, Tegner MJ (1984b) Catastrophic storms, El Nino, and patch stability in a Southern California kelp community. Science 224:283–285

Dean TA, Schroeter SC, Dixon JD, (1984) Effects of grazing by twospecies of sea urchins (*Strongylocentrotus franciscanus* and *Lytechinus anamesus*) on recruitment and survival of two species of kelp (*Macrocystis pyrifera* and *Pterygophora californica*). Mar Biol 78:301–313

De Ridder C, Lawrence JM (1982) Food and feeding mechanisms: Echinoidea. In: Echinoderm Nutrition, Jangoux M, Lawrence JM (eds), AA Balkema, Rotterdam

Dix TG (1970) Biology of *Evechinus chloroticus* (Echinoidea: Echinometridae) from different localities: General. N Z J Mar Freshw Res 4:91–116

Dunton KH, Schell DM (1986) Seasonal carbon budget and growth of *Laminaria solidungula* in the Alaskan High Arctic. Mar Ecol Prog Ser 31:57–66

Ebert TA (1968) Growth rates of the sea urchin *Strongylocentrotus purpuratus* related to food availability and spine abrasion. Ecology 49: 1075–1091

Ebert TA (1980) Relative growth of sea urchin jaws: an example of plastic resource allocation. Bull Mar Sci 30:467–474

Ebling FJ, Hawkins AD, Kitching JA, Muntz L, Pratt VM, (1966) The Ecology of Lough Ine. XIV. Predation and diurnal migration in the *Paracentrotus* community. J Animal Ecol 35:559–566

Elner RW, Vadas RL (1990) Inference in ecology: the sea urchin phenomenon in the Northwest Atlantic. Amer Nat (in press)

Estes JA, Smith NS, Palmisano JF (1978) Sea otter predation and community organization in the western Aleutian Islands, Alaska. Ecol 59:822–833

Foreman RE (1977) Benthic community modification and recovery following intensive grazing by *Strongylocentrotus droebachiensis*. Helgolander wiss. Meeresunters 30:468–484

Fuji A (1962) Studies on the biology of the urchin. V. food consumption of *Strongylocentrotus intermedius*. Jap J Ecol 12:181–186 *

Fuji A (1967) Ecological studies on the growth and food consumption of Japanese common littoral sea urchin, *Strongylocentrotus intermedius* (A. Agassiz). Mem Fac Fish 15:83–160

Gaines SD, Lubchenco J (1982) A unified approach to marine plant–herbivore interactions 2. Biogeography. Annu Rev Ecol Syst13:111–38

Garnick E (1978) Behavioral ecology of *Strongylocentrotus droebachiensis* (Muller) (Echinodermata: Echinoidae) aggregating behavior in chemotaxis. Oecologia 37:77–84

Grunbaum H, Bergman G, Abbott DP, Ogden JC (1978) Intraspecific agonistic behavior in the rock–boring sea urchin *Echinometra lucunter* (L.) (Echinodermata: Echinoidea). Bull Mar Sci 28:181–188

Guerinot ML, Patriquin DG (1981) The association of N2–fixing bacteria with sea urchins. Mar Biol 62:197–207

Guida VG (1976) Sponge predation in the oyster reef community as demonstrated with *Clione gelata* Grant. J Exp Mar Biol Ecol 25:109–122

Hagen NT (1987) Sea urchin outbreaks and nematode eqizootics inVestfjorden, Northern Norway. Sarsia 72:213–229

Harrold C, Reed DC (1985) Food availability, sea urchin grazing,and kelp forest community structure. Ecology 66:1160–1169

Hatcher BG, Larkum AWD, (1983) An experimental analysis of factors controlling the standing crop of the epilithic algal community on a coral reef. J Exp Mar Biol Ecol 69:61–84

Hay ME (1984) Patterns of fish and urchin grazing on Caribbean coral reefs: are previous results typical? Ecology 65:446–454

Hay ME, Fenical W, (1988) Marine plant–herbivore interactions: the ecology of chemical defence. Ann Rev Ecol Syst 19:111–145

Hay ME, Fenical W, Gustafson K (1987) Chemical defense against diverse coral–reef herbivores. Ecology 68:1581–1591 **

Hay ME, Paul VJ, Lewis SM, Gustafson K, Tucker J, Trindell RN,(1988) Can tropical seaweeds reduce herbivory by growing at night? Diel patterns of growth, nitrogen content, herbivory, and chemical versus morphological defenses. Oecologia 75:233–245

Herring PJ (1972) Observations on the distribution and feeding habits of some littoral echinoids from Zanzibar. J Nat Hist 6:169–175 **

Himmelman JH (1980) The role of the green sea urchin, *Strongylocentrotus droebachiensis*, in the rocky subtidal region of Newfoundland. In: Pringle JD, Sharp JD, Caddy JF (eds) Can.Tech Rep Fish Aquat, Sci 954:92–119 *

Himmelman JH, Steele DH (1971) Food and predators of the green sea urchin *Strongylocentrotus droebachiensis* in Newfoundland waters. Mar Biol 9:315–22

Himmelman JH, Cardinal A, Bourget E (1983) Community development following removal of urchins, *Strongylocentrotus droebachiensis*, from the rocky subtidal zone of the St. Lawrence Estuary, Eastern Canada. Oecologia 59:27–39

Hughes RN (1980) Optimal foraging theory in the marine context. Oceanogr Mar Biol Ann Rev 18:423–81

Hughes RN (1988) Optimal foraging in the intertidal environment: Evidence and constraints. In (eds) Chelazzi G, Vannini M. Behavioral Adaptation to Intertidal Life, Plenum Press

Jangoux M, Lawrence JM (eds) (1982) Echinoderm Nutrition, AA Balkema, Rotterdam 654p

Johansen K, Vadas RL (1967) Oxygen uptake and responses to respiratory stress in sea urchins. Biol Bull 132: 16–22

Johnson CR, Mann KH (1982) Adaptations of *Strongylocentrotus droebachiensis* for survival on barren grounds in Nova Scotia. In(ed) Lawrence JM, Internl Echinoderm Conf, AA Balkema Rotterdam

Jones NS, Kain JM (1967) Subtidal algal colonization following the removal of *Echinus*. Helgo. Meeres 15:460–466

Kawamura K (1973) Fishery biological studies on a sea urchin, *Strongylocentrotus intermedius* (A. Agassiz). Sci Rep Hokkaido Fish Exp Sta 16:1–54 *

Keats DW, Steele DH, South GR (1984) Depth–dependent reproductive output of the green sea urchin, *Strongylocentrotus droebachiensis* (O.F. Muller), in relation to the nature and availability of food. J Exp Mar Biol 80:77–91 *

Keats DW, Steele DH, South GR (1987) Ocean pout (*Macrozoarces americanus* (Bloch and Schneider) (Pisces: Zoarcidae) predation on green sea urchins (*Strongylocentrotus droebachiensis* (O.F.Mull.) (Echinodermata: Echinoidea) in eastern Newfoundland. Can J Zool 65:1515–1521

Keller BD (1983) Coexistence of sea urchins in seagrass meadows: an experimental analysis of competition and predation. Ecol 64:1581–1598 **

Kirkman H (1984) Standing stock and production of *Ecklonia radiata* (C.Ag.) J. Agardh. J Exp Mar Biol Ecol 76:119–130

Kirkman H, Young PC (1981) Measurement of health, and Echinoderm grazing on *Posidonia oceanica* (L.) Delile. Aquatic Bot 10: 329–338

Klinger TS, Lawrence JM (1985) Distance perception of food and the effect of food quantity on feeding behavior of *Lytechinus variegatus* (Lamarck) (Echinodermata: Echinoidea). Mar Behav Physiol 11:327–344

Larson BR, Vadas RL, Keser M (1980) Feeding and nutritional ecology of the sea urchin *Strongylocentrotus droebachiensis* in Maine, USA. Mar Biol 59:49–62 *

Lawrence JM (1975) On the relationship between marine plants and sea urchins. Annu Rev Oceanogr Mar Biol 13:213–86

Lawrence JM (1987) A Functional Biology of Echinoderms. Croom Helm, London & Sydney, p. 340

Lawrence JM, Sammarco PW (1982) Effects of feeding on the environment Echinoidea. In: Echinoderm Nutrition, Jangoux M, Lawrence JM (eds) AA Balkema, Rotterdam

Leighton DL, (1966) Studies of food preference in algivorous invertebrates of southern California kelp beds. Pac Sci 20:104–113 *

Lessios HA, Robertson DR, Cubit DJ (1984) Spread of *Diadema* mass mortality through the Caribbean. Science NY 226:335–337

Levitan DR (1988) Density–dependent size regulation and negative growth in the sea urchin *Diadema antillarum* Phillipi. Oecologia 76:627–629
Lewis JB, Storey GS (1984) Differences in morphology and life history traits of the echinoid *Echinometra lacunter* from different habitats. Mar Ecol Prog Ser 15:207–211
Lissner AL (1980) Some effects of turbulence on the activity of the sea urchin *Centrostephanus coronatus* Verrill. J Exp Mar Biol Ecol 48:185–93
Littler MM, Taylor PR, Littler DS (1983) Algal resistance to herbivory on a Caribbean barrier reef. Coral Reefs 2:111–18
Lowe EF, Lawrence JM (1976) Absorption efficiences of *Lytechinus variegatus* (Lamarck) (Echinodermata, Echinoidea) for selected marine plants. J Exp Mar Biol Ecol 21:223–34
Lowry LF, Pearse JS (1973) Abalones and sea urchins in an area inhabited by sea otters. Mar Biol 23:213–19
Lubchenco J, Gaines SD (1981) A unified approach to marine plant–herbivore interactions. 1. Populations and communities. Annu Rev Ecol Syst 12:405–37
Maier D, Roe P (1983) Preliminary investigations of burrow defense and interspecific aggression in the sea urchin, *Strongylocentrotus purpuratus*. Pac Sci 37: 145–149
Mann KH (1973) Seaweeds: Their productivity and strategy for growth. Science NY 182:975–981
Mann KH, Wright JLC, Welsford BE, Hatfield E (1984) Responses of the sea urchin *Strongylocentrotus droebachiensis* (O. F. Muller) to water–borne stimuli from potential predators and potential food algae. J Exp Mar Biol Ecol 79:233–244
Mattison JE, Trent JD, Shanks AL, Akin TB, Pearse JS (1977) Movement and feeding activity of red sea urchins (*Strongylocentrotus franciscanus*) adjacent to a kelp forest. Mar Biol 39:25–30
McClanahan TR, Muthiga NA (1989) Patterns of predation on a sea urchin, *Echinometra mathaei* (de Blainville), on Kenyan reefs. J Exp Mar Biol Ecol 126:77–94

McConnell OJ, Hughes PA, Targett NM, Daley J (1982) Effects of secondary metabolites from marine algae on feeding by the sea urchin, *Lytechinus variegatus*. J Chem Ecol 8:1437–1453 +

McFarland D (1981) The Oxford Companion to Animal Behavior. Oxford Univ. Press, Oxford, p 657

McPherson BF (1968) Contributions to the biology of the urchin *Eucidaris tribuloides* (Lamarck) Bull Mar Sci 18:400–443 **

McPherson BF (1969) Studies on the biology of the tropical sea urchins, *Echinometra lucunter* and *Echinometra viridis*. Bull Mar Sci (19):195–213 **

Menge BA (1972) Foraging strategy of a starfish in relation to actual prey availability and environmental predictability. Ecol Monog 42:25–50

Miller RJ, Colodey AG (1983) Widespread mass mortalities of the green sea urchin in Nova Scotia, Canada. Mar Biol 73:263–267

Moore HB, Jutare T, Bauer JC, Jones JA (1963) The biology of *Lytechinus variegatus*. Bull Mar Sci Gulf Carib 13:23–53 **

Moore HB. McPherson BF (1965) A contribution to the study of the productivity of the urchins *Tripneustes esculentus* and *Lytechinus variegatus*. Bull Mar Sci Gulf Carib 15(4):856–871 **

Nedelec H, Verlaque (1984) Alimentation de L'oursin *Paracentrotus lividus* (Lamarck) Dans un herbier a *Posidonia oceanica* (L) Delile en Corse (Mediterranee–France). In: Intern workshop on *Posidonia oceanica* beds, Boudouresque C, DeGrissac, AJ, Olivier J (eds) GIS Posidonie, Fr:350–363 *

Neil SR St J, Larkum H (1965) Ecology of some echinoderms in Maltese waters Symp Underwat Assoc Malta p51–55

Neill JB (1988) Experimental analysis of burrow defense in *Echinometra mathaei* (de Blainville) on Indo–West Pacific reef flat. J Exp Mar Biol Ecol 115: 127–136

Nelson BV, Vance RR (1979) Diel Foraging Patterns of the sea urchin *Centrostephanus coronatus* as a predator avoidance strategy. Mar Biol 51:251–258

Norton TA (1978) The factors influencing the distribution of *Saccorhiza polyschides* in the region of Lough Ine. J mar biol Ass UK 58:527–536 *

Odintsov VS (1981) Nitrogen fixation (acetylene reduction) in the digestive tract of some echinoderms from Vostok Bay in the Sea of Japan Mar Biol Lett 2:259–263
Ogden JC (1976) Some aspects of herbivore–plant relationships on Caribbean reefs and seagrass beds. Aquatic Bot 2:103–116 **
Ogden JC, Abbott DP, Abbott I (1973a) Studies on the activity and food of the echinoid *Diadema antillarum* philippi on a West Indian patch reef. West Indies Lab, Spec Pub 2: p 96 **
Ogden J, Brown RA, Salesky N (1973b) Grazing by the echinoid*Diadema antillarum* Philippi: Formation of halos around West Indian Patch Reefs. Science NY 182:715–717 **
Ogden J, Lobel PS (1978) The role of herbivorous fishes and urchins in coral reef communities. Env Biol Fish 3:49–63 **
Ott J, Mauer L (1977) Strategies of energy transfer from marine macrophytes to consumer levels: the *Posidonia oceanica* example. In: Keegan PO, Ceidigh PO, Boaden PJS (eds) Biology of Benthic Organisms, 11th European Symp Mar Biol, Galway
Paine RT (1966) Food web complexity and species diversity. Amer Nat 100:65–75
Paine RT, Vadas RL (1969) The effects of grazing by sea urchins, *Strongylocentrotus* spp., on benthic algal populations. Limnol Oceanogr 14:710–19
Patridge L, Green P (1985) II. Intraspecific feeding specializations and population dynamics. In: Behavioural Ecology, Sibly RM, Smith RH (eds), Blackwell Sci Pub, Oxford:207–226
Paul O, Verlaque M, Boudouresque C (1984) Etude du contenu digestif de l'oursin regulier *Psammechinus microtuberculatus* Dans l'herbier a *Posidonia oceanica* de la baie De Port–cros (Var–France). In: Intern workshop *Posidonia oceanica* beds. Boudouresque CF, de Grissac A, Olivier J (eds), GIS Posidonie, Fr.:365–371 *
Paul VJ, Hay ME (1986) Seaweed susceptibility to herbivory: chemical and morphological correlates. Mar Ecol Prog Ser 33:255–264

Pearse JS, Hines AH (1979) Expension of a central California kelp forest following the mass mortality of sea urchins. Mar Biol 51:83–91
Pearse JS, Pearse VB (1973) Removal of glycine from solution by the sea urchin *Strongylocentrotus purpuratus*. Mar Biol 19: 281–284
Peterson CH, Renaud PE (1989) Analysis of feeding preference experiments. Oecologia 80:82–86
Randall JE (1965) Grazing effects on seagrasses by herbivorous reef fishes in the West Indies. Ecology 46:255–260
Randall JE, (1967) Food habits of reef fishes of the West Indies. Stud Trop Oceanogr 5:665–847
Rivera JA (1978) Aspects of the biology of *Lytechinus variegatus* (Larmarck,1816) at Jobos Bay, Puerto Rico (Echinoidea: toxopneustidae). MSc thesis Univ Puerto Rico, Mayaquez 192p **
Rogers CS, Salesky NH (1981) Productivity of *Acropora palmata* (Lamarck), macroscopic algae, and algal turf from Tague Bay reef, St. Croix, US Virgin Islands. J Exp Mar Biol Ecol 49:179–187
Sammarco PW (1980) *Diadema* and its relationship to coral spat mortality: grazing, competition, and biological disturbance. J Exp Mar Biol Ecol 45:245–72
Sammarco PW, Levinton JS, Ogden JC (1974) Grazing and control of coral reef community structure by *Diadema antillarum* Philippi (Econinodermata, Echinoidea): a preliminary study. J Mar Res 32:47–53
Schiel DR (1982) Selective feeding by the echinoid, *Evechinus chloroticus*, and the removal of plants from subtidal algal stands in Northern New Zealand. Oecologia 54:379–388 *
Sloan NA, Campbell AC (1982) Perception of food. In: Echinoderm Nutrition, Jangoux M, Lawrence JM, AA Balkema, Rotterdam
Steneck RS (1983) Escalating herbivory and resulting adaptive trends in calcareous algal crusts. Paleobiology 9:44–61

Tegner MJ, Dayton PK (1981) Population structure, recruitment and mortality of two sea urchins (*Strongylocentrotus franciscanus* and *S. purpuratus*) in a kelp forest near San Diego, California. Mar Ecol Prog Ser 5:255–268

Tertschnig WP (1989) Diel activity patterns and foraging dynamics of the sea urchin *Tripneustes ventricosus* in a tropical seagrass community and a reef environment (Virgin Islands). Mar Ecol 10(1):3–21

Traer K (1979) The consumption of *Posidonia oceanica* Delile by Echinoids at the Isle of Ischia. Proc European Colloq Echinoderm Brussels p 241–244 **

Tsuchiga M, Nishihera M (1985) Agonistic behavior and its effect on the dispersion pattern in two types of the sea urchin, *Echinometra mathaei* (Blainville). Galaxea 4:37–48

Vadas RL (1968) The ecology of *Agarum* and the kelp bed community. Ph.D thesis Univ of Washington, Seattle *

Vadas RL (1977) Preferential feeding: an optimization strategy in sea urchins. Ecol Monogr 47:337–71 *

Vadas RL (1985) Herbivory. In: Handbook of phycological methods ecological field methods: Macroalgae, Littler MM, Littler DS, Cambridge Univ Press: 531–572

Vadas RL, Fenchel T, Ogden JC (1982) Ecological studies on the sea urchin, *Lytechinus variegatus*, and the algal–seagrass communities of the Miskito Cays, Nicaragua. Aquat Bot 14:109–25**

Vadas RL, Elner RW, Garwood PE, Babb IG (1986) Experimental evaluation of aggregation behavior in the sea urchin *Strongylocentrotus droebachiensis*. Mar Biol 90:433–448

Vance RR (1979) Effects of grazing by the sea urchin, *Centrostephanus coronatus* on prey community composition. Ecology 60:537–46

Vance RR, Schmitt RJ (1979) The effect of the predator–avoidance behavior of the sea urchin (*Centrostephanus coronatus*) on the breadth of its diet. Oecologia 44:21–25

Vasquez JA, Castilla JC, Santelices B (1984) Distributional patterns and diets of four species of sea urchins in giant kelp forest (*Macrocystis pyrifera*) of Puerto Toro, Navarine Island,Chile. Mar Ecol Prog Ser 19:55–63 *

Verlaque M (1984) Biologie des juveniles de l'oursin herbivore *Paracentrotus lividus* (Lamarck): selectivite du broutage et impact de l'espece sur les communautes algales de substrat rocheux en Corse (Mediterranee, France). Bot Mar XXVII:401–24 *

Vermeij GJ 1978) Biogeography and Adaptation: Patterns of Marine Life, Harvard Univ Press, Cambridge, Mass

Webster SK (1975) Oxygen consumption of the purple sea urchin with special reference to the reproductive cycle. Biol Bull 148:157–164

Witman JD (1985) Refuges, biological disturbance, and rocky subtidal community structure in New England. Ecol Monogr 55:421–445

Zieman JC, Wetzel RG (1980) Productivity in seagrasses: methods and rates. In Phillips RC, McRoy CP (eds) Handbook of Seagrass Biology: An Ecosystem Perspective, Garland STPM Press, NY 87–116

ORAL DISCUSSION – R.L. VADAS

Q (DeMott) It is often found that where selectivity is low, there are very few grazers, and so you have to take into account grazer density.

A. I agree. Where urchin densities are high, there seems to be little selectivity. We have noticed this in Maine, for example.

Q (Goss) How do urchins react to each other ?

A. Sedentary burrowers show agonistic behavior and in some cases bite the spines of the intruding animals. They also appear to "fence" with their spines. In other urchins there are considerable spine and podial movements initially but no apparent aggressive movements.

Q (Caldwell) What have we learned from the epidemic destruction of *Diadema* in the Caribbean ?

A. That sea urchins have a major impact not only on algal communities but also on the structure of coral reefs.

ORAL DISCUSSION – R.L. Vadas

Q. (DeMott) It is often found that where selectivity is low, there are very few grazers, and so you have to take into account grazer density.
A. I agree. Where urchin densities are high, there seems to be little selectivity.

Q. (Goss) How do the urchins react to each other ?
A. Sedentary burrowers show agonistic behavior and actually fence with their spines. In other urchins there are considerable spine and podial movements initially but no apparent aggressive movements.

Q. (Cladwell) What have we learned from the epidemic destruction of *Diadema* in the Caribbean ?
A. That sea urchins have a major impact not only on algal communities but also on the structure of the coral reefs.

EFFECTS OF FOOD VALUE OF ARTIFICIAL AND NATURAL SEDIMENTS ON FUNCTIONAL RESPONSE AND NET RATE OF ENERGY GAIN BY A DEPOSIT–FEEDING POLYCHAETE

Gary L. Taghon, Ron R. Greene & David Bard, Oregon State University

Introduction

Deposit feeders are a ubiquitous trophic group in aquatic sediments. While many of the details of the nutrition of deposit feeders remain under debate (see Lopez *et al.* 1989 for a recent review), it is clear that some mixture of organic matter produced elsewhere (e.g. sedimenting planktonic debris), as well as organic matter produced *in situ* (e.g. bacteria, benthic diatoms) serves as food. Deposit feeders have high feeding rates, typically ingesting 10– 100 times their own mass of sediments each day (Lopez & Levington 1987, Taghon 1988). Such prodigious feeding rates are attributed, not to unusually high energy demands, but to the low concentrations of nutritionally useful organic matter in sediments.

An alternative way of expressing feeding rates is in terms of encounter rates with potential food items, or decision frequencies. Encounter rates for several species of deposit feeders, based on the physical characteristics of sediment particles and typical feeding rates, are given in Table 1. These are minimum encounter rates since they are based on the amount of sediment actually ingested; any particles that were rejected after being encountered are not included. Following a basic tenet of optimal foraging theory, that an animal makes a decision whether to accept or reject an encountered food item, these high encounter rates translate to high decision frequences. *A priori*, we might expect that the evolution of behavioral or neural mechanisms capable of dealing with such high decision frequencies would be unlikely, especially when the food source is relatively dilute. Indeed, experimental data indicate that, while general trends exist for selection of sediment particles on the basis of characteristics such as size, surface texture, specific gravity, and the presence of organic coatings (reviewed in Taghon 1989), selection by many deposit feeders can be explained more parsimoniously by mechanical processes (Jumars *et al.* 1982, Taghon 1982). For example, small particles, particles of low specific gravity, and particles with an organic coating will adhere more readily to mucus–coated feeding appendages and thus will tend to be ingested disproportionately.

Alteration of feeding rate is a more likely candidate for behavioral control of energy acquisition in these animals. Changes in feeding rate in response to the quantity and quality

NATO ASI Series, Vol. G 20
Behavioural Mechanisms of Food Selection
Edited by R. N. Hughes

Table 1. Estimates of encounter rates (Hz) with potential food items for some deposit feeders, based on feeding rates (by mass or volume) and physical properties of particles. C = crustacean, P = polychaete.

Species	Particle Size	Encounter Rate	Reference
Abarenicola pacifica (P)	125 μm	22	Taghon, unpublished data
Amphicteis scaphobranchiata (P)	28	200	Taghon and Jumars, 1984
Corophium spp. (C)	9	23	Miller, 1984
Hobsonia florida (P)	28	3.6	Taghon and Jumars, 1984
Pseudopolydora japonica (P)	53	3.2	Taghon and Jumars, 1984

of organic matter in sediments are known to occur over short time intervals. For example, some deposit feeders will initially eat particles devoid of food value, but within minutes feeding rate slows and eventually stops. Furthermore, individual animals will change feeding rates over a few hours when presented with sediments of different organic content (e.g., Taghon and Jumars 1984). Such observations indicate that behavioral control of feeding, in the sense of short-term response to variations in food resources, is often manifested through changes in feeding rate. Consequently, modeling efforts have focused on the functional response (changes in feeding rate) of deposit feeders to food value of sediments.

Models to predict how feeding rate should change in response to variable food content of sediments fall into two classes. Those based on optimal foraging theory (OFT) assume an animal's behavior maximizes net rate of energy gain (Taghon 1981). Alternatively, models based on the principle of homeostasis assume feeding behavior aims to maintain a relatively constant net rate of energy gain despite fluctuations in food supply or food value (Calow 1982). A major difference between these alternatives is that the OFT-based model predicts a positive relation between feeding rate and food value, while the homeostatic model predicts an inverse relation. The fact that both of these relations have been observed does not allow a choice between the models. In an important paper, Phillips (1984) merged these two approaches. He showed that a more physiologically realistic OFT-based model could accommodate both of the observed relations. Specifically, when an upper limit on absorption rate or net rate of gain (i.e., growth rate) is imposed in the OFT model, feeding rate should reach a maximum at some intermediate food value and decline at both lower and higher food values. We set out to test this prediction.

Experimental approach

Our experimental animal was *Abarenicola pacifica*, a subsurface deposit-feeding polychaete. Our general approach was to measure feeding and growth rates of juvenile animals offered sediments of different food value. We used two methods of preparing these sediments. In one, we made artificial sediments by mixing clean, well-sorted quartz sand (passing a 125-µm mesh) with known amounts of a commercial cereal product (Gerber's Mixed Cereal® for babies). The other approach was to use natural sediments which had been diluted to varying degrees with the clean quartz sand to produce a series of experimental sediments with progressively lower concentrations of organic matter. Sediment processing rates of individual worms fed the experimental sediments were measured by collecting, drying and weighing fecal castings. Relative growth rate was chosen as a measure of net rate of gain, and was computed as the percentage change in ash-free dry body weight per day based on estimates of the initial and final ash-free dry weights of worms (see legend of Table 2):

$$\text{Relative growth rate} = \frac{[\ln(\text{Final AFDW}) - \ln(\text{Initial AFDW})]}{t_{final} - t_{initial}} \times 100$$

Details on food value of the sediments and other experimental conditions are given in Table 2.

Because our methods involved using artificial sediments or altered natural sediments, we conducted two preliminary experiments to determine whether particle selectivity was occurring and, if so, whether selectivity patterns were affected by the experimental treatments. Treatment-dependent changes in particle selectivity would invalidate our simple method of weighing fecal castings to measure feeding rates. To assess particle size selection

Table 2. Summary of experimental conditions. Means of various parameters based on **n** replicates, S.D. = standard deviation, C.V. = coefficient of variation. For experiments with artificial sediments, protein content of sediments is calculated value, based on amount of cereal added to clean sand and the manufacturer's data on proximate composition of the cereal. In this series of experiments, initial ash-free dry weight of animals was estimated from a regression equation of AFDW on wet weight developed from a sample of 118 worms (r^2 = 0.986). For experiments with natural sediments, protein content was measured using the procedure of Mayer et al. (1986). In these experiments, initial AFDW of animals was estimated from a regression of AFDW on body volume, as determined with an image analysis system, for a sample of 62 worms ($r^2 = 0.966$).

Experiment	Date animals collected	Duration of experiment	Temperature			Initial Ash-Free Dry Weight (g)			mg Protein (g Sediment)$^{-1}$		
			Mean (n)	S.D.	C.V.	Mean (n)	S.D.	C.V.	Mean (n)	S.D.	C.V.
Experiments with artificial sediments											
1A	1 August 1984	48 d	15.0 (47)	0.9	6%	0.0155 (15)	0.0026	17%	0.0051		
1B	"	"	"	"	"	0.0182 (15)	0.0038	21%	0.026		
2A	8 July 1985	22 d	15.8 (62)	1.0	6%	0.0105 (10)	0.0032	30%	0.051		
2B	"	"	"	"	"	0.0086 (10)	0.0021	24%	0.10		
2C	"	"	"	"	"	0.0103 (10)	0.0046	45%	0.15		
Experiments with natural sediments											
3A	6 July 1988	18 d	13.5 (25)	0.9	7%	0.0093 (18)	0.0045	48%	0.038 (6)	0.022	57%
3B	"	"	"	"	"	0.0110 (17)	0.0066	60%	0.086 (6)	0.040	46%
3C	"	"	"	"	"	0.0140 (17)	0.0097	69%	0.145 (6)	0.021	14%
3D	"	"	"	"	"	0.0120 (17)	0.0075	62%	0.635 (3)	0.068	11%
4A	4 August 1988	18 d	12.2 (21)	0.6	5%	0.0228 (20)	0.0067	29%	0.014 (12)	0.002	16%
4B	"	"	"	"	"	0.0218 (20)	0.0059	27%	0.021 (11)	0.004	20%
4C	"	"	"	"	"	0.0225 (20)	0.0060	27%	0.064 (12)	0.021	33%
4D	"	"	"	"	"	0.0190 (19)	0.0070	37%	0.140 (12)	0.052	37%
4E	"	"	"	"	"	0.0189 (19)	0.0047	25%	0.536 (12)	0.245	46%
4F	"	"	"	"	"	0.0168 (15)	0.0029	17%	0.676 (12)	0.299	44%

over the size range of sediments we used, a small amount of clean glass microbeads was mixed with natural sediments and fed to worms. The size-frequency distribution of beads in the sediment and fecal castings is shown in Figure 1. A two-sample Kolmogorov-Smirnov test showed no significant difference. This result agrees with Self and Jumars (1988), who also found that *A. pacifica* shows no strong selection over this size range of particles. These results suggest that adding clean particles in this size range is an effective method for changing the food concentration of natural sediments.

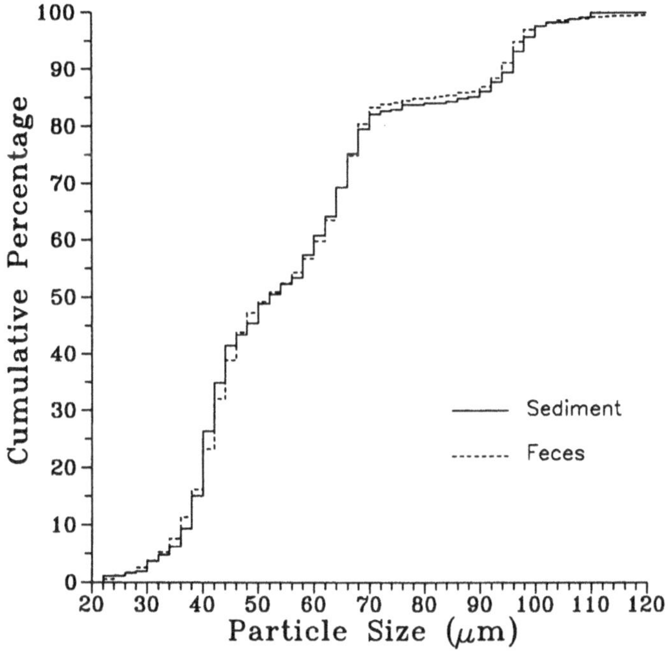

Figure 1. The cumulative percentages of glass microbeads, in 2-μm size intervals, in sediments (solid line, n = 3) and fecal castings from *A. pacifica* (dashed line, n = 12). The sizes of microbeads in sediment or fecal samples dispersed on microscope slides were measured at 400x using a compound microscope interfaced with a Zeiss Videoplan® image analysis system.

The feces of deposit feeders commonly have a higher organic content than the "bulk" sediment they feed on, reflecting selection for some organically enriched subset of the total sediment (e.g., Hylleberg (1975) reported increases in total organic matter and protein in the feces of *A. pacifica*). To determine whether adding variable amounts of organic-free sand to natural sediments altered selectivity on the basis of particle organic content, we fed worms natural sediments and natural sediments diluted with 75% (on a volumetric basis) clean sand. The results (Table 3) show that feces were enriched on the basis of both total organic matter and enzymatically available protein, but the enrichment factors were not substantially different between the two sediment types; disregarding digestion, the same proportional amount of organic matter was ingested despite changes in the absolute amount of organic matter in sediments. Thus, selection for organically enriched particles was not markedly changed by adding large amounts of organic-free particles.

Based on these pre-tests, we concluded that our techniques for altering organic contents of sediment did not lead to significant alterations in particle selectivity by *A. pacifica*.

Table 3. Total organic matter and enzymatically available protein (both in mg g^{-1}) in two experimental sediments and fecal castings of *A. pacifica* feeding on those sediments. Total organic matter was measured as weight loss of sediments after ignition at 500° C for 12 h. Enzymatically available protein content was measured using the procedure of Mayer et al. (1986). Feces enrichment factor calculated as value of total organic matter or protein in fecal castings divided by value in sediments fed to worms.

	Total Organic Matter		Protein	
	Mean (n)	S.D.	Mean (n)	S.D.
100% natural sediments				
Pre-ingested	9.2 (8)	0.8	0.153 (11)	0.046
Fecal castings	15.3 (8)	1.2	0.250 (12)	0.046
Feces enrichment	1.66		1.63	
25% natural sediments: 75% organic-free sand				
Pre-ingested	3.4 (2)	0.3	0.025 (4)	0.004
Fecal castings	5.6 (2)	0.8	0.033 (4)	0.006
Feces enrichment	1.65		1.32	

Feeding and growth rates on artificial sediments

We conducted two series of experiments using clean sand mixed with varying amounts of cereal (Table 2). Organic contents of the sediments are expressed in terms of protein level because protein is generally thought to limit the growth of detritivores (Tenore and Rice 1980). Daily sediment processing rates for Experiments 1 and 2 are plotted in Figures 2 and 3, respectively. Because the size range of animals used in any one treatment within these experiments was fairly narrow (Table 2), sediment processing rates were not normalized to animal size (such normalization did not, in any event, reduce variation in feeding rates).

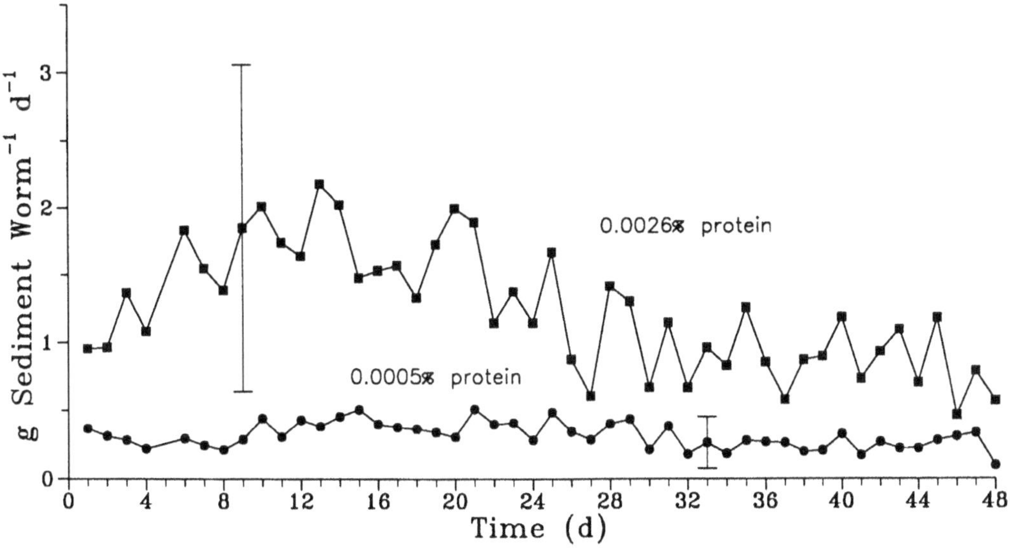

Figure 2. Mean sediment processing rates (± 1 standard deviation) over time for *A. pacifica* in Experiment 1. For clarity, standard deviations are only plotted for day 33 for Experiment 1A and day 9 for Experiment 1B. These are typical of the average coefficients of variation for Experiment 1A (74%) and 1B (65%).

Sediment processing rates, averaged over the duration of these experiments, are plotted against protein content of the sediments in Figure 4A. A Kruskal-Wallis nonparametric analysis of variance revealed significant differences in sediment processing rate and growth rate among the five levels of sediment protein content (Table 4). Sediment processing rate was greatest for sediment with ≈0.05 mg protein (g sediment)$^{-1}$, decreasing

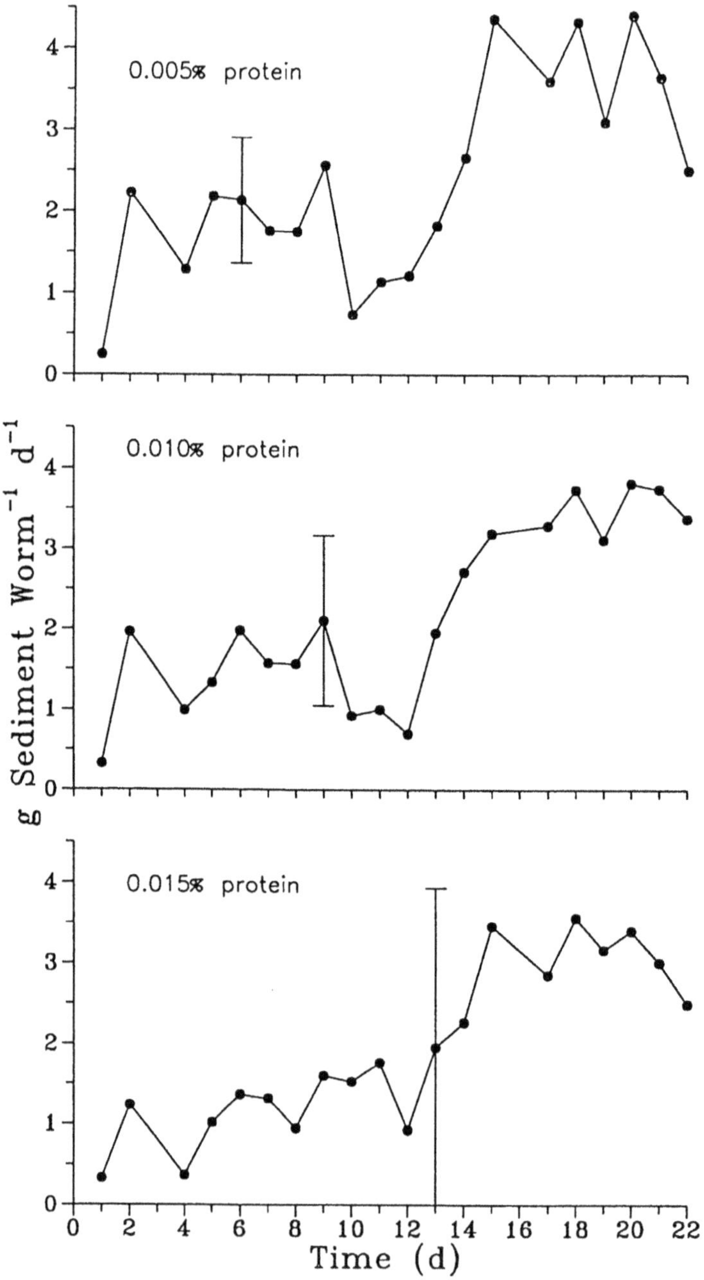

Figure 3. Mean sediment processing rates (± 1 standard deviation) over time for *A. pacifica* in Experiment 2. For clarity, standard deviations are only plotted for day 6 for Experiment 2A, day 9 for Experiment 2B, and day 13 for Experiment 2C. These are typical of the average coefficients of variation for Experiment 2A (36%), 2B (49%), and 2C (91%).

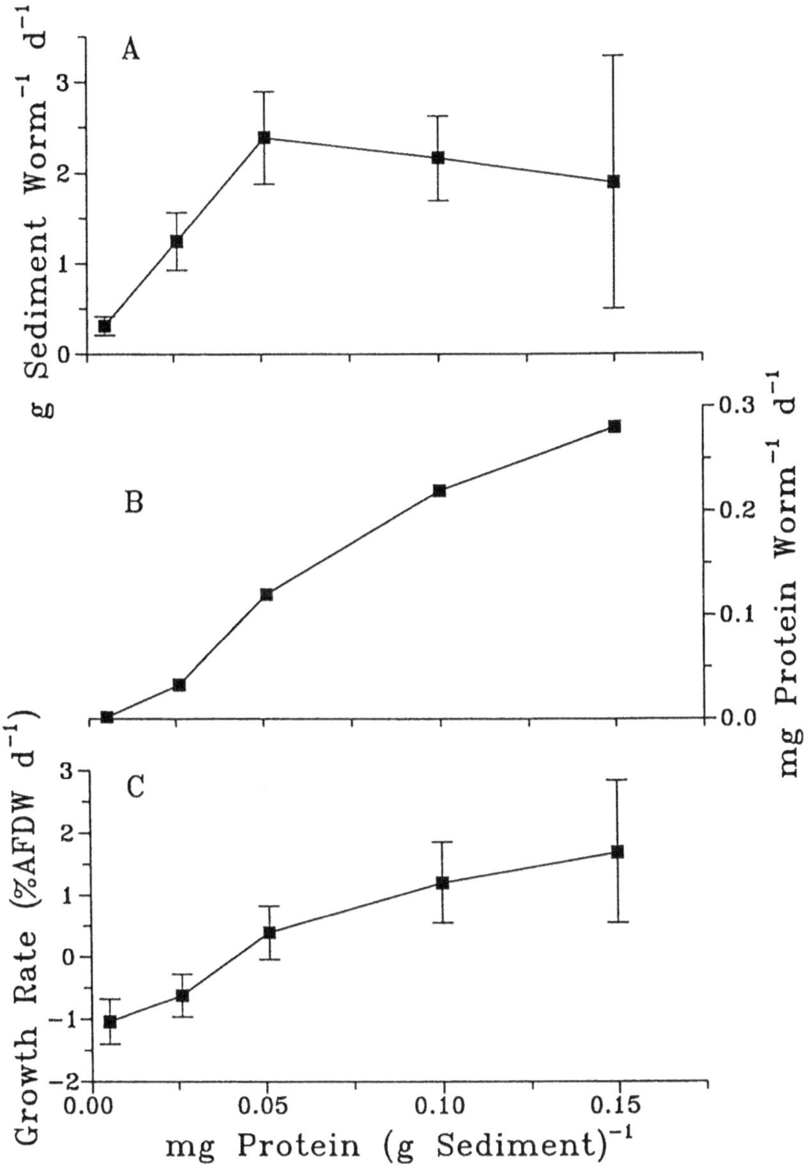

Figure 4. Combined results from Experiments 1 and 2. (A) Sediment processing rate (mean ± 1 standard deviation) versus protein content of artificial sediments. (B) Protein ingestion rates, calculated on basis of means in (A) and protein concentration in sediment. (C) Relative growth rate (mean with standard deviation) versus sedimentary protein content. Growth rates expressed as daily percentage change in ash-free dry weight.

significantly on sediments with lower protein contents. Sediment processing rate also decreased as protein content increased, but the decrease was not significant using the conservative nonparametric multiple comparison test described by Hettmansperger (1984).

Figure 4B shows protein ingestion rates, calculated as the product of sediment processing rate and protein concentration of the sediment. Although sediment processing rate peaks at an intermediate value of sediment protein content (Fig. 4A), the ingestion rate of protein continues to increase over the range of sedimentary protein levels used in these experiments. Growth rates show the same pattern (Fig. 4C), with significantly greater growth rates occurring in worms fed the two sediments highest in protein (Table 4).

Feeding and growth rates on natural sediments

By diluting natural sediments with clean sand, we obtained sediments with a wider range of protein content than in the experiments with artificial sediments. Nonetheless, sediment consumption rate showed a similar pattern of variation with protein content (Fig. 5A), peaking at 0.05-0.1 mg protein (g sediment)$^{-1}$. The decline in sediment consumption rate at higher protein levels was much more evident in the experiments with natural sediments.

Protein ingestion rate (Fig. 5B) and growth rate (Fig. 5C) increased as the protein content of the sediment increased (see Table 4 for significance testing). These patterns are similar to those seen for worms fed on artificial sediments (Fig. 4).

Discussion

The consumption rate of sediment by juvenile *A. pacifica* varied non-monotonically with protein content of sediment, our quantitative measure of food value. Sediment consumption rates peaked at a food value of 0.05-0.1 mg protein (g sediment)$^{-1}$ in all experiments, whether the protein originated from a well-defined, extremely labile source (Experiments 1 and 2) or from the more complex mixture of detrital and living sources

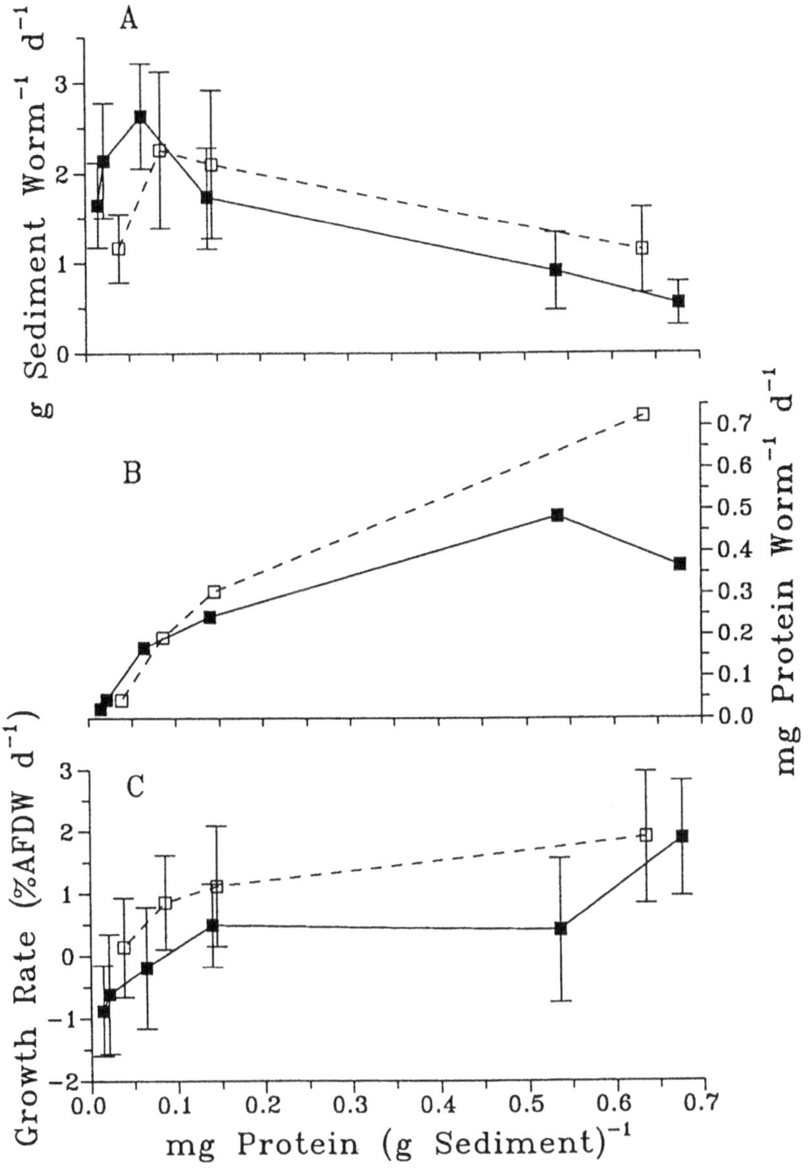

Figure 5. As in Figure 4, but for worms feeding on natural sediments diluted with organic-free sand. Open squares and dashed lines are data from Experiment 3, solid squares and continuous lines are data from Experiment 4.

Table 4. Results of Kruskal-Wallis nonparametric analyses of variance on sediment processing rates and growth rates. All K-W statistics highly significant ($p < 0.0005$). Nonparametric multiple comparison tests performed according to Hettmansperger (1984) with individual α levels corrected to account for multiple testing. Ordering of treatments (refer to Table 2 for experimental conditions) reflects ranking of median values for the variable, from lowest to highest. Treatments sharing a common underline are not significantly different at the indicated α level. ** = significant at overall $\alpha = 0.05$. * = significant at $0.05 <$ overall $\alpha < 0.10$.

Variable	Kruskal-Wallis Statistic	Multiple Comparison Test
Artificial sediments - Experimental series 1 and 2		
Sediment processing rate	43.5	1A 1B 2C 2B 2A **
Growth rate	48.5	1A 1B 2A 2B 2C **
Natural sediments		
Experimental series 3		
Sediment processing rate	18.0	3D 3A 3C 3B **
Growth rate	20.4	3A 3B 3C 3D *
Experimental series 4		
Sediment processing rate	57.9	4F 4E 4A 4D 4B 4C *
Growth rate	49.0	4A 4B 4C 4D 4E 4F *

present in aquatic sediments (Experiments 3 and 4). These results indicate that the method of Mayer et al. (1986) is indeed measuring a fraction of the total organic matter present in natural sediments which is more relevant to the nutrition of deposit feeders than usual bulk measures of organic carbon.

Clearly, worms alter their feeding rate in response to food value of sediments, but how does the pattern of variation compare with the predictions of the OFT-based and homeostatic models? On the basis of observed patterns in sediment processing rate with variations in food concentration, both models can be rejected because neither predicts an intermediate peak in sediment processing rate. Following the prediction of Phillips' (1984) melding of these two models, feeding rate on sediments first increases, peaks, and then declines as food value increases. The decline, however, is not steep enough to maintain a constant rate of protein intake, as the homeostatic part of the combined model predicts. Indeed, maintaining such a constant rate of gross gain would be "suboptimal," since worms grow little if at all at the sedimentary protein level where sediment consumption rate is maximal (Figs. 4 and 5). Because growth rate continues to increase with food value, it is clear that juvenile *A. pacifica* have not reached their physiologically maximum net rate of gain within the range of food values used in these experiments. This result is particularly interesting since the sedimentary protein levels in these laboratory experiments were equal to or less than field levels.

While Phillips' (1984) model predicts that net rate of gain should remain constant as sediment processing rate declines at higher food concentrations, the data in Figures 4 and 5 indicate that growth rate is continuing to increase. Thus, we tentatively reject Phillips' model. Our data do fit quite well the predictions of Dade et al. (this volume) for Michaelis-Menten kinetics of digestion and absorption in the case where multiple substrates are being utilized (i.e., $n > 1$ in their formulation). Although we have no direct measures of absorption rate, our data are consistent with the premise of Dade et al. that *A. pacifica* is adjusting its feeding rate in order to maximize absorption rate despite fluctuations in food concentration.

Alteration of feeding rate represents an important behavioral response of *A. pacifica* to changes in food value of the sediments it inhabits. It is clear that changes in feeding rate have major effects on net rate of gain, but we know little about the natural spatial and temporal variations in sediment food value and how these affect *A. pacifica* at the individual and population level. Furthermore, the behavioral responses of only a few species of deposit feeders have been studied; thus we can not yet generalize our present results.

Acknowledgments

This research was supported in part by Grant OCE-8214708 from the National Science Foundation. D. Bard was supported by the NSF Research Experience for Undergraduates program.

References

Calow P (1982) Homeostasis and fitness. Am Nat 120:416-419

Hettmansperger TP (1984) Statistical inference based on ranks. John Wiley and Sons, New York

Hylleberg J (1975) Selective feeding by *Abarenicola pacifica* with notes on *Abarenicola vagabunda* and a concept of gardening in lugworms. Ophelia 14:113-137

Jumars PA, Self RFL, Nowell ARM (1982) Mechanics of particle selection by tentaculate deposit feeders. J Exp Mar Biol Ecol 64:47-70

Lopez G, Levinton J (1987) Ecology of deposit-feeding animals in marine sediments. Quart Rev Biol 62:235-260

Lopez G, Taghon G, Levinton J (eds) (1989) Ecology of marine deposit feeders. Springer-Verlag, New York Berlin Heidelberg London Paris Tokyo

Mayer LM, Schick LL, Setchell FW (1986) Measurement of protein in nearshore marine sediments. Mar Ecol Prog Ser 30:159-165

Miller DC (1984) Mechanical post-capture particle selection by suspension- and deposit-feeding *Corophium*. J Exp Mar Biol Ecol 82:59-76

Phillips NW (1984) Compensatory intake can be consistent with an optimal foraging model. Am Nat 123:867-872

Self RFL, Jumars PA (1988) Cross-phyletic patterns of particle selection by deposit feeders. J Mar Res 46:119-143

Taghon GL (1981) Beyond selection: optimal ingestion rate as a function of food value. Am Nat 118:202-214

Taghon GL (1982) Optimal foraging by deposit-feeding invertebrates: roles of particle size and organic coating. Oecologia (Berlin) 52:295-304

Taghon GL (1988) The benefits and costs of deposit feeding in the polychaete *Abarenicola pacifica*. Limnol Oceanogr 33:1166-1175

Taghon G (1989) Modeling deposit feeding. In: Lopez G, Taghon G, Levinton J (eds) Ecology of marine deposit feeders. Springer-Verlag, New York Berlin Heidelberg London Paris Tokyo, p 223

Taghon GL, Jumars PA (1984) Variable ingestion rate and its role in optimal foraging behavior of marine deposit feeders. Ecology 65:549-558

Tenore KR, Rice DL (1980) A review of trophic factors affecting secondary production of deposit-feeders. In: Tenore KR, Coull BC (eds) Marine benthic dynamics. University of South Carolina, Columbia, p 325

ORAL DISCUSSION - G.L. Taghon

Q. (Belovsky): Your animals are making ingestion decisions over a long period of time, as incorporated in linear programming models for herbivores. You're talking about short–term effects on long–term ingestion.
A. Our data show it takes about two days from the time animals are first offered different types of sediment for differences in feeding rates to become evident. Part of that time is probably due to the disturbances caused by our experimental manipulations; the animals must spend some time rebuilding their burrows. We don't know the time scales over which the animals deplete their local foraging area in the field and move to new, nearby areas where the organic content of the sediment may differ.

Q. (Peterson) Does the worm have no cost in feeding ?
A. It has a very small mechanical cost associated with moving sediment through the gut, amounting to only 1 – 2% of the standard or resting metabolic cost. I have some estimates for the "cost of growth" and they are in line with reported values for specific dynamic action for various other organisms, about 50 – 60% of the absorbed organic matter. Another cost of feeding involves secretion of the mucus used to lubricate the gut lining and prevent abrasion from the sediment particles. Much of this mucus is lost in the feces.

Q. (Gilliam): Your animals survive starvation for a long time. Perhaps their behaviour is selected not for Optimal Foraging, but for reducing risks of mortality ? For example, they might slow down or hide when at risk ?
A. This animal really cannot do much about reducing its risk of mortality. It lives beneath the sediment surface and is relatively sedentary as part of its normal mode of life.

Q. (Jørgensen) The growth rates are low and there appears to be a lot of variability in the data.
A. We typically measure growth rates of 3 – 4% per day for worms in the laboratory and animals transplanted back into the field. There is much variability in the data. I suspect it is real and may be due to genetic differences among individuals but I have not yet examined this possibility quantitatively.

SUPPLY-SIDE OPTIMIZATION: MAXIMIZING ABSORPTIVE RATES

W. Brian Dade, Peter A. Jumars and Deborah L. Penry
School of Oceanography, WB-10
University of Washington
Seattle, WA 98195, U.S.A.

> ***... Pooh and Piglet walked home thoughtfully together in the golden evening, and for a long time they were silent.***
> ***"When you wake up in the morning, Pooh," said Piglet at last,"what's the first thing you say to yourself ?"***
> ***"What's for breakfast ?" said Pooh. "What do you say, Piglet ?"***
> ***"I say, I wonder what's going to happen exciting today ?" said Piglet.***
> ***Pooh nodded thoughtfully.***
> ***"It's the same thing," he said.***
>
> **A. A. Milne**
> ***The World of Pooh***

INTRODUCTION

The complexities of feeding behavior extend far beyond the simple priorities of life in The Hundred Acre Wood. Vermeij (1987), for example, has suggested that predator-prey relations have continually escalated as predators and prey co-evolved. A consequence of such escalation is that prey defences and predator weapons against them both continue to become more effective. The ultimate source of the material and energy for this escalation in heterotrophs with digestive tracts is absorption across the gut wall. Thus it is arguable that the lowest common denominator in foraging theory should be absorption of digestive products; the ultimate evolutionary escalation must be in rate of gain from food. Most foraging theory, however, focuses on choice of diet and feeding location (Stephens and Krebs, 1986) and stops with prey acquisition and pre-ingestive handling -- despite the fact that digestive handling times often far exceed pursuit and ingestive handling times.

We have adapted a body of theory within chemical engineering aimed at the analysis and design of chemical reactors to analyze the process of digestion in animals possessing guts (Penry and Jumars 1986, 1987; Jumars and Penry 1989). In so doing, we address the scale on which optimization of foraging behavior must ultimately take place: that of anabolic supply to

NATO ASI Series, Vol. G 20
Behavioural Mechanisms of Food Selection
Edited by R. N. Hughes

fuel life's activities. For example, the most frequent pattern of digestive processing seen in animals is the combination of plug flow (little axial mixing in the gut) with enzymatic digestion that follows Michaelis-Menten kinetics. Penry and Jumars (1987) demonstrated within the context of chemical reactor theory that this combination of digesta mixing and reaction kinetics should maximize rate of production of digestive products. With this combination of characteristics, they noted, digestive product formation rate rises continuously with ingestion rate; given Michaelis-Menten kinetics, increasing residence time of material in the gut decreases substrate concentration and thus decreases overall reaction rate. Penry and Jumars (1987) also suggested that optimal throughput rates arise as increasing gross gains with increasing feeding (and thus digestion) rate become offset by steeply-rising costs that are also proportional to feeding rate. The need to invoke ill specified and thus unmeasurable costs has been a chronic problem in optimal foraging theory. Further, digestive costs measured to date are apparently too small to be the prime factors limiting the rate at which material is moved through guts (e.g.: Bohrer and Lampert, 1988; Taghon, 1988). Further still, Phillips (1984) noted that physiological limitation of an animal's capacity for assimilation could confound predictions from simple cost-benefit optimization of the ingestion process alone.

We have continued to develop and modify chemical reactor theory for application to heterotrophy because our prior applications have stopped short of the key step of absorption. Our continuing work on digestion has been inspired in large measure by the evocative graphical arguments of Sibly (1981, Fig. 5.2) and can be viewed as an attempt to quantify the arguments inherent in his graphical procedure. Our primary purpose in this paper is to derive a coupled model of digestion and absorption and to give preliminary results from it. A summary of notation used in this derivation appears in Table 1. What we realized when we progressed to the stage of modeling absorption of digestive products is that maximization of absorption rate sets an optimal ingestion rate. Faster feeding produces digestive products at a greater rate, but one that results in undigested food and digestive products being defecated without being used. Bayne et al. (1987) recognized this trade-off in assimilation efficiency versus ingestion rate in their experiments with the suspension-feeding mussel *Mytilus edulis*. Moreover, an ingestion rate below optimal yields an incremental absorptive rate not worth the incremental time required to obtain it relative to what could be realized with infusion of new food substrate at a slightly faster ingestion rate. The implicit trade-offs in digestive product formation rate and absorption rate thus prove to be special cases of the principle of lost opportunity (Stephens and Krebs, 1986). Below the optimal ingestion rate an animal loses the opportunity for greater gut-averaged absorption rate that is driven by concentration of digestive products, that is, ingesta is over utilized, while above the optimal ingestion rate an animal loses the opportunity to absorb products because undigested and digested materials are under utilized. Given the ability to predict an optimal absorption rate, it is possible for us to predict an optimal absorption

TABLE 1 Notation used in the model of coupled digestion and absorption. In addition, the subscript $_0$ refers to input (feed or time 0) values, while the subscript $_f$ refers to final (output or fecal) values, and (with the exception of ER') dimensionless quantities that we derive are indicated with a subscripted asterisk (*). Because of its common prior usage in chemical engineering, however, we use X without the subscripted asterisk. Energy can be substituted for mass by converting mol to their equivalents in cal throughout.

Notation	Meaning
A	absorbed digestive products (mol)
AR	absorption rate (mol volume^{-1} time^{-1})
AR_*	AR/W_{max}
C_F	input concentration of food resource or digestive substrate (mol volume^{-1})
C_{F*}	C_F/K_m
C_P	concentration of digestive product (mol volume^{-1})
C_{P*}	C_P/M_m
∂C_F	incremental change in C_F (mol volume^{-1})
∂C_P	incremental change in C_P (mol volume^{-1})
∂V	incremental change in volume (volume)
ER_*	egestion ratio, OM output/OM input
ER'	renormalized egestion ratio = $C_{Pf}/M_m + (C_{Ff}/K_m)/(C_{F0}/K_m)$
F	input or feed (mol)
K_m	half-saturation constant for digestion (mol volume^{-1})
M_m	half-saturation constant for absorption (mol volume-1)
n_1	conversion rate exponent
n_2	absorption rate exponent
OM	organic matter, F + P (mol volume^{-1})
P	intermediate digestive products (mol)
q_0	volumetric input flow rate (volume time^{-1})
r_{FP}	rate of F to P conversion (mol volume^{-1} time^{-1})
r_{PA}	rate of product absorption (mol volume^{-1} time^{-1})
s	specific dynamic action coefficient $(0.0 \leq s \leq 1.0)$
τ	throughput time (time)
τ_*	$\tau V_{max}/K_m$
V_{max}	maximal digestive rate (mol volume^{-1} time^{-1})
W_{max}	maximal absorptive rate (mol volume^{-1} time^{-1})
X	fractional conversion of F to P

efficiency. We show that explicit application of the principle of lost opportunity to digestion and absorption has many additional uses.

One controversial implication of our applications of the principle of lost opportunity to digestion and absorption is that increasing weight-specific respiration with decreasing absolute weight is more a function of anabolism than catabolism. Small organisms *can* have a higher weight-specific metabolic demand because they can have a higher weight-specific rate of metabolic gain. Since measured demand (O_2 utilization) is based on prior gain, it should be more obvious than it has been that anabolism determines catabolic potential, not vice versa. Less controversial but no less important are implications for mass and energy flow in ecosystems. There is increasing realization that many empirically determined characteristics of communities are direct consequences of individuals acting in a manner that maximizes individual fitness (Huston et al.,1988; Jumars et al. 1989). Material cycling in aquatic systems is especially sensitive to the fraction of digestive substrates and products released in dissolved form because solutes are rapidly dispersed by physical mechanisms in streams, lakes and oceans, while particulate remains can follow very different pathways.

A MODEL FOR PLUG-FLOW DIGESTION AND ABSORPTION

Simple relationships

Here we extend the earlier treatment of digestion as an optimized process (Penry and Jumars 1986, 1987; Jumars and Penry, 1989) to predict trends in gut throughput rates, throughput compositions, and egesta compositions associated with maximal absorptive rates as food resources and reaction kinetics vary. In development of the model presented here we assume that an animal can and will modify its feeding behavior (i.e., throughput rate) instantaneously and that digestion and absorption proceed as a single two-step reaction series with each step showing saturation kinetics. We further assume that food encounter rate does not limit the rate of ingestion. The latter assumption has two clear benefits: It allows us to characterize food simply via the concentration of some growth-limiting component, and it is easily relaxed. We treat this concentration together with the kinetics of digestion and absorption as the determinants of food quality. If encounter rate does limit ingestion rate (food quantity), then it is a simple procedure (easily carried out by the reader in the graphs we present) to bound the optimization with the encounter rate-limited ingestion rate.

Digestion of proteins and complex carbohydrates in a raw feed occurs by way of enzymatic conversion to intermediate products such as small peptides, free amino acids, and monosaccharide sugars. These smaller molecules, in turn, are taken up at the gut wall by carrier-mediated transport (Karasov and Diamond, 1988). Within the simple tubular gut of many animals, such reactions can be considered to occur in an ideal plug flow reactor (PFR; Fig. 1). Aspects of PFR performance are discussed in considerable detail in Levenspiel (1972). In short, a PFR is characterized by equal residence time for all mass elements passing through the vessel; that is, there is no mixing along the flow path. We assume that there is perfect mixing of digestive (= conversion) products, however, across the flow path to facilitate absorption (= assimilation) at the gut wall. Because reactions continue along the flow path, concentration gradients in both food substrate and intermediate product are established, and a mass balance must be defined for a differential element of volume ∂V. The problem thus becomes one of performance in an ideal PFR that hosts the single two-step reaction series,

$$F \xrightarrow{r_{FP}} P \xrightarrow{r_{PA}} A$$

(representing enzymatic conversion of food F to product P at rate r_{FP}, and absorption of P across the gut wall at rate r_{PA}). Enzymatic conversion can be described by the mass balance within ∂V:

accumulation of F + disappearance of F by conversion to P = input of F - output of F.

Absorption of P across the gut wall can be described by the mass balance within ∂V:

accumulation of P + production of P from F + absorption of P across gut wall = input of P - output of P

We assume for this initial analysis that absorptive sites are uniformly distributed along the entire length of the gut, just as we assume that digestive kinetics are uniform along the entire length of the gut (i.e. C_F and C_P are changing due to digestion and absorption, but rate parameters from relevant reaction rate equations are invariant with location). We set $C_{P0} = 0$ for all the model runs presented here. Under steady state, accumulation rates of both F and P are equal to zero, and reactor performance can then be described in terms of mass balances per unit of time in the differential volume ∂V (referring to the notation summarized in Fig. 1 and Table 1):

$$(-r_{FP})\partial V = q_0 C_F - q_0(C_F + \partial C_F),$$

disappearance rate of F by conversion to P = (volumetric flow rate) (input of F - output of F),

$$(r_{FP} - r_{PA})\partial V = q_0 C_P - q_0(C_P + \partial C_P),$$

production rate of P from F + absorption rate of P across gut wall = (volumetric flow rate) (input of P - output of P).

With some rearrangement,

$$\frac{\partial V}{q_0} = \frac{\partial C_F}{r_{FP}} = \frac{\partial C_P}{r_{PA} - r_{FP}} \tag{1}$$

where C denotes the concentration of F or P and q_0 is volumetric flow rate. Note that r_{FP} and r_{PA} are positive in the direction of digestion and absorption. This notation has been modified

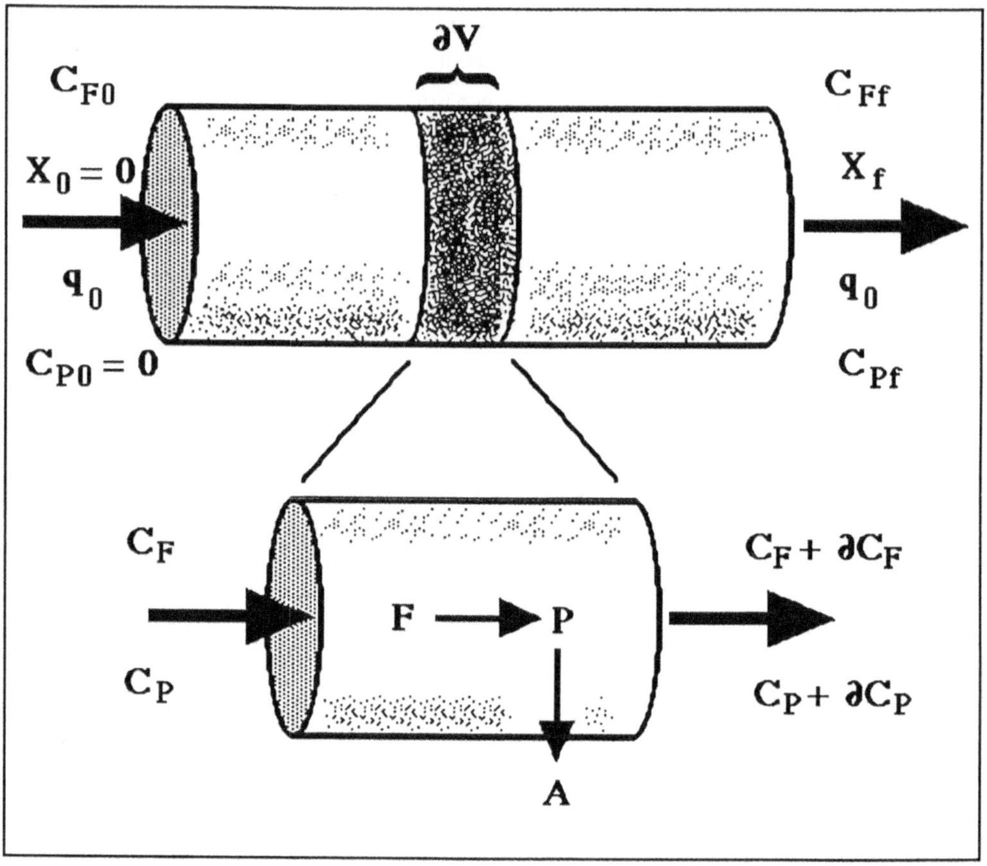

FIGURE 1. Schematic of digestion and absorption in a plug-flow gut. See text and Table 1 for explanation and notation.

from Penry and Jumars (1986, 1987) to accommodate the additional terms needed to describe absorption.

Treating absorption as a seemingly homogeneous reaction in the fluid phase is not unrealistic in the presence of perfect radial mixing. With perfect mixing, a fluid element in any unit of length of gut is in contact instantaneously with the same unit of length of gut lining. Thus a given areal concentration of absorptive sites on the gut lining is readily converted to a volumetric concentration of absorptive sites simply by dividing the number of absorptive sites per unit of length of gut by the volume of digesta per unit of length of the gut. The presence of villi or other ramifications of the gut wall is easily incorporated into this interpretation as well.

The two leftmost terms of Eq. 1 describe the relationships governing the fate of food resource F. In constant-density systems $C_F = C_{F0}(1\text{-}X)$, where X is the fraction of F converted to intermediate product P. Accordingly, $-\partial X = \partial C_F/C_{F0}$, and the two leftmost terms can be integrated over reactor volume to calculate throughput time, τ, required for final conversion X_f at either a fixed throughput flow rate q_o or specified reaction rate r_{FP}:

$$\tau = \int_0^V \frac{dV}{q_0} = \int_{C_{F0}}^{C_{Ff}} \frac{dC_F}{r_{FP}} = C_{F0} \int_0^{X_f} \frac{dX}{-r_{FP}} \qquad (2).$$

For constant-volume, constant-density systems, distance along the gut can be expressed in terms of τ. We explicitly assume that digestion causes no appreciable change in volume or density of digesta. This assumption can be relaxed fairly easily, but it is more widely applicable than one might suppose. We work primarily with deposit feeders, the bulk of whose ingested food is completely indigestible. Moreover, in terrestrial animals (e.g., *Homo sapiens*) solid food often is of order of 10% of gut throughput, with the rest being added fluids that are resorbed in the hindgut (e.g., Drasar and Barrow, 1985).

The two rightmost terms in Eq. 1 can be rearranged to yield the relationship:

$$\frac{\partial C_P}{\partial C_F} = \frac{r_{PA}}{r_{FP}} - 1 \qquad (3).$$

Eq. 3 represents the steady-state balance between production and absorption of digestive product P. Not surprisingly, Eq. 3 indicates that this balance is a function of the ratio of absorption rate to conversion rate. For example, a small decrease in food resource concentration C_F due to conversion to product P will result in a net increase in product concentration C_P if the ratio of absorption rate to conversion rate is less than unity (i.e., new P is produced more rapidly than it is absorbed). Conversely, a small decrease in C_F due to conversion will result in an incremental decrease in C_P if the ratio of absorption rate to conversion rate is greater than unity. To evaluate PFR performance as well as optimal absorption rate and distributions of food and digesta through the gut, Eq. 3 must be solved to evaluate C_P as an explicit function of C_{F0} and r_{PA}/r_{FP}.

Absorption in an ideal series-reaction PFR is evaluated by noting that over the entire reactor volume V a steady-state mass balance requires that:

total absorption = amount of P produced - amount of P egested = $V(C_{F0}X_f - C_{Pf})$.

Thus the gut-averaged absorption rate per gut volume is:

$$AR = \frac{\text{total absorption}}{\text{(volume) (throughput time)}} = \frac{C_{F0}X_f - C_{Pf}}{\tau} \qquad (4A).$$

Costs of digestion can be incorporated into this approach as well. For example, increasing throughput rates could result in increased mechanical costs of material handling to the animal. The rate at which these costs are incurred may be expressed as a fraction of substrate throughput rate or: rate of mass loss due to digestion $= \dfrac{sC_{F0}}{\tau}$, $0 < s < 1$.

Thus net absorption rate is:

$$AR = \frac{\text{total absorption}}{\text{(volume) (throughput time)}} = \frac{C_{F0}(X_f - s) - C_{Pf}}{\tau} \qquad (4B).$$

A non-zero s establishes a minimal conversion factor X_{min} for net positive absorption rate.

Utilizing Eq. 2 and 3 to evaluate Eq. 4 for $0 < X < 1$ enables us to predict distributions of component F and product P though the gut, throughput rates (τ^{-1}) and relative amounts of unconverted F and unabsorbed P egested at optimal absorption rates for given input concentrations of F and P, cost coefficient s, and specified digestive reaction kinetics.

Kinetics of digestion and absorption

Hydrolysis of ingested organic material by digestive enzymes falls into a class of saturation-limited, catalytic reactions whose rates can be described by the Michaelis-Menten equation (Briggs and Haldane, 1925):

$$-r_{FP} = \frac{V_{max} C_F}{K_m + C_F} \qquad (5A),$$

where V_{max} is the maximal attainable conversion rate at saturation and K_m is the concentration yielding conversion rates of $1/2(V_{max})$. [Note that $-r_{FP} = 0.5V_{max}$ at $K_m = C_F$; $-r_{FP} = V_{max}$ for $K_m << C_F$; and $-r_{FP} = V_{max}/K_m$ for $K_m >> C_F$.] A modified form of the Michaelis-Menten equation to accommodate conversion-absorption rates that are other than first order is

$$-r_{FP} = \frac{V_{max} C_F^{n_1}}{K_m^{n_1} + C_F^{n_1}} \qquad (5B).$$

Although the form of Eq. 5B is somewhat arbitrary, it conveniently preserves the physical meaning of reaction rate at saturation, V_{max}, and half-saturation concentration, K_m, while allowing characterization of more complex reactions at low C_F.

Reactions governing uptake of small-molecular intermediates at the gut wall occur at rates given by a similar form (e.g. Fisher and Parsons, 1953; Wilson, 1962; Davenport, 1972):

$$-r_{PA} = \frac{W_{max} C_P}{M_m + C_P} \qquad (5C),$$

where W_{max} and M_m are analogous to V_{max} and K_m as noted above. Again, a modified form of the Michaelis-Menten equation for reactions of varying order in C_P is

$$-r_{PA} = \frac{W_{max} C_P^{n_2}}{M_m^{n_2} + C_P^{n_2}} \qquad (5D).$$

Digestive or absorptive kinetics may differ from simple Michaelis-Menten form (Fig. 2) for many reasons. Exponents n_1 and n_2 less than one cause the curve of reaction rate versus substrate concentration to deviate from the classic Michaelis-Menten hyperbola by being more nearly linear. Exponential values less than unity are thus used to represent either of two relatively common situations. One is incorrect identification of the limiting nutrient. Measurements may be made of some bulk ingesta constituent whose concentration is correlated inversely with that of the rate-limiting ingredient; one might, for example, measure total organic carbon, while the ingredient driving the digestive reaction is some much more labile component that is proportionately scarce in foods characterized by high total organic carbon. The other situation is passive absorption of products (e.g., of volatile fatty acids in ruminants). Exponents n_1 and n_2 greater than one, on the other hand, give the curve of reaction rate versus substrate concentration greater nonlinearity than the hyperbolic form, making it sigmoidal. By analogy with electromagnetic resistance laws, exponents greater than one are appropriate in description of overall kinetics controlled by parallel two-step reaction series involving multiple food substrates and digestive products. This description would appear the most generally compatible with the energy maximization premise when rate of energy gain limits growth and energy is extracted from a wide array of food molecules. Alternatively, the comparatively slow reaction rate at low concentration (Fig. 2) for these higher exponents could be used to represent either a reaction that shows diffusional rate limitation when food is of low quality and the high bulk restricts diffusion or a situation in which enzyme is induced in proportion to substrate concentration. The latter case would be treated more accurately by varying V_{max} or W_{max}, but our simpler formulation gives some insight nonetheless into the kinetic consequences. We thus give results with varying n as a sensitivity analysis and to achieve generality.

Governing equations for optimal PFR deposit feeding behavior

Substitution of Eq. 5B into Eq. 2 and performing the integration yields throughput time (τ) as a function of input substrate concentration (C_{F0}) and final conversion fraction (X_f):

$$\tau = \frac{C_{F0}}{V_{max}} X_f - \frac{K_m}{V_{max}} \ln\left(1 - X_f\right) \qquad n_1 = 1.0 \qquad (6A)$$

and

$$\tau = \frac{C_{F0} X_f}{V_{max}} - \frac{K_m^{n_1}}{V_{max}} \left[\frac{\left(C_{F0}(1-X_f)\right)^{1-n_1} - \left(C_{F0}\right)^{1-n_1}}{1 - n_1} \right] \qquad n_1 \neq 1.0 \qquad (6B).$$

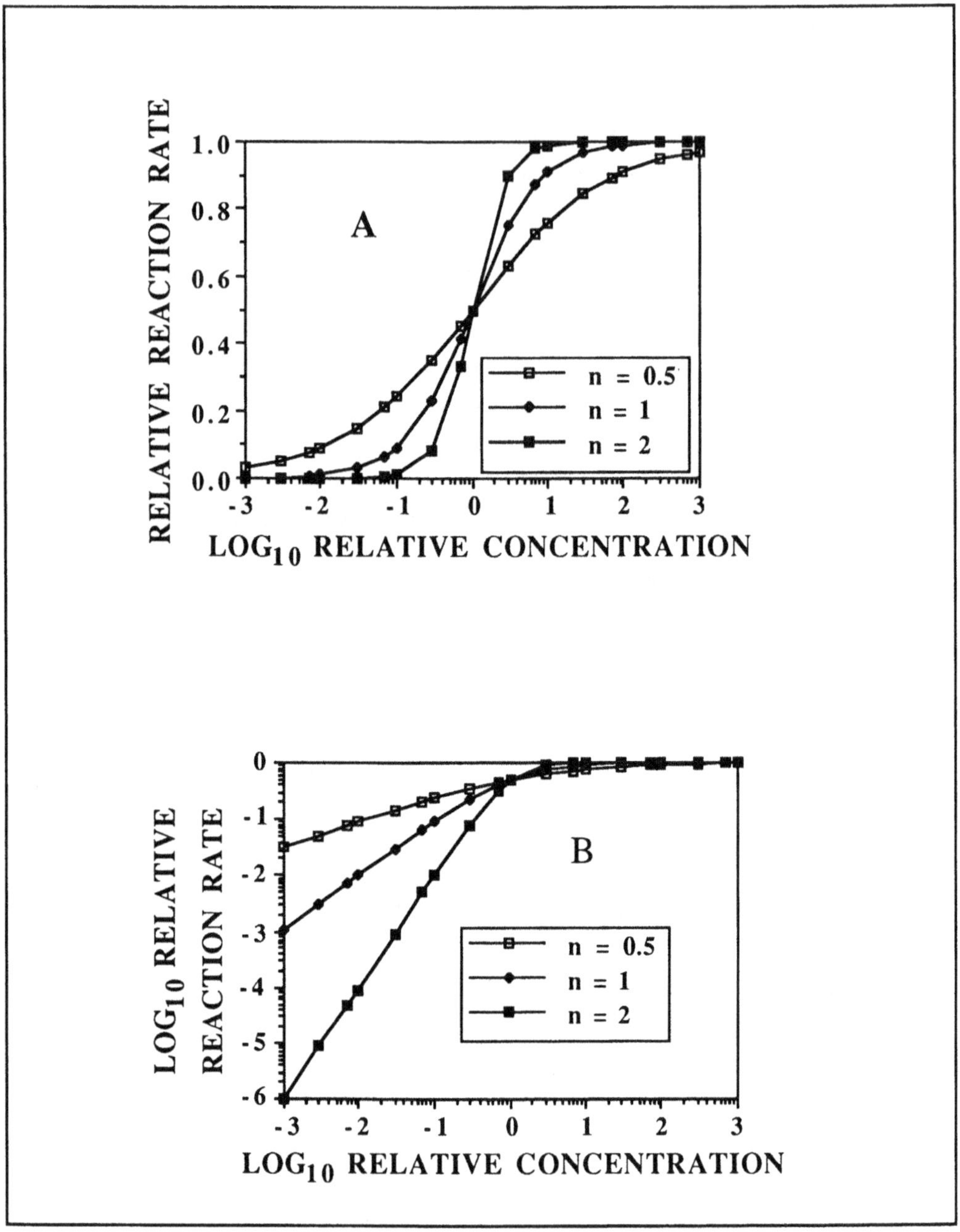

FIGURE 2. Proposed saturation-limited kinetics (r/r_{max}) vs relative reactant concentration (C_*) for n^{th} order digestive and absorptive reactions; n = 0.5, 1.0 and 2.0. (A) Semi-log plot and (B) log-log plot to demonstrate strongly nonlinear behavior.

Eq. 6A and 6B, in turn, can be rearranged to calculate relative throughput time:

$$\tau_* = \frac{\tau V_{max}}{K_m} = C_{F0*} X_f - \ln(1 - X_f) \qquad n_1 = 1.0 \qquad (6C),$$

and

$$\tau_* = C_{F0*} X_f - \frac{\left(C_{F0*}(1-X_f)\right)^{1-n_1} - \left(C_{F0*}\right)^{1-n_1}}{1 - n_1} \qquad n_1 \neq 1.0 \qquad (6D),$$

where $C_{F0*} = C_{F0}/K_m$. Defining relative throughput time in this way is helpful in predicting PFR performance under different input conditions. For example, under very low concentrations of substrate in the ingested food ($C_{F0*} << 1.0$), τ_* approaches [-ln (1-X_f)] when $n_1 = 1$, and throughput time is relatively independent of input substrate concentration (C_{F0}). Moreover, at low relative feed concentrations ($C_{F0*} << 1.0$) throughput time required for feed conversion X_f is a function of conversion rate order as well (Fig. 3). Under these food quality-limiting conditions, relative throughput time required for conversion X_f increases with increasing conversion rate order (n_1). At or near saturation levels of input concentration ($C_{F0*} >> 1.0$), however, throughput time becomes linearly dependent on input concentration and can become quite long at relatively high conversions. This trend is not reaction-order dependent (Fig. 3). Substitution of Eq. 5B and 5D into Eq. 3 yields:

$$\frac{\partial C_P}{\partial C_F} = \frac{W_{max}}{V_{max}} \frac{C_P^{n_2}}{C_F^{n_1}} \frac{\left(K_m^{n_1} + C_F^{n_1}\right)}{\left(M_m^{n_2} + C_P^{n_2}\right)} - 1 \qquad (7A).$$

Solutions obtained for Eq. 7A can be made general by multiplying both sides of the equation by K_m/M_m and factoring K_m and M_m out of the terms in parentheses. The dimensionless form of the equation then becomes:

$$\frac{\partial C_{P*}}{\partial C_{F*}} = \left(\frac{W_{max}}{V_{max}} \frac{C_{P*}^{n_2}}{C_{P*}^{n_1}} \frac{\left(1 + C_{F*}^{n_1}\right)}{\left(1 + C_{P*}^{n_2}\right)} - 1 \right) \frac{K_m}{M_m} \qquad (7B),$$

where as above $C_{F*} = C_F/K_m$ and $C_{P*} = C_P/M_m$. As with throughput time, expressing in this way the balance between production and absorption of product P is helpful in analysis of PFR performance under food concentration-limiting conditions and saturated, rate-limiting conditions.

Substitution of Eq. 6C and 6D with solutions for Eq. 7B into Eq. 4B enables direct calculation of relative absorption rate:

$$AR_* = \frac{AR}{W_{max}} = \frac{V_{max}}{W_{max}} \frac{\left(C_{F0*}(X_f - s) - C_{Pf*}\frac{M_m}{K_m}\right)}{\tau_*} \qquad (8),$$

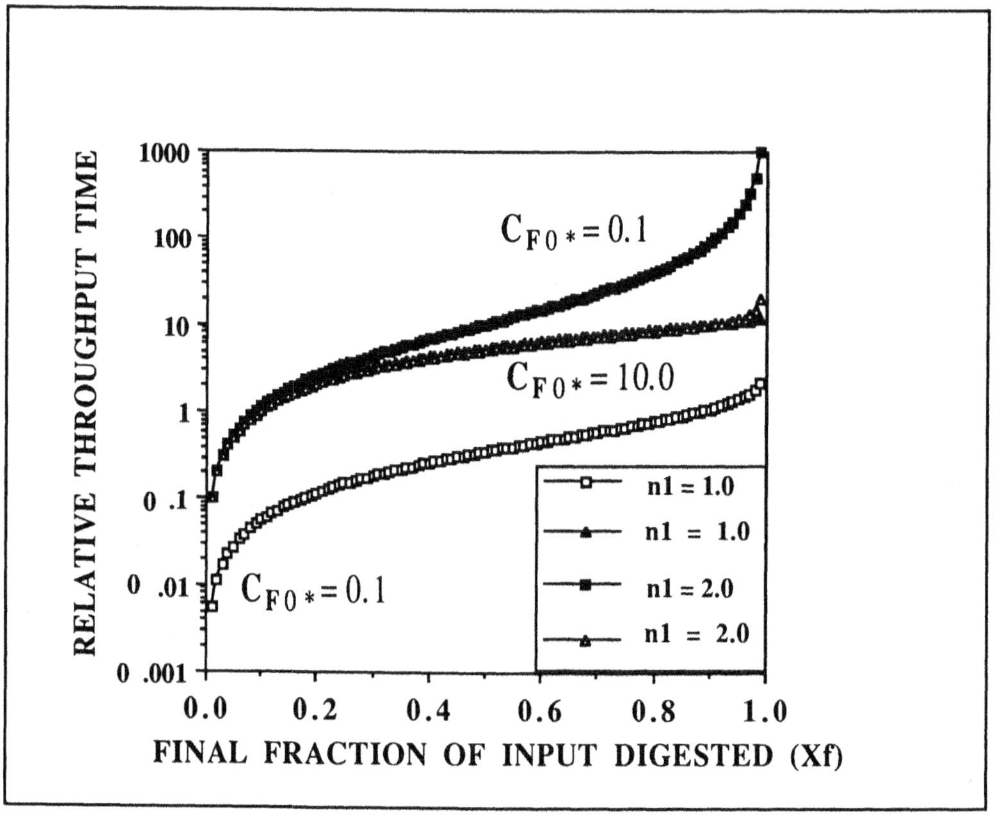

FIGURE 3. Predicted relative throughput times (τ_*) required to convert fraction X_f of initial relative food resource concentrations (C_{F0*} = 0.1, 10) by conversion to digestive product C_P for varying reaction order (n_1 = 1.0, 2.0). Note the near overlap in τ_* near saturation ($C_{F0*} \geq 10$) for both reaction orders in contrast to the increase in throughput time with reaction order at sub-saturating food input concentrations.

where $0 < AR_* < 1.0$ and as above $C_{F*} = \frac{C_F}{K_m}$, $C_{P*} = \frac{C_P}{M_m}$. Marginal analysis of this function suggests that optimal absorptive rates exist for the condition $\frac{\partial AR_*}{\partial \tau_*} = 0$, or with differentiation of Eq. 8 and rearrangement:

$$(AR_*)_{opt} = \frac{V_{max}}{W_{max}} \frac{\partial \left\{ C_{F0*}\left[(X_f)_{opt} - s \right] - (C_{Pf*})_{opt} \frac{M_m}{K_m} \right\}}{\partial (\tau_*)_{opt}} \qquad (9).$$

Eq. 9 indicates that optimal gut-average absorptive rate is achieved when gut-average absorptive rate just equals the marginal gain in absorption at the end of the gut. The situation is analogous to the application of the Marginal Value Theorem to duration of stay in food patches (Stephens and Krebs, 1986). This condition thus defines the optimal throughput rate: Shorter throughput times result in under-utilization of available food resource; longer throughput times suffer "diminishing returns" that, while increasing the total amount absorbed, serve to decrease gut-average absorptive rate.

To recapitulate briefly, our approach treats the digestive process as a single two-step reaction series in which conversion of input substrate to intermediate, small-molecular weight products and subsequent absorption of those products each is governed by spatially uniform rate equations of Michaelis-Menten form. The equations governing gut performance are derived from steady-state mass balances and are made dimensionless with respect to Michaelis-Menten parameters. A variety of potential digestive and absorptive reaction orders are incorporated to assess robustness of model predictions over a broad range of input food concentrations and reaction rates. To calculate substrate and product distributions through the gut as well as relative absorption rates for varying input substrate concentrations, Eq. 7B is solved numerically with a fourth-order Runge-Kutta algorithm for varying relative input food concentrations (C_{F0*}), for varying degrees of food conversion ($0 < X_f < 1$), and for specified reaction orders (n_1, n_2) and kinetic coefficient ratios (W_{max}/V_{max}, M_m/K_m, s). Errors incurred in solutions for C_{Pf} are within about 1% or less of analytical solutions obtained for linearized forms of Eq. 5 under end-member conditions (i.e., $C_{F*}, C_{P*} << 1$; $C_{F*}, C_{P*} >> 1$).

Relative absorption rate, AR_*, for varying input substrate concentrations is then evaluated from Eq. 8. Next, optimal absorption rate, $(AR_*)_{opt}$, its associated optimal throughput rate, $(\tau_*^{-1})_{opt}$, and optimal egestion ratio, $(ER_*)_{opt}$, are calculated. ER_* can be considered as (1 - absorptive efficiency) and is calculated from:

$$ER_* = \frac{\text{OM output}}{\text{OM input}} = \frac{\text{unconverted feed + unabsorbed intermediate}}{\text{input feed}} =$$

$$\frac{C_{Ff} + C_{Pf}}{C_{F0}} = (1 - X_f) + \frac{C_{Pf*} M_m}{C_{F0*} K_m}.$$

Enzymatic secretions into the gut could be added to the numerator to the extent that they are not subsequently resorbed. We do not add this complication but do note that with its addition ER_* values greater than one are possible (Bayne et al., 1987). We also evaluate the distribution of throughput components under the optimal operating strategy and do so for a range of input conditions (C_{F0*}) and reaction orders (n_1, n_2). To demonstrate relatively robust trends in digestive and absorptive behavior of simple, tubular guts, only results for rate coefficient ratios (W_{max}/V_{max}, M_m/K_m) of unity and zero digestive costs (s) are shown. With information available on a given species, it would be straightforward to tailor the results further.

RESULTS

"Performance" curves (Fig. 4) reveal that possible digestive-absorptive alternatives do exhibit maximal rates of absorption per gut volume. The curves all show maxima, but those maxima are sharper for lower input food concentrations (C_{F0*}) and higher reaction order (n). We caution that the plots are log-linear and thus understate the sharpness of the optima. Relative absorption rate, not surprisingly, approaches saturation with increasing degree of digestive saturation (i.e., food substrate concentration) for any digestive-absorptive reaction order (Fig. 5A). For digestive kinetics of first order or less, relative throughput rates are predicted to be decreasing monotonically with increasing degree of digestive saturation (Fig. 5B). Optimal relative throughput rates, however, exhibit maxima at intermediate degrees of digestive saturation for digestive reaction kinetics of order greater than unity (Fig. 5B), even when absorption kinetics are of simple Michaelis-Menten form ($n_2 = 1$). Similar trends are observed for other values of W_{max}/V_{max}, M_m/K_m, and C_{P0*}. Of particular interest in microbial and detrital food webs is the fraction of organic matter predicted to be passed without absorption (Fig. 6).

Total ER* in Fig. 6A represents the quantity (organic matter out)/(organic matter in) vs. degree of digestive saturation and thus represents the quantity (1 - absorption efficiency), with areas under respective curves indicating relative importance of undigested food input (C_{Ff}) and unabsorbed digestive product (C_{Pf}). Somewhat counter-intuitively, absorptive efficiency is predicted to increase with increasing degree of digestive saturation; the reason is that there is no marginal gain from gut passage when the material contained in the gut is still capable of saturating absorption kinetics. While presentation of absorptive efficiency is useful in the context of the published literature, Fig. 6A can be misleading because it represents a plot of ($1/C_{F0}$) vs C_{F0}. An alternative presentation (Fig. 6B; ER' = ($C_{Pf}/M_m + C_{Ff}/K_m$) leaves the ordinate independent of food resource concentration. Increasing total ER' with increasing degree of digestive saturation reflects, not surprisingly, the effects of the quality "richness" of food resources available; rate of organic matter egestion in feces clearly rises with food richness (C_{F0}). Inspection of both Figures 6A and 6B suggests that unutilized digestive product P can make up a significant fraction of the egesta. Without the explicit model it would not be obvious that optimal rate of absorption by an animal coincides with behavior that increases both digestive efficiency (%) and egestion rate (mol volume^{-1} time^{-1}) of undigested food and unabsorbed product with increasing degree of digestive saturation (C_{F0*}). Relatively higher throughput rates and concentrations of unabsorbed digestive product at low C_{F0*} may seem counterintuitive, but are necessary to sustain optimal absorptive rates via maintenance of C_P gradients across the gut wall in the face of dilute food resources.

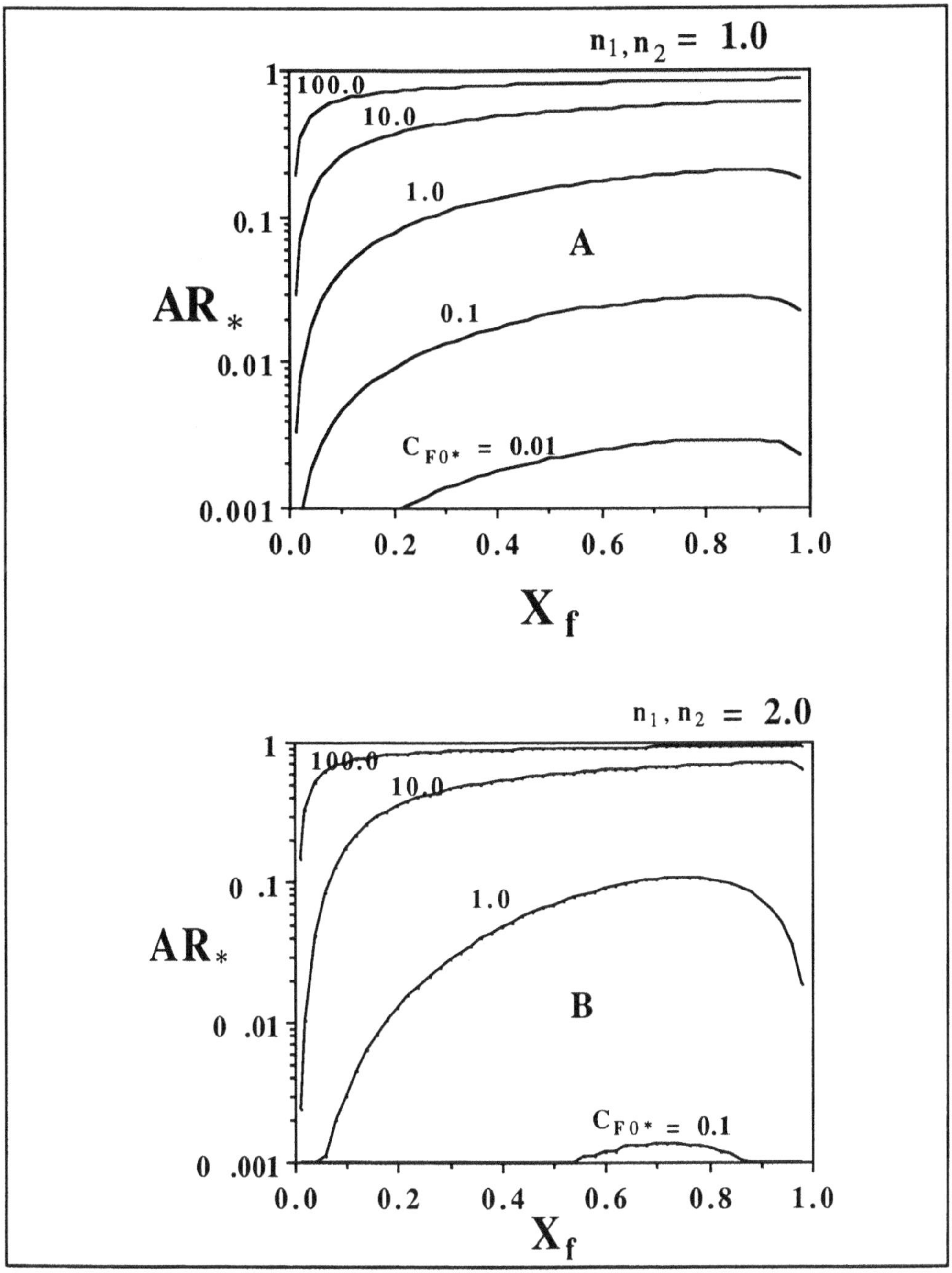

Figure 4. Predicted gut-averaged relative absorption rate (AR_*) as a function of final conversion fraction (X_f) and of varying initial food concentration (C_{F0*}), where $W_{max}/V_{max} = M_m/K_m = 1$ for uniformly (A) first-order and (B) second-order kinetics. Patterns are similar for other values of W_{max}/V_{max} and M_m/K_m.

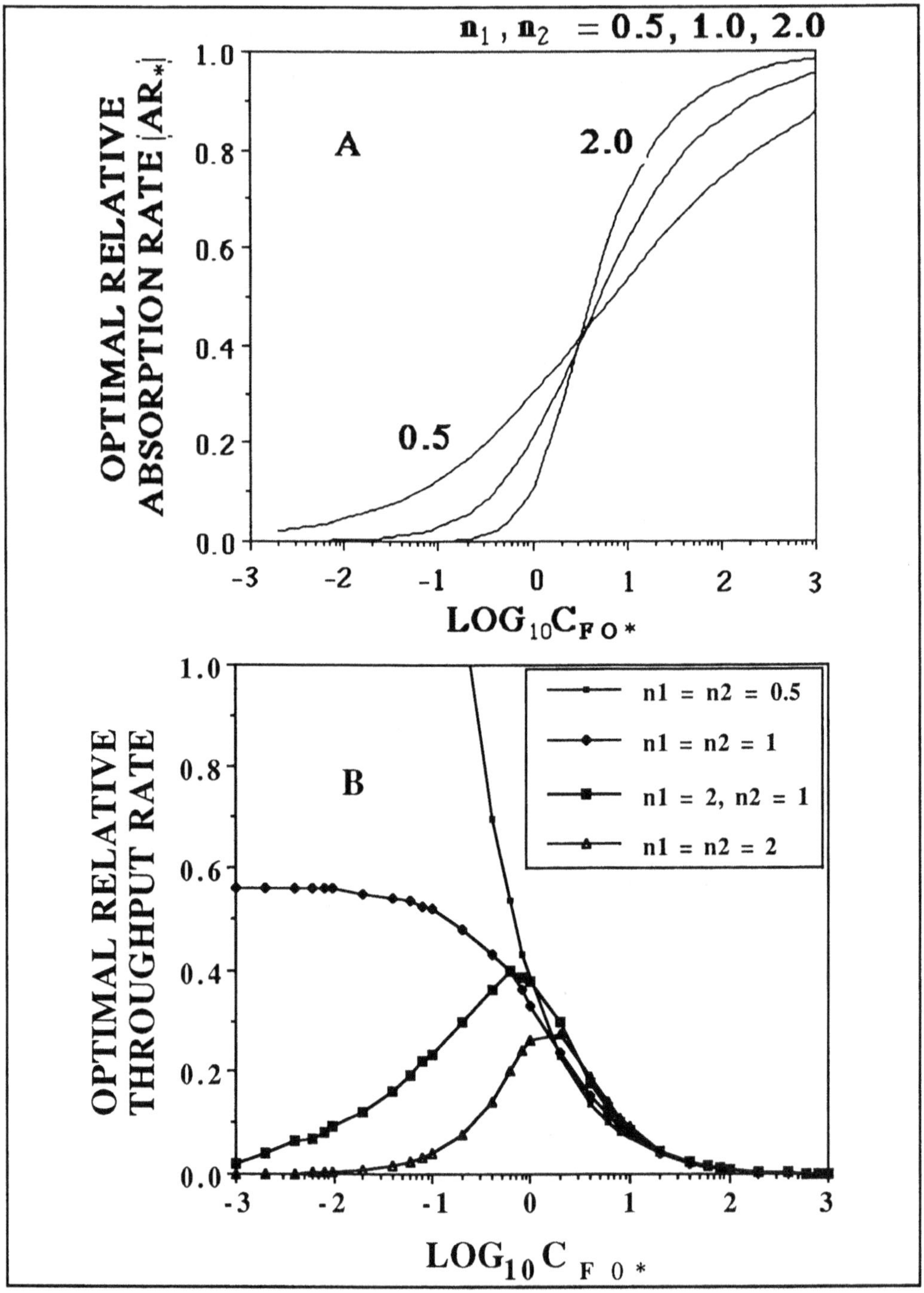

FIGURE 5. Predicted trends in optimal absorption (A) and throughput (B) rates versus C_{F0*} for reactions of varying orders. $W_{max}/V_{max} = M_m/K_m = 1.0$, but patterns are similar for other values. Optimal relative throughput rate $=1/\tau_{*opt}$.

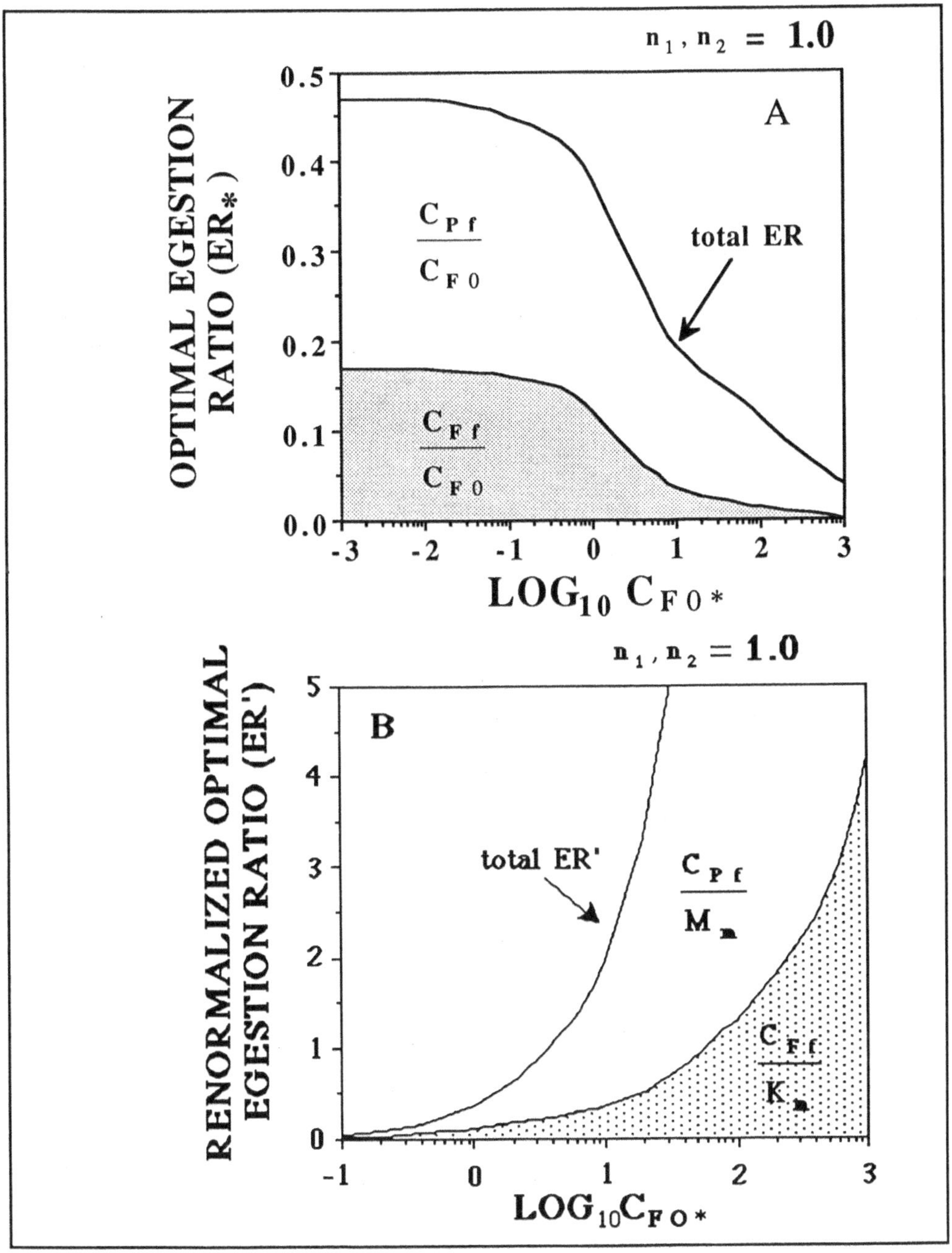

FIGURE 6: (A) Optimal egestion ratio (ER*) and (B) renormalized egestion ratio (ER') vs. relative food resource concentration (C_{F0*}). Similar trends are observed for varying reaction-rate order, Wmax/V_{max}, and M_m/K_m.

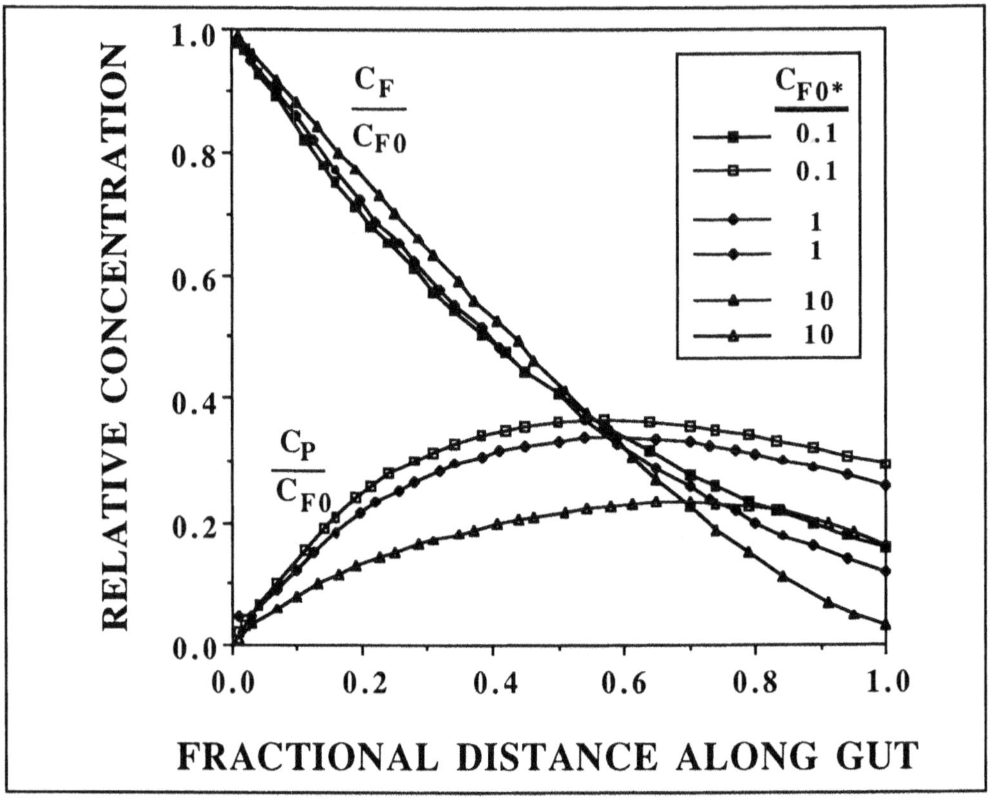

FIGURE 7: Predicted distribution of relative concentration of unconverted substrate (C_F/C_{F0}) and unabsorbed product (C_P/C_{F0}) for various input food concentrations (C_{F0*}) under the optimal operating policy (maximal absorption rate) as a function of fractional distance along the gut for $W_{max}/V_{max} = M_m/K_m = 1.0$, $n_1 = n_2 = 1.0$. Similar trends are observed for varying reaction rate constants and order.

The axial pattern of relative concentrations of unconverted food resource and unabsorbed digestive product through the gut under the optimal operating policy (Fig. 7) are not as surprising; relative concentrations of food resource are predicted to diminish with distance through the gut, while undigested products peak in concentration before the end of the gut is reached. As with egestion ratios, similar trends are observed for varying reaction rate constants and orders.

DISCUSSION

Relationships to prior processing rate models and observations

This coupled digestion-absorption model is the logical extension of optimal digestion approaches to predict aspects of feeding behavior (Penry and Jumars 1986, 1987; Jumars and Penry, 1989). For a range of input food concentrations (C_{F0}), Jumars and Penry (1989) solved the easier problem of predicting conversion (X_f) and throughput times (τ) needed to maintain a fixed rate of digestive product formation. For standard ($n_1 = 1$) Michaelis-Menten kinetics, optimal digestion modeling led to the conclusion that shorter throughput times always yielded greater digestive production rates. Thus Jumars and Penry (1989) were forced to conclude that throughput rate in an animal acting to maximize net rate of gain must be set at some optimum by unidentified costs that rise rapidly above some as yet undetermined rate of material processing. The present approach circumvents the unsatisfying conclusion that unknown costs determine throughput time by allowing the balancing kinetics of digestion and absorption, coupled through the concentration of digestive products, to set the optimal rate of material processing. This approach thus represents an important extension of the intuitive arguments of Phillips (1984), in which limitations to assimilation were invoked within the context of a cost-benefit balance in feeding behavior.

Kofoed et al. (1989) started at the other end of the digestion-absorption couple. Their model includes linear absorption kinetics, but it does not include digestion. Thus they implicitly assumed that either food is ingested in a form that can be absorbed directly without conversion or that the kinetics of digestion are so much faster than those of absorption that digestion can be ignored kinetically. Consideration of their experimental protocol suggests that this assumption is true for the experimental conditions provided but that it may not be true in nature. In their experiments Kofoed and coworkers gave the animal pure, rich foodstuffs (bacteria or diatoms) to conduct absorption measurement; this would place the animal under high C_{F0*} in terms of the digestion-absorption model; hence the kinetics of both digestion and absorption would be saturated, and the implicit assumption that digestion kinetics are unimportant would hold. Thus their protocol might provide a valuable, non-destructive alternative to the techniques of Karasov and Diamond (1983) for getting at maximal absorption rate (W_{max}) freed of the effects of digestive kinetics. The form of the equation that Kofoed et al. (1989) used further implicitly assumed that the entire volume of the gut operates as a continuously stirred tank reactor (or CSTR in the treatment by Penry and Jumars, 1987). We suspect that plug flow is the more frequently accurate representation of digesta mixing (Penry and Jumars, 1987), especially in the absorptive portions of animal guts, and that Michaelis-Menten kinetics more accurately represent absorption (Davenport, 1977) than do linear absorption coefficients. Kofoed et al.'s (1989) preliminary results do fall on curves compatible with their model, but the trends could also be fitted with plug flow and Michaelis-Menten kinetics.

In contrast to the absorption focus of Kofoed et al. (1989), it might be reasonable to do the opposite and focus on digestion alone if absorption were never rate limiting. The presence of an optimum in the coupled model, however, suggests that natural selection will drive ingestion rate away from either extreme. There is an optimum precisely where neither digestion nor absorption rate can be ignored. The coupled digestion-absorption model, however, clearly needs the scrutiny of tests with full closure on the kinetic parameters of digestion and absorption. Net absorption rate is the key measurement and whether it is maximized at the ingestion rates observed under *ad libitum* food supply is the key question. Fortunately, reasonably artifact-free methods for characterizing absorption kinetics have been developed (Karasov and Diamond, 1983).

At first sight the existence of an optimal gut retention time without the need for inclusion of explicit costs appears suspicious. Implicit costs -- the lost opportunity to absorb digestive products at a greater rate -- are exceedingly important ones, however. They represent true costs in evolutionary terms: reductions in the ability to respond to the pressures of natural selection. The search for explicit costs of digestion, moreover, has found them to be small in comparison with the subsequent costs of constructing new molecules from digestive products, the so-called "specific dynamic action" (Kiørboe et al., 1985; Bohrer and Lampert, 1988; Taghon, 1988). Rather than conclude that a more thorough search for explicit digestive and absorptive costs is needed, we suggest instead that evolution is likely to have acted to poise animals at the processing-rate optima we have identified. What needs evaluation in order to test this idea is whether explicit costs, e.g. of enzyme production, are large or trivial in comparison with potential gains lost in small departures from the otherwise optimal processing rates.

Our model adds fuel and some light to the controversy over whether volumetric ingestion rates should decrease or increase with increasing food quality (Cammen, 1989; Taghon, 1989). Specifically, it demonstrates that an animal operating to maximize its rate of absorption may uniformly decrease feeding rate with increasing food quality if the digestive reaction shows simple Michaelis-Menten kinetics. If, on the other hand, digestion consists of a number of reactions in parallel, each described by hyperbolic kinetics ($n_1 > 1$), then a peak feeding rate at intermediate food quality is to be expected. Therefore measurements of feeding rate against some measure of food quality alone (e.g., Taghon and Jumars, 1984) are not sufficient to determine whether an animal is processing ingested material in a manner that maximizes rate of gain from absorption. Taghon and Jumars' (1984) findings of higher volumetric ingestion rate with higher food protein content to be consistent with the maximization of absorption would imply (cf. Fig. 5B) that $n_1 > 1$ and $C_{F0} < 1$ for the protein ingested as bovine serum albumin. In any event, neither their nor our present results are easily reconciled with the idea of digestive homeostasis (Calow, 1982; Cammen, 1989). An animal seeking to achieve a constant rate of absorptive gain at any fixed C_{F0*} would be faced with the quandary of having two alternative solutions at any rate below the maximal achievable AR* (Fig. 4).

The coupled model does help to refine the concept of food quality. Information on quality is contained within C_{F0} as well as within all four kinetic parameters of the digestive and absorptive reactions (V_{max}, W_{max}, K_m and M_m). This five-parameter definition of food quality is consistent with the observation that animals select against foods that they do not have the enzymatic capability to digest or absorb (Martinez and Stevens, 1989) Penry and Jumars (1987) point out that physical form and diffusional constraints may complicate the kinetics of digestion, and they certainly can complicate absorption as well. Before testing of the present model, however, we hesitate to add this degree of complexity. Some degree of phase heterogeneity in the reactions and some degree of diffusional control of kinetics can be accommodated in the Michaelis-Menten parameterization without serious violation of its interpretation.

Quantity of food enters the model most directly as throughput rate, q_0, which can be converted easily to τ in a fixed-volume system: τ = (gut volume)/q_0. Throughput time also is directly related to X_f (Eq. 6). If the quantity of food available is unlimited, then optimization would argue for the solutions depicted in Fig. 4. Solving for the optimal processing rate under an ingestion rate set lower than the food quantity-unlimited optimum is a simple matter of truncating the range of q_0, τ, and X_f over which the maximal AR is sought. Just as simple intuition would suggest, food shortage should drive τ and X_f up; intuition, however, might not predict the degree of nonlinearity shown above $X_f = 0.9$ (Figs. 3 and 4).

Whether or not the details of this model hold up under scrutiny, the focus it provides on net absorption rate as the determinant of growth potential suggests a critical re-appraisal of fundamental metabolic scaling arguments. The observation that net growth of animals scales with body mass raised to an exponent between 2/3 and 1 is usually explained as indicative of decreased metabolic demand per unit of body weight with increasing body weight. We find this explanation inconsistent when metabolic components are placed in their logical sequence. Prior gain clearly determines whether an animal has material to oxidize at present; present oxidation much less clearly indicates future gain (e.g., Kristensen, 1989). There are many ways to reduce metabolic demand, especially in small individuals (spores, seeds, resting stages); in heterotrophs there are no ways to oxidize material that was not absorbed. The very fact that smaller animals show more rapid population growth (e.g., Fenchel, 1974) is inconsistent with the overriding importance of metabolic "demand." Since they supposedly have a higher weight-specific metabolic demand, smaller organisms should fare more poorly.

The greater growth potential of smaller organisms is far more consistently explained by the supply side of metabolism. The data clearly demonstrate that smaller organisms can grow faster, not that small organisms must grow faster. Isometric growth increases absorptive sites only as body volume raised to a power of 2/3. Allometries such as increasing gut length in proportion to volume or increasing ramification of gut linings with animal size can raise the exponent to some extent, but the diffusion advantage of small individuals in gaining nutrients at a greater rate is physically irrevocable. At some point further ramification of absorptive sites into the gut must displace gut contents to an extent that decreases net rate of gain. For these reasons we hope to see the approaches of Karasov and Diamond (1988) extended to examine the issue of the scaling of absorption rate with body size within species.

Model extended

That we work on deposit feeders helps explain many of the assumptions of the model. These animals characteristically spend most of their time feeding and appear far more rarely limited in growth potential by food quantity than are most other trophic groups. They also are incapable of digesting much of their food on a volumetric basis.

Our reasons for not starting out, say, with the more complicated and possibly more generally applicable assumption that volume of gut contents changes significantly during digestion goes beyond a predilection for animals that eat sand and mud. Volume reduction during digestion can have two important effects. One is to permit input rate to the gut to exceed output rate ($q_0 > q_f$) from the gut by a significant amount; thus a gut of a given size can allow greater ingestion rate when a significant fraction of the volume ingested can be digested and absorbed away (e.g., Lehman, 1976). A more subtle but possibly equally important effect of volume change during digestion and absorption will be to vary the concentration of digestive enzymes and thereby V_{max}. Both effects are reasonably straightforward to model, but are likely to be highly species specific and diet specific in magnitude. Here we have endeavored to provide a template for the parts of the system that are common to most species with the hopes that it will be modified as applications demand. Specifying the extent to which a food parcel can be digested and absorbed further allows one to use the digestion-absorption theory and the principle of lost opportunity to set an upper bound on selection costs that an animal should bear. Such costs should be no larger than the added rate of absorptive gain stemming from selection.

Along similar lines, our assumption that digestion and absorption kinetics are constant along the full length of the gut need to be modified for applications to particular species. In several kinds of invertebrates, however, digestive and absorptive functions are broadly distributed axially (e.g.: Michel, 1988; Féral, 1989). A useful and fairly general elaboration of the model might divide the gut into three segments with successive functions of primarily digestion (foregut and stomach), digestion plus absorption (midgut and hindgut), and neither (rectum). The purely digestive portion may be treated with the equations of Penry and Jumars (1987). The intestine may be treated with the present equations by modifying the input stream to have a non-zero C_{P0} determined by output from the foregut. Digestion could be allowed to continue beyond the actively absorbing midgut by taking the midgut's output as input for the digestive equations. Modifying the equations in this way requires specification of gut geometry and thus is less general than the template we provide, but it raises the very exciting possibility of determining an optimal geometry of digestion and absorption, linking the study of histology and internal morphology with the ecology of foraging. Another interesting complication would be to calculate an optimal retention time and diet choice for a mixture of food items varying in quality. An obviously needed further addition is to allow digestive and absorptive kinetics to vary with time -- with adaptive state of the individual (e.g., Karasov et al., 1983)

An added benefit of these extensions would be to better define the chemical environment of an animal's gut and the chemical output from it. The latter has obvious advantage in studies of material processing (e.g., Jumars et al., 1989). The former provides a context for interpreting the results of past observations of microbial growth in animal guts (e.g.: Deming and Colwell, 1982; Drasar and Barrow, 1985; Plante et al., 1989) and for designing new experiments. Our demonstration that a certain amount of digestive inefficiency is optimal in terms of the gut owner's absorptive rate of gain explains the existence of hindgut and fecal resources for microbial growth and for utilization by detritivores.

To extend the essence of Pooh's comment to Piglet beyond the joy of throughput, there are many aspects of feeding that can be exciting. The model presented here is for us a truly exciting advance in ongoing efforts to understand and to predict the influence of digestive and absorptive processes on foraging behavior. Indeed, the insight of supply-side optimization has whetted anew our appetite for inquiry into the ecology of feeding.

SUMMARY

We extend a model of optimal digestion to the step of absorption by way of a generalized performance analysis of guts behaving like continuous, plug-flow chemical reactors hosting series reactions at steady state. The model predicts an optimal rate of ingestion or gut residence time without inclusion of explicit costs. Costs are included implicitly via the principle of lost opportunity: processing faster or slower than the optimum leads to lesser rates of absorption. This criterion for optimality can be evaluated explicitly by noting that maximal absorptive rates of digestive products in simple tubular guts are achieved when marginal gain in absorption at the end of the gut just equals gut-averaged absorptive rate. In simple tubular guts hosting classical Michaelis-Menten digestive and absorptive kinetics, optimal throughput rates of food in unlimited quantities are predicted to <u>decrease</u> with increasing food quality. Higher-order digestive reaction kinetics, however, lead to throughput rate maxima at intermediate food qualities. Animals acting to maximize individual rates of absorption of digestive products are predicted to egest substantial components of undigested food and unabsorbed digestive products, with the efficiency of absorption and rate of egestion of usable organic matter both rising with food quality. Perspectives advanced by this model highlight the role of digestion-absorption reaction kinetics in foraging behavior.

ACKNOWLEDGEMENTS

Our modeling efforts and their exposition were supported by the U.S. Office of Naval Research Contract N00014-87-K-0160 and National Science Foundation Grant OCE 86-08157.

REFERENCES

Bayne BL, Hawkins AJS, Navarro E (1987) Feeding and digestion by the mussel *Mytilus edulis* L. (Bivalvia: Mollusca) in mixtures of silt and algal cells at low concentrations. J Exp Mar Biol Ecol 111:1-22

Boher RN, Lampert W (1988) Simultaneous measurement of the effect of food concentration on assimilation and respiration in *Daphnia magna* Straus. Funct Ecol 2:463-472

Briggs GE, Haldane JBS (1925) A note on the kinetics of enzyme action. Biochem J 19:338-339

Calow P (1982) Homeostasis and fitness. Am Nat 120:416-419

Cammen LM (1989) The relationship between ingestion rate of deposit feeders and nutritional value. In: Lopez GR, Taghon GL, Levinton JS (eds) Ecology of marine deposit feeding. Springer-Verlag, New York, p. 201

Davenport HW (1977) Physiology of the digestive tract, 4th edn. Year Book Medical Publishers, Inc., New York

Deming JW, Colwell RR (1982) Barophilic bacteria associated with digestive tracts of abyssal holothurians. Appl Environ Microbiol 44:1222-1230

Drasar BS, Barrow PA (1985) Intestinal microbiology. Am Soc Microbiol, Washington, D.C.

Fenchel T (1974) Intrinsic rate of natural increase: the relationship with body size. Oecol 14:317-326

Féral J-P (1989) Activity of the principal digestive enzymes in the detritivorous apodous holothuroid *Leptosynapta galliennei* and two other shallow-water holothuroids. Mar Biol 101:367-379

Fisher RB, Parsons DS (1953) Glucose movements across the gut wall of the rat small intestine. J Physiol 119:210-223

Huston M, DeAngelis D, Post W (1988) New computer models unify ecological theory. BioSci 38:682-691

Jumars PA, Penry DL (1989) Digestion theory applied to deposit feeding. In: Lopez GR, Taghon GL, Levinton JS (eds) Ecology of marine deposit feeding. Springer-Verlag, New York, p. 114

Jumars PA, Penry DL, Baross JA, Perry MJ, Frost BW (1989) Closing the microbial loop: dissolved carbon pathway to heterotrophic bacteria from incomplete ingestion, digestion and absorption in animals. Deep-Sea Res 36:483-495

Karasov WH, Diamond JM (1983) A simple method for measuring intestinal solute uptake in vitro. J Comp Physiol 152:105-116

Karasov WH, Diamond JM (1988) Interplay between physiology and ecology in digestion. BioSci 38:602-611

Karasov WH, Pond RS III, Soberg DH, Diamond JM (1983) Regulation of proline and glucose transport in mouse intestine by dietary substrate levels. Proc Nat Acad Sci USA 80:7674-7677.

Kiørboe T, Möhlenberg F, Hamburger K (1985) Bioenergetics of the planktonic copepod *Acartia tonsa:* relation between feeding, egg production and respiration, and composition of specific dynamic action. Mar Ecol Progr Ser 26:85-97

Kofoed L, Forbes V, Lopez GR (1989) Time-dependent absorption in deposit feeders. In: Lopez GR, Taghon, GL, Levinton JS (eds) Ecology of marine deposit feeding. Springer-Verlag, New York, p. 129.

Kristensen E (1989) Oxygen and carbon dioxide exchange in the polychaete *Nereis virens:* influence of ventilation activity and starvation. Mar Biol 101:381-288

Lehman JT (1976) The filter-feeder as an optimal forager, and the predicted shapes of feeding curves. Limnol Oceanogr 21:501-516

Levenspiel O (1972) Chemical reaction engineering, 2nd edn. Wiley, New York

Maritinez del Rio C, Stevens BR (1989) Physiological constraint on feeding behavior: intestinal membrane disaccharides of the starling. Science 243:794-796

Michel C (1988) X. Intestine and digestive glands. In: Westheide W, Hermans CO (eds) The ultrastructure of Polychaeta, Microfauna marina, vol 4. Gustav Fischer Verlag, Stuttgart. p. 157

Penry DL, Jumars PA (1986) Chemical reactor analysis and optimal digestion theory. BioSci 36:310-315

Penry DL, Jumars PA (1987) Modeling animal guts as chemical reactors. Am Nat 129:69-96

Phillips NW (1984) Compensatory intake can be consistent with an optimal foraging model. Am Nat 123:867-872

Plante CJ, Jumars PA, Baross.JA (1989) Rapid bacterial growth in the hindgut of a marine deposit feeder. Microbial Ecol, in press.

Sibly, RM (1981) Strategies of digestion and defecation. In: Townsend CR, Calow P (eds) Physiological ecology: an evolutionary approach to resource use. Sinauer Associates, Sunderland, MA

Stephens DW, Krebs JR (1986) Foraging theory. Princeton University Press, Princeton, New Jersey

Taghon GL (1988) The benefits and costs of deposit feeding in *Abarenicola pacifica.* Limnol Oceanogr 33:1166-1175

Taghon GL (1989) Modeling deposit feeding. In: Lopez GR, Taghon GL, Levinton JS (eds) Ecology of marine deposit feeding. Springer-Verlag, New York, p. 223

Taghon GL, Jumars PA (1984) Variable ingestion rate and its role in optimal foraging behavior of marine deposit feeders. Ecol 65:549-558

Vermeij GJ (1987) Evolution and escalation: an ecological history of life. Princeton University Press, Princeton, New Jersey.

Wilson TH (1962) Intestinal absorption. W. B. Saunders Company, Philadelphia, Pennsylvania

ORAL DISCUSSION – P.A. Jumars

Q. (Real) How would you go about measuring the kinetic parameters in your model ?
A. The tough one in our system, because of the small sizes of the animals, is knowing the characteristics of the digestive enzymes in situ. The usual techniques of grinding whole animals release stored enzymes and so cannot be used to infer activities in the gut lumen. Karasov and Diamond's technique, on the other hand, is a straightforward way of estimating the absorption kinetics. It might be possible from net changes plus absorption kinetics to back out digestive kinetics.

Q. (Hobbs) Have you examined the effects of gut fullness ?
A. In the model as written, there is no change in gut fullness due either to uncoupled variations in ingestion and egestion or to absorption. In deposit feeders such changes are minimal. I have modeled (separately from the present text) the effects of absorptive change in volume; as you would expect absorption of a significant volumetric fraction of the digesta effectively increases the ingestion rate that a gut of given volume can accommodate.

Q. (Hobbs) What is the meaning of the exponent, n, in the hyperbolic model ?
A. Values of n less than one indicate more nearly linear kinetics of uptake and values of n above one indicate more sigmoidal uptake kinetics (more nonlinear) than the classic Michaelis–Menten form. Values of n less than one are suggestive of measurement artifacts or of passive absorption (e.g. of volatile fatty acids in ruminants). Values of n above one may be quite common in nature if digestion and absorption are rate limited by reactions running in parallel.

Q. (DeMott) Can you offer the animal two different particles of different digestibilities ?
A. Your question can be rephrased in two different ways that both can be addressed with the present model by changing its boundary conditions: How much is selection worth (in absorption rate per time) and what is the optimal gut retention time for an animal constrained to eat a mixture of two food types of differing kinetics ? Both are very interesting questions well worth pursuing.

FORAGING STRATEGY OF A DEPOSIT FEEDING CRAB

R. K. Zimmer-Faust
Marine Environmental Sciences Consortium
and Aquatic Biology Program
University of Alabama
P.O. Box 369-370
Dauphin Island, Alabama 36528
U.S.A.

Animals living in complex natural habitats commonly exhibit simple, stereotyped behavior that is highly efficient in acquiring resources (*e.g.* Breck, 1978; Waage, 1979; Hanski, 1980; Ydenburg, 1984; Zimmer-Faust, 1987). Such behavior reduces central nervous system processing and simplifies decision making (Oaten, 1977; Iwasa et al., 1981; Anderson, 1983). Bookstaber and Langsam (1985) have demonstrated that stereotyped behavior can evolve under conditions of extended uncertainty, where high costs in neural investment or energetic expenditures may make it optimal for an animal not to be fully responsive to fine-grain environmental cues. Given these findings, it is important for investigators to consider the temporal and spatial scales over which foragers integrate sensory information in discriminating food resources. By sampling distributions of food on an inappropriately large scale, or by averaging measurements over time or space, investigators may conclude that one habitat yields higher payoffs than another, even though on a finer scale, there may be significant overlap in distributions of food that obscure a forager's discrimination ability. The time and energy that a forager needs to discriminate high quality patches in complex natural habitats can become so large that the most efficient response is simply to harvest each patch encountered. Patches may then be harvested at random with respect to quality, though it might be expected that forager movements between patches will not be random in order to minimize travel and to avoid patches already visited.

This paper focuses on foraging by the deposit-feeding crab, *Scopimera inflata*, for which field observations and experiments can be performed to tease apart effects of food gradients, predation, and intraspecific competition on foraging efficiency. I summarize results that describe the natural history of grazing by crabs which I demonstrate is highly stereotyped yet well-adapted to take advantage of natural distributions of food. I briefly discuss effects of predation and intraspecific competition on crab foraging and show that: (1) contests for space among crabs are quickly and non-injuriously resolved to maximize time spent feeding, and (2) contest winners and losers forage with nearly equal efficiency, but contest losers feed further away from burrow entrances and may be exposed to higher risks of predation. *Scopimera* is an energy maximizer (Schoener, 1971; Hixon, 1982) that lives in extreme, food-poor habitats.

NATO ASI Series, Vol. G 20
Behavioural Mechanisms of Food Selection
Edited by R. N. Hughes

Scopimera inflata: A model organism for investigating foraging strategy

The sand bubbler crab, *Scopimera inflata*, is a member of the family Ocypodidae, which includes *Uca* (fiddler crabs) and *Ocypode* (ghost crabs) among other genera. The reproductive (*e.g.* Christy, 1978; 1983; Salmon and Hyatt, 1983; Christy and Salmon, 1984) and foraging ecologies (*e.g.* Robertson et al., 1980, 1981; Murai et al., 1982, 1983; Robertson and Newell, *1982a*, *b*; Zimmer-Faust, 1987) of this family have received considerable attention recently. However, strategies employed by foraging crabs have been largely unexplored.

Scopimera employs a flotation feeding mechanism similar to that of other ocypodid crabs (Fielder, 1970). Water, originally stored in the gill chambers, is used to separate and suspend organic materials from sand (Miller, 1961; Fielder, 1970; Wolcott, 1976; Quinn, 1980; Robertson and Newell, 1982*a*). Particulate organics adhere to specialized spoon-tipped setae on the second maxillipeds and are then ingested. Spherical pellets of rejected inorganic materials accumulate in the dorsal buccal region; they are periodically removed by the chelae, swept underneath the body and kicked to the rear by the animal (Fig. 1). Feeding crabs move sideways along a radius from the burrow. They alternatively lift substrates to the mouth with chelae, feed, and then deposit a pellet whose mass is directly proportional to the size of the crab that made it (Fig. 2). This process is repeated and a row of pellets is formed tangential to

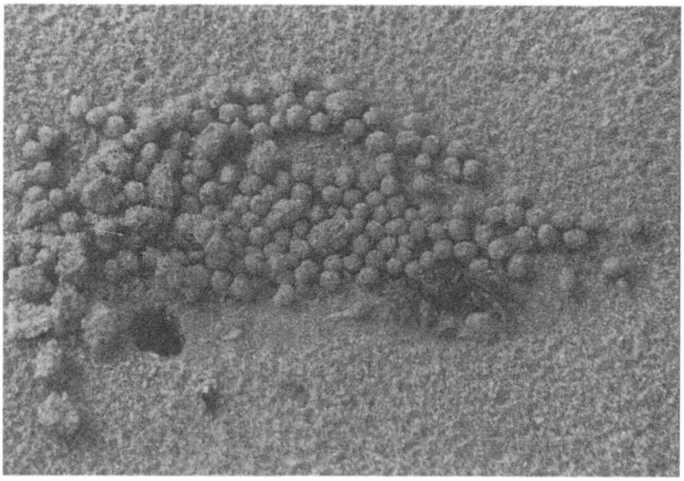

Figure 1. Circular path foraged by *Scopimera*. Feeding is initiated by a crab when it emerges from a burrow and begins scraping the top layer of substrates, lifting materials to its mouth. Separation is made of the lighter organics from heavier inorganics, which accumulate as a spherical pellet. Pellets are deposited on the sand. Small, spherical pellets are produced in feeding, while larger cylindrical pellets (lower left) are produced in burrow maintenance and construction. Movement in this figure is clockwise with respect to the burrow entrance. Crab body size (carapace width) is about 10 millimeters.

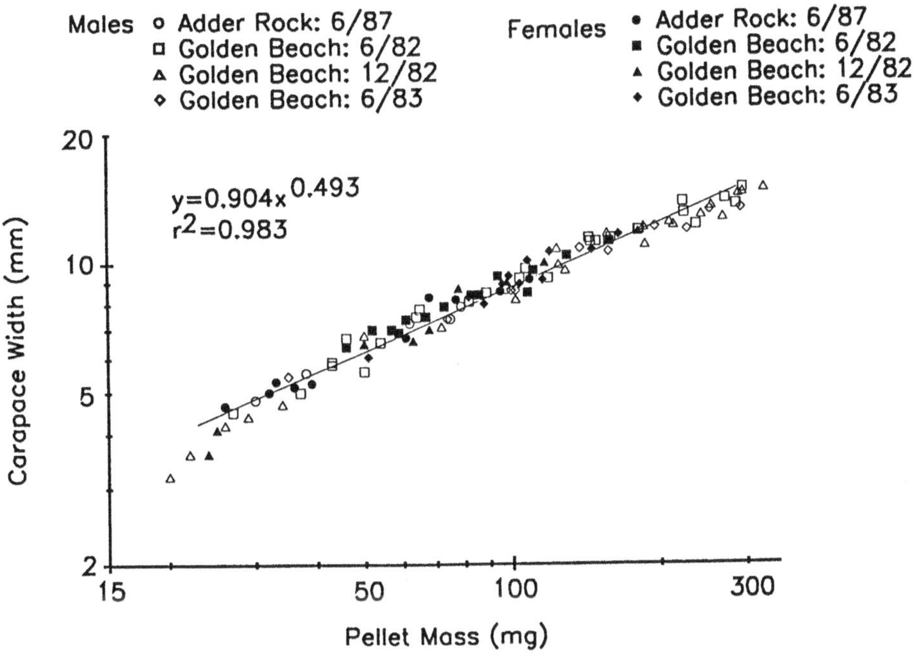

Figure 2. Pellet mass can be used to predict crab size (carapace width) when capture is impossible. Pellets produced by male and female crabs were sampled at Golden Beach, Caloundra, in June and December, 1982, and in June, 1983, and at Adder Rock Beach, North Stradbroke Island, in June, 1987. There are no effects on pellet mass associated with crab gender, study site, or time at which pellets were collected.

the burrow entrance. When a crab reaches a position, usually $\leq$ 30 cm away from the burrow entrance, it momentarily stops feeding, takes one step forward, and then forages back to the burrow. The crab next steps forward and continues foraging outward from the burrow. Confluent rows of pellets become radially arranged around the burrow entrance as feeding continues, and a trough becomes visible on the edge most recently fed where surface substrates (0-3 mm depth) have been lifted, harvested, and removed to form pellets (Fig. 1).

The foraging behavior of *Scopimera* is remarkably consistent in time and space. It was first described by Fielder (1970) based on observations he made of crabs grazing on sandy beaches near Townsville, Queensland, Australia, in the year, 1965. I observed thousands of crabs grazing on beaches near Brisbane, Queensland, Australia, about 900 km south of Townsville, during June, 1982 - June, 1983, and again during June and July, 1987. My observations were essentially identical to those reported by Fielder (1970). The unique but stereotyped behavior of *Scopimera*, and the ease with which this crab can be observed and its habitat experimentally manipulated make *S. inflata* a model organism for study.

Natural history of *Scopimera*

Scopimera lives on sandy beaches in tropical and sub-tropical eastern Australia where it establishes burrows in the high inter-tidal zone (1.5-2.1 MLLW). Crabs emerge from burrows to feed only on the ebb tide during daylight and they feed only on surface sediments (0-3 mm depth), because food (= biomass) concentrations are significantly higher than in sub-surface sediments (Zimmer-Faust, 1987). The beaches on which crabs live and feed are characterized by well-sorted, intermediate grain sand (median particle diameter: 265-320 μm) having chlorophyll *a* values of 1.25-2.50 μg/g substrate (dry weight) and adenosine triphosphate values of 0.3-0.8 μg/g substrate over a yearly cycle. Organic carbon and nitrogen concentrations of 0.3-0.9 mg/g substrate and 0.01-0.08 mg/g substrate are common, respectively. These biomass concentrations are lower than values reported for sediments in other littoral and in deep-sea environments (P.A. Jumars, *pers. comm.*), and they are 3-20 times lower than concentrations measured in substrates harvested by other species of deposit-feeding crabs (Ono, 1965; Icely and Jones, 1978; Robertson et al., 1980; Quinn, 1983).

Foraging strategy

Scopimera begins foraging anywhere around its burrow entrance. The starting position and direction of movement then determine where crabs subsequently feed. Each crab moves either clockwise or anticlockwise, though once a direction is chosen it is maintained constant for an entire day. Systematic grazing of this type reduces the chance that a crab feeds in areas previously grazed and, for this reason, it is a commonly observed phenomenon among other invertebrate and vertebrate grazers (Janzen, 1971; Heinrich, 1976; Gill and Wolf, 1977; Kamil, 1978; Corbet et al., 1981). When resources are distributed homogeneously or so complexly that animals cannot efficiently sample or interpret food gradients, circular paths optimize net resource yields (Covich, 1976; Andersson, 1978; Hixon, 1980). By grazing a circular path *Scopimera* minimizes distances traveled from burrows while it nearly maximizes use of substrates immediately adjacent to burrows.

Food gradients

Is *Scopimera* an "optimal" forager? To answer this question, it must be determined how food resources distribute in habitats occupied by crabs. I sampled substrates from the habitats of 27 crabs (nine crabs chosen at random every other month, for five months; *see* methods in Zimmer-Faust, 1987). Crabs were allowed to graze freely for 15 minutes to establish both a clear foraging position and a direction of movement around a burrow entrance. A circle with the burrow at the center and a radius extending to the most distant pellet was taken as the foraging range. Circular foraging ranges were divided into four equal sectors and substrates from each sector were collected then analyzed for chlorophyll *a*, ATP, and ash-free dry mass (AFDM) as indicators of biomass. Sectors around each burrow were ranked one through four according to biomass concentration and ratios were calculated to determine the relative change in biomass for sectors of consecutive ranks and for sectors of highest and lowest ranks. Ratios were computed by dividing the value of biomass in the higher ranked sector by that in the lower ranked sector. Sectors of consecutive rank differed little in biomass (ratios: 1.07-1.18), while sectors of highest and lowest ranks differed greatly (ratios: 1.32-1.51). This last result indicates that biomass was not homogeneously distributed. To determine whether measurements were affected by sector size, I divided the foraging ranges of six additional animals into 16 equal sectors for which biomass determinations were made. Consecutively ranked sectors differed by a mean ratio of 1.06 ($\pm$ 0.01 *sem* AFDM), while sectors of highest and lowest ranks differed by a mean ratio of 2.56 ($\pm$ 0.82 *sem* AFDM). Sector size quantitatively affected measurements, but the qualitative conclusions were unchanged; biomass is not homogeneously distributed though differences are slight between sectors of consecutive rank.

Were distributions of biomass predictably or randomly arrayed around burrow entrances? To answer this question, I constructed a model for comparison with empirical data. The range foraged by a single crab was assumed to be circular and divided into any number (n) of sectors. It was further assumed that within the habitat of each individual crab, sectors could be ranked in descending order of biomass. Ranks of adjacent sectors were taken to covary at random. For these conditions, the sum of absolute differences between ranks of adjacent sectors was:

$$\sum_{x=1}^{n-1} x(n-x) \qquad (1)$$

with a mean difference of

$$\frac{\sum_{x=1}^{n-1} x(n-x)}{\sum_{x=1}^{n-1} n-x} \qquad (2)$$

where x is the frequency of a given difference in rank (1,2,3,... n-1).

When foraging ranges of single crabs were divided into four or eight sectors, empirical differences in ranks of adjacent sectors were nearly identical to those predicted by the model (random assortment; Fig. 3A and B; G test for goodness-of-fit with Williams' correction; four sectors: G = 0.165, df = 2, P > 0.90; eight sectors: G = 3.54, df = 5, P > 0.50). Slightly different results were obtained when 16 sectors were considered; empirical values then differed significantly from those predicted (Fig. 3C; G test: $P < 0.05$). However, the average difference observed of 4.695 between ranks of

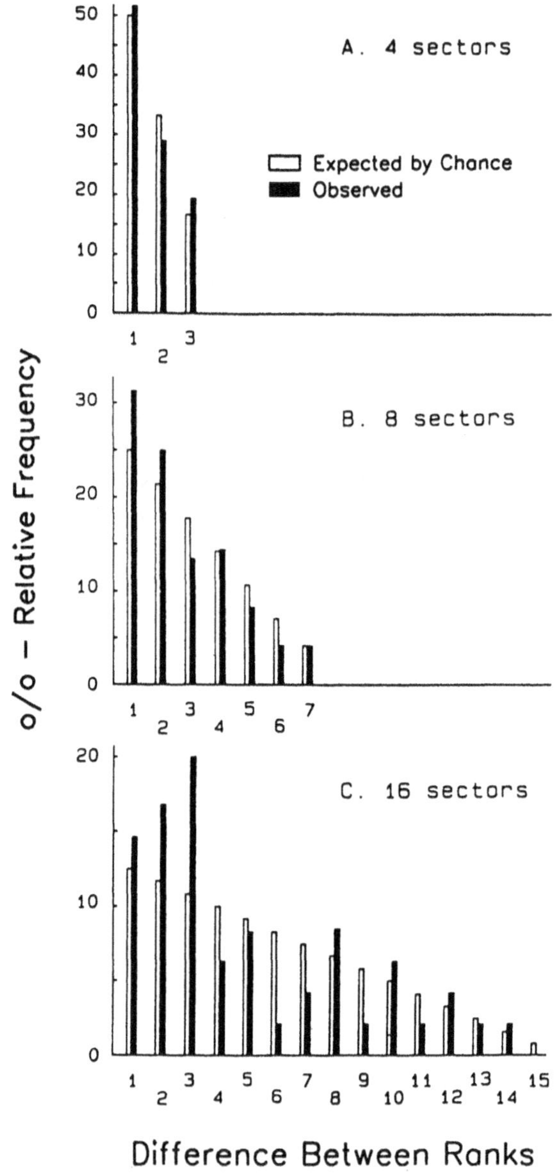

Figure 3. Relative frequency distributions comparing observed with expected differences in rank between adjacent sectors. Ranks were assigned in descending order of biomass concentration. Determinations using eight sectors were made by sampling biomass from 16 sectors, averaging values for adjacent sectors, then assigning ranks. Data are combined for (A) 27 animals and 4 sectors (N = 108), (B) 6 animals and 8 sectors (N = 48) and (C) 6 animals and 16 sectors (N = 96). Expected values were generated by assuming that the ranks of adjacent sectors covaried at random (after Zimmer-Faust, 1987).

adjacent sectors was similar to the 5.667 predicted if ranks covaried at random. This last finding would be expected if spatial distributions of biomass deviated slightly from randomness, but also if sector size had become smaller than microhabitat patches. I conclude that biomass distributions are complex and nearly distributed at random within circular habitats of individual crabs.

If substrates near burrows are food-poor and food distributions are complex, why do crabs not leave and feed elsewhere? One reason is that burrows act as the only refuges guarding crabs against desiccation and avian predation and *Scopimera* is unable to dig new burrows in substrates exposed on sandy shores (Zimmer-Faust, *unpubl. data*). Another reason possibly is there may be little or no added benefit in feeding away from burrows. To test this latter hypothesis, I collected surface substrates (0-3 mm depth) from twelve, 1-m^2 quadrats chosen at random on each of three transects (50-m length) positioned parallel to shore in "high" (1.6-2.1 m MLLW), "high-intermediate" (1.2-1.6 m MLLW) and "low-intermediate" (0.8-1.2 m MLLW) inter-tidal zones. Crabs occupied sediments only in the "high" zone which, on its landward margin, bordered a supra-littoral zone that was dry, irregularly flooded and clearly unsuitable for feeding. Substrate samples were collected at Golden Beach once every other month over a one year period (June, 1982 - June, 1983) and at Adder Rock Beach and at Amity Point, North Stradbroke Island, periodically during 1982 and 1983, and again once in 1987. I found that the difference in mean biomass concentration between transects was only occasionally significant ($P < 0.05$, in 3 of 10 days sampled; one-way ANOVA's with replicates), and where significance occurred, the zone inhabited by crabs always was highest in biomass. These results demonstrate that crabs gain no additional energetic benefit by foraging away from burrows.

Effects of conspecific interactions on foraging

Fielder (1970) noted that crabs visually displayed to one another by waving claws and he attributed this behavior to intraspecific competition for food or space. This conclusion seems plausible given that feeding ranges of crabs frequently overlap and that food abundances are low in habitats occupied by crabs. In studying how conspecific interactions affect foraging efficiency, my goals were to: (1) repeat Fielder's (1970) observations to determine whether crabs interfere with one another, (2) determine whether, or not, interference is associated with contests for space or food, (3) determine the amount of resource won or lost in contests, and (4) determine how time allocation differs between contestants and non-contestants.

I watched a total of 116 crabs from the time they first emerged on an ebb tide until they finished feeding as flood tide approached. Thirty-five of these crabs engaged in contests where they waved claws, or combated, as they contacted each other and disputed areas where feeding ranges overlapped. "Winning" contestants foraged areas of dispute while "losing" contestants moved away to graze other areas. Larger crabs won 85.9% of the contests although mean body size (carapace width) of winning contestants, 7.36 mm ($\pm$ 1.61 mm *SD*), was less than one millimeter larger than mean body size (carapace width) of losing contestants, 6.52 mm ($\pm$ 1.51 mm *SD*).

Interestingly, time-activity budgets for contestants and non-contestants were nearly identical because crabs spent most of the time feeding (Fig. 4) and agonistic displays were short, usually lasting less than two minutes.

I tested the hypotheses that contest winners extend feeding ranges at the expense of ranges grazed by contest losers. Because surface-area and depth grazed by crabs depend on body size (Fig. 5), I limited analyses to the 12 conflicts in which contestants differed by less than one millimeter in carapace width. For these conditions, I found no significant difference in either surface-area, depth, travel time, or substrate biomass harvested by contest winners and losers (Wilcoxen signed rank test: $T_x \geq 33$, $N \geq 12$, $P \geq 0.30$, for all comparisons). These results demonstrate that competition does not affect feeding or foraging efficiency. I also found that contest losers harvested

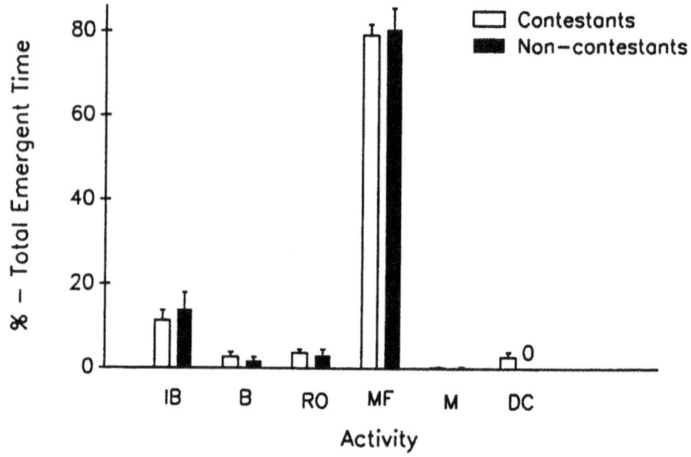

Figure 4. Daily time-activity budgets for 116 crabs measured during July, 1987, at Adder Rock Beach, North Stradbroke Island; 35 of these crabs contested space. Habitats were exposed by ebb tides for periods of 5-6 hours during daytime. There are no significant differences in the activity budgets of contestants and non-contestants, except for "display and combat" (Student's *t*-test: $P < 0.01$, using arc-sin transformed data). IB: crab was in burrow, B: crab was burrowing, RO: crab was resting outside of burrow usually at the entrance, MF: crab was moving and feeding, M: crab was moving but not feeding, DC: crab was displaying or combating. Bars are ± one standard deviation.

substrates situated, on average, several centimeters further away from burrow entrances than did contest winners (Wilcoxen test: $T_x = 16$, $N = 12$, $P < 0.05$). This finding leads me to believe that competition may affect crab exposure to predation, because *Scopimera* suffers high mortality at some study sites where shorebirds eat as much as 0.6% of total, emerged crab populations per day. Many shorebirds, including sandpipers, exhibit short, high acceleration runs that are effective in catching surface-feeding crabs, including *Scopimera*, even as crabs flee to burrows (Boshe, 1982; Zwarts, 1985; Piersma, 1986*a,b*). Whether contest winners and losers suffer different rates of mortality has yet to be determined, consequently, unmistakable connections between competition and predation risk in *Scopimera* have still to be established. It is believed that avian predators are important elements that structured the central-place foraging and mating behaviors of deposit feeding crabs (Christy and Salmon, 1984). Therefore,

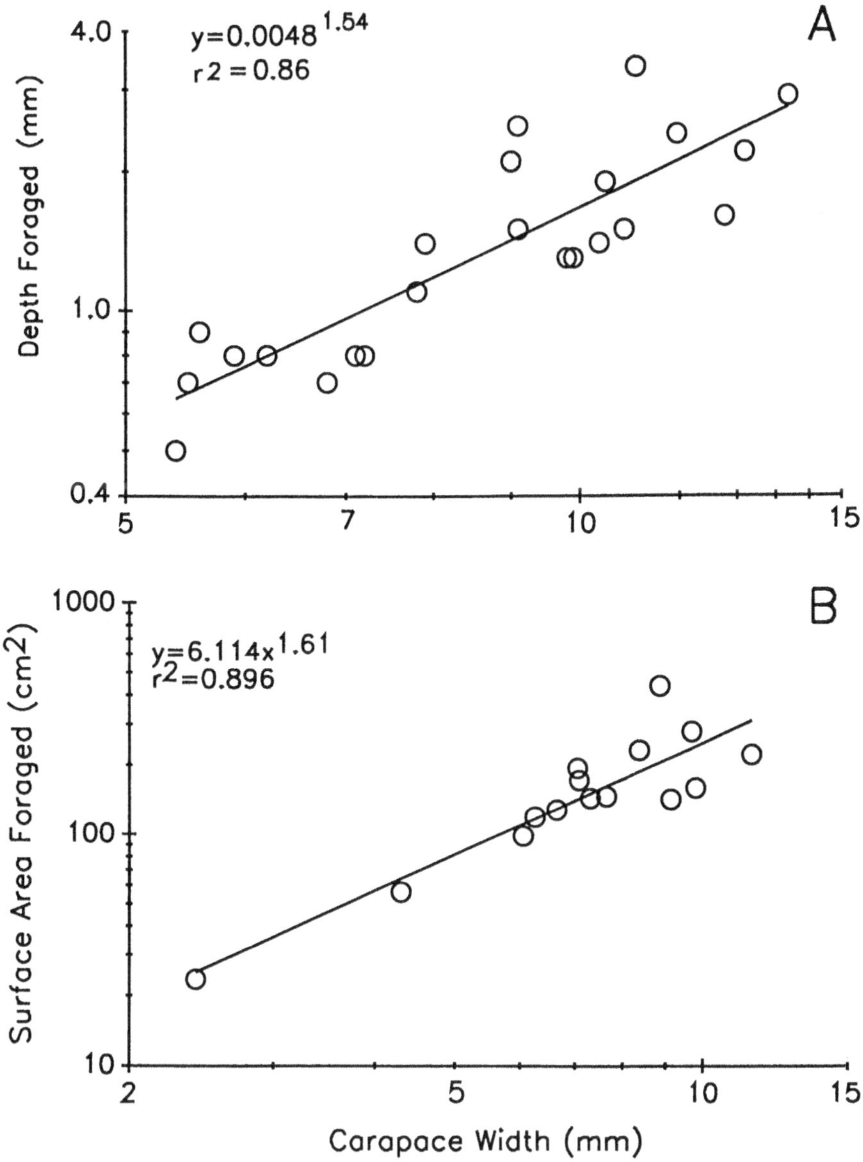

Figure 5. (A) Depth and (B) total surface-area grazed during ebb tide depend on crab body size (carapace width). Data were collected for crabs feeding at Golden Beach, Caloundra, during July, 1982; crabs did not have overlapping feeding ranges. Depth determinations were made using dental impression material (polysiloxane) that I cast in templates, while surface-area determinations were made using photographic records and using a computer digitizing pad.

determinations of the roles played by food and predators in mediating foraging by *S. inflata*, and in limiting crab abundances, will be exciting avenues for future research.

Summary

The deposit-feeding crab, *Scopimera inflata*, exhibits foraging behavior that is remarkably consistent in time and space. Crabs live in food-poor habitats and they graze circular paths around burrow entrances. The behavior of *Scopimera* minimizes path lengths and distances traveled away from burrows yet nearly maximizes use of substrates immediately adjacent to burrows. Crabs forage near burrows because there is no added benefit in feeding away from burrows (although there are added costs). Crabs frequently are aggregated in space and they contest areas where feeding ranges overlap. Contest winners and losers forage with nearly equal efficiency, but contest losers graze significantly further from burrow entrances. I hypothesize that, for a crab of given size, probability of being eaten by shorebirds covaries positively with crab distance from a burrow, consequently, the resource won or lost in competition appears not to be food but rather exposure to predation. I have yet to determine whether contest winners and losers suffer different rates of mortality, therefore, unmistakable connections between intraspecific competition and predation risk in *Scopimera* have still to be established.

Acknowledgements

This work was supported by a post-doctoral research fellowship and a faculty research grant from the University of Queensland, Australia, and by a grant from the University of California Research Expeditions Program. I thank D. Fielder and J. Kikkawa for logistical support in field research and for use of laboratory facilities. M. Tamburri assisted with data analysis, while G. Crozier provided encouragement and computer facilities. C. Wood, R. Horton and L. Bryant graciously assisted with figures and word processing. Thanks go to T. Piersma for his suggestions on literature pertaining to bird predation on deposit feeding crabs and to C. Jorgensen for comments that greatly improved earlier drafts of this manuscript. Travel awards by the University of Alabama and by NATO allowed me to attend this conference. I am especially grateful to R. Hughes for convening this workshop and for acting as editor, and to H. Hughes for assisting as conference organizer. This work is contribution number 164 from the Marine Environmental Sciences Consortium.

Literature Cited

Anderson DJ (1983) Optimal foraging and the traveling salesman. Theor Pop Biol 24:145-159

Andersson M (1978) Optimal foraging area: size and allocation of search effort. Theor Pop Biol 13:397-409

Bookstaber R, Langsam J (1985) On the optimality of coarse behavioral rules. J

Theor Biol 116:161-193
Boshe JI (1982) Predation of fiddler crabs *Uca stenodactyla* (Ocypodidae) by the common shore birds in Pangani Beach, Tanzania. Afr J Ecol 20:237-240
Breck JE (1978) Suboptimal foraging strategies for a patchy environment. Dissertation Michigan State University, East Lansing, Michigan, USA
Christy JH (1978) Adaptive significance of reproductive cycles in the fiddler crab *Uca pugilator*: a hypothesis. Science 199:453-455
Christy JH (1983) Female choice in the resource-defense mating system of the sand fiddler crab, *Uca pugilator*. Behav Ecol Sociobiol 12:169-180
Christy JH, Salmon M (1984) Ecology and evolution of mating systems of fiddler crabs (Genus *Uca*). Biol Rev 59:483-509
Corbet SA, Cuthill I, Fallows M, Harrison T, Hartley G (1981) Why do nectar-feeding bees and wasps work upward on inflorescences? Oecologia 51:79-83
Covich AP (1976) Analyzing shapes of foraging areas: some ecological and economic theories. Annu Rev Ecol Syst 7:235-257
Fielder DR (1970) The feeding behaviour of the sand bubbler crab *Scopimera inflata* (Decapoda, Ocypodidae). J Zool 160:35-49
Gill FB, Wolf LL (1977) Nonrandom foraging by sunbirds in a patchy environment. Ecology 58:1284-1296
Hanski I (1980) Movement patterns in dungbeetles and in the dung fly. Anim Behav 28:953-964
Heinrich B (1976) The foraging specializations of individual bumblebees. Ecol Monogr 46:105-128
Hixon M A (1980) Food production and competitor density as the determinants of feeding territory size. Am Nat 115:510-530
Hixon MA (1982) Energy maximizers and time minimizers: theory and reality. Am Nat 119:596-599
Icely JD, Jones DA (1978) Factors affecting the distribution of the genus *Uca* (Crustacea: Ocypodidae) on an East African shore. Estuar Coast Mar Sci 6:315-332
Iwasa I, Higasi M, Yanamura N (1981) Prey discrimination as a factor determining the choice of optimal foraging strategy. Am Nat 117:710-723
Janzen DH (1971) Euglossine bees as long-distance pollinators of tropical plants. Science 171:203-205
Kamil AC (1978) Systematic foraging by a nectar-feeding bird, the Amakihi (*Loxops virens*). J Comp Physiol Psychol 92:388-396
Miller DC (1961) The feeding mechanism of fiddler crabs, with ecological considerations of feeding adaptations. Zoologica 46:89-100
Murai M, Goshima S, Nakasone Y (1982) Some behavioral characteristics related to food supply and soil texture of burrowing habitats observed on *Uca vocans vocans* and *U. lactea perplexa*. Mar Biol 66:191-197
Murai M, Goshima S, Nakasone Y (1983) Adaptive droving behavior observed in the fiddler crab *Uca vocans vocans*. Mar Biol 67:159-164
Oaten A (1977) Optimal foraging in patches: a case for stochasticity. Theor Pop Biol 12:263-285
Ono Y (1965) On the ecological distribution of ocypodid crabs in the estuary. Mem Fac Sci Kyushu Univ Ser E 56:165-177
Piersma T (1986*a*) Foraging behaviour of Terek Sandpipers *Xenus cinereus* feeding on sand-bubbling Crabs *Scopimera globosa*. J Orn 127:475-486
Piersma T (1986*b*) Eastern curlew *Numenius madagascariensis* feeding on *Macrophthalmus* and other ocypodid crabs in the Nakdong estuary, South Korea. Emu 86:155-160
Quinn RH (1980) Water uptake and feeding in the soldier crab, *Mictyris longicarpus* Laetrille, 1806 (Decapoda, Mictyridae). Dissertation, University of Queensland, Brisbane, Australia
Robertson JR, Bancroft K, Vermeer G, Plaisier K (1980) Experimental studies on the

foraging behavior of the sand fiddler crab *Uca pugilator* (Bosc, 1802). J Exp Mar Biol Ecol 44:67-83
Robertson JR, Fudge JA, Vermeer GK (1981) Chemical and live feeding stimulants of the sand fiddler crab *Uca pugilator* (Bosc). J Exp Mar Biol Ecol 53:47-64
Robertson JR, Newell SY (1982*a*) Experimental studies of particle ingestion by the sand fiddler crab *Uca pugilator* (Bosc). J Exp Mar Biol Ecol 59:1-21
Robertson JR, Newell SY (1982*b*) A study of particle ingestion by three fiddler crab species foraging on sand sediment. J Exp Mar Biol Ecol 65:11-17
Salmon M, Hyatt GW (1983) Spatial and temporal aspects of reproduction in North Carolina populations of fiddler crabs (*Uca pugilator* Bosc). J Exp Mar Biol Ecol 70:21-43
Schoener TW (1971) Theory of feeding strategies. Annu Rev Ecol System 11:369-404
Waage J K (1979) Foraging for patchily-distributed hosts by the parasitoid, *Nemeritis canescens*. J Anim Ecol 48:353-371
Wolcott TC (1976) Uptake of capillary water by ghost crabs. Nature 264:756-757
Ydenberg RC (1984) Great Tits and giving-up times: decision rules for leaving patches. Behaviour 90:1-24
Zimmer-Faust RK (1987) Substrate selection and use by a deposit-feeding crab. Ecology 68:955-970
Zwarts L (1985) The winter exploitation of fiddler crabs *Uca tangeri* by waders in Guinea-Bissau. Ardea 73:3-12

ORAL DISCUSSION – R.K. Zimmer–Faust

Q. (Godin) Is there a relationship among temperature, foraging range and extraction efficiency ?
A. Temperature doesn't affect extraction efficiency but does affect time in burrow.

Q. (Ansell) Do the crabs increase distance from their burrow with increased foraging time ?
A. Yes, there is more vigilance behaviour at the beginning of a feeding bout, but after half an hour they are out, foraging full circle.

Q. (Jørgensen) How do they stabilise the sand balls ?
A. The crabs use some organic material, but I don't know exactly what it is.

Q. (Lucas) How much are they moving around from day to day ?
A. In a low energy environment, very little. But I cannot follow individuals easily from day to day.

Q. (Gilliam) Is this an intertidal region ?
A. Very high intertidal, but there are dynamic processes driving the renewal of nutrients in the sand from day to day.

Q. (Sih) Have you measured optimal patch use by the crabs ?
A. No, but the crabs extract 20 – 30% of the nutrients within the biomass level to which they are constrained. They seem to stay in an area for a given period of time before moving on. They don't seem to compare where they are feeding now with where they fed before. Crabs may never face a tractable choice situation in the natural habitat because distributions of biomass are too complex, given that crabs must contact substrates to assess their worth. Selective grazing may be inefficient due to high sampling costs and neural investment needed to interpret complex food gradients.

RETENTION EFFICIENCY, PERCEPTUAL BIAS, AND ACTIVE CHOICE AS MECHANISMS OF FOOD SELECTION BY SUSPENSION-FEEDING ZOOPLANKTON

William R. DeMott
Department of Biological Sciences and
Crooked Lake Biological Station
Indiana University-Purdue University at Fort Wayne
Fort Wayne, Indiana 46815
U.S.A.

Suspension-feeding zooplankton live in environments in which nutritious algae are mixed with particles of low food value, such as silt, detritus, and indigestible or toxic algae (Porter 1977). In this situation an ability to select nutritious particles should be advantageous, provided that the cost of handling and rejecting individuals particles is low. Nonetheless suspension feeding is often considered synonymous with "filter feeding", a term which suggests a passive, mechanical mode of food collection. For a filter feeder, food selection is largely a function of the range of particle sizes which can be retained by the feeding apparatus and successfully ingested (Boyd 1976). Within the past 10 years, however, a rapidly growing number of studies have shown that many taxa of zooplankton are not mechanical filter feeders but are able to use complex behaviors to select between individual particles which differ size or nutritional value (reviewed by Price 1988).

With only a few exceptions among predatory species, zooplankton are nonvisual feeders; taxa which actively select between individual particles must use chemical and mechanical cues to detect and handle particles and to assess their nutritional value. Experiments with artificial particles provide evidence for a sharp dichotomy between mechanical filter feeders and taxa which use chemical cues to assess the nutritional value of individual food items. For example, experiments with algal-flavored and untreated polystyrene spheres revealed three kinds of responses among diverse taxa of freshwater zooplankton (DeMott 1986). Some species fed nonselectively (cladocerans *Daphnia*, *Ceriodaphnia*, *Diaphanosoma*, *Chydorus*, and rotifer *Brachionus*), others fed preferentially on flavored spheres (calanoid and cyclopoid copepods, cladoceran *Bosmina* and rotifer *Filinia*), whereas a third group infrequently ingested any spheres, even though they readily feed on algae of similar sizes (rotifers *Polyarthra*, *Keratella*, and *Synchaeta*).

NATO ASI Series, Vol. G 20
Behavioural Mechanisms of Food Selection
Edited by R. N. Hughes

Presumably these rotifers were not "fooled" by the flavor treatment.

Since particle size is not a consistent indicator of nutritional value, mechanical filter feeding provides little opportunity for specializing on high-quality particles. In specific circumstances, however, restricting the size range of particles collected may be advantageous. For example, small cladocerans may benefit from their inability to collect and ingest large filaments or colonies of toxic cyanobacteria (Fulton and Paerl 1987a, b). Large cladocerans such as *Daphnia* can respond to the presence of large toxic or interfering particles by narrowing the gap between the valves of the carapace (Gliwicz and Siedlar 1980). This behavior restricts the entry of large particles but also reduces feeding rates for small high-quality particles.

The ability of filter-feeding cladocerans to capture very small bacteria-sized particles is not related to body size but is correlated with the "mesh-size" of the fine setules on the filtering limbs (Gophen and Geller 1984; Brendelberger 1985; DeMott 1985; Hessen 1985). While an ability to eat bacteria is considered a significant advantage (Geller and Müller 1981), the potential benefits of a coarse mesh-size have received little attention. It could be advantageous to avoid eating fine particles in lakes and reservoirs with high loads of suspended clay. Alternatively, a coarse filter provides less resistance to water flow, and might, therefore, significantly reduce energy expenditure or allow higher maximal feeding rates. A current controversy focuses on energetic costs and even whether water actually flows through the "filtering limbs". Using models of fluid mechanics, Brendelberger et al. (1986) calculated that the cost of sieving water through the fine setules of *Daphnia* would be only a few percent of the total energy budget. In contrast, Gerritsen et al. (1988) calculated that *Daphnia* must use 20-60% of its total metabolism in sieving but used their results to support the hypothesis that water does not flow through the "filtering limbs".

Because of the importance of calanoid copepods in oceans and lakes, most of our knowledge of the behavioral mechanisms by which zooplankton select high-quality diets comes from studies of this group (reviewed by Koehl 1984). In addition to experiments with artificial particles (reviewed by DeMott 1988a), food selection experiments have shown that calanoid copepods can discriminate between live and dead algae of the same species (Starkweather and Bogdan 1980; Paffenhöfer and Van Sant 1986; DeMott 1988b), readily digested and digestion-resistant algae (DeMott 1989), nontoxic and toxic algae (Huntley et al. 1986; Fulton and Pearl 1987a; DeMott 1989), and even nitrogen-rich and nitrogen-limited algae of the same species (Cowles et al. 1988; Butler et al.

1988). Despite these capabilities, analyses of gut contents and fecal pellets of copepods from natural populations provide evidence of rather broad diets, often including substantial amounts of detritus, silt, and other low-quality particles (e.g. Turner 1984; Hart 1987). Thus, predicting diets in nature will require a detailed knowledge of both the behavioral mechanisms of selection and how the degree of discrimination is influenced by internal and environmental variables.

Direct visual observations using high-speed cinematography reveal that calanoid copepods can detect individual particles at a distance and may orient their body position or alter feeding currents to increase the probability of capture (Strickler 1982). Larger algae are detected at greater distances and are more likely to elicit precapture responses (Price et al. 1983; Price and Paffenhöfer 1984; Vanderploeg and Paffenhöfer 1985). Thus, preferences for larger algae may simply reflect an increased probability of detection, a perceptual bias, rather than a decision-based process. Decision-based selection ("active choice") appears to be important in selection based on nutritional value rather than particle size. High-speed movies show that captured particles are manipulated and often appear to be tasted at the mouth before being ingested or rejected. Low-quality particles such as fecal pellets or plastic beads are more likely to be rejected after capture than are living algae (Paffenhöfer and Van Sant 1985; Vanderploeg and Paffenhöfer 1985).

A current controversy concerns whether chemical or mechanical signals are used in remote detection. Because crustaceans are known to have very sensitive and finely-tuned chemosensory capabilities (reviewed by Atema 1987a), most workers have assumed that distance chemoreception ("smell") is used in remote detection. A living alga is surrounded by a "phycosphere" within which the concentration of sugars and amino acids and other excreted matter is thought to exceed background levels (Bell and Mitchell 1972). According to a model developed by Andrews (1983), as a particle in the feeding current is moved toward a copepod, the phycosphere associated with the alga is sheared and, as a result, is deformed and elongated. This leading edge of the phycosphere could then be detected by chemosensory cells on the antennae and mouthparts. Larger algae presumably excrete more chemicals and are therefore detected at greater distances.

The feasibility of particle detection via distance chemoreception depends on both the sensitivity of the animal's sensory organs and the "signal to noise ratio" in the environment (Atema 1987b; Zimmer-Faust in press). Legier-Visser et al (1986) point out that zooplankton loose considerable amounts of dissolved organic matter as algae are handled and sometimes broken during feeding. They argue that elevated concentrations of algal products in the vicinity of a feeding copepod would create a chemical noise which would be elevated far above the signal produced by the exudates of an approaching alga. As an alternative, Legier-Visser et al. (1986) suggest that copepods can use mechanoreception to detect pressure disturbances created by suspended particles. Larger particles would create a greater disturbance and would therefore be detected at greater distances. Whereas the importance of mechanoreception in the detection of swimming prey by predatory zooplankton has been demonstrated (e.g. Kerfoot 1978), the use of mechanoreception to detect small, nonmotile particles has not been established. Since cyclopoid copepods detect animal prey via swimming disturbances, Price (1988) suggested that this form of mechanoreception might explain the reported preferences of certain small, marine cyclopoids for motile algae (e.g. Uchima and Hirano 1986).

The chemoreception and mechanoreception hypotheses of distance detection imply different patterns and mechanisms of selection, especially between living algae and nonliving particles. According to the mechanoreception hypothesis of Legier-Visser et al. (1986), essentially all selection based on nutritional quality occurs after capture as copepods decide whether to reject or ingest particles after tasting at the mouth. Distance chemoreception adds two additional possibilities, one involving decision-based selection and the other a perceptual bias. In the first, termed "passive rejection" by Strickler (1984), particles are detected and identified but no capture responses are initiated. In this way the costs of capturing and rejection low-quality particles could be avoided. Copepods would also show reduced feeding on detritus and inert particles because these would lack the chemical odors necessary for detection.

Understanding the underlying mechanisms of food selection is crucial in the development and application of predictive models to zooplankton food selection (reviewed by Hughes 1980). Because zooplankton have been considered mechanical filter feeders, some models have focused on optimal ingestion rates without considering selection (e.g. Taghon 1981) or have assumed that food selection would be a function of particle size and feeding limb morphology (e.g. Lam and Frost 1976). Vanderploeg et al. (1984) argued that perceptual biases should result in fixed ("invariant") size preferences among high-quality algae. Unlike the invariant preferences predicted by models of mechanical retention efficiency, these size preferences would not be expected to correlate with feeding limb morphology. Lehman (1976) presented three models; one predicting optimal ingestion rates, a second which assumed mechanical selection based on size, and a third based on active choice between particles of differing digestibility.

Unlike conventional optimal diet models (reviewed by Stephens and Krebs 1986) theoretical models for suspension-feeding zooplankton have assumed that handling time is negligible and that gut processing places upper limits on ingestion rates (Lehman 1976; Lam and Frost 1976; Taghon 1981). It follows from these assumptions that nutritional benefit can be maximized by food selection based on particle quality rather than particle size. Lehman's (1976) optimal diet model assumed that the "filter feeder" first captures particles and rejects or ingests individual items after assessing their nutritional value. If rejection costs are not significant, acceptance of low-quality particles depends solely on the abundance of high-quality particles. The proportion of low-quality particles in the diet should decline to zero when the abundance of better particles is sufficient to saturate gut processing. On the other hand, if rejection costs are significant, the optimal diet depends on the densities of all particles. When poorer particles become abundant, they should be included in the diet because their total rejection would become too costly. Thus, in common with other optimal diet models, Lehman's (1976) model predicts that discrimination against low ranking foods should be strong when preferred foods are abundant and weak when preferred foods are scarce.

In a recent test, the calanoid copepod *Eudiaptomus* spp. exhibited shifts in selectivity which agreed with the qualitative predictions of Lehman's (1976) optimal diet model (DeMott 1989). Low-quality particles including digestion-resistant algae and dead algae were weakly discriminated against at low food concentrations and strongly discriminated against at high food concentrations. Strong, invariant discrimination against a toxic blue-green was considered consistent with optimal foraging, since ingestion would be deleterious even when other food sources are scarce or absent. In a second recent study, Vanderploeg et al. (1988) showed that *Diaptomus sicilis* discriminates against large, difficult-to-handle net diatoms when *Chlamydomonas* is abundant. In this situation, handling times and handling costs could be important.

Direct visual observations of zooplankton feeding often require expensive filming equipment and tedious analyses. Unlike the situation with terrestrial vertebrates (Zach and Smith 1981), however, the small size of zooplankton and the relative homogeneity of pelagic environments makes it possible to conduct simple but realistic food selection experiments under controlled laboratory conditions. This paper has used six series of laboratory experiments as a framework for examining a wide range of mechanisms for food selection by suspension-feeding, crustacean zooplankton:

1. Previous studies have shown a correlation between filter mesh-size and the ability of cladocerans to feed on very small particles but have not directly considered the advantages of a coarse filter. This question is considered in experiments with *Sida* and *Diaphanosoma*, closely related cladocerans with coarse and fine filter setules, respectively.

2. A preliminary experiment investigated the effects of dissolved algal extracts on selection between large and small high-quality algae by a calanoid copepod. Assuming that distance chemoreception is important in the detection of large algae, I predicted that the addition of algal extracts would interfere with detection, thereby decreasing size selectivity.

3. An experiment tested whether a variety of freshwater zooplankton would discrimination between high-quality motile and nonmotile algae. Preferential feeding on motile algae could indicate that the "swimming disturbance" is used in prey detection.

4. Both perceptual biases for large particles and active discrimination based on food quality need to be considered in copepod feeding. I examined the interaction between these two factors in selection experiments pairing large, low-quality particles with small, high-quality particles.

5. Lehman's (1976) optimal diet model predicts that gut fullness and hunger rather than food concentration *per se* control feeding rates and food selection behavior. This hypothesis was tested by investigating the effects of starvation on selection between high- and low-quality particles by calanoid copepods.

6. Responses to algal toxicity were tested in experiments pairing a toxic and a presumably nontoxic species of the cyanobacterium *Oscillatoria*.

Methods

Experimental animals and food resources--Food selection experiments were run at three laboratories using cladocerans and copepods collected with plankton net tows at nearby lakes: the Franz Theodore Stone Laboratory, Put-In-Bay, Ohio (Lake Erie); the Max Planck Institute of Limnology, Plön, West Germany (Schöhsee); and the Crooked Lake Biological Station, near Columbia City, Indiana (Little Crooked Lake). Two additional cladocerans, *Daphnia pulicaria* and *Bosmina longispina*, were cultured at the Max Planck Institute. Animals from the field were used in experiments on the same day as collection or maintained in 2-L jars with filtered lake water and fed cultured algae for 1-10 days prior to use in an experiment.

Table 1 lists the sources and characteristics of the eight species of algae used in feeding experiments. Exponentially growing cultures were prepared for experiments. Algae for experiments conducted at the Max Planck Institute were cultured in a modified Chu 12 medium as described in DeMott (1989). Algae for experiments conducted in the U.S.A. were grown in batch cultures in Erlenmeyer flasks with a modified MBL medium (Stemberger 1981). Algal dimensions were estimated by measuring 50 cells or colonies of each taxon with a microscope.

Readily digested, nontoxic algae were classified as "high quality" without regard to size or the presence of spines. I also used a strain of *Oscillatoria agardhii* from Lake Washington which has been shown to inhibit growth and to increase mortality in two species of *Daphnia* (Infante and Abella 1985). The nutritional value and toxicity of a morphologically similar congener (*Oscillatoria* sp.) has not yet been tested. Algae with gelatinous sheaths are often poorly assimilated (Porter 1976) and can survive gut passage (Kerfoot et al. 1985). Even when these algae are digested, the presence of a gelatinous sheath should result in low nutritional gain per volume ingested. Algae which were heat-killed 2 d prior to experiments were considered models of detritus, a low-quality food type (see DeMott 1988b, 1989).

Experimental methods and design--Two basic kinds of laboratory feeding experiments were conducted. The first tested selection between polystyrene microspheres which differed in size. The second examined selection between pairs of radioactively labeled algae (live or dead) which differed in size, shape, motility or nutritional value. In each experiment animals were acclimated to unlabeled algae and then allowed to feed on microspheres or radioactively labeled algae for a period shorter that the gut passage time. Sensitive and precise estimates of selectivity and feeding rates were made by counting microspheres within the guts of individuals or by using dual-label liquid scintillation counting to determine the radioactivity of ingested algae.

Experiments testing selection between 2- and 6-μm microspheres by cladocerans *Diaphanosoma birgei* and *Sida crystallina* followed methods of DeMott (1986). Previous studies have shown that *Diaphanosoma* does not discriminate between algal-flavored and untreated microspheres (DeMott 1986) and that *Sida* does not discriminate between algae and microspheres of the same size (DeMott 1988). Thus, microspheres should provide a realistic indication of particle size selection by these taxa. Animals were

Table 1. Sources and characteristics of algae used in feeding experiments. Size is expressed as mean length and mean width (greatest axial linear dimension).

Taxa	Source[1]	Food Quality	Length (μm)	Width (μm)
Chlamydomonas reinhardi	UT90	high	7	6
Chlamydomonas sphaeroides	UG58.72	high	12	12
Cartaria sp.	CBS	high	28	24
Scenedesmus acutus	MPI	high	11	5
Scenedesmus quadricaudus	CBS	high	45[2]	30
Pandorina morum	CBS	gelatinous	40	36
Oscillatoria agardhii	UW	toxic	1060	5
Oscillatoria sp.	CBS	?	760	5

[1]UT-University of Texas Culture Collection; UG-University of Göttingen Culture Collection; MPI-Max Planck Institute of Limnology; CBS-Carolina Biological Supply; UW-University of Washington Depart. of Botany

[2]Does not include 4 spines with mean length of 24 μm.

collected from Lake Erie and acclimated for 2 h to a low concentration of *C. reinhardi* in filtered lake water before the addition of low, equal concentrations (by number) of the two sizes of microspheres. Due to differences in feeding rates, the smaller cladoceran was given a higher concentration of microspheres (*Diaphanosoma*, 1,000 microspheres/mL of each size; *Sida*, 200 or 500 microspheres/mL of each size). After a 10-min feeding trial, animals were anesthetized in carbonated water and preserved in a formalin-sucrose mixture. Animals were placed on microscope slides, cleared with tissue solubilizer, and all microspheres in guts were counted with a microscope.

Experiments in which one cell type was labeled with [^{14}C]bicarbonate and the other with [^{32}P]orthophosphate were used to quantify selection between pairs of algal foods. A single preliminary experiment tested whether the addition of soluble algal extract would interfere with feeding on a mixture of large and small high-quality algae (*Cartaria* sp. and *Chlamydomonas reinhardi*). To prepare a soluble extract, approximately equal density mixtures (by cell volume) of the two algae were filtered onto glass fiber filters (GF/C), frozen, thawed, sonicated, refiltered and centrifuged. The concentration of dissolved organic carbon (DOC) in the supernate was estimated by subtracting the amount of particulate organic carbon (POC) collected after extraction from the amount of POC measured in untreated samples. POC was estimated in replicate treated and untreated samples using a wet oxidation technique (Wetzel and Likens 1979). Animals were acclimated to a low concentration of unlabeled algae and then allowed to feed for 10 min in a suspension containing radioactively labeled algae and dissolved extract.

As in DeMott (1989), experiments designed to test the effects of food concentration on selection behavior consisted of four treatments for each pair of particles: low quantities (0.125 ppm) of one particle alone, low quantities (0.125) of the alternative particle alone, mixtures of both particles in low but equal mixtures (0.25 ppm total), and mixtures of both particles in high but equal volumes (2.0 ppm total; 1.0 ppm = $10^6 \mu m^3$ mL^{-1}). Volumes of 0.125 and 1.0 ppm represent about 25 and 200 µgC/liter of high-quality algae. Animals were acclimated to experimental concentrations of unlabeled algae for 2-6 h and then allowed to feed on labeled algae for 10 min.

Experiments to test the effects of hunger on the food selection behavior of calanoid copepods involved either starving animals in filtered (GF/C glass fiber) lake water for 24 h and measuring selection at various times after the introduction of high but equal concentrations of two food types (2.0 ppm total), or acclimating animals to high but equal concentrations of two food types (2.0 ppm total) for 24 h and then measuring selection after varying periods in filtered lake water.

Methods for preparing radioactively labeled algae followed DeMott (1988a) with one exception. In most experiments labeled algae were concentrated by centrifugation and rinsed with isotope-free culture media to remove dissolved isotope. In the one exception, radioactively labeled filaments of *Oscillatoria* spp. were filtered through a 500-µm mesh screen to remove clumps and excessively long filaments and collected and rinsed on a 100-µm mesh screen. Radioisotope experiments were run in beakers with 300, 500, or 600 mL of feeding suspension following the protocol of DeMott (1989). Clearance rates were calculated by comparing the specific activities (dpm/ind.) of animals with the specific activities (dpm/mL) of suspended particles. Basic methods for dual-label scintillation counting and the calculation of feeding rates followed Lampert and Taylor (1985).

Data analysis--Food selection was quantified by the selectivity coefficient α (Chesson 1983). All experiments involved pairs of particles and particle depletion was negligible due to short duration. In this simple case, α is the ratio of the clearance rate on one food to the sum of the clearance rates on both food types. The index ranges from 0 to 1.0 with a value of 0.50 indicating nonselective feeding. Statistical comparisons between experimental treatments involved a two-step process. First a t-test or one-way ANOVA was used to determine whether selectivity varied between treatments (food concentration, algal exudates, or acclimation time). When selectivity did not vary between treatments, values of α were pooled and a t-test determined whether feeding

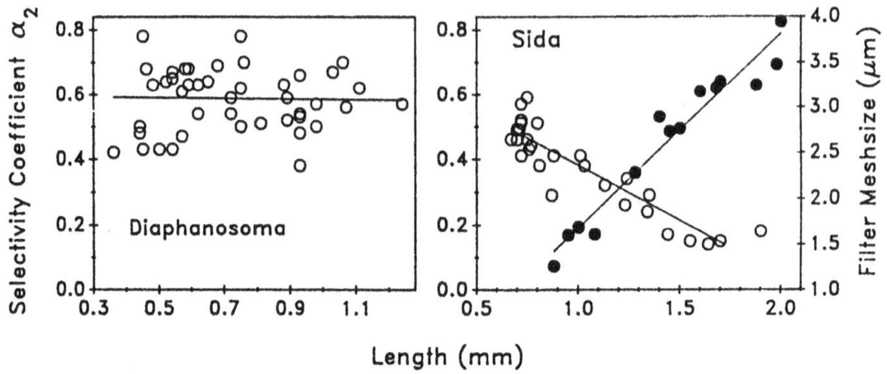

Fig. 1. Relations between cladoceran body size and selection between 2- and 6-μm polystyrene spheres. Each open circle represents a selectivity coefficient for individual *Diaphanosoma birgei* and *Sida crystallina* from Lake Erie. Values greater or less than 0.50 indicate preferences for 2- or 6-μm spheres, respectively. Filter mesh-size of the endites of the third limbs of *Sida* (filled circles) was measured as described in DeMott (1985).

was selective or nonselective (α different from 0.50). Statistical tests are based on three replicate beakers per treatment. Error bars in figures show ± 1 SE.

Results

Selection between particles of similar nutritional value--Diaphanosoma and *Sida* belong to the same cladoceran family (Sididae) and exhibit striking similarity in the placement and gross morphology of the five pair of filtering limbs which each uses to collect food particles (Cannon 1933). They represent opposite extremes, however, in the spacing of the filter setules on these limbs. *Diaphanosoma* possesses the finest filter setules reported among Cladocera (intersetular distance <0.5 μm; Geller and Müller, 1981), whereas *Sida crystallina* possesses the coarsest filter setules. In *Sida*, filter mesh-size increases linearly with body length (Fig. 1, filled circles; mesh-size (μm) = 2.13*body length (mm) - 0.45, $r^2 = 0.95$; $p<0.001$). Experiments testing selection between 2 and 6μm polystyrene microspheres revealed differences between *Diaphanosoma* and *Sida* which correspond to differences in the gaps between their filter setules. *Diaphanosoma* exhibited no relation between selectivity and body size (correlation coefficient r = -0.024) but showed a weak, yet significant preference for 2-μm spheres (α_2, mean$\pm$SE, 0.59$\pm$0.02, t-test, $p<0.001$). In contrast, *Sida* exhibited a strong decline in the

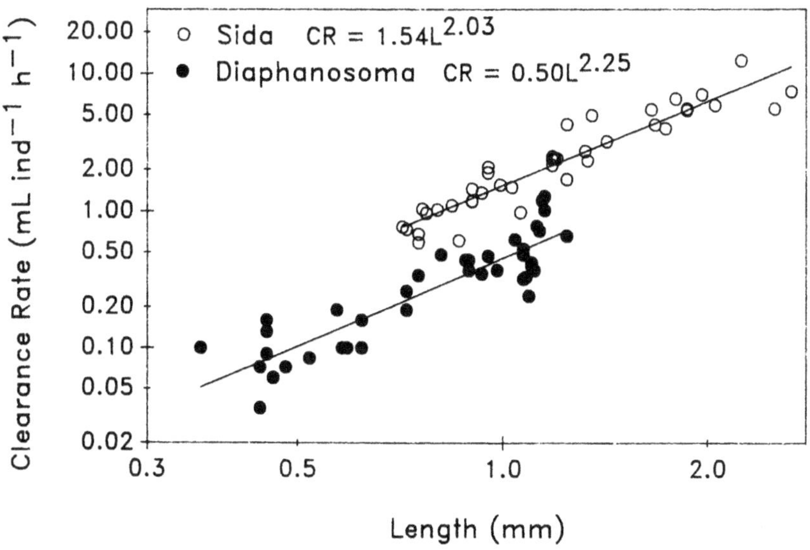

Figure 2. Relation between body length and the clearance rates of two cladocerans for polystyrene spheres. Each circle represents an individual; lines are least-squares regression lines. Clearance rates were calculated for the more efficiently collected particle size for each species (2-μm spheres for *Diaphanosoma* and 6-μm spheres for *Sida*).

proportion of 2-μm spheres ingested with increases in body size ($r = -0.91$, $p<0.0001$) corresponding to increases in the porosity of its filtering limbs.

The clearance rates of *Diaphanosoma* and *Sida* both increased with approximately the square of body length (Fig. 2). The clearance rate of *Sida* was, however, about three times higher than that of *Diaphanosoma* (ANCOVA comparing the two regressions, $p<0.001$). For example, the predicted clearances rates for individuals 1 mm in length are: *Sida*, 1.54 mL/h and *Diaphanosoma*, 0.50 mL/h. The highest clearance rate measured was 12.4 mL/h by a *Sida* 2.2 mm in length.

Figure 3 shows the results of a test of the effects of food concentration and a soluble algal extract on selection by a calanoid copepod between two high-quality flagellates. The concentration of DOC added was 2.6 mgC/L, about 50 times higher than the concentration of algal POC. Despite very high precision in estimates of clearance rates and selectivity neither algal extracts nor food concentration influenced selectivity. In each treatment, copepods exhibited a strong invariant preference for the larger flagellate (*Cartaria* sp.) over the smaller flagellate (*C. reinhardi*; t-test on pooled α values, selective feeding, $p<0.0001$).

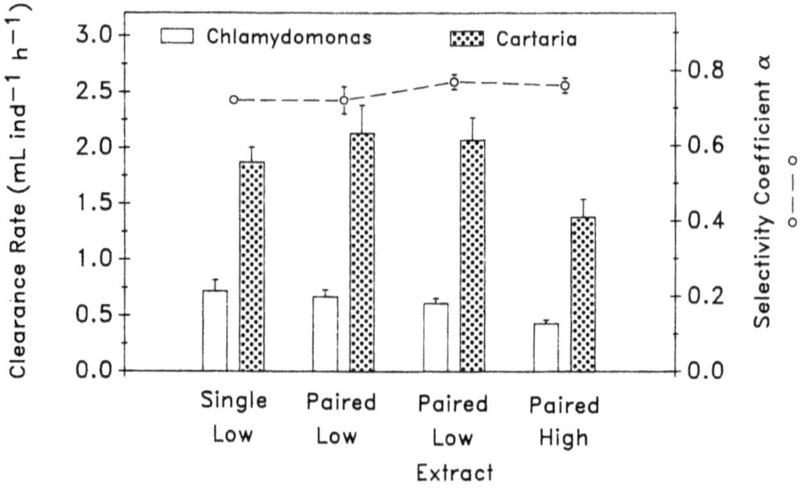

Figure 3. Effects food concentration and dissolved algal extract on selection between large (*Cartaria* sp.) and small (*C.reinhardi*), high-quality flagellates by adult female *Diaptomus birgei* from Little Crooked Lake.

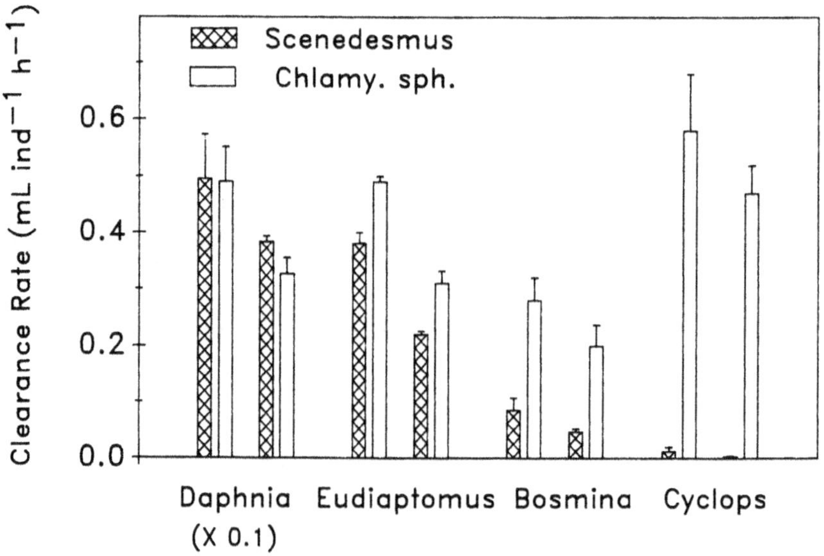

Fig. 4. Selection between a motile green alga (*C. sphaeroides*) and a slightly smaller nonmotile green alga (*S. acutus*). Experimental animals included adult Cladocera from laboratory cultures (*Daphnia pulicaria* and *Bosmina longispina*) and copepods from Schöhsee (adult female *Cyclops kolensis* and mixtures of adult and late copepodites of *Eudiaptomus gracilis* and *E. graciloides*). The first and second pairs of bars for each species represent feeding rates at low and high food concentrations, respectively. Feeding rates of large *Daphnia* (2.0-2.5 mm length) were multiplied by 0.1.

Feeding rates in respective single-species suspensions were good predictors of feeding rates and selectivity in mixtures.

Two cladocerans and two copepods feeding within the same beakers showed marked differences in selection between a motile green alga and a slightly smaller nonmotile green alga (Fig. 4). Although none of the four taxa exhibited a change in selectivity with changing food concentration, only *Daphnia* fed nonselectively on the two food types (α, mean $\pm$ SE, 0.52 ± 0.02). *Eudiaptomus*, *Bosmina*, and *Cyclops* each showed significant preferences for larger, motile species (t-tests on α's, each $p < 0.001$). Selectivity was weak for *Eudiaptomus* (α, 0.43 ± 0.01), intermediate for *Bosmina* (0.21 ± 0.02) and very strong for *Cyclops* (0.012 ± 0.008).

Selection between particles which differ in nutritional quality--Two experiments were designed to test for interactions between particle size, nutritional value, and food concentration. In these experiments, the small, high-quality flagellate *C. reinhardi* was paired large low-quality particles, either dead *Cartaria* or gelatinous *Pandorina morum* (Fig. 5, upper and middle panels). In each experiment *Diaptomus birgei* fed selectively on large low-quality particles at low food concentrations but exhibited a preference for the small, high-quality alga at high food concentrations (t-tests on selectivity coefficients, each $p < 0.01$). A similar shift in selectivity was observed when live *Cartaria* sp. was paired with an alga which was assumed, *a priori*, to be high in nutritional value but difficult to handle and ingest. *Scenedesmus quadricaudus* consisted primarily of rectangular colonies bearing stout spines at each corner (Table 1). Both food types were eaten at high rates when offered at low concentrations, alone or together, but *S. quadricaudus* was strongly discriminated against in high-concentration mixtures (Fig. 5, bottom, $p < 0.01$).

Two experiments were designed to separate the effects of food concentration and hunger on selection between high- and low-quality foods. When *D. birgei* was starved for 24 h and then fed high-concentration mixtures of *C. reinhardi* and dead *Cartaria* it initially fed selectively on the large dead alga but showed a rapid shift in selectivity during the first 30 min of acclimation to the high density of food (Fig. 6; ANOVA, change in selectivity, $p < 0.00001$). There was no difference in selectivity between high- and low-concentration mixtures offered without prefeeding (t-test comparing time = 0 treatments). In the second experiment, *Diaptomus birgei* fed on a high concentration mixture of *C. reinhardi* and dead *Cartaria* for 24 h and then was allowed to feed on low

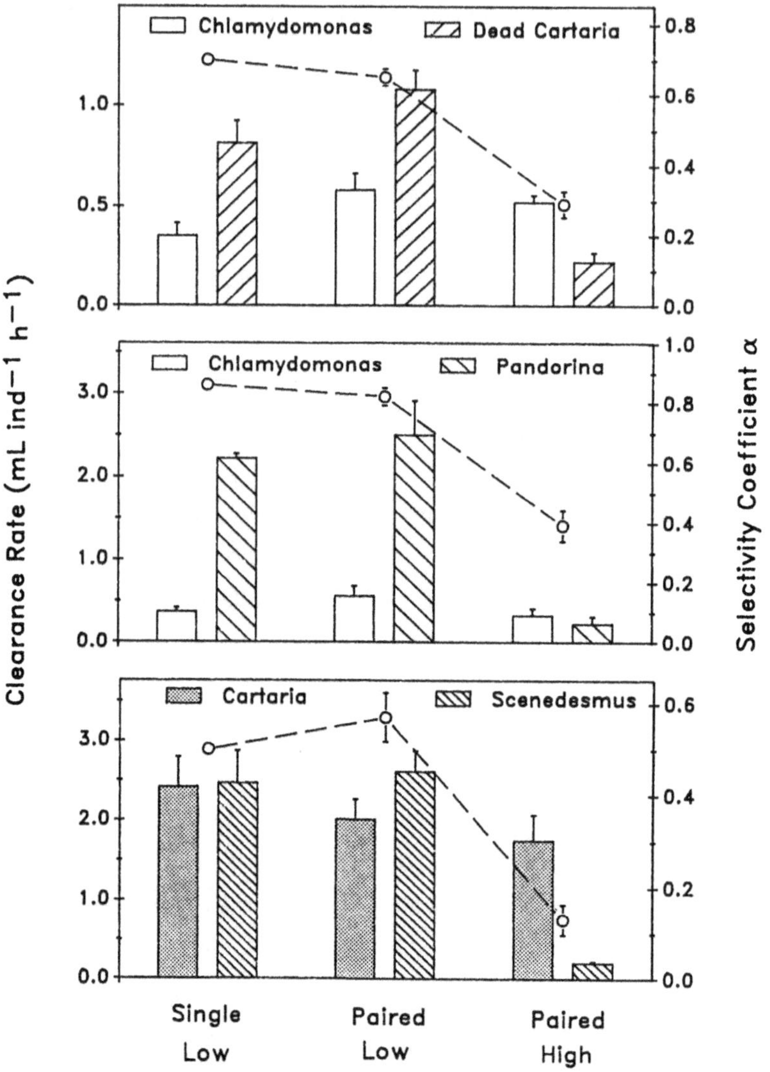

Fig. 5. Effects of food concentration on discrimination between algae which differ in size, quality, and manageability by adult female *Diaptomus birgei* from Little Crooked Lake.

concentrations of labeled particles after various periods of starvation. Here the copepods showed a gradual increase in the proportion of dead *Cartaria* in the diet over about 6 h (Fig. 7; ANOVA on α values, $p<0.00001$).

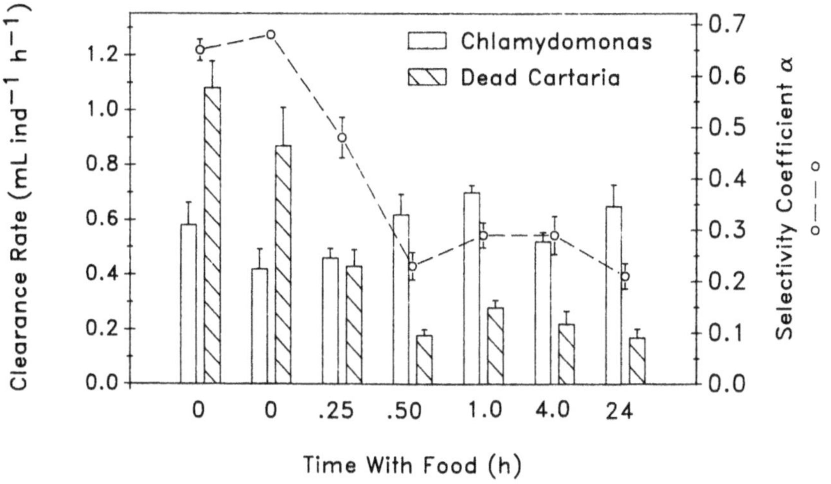

Fig. 6. Time course for the starvation effect on the selective feeding of *D. birgei* on *C. reinhardi* and dead *Cartaria*. The first and second time = 0 treatments represent trials with low and high concentrations, respectively, of labeled algae (see text for further explanation).

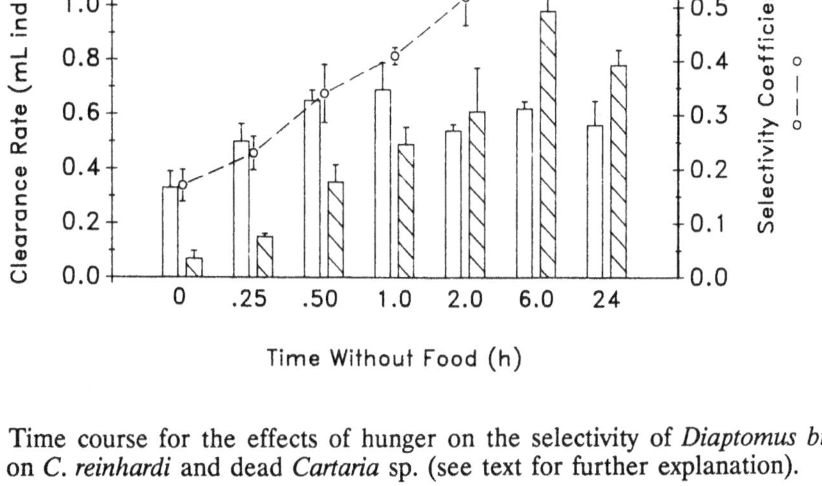

Fig. 7. Time course for the effects of hunger on the selectivity of *Diaptomus birgei* feeding on *C. reinhardi* and dead *Cartaria* sp. (see text for further explanation).

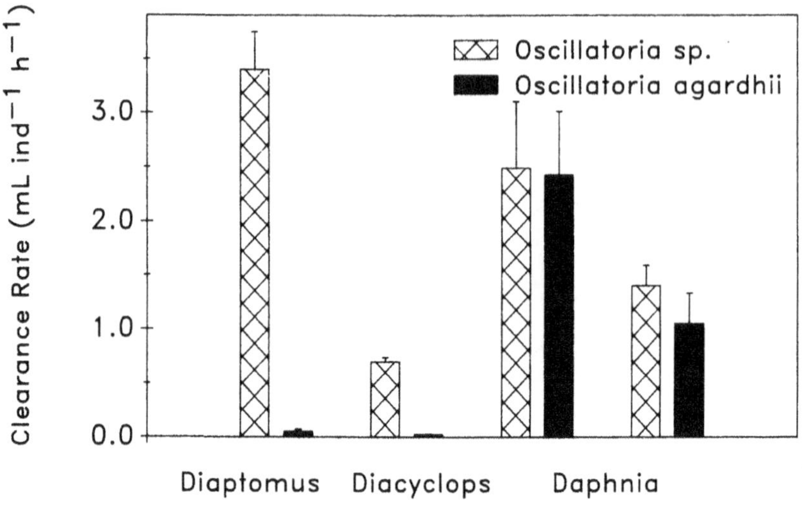

Fig. 8. Discrimination between toxic *Oscillatoria agardhii* and presumably nontoxic *Oscillatoria* sp. in low-concentration mixtures. Experimental animals were adult female *Diaptomus birgei*, CIII-CIV copepodites of *Diacyclops thomasi*, and adult (1.8 mm) and juvenile (1.0 mm) *Daphnia pulicaria* from Little Crooked Lake.

The final experiment tested whether three species of zooplankton from Little Crooked Lake would discriminate between toxic *Oscillatoria agardhii* and a presumably nontoxic congener of very similar morphology (*Oscillatoria* sp.). Both *D. birgei* and *D. thomasi* exhibited very strong discrimination against toxic *O. agardhii* (α's of 0.023 ± 0.008 and 0.035 ± 0.004, respectively; selective feeding, t-tests, both $p < 0.0001$). In contrast, both adult and juvenile *Daphnia pulicaria* fed nonselectively (α's, 0.49 ± 0.04 and 0.41 ± 0.06, respectively).

Discussion

Retention efficiency--As expected for mechanical filter feeders, comparisons between *Sida* and *Diaphanosoma* show a close relationship between the physical structure of the feeding appendages and "retention efficiencies" for small particles. Results presented here are also in good agreement with previous studies of size-selective feeding which used natural foods rather than artificial particles. These studies have shown that *Diaphanosoma* feeds very effectively on bacteria whereas *Sida* has zero feeding rates on solitary bacteria, low feeding rates on picoplankton (1-3 μm diameter) and very high feeding rates on slightly larger algae (5-7 μm width; DeMott 1985; Brendelberger 1985).

Maximal clearance rates of *Sida* were markedly higher than those of *Diaphanosoma* and also higher than the highest reported rates for other fine-meshed taxa, including *Daphnia* (e.g. Haney 1985). Other studies have also reported maximal clearance rates of *Sida* higher than those of daphnids of similar size (Downing and Peters 1980; DeMott 1985). *Holopedium gibberum*, a planktonic member of the Sididae with coarse filter setules also exhibits exceptionally high maximal clearance rates (Haney 1985). Thus, these results suggest that coarse filters may allow higher maximal clearance rates and tend to contradict the conclusion of Brendelberger et al. (1986) that cost of sieving is negligible. On the hand, contrary to the hypothesis of Gerritsen et al. (1988), the apparent advantage for coarse setules and zero and near-zero feeding rates for *Sida* on very fine particles support the hypothesis that particles are collected primarily by sieving water through the filter setules.

The hypothesis that coarse filters provide an advantage when very fine, low-quality particles are exceptionally abundant is not supported by field evidence on species distributions. For example, although copepods fared better than cladocerans during periods of exceptionally high concentrations of suspended solids in a South African reservoir, the cladoceran community consisted primarily of daphnids, *Diaphanosoma*, and *Moina*, taxa with fine filter meshes (Hart 1986).

Perceptual bias--In contrast to sidids and daphnids, size-selective feeding by diaptomid copepods is poorly predicted by the physical structure of the feeding limbs. For example, Vanderploeg and Ondricek-Fallscheer (1982) reported that the mean intersetular distance on the second maxillae of adult *Diaptomus sicilis* is about 3 μm. Nonetheless, feeding rates on particles <6-μm diameter were very low and size preferences increased continuously to a size of 15-20 μm. The adult female *Diaptomus birgei* used in this study (~1.5 mm total length) are similar in size to *D. sicilis* and possess second maxillae with intersetular distances of about 4 μm (DeMott unpublished). Nonetheless, *D. birgei* fed on large *Cartaria* sp. (range 13-40 μm diameter) at 3-5 times higher rates than on *C. reinhardi* (5-9 μm; Figs. 3 and 5). Additional experiments have shown a continuous increase in selectivity with algal size with maximal clearance rates on *Pediastrum* (60 μm diameter) and long filaments of *Oscillatoria* sp. (DeMott, unpublished). Such results are best explained by an increases in the probabilities of remote detection and active capture with increases in algal size (Price and Paffenhöfer 1985; Vanderploeg and Paffenhöfer 1985).

The relative importance of the passive capture of small particles and the active capture of large particles appears to vary dramatically among marine calanoid taxa (reviewed by Donaghay 1988). While most taxa show strong preferences for larger particles, *Neocalanus plumchrus*, a very large species which lives in north Pacific waters dominated by nannoplankton, is a marked exception. It exhibits nonselective, apparently passive feeding on particles ranging from at least 2-30 μm diameter (Landry and Lehner-Fournier 1988). Perhaps the success of filter-feeding daphnids in freshwater has prevented freshwater calanoids from occupying a passive filter-feeding niche.

Whether copepods use chemical or mechanical cues or some combination of the two in distance detection has been an open question; current evidence is indirect and somewhat contradictory. For example, remote detection of chemical exudates is favored by Paffenhöfer and Van Sant's (1985) observation that *Eucalanus pileatus* actively captured algal cells but not plastic spheres of similar size. Observations by Gill and Poulet (1988) with *Temora longicornis*, however, indicate that both plastic spheres and algal cells were detected at a distance and provoked oriented capture responses. Several studies involving additions of amino acids, algal exudates, or algal extracts have shown that starved copepods respond to point sources by swarming in the vicinity of the source (Poulet and Marsot 1980; Poulet and Ouellet 1982), and they respond to homogeneous mixtures by increasing the rate of movement of their feeding appendages (Gill and Harris 1987; Gill and Poulet 1988). These studies provide direct evidence of abilities to detect "algal scents" but appear more pertinent to the detection of patches of algae than the detection of individual cells.

My experiments provide two lines of evidence which run counter to the chemoreception hypothesis. Assuming that the chemoreception hypothesis is correct, I predicted that the addition of algal extracts would create a background "noise" which would interfere with remote detection. Since remote detection is more important for larger algae, I expected that selectivity between *Cartaria* sp. and *C. reinhardi* would decline. Contrary to my prediction, the addition of algal extract had no effect on selectivity or feeding rates (Fig. 3). Cowles et al. (1988) reported that additions of algal exudates lead to higher ingestion rates for *Acartia tonsa*, perhaps by stimulating greater feeding efforts. Such confounding effects should have been avoided in my experiments since comparisons were based on selectivity coefficients rather than feeding rates on a single cell type. Whole-cell extracts insured that all signal chemicals were initially present. However, the possibilities that concentrations were too low or that more complex chemicals were degraded by the extraction process cannot be ruled out.

High feeding rates on detrital matter would also seem contrary to the chemoreception hypothesis of distance detection, since nonliving matter, including the 2 day-old heat-killed *Cartaria* used in this study would not be expected to leak significant amounts of dissolved chemicals. Nonetheless, hungry *D. birgei* "selected" large dead *Cartaria* over live *C. reinhardi* (Figs. 5, 6, 7). Similar experiments with *Eudiaptomus* spp. and three species of marine copepods demonstrated selective feeding on clumps of dead algae (DeMott 1988b). Interpretations are complicated by comparisons between experiments, which show that feeding rates on low concentrations of dead *Cartaria* by *D. birgei* are only about one-half those recorded for live *Cartaria* (Figs. 3, 5, 6). Without direct visual observations, it is not possible to determine whether this difference is due to better detection of live cells or increased rejection of dead cells. Using high speed films, Vanderploeg and Paffenhöfer (1985) observed that fecal pellets and other debris were more likely to be actively captured and more likely to be rejected than were smaller clumps and individual cells of living algae. Vanderploeg et al. (this volume) used high-speed filming to demonstrate that *D. sicilis* actively captures large plastic spheres, nylon rods and toxic *Anabaena*. This direct evidence for distance mechanoreception is consistent with my indirect evidence against distance chemoreception. These results also support Lehman's (1976) assumption that selection against digestion-resistant particles occurs via post-capture rejection.

Since cyclopoid copepods, unlike calanoids, do not create feeding currents, cyclopoids should be especially dependent on remote detection of algal prey and yet would not be able to shear the phycosphere as required in the Andrews (1983) model. In an experiment with motile and nonmotile, high-quality algae, *Cyclops kolensis* exhibited a more than 50 times higher feeding rate on the slightly larger, spherical motile alga, *Chlamydomonas sphaeroides* than on the elongate, nonmotile alga *Scenedesmus acutus*. Although the effects of size cannot be completely excluded, this result provides some support for the notion that freshwater cyclopoid copepods use swimming disturbances in detecting motile algae. Interestingly, the raptorial cyclopoid exhibited higher clearance rates on the motile alga than did larger calanoid copepods. Particle size differences probably account for the weak selection between *C. sphaeroides* and *S. acutus* by *Eudiaptomus* spp. *Bosmina longispina* (0.6-0.8 mm length) exhibited an intermediate preference for the larger, motile alga. Previous studies have shown that other bosminids select motile algae over nonmotile algae (Gilbert and Bogdan 1981) and large algae over small algae (e.g. Bleiwas and Stokes 1985; DeMott 1985). Thus, bosminids may use both size and motility in active capture of algae (see DeMott and Kerfoot 1982).

Active choice--In contrast to experiments with particles of similar nutritional value, *Diaptomus birgei* exhibited marked changes in selectivity with changes in food concentration when fed mixtures of high- and low-quality food particles. In agreement with optimal foraging theory and previous experiments with *Eudiaptomus* spp. (DeMott 1989), *D. birgei* discriminated against low-quality foods (i.e. dead *Cartaria* and *Pandorina*) at high food concentrations. Such examples of behavioral flexibility are strong evidence for active choice in food selection by calanoid copepods. These results, which are predicted by Lehman's (1976) optimal diet model, also help explain why animals which are capable highly selective feeding often ingest substantial amounts of detritus and other low-quality particles in nature. Experiments demonstrating selection between nitrogen-rich and nitrogen-limited algae of the same species were run at relatively high food concentrations (Cowles et al. 1988; Butler et al. 1988). Foraging theory predicts that active selection between nitrogen-rich and nitrogen-limited algae would be negligible under low food conditions.

At low food concentrations *Diaptomus birgei* fed size-selectively on large, low-quality particles (i.e. dead *Cartaria*, gelatinous *Pandorina*) whereas at high food concentrations the copepod selected a small, high-quality flagellate (*C. reinhardi*). Presumably, the probability of detection is primarily a function of size, and the probability that low-quality particles will be rejected increases with increasing abundance of high-quality particles and declining hunger. This interaction between size and quality emphasizes the need for considering both of these variables in models of copepod feeding. The low and high food concentrations used here (25-200 μgC/L of high-quality algae) are well within the range experienced by zooplankton in moderately productive environments, although the "high concentration" may often be restricted to "spring bloom" conditions. The tendency for hungry copepods to feed size-selectively on low-quality algae helps explain why particle size is a reasonably consistent predictor of feeding selectivity, even in environments with a high proportion of low-quality particles (e.g. Vanderploeg 1981).

An experiment showing discrimination against *Scenedesmus quadricaudus* suggests that handling costs or handling times can be significant for large algae with spines (see also Vanderploeg et al. 1988). Handling times for readily ingested algae are very short.

For example, Gill and Poulet (1988) reported that *Temora longicornis* handled a 25-μm centric diatom for 0.06 s prior to ingestion. Plastic spheres, which were usually rejected, required handling times of 0.31 s for small spheres (10-25 μm) and "several seconds" for large spheres. Thus, "recognition time" (Hughes 1979) could be important for low-quality particles. In contrast to the rapid ingestion of small spherical algae, Vanderploeg et al. (1988) observed that *D. sicilis* required 2.2 s to ingest a 120-μm-long segment of the filamentous diatom *Melosira islandica*. Movies also show that several particles can be handled simultaneously and that additional particles can be actively captured and ingested while others continue to be manipulated (Strickler 1984; Vanderploeg and Paffenhöfer 1985). Difficulties in quantifying handling and rejection times and costs are factors which prevent quantitative tests of zooplankton foraging models. Quantitative tests would also require data on relations between gut passage times and assimilation efficiency.

Experiments testing the response of *D. birgei* to rapid changes in food concentration clearly demonstrate that hunger rather than food concentration *per se* control the degree of discrimination between low- and high-quality particles. This result is consistent with Lehman's (1976) assumption that gut processing rather than handling time usually determines the maximal rate of energy gain. Previous studies of the responses of zooplankton to rapid changes in food concentration have often measured elevated ingestion rates on high-quality particles following starvation. The time course of the food-selection response observed in my experiments is similar to the ingestion-rate responses observed for freshwater copepods and cladocerans (e.g. Lampert et al. 1988). Larger marine copepods appear to exhibit somewhat slower responses (e.g. Hassett and Landry 1988). In nature, rapid changes in food concentration are most likely to be experienced by zooplankton undergoing vertical migration. The shifts in selectivity observed in my experiments suggest that vertically migrating copepods would rapidly become more discriminating upon arriving in food-rich surface waters at night. Following descent, increasing hunger would result in increased ingestion of the detritus which tends to dominate at greater depths.

In contrast to the variable selectivity observed with nontoxic, low-quality particles, calanoid copepods show strong, invariant selection against toxic algae. Previous studies with freshwater copepods examined selection between green algae and toxic blue-greens (Fulton and Paerl 1987b; DeMott 1989). Since even nontoxic blue-greens are considered poor-quality foods, my experiment with two species of *Oscillatoria* provides better evidence for active discrimination against toxicity *per se*. Similar experiments have shown strong, invariant discrimination between toxic and nontoxic dinoflagellates by marine copepods (Huntley et al. 1986).

The very high feeding rates of both *D. birgei* and *Diacyclops bicuspidatus* on *Oscillatoria* sp. emphasize the importance of using particles of similar morphology when testing the effects of nutritional characteristics on selectivity. Hartmann (1985) reported only weak selection for the green alga *Ankistrodesmus* over *Oscillatoria agardhii* by *Diaptomus ashlandi* but this result was due to very low feeding rates on both particle types. Other experiments have shown that the needle-like cells of *Ankistrodesmus* are eaten at low rates by *D. birgei* (DeMott unpublished). The relatively high swimming speeds of cyclopoids may result in many direct contacts with long filaments. Thus, high feeding rates by cyclopoids on filaments may be possible without remote detection.

Conclusions--Until recently, aquatic ecologists had grossly underestimated the sensory capabilities and behavioral flexibility of many taxa of suspension-feeding zooplankton. Failure to take into account the possible effects of perceptual biases and active choice on food selection can result in misinterpretation of feeding experiments and, ultimately, an incorrect understanding of zooplankton-food resource interactions. Thus far, models of zooplankton-resource interactions have focused on the effects of grazer size and particle size (e.g. Steele and Frost 1977; Carpenter and Kitchell 1984). My results show how the differing feeding mechanisms of certain cladocerans, and calanoid and cyclopoid copepods result in differing responses not only to algal size, but also to motility, toxicity, and digestibility.

This paper used six series of laboratory experiments as a framework for examining a wide range of behavioral mechanisms for food selection by zooplankton. Inferences concerning the possible advantages of coarse filter setules and possible mechanisms of distance detection were based on very limited comparative data and are necessarily speculative. This is especially true of mechanoreception hypothesis of remote detection, since this hypothesis was supported only indirectly by a lack of evidence for the chemoreception hypothesis. On the other hand, models combining a perceptual bias for larger particles and active choice between low- and high-quality particles should successfully predict a wide range of copepod feeding behaviors. Although long-term learning could be a factor in some cases (Price and Paffenhöfer 1984; Huntley 1988), hunger appears to be the proximal factor controlling the degree of discrimination. The one certainty is that these "simple" animals will continue to surprise us.

Acknowledgements

This study was supported by a fellowship from the Max Planck Society for the Advancement of Science and NSF grant BSR 8717564. Purdue University provided travel funds. I thank Julie DeMott, Felix Moxter, and Desiree Watson for assistance and encouragement and Pete Jumars and Hank Vanderploeg for suggesting improvements in the manuscript. Dr. W.T. Edmondson kindly provided a culture of *Oscillatoria agardhii.*

References

Andrews, JC (1983) Deformation of the active space in the low Reynolds number feeding current of calanoid copepods. Can J Fish Aquatic Sci 40:1293-1302

Atema J (1987a) Chemoreception in the sea: Adaptations of chemoreceptors and behaviour to aquatic stimulus conditions. In: Atema J, Ray RR, Popper AN, Tavolga (eds) Sensory biology of aquatic animals, Springer, New York, pp 387-42

____ (1987b) Distribution of chemical stimuli. In: Atema J, Ray RR, Popper AN, Tavolga (eds) Sensory biology of aquatic animals, Springer, New York, pp 29-56

Bell W, Mitchell R (1972) Chemotactic and growth responses of marine bacteria to algal extracellular products. Biol Bull 143:265-277

Bleiwas AH, Stokes PM (1985) Collection of large and small particles by *Bosmina.* Limnol Oceanogr 30:1090-1092

Boyd CM (1976) Selection of particle sizes by filter-feeding copepods: A plea for reason. Limnol Oceanogr 21:175-180

Brendelberger H (1985) Filter mesh-size and retention efficiency for small particles: comparative studies with Cladocera. Arch Hydrobiol Beih Ergebn Limnol 21:135-146

____, Herbeck M, Lang H, Lampert W (1986) *Daphnia*'s filters are not solid walls. Arch Hydrobiol 107:202

Butler NM, Suttle CA, Neill WE (1989) Discrimination by freshwater zooplankton between single algal cells differing in nutritional status. Oecologia 78:368-372

Cannon HG (1933) On the feeding mechanism of the Branchiopoda. Phil Trans Roy Soc Lon Ser B 222:267-352

Carpenter SR, Kitchell JF (1984) Plankton community structure and limnetic primary production. Am Nat 124:159-172

Chesson J (1983) The estimation and analysis of preference and its relationship to foraging models. Ecology 65:1297-1304

Cowles TJ, Olson RJ, Chisholm SW (1988) Food selection by copepods: discrimination on the basis of food quality. Mar Biol 100:41-49

DeMott WR (1985) Relations between filter mesh-size, feeding mode, and capture efficiency for cladocerans feeding on ultrafine particles. Arch Hydrobiol Beih Ergeb Limnol 21: 125-134

____ (1986) The role of taste in food selection by freshwater zooplankton. Oecologia 69: 334-340

____ (1988a) Discrimination between algae and artificial particles by freshwater and marine copepods. Limnol Oceanogr 33:397-408

____ (1988b) Discrimination between algae and detritus by freshwater and marine zooplankton. Bull Mar Sci 43:486-499

____ (1989) Optimal foraging theory as a predictor of chemically mediated food

selection by suspension feeding copepods Limnol Oceanogr 34:140-154
____, Kerfoot WC (1982) Competitive interactions among cladocerans: nature of the interaction between *Bosmina* and *Daphnia*. Ecology 63:1949-1966
Donaghay PE (1988) Role of temporal scales of acclimation, food quality and trophic dominance inn controlling the evolution of copepod feeding behavior. Bull Mar Sci 43:469-485
Downing JA, Peters RH (1980) The effect of body size and food concentration on the *in situ* filtering rate of *Sida crystallina*. Limnol Oceanogr 25:883-895
Fulton RS, Paerl HW (1987a) Effects of colonial morphology on zooplankton utilization of algal resources during blue-green algal (*Microcystis aeruginosa*) blooms. Limnol Oceanogr 32: 634-644
___, ___ (1987b) Toxic and inhibitory effects of the blue-green alga *Microcystis aeruginosa* on herbivorous zooplankton. J Plankton Res 9: 837-855
Geller W, Müller H (1981) The filtration apparatus of Cladocera: filter mesh-sizes and their implications for food selectivity. Oecologia 49:316-321
Gerritsen J, Porter KG, Strickler JR (1988) Not by sieving alone: observations of suspension feeding in *Daphnia*. Bull Mar Sci 43:377-394
Gilbert JJ, Bogdan KG (1981) Selectivity of *Polyarthra* and *Keratella* for flagellate and aflagellate cells. Verh Internat Verein Limnol 21:1515-1521
Gill CW, Harris RP (1987) Behavioral responses of the copepods *Calanus helgolandicus* and *Temora longicornis* to dinoflagellate diets. J Mar Biol Ass UK 67:785-801
____, Poulet SA (1988) Impedance traces of copepod appendage movements illustrating sensory feeding behavior. Hydrobiologia 167/168:303-310
Gliwicz ZM, Siedlar E (1980) Food size limitation and algae interfering with food collection in *Daphnia*. Arch Hydrobiol 88:155-177
Gophen M, Geller W (1984) Filtering mesh size and food particle uptake by *Daphnia*. Oecologia 64:408-412
Haney JF (1985) Regulation of cladoceran filtering rates in nature by body size, food concentration and diel feeding patterns. Limnol Oceanogr 30:397-411
Hart RD (1986) Zooplankton abundance, community structure and dynamics in relation to inorganic turbidity, and their implications for a potential fishery inn subtropical Lake le Roux, South Africa. Freshwater Biol 16:351-371.
____ (1987) Observations on calanoid diet, seston, phytoplankton-zooplankton relationships, and inferences on calanoid food limitation in a silt-laden reservoir. Arch Hydrobiol 111:67-82
Hartmann HJ (1985) Feeding of *Daphnia pulicaria* and *Diaptomus ashlandi* on mixtures of unicellular and filamentous algae. Verh Internat Verein Limnol 22:3178-3183
Hassett RP, Landry MR (1988) Short-term changes in the feeding and digestion by the copepod *Calanus pacificus*. Mar Biol 99:63-74
Hessen DO (1985) Filtering structures and particle size selection in coexisting Cladocera. Oecologia 66:368-372
Hughes RN (1979) Optimal diets under the energy maximization premise: the effects of recognition time and learning. Am Nat 113:209-221
____ (1980) Optimal foraging in the marine context. Oceanogr Mar Biol A Rev 18: 428-449
Huntley M (1988) Feeding biology of *Calanus*: a new perspective. Hydrobiologia 167/168:83-99
____, Sykes P, Rohan S, Marin V (1986) Chemically-mediated rejection of dinoflagellate prey by the copepods *Calanus pacificus* and *Paracalanus parvus*: mechanism, occurrence, and significance. Mar Ecol Prog Ser 28:105-120

Infante A, Abella SEB (1985) Inhibition of *Daphnia* by *Oscillatoria* in Lake Washington. Limnol Oceanogr 30:1046-1052

Kerfoot WC (1978) Combat between predatory copepods and their prey: *Cyclops*, *Epischura*, and *Bosmina*. Limnol Oceanogr 23:1089-1102

___, DeMott WR, DeAngelis DL (1985) Interactions among cladocerans: Food limitation and exploitative competition. Arch Hydrobiol Ergebn Limnol 21:431-452

Koehl MAR (1984) Mechanisms of particle capture by copepods at low Reynolds numbers: possible modes of selective feeding. In: Meyers DG, Strickler JR (eds) Trophic interactions within aquatic ecosystems. AAAS Select Symp Ser 85, Westview, Boulder, pp 135-166

Lam RK, Frost BW (1976) Model of copepod feeding response to changes in size and concentration of food. Limnol Oceanogr 21:490-500

Lampert W, Taylor BE (1985) Zooplankton grazing in a eutrophic lake: implications of diel vertical migration. Ecology 66:68-82

___, Schmitt, R-D, Muck P (1988) Vertical migration of freshwater zooplankton: test of some hypotheses predicting a metabolic advantage. Bull Mar Sci 43:620-640

Landry MR, Lehner-Fournier JM (1988) Grazing rates and behaviors of *Neocalanus plumchrus*: implications for phytoplankton control in the subarctic Pacific. Hydrobiologia 167/168:9-19

Legier-Visser MF, Mitchell JG, Okubo A, Fuhrman JA (1986) Mechanoreception in calanoid copepods. A mechanism of prey detection. Mar Biol 90:529-535

Lehman JT (1976) The filter feeder as an optimal forager, and the predicted shapes of feeding curves. Limnol Oceanogr 21:501-516

Paffenhöfer G-A, Van Sant KB (1986) The feeding response of a marine planktonic copepod to quantity and quality of particles. Mar Ecol Prog Ser 27: 55-65

Porter KG (1976) Enhancement of algal growth and productivity by grazing zooplankton. Science 192:1332-1334

___ (1977) The plant-animal interface in freshwater ecosystems. Am Sci 65:159-170

Poulet SA, Marsot P (1980) Chemosensory feeding and food-gathering by omnivorous marine copepods. In: Kerfoot WC (ed) The evolution and ecology of zooplankton communities. University Press of New England, Hanover, pp 198-218

___, Ouellet G (1982) The role of amino acids in the chemosensory swarming and feeding of marine copepods. J Plankton Res 4:341-361

Price HJ (1988) Feeding mechanisms in marine and freshwater zooplankton. Bull Mar Sci 43:327-343

___, Paffenhöfer G-A (1984) Effects of feeding experience in the copepod *Eucalanus pileatus*: a cinematographic study. Mar Biol 84:35-40

___, ___ (1985) Perception of food availability by calanoid copepods. Arch Hydrobiol Beih 21:115-124

___, ___, Strickler JR (1983) Modes of cell capture in calanoid copepods. Limnol Oceanogr 28:116-123

Starkweather PL, Bogdan KG (1980) Detrital feeding in natural zooplankton communities: discrimination between live and dead algal foods. Hydrobiologia 73:83-85

Steele JH, Frost BW (1977) The structure of plankton communities. Phil Trans R Soc London 280:485-534

Stemberger RS (1981) A general approach to the culture of planktonic rotifers. Can J Fish Aquat Sci 38:721-724

Stephens DW, Krebs JR (1986) Foraging theory. Princeton

Strickler JR (1982) Calanoid copepods, feeding currents, and the role of gravity.

Science 218:158-160

___ (1984) Sticky water: a selective force in copepod evolution. In Meyers DG, Strickler JR (eds) Trophic interactions within aquatic ecosystems. AAAS Select Symp Ser 85, Westview, Boulder, pp 135-166

Taghon GL (1981) Beyond selection: optimal ingestion rate as a function of food value. Am Nat 118:202-214

Turner JT (1984) Zooplankton feeding ecology: contents of fecal pellets of the copepods *Eucalanus pileatus* and *Paracalanus quasimodo* from continental self water of the Gulf of Mexico. Mar Ecol Prog Ser 15:27-46

Uchima M, Hirano R (1986) Food of *Oithona davisae* (Copepoda: Cyclopoida) and the effect of food concentration at first feeding on the larval growth. Bull Plankton Soc Japan 33:21-28

Vanderploeg HA (1981) Seasonal particle-size selection by *Diaptomus sicilis* in offshore Lake Michigan. J Fish Res Board Can 38:504-517

___, Paffenhöfer GA (1985) Modes of algal capture by the freshwater copepod *Diaptomus sicilis* and their relation to food-size selection. Limnol Oceanogr 30:871-885

___, ___, Liebig JR (1988) *Diaptomus* vs. net phytoplankton: Effects of algal size and morphology on selectivity of a behaviorally flexible, omnivorous copepod. Bull Mar Sci 43:377-394

___, ___, ___ (this volume) New cinematographic observations and hypotheses on the concentration-variable selectivity of calanoid copepods for particles of different food quality. In: Behavioural mechanisms of food selection. Hughes RN (ed) Springer, New York

___, Ondricek-Fallscheer RL (1982) Intersetular distances are a poor predictor of particle-retention efficiency in *Diaptomus sicilis*. J Plankton Res 4:237-244

Wetzel RG, Likens GE (1979) Limnological analyses. WB Saunders Company, Philadelphia

Zach R, Smith JMN (1981) Optimal foraging in wild birds? In: Kamil AC, Sargent TD (eds) Foraging behavior: ecological, ethological and psychological approaches. Garland STPM Press, New York, pp 95-107

Zimmer-Faust RK (in press) The relationship between chemoreception and foraging behavior in crustaceans. Limnol Oceanogr

ORAL DISCUSSION – W.R.Demott.

Questions were raised about the ability of copepods to use chemical cues to detect suspended prey particles from a distance, and how this might be related to particle size.

The observed strong selection of prey is what you would expect for distance detection, but the chemical cues could degrade very rapidly. Perhaps the size effects in the perception model could be tested on a species of alga consiting of 2,4, or 8–cell morphs, but bigger algae might secrete more chemicals and emit a stronger signal; so to test this critically we might need to alter the metabolism of the alga without killing it. Signal molecules are broken down by exogenous enzymes near to receptor organs in moths and in lobsters. This enables the animal to cope with acclimation to constant high concentrations of signal in the environment, by attenuating the chemical at the receptor.

CONCENTRATION-VARIABLE INTERACTIONS BETWEEN CALANOID COPEPODS AND PARTICLES OF DIFFERENT FOOD QUALITY: OBSERVATIONS AND HYPOTHESES*

Henry A. Vanderploeg, Gustav-Adolf Paffenhöfer[1], and James R. Liebig
U.S. Department of Commerce, NOAA
Great Lakes Environmental Research Laboratory
2205 Commonwealth Blvd.
Ann Arbor, Michigan, USA 48105-1593

Introduction

The two major issues of research in feeding behavior of calanoid copepods and, indeed, in all research on zooplankton feeding have been concentration-variable selection and food quality (Vanderploeg in press). The concentration-variable selectivity hypotheses in its earliest form stated that copepods would track peaks in natural particle-size spectra, that is, focus their efforts on the most abundant food (Vanderploeg 1981a; Vanderploeg et al. 1984). The counter hypothesis is that selectivity, when expressed as W' or other appropriate measures, remains invariant no matter what the relative proportions or total concentration of the various food types (Vanderploeg and Scavia 1979a,b; Vanderploeg 1981a; Vanderploeg et al. 1984).

The direct observations of Strickler and colleagues (e.g., Alcaraz et al. 1980; Koehl and Strickler 1981; Strickler 1982; Paffenhöfer et al. 1982), using the newly developed technique of high-speed microcinematography, showed that certain calanoid copepods can create a double-shear scanning current that focuses water near their bodies, and that they respond with coordinated movements of their mouthparts to capture large algae they detect in this current. At this time, olfaction was described as the stimulus (Andrews 1983), and this is still the accepted paradigm, although mechanoreception has been also hypothesized (Legier- Visser et al. 1986). Invariant selectivity became identified with the passive-mechanical selection of filtering since this was one obvious means of obtaining invariant selection (Vanderploeg 1981a; Vanderploeg et al. 1984). At the same time, a number of studies showed that copepods preferentially ingest algae over plastic microspheres and algal-exudate-flavored

[1]Skidaway Institute of Oceanography, P.O. Box 13687, Savannah, Georgia USA 31416

NATO ASI Series, Vol. G 20
Behavioural Mechanisms of Food Selection
Edited by R. N. Hughes
© Springer-Verlag Berlin Heidelberg 1990

microcapsules over unflavored microcapsules (e.g., Poulet and Marsot 1978; Donaghay 1980) and that copepods have numerous mechanoreceptors and chemoreceptors on the mouth parts to make choices (e.g., Friedman 1980). These cinematographic observations and ability of copepods to select particles on the basis of a particle's nutritional value ushered in a new era zooplankton feeding biology emphasizing the importance of nutritional value of particles and the optional foraging narrative, since peak tracking is one facet of the optional foraging story (Pyke et al. 1977; Hughes 1980; Vanderploeg et al. 1984, 1988).

Our goals in this paper are to present new observations and hypotheses on the food selection mechanisms of calanoid copepods and relate these mechanisms to the many new observations on involving food quality (e.g., DeMott 1986, 1988, 1989; Fulton 1988; Fulton and Paerl 1988). We will focus our efforts on *Diaptomus*, a freshwater copepod that creates a scanning current. *Diaptomus* and other scanners like the marine copepods *Paracalanus* and *Eucalanus* are dominant grazers in lakes and in coastal and offshore waters (Vanderploeg and Paffenhöfer 1985; Paffenhöfer and Stearns 1988).

The Current Paradigm on Mechanisms

Selection for High Quality Food -- The current paradigm of feeding mechanisms and its connection to selectivity in calanoid copepods come in large part from high-speed microcinematography of tethered copepods feeding on live algae and microzooplankton of various sizes (e.g., Alcaraz et al. 1980; Koehl and Strickler 1981; Paffenhöfer et al. 1982; Price et al. 1983; Price and Paffenhöfer 1985; Vanderploeg and Paffenhöfer 1985; Vanderploeg et al. 1988; Williamson and Vanderploeg 1988). Figure 1 summarizes our present knowledge on capture modes of *Diaptomus* and their relation to selectivity from experiments with easily ingested high-quality algae and an easily ingested soft bodied rotifer. Selectivity is expressed here as $W_i'=F_i/F_{pref} = m_i/m_{pref,}$ where F_i and F_{pref} are, respectively, clearance rates on prey i and the preferred prey, and m_i and m_{pref} are respective mortality rate coefficients (Vanderploeg and Scavia 1979a). That is, W' is just the clearance rate or mortality rate normalized to that of the preferred prey. We prefer the selectivity coefficient, W', to Vanderploeg and Scavia's (1979b) electivity indices W(=α of Chesson 1983), which is proportional to W', and E* because the values of W and E* depend on number of kinds of prey, and because W' is the coefficient used in the effective food concentration model (Vanderploeg and Scavia 1979a; Vanderploeg et al. 1984). Moreover, W' can be

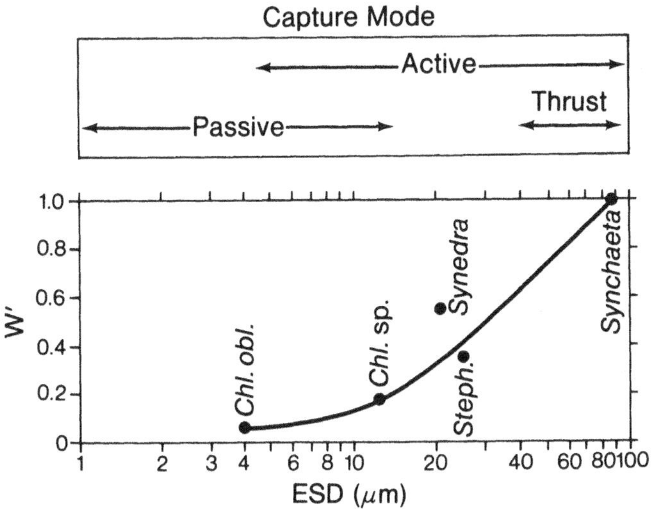

Figure 1. Lower panel: The selectivity (W') vs. equivalent spherical diameter (ESD) curve for *Diaptomus* feeding on high quality food. The curve is a composite of results from experiments of *D. sicilis* feeding on the algae *Chlamydomonas* sp., *C. oblonga*, *Stephanodiscus niagare*, and *Synedra* sp. (Vanderploeg et al. 1984; Vanderploeg et al. 1988) as well as those for *D. pallidus* feeding on the rotifer *Synchaeta oblonga* (Williamson and Butler 1986). Upper panel: The capture modes used for these prey (Vanderploeg and Paffenhöfer 1985; Vanderploeg et al. 1988; Williamson and Vanderploeg 1988).

broken down into a sequence of conditional selectivities at each step of the behavioral chain leading to ingestion in a fashion analogous to Holling's (1966) conditional probabilities for his components-of-predation approach (Vanderploeg in press).

The selectivity curve shown in Figure 1 represents the perceptual bias of *Diaptomus* for the different prey. We can say this because factors including handling and taste that affect the rest of the behavioral chain leading to ingestion do not apply to this high quality food at the low food concentrations at which these experiments were performed. As seen in Figure 1, there is an overlap in size range over which the various capture modes operate. Small particles are captured passively without detection by the copepod. These passively captured particles carried in the scanning current flow in undetected between the feeding appendages, and are funneled toward the mouth (Vanderploeg and Paffenhöfer 1985; Price and Paffenhöfer 1986). As particles get larger an increasing proportion of them are captured actively, that is, they are detected in the scanning current, and coordinated motions of the mouthparts are used to bring the particle toward the median and ultimately ingest it (Koehl and Strickler 1981). As algal size increases, it is probable that the algae are also sensed at greater distances (Vanderploeg and Paffenhöfer 1985). Thus the perceptual bias explains the increasing selectivity with increase in algal size (Price and Paffenhöfer 1985; Vanderploeg and Paffenhöfer 1985) .

The rotifer *Synchaeta*, the most preferred of *Diaptomus*' prey, was captured actively like the algae or with a thrust response that has only been observed for microzooplankton. The thrust response is a vigorous pouncing motion created by a sweeping back of the antennae and thrust of the swimming feet (Williamson and Vanderploeg 1988). This vigorous thrust response is thought not just to be an enthusiastic response to a distantly perceived large particle but probably a directed attack at potentially highly mobile prey that *Diaptomus* recognizes as animal prey (Williamson and Vanderploeg 1988).

In feeding experiments with mixtures of the passively captured alga *Chlamydomonas oblonga* (diameter = 4 μm) and the actively captured *Chlamydomonas* sp. (12 μm), Vanderploeg et al. (1984) showed that selectivity (W') for the small alga remained at approximately 0.3 relative to the large alga no matter what the relative proportions or total concentrations of the algae offered. In contrast, Vanderploeg et al. (1988) showed that at high concentrations of *Synedra* sp. (a rod 240 μm long) in mixtures with *Chlamydomonas* sp., selectivity for *Chlamydomonas* increased (Table 1). They argued that this concentration-variable selectivity could be viewed properly in the classical ethological framework of motivation and excitability of different motor patterns used to capture, handle, and ingest different kinds of algae. Specifically they

Table 1. A summary of *Diaptomus sicilis* clearance rate ratios (F_c/F_s), a measure of selectivity, and proportion (Pr) of attacked *Synedra* only partially ingested (<50%) as a function of relative concentrations (X_c/X_s) and total concentrations of *Chlamydomnas* (C) and *Synedra* (S) in experiments of Vanderploeg et al. (1988).[a]

Total conc.	X_c/X_s	F_c/F_s	Pr	Grouping[b]
Low conc.[c]	10.2	21/35=0.60±0.02	0.18	A
Low conc.[c]	0.52	19/30=0.63±0.04	0.34	A
High conc.[d]	6.0	9.4/7.0=1.3±0.03	0.32	B
High conc.[d]	0.41	10.5/5.1=2.0±0.12	0.61	C

[a] See Vanderploeg et al. for full statistical details.
[b] A different letter indicates a statistically different result.
[c] Total concentration = 0.04-0.06 $mm^3 \cdot liter^{-1}$.
[d] Total concentration = 0.6 $mm^3 \cdot liter^{-1}$.

argued that the complex motor patterns required to handle and ingest *Synedra* were less excitable than the simpler patterns used to ingest small spherical algae. More will be said about the utility of the behavioral approach for understanding food selection below.

Mechanisms and Food Quality - Typically food quality of a particle has been regarded as its nutritional content for the predator. A nutritious particle would be one that was non-toxic, digestible, and contained a balanced composition of proteins, lipids, carbohydrates, and micronutrients to allow growth and reproduction of the suspension feeder. Also, it is necessary to include any factors that would diminish the predator's ability to detect, capture, and ingest a particle. Non-nutritional factors would include, for example: size and other qualities as they affect detection and "filtering efficiency", size and shape as they affect handling and ingestion, and escape abilities of motile prey. For copepods, both nutritional and non-nutritional factors are reflected in their selectivity coefficients. We are emphasizing nutritional quality in this paper because of the strong linkage between selectivity and nutritional quality. The current paradigm argues that the combination of olfaction for distance perception and short-range olfaction, or taste, before ingestion implies the copepod has a sophisticated two-step process that encourages ingestion of large high-quality food.

The sensitivity and possible operation of this two-step olfactory system can be appreciated from the study of Paffenhöfer and Van Sant (1985) on *Eucalanus* feeding on particles of varying nutritional quality. It is the only study in which directly observed feeding mechanisms have been coupled to selectivity data derived from traditional feeding experiments. Results of their separate experiments run with only one or two particle types at a time were combined (Vanderploeg in press) to produce these overall results shown in Table 2. The dead *Rhizosolenia alata* and fecal pellets, which are the same size or nearly the same size as live *R. alata*, had lower selectivities. The fecal pellets had a selectivity about the same as the small alga, *Thalassiosira weissflogii*. The percentage of captures that were active captures for pellets and *T. weissflogii* were 66 and 63, values less than the ~100% observed for *R. alata*. They argued that these results are consistent with a smaller long-range olfactory cue arising from this lower quality food. Moreover, they argued that lack of both long-range and short-range olfactory cues were responsible for the complete lack of ingestion of microspheres when offered alone. Very few microspheres were captured and only 1 of 15 observed captures was an active capture. None of the

Table 2. Selectivity (W') of the marine calanoid copepod *Eucalanus pileatus* in experiments of Paffenhöfer and Van Sant (1985) for particles of different nutritional quality and equivalent spherical diameter (ESD).

Particle	ESD (μm)	Active Capture (%)	W' Offered alone	W' Offered in pairs
Rhizosolenia alata (live)	59	~100	1.0	1.0
R. alata (dead, heat killed)	59	-	0.63	0.83
Thalassiosira weissflogii (live)	14	63	0.36	0.30
Fecal pellets	51	66	0.43	0.44
Polystyrene spheres	20	7	0	0.02

captured beads was ingested. In a mixtures of *T. weissflogii* and microspheres, 42% of the captured microspheres were ingested incidentally with the algae.

How the incidental ingestion occurs is available from observations of the rejection process. Large algae or large particles like fecal pellets are usually ingested or rejected very soon after being brought to the mouth (Price et al. 1983; Paffenhöfer and Van Sant 1985; Vanderploeg and Paffenhöfer 1985), whereas small algae or plastic microspheres captured passively are ingested or rejected as a group after several have accumulated near the mouth. Small particles of low nutritional quality can be hidden among a larger mass of high quality food and be ingested (Paffenhöfer and Van Sant 1985). Thus, incidental ingestion explains the low but non-zero W' for microspheres offered with algae (Table 2).

Specific Questions

Recent experimental work has raised questions about selection and nutritional quality that were not answered by Paffenhöfer and Van Sant's (1985) study since only inert microspheres and fecal pellets were used as the less preferred foods. Experiments of Fulton (1988), Fulton and Pearl (1988), and DeMott (1989) show that selectivity for toxic bluegreen algae is very low. Where in the behavioral chain leading to ingestion or rejection is it that bluegreen algae are rejected? Also, we were intrigued by the experiments of DeMott (1986, 1988) with mixtures of algae, microspheres, and algal-exudate flavored microspheres. A summary of his results for the low concentration experiments are shown in Table 3. Notice there is a range of selectivities for microspheres depending on whether they are flavored and on their size. What are the reasons for this? This question is relevant because *Diaptomus*, although cabable of very selective feeding, can ingest significant quantities of inert suspensoids like calcite in nature (Vanderploeg et al 1987). In addition we wondered whether olfaction was the complete explanation for perceptual bias for larger prey. We came to suspect physical cues (distortion of the flow field caused by a particle in the double shear scanning current) suggested by Legier-Visser et al. (1986) because the elongated alga *Synedra* was always preferred to spherical algae of the same equivalent spherical diameter (1988), as can be seen for one example in Figure 1. We hypothesized that an elongated alga creates a greater deformation in the flow field than a sphere, especially before it becomes aligned in the scanning current (Vanderploeg et al. 1988). We thought also that a detectable signal (change in pressure) might be generated by a randomly oriented alga as it swings to alignment in the current field (Vanderploeg et al. 1988). Also, Barrientos Chacon (1980) showed that the first antennae, the organs thought responsible for distance perception of large particles, have both mechano- and chemoreceptors. Scanners like *Diaptomus* have a greater proportion of

Table 3. Selectivity (W') of *Eudiaptomus* in experiments of DeMott (1988) for algae, microspheres, and algal-exudate-flavored microspheres.[a]

Particle	**Diameter (μm)**	**W'**
Chlamydomonas reinhardti	6	1.00
Flavored microsphere	6	0.19
Unflavored microsphere	6	0.025
Flavored microsphere	12	0.071
Unflavored microsphere	12	0.0061

[a]Data from Tables 1 and 3 of DeMott (1988) were combined by the technique of Vanderploeg (in press) to produce these results. Results are for mixture experiments with a low concentration of algae ($\leq$ 1000 cells•ml^{-1}). Percent standard errors of the mean for the W' data for microspheres are approximately 25%. Standard error of W'of *C. reinhardti* is zero since it was always the preferred particle.

chemoreceptors than the estuarine copepod *Acartia*, which does not create a scanning current.

We attempted to answer all these questions by using high-speed microcinematography to make direct observations on the entire behavioral sequences from capture to rejection or ingestion for particles of different kinds: (1) toxic and non-toxic bluegreen algae, (2) algae and flavored and unflavored microspheres of various sizes, and (3) large nylon rods (13x200 μm). We also discuss the utility of classic ethological approaches for understanding concentration-variable selectivity that DeMott (1989) has reported for mixtures of particles of high and low nutritional quality, as for example the *Planktosphaeria* vs. *Chlamydomonas* pair and the dead vs. live *Scenedesmus* pair shown here in Table 4. His results for particles of low nutritional quality parallel ours for the difficult to handle *Synedra* (Vanderploeg et al. 1988).

Table 4. Selectivity (W') of *Eudiaptomus* for the less preferred alga in indicated pairs of algae of different food quality (H=High, T=Toxic, G=Gelatinous sheath) offered alone and at high and low concentrations in experiments of DeMott (1989).[a]

Food Pair	ESD (μm)	Food quality	W' Alone	W' Low	W' High
1. *Scenedesmus*	5	H	0.67	0.79	0.72
Chlamydomonas	12	H			
2. *Microcystis*	5	T	0.060	0.18	0.15
Monoraphidium	5	H			
3. *Planktosphaeria*	23	G	0.82	0.54	0.16**
Chlamydomonas	12	H			
4. Dead (sterile) vs.			0.025	0.014	0.011
Live *Staurastrum*	16	H			
5. Dead (sterile) vs.			0.27	0.45	0.19**
Live *Scenedesmus*	5	H			

[a]W' is given only for less preferred alga in pair; by definition, W' of the preferred alga = 1. Concentrations of each alga offered alone or in pairs at the low concentration were each 0.125 $mm^3 \cdot liter^{-1}$. The concentration of algae offered in pairs at the high concentration were each 1.0 $mm^3 \cdot liter^{-1}$. Standard errors of the mean of the W' data are on the order of 5%.
**indicates highly significant differences between low and high concentration results.

Methods

Filming Techniques -- All filming was done at the Great Lakes Environmental Research Laboratory using a microcinematography apparatus identical to the one described by Alcaraz et al. (1980). All films were made at rates between 250 and 500 frames•s^{-1}. Adult female *Diaptomus* sicilis were used, and they were tethered to a fine cat hair (Vanderploeg and Paffenhöfer 1985; Vanderploeg et al. 1988) to hold the animal in place inside of the 35-ml aquarium of 2.1-cm-W x 4.0-cm-L x 4.2-cm-H dimensions. Films were made through the width of the aquarium to observe a lateral view of the animal. Recently, an addition was made to the system to allow an experimental feeding suspension to be replaced with filtered lake water with the animal still in the aquarium. A small glass inlet tube connected with rubber tubing to a reservoir (2 l) of water and a small glass outlet tube were used to flush out the cuvette with filtered lake water. This flow-through system keeps the animal under water while purging the system of particles. A clamp turns the flow on and off and gravity controls the rate. The flow rate (25-50 mls•min^{-1}) is slow enough so the animal is not harmed, yet fast enough to purge the system in 6 to 15 min using a total water volume of 300-500 ml. This makes it possible to film the same animal in many different feeding suspensions in the same day without having to remove the animal. Removing and replacing the animal is difficult, time consuming, and potentially harmful to the animal. In the past we often used different animals for each suspension. The flow aquarium allows us to minimize problems associated with individual variation as well. During the purging process the animal behaves normally: it continues to scan and does not try to escape. It will even ingest food if food is captured.

Toxic Blue-green Algae -- Non-toxic (UTEX 1444) and toxic (UTEX 2383) strains of the filamentous colonial blue-green alga *Anabaena flos-aquae* were used in the filming experiments to clearly contrast the effect of toxicity with nontoxicity on the feeding behavior; these species have a very similar gross morphology (Fulton 1988). Experiments of DeMott (unpublished) showed that the non-toxic strain is readily ingested, and Fulton (1988) showed that the toxic strain is strongly selected against in mixtures with *Chlamydomonas*. Both algae were cultured in WC medium at 15°C using general methods described by Vanderploeg et al. (1984). For the filming, exponentially growing cultures of the *A. flos-aquae* strains were screened with a 20-µm sieve and resuspended in 0.22-µm-filtered lake water. The cultures were also screened with a 200-µm screen to remove very long colonies. Colonies used in filming generally fell in the size range 300-500 μm.

We preconditioned the animal to a low concentration (0.05 mm^3•$liter^{-1}$) of the toxic and non-toxic *Anabaena* for 1 h. The suspension was then flushed out of the cuvette with 0.22-μm-filtered lake water, and 1 h later the filming was started. The following film sequences were made of *D. sicilis* in 3000 colonies• ml^{-1} *Anabaena* suspensions: non-toxic *Anabaena* (1.0 mm^3•$liter^{-1}$) 45 s after addition of algae, non-toxic *Anabaena* 7.5 min after addition, toxic *Anabaena* (1.4 mm^3• $liter^{-1}$) 45 sec after addition, and toxic *Anabaena* 12.2 min after addition. The animal was in filtered lake water for approximately 25 min between the non-toxic and toxic suspension. The temperature during filming was 14°C.

These films, run at 300 frames-s^{-1}, allowed 53 s of observation for a 122-m-long roll, the longest roll taken by the camera. Filming is expensive, about $130 per roll; therefore, we chose a high *Anabaena* concentration to maximize events on film. Films shot before this experiment showed that toxic *Anabaena* was always rejected, but that non- toxic *Anabaena* was sometimes rejected in films we made after waiting a few minutes after the introduction of the algae. We suspected satiation was responsible for the result (Vanderploeg et al. 1988). Therefore, we chose to make the first films 45 s after introduction of the algae to minimize satiation-driven rejections.

Preliminary Experiments with Flavored and Unflavored 11-μm Microspheres -- The particles used in the feeding suspension were 11-μm polystyrene microspheres (Duke Scientific). The treated microspheres were flavored with *Chlamydomonas reinhardti* (4.5 μm) following the method of DeMott (1986). Our *C. reinhardti* was smaller than his so we adjusted the concentration of algae used to flavor the microspheres upward to match his volumetric concentration of algae used to flavor his microspheres. *D. sicilis* was acclimated to a mixture containing the complete array of *Chlamydomonas* and microspheres (flavored and unflavored), each at a low concentration of 0.05 mm^3s•$liter^{-1}$ for 1 h before filming. Films of *D. sicilis* were made at 20 min intervals using six experimental treatments: control beads at 20,000 ml^{-1}, control beads and *C. reinhardti* (4.5 μm) at 20,000 ml^{-1} each, flavored beads at 20,000 ml^{-1}, flavored beads and *C. reinhardti* at 20,000 ml^{-1} each, flavored beads at 3564 ml^{-1} (3 mm^3•liter-1) and *Chlamydomonas* sp. (11-μm) at 3564 ml^{-1} (3 mm^3•$liter^{-1}$).

Experiments with Large Rods and Spheres -- At the end of the feeding experiment with toxic and non-toxic Anabaena we added enough 13x200-μm nylon rods to make 1400 rod•ml^{-1} suspension. Rods were made by microtoming bundles of nylon yarn as described by Vanderploeg (1981b) for evaluating Coulter counter function. About 4 h before the experiment, a concentrated stock suspension of rods was made by adding filtered lake water to the microtomed bundles of yarn and sonifying them to separate them. No surfactant was added to separate the rods. The flow aquarium was purged, and 1 min after mixing in the rods, a film as made.

A few days later, after viewing the films from the rod experiments, we made films of the same tethered *Diaptomus* in suspensions of large spheres. On the morning of each of the two experiments with the spheres, *Diaptomus* was fed a moderate concentration (< 1 mm^3 liter^{-1}) of *Chlamydomonas* sp. After purging the system and waiting 2 h, the first film was made. Each polystyrene microsphere (Duke Scientific) feeding suspension (1400 ml^{-1}) was made by pipetting a measured amount of a microsphere working stock into the filming cuvette containing filtered lake water and the animal. The working stock was made by placing a number of drops of the original microsphere solution into a measured volume (20 or 25 ml) of 0.22-μm-filtered lake water and sonifying the suspension to separate the microspheres. The number of drops added was determined by Coulter Counter measurement of the concentration of the microspheres. Five different sizes of microspheres, having diameters of 14, 29, 42, 48 and 102 μm, were used. Filming was started about 2 minutes after mixing in the microspheres. The cuvette was purged of microspheres immediately after the completion of each film, and there was a 20-minute waiting period between films.

To determine if there was an effect of any odors associated with the microspheres, another film was made a few days later of 48-μm-diameter microspheres which were thoroughly rinsed. These microspheres were rinsed four times in distilled water and left in distilled water overnight. They were then resuspended in 0.22-μm-filtered lake water and prepared for filming as above. Similar results were obtained for these washed microspheres.

Results

Toxic and Non-Toxic Anabaena -- The summary of results in Table 5 shows that both toxic and non-toxic *Anabaena* are actively captured, indicating that *Diaptomus* did not use long-range olfactory cues to distinguish between toxic

Table 5. Number of active and passive captures, rejections, and times between capture and rejection per 53-s film sequence by *Diaptomus sicilis* for non-toxic (UTEX 1444) and toxic (UTEX 2383) strains of *Anabaena flos-aquae* offered at 3000 colonies•ml^{-1}.

Strain	Time after addition (min)	No. of captures		No. of rejections	Rejection response time (ms)
		Active	Passive		
Non-toxic	0.8	6	0	1	367
Non-toxic	7.5	11	0	8	207 ±49(7)
Toxic	0.8	10	2[a]	12	138 ±40(7)
Toxic	12.2	6	0	5	186 ±101(5)

[a]Rejected before entering maxillae by flap of swimming feet.

and non-toxic strains. However, 2 toxic *Anabaena* colonies entering on a perfect line for passive capture were avoided before entry into the second maxillae. All captured colonies of the toxic strain in these films and in all other films made were rejected. In the case of the non-toxic strain, 5 out of 6 captures resulted in ingestion during the first film, but 8 of 11 were rejected in the second film. This demonstrates the tendency of *Diaptomus* to capture more prey than it ingests when prey are abundant. In all cases here, rejections occurred relatively quickly after capture for both toxic and non-toxic strains.

Experiments with Flavored and Unflavored Microspheres -- Some of the algal-exudate flavored spheres were captured actively. The 11-µm spheres when offered alone were definitely only captured passively (Table 6). In all cases the microspheres appeared to be rejected after capture. We cannot make quantitative statements beyond this because captures and rejections were hard to evaluate at the high concentration of microspheres used.

Large Nylon Rods and Spheres -- Unmistakable, vigorous active captures like those reported for algae (Vanderploeg and Paffenhöfer 1985) were made for both nylon rods and large microspheres. Flaps of the maxillipeds and swimming feet often aided the second maxillae in a manner similar to patterns observed for large algae (Vanderploeg and Paffenhöfer 1985). These experi-

Table 6. Number of passive and active captures per film sequence by *Diaptomus sicilis* for inert particles of various shapes and sizes. Each entry represents one film sequence of the indicated length for the same animal. The nylon-rod results are for 15°C, and those for microspheres are for 10°C. Concentrations were always 1400 particles•ml^{-1} except for the 11-μm micropheres of June 21 offered at 20,000•ml^{-1}.

Date	Particle Type	Particle Dimensions (μm)	Sequence length (s)	Number of Captures Passive	Number of Captures Active
June 21	Microsphere	11	64	Many	0
June 22	Nylon rod	13x200	53	0	25
June 28	Microsphere	48	13	1	6
June 28	Microsphere	102	13	0	5
June 29	Microsphere	14	13	1	2
June 29	Microsphere	29	13	0	6
June 29	Microsphere	42	13	0	6
July 6	Microsphere	48	13	0	10

ments confirm the hypothesis of Legier-Visser et al. (1986) that suspension-feeding copepods can use physical cues to detect and capture large particles. A particle of about 14 μm in diameter is apparently the smallest that *Diaptomus* can detect by mechanoreception (Table 6).

Once captured, the typical behavioral sequence for handling large algae was followed until rejection at or near the mouth. We could clearly see the spheres being passed from the first maxillae to the labial palp area, where rejection occurred. Many of the spheres were too large (> 30 μm) to be ingested. An exception to this sequence was results for the 102-μm microspheres: they were too large to fit well between the second maxillae and were not passed forward efficiently. They were usually rejected while still in the second maxillae. Ingestion of an elongated alga requires complicated manipulation of the alga so that it can be inserted into the mouth perpendicular to the long axis of the body (Paffenhöfer et al. 1982; Vanderploeg et al. 1988). Typically the captured nylon rods were handled by this same sequence and inserted into the mouth, where rejection occurred. There are contact chemoreceptors on all the appendages used for handling the spheres and rods (Friedman 1980). Apparently, they were not used to reject these large inert particles.

Discussion: New Paradigms

Importance of Mechanical and Chemical Cues -- Once particles get larger than some threshold size, they can be detected by mechanoreception and captured actively. The capture response for the 102-μm microspheres appeared not to be as enthusiastic and appeared to occur at a shorter distance. Perhaps *Diaptomus* recognized this particle was too large to ingest.

What are the relative contributions of mechano- and chemoperception to the W' vs. equivalent spherical diameter curve (Figure 1)? *Diaptomus* can capture live *Chlamydomonas* as small as 5 μm. Flavored 11-μm microspheres can be actively captured, but not unflavored microspheres. Paffenhöfer and Van Sant (1985) observed a high proportion (63%) of active captures with a 14-μm diatom (*T. weissflogii*) but only 7% (1 of 15) with 20-μm microspheres for *Eucalanus pileatus* (Table 2). *E. pileatus* (prosome length = 1.9 mm) being larger than *D. sicilis* (prosome length = 1.2 mm) may explain why it did not actively capture more 20-μm microspheres. All of these results lead to this hypothesis: With increasing particle (algal) size, chemoperception by the calanoid increases up to a point; with further increases in size chemoperception does not increase, but mechanoperception does. The relative contributions of mechano- and chemoperception to the W' vs. equivalent spherical diameter curve (Figure 1) need to be worked out.

There are probably many interesting synergies of chemical and mechanical cues related to capture of particles. In the studies of Williamson and Vanderploeg (1988), the thrust response was often used to capture small microzooplankton (rotifers). We have not seen this pouncing behavior with algae or any of the large inert spheres or rods. Perhaps both chemical and mechanical cues are necessary to elicit this response. Another possibility is that the prey's movement may inform *Diaptomus* that there is a microzooplankter nearby; however, Williamson and Vanderploeg (1988) discount this possibility because they observed thrust responses for non-moving (non-escaping) rotifers. We cannot rule out the possibility that hydrodynamic signals produced by moving prey in the very far field alerted *Diaptomus* of their presence.

It would appear at first glance that use of mechanoreception for the capture of large particles is maladaptive. However, consider that large inert particles like nylon rods or plastic microspheres are not common in nature. Large mineral particles do not stay suspended for long. Objects like fecal pellets may have some nutritional value, yet not give off much of a scent because of the membrane that surrounds them. The remaining large objects are algae and microzooplankton. Perhaps the mechanosensory system can perceive large objects at greater distances than the chemosensory system. Perhaps it is faster responding, which would be important for capture of microzooplankton with high escape abilities.

Sensory Modes and Nutritional Quality -- The cinematographic observations on toxic and non-toxic strains of Anabaena help explain the low selectivities observed in *Diaptomus* by Fulton (1988) for the same toxic strain of *Anabaena* used here and by Fulton and Pearl (1988) and DeMott (1989; results shown here in Table 4) for *Microcystis*. With both *Anabaena* and *Microcystis*, post-capture rejection is the explanation. It is possible that, like the nylon rods, both toxic and non-toxic strains are captured in response to mechanical cues. Fulton (1988) showed that the presence of *Anabaena* in mixtures with *Chlamydomonas reinhardti* (6 mm) did not depress clearance rates on *C. reinhardti*. This is explained by the immediate rejection of *Anabaena* that we observed in our films. In two cases, this rejection occurred before entry into the second maxillae.

Our collective experience (Paffenhöfer et al. 1982; Price and Paffenhöfer 1985; Vanderploeg and Paffenhöfer 1985; Vanderploeg et al. 1988) can be used to generate hypotheses that explain *Diaptomus*' selectivities for non-toxic particles of low nutritional quality (Tables 3 and 4). We hypothesize that the probability of ingestion a single particle or group of particles collected at the mouth depends on the strength of the appropriate chemical signal given off per unit volume (or area) for the particle or group of particles being tasted. This implies there is an interaction between selection and size of the low-quality particle, because a small low-quality particle, captured passively, can be hidden in a mass of high quality particles at time of tasting, as suggested by Paffenhöfer and Van Sant's (1985) data. This explanation clearly applies to DeMott's (1988) results for *Chlamydomonas* and flavored and unflavored microspheres shown here in Table 3: the selectivities for flavored and unflavored large microspheres (12 μm) are lower than respective selectivities for flavored and unflavored small (6 μm) microspheres.

We believe the same signal-strength argument applies to DeMott's (1989) results for algae of varying nutritional quality, as for example, food pairs 3, 4, and 5 in Table 4. In the case of pairs 4 and 5, the dead *Staurastrum* would be selected against more stringently than the dead *Scenedesmus* because of the former's larger size. Distance olfaction could also play a role in the case of *Staurastrum* since less algal exudate associated with the dead cells could lead to a smaller perceptive volume for *Diaptomus* scanning for food. In the case of food pair 3, the *Planktosphaeria* by virtue of its large size could result in a large perceptive volume caused by mechanoreception. Its gelatinous sheath, however, could inhibit signal strength and hence lower selection for this large cell.

In algal pairs 3 and 5 in Table 4, selectivity for the lower quality food dropped as food concentration increased. We believe this is a satiation-driven motivational response. At high food concentrations, where captures typically greatly exceed ingestions (Vanderploeg and Paffenhöfer 1985; Vanderploeg et al. 1988), low motivation may require greater signal strength to initiate the ingestion reflex. Like Leyhausen's (1973) cat in a room full of mice, *Diaptomus* will capture, handle, and even partially ingest more particles than it can ingest. For example, at high concetrations of Synedra, 61% of the (obviously) attacked *Synedra* were only partially ingested (Table 1). Motivation will be fluctuating in high concentrations, and if a high-strength chemical cue is available as motivation is increasing the ingestion response will be triggered. This explanation is analogous to the motivation argument we presented for the difficult-to-handle *Synedra* (Table 1) that follows the same concentration- variable selectivity pattern (Vanderploeg et al. 1988). This result for low-nutritional-quality food is also consistent with another aspect of signal strength: As the proportion of low quality particles increases, the proportion of low quality particles increases in the group of particles to be tasted at the mouth, resulting in a higher probability of rejection. This argument would be most relevant for small particles.

Summary

Direct high-speed microcinematographic observations of the suspension-feeding copepod *Diaptomus* showed that it uses mechanoreception in addition to olfaction to detect particles in its scanning current. Mechanoreception may explain part of the reason for the increase in selectivity that occurs with increasing particle size. Much of the selection for particles of varying nutritional quality is determined by taste at the mouth. Particles of low nutritional quality as well as toxic blue-green algae are rejected after capture. Concentration-variable selection of particles of low nutritional quality is a satiation-driven response that can be understood from classical ethological principles.

Acknowledgements

We thank W.R. DeMott for useful discussions during the course of this research. We thank both W.R. DeMott, and C.B. Jorgensen for their comments on the manuscript. We are appreciative of Roger and Helen Hughes efforts in organizing this productive and enjoyable workshop. This is GLERL Contribution No. 686. NATO provided funds to help defray expenses of attending this workshop.

References

Alcaraz M, Paffenhöer G-A, Strickler JR (1980) Catching the algae: a first account of visual observations on filter feeding calanoids. Am Soc Limnol Oceanogr Spec Symp 3:241-248. New England

Andrews JC (1983) Deformation of the active space in the low Reynolds number feeding current of calanoid copepods. Can J Fish Aquatic Sci 40:1293-1302

Barrientos Chacon, Y (1980) Ultrastructure of sensory units of the first antennae of calanoid copepods. MS Thesis, University of Ottawa, Ontario, Canada, 81 pp

Chesson J (1983) The estimation and analysis of preference and its relationship to foraging models. Ecology 64:1297-1304

DeMott WR (1986) The role of taste in food selection by freshwater zooplankton. Oecologia 69:334-340

DeMott WR (1988) Discrimination between algae and artificial particles by freshwater and marine copepods. Limnol Oceanogr 33:397-408

DeMott WR (1989) Optimal foraging theory as a predictor of chemically mediated food selection by suspension-feeding copepods. Limnol Oceanog 34:140

Donaghay PL (1980) Grazing interactions in the marine environment. Am Soc Limnol Oceanogr Spec Symp 3:234-240. New England

Friedman MM (1980) Comparative morphology and functional significance of copepod receptors and oral sturctures, Am Soc Limnol Oceanogr Spec Symp 3:185-197. New England

Fulton, RS III (1988) Grazing on filamentous algae by herbivorous zooplankton. Freshwater Biology 20:263-271

Fulton RS III, Paerl HW (1988) Zooplankton feeding selectivity for unicellular and colonial *Microcystis aeruginosa*. Bull of Mar Sci 43(3):500-508

Holling, CS (1966) The functional response of predators to prey density and its role in mimicry and population regulation. Mem Entomological Soc Canada

45:3-60
Hughes RN (1980) Optimal foraging in the marine context. Oceanogr Mar Biol Annu Rev 18:428-449
Koehl MAR, Stricker JR (1981) Copepod feeding currents: food capture at low Reynolds number, Limnol Oceanogr 26:1062-1073
Légier-Visser J, Mitchell, Okubo A, Fuhrman JA (1986) Mechanoreception in calanoid copepods, a mechanism for prey detection. Mar Biol 90:529-535
Leyhausen P (1973) Verhaltensstudien an katzen, Parey, Berlin and Hamburg. 232 pp
Paffenhöfer G-A, Stearns DE (1988) Why is Acartia tonsa (Copepoda: Calanoida) restricted to the nearshore environment? Mar Ecol Prog Ser 42:33-38
Paffenhöfer G-A, Stricker JR, Alcaraz M (1982) Suspension-feeding by herbivorous calonoid copepods: a cinematographic study. Mar Biol 67:193-199
Paffenhöfer G-A, Van Sant KB (1985) The feeding response of a marine planktonic copepod to quantity and quality of particles. Mar Ecol Prog Ser. 27:55-65
Poulet SA, Marsot P (1980) Chemosensory grazing by marine calanoid copepods (Arthropoda: Crustacea). Science 200:403-405
Price HJ, Paffenhöffer G-A (1985) Perception of food availability by calanoid copepods. Arch Hydrobiol Beih 21:115-124
Price HJ, Paffenhöffer G-A (1986) Capture of small cells by the copepod Eucalanus elongatus. Limnol Oceanogr 31:189-194
Price HJ, Paffenöffer G-A, Strickler JR (1983) Modes of cell capture in calanoid copepods. Limnol Oceanogr 28:156-163
Pyke GH, Pulliam HR, Charnov EL (1977) Optimal foraging: a selective review of theory and test, Q. Rev Biol 52:137-154.
Strickler JR (1982) Calanoid copepods, feeding currents, and the role of gravity. Science 218:158-160
Vanderploeg HA (1981a) Seasonal particle-size selection by *Diaptomus sicilis* in offshore Lake Mich. Can J Fish Aquat Sci 38:504-517
Vanderploeg HA (1981b) Effect of the algal length/aperture length ratio on Coulter analyses of Lake seston. Can J Fish Aquat Sci 38:912-916
Vanderploeg HA (in press) Feeding mechanisms and their relation to particle selection and feeding in suspension-feeding zooplankton. In: Wotton R (ed) The Biology of Particles in Aquatic Systems
Vanderploeg HA, Paffenhöfer G-A (1985) Modes of algal capture by the freshwater copepod *Diaptomus sicilis* and their relation to food-size selection. Limnol Oceanogr 30:871-885
Vanderploeg HA, Paffenhöfer G-A, Liebig JR (1988) *Diaptomus* vs. net phytoplankton: effects of algal size and morphology on selectivity of a behaviorally flexible, omnivorous copepod. Bull Mar Sci 43:377-394
Vanderploeg HA, Scavia D (1979a) Calculation and use of selectivity coefficients of feeding: zooplankton grazing. Ecol Modeling 7:135-149
Vanderploeg, HA, Scavia D (1979b) Two electivity indices for feeding with special refernce to zooplankton grazing. J Fish Res Board Can 36:362- 365
Vanderploeg HA, Scavia D, Liebig JR (1984) Feeding rate of *Diaptomus sicilis*

and its relation to selectivity and effective food concentration in algal mixtures and in Lake Michigan. J Plankton Res 6:919-941

Vanderploeg HA, Eadie BJ, Liebig JR, Tarapchak SJ, Glover, RM (1987) Contribution of calcite to the particle-size spectrum of Lake Michigan seston and its interactions with the plankton. Can J Fish Aquat Sci 44:1898-1914

Williamson CE, Butler NM (1986) Predation on rotifers by the suspension-feeding copepod *Diaptomus pallidus*. Limnol Oceanogr 31:393-402

Williamson CE, Vanderploeg HA (1988) Predatory suspension-feeding in *Diaptomus*: prey defenses and the avoidance of cannibalism. Bull Mar Sci 43:561-572

WATER PROCESSING IN FILTER–FEEDING BIVALVES

C.Barker Jørgensen,
University of Copenhagen

Introduction

Filter feeding is an adaptation allowing animals to feed on suspended microscopic food particles that cannot be sensed and seized individually. Such food particles are primarily the unicellular phytoplankton that constitute the major primary producers in the sea and all but the smallest and shallowest lakes. Filter feeders process the ambient water by pumping it through filters that retain suspended matter, usually irrespective of whether this matter is of food value or not. Water processing in bivalves is usually described in terms of physiological regulation. However, such interpretation of observations on capture, transport and sorting of food and other particles are often inconsistent with the physical and other conditions applying.

Functional Morphology of Bivalve Feeding

Feeding mechanisms in filter feeding bivalve have traditionally been investigated on preparations of animals with one valve and mantle removed, on isolated gills and gill fragments, as well as on excised labial palps. Ciliary tracts and their function were studied by means of suspended particles of various kinds dropped onto the surface of gills and palps. Some outstanding early contributions to this field of study of bivalve feeding are those of Orton 1912, Kellogg 1915, and Yonge 1923–24, 1926–27. The tradition culminated in the work of Atkins 1936,1937a,b, who described the ciliary tracts on the frontal surface of the gill filaments in some 90 species of bivalves. She established the classification of gill types, manily based on patterns of ciliary tracts, that is still extensively used. To Atkins and her successors, sorting of particles was the key concept to the understanding of the feeding mechanism in filter feeding bivalves. It was assumed that sorting mechanisms were best demonstrated by the use of dense particles, such as carborundum powder, because they ensured that the particles were "actually in contact with the cilia, and thus moved by them and not by some superficial water current" (Atkins,

NATO ASI Series, Vol. G 20
Behavioural Mechanisms of Food Selection
Edited by R. N. Hughes

1936).

Atkins emphasised the importance of distinguishing between two main types of frontal ciliary tracts, the lateral tracts of fine cilia that were thought to preferentially carry small and light particles, and the median tract of coarse cilia, believed to carry large and heavy particles (Fig. 1). The fine tracts terminate in food grooves along the dorsal and/or ventral margins of the gills, and they constitute acceptance tracts that select potential food particles in the processing

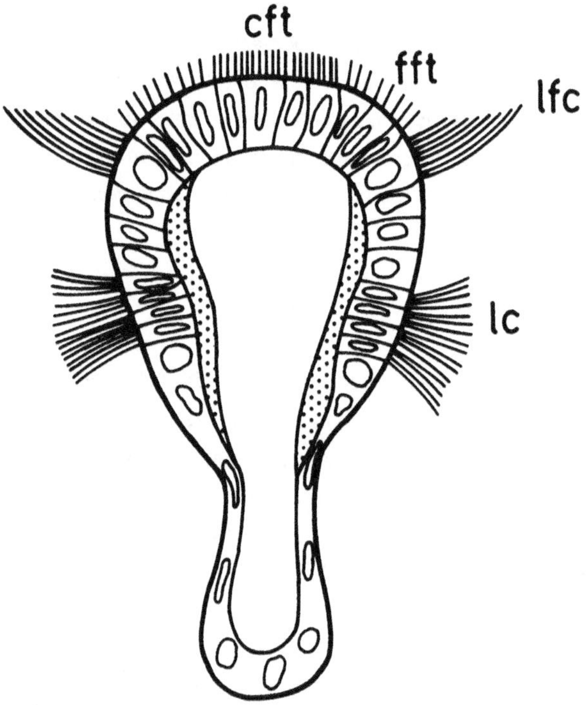

Fig. 1. Diagram of transverse section of a gill filament showing the differentiation of the frontal cilia into the coarse median tract (cft) and fine lateral tracts (fft). lfc indicates the laterofrontal cirri and lc the bands of lateral cilia (redrawn from Owen and McCrae, 1976).

of the ambient water. The coarse tracts lead to the ventral margin of the gills where they continue into other coarse tracts, the terminal frontal cilia, situated outside the food groove (Fig. 2). Mostly, the fine ciliary tracts were observed to be constantly active, whereas the coarse tracts were only active when stimulated by the presence of coarse or heavy particles. Moreover, heavy particles stimulated profuse secretion of mucus that consolidated the particles. But fine particles were also assumed to be embedded in mucus to ensure their transport along the tracts of fine cilia.

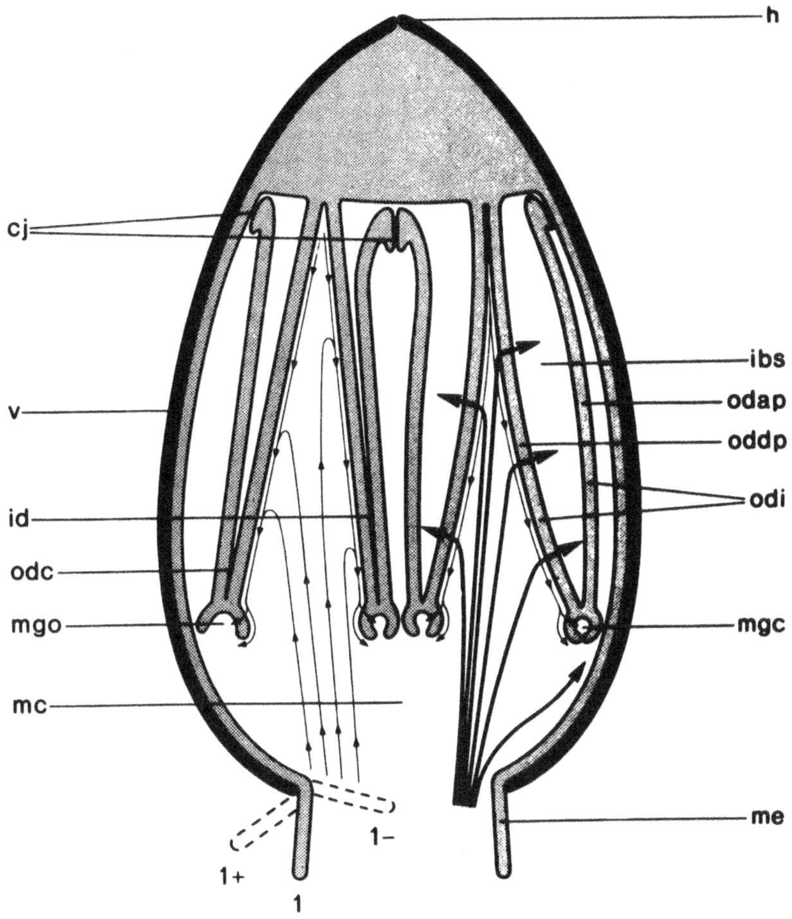

Fig. 2. Diagram of cross section of mussel, indicating main water currents when the gill transports water both through the interfilament canals and along the frontal surface of the gill filaments (right side of diagram, demibranchs are inflated) and when only the frontal surface of the filaments transports water (left side of diagram, demibranchs collapsed). Heavy lines indicate strong water currents; fine lines, weak currents. Currents are only shown between an outer and an inner demibranch. The diagram also illustrates shapes of the marginal groove from closed to widely open, as well as the varying orientation of the mantle edges in fully open animals. cj, Ciliary junctions; h, hinge; id, inner demibranch; ibs, infrabranchial space; mc, mantle cavity; me, mantle edge; mgc, marginal food groove, closed; mgo, marginal groove, open; odc, outer demibranch, collapsed; odi, outer demibranch, inflated; odap, outer demibranch, ascending part; oddp, outer demibranch, descending part; v, valve; 1, mantle edges parallelly oriented; 1+ mantle edges diverging; 1-, mantle edges converging (from Jørgensen. 1975).

The inner side of the labial palps is folded and the folds are furnished with complex systems of ciliary tracts, assumed to sort particles according to size, so that small particles are preferentially carried towards the mouth and large particles along rejection tracts. Allen (1958) attempted a synthesis of features that were common to all bivalves, and he distinguished between acceptance tracts running across the ridges of the palp folds towards the mouth, rejection tracts running in the bottom of the troughs between the folds, and resorting tracts along the aboral sides of the folds.

From the large number of studies of how preparations of gills and palps deal with particles of various kinds and sizes, a mucociliary mechanism of bivalve feeding emerged that may be summarized as follows. A flow of water is maintained through the gills by means of the activity of the lateral cilia (Fig. 1). The water is filtered at the entrance to the interfilament canals by the laterofrontal cirri, which both strain particles from the water and throw them onto the frontal surface of the filaments. The particles adhere to mucus, secreted by numerous gland cells on the filaments. The mucus-embedded particles are carried by frontal ciliary tracts to the ventral and/or dorsal margins of the gills and farther towards the mouth. The material transported along dorsal ciliary tracts reaches the mouth directly whereas material transported along the ventral margins is taken up by the palps on the inner surfaces of which ciliary tracts carry out the final sorting of material for acceptance or rejection.

The tradition in the studies of the functional morphology of bivalve feeding structures, established in the thirties, has continued up to the present time (Owen, 1953; Purchon, 1954, 1955a,b; Clausen, 1958; Allen, 1958, 1968; Nelson, 1960; Stasek, 1962; Morton, 1969, 1970b, 1976; Fankboner, 1971; Judd, 1971; Narchi, 1972; Hughes, 1975; Lam, 1977; Ansell, 1981).

Ciliary Feeding and Fluid Mechanics

The mucociliary concept of feeding in bivalves has been developed without consideration of physical constraints imposed by the low Reynolds-number fluid-dynamics that apply to suspension-processing in filter feeders (Jørgensen, 1983).

The Reynolds number Re is given by the relation

$$Re = \rho l v / \mu,$$

where ρ is the density of the fluid, l is a characteristic length, e.g., a diameter of a tube or dimension of a particle, v is the velocity and μ is the dynamic viscosity. Re is thus a dimensionless number that expresses the ratio between inertial forces and viscous forces. This implies that a flow is laminar, with orderly streamlines, when the Reynolds number is less than a certain critical value, e.g., ≈ 2000 for pipe flow. Above this value the flow is unstable and becomes turbulent. Furthermore, when Re < 1 inertia effects are negligibe. Physical dimensions and fluid velocities are normally so small in filter feeding animals that we always deal with laminar flow, and in ciliary feeding the flow is creeping, R << 1.

The parameter characterizing the motion of a suspended particle is the Stokesian time constant,

$$tp = (dp^2/18\nu) \cdot (\rho p/\rho),$$

where dp denotes the equivalent spherical diameter of the particle, $\nu = \mu/\rho$ the kinematic viscosity of the fluid, ρp the density of the particle, and ρ the density of the fluid.

tp is a measure of the time it takes the particle, if initially at rest, to be accelerated to the fluid velocity. Similarly, tp is a measure for the time it takes a particle to come to rest if the fluid is suddenly brought to rest. For a 10 μm-diameter particle in water, tp ≈ 10^{-6} s, hence such particles will closely follow the intantaneous fluid motion as long as this is characterized by times that are orders of magnitude greater, which is normally the case.

Investigators of filter feeding do not usually directly refer to the physical conditions applying, but mechanisms suggested reveal that observations made on particle movements have been interpreted in terms of "macroscopic world" physics, dominated by inertial forces (Purcell, 1977; Vogel, 1981). This is evident, e.g., when it is implied that inertial momentum governs particle movements. Thus, particles are said to "impinge" on ciliated surfaces when the flow directed towards the surface is deflected, leaving the water depleted of particles (Atkins, 1936; Werner, 1955; Bullivant, 1968; Fretter & Montgommery, 1968; Stebbing & Dilly, 1972; Bernard, 1974; Foster-Smith, 1975; Gilmour, 1979). Other cilia, such as, the laterofrontal cirri of the bivalve gill filaments, are said to intercept particles in the through current and to "throw" or "flick" particles onto the frontal surface of the filaments (Yonge, 1926-27; Morton, 1969, 1983).

Such statements imply that particles come into physical contact with the cilia and that the stroke that hits a particle conveys inertial momentum to the particle, momentum that carries it across streamlines to land on and thus to obtain contact with the cilia of the frontal tracts. It is, moreover, implied, and often explicitly stated, that the physical impact of particles hitting the frontal surface of the gill filaments stimulates the secretion of mucus that is needed to bind the particles.

When large particles of high density enter through the inhalant opening of oysters and other bivalves they have been assumed to settle on the inner mantle surface before they reach the gill, a gravimetric mechanism that Bernard (1974) considered to be the major sorting mechanism in filter feeding bivalves.

Thus, according to prevailing mucociliary concepts of filter feeding in bivalves, inertial forces are incorrectly assumed to be active from when ambient water enters the mantle cavity until it passes through the interfilament canals, depleted of particles.

Observations of Undisturbed Feeding

Usually it is understood, often implicitly, that observations made on preparations also apply in the intact, undisturbed animal. MacGinitie (1941) questioned this, and he attempted to observe feeding in undisturbed bivalves through a glass window covering a hole drilled in a valve and cut through the underlying mantle. MacGinitie found that the operation resulted in abnormal feeding, and it took some weeks in fresh sea water before the operated bivalves resumed normal feeding. During this a mucus sheet was assumed to cover the frontal surface of the gill, acting as a filter to strain particles from the through current. MacGinitie inferred the presence of a mucus sheet from the observation that the course of particles retained on the gill surface in undisturbed bivalves might cross the gill filaments, independently of existing ciliary tracts.

Foster-Smith (1975, 1978) resumed observations of feeding in bivalves furnished with windows. The animals, Mytilus edulis, Cerastoderma edule and Venerupis pullastra, were kept in running sea water where they might remain healthy for many months. The experimental conditions, therefore, seemed compatible with normal feeding. Foster-Smith did, however, not observe mucus sheets but mucus strings, extending across the gill filaments. Algal cells used as food were observed embedded in the mucus strings, whereas single cells were rarely seen even at low concentrations. Foster-Smith was thus unable to confirm the filtering mechanism proposed by MacGinitie, but agreed with the final result that captured food particles are embedded in mucus.

Bernard (1974) made an extensive study of the feeding process in the Pacific oyster, Crassostrea pacifica. Movements of particles, colloidal graphite and sand grains, on the gills of undisturbed oysters were recorded and measured after careful removal of the anterior portion of a valve. Animals survived several months after the operation and appeared to function normally. This function varied with the concentration of suspended particles. At low concentrations the frontal surface of the gill filaments is stated to be covered by bands of a serous mucus that entraps prospective food particles. At high concentrations another type of mucus, produced by the goblet cells, was secreted to entangle particles destined for rejection.

Available reports on the function of the gills in supposedly undisturbed, feeding bivalves are difficult to reconcile. However, the descriptions agree in food particles being entrapped in mucus on the gill surface and they are thus in accordance with the conventional mucociliary concept of the feeding mechanism.

Ciliary Filter Feeding without Mucus

Transport of particles in water currents along the surface of the bivalve gill or along other ciliated tracts has been described repeatedly over the years (e.g., Bronn, 1862; Stenta, 1903; Segerdahl, 1922; Gray, 1929; Nicol, 1930; Gosselin & O'Hara, 1961; Gilmour, 1979). Any observer of veliger larvae in a fresh sample of zooplankton will be familiar with the free rotation and intermingling of particles within the stomach, witnessing the absence of mucus from ingested food particles. Accordingly, feeding in veligers and other ciliary-feeding planktonic larvae has been described without involving mucus as part of the feeding mechanism (Millar, 1955; Thompson, 1959; Fretter & Montgommery, 1968; Strathmann et al., 1972). Yonge (1926-27), however, does state that veligers embed food particles in mucus.

Descriptions of transport of food particles in suspension in currents produced by ciliary tracts usually pass without comment, presumably because of a lack of awareness of any problem being inherent in such transport. To those who did recognize a problem, mucus secreted by the ciliated epithelium was the only means available to keep captured particles from escaping the ciliary tracts responsible for the transport of the particles to the mouth, and Morton (1967) probably expressed a general feeling by stating that unbound with mucus "a multitude of small unruly particles would be no sooner collected than dispersed".

The persistence on mucus being needed to keep collected particles on the ciliary tracts is remarkable in view of the easy demonstration that it is not (Fretter & Montgommery, 1968; Jørgensen, 1975, 1976, 1981a; Owen & McCrae, 1976; Owen, 1978). Thus, if gills of a bivalve are exposed to a suspension of particles, such as cultured flagellates, it can be observed that flagellates are collected and concentrated in water currents along gill filaments and dorsal or ventral food grooves. As long as the flagellates move suspended in the surface currents, their own movements do not suffice to help them escape from the currents. However, at the cut ends of a tract, where the flow of water stops, concentrated suspensions of flagellates immediately redisperse in the ambient medium and they continue to swim normally, indicating that they have not become smeared with mucus during the process of retention and transport by the gill (Jørgensen, 1976). Thus, forces must act upon the particles to keep them in suspension within the water currents produced by the ciliary tracts. These forces may be the same as those that maintain particles moving in laminar flow along a wall at a certain distance from the wall (Jørgensen, 1981a).

Fluid mechanical forces acting on particles suspended in laminar ciliary currents thus constitute an alternative mechanism for the entrapping of food particles. It, therefore, became crucial to know in what state the particles arrive in the stomach.

An extensive literature deals with the stomach content of filter feeding bivalves and with the ways the stomach deals with ingested food. This literature consistently refers to food as particles embedded in mucus (e.g., List, 1902; Vonk, 1924; Yonge, 1935; Morton, 1960; Reid, 1965; Kristensen, 1972; Bernard, 1974). Ingestion of particles in mucus has also been directly observed. Thus, Allen (1958) recorded that most of the material accepted after sorting on the palps entered the oesophagus within a string of mucus that was subsequently wrapped around the tip of the crystaline style. The observations were, however, made on preparations and may not apply to undisturbed feeding. In case of stomach analysis the filling of the stomach with mucus may have resulted from the strongly increased mucus secretion induced by the opening of the animal (Jørgensen, 1981b). In order to determine the normal state of the stomach content this should be sampled immediately. In intact mussels, Mytilus edulis, that were clearing suspensions of graphite, samples of the stomach content, taken by pipette within seconds after a mussel had been opened by cutting of the adductor muscle, showed that practically all graphite ingested was present in a freely suspended state. Particles retained by the gills are thus predominantly ingested in suspension. Moreover, ingestion of particles can occur simultaneously with the elimination of mucus-embedded particles as pseudofaeces (Jørgensen, 1981b). Also Kiørboe and Møhlenberg (1981) found that algal food particles, Phaeodactylum tricornutum, fed to M. edulis below and above the threshold for pseudofaeces production, were freely suspended in samples of the oesophagus content.

Mucus and Sorting of Particles: Cleaning and Feeding

The suspensions to which filter feeding bivalves are exposed in nature are mixtures of particles of food value, particularly phytoplankton, and particles without food value, particularly silt, which may be present at concentrations that are many times higher than those of food particles. High concentrations of silt reduce the overall food value of the suspended particulate matter by a dilution effect. Above threshold concentrations for the production of pseudofaeces this dilution effect would be counterbalanced if mechanisms existed by which the bivalves were able to sort food particles from silt for preferential ingestion of food. Such sorting has been demonstrated in a number of filter feeding bivalves (Kiørboe et al., 1980; Kiørboe & Møhlenberg, 1981; Newell & Jordan, 1983). Kiørboe & Møhlenberg (1981) compared the efficiency with which 10 species of filter feeding bivalves separated algae from silt presented as a suspension of 10-$30 \cdot 10^3$ cells ml^{-1} and 10-20 mg l^{-1} of bottom material. The selection efficiency, expressed as the ratio of the chlorophyl a (mg dry wt)$^{-1}$ in the suspension offered and in the pseudofaeces rejected, varied from about 2 in Arctica islandica to about 16 in Spisula subtruncata. The efficiency was highly correlated with the size of the labial palps, expressed by the ratio palp area (mm^2)/clearance (ml min^{-1}).

The efficiency of selection could not be related to gill type, and there was no clear correlation between selection efficiency or palp size and the turbidity characteristic of the habitat of the species. However, such correlation was striking in the two populations of Mytilus edulis investigated. The Wadden Sea mussels, which are exposed to concentrations up to several hundreds of mg l^{-1}, have larger palps and more efficient particle selection than the mussels from Øresund, where the maximum concentration of particulate matter recorded in bottom-near water is about 10 mg l^{-1} (Kiørboe & Møhlenberg, 1981; Theisen, 1982). Also selection among particles of food value, such as various, types of algae, has been reported (Shumway et al., 1985), but the evidence seems inconclusive (Jørgensen, 1989).

The close correlation between palp size and selection efficiency suggested the labial palps to be the site of sorting, but mechanisms remained obscure. If sorting of particles for acceptance or rejection does occur on the palps the sorting mechanism presumably must be operating within a matrix of mucus entangling the particles. This is because only the mucus-embedded material that travels along the ventral margin of the demibranchs outside the food groove seems to be transferred onto the palps where these envelope the margins of the demibranchs (Jørgensen, 1981b). Particulate matter in suspension is transported inside the food groove directly to the mouth. No tracts are known to lead from the food grooves onto the palps. Ciliary currents on the palp surface are capable of separating smaller fragments or even single particles or cells from larger masses of mucus-embedded material but the separation does not free the particles of mucus. This can be inferred from the stickiness the particles have acquired (unpublished observations).

Newell & Jordan (1983) attempted to explain sorting of particles on the palps by suggesting that the mucus upon passing between the opposing palps changes from a mucous to a serous state, by which change the particles should become free to move and thus accessible for sorting. Mucus does in fact change its rheological properties when exposed to shearing stress. Mucus is secreted as a viscoelastic substance, a property of fundamental importance for its function in clearing ciliated surfaces (Litt, 1971). At high shear rates mucus looses its elastic properties with a concurrent decrease in viscosity, and it assumes Newtonian properties (Ronkin, 1955; Reid, 1970; Simkiss & Wilbur, 1977; Denny & Gosline, 1980; Grenon & Walker, 1980). It has been hypothesized that shearing stress may be responsible, at least partially, for the establishment of serous sublayers surrounding the cilia in mucociliary systems, of widespread occurrence among animals (Blake, 1973; Blake & Sleigh, 1974; Simkiss & Wilbur, 1977). However, if ciliary tracts do exert shear forces high enough to affect the rheological properties of mucus, such effects seem to be restricted to the ciliary sublayer, and not to extend beyond the ciliary tips (Blake, 1973), that is, to the level of a possible sorting. It, therefore, seems unlikely that mechanisms exist on the palps that can re-establish conditions compatible with sorting of individual particles.

The lack of sorting of particles in suspension on the palps is further supported by Kiørboe et al.´s (1980) finding of a threshold value for the addition of silt before the selection mechanism becomes operative and pseudofaeces production starts. In Mytilus edulis from Øresund, Denmark, this threshold was 1 mg silt l^{-1}. If exposure of mixtures of algae and silt in suspension to sorting mechanisms on the palps is obligatory, it is difficult to understand how such sorting can remain inoperative below certain concentrations of silt. Kiørboe et al. also found a linear relationship between the rate of production of pseudofaeces and the material retained by the gills. As the rate at which the material was retained increased from 0.057 mg min^{-1} to about 3 mg min^{-1}, pseudofaeces production increased from zero to close to 3 mg min^{-1}, corresponding to only a moderate increase in ingestion rate from about 32 µg min^{-1} at a silt concentration of 1 mg l^{-1} to 50 µg min^{-1} at a 60 times higher concentration of silt. Such a regulation of the ingestion rate concurrently with the rejection of increasing proportions of retained material after sorting is consistent with the concept of a restricted capacity of the food grooves for transporting particles in suspension, any surplus being taken care of by mucus produced by the gills and eventually transferred to the palps. This would imply a re-location of the sorting from the palps to the gills.

Sorting of particles presumably depends upon the ability simultaneously to ingest particles in suspension and to eliminate mucus-bound particles as pseudofaeces. This ability again depends upon spatial separation of acceptance and rejection tracts on the gills. How such separation ensues may be elucidated from observations of the exposed gill surface in, e.g., Mytilus edulis (Jørgensen, 1981b).

As mentioned, cutting of the adductor muscle to expose the gill surface strongly stimulates the secretion of mucus, but the secretion gradually abates to reach low levels within some hours. The secretion does not, however, stop as in the intact mussel not producing pseudofaeces. The mussels with cut adductor muscle responded with profuse secretion of mucus to low concentrations of particles in the water. Mutilation in Mytilus edulis thus enhanced the sensitivity of the mechanism that controls the secretion of mucus in response to particulate matter in the water passing through the gill. However, the close relationship persisted between the concentration of particulate matter and the rate of secretion of mucus.

Mucus is secreted in the form of tiny threads that tend to coalesce with secretions from other mucus cells and to form longer and longer threads or strings traversing many filaments. These mucus filaments are carried directly on the ciliary tracts and they advance slowly compared with the water currents that flow distally to the mucus strings. Particles suspended in the currents along the frontal surface of the filaments thus pass the mucus strings at close distance and may even make contact with and stick to a mucus string, that is, pass from an ingestion pathway to a rejection pathway. The chance of being caught in a mucus string must depend upon the rate of secretion of mucus, which determines the thickness and spacing of the mucus strings traversing the filaments. Presumably, capture of particles by the mucus strings also depends upon size, shape and surface properties of the particles. The spatial and temporal relationships between the surface currents along the gill filaments and the mucus strings carried across the filaments provide a basis for the selective rejection of particles from mixtures when the total concentration of particulate matter in suspension exceeds the capacity of the digestive tract.

This hypothesis for the mechanism of particle sorting is based primarily on observations of the filibranch bivalve, Mytilus edulis, but presumably it may apply equally well to the plicate gills which exhibit a more pronounced spatial separation of rejection tracts and acceptance tracts. The ciliary tracts on the plical crests transport mucus strings that may span across several plicae (Foster-Smith, 1975). In most species this transport is ventral to the margin and further along to the palps. Simultaneously, suspended particles are carried in currents, created along the filaments that constitute the throughs of the plicae, to the currents within the dorsal and/or ventral food grooves leading directly to the mouth. As mentioned, there was nothing to indicate, though, that plicate gills were correlated with more efficient separation of algae from silt (Kiørboe & Møhlenberg, 1981).

If the complex system of folds and ciliary tracts on the inside of the palps does not act to sort particles the function remains to be ascertained. Such a function may be to form the particle-loaded mucus-strings into pseudofaeces. Observations of exposed palps in preparations of mussels with cut adductor muscles show that the activity of the ciliary tracts tends to rotate and wind up mucus strings. This winding-up presumably constitutes the major feature in the formation of pseudofaeces before they are dropped onto rejection tracts along the mantle edges. Such a function of the palps in the formation of pseudofaeces would be consistent with the finding that Mytilus edulis from habitats characterized by high particle loads, correlated with high levels of pseudofaeces formation, also possess larger palps than do mussels from populations exposed to only low particle loads (Kiørboe et al., 1980, Theisen, 1982).

Reconciliation of Views

Early descriptions of feeding mechanisms in filter feeding bivalves did not clearly distinguish between transport of particles in suspension or bound in mucus. However, at the turn of the century the ambiguity was resolved in the statement that retained food particles are bound to mucus strings that are transported by physical contact with the beating cilia (Stenta, 1903; Wallengren, 1905). After mucus had been generally accepted as an integral element in the feeding process, descriptions often omitted mentioning its presence, and its involvement in the treatment of particles on gills and palps can only be inferred, e.g., from a casual reference to the production of pseudofaeces (Stasek, 1965). The vast amount of observations published on particle transport and sorting on the various types of bivalve gills seems, however, to fit the concept of two distinct mechanisms of ciliary particle transport: particles may be carried along ciliary tracts free in suspension within the laminar currents produced by the ciliary tracts or bound to mucus that tends to be transported by direct ciliary contact.

This distinction between two separate physical states for particle transport along gill filaments may also resolve the widely accepted paradox that tracts of fine cilia selectively transport small and light particles and that tracts of coarse cilia selectively transport large and heavy particles. There seems to be no physical basis for a preference of "fine" cilia for small particles and of "coarse" cilia for large particles. Moreover, the distinction between "fine" and "coarse" cilia is misleading because the diameter of cilia forming the fine and coarse ciliary tracts is the same, viz. 0.2 µm, the standard diameter of cilia and flagella. But the cilia are sparsely and evenly distributed along the lateral "fine" tracts as compared with the median "coarse" tract, where the densely spaced cilia tend to form cirral-like tufts (Owen & McCrae, 1976; Ribelin & Collier, 1977).

The interpretations in terms of particle selection by means of "fine" and "coarse" ciliary tracts are based on observations made on exposed gills under conditions where mucus is produced and rejection tracts thus activated. Under such conditions we now know that the gill surface presents two different environments to particles, laminar surface currents along the filaments and mucus strings passing perpendicularly to these currents. Median frontral ciliary tracts on the gill surface constitute prominences, both on the filibranch gills and on the plicate gills where the dense tracts are preferentially found on the crests. These structural features result in the mucus strings primarily being formed and transported by the activity of the tracts of "coarse" cilia. Given that large particles in suspension in the surface currents are more likely to be caught than small particles when the particles pass a mucus string, the overall pattern of events appears to be that large particles are preferentially transported by tracts of "coarse" cilia and small particles by tracts of "fine" cilia.

Finally, an attempt should be made to reconcile Bernard´s (1974) description of bivalve suspension feeding with that presented here. As mentioned, Bernard states that prospective food particles are entrapped in watery, serous mucus overlying the frontal cilia. The thickness of the layer is given as 17 µm, or about three times the length of the frontal cilia. This feeding mucus is stated to constitute a continuous stream moving from the gills towards the mouth whether food particles are present or not. From the dimension of the stream and its travelling speed Bernard calculated that the rate of mucus production in an 8.2 cm long oyster would amount to 7.9 cm^3 h^{-1}. He made the important observation that the mucus from the food grooves contained about 3 % of dry material against 12% in the mucus from the mouth region. According to Bernard this concentration of matter takes place on the palps, which reduce the mucus so that the particles enter the mouth in one fourth the original volume.

At a rate of secretion of 7.9 ml of mucus per h containing 3 % of dry algal matter, 7.9 x 0.03 = 0.24 g algal matter would be ingested per h. It is not stated at which concentration of algae the oysters were feeding, but since they did not secrete "rejection" mucus the concentration was below the threshold for the production of pseudofaeces. Neither is it stated at what rate the oysters filtered the water. If we assume a filtration rate of 7 l h^{-1} and a food concentration of 1 mg l^{-1} this corresponds to an ingestion rate of 7 mg h^{-1}, or $(7/240)\cdot 100 = 3$ % of the rate calculated by Bernard. This would imply that mucus flow in the food grooves amounts to only a small fraction of the flow that transports the food particles along the frontal surface of the gill filaments. Such a reduction in volume of secreted mucus at the transition from the filaments to the ventral food groove is not easily explained. But Bernard´s statements concerning the secretion of serous mucus along the frontal filaments for the confinement of food particles may constitute an interpretation of observations he felt to imply such secretion. It thus seems that the observations and data available are consistent with the fluid mechanical mechanism for the transport of particles suggested above. This mechanism implies stepwise and steep increases in concentration of suspended particles at the transition from the surface currents along the gill filaments to the ventral food grooves and again from these grooves to the oesophagus, the increases in concentration corresponding to the reduction in volume of the currents containing the particles (Jørgensen, 1981b).

Conclusions

Filter feeding bivalves process the ambient water at rates that are basically independent of the concentration of suspended particulate matter, including food particles as well as silt. Above threshold concentrations for the production of pseudofaeces, sorting of food particles, algae, from silt for preferential ingestion of algal cells may obtain. The sorting of particles seems to depend upon the ability simultaneously to ingest particles in suspension and to eliminate mucus-bound particles as pseudofaeces. Presumably, differential capture of suspended particles in mucus arises from differences in physical properties of the particles. The preferential ingestion of food particles at high loads of suspended matter in the environment, therefore, appears to be incidental.

Acknowledgements: Supported by a grant from the Carlsberg Foundation

References

Allen JA (1958) On the basic form and adaptations to habitat in the Lucinacea (Eulamellibranchia). Phil Trans Roy Soc Series B 241: 421-484

Allen JA (1968) The functional morphology of Crassinella mactracea (Linsley) (Bivalvia: Astartacea). Proc Malacol Soc, London 38: 27-40

Ansell AD (1981) Functional morphology and feeding of Donax serra Röding and Donax sordidus Hanley (Bivalvia: Donacidae). J Mollusc Stud 47: 59-72

Atkins D (1936) On the ciliary mechanisms and interrelationships of lamellibranchs. Part I. Some new observations on sorting mechanisms in certain lamellibranchs. Quart J Microsc Soc 79: 181-308

Atkins D (1937a) On the ciliary mechanisms and interrelationships of lamellibranchs. Part II. Sorting devices on the gills. Quart J Microsc Soc 79: 339-373

Atkins D (1937b) On the ciliary mechanisms and interrelationships of lamellibranchs. Part III. Types of lamellibranch gills and their food currents. Quart J Microsc Soc 79: 375-421

Bernard FR (1974) Particle sorting and labial palp function in the Pacific oyster Crassostrea gigas (Thunberg, 1795). Biol Bull Mar Biol Lab, Woods Hole 146: 1-10

Blake J (1973) A note on mucus shear rates. Resp Physiol 17: 394-399

Blake JR, Sleigh MA (1974) Mechanics of ciliary locomotion. Biol Rev 49: 85-125

Bronn (1862) Classen und Ordnungen des Thierreichs. Bd III, Abth. I. Kopflose Weichthiere (Malacozoa Acephala). Leipzig

Bullivant JS (1968) A revised classification of suspension feeders. Tuatara 16: 151-160

Clausen C (1958) On the anatomy and histology of the eulamellibranch Kelliella miliaris (Philippi) with observations on the ciliary mechanisms in the mantle cavity. Nytt Mag Zool 6: 144-175

Denny MW, Gosline JM (1980) The physical properties of the pedal mucus of the terrestrial slug, Ariolimax columbinus. J Exp Biol 88: 375-393

Fankboner PV (1971) The ciliary currents associated with feeding, digestion, and sediment removal in Adula (Botula) falcata Gould 1851. Biol Bull Mar Biol Lab, Woods Hole 140: 28-45

Foster-Smith RL (1975) The role of mucus in the mechanism of feeding in three filter-feeding bivalves. Proc Malacol Soc, London 41: 571-588

Foster-Smith RL (1978) The function of the pallial organs of bivalves in controlling ingestion. J mollusc Stud 44: 83-99

Fretter V, Montgomery MC (1968) The treatment of food by prosobranch veligers. J Mar Biol Ass UK 48: 499-520

Gilmour THJ (1979) Feeding in pterobranch hemichordates and the evolution of gill slits. Can J Zool 57: 1136-1142

Gosselin RE, O'Hara C (1961) An unsuspected source of error in studies of particle transport by lamellibranch gill cilia. J Cell Comp Physiol 58: 1-9

Gray J (1929) The mechanism of ciliary movement. Amer Natur 63: 68-81

Grenon J-F, Walker G (1980) Biochemical and rheological properties of the pedal mucus of the limpet, Patella vulgata L. Comp Biochem Physiol 66B: 451-458

Hughes TG (1975) The sorting of food particles by Abra sp. (Bivalvia: Tellinacea). J exp mar Biol Ecol 20: 137-156

Judd W (1971) The structure and habits of Divariscintilla maoria Powell (Bivalvia: Galeommatidae). Proc malacol Soc, London 39: 343-353

Jørgensen CB (1975) On gill function in the mussel Mytilus edulis L. Ophelia 13: 187-232

Jørgensen CB (1976) Comparative studies on the function of gills in suspension feeding bivalves, with special reference to effects of serotonin. Biol Bull Mar Biol Lab, Woods Hole 151: 331-343

Jørgensen CB (1981a) A hydromechanical principle for particle retention in Mytilus edulis and other ciliary suspension feeders. Mar Biol 61: 277-282

Jørgensen CB (1981b) Feeding and cleaning mechanisms in the suspension feeding bivalve Mytilus edulis. Mar Biol 65: 159-163

Jørgensen CB (1983) Fluid mechanical aspects of suspension feeding. Mar Ecol - Progr Ser 11: 89-103

Jørgensen CB (1989) Bivalve Filter Feeding: Hydrodynamics, Physiology and Ecology (in manuscript)

Kellogg JL (1915) Ciliary mechanisms of lamellibranchs with descriptions of anatomy. J Morphol 26: 625-701

Kiørboe T, Møhlenberg F (1981) Particle selection in suspension-feeding bivalves. Mar Ecol - Progr Ser 5: 291-296

Kiørboe T, Møhlenberg F, Nøhr O (1980) Feeding, particle selection and carbon absorption in Mytilus edulis in different mixtures of algae and resuspended bottom material. Ophelia 19: 193-205

Kristensen JH (1972) Structure and function of crystalline style of bivalves. Ophelia 10: 91-108

Lam VWW (1977) Shell form and diagnostic differences in the structure of the siphons and ciliary currents of the ctenidia in coastal species of the Tapetinae (Bivalvia: Veneracea) in Hong Kong. Proceedings I. International Workshop on the Malacofauna of Hong Kong and Southern China 1977, pp 11-31. Hong Kong.

List T (1902) Die Mytiliden des Golfes von Neapel und der angrenzenden Meeres-Abschnitte. Friedländer, Berlin.

Litt M (1971) Flow behavior of mucus. Ann Otol 80: 330-335

MacGinitie GE (1941) On the method of feeding in four Pelecypods. Biol Bull Mar Biol Lab, Woods Hole 80: 18-25

Millar RH (1955) Notes on the mechanism of food movement in the gut of the larval oyster, Ostrea edulis. Quart J Microsc Sci 96: 539-544

Morton B (1969) Studies on the biology of Dreissena polymorpha Pall. I. General anatomy and morphology. Proc malacol Soc, London 38: 301-321

Morton B (1970) The functional anatomy of the organs of feeding and digestion of Teredo navalis Linnaeus and Lyrodus pedicellatus (Quatrefages). Proc malacol Soc, London 39: 151-167

Morton B (1976) The biology, ecology and functional aspects of the organs of feeding and digestion of the S.E. Asian mangrove bivalve, Enigmonia aenigmatica (Mollusca: Anomiacea). J Zoöl, London 179: 437-466

Morton B (1983) Feeding and digestion in Bivalvia. In: Wilbur KM (ed) The Mollusca vol. 5 Academic Press, New York, London, p 65
Morton JE (1960) The functions of the gut in ciliary feeders. Biol Rev 35: 92-140
Morton JE (1967) Guts. Arnold, London
Narchi W (1972) On the biology of Iphigenia brasiliensis Lamarck, 1818 (Bivalvia, Donacidae). Proc malacol Soc, London 40: 79-91
Nelson TC (1960) The feeding mechanism of the oyster. II. On the gills and palps of Ostrea edulis, Crassostra virginica and C. angulata. J Morphol 107: 163-203
Newell RIE, Jordan SJ (1983) Preferential ingestion of organic material by the American oyster Crassostrea virginica Mar Ecol - Progr Ser 13: 47-53
Nicol EAT (1930) The feeding mechanism, formation of the tube, and physiology of digestion in Sabella pavonina. Trans Roy Soc Edinburgh 56: 537-598
Orton JH (1912) The mode of feeding of Crepidula, with an account of the current-producing mechanism in the mantle cavity, and some remarks on the mode of feeding in Gastropods and Lamellibranchs. J Mar Biol Ass UK 9: 444-478
Owen G (1953) On the biology of Glossus humanus (L.) (Isocardia cor Lam.). J Mar Biol Ass UK 32: 85-106
Owen G (1978) Feeding and digestion in the bivalvia. Adv Comp Physiol Biochem 5: 1-35
Owen G, McCrae JM (1976) Further studies on the latero-frontal tract of bivalves. Proc Malacol Soc, London B. 194: 527-544
Purcell EM (1977) Life at low Reynolds number. Amer J Phys 45: 3-11
Purchon RD (1954) A note on the biology of the lamellibranch Rocellaria (Gastrochaena) cuneformis Spengler. Proc Zool Soc, London 124: 17-33
Purchon RD (1955a) The functional morphology of the rock-boring lamellibranch Petricola pholadiformis Lamarck. J Mar Biol Ass UK 34: 257-278
Purchon RD (1955b) The structure and function of the British Pholadidae (rock-boring lamellibranchia). Proc Zool Soc, London 124: 859-911
Reid L (1970) Chronic bronchites - A report on mucus research. Proc R Instn Gt Br 43: 438-463
Reid RGB (1965) The structure and function of the stomach in bivalve molluscs. J Zool, London 147: 156-184
Ribelin BW, Collier A (1977) Studies on the gill ciliation of the American oyster Crassostrea virginica (Gmelin). J Morphol 151: 439-449
Ronkin RR (1955) Some physicochemical properties of mucus. Archs Biochem Biophys 56: 76-89
Segerdahl E (1922) Investigations on the effect of a direct electric current on the ciliary motion of the Anodonta gill. Skand Arch Physiol 42: 62-76
Shumway SE, Cucci RC, Yentsch CM (1985) Particle selection, ingestion, and absorption in filter-feeding bivalves. J exp mar Biol Ecol 91: 77-92
Simkiss K, Wilbur KM (1977) The molluscan epidermis and its secretions. Symp Zool Soc London 39: 35-76
Stasek CR (1962) Aspects of ctenidial feeding in immature bivalves. The Veliger 5: 78-79

Stasek CR (1965) Feeding and particle-sorting in Yoldia ensifera (Bivalvia: Protobranchia) with notes on other nuculanids. Malacologia 2: 349-366

Stebbing ARD, Dilly PN (1972) Some observations on living Rhabdopleura compacta (Hemichordata). J Mar Biol Ass UK 52: 443-448

Stenta M (1903) Zur Kenntniss der Strömungen im Mantelraume der Lamellibranchiaten. Arb Zool Inst Univ Wien 14: 211-240

Strathmann RR, Jahn TL, Fonseca JRC (1972) Suspension feeding by marine invertebrate larvae: clearance of particles by ciliated bands of a rotifer, pluteus, and trochophore. Biol Bull Mar Biol Lab, Woods Hole 142: 505-519

Theisen BF (1982) Variation in size of gills, labial palps, and adductor muscle in Mytilus edulis L. (bivalvia) from Danish waters. Ophelia 21: 49-63

Thompson TE (1959) Feeding in nudibranch larvae. J Mar Biol Ass UK 38: 239-248

Vogel S (1981) Life in Moving Fluids. Willard Grant, Boston

Vonk HJ (1924) Verdauungsphagocytose bei den Austern. Z vergl Physiol 1: 607-623

Wallengren H (1905) Zur Biologie der Muscheln. II. Die Nahrungsaufnahme. Lunds Univ Årsskr N.F. Afd. 2, Bd. 1, Nr. 3, p. 1-58

Werner B (1955) Über die Anatomie, die Entwicklung und Biologie des Veligers und der Veliconcha von Crepidula fornicata L. (Gastropoda Prosobranchia). Helgoländer Wissens Meeresunters 5: 169-217

Yonge CM (1923-24) Studies on the comparative physiology of digestion. I. The mechanism of feeding, digestion and assimilation in the lamellibranch Mya. J Exp Biol 1: 15-63

Yonge CM (1926-27) Structure and physiology of the organs of feeding and digestion in Ostrea edulis. J Mar Biol Ass UK 14: 295-386

Yonge CM (1935) On some aspects of digestion in ciliary feeding animals. J Mar Biol Ass UK 20: 341-346

ORAL DISCUSSION – C.B. Jørgensen.

The question was raised about the possibility of electrical charges affecting the movement of particles. CBJ responded that charged particles behaved like neutral particles. Also a comment from the audience indicated that we wouldn't expect a charge to have any effect in seawater. The problem of sorting different quality silt by strictly mechanical means was questioned from an evolutionary perspective. CBJ thought the sorting mechanism acted as an adaptation to simulataneously exploit low densities of suspended algae and clean the gills of excess silt. It was stressed that the mucous strands merely served to remove silt via pseudofaeces and that the two processes, feeding and cleaning, should be considered as two systems. The relationship between palp size and the amount of suspended matter in the water is probably related in causal manner. When the amount of suspended matter in the water exceeds the carrying capacity of the digestive tract, the pseudofaeces are produced from long mucous strands, formed into small balls on the palps and rejected.

BEHAVIOURAL PLASTICITY IN THE SUSPENSION FEEDING OF BENTHIC ANIMALS

B. Okamura
Department of Zoology
South Parks Road
Oxford OX1 3PS
United Kingdom

INTRODUCTION

Because the world is heterogeneous, consumers typically must acquire food that is distributed patchily. Many consumers solve this problem by moving through the environment to locate and exploit patches of food, and a large literature has arisen concerning various aspects of the behaviour of mobile consumers. Much of this literature has taken an optimality approach, considering questions such as how much time foragers should allocate to patches (or prey) of differing profitability so as to maximize their profits (or fitness) before moving on (see Townsend & Hughes 1981 and Krebs & McCleery 1984 for review). However, not all consumers are mobile; some are fixed in one place. Such organisms include plants and a large number of benthic invertebrates. In the latter, habitat selection may be effected once in the lifetime by habitat-related preferences at larval settlement: for the rest of their lives such organisms have little control over resource availability. (However, in clonal organisms, growth and fragmentation may provide additional means of sampling different patches within the life cycle.)

Many sessile benthic invertebrates feed on material suspended in the water column. Morphological correlates of this lifestyle include radial or planar feeding structures (a morphology inferred to maximize the amount of food intercepted from unpredictable or from uni- or bidirectional currents), and the loss or lack of development of highly-cephalized nervous systems associated with the sensory processing of active animals. Suspension feeding is one of the major feeding modes in aquatic environments, and nearly every animal phylum possesses members that ingest suspended particulate matter. Being relatively immobile, benthic suspension feeders depend on the general movement of water to introduce food to their local habitats. Thus, rather than searching out food-rich patches, benthic suspension feeders must make do with what is delivered.

Research on suspension feeders during the past several decades makes it clear that our old simplistic notions of suspension feeding need to be revised. Many of these animals actually employ a variety of feeding techniques and capture food by mechanisms other than traditional sieving. My aim is to review the growing evidence that feeding plasticity is common

NATO ASI Series, Vol. G 20
Behavioural Mechanisms of Food Selection
Edited by R. N. Hughes

amongst benthic suspension feeders despite their apparent simplicity and to argue that this plasticity allows these animals to act opportunistically and thereby exploit the variable food resources they encounter. Such behavioural plasticity may be a particularly important resource acquisition strategy of immobile organisms facing spatially and temporally heterogeneous food resources and thus may represent a different approach to foraging than that employed by more traditionally-studied mobile animals. As background to the topic I briefly describe some general aspects of the biology of suspension feeding and the nature of the food resources of suspension feeders before going on to review feeding plasticity within groups. The scope of the paper is restricted to benthic marine suspension feeders.

GENERAL ASPECTS OF SUSPENSION FEEDING

Suspension feeders feed on suspended material that is often much smaller than themselves and is present in low concentrations; that is, they live in a nutritionally dilute environment (Conover 1968). They must therefore process relatively large quantities of water and can do so effectively (Glynn 1973, Buss & Jackson 1981, Mook 1981, Jordan & Valiela 1982, Carlson et al. 1984, Dame et al. 1984, Wildish & Kristmanson 1984, Dame & Dankers 1988, Fréchette et al. 1989). Traditionally, suspension feeders are viewed as being either active or passive feeders (Jørgensen 1966). Active suspension feeders expend energy to create currents past their feeding surfaces (e.g.sponges, brachiopods, bryozoans, phoronids, bivalve molluscs, some polychaetes, and ascidians). Passive suspension feeders rely entirely on ambient water currents to introduce food to their feeding surfaces (e.g. hydroids, gorgonians, corals, anemones, crinoids, ophiuroids, and holothurians). However, it has gradually become evident that these categories are not mutually exclusive. Many suspension feeders capable of creating their own feeding currents will take advantage of ambient currents and feed passively, thereby presumably reducing the costs of the feeding process. LaBarbera (1984) has termed such animals facultatively active suspension feeders.

For a long time animals were believed to collect suspended particulate matter by retaining it on sieve elements. Hence they were termed filter feeders. However, a consideration of filtering mechanisms analogous to mechanisms of particle capture by aerosol filters suggested strict sieving need not always apply (Rubenstein & Koehl 1977). This clarification has sparked a number of investigations that have revealed mechanisms other than sieving are important (Koehl 1977, LaBarbera 1978, Meyer 1979, Ross & Craig 1980, Patterson 1984, Braimah 1987, Best 1988, Leonard 1989). Suspension feeders are thus not always animals that sieve particles out of suspension, nor are they invariably active or passive in their approach. Furthermore, in some kinds of suspension feeding food items may be so large as to be individually sensed and seized while in other cases particles are so small that they are only obtained in bulk by processing the surrounding water. These cases have been termed macrophagy and microphagy respectively (Jørgensen 1975, Fauchald & Jumars 1979).

THE FOOD OF SUSPENSION FEEDERS

The potential food of suspension feeders can be classified roughly into phytoplankton, algal spores, organic detritus, dissolved organic matter, and heterotrophic organisms (bacteria, fungi, yeasts, zooplankton, gametes, and larval forms) (Jørgensen 1966, Sebens & Koehl 1984, Young & Chia 1987). In general, the food of suspension feeders is patchy on both spatial and temporal scales and is highly variable in composition. This variability occurs throughout a hierarchical spectrum of space and time scales, thus a single model of variability will resolve only a fraction of the total variability (Mackas et al. 1985). The intensity, composition, and scale-dependence of plankton spatial patchiness is strongly regulated by physical oceanographic processes (e.g. turbulent advection, upwelling, convergence, vertical mixing, turbulent diffusion, and topographic influences on circulation patterns) (e.g. Denman & Powell 1984, Mackas et al. 1985, Bennett & Denman 1989). Local plankton-dense patches can arise because currents and tidal fronts may concentrate suspended material, upwelling and other physical processes make available rich supplies of some limiting factor, and/or through behavioural (e.g. swimming) or physiological (e.g. buoyancy) mechanisms that result in aggregation (Mackas et al. 1985). Zooplankton may show peaks of abundance that lag behind the peaks of their phytoplankton prey, although purely biological forcing mechanisms (e.g. predator/prey oscillations) do not appear to be as important as biological-physical interactions at least at larger spatial scales (Mackas et al. 1985). Variability in diurnal patterns of photosynthesis has been attributed to temporal and spatial differences in irradiance, photoperiod, nutrient concentrations, temperature, and/or species composition of the community (Putt et al. 1988). In addition, abundances of zooplankton and larvae can show diel variation, reflecting vertical migration, the propensity of benthic animals to leave the substratum and swim during periods of calm (demersal plankters), and rhythmic patterns of larval release (e.g. Sebens & Koehl 1984, Jacoby & Greenwood 1988). Marked seasonality in the abundance of both phytoplankton and larvae is common except in some subtropical and tropical areas (see Valiela 1984 for general discussion of seasonal production in the oceans). Finally, primary productivity can show interannual variation on global scales due to atmospheric forcing events (Philander 1983, Shannon et al. 1986). The best known example is the El Niño in the Eastern Pacific in which a cessation or relaxation of upwelling or alternatively a depression of the seasonal thermocline occurs on average every 4 years (range = 2-10 yrs). The effect is that local equatorial winds do not bring nutrient rich waters to the surface and there can be major and widespread climatological and biological consequences

(Barber & Chavez 1983, Cane 1983). These periodic events vary in amplitude, spatial characteristics, and biological consequences (Cane 1983).

Waves, ambient currents and tidal flux influence the general pattern of flow and hence the movement of suspended particles through the habitats of benthic suspension feeders. Benthic animals experience variability in suspended material at a smaller scale as well. As water flows across a surface a velocity gradient or boundary layer develops above the surface due to the no-slip condition at the fluid/surface interface (see reviews by Jumars & Nowell 1984 and Nowell & Jumars 1984). Suspended material will be sorted differentially within that velocity gradient depending on its specific gravity: materials of higher specific gravity occur nearer the substratum in regions of slower flow due to their greater tendency to settle out of suspension than lighter, more neutrally-buoyant material such as organic particles or living organisms (Muschenheim 1987a). In addition, resuspension of material from the bottom can result in greater abundances of benthically-derived material relatively close to the substratum (Sebens & Koehl 1984, Johnson 1988).

The food of suspension feeders varies qualitatively. Suspended particulate material includes planktonic organisms from different taxonomic and trophic groups and that range in size from 0.02 mm to 200 cm (Sieburth et al. 1978) and organic aggregates resulting from a variety of sources (Wotton 1984). Most suspension feeders ingest particles ranging in size from several mm to several mm in length, and there is a large overlap in the sizes ingested (Jørgensen 1966). The types of particles captured by suspension feeders are dictated by properties of both the feeding apparatus and the particles (Rubenstein & Koehl 1977), and post-capture sorting may select certain particles for ingestion and others for rejection. Variation in particle shape, charge, surface characteristics, and nutritional quality may also influence suspension feeding. Benthic suspension feeders thus typically face a heterogeneous array of suspended material that varies in quantity, quality and rates of movement in time and space.

THE FEEDING BEHAVIOURS OF SUSPENSION FEEDERS

In this section I will discuss specifically the kinds of feeding behaviours observed in a variety of marine benthic suspension feeders pointing out the factors (when known) promoting these different behaviours. There is a corresponding literature on the suspension-feeding behaviour of zooplankton (e.g. see Cowles et al. 1988 for a summary of feeding techniques employed by copepods and cladocerans). When considering behaviour I shall take a broad view: I will discuss examples where organisms utilize different suspension-feeding techniques and cases in which suspension feeding exists as one kind of feeding mode in association with others. However I will not discuss patterns of feeding behaviour that result from the utilization of one feeding mechanism. For instance, varying particle characteristics can elicit differential feeding rates in many suspension feeders. Such cases will not be discussed except in instances where different feeding rates are associated with different feeding techniques. The review is not meant to be exhaustive, but rather illustrative of the kinds of techniques and feeding modes employed by a broad range of animals. In most cases, the generality of these behaviours in each group remains to be determined.

Amoeboid Protists

Foraminiferans employ fine pseudopodia to capture their food. Individual pseudopodial threads constantly change their extent and configuration probably via microtubule assembly and disassembly and active sliding (Sleigh 1989). A food item impacting on a sticky pseudopodial thread will be engulfed by a food vacuole. Benthic foraminiferans will engulf e.g. bacteria, diatoms, other algae, and similar small prey encountered by the wandering pseudopodial threads. Suspension feeding foraminiferans produce pseudopodial nets for food capture.

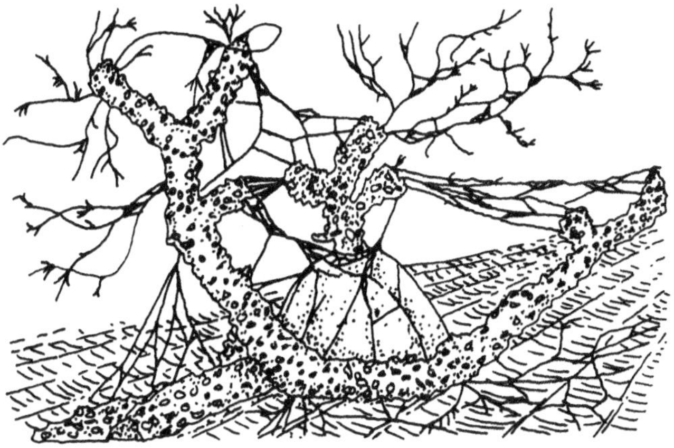

Figure 1. *Cibicides refulgens* growing on bivalve shell (see text for details) (after Alexander & DeLaca 1987). (Not to scale.)

The calcareous foraminiferan *Cibicides refulgens* living epizooically on scallops in McMurdo Sound, Antarctica utilizes a combination of morphological and physiological feeding adaptations (Alexander & DeLaca 1987). This organism erects a superstructure of agglutinated tubes as extensions from its calcareous test and from which a pseudopodial net is deployed (see Figure 1). Pseudopodia stream in the conduits of the agglutinated tubes and emerge from the tips some 5 mm above the scallop shell. The pseudopodia can extend another 3 mm from the tips of the tubes to form nets that capture suspended diatoms and bacteria. In addition to this suspension feeding mode *C. refulgens* employs two others. It grazes algae and bacteria living on the scallop shell, and it erodes through the scallop shell and utilizes free amino acids present in high concentrations in the mantle cavity of the scallop.

Alexander and DeLaca (1987) suggest that this combination of feeding modes enhances the ability of *Cibicides refulgens* to acquire nutrients in an otherwise oligotrophic and seasonal environment. The period of pronounced primary productivity in McMurdo Sound is of limited duration. Thus *C. refulgens* appears to benefit nutritionally by augmenting the year-round uptake of highly-concentrated mantle cavity amino acids with nutrients gained through suspension feeding and grazing activities. The rarity of loosely attached or roaming individuals would support this view.

Cnidaria

Cnidarians are passive suspension feeders using their nematocysts to trap and ensnare zooplankton and other food items that contact their tentacles. The feeding behaviours of a variety of reef corals have been observed (Lewis & Price 1975, Lewis 1976). Many, in addition to tentacular feeding, produce mucus nets and strands that capture suspended material ranging from fine particulate matter to larger zooplankton. Ciliary currents on the oral disk, tentacles and stalk facilitate particle-gathering of the mucus by lifting strands and nets off the epidermal surface and exposing them to water movement. These mucus strands are employed in some corals both at night when polyps are expanded and tentacular feeding occurs, and in the day when polyps are contracted. (In the latter case strands are eventually ingested.) Two additional feeding techniques have also been observed: 1) food capture by ciliary currents on the oral disk (such currents normally serve a cleansing function), and; 2) feeding by means of mesenterial filaments extruded from the gastrovascular cavity. However, these latter behaviours have been observed less frequently and so are thought to be relatively unimportant. Lewis and Price argue that the ability of reef corals to feed by mucus filaments greatly increases the potential food sources available to them. This is thought to be significant given the very low abundance of zooplankton in tropical waters (Johannes et al. 1970). The contribution to coral nutrition by symbiotic zooxanthellae will be discussed in a later section.

Mollusca

In bivalve molluscs the basic molluscan ctenidium has been elaborated into an organ of large surface area used to process particulate material. Suspension feeding bivalves possess an incurrent region, sometimes developed as a siphon, into which water is pumped into the mantle cavity. There particulate matter is screened by the ctenidia and rejected material and wastes are expelled through an excurrent area (or siphon).

A combination of suspension feeding and deposit feeding have been observed in both heterodont and palaeotaxodont bivalves. Brafield and Newell (1961) observed *Macoma balthica* (a heterodont bivalve with eulamellibranch ctenidia) deposit feeding when the tide was out and suspension feeding when the tide was in, and Hughes (1969) found that *M. balthica* would readily feed on suspended algae in the laboratory. Ōlafsson (1986) found that the switch from deposit to suspension feeding in *M. balthica* occurs with changes in flow velocity: clams would suspension feed with the advent of flow and deposit feed when flow ceased. He found variation in such feeding behavior depended on habitat. Clams taken from sandy habitats would readily suspension feed from flow and deposit feed in the absence of flow, while clams from muddy habitats would deposit feed regardless of flow. Since flow regimes and sediment composition are linked, suspension feeding should predominate in *M. balthica* in sandy sites and deposit feeding should predominate in muddy sites. Ōlafsson suggests that suspension feeding is a response to greater amounts of suspended matter present at high flow rates while deposit feeding is a response to a decrease in the concentration of suspended material. This view has been proposed to explain the alternate feeding behaviors observed in other animals (see annelid section).

Yoldia eightsi (a palaeotaxodont form with protobranchiate ctenidia) is the most common bivalve over large areas of the Antarctic and Subantarctic seabed. Davenport (1988) argues that this animal is so successful because it exploits all energy sources available to it by both deposit and suspension feeding. Suspension feeding is probably most important during diatom blooms in summer periods while deposit feeding is likely to contribute the main source of organic material for much of the year.

Thiesen (1972) provides evidence that in addition to suspension feeding, young mussels (*Mytilus edulis*) gain food through deposit feeding. Young mussels are very active and use their foot to creep along the substratum and to clean their valves. Thiesen found that stained yeast deposited on the valves and on the substratum was picked up, transported by ciliary action along the foot to the mantle cavity, and ingested. Thus in young mussels deposit feeding may augment suspension feeding. However, Thiesen points out that the nutritional gains associated with the activity of the foot are probably incidental to the primary functions of the foot's activity, these being movement and the prevention of fouling.

Bryozoa

Bryozoans are active suspension feeders creating feeding currents with their crown of tentacles (the lophophore). Coordinated action of the cilia lining the tentacles draws in water from above the lophophore and suspended material is ingested by several techniques. Strathmann (1982) has demonstrated that the lateral cilia of the lophophores detect individual particles and undergo a reversal in beat which tends to throw particles onto the frontal ciliary tracts of the tentacles where food is conveyed to the mouth at the base of the lophophore (Figure 2A). Particles can also be transported directly to the mouth by feeding currents (Best & Thorpe 1986). This may be particularly significant when particle concentrations are high (Okamura 1987).

In addition to the above ciliary techniques, tentacular activity also appears to be important in the feeding of bryozoans. This activity can range from simple tentacular flicking in which a particle is directed by an individual tentacle towards the mouth (Figure 2B) to the coordinated activity of all tentacles resulting in the encagement of larger particles such as zooplankton and larvae (Figure 2C) (Winston 1978, pers. obs.). In a recent study I found evidence that tentacular activity becomes more important for the arborescent cheilostome *Bugula neritina* as both particle size and ambient flow velocity increase (Okamura 1987). Such flow-induced behaviour is reminiscent of the flow-induced switch from deposit to suspension feeding in bivalves and annelids (see relevant sections).

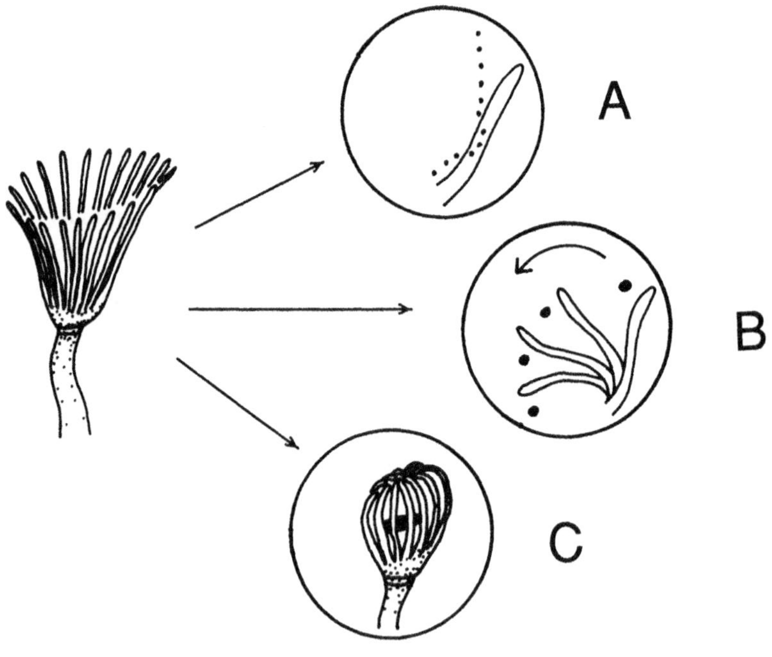

Figure 2. Schematic representation of bryozoan feeding techniques: lateral cilia lining tentacle of lophophore reverse beat to direct particle onto frontal ciliary tract (cilia not depicted) (A); tentacular flicking directs particle towards mouth (B); tentacular encagement of large particle (C). Extended lophophore to left shows relationship of structures.

Echinodermata

Ophiuroids can be scavengers, carnivores, deposit feeders and suspension feeders, and some species employ at least several of these approaches (see Warner 1982 for review). Suspension feeding is included in the feeding behavioural repertoire of many ophiuroids (Pentreath 1970, Warner 1982) and can be achieved by various techniques which include the use of especially sticky mucus, active tube feet, and mucus-covered spines to entrap particles. Fontaine (1965) found *Ophiocomina nigra* to be particularly opportunistic in its feeding behaviour. This species is capable of feeding microphagously on relatively small particles by employing both suspension feeding and deposit feeding techniques. When deposit feeding, *O. nigra* picks up particles from the surface of the substratum with its tube feet which then transfer the particles to the mouth. Water movement stimulates suspension feeding, and copious mucus is secreted from glands on the arm spines. The mucus gets strung out between the spines forming a net that entraps particles. In addition, the sticky arms are moved to and fro through the water presumably intercepting particles. *O. nigra* feeds macrophagously on larger particles as well. Bits of carrion are ingested, macroalgae are grazed, and larger zooplankters (e.g. some copepods) are captured by the arm-loop technique. In the latter instance the arm bends laterally around the food so as to grip it within a coil. Further coiling then proceeds up the arm carrying the food with it while the entire arm simultaneously bends towards the mouth.

The sand dollar *Dendraster excentricus* has been observed to capture actively swimming prey (nauplii, mysids, copepods) with spines, tube feet and large bidentate pedicellaria (Timko 1976). Spines form cages trapping zooplankters which are then captured and masticated by the open jaws of the pedicellaria. The prey is then moved to the food groove by the spines and tube feet. Small, immobile food items (> 50 mm in diameter) can be grasped by tube feet and pushed toward the food groove, while even smaller particles (< 50 mm in size) very close to the test surface can be swept directly into the food grooves by ciliary currents. *D. excentricus* uses mucus strings in food grooves to convey small particles but dispenses with mucus transport and uses its tube feet when food items are larger. Gut contents suggest mucus ciliary feeding is of minor importance as most gut particles are too big to be moved by ciliary currents. *D. excentricus* is also capable of deposit feeding.

Crinoids feed on particles that directly impinge on their modified tube feet with the aid of adhesive mucus. In addition to particle interception by tube feet, the comatulid *Florometra serratissima* has been seen to extend mucus threads that intercept suspended material (Byrne & Fontaine 1981), while the orientation (oral side up, arms spread out and contacting the substratum) of several other comatulids suggests they may feed on deposited material in addition to suspension feeding (Meyer 1982).

Microphagous feeding on either deposited or suspended particles is considered to be characteristic of a number of polychaete families. However, it is now evident that many polychaetes employ a variety of feeding modes and techniques (see review by Fauchald & Jumars 1979).

Spionid polychaetes were once considered to be strict deposit feeders using their two palps to browse by sweeping and flailing over the sediment to dab or collect particles with mucus (Self & Jumars 1978). Recent observations show that spionids are also suspension feeders. *Pygospio elegans* can filter particles through a mucus net within its tube, it can catch plankton with its palps in a raptorial fashion, and it can use its palps to intercept smaller suspended material (Fauchald & Jumars 1979, Taghon et al. 1980, Jumars et al. 1982). In the latter case worms lift their palps into the water column and twist them in gradually tapering helices (Taghon et al. 1980). Observations of other spionids indicate their palps can employ a range of behaviours: 1) both can deposit feed; 2) both can lash rhythmically through the water while suspension feeding; 3) both can lash irregularly while suspension feeding; 4) both may be held rigid while suspension feeding; 5) one may lash the water while the other deposit feeds; 6) one may be held rigid while the other deposit feeds (Dauer et al. 1981). Suspension feeding activity in spionids can be invoked by high particle fluxes in currents (Taghon et al. 1980, Muschenheim 1987b) or by the addition of particles to suspension (Dauer et al. 1981).

Polychaetes long considered to be strict suspension feeders also show plasticity in feeding behaviour. The Chaetopteridae live in tubes and secrete mucus bags or strings that strain particles from currents set up by fan-shaped notopodia. However, *Spiochaetopterus* sp., under conditions of low particle concentration, will search the sediment surface with its palps, while *Phyllochaetopterus prolifica* can also suspension feed by catching particles impacting its erect palps and has been seen to use its palps deposit feeding as well (Fauchald & Jumars 1979). Sabellid polychaetes possess large tentacular crowns that create feeding currents and capture suspended particles. In addition to normal suspension feeding, the sabellid *Manayunkia* sp. employs several branchial filaments of the tentacular crown to deposit feed (Lewis 1968). Perhaps the most bizarre combination of feeding modes is seen in the sabellid subfamily Fabriciinae. These worms are common in deep water and are capable of moving around with their posterior end first and trailing their tentacular crown. When moving they frequently drag along a tangle line of mucus which is periodically retrieved and ingested. When sitting in their tubes, the worm's tentacles not only capture suspended material in the normal sabellid style, but also sweep the surface of the substratum to collect deposited material. The tentacular crown of these worms is simple and proportionately smaller than in other sabellids. Fauchald and Jumars (1979) argue that their mobility and reduced tentacular crowns suggest the Fabriciinae are becoming deposit feeders secondarily and that this, in addition to a moderate suspension feeding capacity, is an adaptation to life in nutrient poor or physically unstable environments.

Nereids are regarded as errant polychaetes that search out their food. *Nereis diversicolor* feeds omnivorously on plant, animal, and detrital material. It possesses strong jaws that grab such items and drag them into its burrow. It also spends time hunting for live prey such as amphipods and small oligochaetes, and it will browse on live algae (Pashley 1985 cited in Barnes et al. 1988). In addition, *N. diversicolor* is capable of suspension feeding by irrigating water through a mucus cone secreted in its tube and deposit feeding on surface sediment particles (Harley 1950, Pashley 1985 cited in Barnes et al. 1988). Undulations of the worm drive water through the cone which is periodically ingested. It is not known if this behaviour is unique to *N. diversicolor* or occurs in other nereids (Harley 1950).

I have discussed some of the more well-described instances of behavioural plasticity in suspension feeding polychaetes. For further discussion the reader is referred to Fauchald and Jumar's review (1979).

Arthropoda

Barnacles can suspension feed on particles ranging in size from several micrometers to several millimeters in diameter (Crisp & Southward 1961). They do so by retaining particles on the cirral net and on the setae of the smaller cirri surrounding the mouth. Balanoid barnacles can actively sweep the cirri at varying rates through the water to create their own feeding currents. However, under conditions of increased flow the cirral net may be extended and held open, facing into the flow, and will be periodically withdrawn to ingest captured food. (See Figure 3 .) The fast cirral beating of balanoid barnacles appears to be particularly suited to the capture of relatively large particles in still or slowly-moving water (Crisp & Southward 1961), while passive extension into flow allows the exploitation of external currents for feeding.

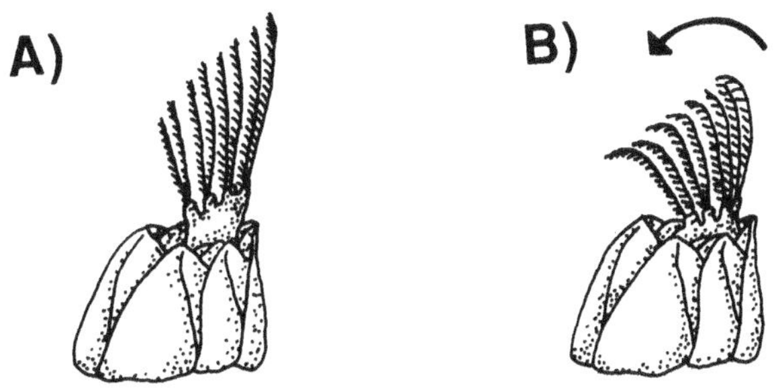

Figure 3. Barnacle with cirri held passively into current and straining particles (A), and actively sweeping cirri through water (B) to capture food items.

Several investigators have noted that tube-dwelling gammarid amphipods will ingest particles obtained through both suspension and deposit feeding (Hart 1930, Meadows & Reid 1966, Mills 1967, Miller 1984). In suspension feeding, pleopods of the second antennae create feeding currents, and particles in suspension are trapped on the setal baskets of the second gnathopods and collected periodically. In some cases one of the whirling second antennae occasionally scrapes the bottom to resuspend detritus from the sediment surface. In deposit feeding, the second antennae are used to collect material from the substratum.

ADDITIONAL FEEDING MODES AND MECHANISMS OF SUSPENSION FEEDERS

The foregoing discussion reviewed aspects of suspension feeding behaviours in different groups. I have left consideration of several additional nutritional modes and feeding mechanisms until this point because they are scattered or ubiquitous across groups, and/or because they do not really fall within the category of behaviour. However, it is necessary to discuss them briefly as they shall be relevant to the following discussion.

Association with Symbionts

A number of suspension feeders are hosts to symbiotic microorganisms (cyanobacteria, chemoautotrophic bacteria, and various photosynthetic algae). Such animals include pogonophorans and some corals, anemones, bivalves, polychaetes, oligochaetes, ascidians, and sponges. Investigations show that nutrients are transported from the symbionts to the hosts (see Smith & Douglas 1987 and Southward 1987 for review), and compounds produced by symbiotic chemoautotrophic bacteria appear to represent the only food source for gutless animals (e.g. pogonophorans). The contribution to host nutrition by symbionts in other associations can also be substantial (e.g. Wellington 1982). Indeed in coral reef habitats much of the primary productivity is carried out within animal tissues (Wilkinson 1983, Valiela 1984, Hatcher 1988).

Uptake of Dissolved Organic Matter

Uptake of dissolved organic matter (DOM) occurs in a wide range of invertebrates (see reviews by Jørgensen 1976, Stewart 1979, Stephens 1982). However, uptake has generally been studied at exceedingly high substrate concentrations and in the presence of bacteria which themselves have very high uptake rates thereby confounding measurements. Hence the ecological importance of uptake is largely unclear. The mussel *Mytilus edulis* and the ascidian *Styela montereyensis* have been shown to take up DOM from natural seawater (Manahan et al. 1982), but the relative contribution this makes to their nutritional requirements is not known. DOM may provide a significant source of nutrition for adult animals lacking a functional gut (Southward & Southward 1982) (although recent evidence of the potential importance of chemoautotrophic bacterial symbionts in gutless animals diminishes the nutritional role of DOM (e.g. Cavanaugh 1985)), for developing invertebrate larvae whose guts are not functional (or to nonfeeding larvae) (Manahan & Crisp 1982), and for organisms regenerating body parts (Clements et al. 1988). It is generally felt that DOM represents a small source of dietary requirements for most marine invertebrates (Valiela 1984), however the importance of DOM as an alternate or additional food source for animals such as suspension feeders that experience fluctuations in the quality or quantity of their food supply remains to be determined.

Flow Induction: Sponges, Brachiopods, and Ascidians

Flow can be induced in biological systems by three transducing mechanisms: dynamic force, the Bernoulli effect, and viscous entrainment (Vogel 1978). Among active benthic suspension feeders, such passive induction of flow may reduce the cost of moving water over feeding surfaces. There is evidence that three different types of active suspension feeders with internalized feeding surfaces make use of this phenomenon. These are sponges, brachiopods, and ascidians.

Sponges are able to make use of induced flow due to the arrangement of their plumbing (Vogel 1974). The lower incurrent pores (ostia) of sponges tend to occur in the more slowly-moving regions of the boundary layer than will the higher excurrent pores (oscula). The result of this velocity differential between the ostia and oscula means that water is passively induced to flow through the body of the sponge by viscous entrainment. Due to its viscosity, fluid flowing over a surface tends to draw fluid out of any hole in that surface. The faster the fluid flows past the hole, the greater the viscous entrainment of fluid from the hole. While viscous entrainment tends to draw water out of the ostia and the oscula, the effect is much greater at the oscula because the flow past them is faster. Thus water enters the ostia and flows through the oscula. Flow from ostia to oscula may also arise by the Bernoulli effect which can occur when fluid flows over a moundlike projection. The divergence of streamlines over the mound results in faster flows over the top. By Bernoulli's Principle, pressure is lower where velocity is higher, thus fluid flows from ostia to oscula following the pressure gradient. Finally flow can also be induced by the dynamic pressure of the current on the ostia. Since many of the ostia are oriented more or less directly into the current the dynamic pressure on these exceeds the static pressure exerted on the perpendicularly-oriented oscula and flow from ostia to oscula is induced. The relative importance of these mechanisms is not clear, but it is likely that viscous entrainment is more important for tall chimney-like sponges while the Bernouilli effect may be more important for short moundlike forms (Alexander 1979).

Induced flow through sponges was elucidated by observing both model and real sponges exposed to flow (Vogel & Bretz 1972, Vogel 1974). Previous understanding of sponge feeding currents was based on observations of currents set up by sponges in still water, however as most sponges do not flourish in still water (Vogel 1977) induced flows may commonly occur. The extent to which sponge choanocytes vary the rate of flagellar beating in response to flow induction remains to be determined. It may be that feeding currents are actively produced when external currents are weak. When strong currents prevail sponges may rely passively on induced flows. However, flagellar activity of the choanocytes has been suggested to be more important for driving water locally through the choanocyte collars than in the creation of feeding currents through the sponge canals (Alexander 1979). Thus flagellar activity may occur regardless of external flow conditions.

LaBarbera (1977, 1981) observed articulate brachiopod behaviour in a laboratory flume and found that orientation perpendicular to currents forces water into the sides of the gaping valves. Water is then driven through the lophophore and exits via the central excurrent area of the gape. Flow induction appears to occur by dynamic pressure on incurrent areas due to deceleration of ambient flow, lower pressure at the excurrent area which is oriented into the faster flows above the substratum, and viscous entrainment. It is not known whether lophophoral cilia vary their rate of beating in response to induced flows.

The solitary ascidian, *Styela montereyensis*, lives exposed to waves and currents. The flexible stalk of the animal results in passive orientation of the recurved incurrent siphon into currents (Young & Braithwaite 1980). Ambient currents enhance internal flow and feeding by forcing water into the incurrent siphon and through the branchial basket. Viscous entrainment in which some water is drawn out of the excurrent siphon probably reinforces the effect of dynamic pressure. Young and Braithwaite (1980) suggest that the low energy demand for food-gathering resulting from induced flow permits the low metabolic rate of *S. montereyensis* and also provides energy for the attainment of large size, indeterminate growth, and continuous reproduction. As a number of other solitary, stalked ascidians display similar recurved incurrent siphons and some deep-sea forms have unciliated branchial baskets, induced flow may be of general importance in these animals (Young & Braithwaite 1980).

VARIABLE AND INVARIANT FEEDING BEHAVIOURS

Because they are sessile, benthic suspension feeders are unable to track and pursue their food. Rather, they must contend with the food that is delivered to their local vicinity, and they must process large volumes of water to obtain it (Conover 1968). Jørgensen (1975) suggests that the clue to understanding characteristics of suspension feeding may be the generally low concentration of food available. He argues that the water transporting structures of suspension feeders may be dimensioned for continuous feeding at low rates (Jørgensen 1966, 1975). However, an alternate strategy is for suspension feeders to exploit as much of the potentially available suspended material as possible by utilizing different feeding techniques and/or to augment suspension feeding with other feeding modes. As I have shown above, many organisms have adopted this strategy. The variety of feeding modes and techniques employed by these animals allows them to make the most of their heterogeneous food sources. As different kinds of food items are obtained by using different kinds of feeding techniques, these animals are not specialists on one food type but act opportunistically and ingest many food types.

Suspension feeders that do have an invariant approach to feeding may suffer from competition with suspension feeders that are more plastic in their feeding behaviour. Unless the suspended material is relatively uniform in character and the net gain achieved by either approach is similar, the opportunistic approach of flexibility in feeding behaviour may prove superior and invariant suspension feeders may be excluded. Invariant, low feeding rates may be suitable where food supplies are low or relatively constant and hence predictable, but such a strategy may be inappropriate elsewhere. Articulate brachiopods appear to be an example of such suspension feeders: indeed Thayer (1986) has termed them 'minimal organisms'. Articulates possess a number of mechanisms that minimize energetic expenditure such as a low metabolic rate, laminar flow in active pumping, the ability to exploit ambient water currents to induce flows through their lophophore (LaBarbera 1981), relatively weak feeding currents, and the utilization of energetically-costly mucus (Paine 1971, Denny 1980) only during particle rejection (Thayer 1986). Articulate brachiopods are restricted to high latitudes and other regions of seasonally limited food supplies and to places such as cryptic reef habitats. This restricted distribution has been argued to reflect the inability of brachiopods to compete with the bivalves who produce strong feeding currents and can afford to be energetically inefficient in habitats where suspended food levels are higher.

A further complication to understanding the comparative approaches to suspension feeding is that actively produced currents in many suspension feeders may be required to meet respiratory demands. Separating respiratory requirements from nutritional ones is crucial to Jørgensen's proposal (Jørgensen 1966, 1975). Thin suspension feeders whose respiratory requirements are met by diffusion across the body wall may provide examples of animals whose self-generated currents are used only for feeding. However small animals have relatively high metabolic rates and hence large respiratory demands. Furthermore, depending on the flow environment (Nowell & Jumars 1984), small animals may be hidden within deep boundary layers across which diffusion occurs slowly (Vogel 1981), thus even in small organisms teasing apart respiratory and nutritional considerations may be problematic. Minimal levels of current production in suspension feeders that vary their ciliary currents in response to particle concentration (e.g. some bivalves, bryozoans, ascidians) might be reflective of respiratory requirements alone, however by varying their currents these organisms are not feeding at the continuous low rates predicted by Jørgensen.

SUSPENSION FEEDERS AS FORAGERS

The foraging behaviour of animals in stochastic environments has been the focus of recent theoretical and empirical study (e.g. see Real & Caraco 1986 and Stephens & Krebs 1986 for review). When a forager typically encounters unpredictable food sources it must select a strategy for exploiting these resources. In theory the forager must learn the probability distributions associated with different behaviours. Much empirical work has concentrated on visitation rates and optimal patch choice by mobile animals that encounter their food sequentially. However other animals, including immobile suspension feeders, may often face heterogeneous food resources simultaneously (Stephens & Krebs 1986). Many benthic suspension feeders employ a range of foraging behaviours to feed from their unpredictable food sources. Suspension feeders that employ different feeding techniques should therefore be able to perceive and assess prey availability and vulnerability directly or by some correlated cue and subsequently to adopt the appropriate feeding strategy for any given prey. It would be of interest to determine the relative costs and gains associated with alternate feeding techniques employed by suspension feeders to assess whether the techniques chosen tend to maximize their net energy gain. The amount of time suspension feeders allocate to different feeding methods should reflect the availability of appropriate food and the relative costs of obtaining it if they are foragers that maximize their energetic gains as would be predicted in the first instance by optimal foraging theory (e.g. Stephens & Krebs 1986). Attributes of benthic suspension feeders that might make them particularly suitable to studies of foraging behaviour are their sessile nature, the potential ease with which food characteristics (e.g. particle quality, quantity, uniformity) could be manipulated, and the absence of search costs that must be accounted for in the foraging constraints of mobile animals.

WHY SUSPENSION FEEDING BEHAVIOURS HAVE BEEN UNDER-APPRECIATED

As previously mentioned, a morphological correlate of the benthic suspension-feeding lifestyle is the loss or lack of development of highly cephalized nervous systems used for sensory processing by mobile animals. This simplicity of the nervous systems of benthic suspension feeders belies their behavioural plasticity and perhaps has been partially responsible for engendering a view of suspension feeders as strictly active pumpers or passive filterers. This is ironic since it is their sessile nature that may necessitate some degree of behavioural plasticity (see previous section). Furthermore, there has been an erroneous impression arising from the original descriptions of feeding modes that animals adhere to one particular approach to feeding (Fauchald & Jumars 1979).

Another reason behavioural plasticity in suspension feeding has been under-appreciated certainly has to do with the past tendency to study suspension feeding under uniform and artificial conditions. For instance many earlier descriptions of suspension feeding were based on observations of animals in small volumes of still water. Such observations can be fraught with experimental artifacts since these animals live surrounded by large volumes of moving water and generally will not be subject to the wall effects that arise in containers that hold the animals being observed (Vogel 1981). Indeed it should be evident from earlier discussion that moving water often will elicit different kinds of feeding behaviours and can induce flows through the bodies of suspension feeders with internalized feeding surfaces. Studies in still water can thus fail to detect many important aspects of suspension feeding.

Quantification of suspension feeding has most often been done by assessing clearance rates of uniform particles from suspension. Again, such studies can fail to detect important aspects of feeding because the conditions that elicit different feeding behaviours are lacking. This paper describes examples where different feeding techniques are employed to capture particles that vary in size (e.g. by bryozoans, echinoderms). Variation in such particle attributes as shape, surface chemistry, concentration, and food value can often influence feeding rates, and it is now clear that inconsistencies amongst many studies are due in part to differences of these sorts. Such studies are now judged to be inappropriate measures of feeding in nature (Jørgensen 1975) although they may be valuable in providing information on feeding mechanisms and inherent vulnerabilities of prey.

PROBLEMS FOR FUTURE STUDY

The fluid environment experienced by benthic animals can vary during ontogeny. Small individuals may be hidden deep within the boundary layer while large organisms extend further into flow. Hence both the movement of water and the particulate material suspended in it will change as many benthic animals grow (see Statzner & Holm 1989 for discussion of the effects of flow alteration during ontogeny). The foregoing review provided examples of organisms employing alternate feeding techniques depending on flow conditions and particle size. The extent to which alternate modes or techniques may exist at different ontogenetic benthic stages of suspension feeders is unexplored, but it would seem likely that the relative importance of different feeding behaviours may vary with ontogenetic stages of attached individuals. It is already clear that at earlier ontogenetic stages ciliary feeding by the planktotrophic larvae characteristic of many benthic suspension feeders can be very different from the feeding techniques employed by the adult forms. It has also been suggested that uptake of DOM may be important at metamorphosis when tissue reorganization occurs and particle feeding may be arrested for several days (Manahan & Crisp 1982).

Recent work on suspension feeding from flow has helped to elucidate aspects of feeding not observed in still water (e.g. Vogel 1974, LaBarbera 1977, Taghon et al. 1980, Okamura 1984, 1985, 1987, 1988, Patterson 1984, Best 1988). However, these studies have concentrated on feeding in steady, unidirectional flow conditions. Many suspension feeders live in oscillatory flow. In such conditions, turbulent mixing is enhanced and boundary layer thickness diminished. Both effects can be expected to influence the behaviour of suspension feeders, and feeding behaviour in oscillatory flow merits much further study (see Hunter 1989 for discussion of the effects of oscillatory flow on feeding rates in hydroids).

As mentioned earlier, many studies of suspension feeding have focused on ingestion or clearance of uniform particles. When offered an array of particles suspension feeders may be selective, and different particles may elicit different feeding techniques. In some instances, patterns of selectivity may be assessed adequately by extrapolating from single-prey experiments (e.g. Greene et al. 1986). However, in many cases patterns of feeding on single prey types do not relate to patterns of selective feeding when multiple prey types are available. For instance, Jumars et al. (1982) found selectivity of deposit-feeding polychaetes to depend on the nature of the mixture presented, and Colton (1987) demonstrated the failure of one-predator/one-prey functional responses of damselfly naiads to predict correctly the one-predator/two-prey functional responses to copepods and cladocerans. I have found evidence that bryozoan patterns of feeding on single-sized suspensions cannot be used even to make qualitative predictions of selective feeding from mixtures, and furthermore that the patterns of feeding varied with flow (Okamura, in prep.). Moore (1988) showed that selectivity of *Chaoborus* changes with increasing prey density, and DeMott (1989) found selection by copepods against low-quality food particles to be sensitive to food concentration. Suspension feeding on uniform particles can provide information on the range of feeding techniques animals may employ. The next step will be to ascertain the necessary conditions that elicit various behaviours, and the interaction between flow regime and particle characteristics must be addressed explicitly. One might predict that there will be a minimum level of particle concentration below which energetically-costly behaviours will not be invoked and that flow rates will have important effects since they will determine particle fluxes and perhaps vulnerabilities.

Clearly the ability of some organisms to combine feeding modes will need to be taken into account in future views of benthic community structure. This has particular relevance to understanding the maintainence of deposit and suspension-feeding communities of soft sediments (see Probert 1984 for discussion) since a number of important constituents appear to employ both feeding modes (e.g. spionid polychaetes, tellinid bivalves). The view that trophic group interactions between assemblages of deposit and suspension feeders are important in structuring communities of soft sediments may need revision.

This paper provides evidence that feeding plasticity occurs commonly in many suspension feeders from different phyla. Given that plasticity in suspension feeding is distributed so widely across groups, simple policy models (as defined by Cheverton et al. 1985) such as maximizing energy intake, may provide a useful means to compare the behaviours and designs of these diverse organisms.

CONCLUSIONS

The energetics of suspension feeding as dictated by the low availability of suspended material would appear to make this a marginal existence (LaBarbera 1984). Thus any mechanism(s) that reduces the cost of moving water through filtering structures would be expected to be favored by natural selection (LaBarbera 1984). Some organisms are so arranged that flow is passively induced through their bodies. Others may passively feed by holding their filtering elements into flow but will resume active pumping when flow decreases or ceases. Still others don't bother to move water at all but always orient their feeding surfaces into ambient currents. However, this is only part of the story since any mechanism(s) that would maximize the food gathered by suspension feeders may also be advantageous. Many suspension feeders house symbionts from which they receive nutrition in addition to that obtained by other feeding techniques. A number of suspension feeders deposit feed when suspended material is not available. Many suspension feeders are capable of taking up DOM which may augment their feeding. Finally, many suspension feeders are able to exploit the heterogeneous array of suspended particles by using different feeding techniques that enable them to capture different particle types. The combination of feeding modes and techniques no doubt has provided these animals with an escape from the marginal existence of suspension feeders lacking such capabilities and has allowed them wide success in marine environments.

Many benthic suspension feeders, while relatively simple organisms, nonetheless show behavioural plasticity in response to suspended material. The variety of feeding modes and techniques observed in these animals allows them to act opportunistically to exploit the variability so characteristic of suspended particulate matter. Such behavioural plasticity appears to be an important strategy for these immobile animals that rely on spatially and temporally heterogeneous food resources. Future studies that incorporate other aspects of the flow regimes and the nature of particulate matter in the habitats of suspension feeders will probably elucidate further variations in the feeding behaviour of these animals.

ACKNOWLEDGEMENTS

Many thanks to D. Bazely, H.C. Bennet-Clark, P.M. Hammond, C.K. Kelly, M.A.R. Koehl, and J.R. Krebs for kindly reading and providing helpful comments on the manuscript.

REFERENCES

Alexander RM (1979) The Invertebrates. Cambridge University Press, Cambridge

Alexander SP, DeLaca TE (1987) Feeding adaptations of the foraminiferan *Cibicides refulgens* living epizooically and parasitically on the antarctic scallop *Adamussium colbecki*. Biol Bull 173: 136-159

Barber RT, Chavez FP (1983) Biological consequences of El Niño. Science 222: 1203-1210

Barnes RSK, Calow P, Olive PJW (1988) The Invertebrates. A New Synthesis. Blackwell Scientific Publications, Oxford

Bennett AF, Denman KL (1989) Large-scale patchiness due to an annual plankton cycle. J Geophys Res 94: 823-829

Best BA (1988) Passive suspension feeding in a sea pen: Effects of ambient flow on volume flow rate and filtering efficiency. Biol Bull 175: 332-342

Best MA, Thorpe JP (1986) Effects of food particle concentration on feeding current velocity in six species of marine Bryozoa. Mar Biol 93: 255-262

Brafield AE, Newell GE (1961) The behaviour of *Macoma balthica* (L.). J Mar Biol Ass U.K. 41: 81-87

Braimah SA (1987) Mechanisms of filter feeding in immature *Simulium bivattatum* Malloch (Diptera: Simuliidae) and *Isonychia campestris* McDunnough (Ephemeroptera: Oligoneuriidae). Can J Zool 65: 504-513

Buss LW, Jackson JBC (1981) Planktonic food availability and suspension-feeder abundance: Evidence of *in situ* depletion. J Exp Mar Biol Ecol 49: 151-161

Byrne M, Fontaine, AR (1981) The feeding behaviour of *Florometra serratissima* (Echinodermata: Crinoidea). Can J Zool 59: 11-18

Cane MA (1983) Oceanographic events during El Niño. Science 222: 1189-1195.

Carlson DJ, Townsend AR, Hilyard AL, Eaton JF (1984) Effect of an intertidal mudflat on plankton of the overlying water column. Can J Fish Aquat Sci 41: 1523-1528.

Cavanaugh CM (1985) Symbiosis of chemoautotrophic bacteria and marine invertebrates from hydrothermal vents and reducing sediments. Bull Biol Soc Wash 6: 373-388

Cheverton J, Kacelnik A, Krebs JR (1985) Optimal foraging: Constraints and currencies. In: Holldobler B, Lindauer M (eds) Experimental Behavioral Ecology. Fischer Verlag, Stuttgart, p 109

Clements LAJ, Fielman KT, Stancyk SE (1988) Regeneration by an amphiurid brittlestar exposed to different concentrations of dissolved organic material. J Exp Mar Biol Ecol 122: 47-61

Cohen RRH, Dresler PV, Phillips EJP, Cory RL (1984) The effect of the Asiatic clam *Corbicula fluminea* on phytoplankton of the Potomac River Maryland. Limnol Oceanogr 29: 170-180

Colton TF (1987) Extending functional response models to include a second prey type: an experimental test. Ecol 68: 900-912

Conover RJ (1968) Zooplankton - Life in a nutritionally dilute environment. Amer Zool 8: 107-118

Cowles TJ, Olson RJ, Chisholm SW (1988) Food selection by copepods: Discrimination on the basis of food quality. Mar Biol 100: 41-49

Crisp DJ, Southward AJ (1961) Different types of cirral activity in barnacles. Phil Trans Roy Soc B 243: 271-308
Dame RF, Dankers N (1988) Uptake and release of materials by a Wadden Sea mussel bed. J Exp Mar Biol Ecol 118: 207-216
Dame RF, Zingmark RG, Haskin E (1984) Oyster reefs as processors of estuarine materials. J Exp Mar Biol Ecol 83: 239-247
Dauer DM, Maybury CA, Ewing RM (1981) Feeding behavior and general ecology of several spionid polychaetes from the Chesapeake Bay. J Exp Mar Biol Ecol 54: 21-38
Davenport D (1988) The feeding mechanism of *Yoldia (=Aequiyoldia) eightsi* (Courthouy). Proc R Soc Lond B 232: 431-442
DeMott WR (1989) Optimal foraging theory as a predictor of chemically mediated food selection by suspension-feeding copepods. Limnol Oceanogr 34: 140-154
Denman KL, Powell TM (1984) Effects of physical processes on planktonic ecosystems in the coastal ocean. Oceanogr Mar Biol Ann Rev 22: 125-168
Denny M (1980) Locomotion: The cost of gastropod crawling. Science 208: 1288-1290
Fauchald K, Jumars PA (1979) The diet of worms: A study of polychaete feeding guilds. Oceanogr Mar Biol Ann Rev 17: 193-284
Fontaine AR (1965) The feeding mechanisms of the ophiuroid *Ophiocomina nigra*.. J Mar Biol Ass UK 45: 373-385
Fréchette MCA, Butman,Geyer WR (1989) The importance of boundary-layer flows in supplying phytoplankton to the benthic suspension-feeder *Mytilus edulis* L. Limnol Oceanogr 34: 19-36
Glynn PW (1973) Ecology of a Caribbean coral reef. The *Porites* reef-flat biotope: Part II. Plankton community with evidence for depletion. Mar Biol 22: 1-21
Greene CH, Landry MR, Monger BC (1986) Foraging behavior and prey selection by the ambush predator *Pleurobrachia bachei*. Ecology 67: 1493-1501
Harley MB (1950) Occurrence of a filter-feeding mechanism in the polychaete *Nereis diversicolor*. Nature 165: 734-735
Hart TJ (1930) Preliminary notes on the bionomics of the amphipod*Corophium volutator* Pallas. J Mar Biol Ass UK 16: 761-789
Hatcher BG (1988) Coral reef primary productivity : A beggar's banquet. TREE 3: 106-111
Hughes RN (1969) A study of feeding in *Scrobicularia plana*. J Mar Biol Ass UK 49: 805-823
Hunter T (1989) Suspension feeding in oscillating flow: The effect of colony morphology and flow regime on plankton capture by the hydroid *Obelia longissima*. Biol Bull 176: 41-49
Jacoby CA, Greenwood JG (1988) Spatial temporal and behavioral patterns of emergence of zooplankton in the lagoon of Heron Reef Great Barrier Reef, Australia. Mar Biol 97: 309-328
Johannes RE, Coles SL, Kuenzel NT (1970) The role of zooplankton in the nutrition of some scleractinian corals. Limnol Oceanogr 15: 579-586
Johnson AS (1988) Hydrodynamic study of the functional morphology of the benthic suspension feeder *Phoronopsis viridis* (Phoronida). Mar Biol 100: 117-126
Jordan TE, Valiela I (1982) A nitrogen budget of the ribbed mussel *Geukensia demissa* and its significance in nitrogen flow in a New England salt marsh. Limnol Oceanogr 27: 75-90
Jørgensen CB (1966) Biology of suspension feeding. Pergamon Press, Oxford
Jørgensen CB (1975) Comparative physiology of suspension feeding. Ann Rev Physiol 37: 57-79
Jørgensen CB (1976) August Pütter, August Krogh, and modern ideas on the use of dissolved organic matter in aquatic environments. Biol Rev 51: 291-328
Jumars PA, Nowell ARM (1984) Fluid and sediment dynamic effects on marine benthic community structure. Amer Zool 24: 45-55
Jumars PA, Self RFL, Nowell ARM (1982) Mechanics of particle selection by tentaculate deposit-feeders. J Exp Mar Biol Ecol 64: 47-70
Koehl MAR (1977) Effects of sea anemones on the flow forces they encounter. J Exp Biol 69: 87-105

Krebs JR, McCleery RH (1984) Optimization in behavioural ecology. In: Krebs JR, Davies NB (eds) Behavioural Ecology. An Evolutionary Approach. 2nd edn. Blackwell Scientific Publications, Oxford, p 91

LaBarbera M (1977) Brachiopod orientation to water movement. I.Theory, laboratory behavior, and field orientations. Paleobiol 3: 270-287

LaBarbera M (1978) Particle capture by Pacific brittle stars: Experimental test of the aerosol suspension feeding model. Science 201: 1147-1148

LaBarbera M (1981) Water flow patterns in and around three species of articulate brachiopods. J Exp Mar Biol Ecol 55: 185-206

LaBarbera M (1984) Feeding currents and particle capture mechanisms in suspension feeding animals. Amer Zool 24: 71-84

Leonard AB (1989) Functional response in *Antedon mediterranea* (Lamarck) (Echinodermata: Crinoidea): The interaction of prey concentration and current velocity on a passive suspension feeder. J Exp Mar Biol Ecol 127: 81-103

Lewis DB (1968) Feeding and tube-building in the Fabriciinae (Annelida: Polychaeta). Proc Linn Soc Lond 179: 37-49

Lewis JB (1976) Experimental tests of suspension feeding in Atlantic reef corals. Mar Biol 36: 147-150

Lewis JB, Price WS (1975) Feeding mechanisms and feeding strategies of Atlantic reef corals. J Zool Lond 176: 527-544

Mackas DL, Denman KL, Abbott MR (1985) Plankton patchiness: Biology in the physical vernacular. Bull Mar Sci 37: 652-674

Manahan DT, Crisp DJ (1982) The role of dissolved organic material in the nutrition of pelagic larvae: Amino acid uptake by bivalve veligers. Am Nat 22: 635-646

Manahan DT, Wright SH, Stephens GC, Rice MA (1982) Transport of dissolved amino acids by the mussel *Mytilus edulis*: demonstration of net uptake from natural seawater. Science 215: 1253-1254

Meadows PS, Reid A (1966) The behaviour of *Corophium volutator* (Crustacea: Amphipoda). J Zool Lond 150: 387-399

Meyer DL (1982) Food and feeding mechanisms: Crinozoa. In: Jangoux M, Lawrence JM (eds) Echinoderm nutrition. AA Balkema, Rotterdam, p 25

Miller DC (1984) Mechanical post-capture particle selection by suspension- and deposit-feeding *Corophium*. J Exp Mar Biol Ecol 82: 59-76

Mills EL (1967) The biology of an ampeliscid amphipod crustacean sibling species pair. J Fish Res Bd Canada 24: 305-355

Mook DH (1981) Removal of suspended particles by fouling communities. Mar Ecol Prog Ser 5: 279-281

Moore MV (1988) Density-dependent predation of early instar *Chaoborus* feeding on multispecies prey assemblages. Limnol Oceanogr 33: 256-268

Muschenheim DK (1987a) The dynamics of near-bed seston flux and suspension-feeding benthos. J Mar Res 45: 473-496

Muschenheim DK (1987b) The role of hydrodynamic sorting of seston in the nutrition of a benthic suspension feeder *Spio setosa* (Polychaeta: Spionidae). Biol Oceanogr 4: 265-288

Nowell ARM, Jumars PA (1989) Flow events in aquatic benthos. Ann Rev Ecol Syst 15: 303-328

Okamura B (1984) The effects of ambient flow velocity, colony size, and upstream colonies on the feeding success of Bryozoa. I. *Bugula stolonifera* Ryland, an arborescent species. J Exp Mar Biol Ecol 83: 179-193

Okamura B (1985) The effects of ambient flow velocity, colony size, and upstream colonies on the feeding success of Bryozoa. II. *Conopeum reticulum* (Linnaeus), an encrusting species. J Exp Mar Biol Ecol 89: 69-80

Okamura B (1987) Particle size and flow velocity induce an inferred switch in bryozoan suspension-feeding behavior. Biol Bull 173: 222-229

Okamura B (1988) The influence of neighbors on the feeding of an epifaunal bryozoan. J Exp Mar Biol Ecol 120: 105-123

Ólafsson EG (1986) Density dependence in suspension-feeding and deposit-feeding populations of the bivalve *Macoma balthica*: A field experiment. J Anim Ecol 55: 517-526
Paine RT (1971) Energy flow in natural populations of the herbivorous gastropod *Tegula funebralis*. Limnol Oceanogr 16: 86-98
Pashley HE (1985) The foraging behaviour of *Nereis diversicolor* (Polychaeta). PhD Thesis, University of Cambridge
Patterson MR (1984) Patterns of whole colony prey capture in the octocoral, *Alcyonium siderium*. Biol Bull 167: 613-629
Pentreath RJ (1970) Feeding mechanisms and the functional morphology of podia and spines in some New Zealand ophiuroids (Echinodermata). J Zool Lond 161: 395-429
Philander SGH (1983) El Niño southern oscillation phenomena. Nature 302: 295-301
Probert PK (1984) Disturbance, sediment stability, and trophic structure of soft-bottom communities. J Mar Res 42: 893-921
Putt M, Rivkin RB, Prezelin BB (1988) Effects of altered photic regimes on diel patterns of species-specific photosynthesis. I. Comparison of polar and temperate phytoplankton. Mar Biol 97: 435-443
Real L, Caraco T (1986) Risk and foraging in stochastic environments. Ann Rev Ecol Syst 17: 371-390
Ross DH, Craig DA (1980) Mechanism of fine particle capture by larval blackflies (Diptera: Simuliidae). Can J Zool 58: 1186-1192
Rubenstein DI, Koehl MAR (1977) The mechanisms of filter-feeding: Some theoretical considerations. Am Nat 111: 981-994
Sebens KP, Koehl MAR (1984) Predation on zooplankton by the benthic anthozoans *Alcyonium siderium* (Alcyonacea) and *Metridium senile* (Actiniaria) in the New England subtidal. Mar Biol 81: 255-271
Self RFL, Jumars PA (1978) New resource axes for deposit feeders? Bull Mar Sci 35: 440-448
Shannon LV, Boyd AJ, Brundrit GB, Taunton-Clark J (1986) On the existence of an El Niño-type phenomenon in the Benguela System. J Mar Res 44: 495-520
Sieburth JMV, Smetacek, Lenz J (1978) Pelagic ecosystem structure: heterotrophic compartments of the plankton and their relationship to plankton size fractions. Limnol Oceanogr 23: 1256-1263
Sleigh MA (1989) Protozoa and Other Protists. Edward Arnold, London
Smith DC, Douglas AE (1987) The Biology of Symbiosis. Edward Arnold, London
Southward AJ, Southward EC (1982) The role of dissolved organic matter in the nutrition of deep sea benthos. Am Zool 22: 647-659
Southward EC (1987) Contribution of symbiotic chemoautotrophs to the nutrition of benthic invertebrates. In : Sleigh MA (ed) *Microbes in the Sea*. Ellis Horwood Ltd., Chichester
Statzner B, Holm TF (1989) Morphological adaptation of shape to flow: Microcurrents around lotic macroinvertebrates with known Reynolds numbers at quasi–natural flow conditions. Oecologia 78: 145-157
Stephens DW, Krebs JR (1986) Foraging Theory. Princeton University Press, Princeton
Stephens GC (1982) Recent progress in the study of "Die Ernahrung der Wassertiere und der Stoffhaushalt der Gewasser". Am Zool 22: 611-619
Stewart MG (1979) Absorption of dissolved organic nutrients by marine invertebrates. Oceanogr Mar Biol Ann Rev 17: 163-192
Strathmann RR (1982) Cinefilms of particle capture by an induced local change of beat of lateral cilia of a bryozoan. J Exp Mar Biol Ecol 62: 225-236
Taghon GF, Nowell ARM, Jumars PA (1980) Induction of suspension feeding in spionid polychaetes by high particulate fluxes. Science 210: 562-564
Thayer CW (1986) Are brachiopods better than bivalves? Mechanisms of turbidity tolerance and their interaction with feeding in articulates. Paleobiol 12: 161-174
Thiesen BF (1972) Shell cleaning and deposit feeding in *Mytilus edulis* L. (Bivalvia). Ophelia 10: 49-55
Timko PL (1976) Sand dollars as suspension feeders: a new description of feeding in *Dendraster excentricus*. Biol Bull 151: 247-259

Townsend CR, Hughes RN (1981) Maximising net energy returns from foraging. In: Townsend CR, Calow P (eds) Physiological Ecology: An Evolutionary Approach to Resource Use. Blackwell Scientific Publications, Oxford , p 86
Valiela I (1984) Marine Ecological Processes. Springer-Verlag, New York
Vogel S (1974) Current-induced flow through the sponge, *Halichondria.* Biol Bull 147: 443-456
Vogel S (1977) Current-induced flow through sponges in nature. Proc Nat Acad Sci USA 74: 2069-2071
Vogel S (1978) Organisms that capture currents. Sci Amer 239: 128-139
Vogel S (1981) Life in moving fluids. The physical biology of flow. Princeton University Press, Princeton
Vogel S, Bretz WL (1972) Interfacial organisms: passive ventilation in the velocity gradients near surfaces. Science 175: 210-211
Warner G (1982) Food and feeding mechanisms: Ophiuroidea. In: Jangoux M, Lawrence JM (eds). Echinoderm nutrition. AA Balkema, Rotterdam, p 161
Wellington GM (1982) An experimental analysis of the effects of light and zooplankton on coral zonation. Oecologia 52: 311-320
Wildish DJ, Kristmanson DD (1981) Importance to mussels of the benthic boundary layer. Can J Fish Aquat Sci 41: 1618-1625
Wilkinson CR (1983) Net primary productivity in coral reef sponges. Science 219: 410-412
Winston JE (1978) Polypide morphology and feeding behavior in marine ectoprocts. Bull Mar Sci 28: 1-31
Wotton RS (1984) The importance of identifying the origins of microfine particles in aquatic systems. Oikos 43: 217-221
Wright RT, Coffin RB, Ersing CP, Pearson D (1982) Field and laboratory measurements of bivalve filtration of natural marine bacterioplankton. Limnol Oceanogr 27: 91-98
Young CM, Braithwaite LF (1980) Orientation and current-induced flow in the stalked ascidian *Styela montereyensis.* Biol Bull 159: 428-440
Young CM, Chia FS (1987) Abundance and distribution of pelagic larvae as influenced by predation, behavior, and hydrographic factors. In: Giese AC, Pearse JS, Pearse VB (eds) Reproduction of Marine Invertebrates. Vol IX. General Aspects: Seeking Unity in Diversity. Blackwell Scientific Publications, Oxford, p 385

HOW TRAIL LAYING AND TRAIL FOLLOWING CAN SOLVE FORAGING PROBLEMS FOR ANT COLONIES

S. Goss, R. Beckers, J.L. Deneubourg, S. Aron, J.M. Pasteels
Unit of Behavioural Ecology, CP 231
Université Libre de Bruxelles
1050 Bruxelles
Belgium

Introduction

One of the most striking features of an ant colony's behaviour is its capacity for the spatial organisation of foraging activity. The use of trail pheromone to guide fellow workers in the nest to a large food source or rich foraging zone has been extensively studied (e.g. Wilson 1971) and obviously contributes to foraging efficiency. We have recently, however, been able to show that trail laying and trail following behaviour are more than just a means of communicating a food source's location. When more than one trail is present at a time, the interactions between foragers and the trails can lead to the collective selection of the shortest path or the best food source, despite the fact that individual foragers have no means of making such choices. Furthermore, the interactions between the foragers and the food sources can lead to some remarkable foraging patterns, giving the illusion that the foraging force is obeying a complicated global blueprint.

We shall illustrate these ideas by simplifying the ants' behaviour to the extreme. We ignore the recruiting power of a trail, considering only its function as a directional guide for recruits. We consider the individual ants to have no memory (though they know where their nest is) and no sight, being capable only of responding differentially to the pheromone concentation at each antenna. While of course ants are not so simple, this approach lets us see the extent to which "blind" individual mechanisms can lead to "far–sighted" collective choices.

NATO ASI Series, Vol. G 20
Behavioural Mechanisms of Food Selection
Edited by R. N. Hughes

Which way: two equal paths?

This section shows how ants select one of two equal paths rather than using both equally (Deneubourg et al 1989a), and provides the experimental ground-work for the more complicated situations in the rest of the article.

We introduced a diamond-shaped bridge (fig. 1 insert) between a nest of the Argentine ant, *Iridomyrmex humilis*, and a chemically unmarked arena which elicits an exploratory recruitment. Both branches were of equal length, and after a period of initial fluctuation, one branch was rapidly selected (fig. 1).

We know that *I. humilis* workers lay trail pheromone both leaving and returning to the nest, and both during food (Van Vorhis Key and Baker 1986) and exploratory (Aron et al. 1989) recruitment. Initially both branches are chosen equally. However, in marking, each ant that passes modifies the following ant's probability of choosing left or right, a positive feed back system that after initial fluctuation rapidly leads to one of the two forks becoming more or less completely preferred to the other.

We estimated the pheromone quantity at a point by the number of ants that have passed the point. This is possible if we ignore the pheromone's evaporation and consider that each ant lays an average amount of pheromone, whether leaving or returning to the nest. The kinetics of the ants' choice can therefore be represented not only as a function of time but also as a function of the cumulative number of ants passing the point.

Fig. 2 pools the results from 20 different experiments. The dashed curve represents the average and clearly shows how the probability of an explorer choosing what is ultimately the collectively selected branch increases rapidly and non-linearly with the number of explorers that have previously passed on the bridge.

The exploration on the bridge is modelled as follows. After i ants have crossed the bridge there are i pheromone units on the bridge, of which A_i are on branch A, and B_i on branch B. The next ant arriving chooses branch A or B with probabilities $prob_A$ and $prob_B$, depending on A_i and B_i (i.e. on the number of ants that have previously chosen each branch). Having chosen, the ant then adds to the pheromone on the branch chosen. Thus:

(1) $$prob_A = (20 + A_i)^2 / ((20 + A_i)^2 + (20 + B_i)^2) = 1 - prob_B$$

(2) $$A_{i+1} = A_i + \delta \qquad B_{i+1} = B_i + (1-\delta) \qquad (A_i+B_i=i)$$

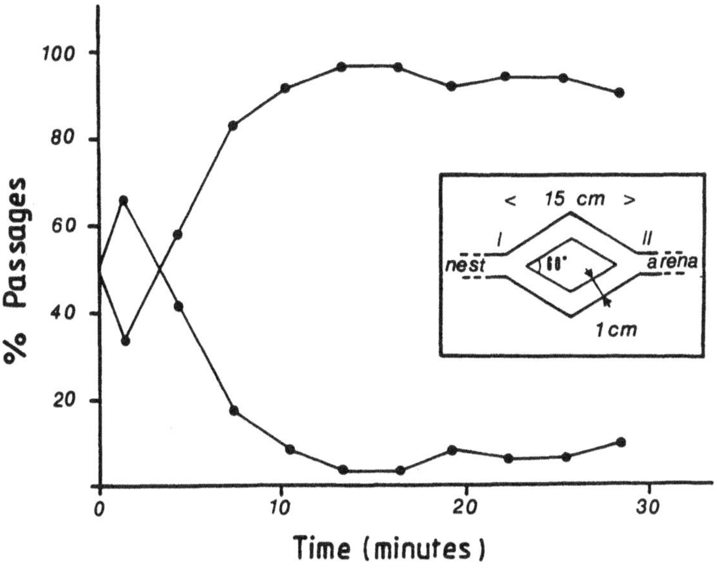

Fig. 1. Percentage of *I. humilis* workers per 3-minute period passing on the two branches of the bridge (insert). From Deneubourg et al. 1989a.

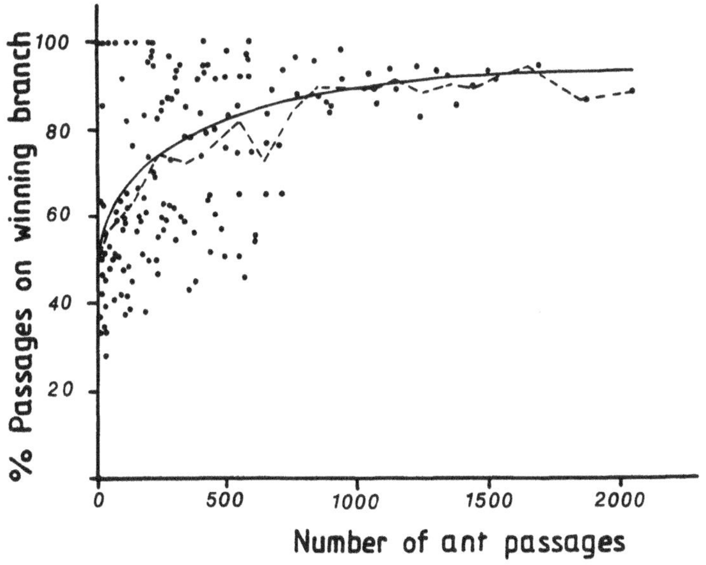

Fig. 2. Percentage of *I. humilis* workers passing on the collectively selected branch of the bridge (fig. 1 insert). 20 experiments, measured every 3 minutes for 30 minutes (dots). The horizontal axis represents the cumulative number of ant passages on both branches of the bridge. The dashed curve represents the average values. The solid curve represents the average of 200 Monte-Carlo simulations, based on eqs. 1 and 2. From Deneubourg et al. 1989a.

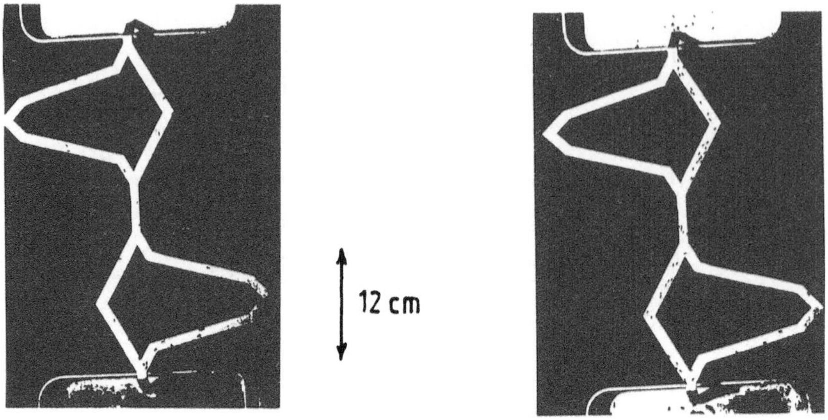

4 min 8 min

Fig. 3. A colony of *I. humilis* selecting the short branches on both modules of the bridge (photos taken 4 and 8 minutes after the bridge was placed). From Goss et al. subm.

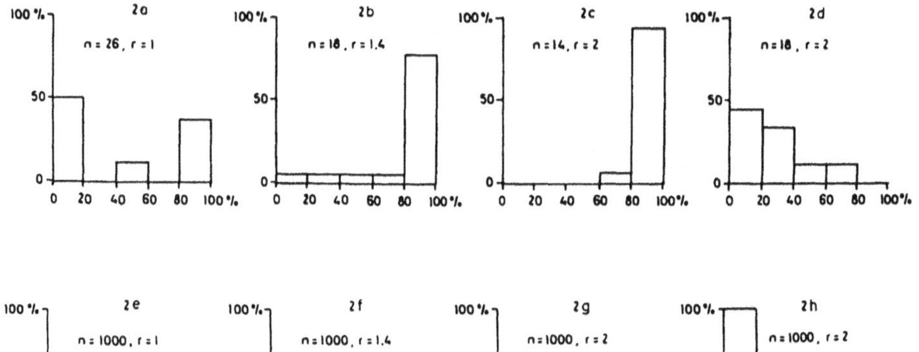

Fig. 4. The distribution of the % of ants that chose the short branch. r = ratio of long branch length to short branch, n = no. of trials. From Goss et al. subm.

Figs. 4a-d: Experiments with *I. humilis*, traffic counted between 30th and 40th minute after the bridge (fig. 3) is placed. Fig 4d: short branch added 30 min. after the experiment began, traffic measured between 20 and 30 minutes later.

Figs. 4e-h: Monte Carlo simulations of eqs. 1, 3 and 4, with ϕ=0.5/sec, traffic counted between the 501st and 1000th ant crossing the bridge. Fig. 4h: short branch added after 1000th crossing, traffic counted between 1501st and 2000th crossing.

where δ is a stochastic variable that takes a value of 1 or 0 with probability $prob_A$ and $prob_B$ respectively. In other words, A is increased by one unit if the ant chooses branch A, otherwise it chooses branch B and adds to B.

The Monte Carlo simulations of the model work by comparing, for each successive ant, a random number generated by computer with the probability of choosing branch A or B as determined by equation 1. The numerical values of 2 and 20 in eq. 1 were fitted empirically by matching the experimentally observed ants' choice of branch on the bridge with Monte-Carlo simulations (fig. 2, dashed vs solid curves).

Note that if only returning explorers marked the bridge, one path would still have been selected, but only as a cumulative choice of returning explorers. Also if you introduce an evaporation of the trail pheromone, especially if there are few explorers, the positive feed-back loop is unable to start up, and statistically the explorers use both paths equally.

Which way: two inequal paths?

This section shows how the mechanisms determined above can lead to the selection of the shortest path between two points (Goss et al. subm.).

The same series of experiments was repeated, except that the recruitment was principally alimentary rather than exploratory. More importantly, the bridge (fig. 3), consisted of two identical modules, each module consisting of two branches of inequal length. Any environmental factors that might bias the workers' choice are neutralised by having one module's short branch on the oppposite side of the bridge as the other's (i.e. the selection of the short branch is not influenced by the fact that it is on the left or the right). Any effect of the explorers' memory is ignored here.

The experiments are summarised in fig. 4a-d. When the two branches were equal ($r=1$) no significant overall preference was noted (in 12 out of 26 experiments, i.e. in 46% of trials, more than 80% of the ants used the left branch; $p > 0.05$) (note that one branch is always selected). With $r=1.4$ and $r=2$, the ants showed a significant and increasing overall selection of the short branch (15/18 and 14/14 respectively; $p < 0.05$). When the short branch was added only after the trail on the long branch was established (fig. 4d, $r=2$) the ants were unable to switch to the shorter branch (2/18; $p < 0.05$).

A similar series with r=2 was performed in red light to which ants are insensitive, more or less eliminating any visual factor. In 11/14 cases more than 80% of the traffic was on the short branch.

The model presented above must be slightly modified to take into account the difference in time taken to cross the short and the long branch. Time is therefore introduced specifically, so we drop the i suffix, but we must introduce the suffix j to distinguish between the pheromone concentration at one end of a branch with that at the other end.

ϕ ants per unit time cross the bridge in each direction and arrive at each of the two choice points (j=1, 2). Each ant chooses the short or the long branch as in eq. 1, and adds to the pheromone on the branch chosen. Those that choose the short branch arrive at the opposite choice point approximately 20 seconds later, given their average speed and the length of the short branch, adding to the pheromone there. Those that choose the long branch do so 20 times r seconds later, where r is ratio of the long branch's length to the short one's. Thus, at time t, the average equations for this stochastic system are:

(3) $$dA_j/dt = \phi prob_{A,j'}(t-20) + \phi prob_{B,j} \qquad (j=1;j'=2 : j=2;j'=1)$$

(4) $$dB_j/dt = \phi prob_{B,j'}(t-20r) + \phi prob_{B,j}$$

The delays in eqs. 3 and 4 are of course the key to which branch is chosen, as they lead to an initial period in which the short branch at both choice points is marked by the ants moving in both directions while each choice point of the long branch is only marked by ants moving in one direction.

Consider the choice point nearest the nest. At time 0 it is unmarked. The first ants arrive, choose randomly and travel along their chosen branch, marking as they go. After 20 seconds, however the short branch will be marked not only by fresh ants arriving from the nest, but also by the first ants returning to the nest that chose the short branch. This happens somewhat later for the long branch, i.e. after 20 times r seconds. Between these two moments, the short branch accumulates an advantage over the long branch at both ends. This is amplified by the choose-and-mark process' autocatalytic nature, and the short branch rapidly becomes preferred unless random noise reverses the advantage. The greater the initial advantage, i.e. the greater ϕ and the difference between the two branches, the more probable it is that the short branch is chosen.

The model's simulations reflect the experimental results. Firstly, the colony's probability of selecting the shortest branch increases with the difference between the two branches (fig. 4e-g). Secondly, if the short branch is only present after the trail on the long branch has been established, the colony is incapable of switching to the short branch, due to the irreversible nature of the positive feed back process involved (fig. 4h).

Finally, if, as for most ant species, the workers only mark when returning to the nest, it is not possible for the colony to select the shortest branch, by the above means, any more often than the long branch, as their is no initial period during which the short branch is marked twice as much as the long. More generally, the greater the difference between the marking made by the ants leaving the nest and that by those returning to the nest, the more the overall choice between the two branches is random. Experiments are under way to confirm this prediction. We do not, however, exclude the existence of other mechanisms.

It is important to note that the selection of the shortest branch is not the result of individual ants comparing the different lengths of each branch, but is an *indirect* consequence of their permanent trail laying and trail following behaviour and the system's physical constraints.

Which source?

This section shows how the mechanisms determined above can lead to the selection of the richer of two sources, involving however a slightly more flexible individual response (Pasteels et al. 1987; Deneubourg et al. subm.). More exactly, it is known for a number of species that the richer the food source, the more the pheromone laid per returning forager.

This series of experiments was performed with another ant species, *Lasius niger*, which lays trail pheromone only when returning to the nest with food. Two large (i.e. unlimited) drops of 1M sucrose solution were presented 60cm apart and 60cm from the entrance to a sand-filled arena. The number of ants at each source was then counted.

Again, after a period of initial fluctuation, one of the sources became abruptly more exploited than the other (fig 5a). If one of the sources is richer than the other (1m vs 0.1M sucrose), then the colony always selects the richer source (fig. 5b), unless the richer source is only introduced well after a recruitment has built up to the poorer source. In this case the colony continues to exploit preferentially the poorer source even though it has discovered the richer one (fig. 5c). Note that similar results to those obtained with *L. Niger* have been obtained with the trail recruiting ants *Iridomyrmex humilis* and *Pheidole pallidula*, all three having very different life-styles. This supports the idea that the trail mechanism is the important factor.

Again eqs. 1 and 2 can form the basis of the model, except that eq. 2 must be modified, becoming:

$$(5) \qquad A_{i+1} = A_i + q\delta \qquad B_{i+1} = B_i + (1-\delta) \qquad (A_i+B_i=i)$$

where q is the ratio of pheromone laid by an average forager returning from the 1M source with respect to that laid by one returning from the 0.1M source. A_i and B_i are the number of pheromone units on the trails leading to the rich and poor source respectively. While the exact form of eq. 2 has not yet been experimentally fitted for *L. niger*, using the same values as for *I. humilis* gives the probability of selecting the richer source as a function of q as shown in fig. 6. Note that small differences in indivisual response are amplified by the trail system into large differences in the collective response.

Once more the selection of the best food source is not the result of foragers making direct comparisons between the sources or of recruits in the nest making direct comparisons between the quality of food offered by the recruiters, but is instead an indirect consequence of the individual foragers laying more pheromone to a richer source.

Finally a weak trail cannot compete with the strong trail, from eq. 2, and so cannot attract enough foragers to become a strong trail (see fig. 5c, and also fig. 4d). A well-established trail leading to a poorer source is thus preferred to weak trail to a better source, and the colony can therefore exhibit behaviour that appears sub-optimal from the foraging benefit point of view.

A rotating foraging trail

Up until now we have stressed the collective choice between two alternatives. We would like now to greatly increase the number of alternatives and to examine the spatial patterns that result. This section shows how the mechanisms so far described can interact with food sources that are small and limited in number to produce a foraging trail that rotates around the nest with a regular period, giving the illusion of systematic and deterministic foraging (Goss and Deneubourg 1989). Note that, as

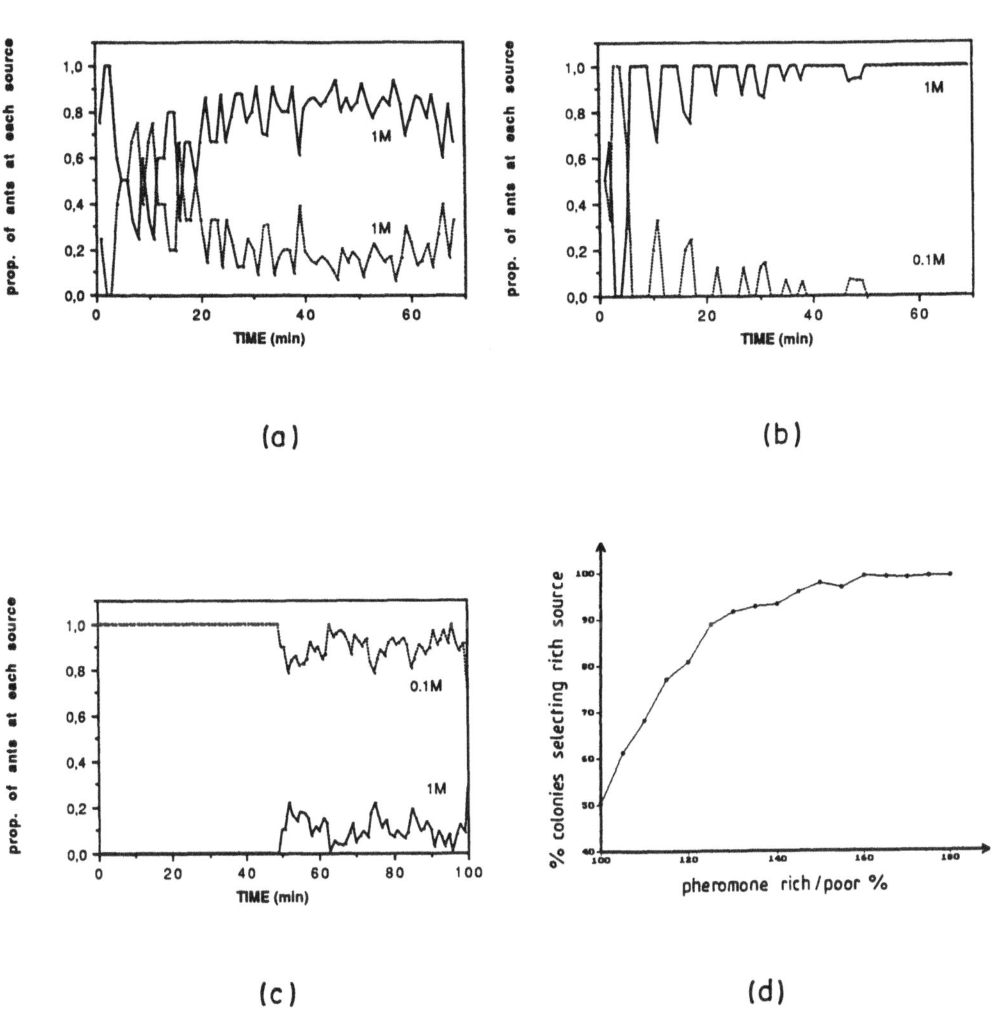

Fig. 5a-c. A hungry colony of *L. niger* was presented with two sugar sources, 60cm apart and both 60cm from the nest in an 80 x 80cm arena. The proportion of ants around each source is given as a function of time. From Deneubourg et al. subm.
Fig. 5a. Two 1M sucrose sources introduced simultaneously.
Fig. 5b. 1M and 0.1M source introduced simultaneously.
Fig. 5c. 1M source introduced 50 mins after a 0.1M source.
Fig. 5d Monte Carlo simulations of eqs. 1 and 5. The % colonies preferring the rich source is presented as a function of the ratio (q) of the quantity of pheromone laid by an ant returning from the rich source to that laid by one returning from the poor source. Each point represents 500 trials in which 1000 ants choose between the two sources.

these examples concern a much longer time-scale than the previous ones, we must take into account the trail pheromone's evaporation.

The idea is based on the field observations of Bernstein (1975) and Rissing and Wheeler (1976) on the harvesting ant *Messor pergandei*. Fig. 6 gives a typical map of the direction of a nest's foraging trail over a score of sequential foraging periods. With a certain degree of irregularity, the trail can be seen to make a complete turn about the nest with a period of several days to 3 weeks or so (the scarcer the food, the faster the rotation).

The model uses the same logic as in equations 1 and 2, but takes into account the arrival and disappearance of food sources. The circular foraging area, with the nest in the centre, is divided into b sectors. ϕ_i seeds arrive per time in each sector, each of which contains S_i seeds. A fraction, r, disappear per unit time by competition, decay, etc.

A trail leads to each sector, characterised by C pheromone units, of which a fraction, e, evaporate per unit time. N foragers leave the nest per unit time. A fraction, f_i, choose sector i, according to the multi-choice equivalent of eq. 1. Of this fraction, a small fraction, q, diffuse into each of the two adjacent sectors (note that the sectors adjacent to sector b are sectors 1 and b-1).

The number of seeds found in a sector is given by the product between a constant, g, the number of foragers, F_i, and a function of the number of seeds in that sector, $S_i/(a + S_i)$, which increases monotonically from 0 to 1 with S_i, the parameter a being a constant. At the end of each step, all the foragers return to the nest. Those that have found a seed add one pheromone unit to the trail leading to the corresponding sector. Those that find no seeds return without marking. The average equations for this process are thus:

(6) $$dS_i/dt = \phi_i - gF_iS_i/(a+S_i) - rS_i \qquad (i=1,\dots,b)$$

(7) $$dC_i/dt = gF_iS_i/(a+S_i) - eC_i$$

(8) $$F_i = N((1-2q)f_i + qf_{i+1} + qf_{i-1})$$

(9) $$f_i = (20 + C_i)^2 / \Sigma(20 + C_j)^2 \qquad (\Sigma f_i=1)$$

Fig. 7 shows a typical Monte Carlo simulation (for simplicity of presentation the foraging area is divided into 4 sectors, but the same results are obtained when it is divided into many more sectors). At time 0, each sector contains ϕ_i/r seeds (the equilibrium value in the absence of ants), and there are no pheromone marks. As previ-

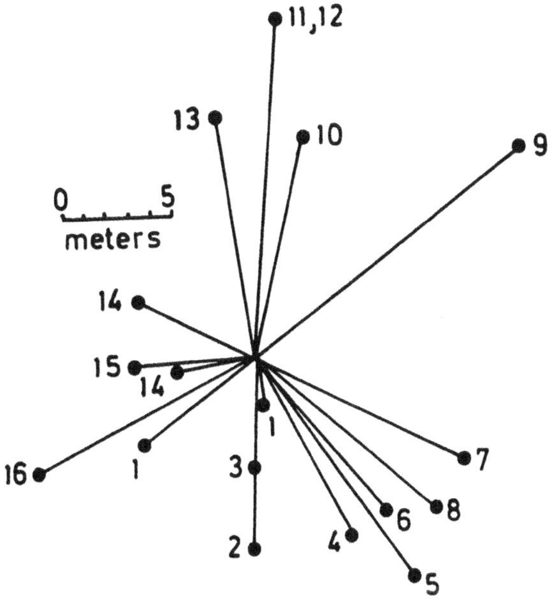

Fig. 6 presents the 16 consecutive foraging columns of the harvester ant *M. pergandei*. Note that two columns coexisted during periods 1 and 14. Redrawn from Rissing and Wheeler 1976.

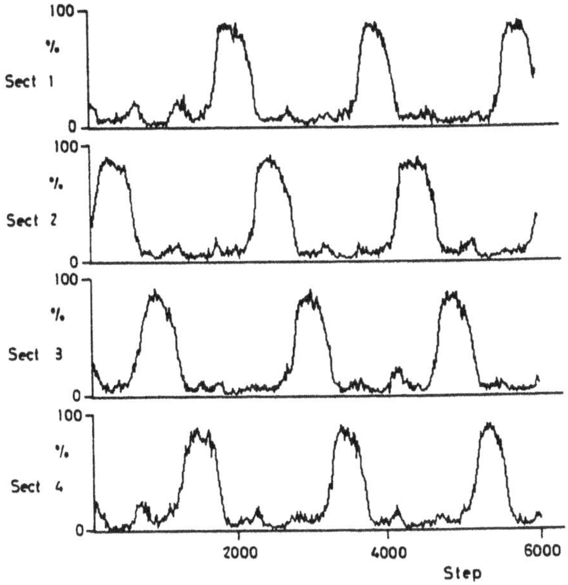

Fig. 7. Monte Carlo simulation of eqs. 6-9 with 4 sectors, showing the % foragers in each sector as a function of time. The foragers form a trail that starts in sector 2 and rotates clockwise to sectors 3, 4, 1, 2, 3, etc., with a regular period. From Goss and Deneubourg 1989.
$N=100$; $e=0.03$; $g=0.1$; $r=0.001$; $q=0.05$; $\phi_i=2$ $(i=1,\ldots,4)$; $a=1000$; $S_0=\phi/r$.

ously on the bridge, one sector, in this case sector 2, is chosen after initial fluctuation, corresponding to the formation of a recruitment trail in real space.

Thereafter, the number of seeds in sector 2 diminishes through the foragers' activity, while they accumulate in the other unfrequented sectors. Eventually, the foragers have difficulty in finding seeds in sector 2, and so the trail to it is no longer reinforced and diminishes by evaporation. At this point, the trails to its two neighbouring sectors 1 and 3 have a higher concentration than sector 4's trail, as they have been exploited at a low rate by foragers diffusing from sector 2. Thus rather than collectively selecting at random the three (b-1) remaining sectors, the foragers collectively choose between sectors 1 and 3, in this case sector 3. When sector 3 is nearly empty, again there are weak trails leading to the two adjacent sectors 2 and 4. However, sector 2 is still more or less empty, not having had enough time to restock. The foragers collectively choose sector 4, by finding seeds more easily there and therefore reinforcing the trail leading to it. Thereafter, they continue to choose the next sector in the same direction as before, and the trail rotates around the nest. By the time it has gone a full circle, the sectors originally exploited have been restocked, and the rotation continues indefinitely.

Once again that there is nothing in an individual ant's behaviour to suggest that it should forage in one sector one day, and the neighbouring one the next. Note also that the rotating trail is not the model's only solution. Depending principally on the number of foragers or food sources the trail rotates slower or faster, but more importantly the model can also generate random foraging or a foraging trail fixed on one sector. This shows how the same individual behaviour in different environmental conditions can generate very different patterns.

Finally, experiments are under way to reproduce this behaviour in other species under laboratory conditions. Besides fitting the model more empirically, we hope to show that this complicated behaviour is not a species specific phenomenon, even though it has not been reported in other ant species, but depends only upon the correct circumstances being available to a common trail-laying and trail-following behaviour.

Army ant foraging patterns

This section shows how the 1-dimensional behaviour corresponding to eqs. 1 and 2 can be extrapolated to 2-dimensions to generate complex foraging patterns when a large number of permanent trail-laying foragers leave the nest together (Deneubourg et al. 1989b).

Army ants (Raignier and Van Boven 1955; Schneirla 1971; Franks 1989) live in colonies of up to 20,000,000 individuals, who are roughly speaking 1 cm long and practically blind. They can form a swarm raiding system made up of 200,000 foragers, covering 100m x 10m in a day, catching some 30,000 prey items which are all transported to the nest. They do this in highly structured and species characteristic patterns, in the total absence of any central organisation (fig. 8).

The traditional approach to these different swarm patterns is to assume that each is optimal for the prey these species hunt, and corresponds to complex species specific individual behaviour. However, simulations of the simple mechanism already described can generate these different patterns (Deneubourg et al. 1989b).

Like *I. humilis*, army ant foragers lay pheromone not only when returning with prey but also to a lesser extent while advancing with the swarm (Schneirla, 1933, 1940). If we consider the swarm to move in a discrete network of points representing continuous 2-D space, at each point each ant chooses ahead left or ahead right, moves and adds to the pheromone at the point chosen. As on the bridge, one of the two points ahead becomes more or less completely preferred to the other. This process repeated at each point along the swarm's path, is the basis of how the swarm forms its trails.

The model is subject to some minor refinements. Firstly, to correspond with observed behaviour (Schneirla 1940; Franks 1985), the probability of each ant's moving at each step increases sigmoidally with the pheromone quantity ahead of it. Secondly, we postulate that there is a maximum number of ants that can occupy one point. Thirdly, a fixed fraction, e, of the pheromone at each point evaporates per step. Only moving ants lay pheromone, at the point at which they arrive, though not if the quantity of pheromone there is greater than a saturation value.

Figure 9a shows a Monte-Carlo simulation of this model. A diffuse front is formed and advances with a concentrated trail extending from its trailing edge to the nest. At the front, the ground is relatively unmarked and the ants move slowly and randomly, thus accumulating and spreading out. At the trailing edge of the front, the number of passages is sufficient for one path, the trail, to have become preferred to all other possibilities.

Let us now introduce the food. We first consider that each point has a fixed probability of containing one non-renewable food item, transportable by one ant. This approximates the situation of *Eciton burchelli*, which feeds largely on scattered arthropods (Franks and Bossert 1983). The first ant that arrives at a point containing a food item takes it and returns towards the nest obeying exactly the same rules as an advancing ant, laying however a greater amount of pheromone per step.

With a low food density, the returning ants do not modify the swarm pattern seen in fig. 9a. However, with higher densities, they cause the central trail to split into lateral trails which themselves branch out, giving a swarm like a river delta (fig. 9b), characteristic of *E. burchelli*.

Finally, *E. hamatum* feeds more on dispersed social insect colonies, and its food distribution can be represented by each point having a very small probability of containing a very large number of food items. *E. rapax*, has an intermediary diet. With this more heterogenous food distribution, the swarm splits up into a number of small columns and deltas (fig. 9c), characteristic of *E. rapax*.

While the individual ants in the simulations have exactly the same qualitative and quantitative behaviour, the different spatial distributions of the foragers returning with food (via the different food distributions) interact with the flow of ants heading away from the nest to produce the qualitatively different patterns.

What about other group-living organisms?

We have seen how one simple individual mechanism can generate a wide range of collective behaviours that appear both complex and adaptive. Just how general is this mechanism?

Basically, the logic may be reduced to the statement: I tend to do as others do, and doing so, influence those that come after me. Such allelomimesis is by no means limited to ants, and behaviour equivalent to that described above may be seen in widely different group-living species and may be mediated by visual or tactile stimuli as well as chemical ones.

Social bacteria use chemical trails to form army-ant-like "swarms" (review in Wilson 1975). Gregarious caterpillars, e.g. *Malacosoma spp.*, also use trails in a manner similar to ants, enabling them both to diffuse information about richer foraging zones (more trail is laid to better food sources) and to find the same collective nest site at the end of a day's activity (Fitzgerald and Peterson 1988).

Trails can also be formed without chemical communication. For example, consider a series of animals crossing a scrubland. The first cross at random. The later animals will find it easier going if they tend to follow some of the previous animals' paths, and eventually this *cumulative* trampling and wearing away of the vegetation

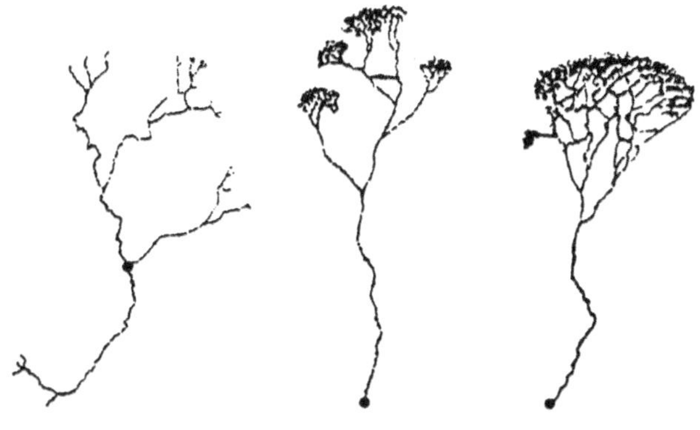

Figure. 8. Foraging patterns of three army ant species, *Eciton hamatum*, *E. rapax*, and *E. burchelli*, from left to right. Redrawn from Rettenmeyer 1963 and Burton and Franks 1985.

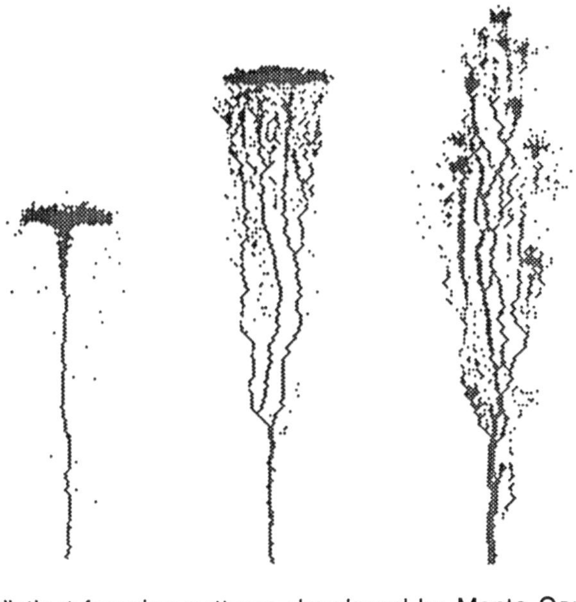

Fig 9. Three distinct foraging patterns developed by Monte Carlo simulations of the same model with three different food distributions. From Deneubourg et al. 1989b.
10 ants leave the nest per step, and advance into the foraging area. If they find food they return to the nest. At each point, the prob. of moving per time step = 0.5 + 0.5 tanh((L+R)/100 -1), where L and R are the quantities of pheromone ahead left and right. Those that move choose between ahead left and ahead right, with the prob. of choosing ahead left = $(5+L)^2 / ((5+L)^2 + (5+R)^2)$. A maximum of 20 ants are allowed at each point. Having moved, the advancing ants lay 1 pheromone unit at the point chosen (with a saturation level of 300 units), and returning ants lay 10 pheromone units (with a saturation level of 1000 units). 1/30th of the pheromone at each point evaporates per time step.
Fig. 9a. No food.
Fig. 9b. Each point has a 1/2 probability of containing 1 food item.
Fig. 9c. Each point has a 1/100 probability of containing 400 food items.

can form a well-defined physical trail, just as the ants chose one branch on the bridge, without the animals necessarily living in groups or even belonging to the same species.

Visual cues can also coordinate foraging activity, and the information centres described in relation to communal roosting or colonial nesting bird species (Ward and Zahavi 1973) show a strong analogy with social insect colonies. Members of a roost are thought to be capable of recognising when other members of the colony return from a successful foraging flight and to choose their next flight direction accordingly, this being the equivalent of trail recruitment. Similar complex decisional or spatial patterns have been observed, and can be understood by similar and even identical analyses of the individual kinetics. For example, pigeons may concentrate on one source, neglecting nearby identical ones (Lefebvre 1983), and Hoffman et al. (1981) describe large mixed flocks of seabirds that form foraging patterns very like those of ants, via a recruitment mechanism.

In conclusion, ants, and also some other group living species, appear to have hit on a simple but effective means of making decisions. By interacting amongst themselves and with the environment, they are able to respond as a group to environmental heterogeneities in spite of their limited and local individual perception. Their system combines robustness and flexiblity, allowing a considerable degree of individual randomness and simplicity appropriate to ant workers, and being functional in a wide range of situations.

Acknowledgements

This work is supported in part by the Belgian program on interuniversity attraction poles, les Instituts Internationaux de Physique et de Chimie, Belgian IRSIA grant #860004, Nato grant #0344/88 and La Fondation Schlumberger pour l'Education et la Recherche.

References

Aron S, Pasteels JM, Deneubourg JL (1989). Trail-laying behaviour during exploratory recruitment in the Argentine ant, *Iridomyrmex humilis* (Mayr). Biol. Behav. in press

Bernstein R (1975) Foraging strategies of ants in response to variable food density. Ecology 56: 213-219.
Burton JL, Franks NR (1985) The foraging ecology of the army ant *Eciton rapax*: an ergonomic enigma? Ecol. Entomol. 10: 131-141.
Deneubourg JL, Aron S, Goss S, Pasteels JM. (1989a) The self-organizing exploratory pattern of the Argentine ant. J Ins Behav in press
Deneubourg JL, Goss S, Franks N, Pasteels JM (1989b) The blind leading the blind: modelling chemically mediated army ant raid patterns. J Ins Behav in press
Deneubourg JL, Beckers R., Goss S, Pasteels JM (subm.) Collective decision making through food recruitment
Fitzgerald TD, Peterson SC (1988) Cooperative foraging and communication in caterpillars. Bioscience 38: 20-25
Franks NR (1985) Reproduction, foraging efficiency and worker polymorphism in army ants. In: Hölldobler B, Lindauer M (eds.) Experimental behavioral ecology. Fortschritte der Zool. BD 31, Fischer Verlag, Stuttgart, p91.
Franks NR (1989). Army ants: a collective intelligence. American Scientist, in press.
Franks NR, Bossert WH (1983). The influence of swarm raiding army ants on the patchiness and diversity of a tropical leaf litter ant community. In: Sutton SL, Chadwick AC, Whitmore TC (eds.) Tropical rain forest: ecology and management. Blackwell, Oxford, p151
Goss S, Deneubourg JL (1989). The self-organising clock pattern of *Messor pergandei* (Formicidae, Myrmicinae). Insectes Soc. in press
Goss S, Aron S, Deneubourg JL, Pasteels JM (subm.) Self-organised short cuts in the Argentine ant
Hoffman W, Heinemann D, Wiens J (1981) The ecology of seabird feeding flocks in Alaska. Auk 98: 437-455
Lefebvre L (1983) Equilibrium distribution of feral pigeons at multiple food sources. Behav. Ecol. Sociobiol. 12: 11-17
Pasteels JM, Deneubourg JL, Goss S. (1987). Self-organization mechanisms in ant societies (I): Trail recruitment to newly discovered food sources. In: Pasteels JM, Deneubourg JL (eds.) From individual to collective behavior in social insects. Birkhäuser, Basel, p 155
Raignier A, Van Boven JKA (1955) Etude taxonomique, biologique et biométrique des *Dorylus* du sous-genre *Anomma* (Hymenoptera: Formicidae). J. Annales du musée Royal du Congo Belge n.s.4 Sciences Zoologiques 2: 1-359
Rettenmeyer CW (1963) Behavioral studies of army ants. Univ. Kansas Sci. Bull. 44: 281-465
Risssing SW, Wheeler J (1976) Foraging responses of *Veromessor pergandei* to changes in seed production. Pan.-Pac. Entomol. 52: 63-72
Schneirla TC (1933) Studies on army ants in Panama. J. Comp. Physiol. 15: 267-299
Schneirla TC (1940) Further studies on the army ant behavior pattern. J. Comp. Psychol. 29: 401-461
Schneirla TC (1971) (Topoff HR, ed.) Army ants: a study in social organization. Freeman, San Francisco
Van Vorhis Key SE, Baker TC (1986) Observations on the trail deposition and recruitment behaviors of the Argentine ant, *Iridomyrmex humilis* (Hymenoptera: Formicidae). Ann Entomol Soc Am 79: 283-288
Ward P, Zahavi A (1973) The importance of certain assemblages of birds as "information centres" for food finding. Ibis 115: 517-534
Wilson EO (1971). The insect societies. Harvard University Press, Cambridge Massachussets
Wilson EO (1975) Sociobiology. Harvard University Press, Cambridge, Massachussets

ORAL DISCUSSION – S.Goss

Q. (Caldwell) What is the importance of the decay time of the pheromone trail ?
A. In the first, short term experiments lasting only half an hour, I ignored decay times. For the circular and later models, I assumed a simple exponential decay. It's important, or the ant would keep going back and forth in the same place. If the decay is very rapid, you need to have large numbers of ants active together.

Q. (Jumars) How do the ants find their way back to the nest when there is no trail ?
A. Army ants are blind. They can only sense polarized light. However, from the trail system, they always know where the nest is. Otherwise it would be too dangerous to leave the nest.

Q. (DeMott) What determines the path diameter of the circular foragers ?
A. The food abundance is important. We've wondered whether the diameter of the circle increases with each successive sweep when the rate of food renewal is low.

Q. (Sih) You note the problem where animals could get stuck in a mediocre area if the trail does not quickly decay. It seems that there should be a relation between the temporal predictability of a spot and the decay rate. If the spot won't be good for long, a rapid decay would allow animals to get away from the spot.
A. We would need to make artificial trails to test this.

Q. (Sih) Do you know what the pheromones are ?
A. Not in general, although you can use chemicals to make trails, for certain species.

INTRASPECIFIC KLEPTOPARASITISM AND FORAGING EFFICIENCY AS CONSTRAINTS ON FOOD SELECTION BY KELP GULLS *LARUS DOMINICANUS*

Philip A.R. Hockey and William K. Steele
Percy FitzPatrick Institute of African Ornithology
University of Cape Town
Rondebosch
South Africa 7700

INTRODUCTION

Kleptoparasitism (food-stealing) is widespread among birds and is an important feeding technique of some families of seabirds, notably Fregatidae (frigatebirds), Chionididae (sheathbills), Stercoraridae (skuas) and Laridae (gulls and terns), being reported for more than 25% of species within these families (Brockmann and Barnard 1979). A useful axis along which to divide the behaviour between species is a 'specialist-opportunist' continuum. No seabird species is known to be an obligate kleptoparasite, but specialists, such as skuas and frigatebirds, obtain a large proportion of their food by kleptoparasitism of heterospecifics. It has been suggested that they may possess structural adaptations associated with these feeding techniques (Brockmann and Barnard 1979), although the evidence for this is weak (Furness 1987). All kleptoparasitic species steal food from other species, but several opportunistic kleptoparasites, such as gulls and terns, also practise this behaviour intraspecifically (e.g. Hopkins and Wiley 1972; Hulsman 1976; Verbeek 1977a,b; Rockwell 1982; Greig *et al.* 1983; Hockey *et al.* in press). In a recent review, Furness (1987) concluded that such opportunistic kleptoparasites were as successful at stealing food as were specialist kleptoparasites.

Brockmann and Barnard (1979) maintain that inter- and intraspecific kleptoparasitism are essentially the same behaviour. However, while the act of food stealing may be the same in both situations, the evolutionary implications of the behaviour are quite different. In the case of interspecific kleptoparasitism, selection for kleptoparasitic skills will begin as soon as the habit provides food for birds which are

NATO ASI Series, Vol. G 20
Behavioural Mechanisms of Food Selection
Edited by R. N. Hughes

food-stressed (Furness 1987). However, when kleptoparasites steal enough food to affect host fitness (reproduction and/or survival), the host should develop defences against the kleptoparasite and an 'interspecific asymmetric arms race' (*sensu* Dawkins and Krebs 1979) will have begun. Under these conditions, benefit accrues to the parasitic species, and cost is incurred by the heterospecific host. Members of the two species each experience a different evolutionary problem, respectively how to raise and lower the frequency and success of parasitic attacks.

The evolution and maintenance of intraspecific kleptoparasitism is intrinsically different, as individuals within one species may experience both evolutionary problems. A specialist kleptoparasite could, in theory, evolve to become an obligate kleptoparasite. However, limits to intraspecific kleptoparasitism are determined by conspecific, rather than heterospecific, host availability, and the behaviour cannot evolve progressively as under Dawkins and Krebs' (1979) 'intraspecific symmetric arms race' scenario.

Barnard and Sibly (1981) developed a general model describing the pay-off between hunting and stealing in single-species flocks. However, this model assumes that an individual follows a fixed, not a mixed strategy, and that the balance between 'producing' and 'scrounging' is related solely to group composition and is independent of differential foraging efficiency within the population. Differences in foraging efficiency are known to exist between age classes in several larids (Verbeek 1977b,c; Ulfstrand 1979; Monaghan 1980; Burger and Gochfeld 1981; Greig *et al.* 1983; Hockey *et al.* in press).

We hypothesised that, in a generalist, opportunistic predator such as the Kelp Gull, not only will several factors influence the incidence of kleptoparasitism, but the relative importance of these factors will vary from site to site. Interpretation of previous studies of this behaviour has been limited by site-specificity (see e.g. Greig *et al.* 1983). To overcome this problem, we examined variations in kleptoparasitic behaviour of Kelp Gulls *Larus dominicanus* under different feeding conditions, and in geographically disparate areas.

The following predictions were examined:

1. Prey type and abundance, and handling techniques of the host will influence the frequency of kleptoparasitism (Brockmann and Barnard 1979).

2. The incidence of kleptoparasitism will be unevenly divided among the gull population (Burger and Gochfeld, 1981) and the behaviour will be of greatest importance to birds

which are relatively inefficient hunters and hence may be more food-stressed than efficient hunters (Furness 1987).

3. Kleptoparasitism will be most frequent when food quality, and/or quantity, hence potential rewards, are high (Brockmann and Barnard 1979).

4. Potential hosts will attempt to reduce their availability to conspecific kleptoparasites by resorting to interspecific kleptoparasitism.

5. Prey choice will be influenced by the risk of losing prey (tested interspecifically by Thompson 1983).

6. Variations in the above factors will lead to differences in the demography of gull populations, dependent on prey type and availability.

STUDY AREAS AND METHODS

The foraging and kleptoparasitic behaviour of Kelp Gulls was observed at four sites in Chile and four sites in South Africa (Table 1). At all sites, the dominant prey items in the diet were recorded. At sites where bivalves were preyed on, size distributions of shells dropped (to break them open) or swallowed whole were determined from beach collections and regurgitated pellets respectively. The following components of foraging behaviour were recorded: active searching time T_s (s); prey handling time T_h (= time from prey capture to completed consumption or until prey was stolen/lost/abandoned (s)); the number of times bivalves were dropped before breaking; the frequency of kleptoparasitic attempts and their outcome (success/failure); and, the age classes of the parasite and host involved in each attack. The demography of the gull population at each site was recorded, and the above parameters of foraging behaviour were recorded by age class. On the basis of plumage pattern we divided gulls into three age classes: AD (adult), 2Y (2-3 year old birds) and 1Y (first year birds) (Hockey *et al.* in press). Extended focal animal observations were avoided to obviate pseudoreplication in the data. Energy values for the sand mussel *Mesodesma donacium* were obtained from Tarifeno-Silva (1980)

Controlled feeding of gulls was carried out at Kalk Bay, South Africa (34° 09'S, 18° 27'E) to determine experimentally the influence of prey attributes on prey choice and kleptoparasitic behaviour. To assess the effects of prey size, independent of prey quality, three strips of fish of three different lengths (5, 10 and 15 cm) were randomly ordered equidistant from a flock of gulls, and the order in which prey were selected, as

Table 1. Summary of study sites: habitat type, main prey of gulls, handling technique and modal prey sizes of 'natural' prey.

Locality	Habitat type	Main prey	Principal Handling Technique	Modal size of bivalve prey (mm)	Obs time (h)
CHILE					
Las Cruces (33 30S, 71 38E)	Sandy Beach	Bivalve: *Mesodesma donacium*	Fly up and drop	70-79	42
Mehuin (39 24S, 73 13W)	Estuarine sand spit	Bivalves: *Mesodesma donacium* *Mulinia* sp.	Swallow whole Fly up and drop	20-29	7
Mar Brava (42 00S, 74 00E)	Sand and shingle beach	*Mesodesma donacium*	Fly up and drop Swallow whole	50-59 30-39	17
Porvenir (53 18S, 70 22E)	Offshore kelp beds	Bivalve: *Gaimardia* sp.	Fly up and drop Swallow whole	50-59 <10	9
SOUTH AFRICA					
Yzerfontein (33 18S, 18 17E)	Sandy Beach	Bivalve: *Donax serra*	Fly up and drop Swallow whole	40-49 <35	46
Hout Bay (34 04S, 18 21E)	Fishing harbour	fish and fish offal	Swallow whole		47
Olifantsbos (34 17S, 18 17E)	Rocky shore	small kelp infauna, crabs *Plagusia chabrus*	Swallow whole		55
Strandfontein (34 05S, 18 30E)	Rubbish dump	domestic waste	Swallow whole/ tear apart		55

well as kleptoparasitic interactions, were recorded (n = 50 trials). To test effects of prey quality independent of prey size, 10 portions each of fish and bread of equal size were randomly positioned in a 5x4 grid spread over 1.5 x 1.0m, and selection was recorded as above (n = 15). To test effects of prey dispersion, independent of prey size and quality, gulls were videotaped while feeding on pilchards *Sardinops ocellata* presented either as one pile of 20 fish (n = 10 replicates), five piles of four fish (n = 8) or as 20 individually spaced fish (n = 8).

RESULTS

Food and feeding techniques

At the different study sites Kelp Gulls fed on a range of prey types (Table 1). On the rocky shore at Olifantsbos most foraging time was spent either among driftline kelp, from which small invertebrates, mainly larvae of kelp flies *Xanthocanace* spp., were dug, or among rocks at the tide's edge where birds preyed on the isopod *Ligia galbrata* and the crab *Plagusia chabrus*. Small prey (*Gaimardia* sp.) also were taken at Porvenir, Tierra del Fuego, where swimming gulls fed by picking bivalves from the fronds and stipes of kelp *Macrocystis pyrifera*. At these sites, prey were swallowed whole. Larger bivalves dominated the diet at all other sites in Chile, and at Yzerfontein in South Africa (Table 1.). Bivalves were captured either by 'paddling' in thixotropic sand at the water's edge, by swimming in shallow water and upending to retrieve swimming or feeding bivalves, or by plunge-diving in deeper (30-50 cm) water. At Mar Brava, some *Mesodesma* were located by walking among shingle patches at the mid tide level. Once captured, small bivalves (modal size 20-40 mm) were swallowed whole. Larger shells (modal size 50-70 mm) were carried into the air and dropped repeatedly until they were sufficiently damaged for gulls to extract the flesh.

At Hout Bay and Strandfontein, where available natural food is greatly supplemented by man's activities, 'unnatural' food is abundant and no specialised techniques are required either to locate or to handle the food. Man-derived foods dominated the diets of Kelp Gulls and Hartlaub's Gulls *Larus hartlaubii* at both sites.

Food type and handling time

When small, abundant prey were eaten, search and handling times were short (Table 2). At Porvenir, ingestion rates of *Gaimardia* ranged between 10 and 51 mussels per minute (Table 2), with a mean handling time of < 1s. At Mehuin, feeding on small *Mesodesma*, between 10 and 14 mussels were eaten per minute with a mean handling time of 1.7 ± 1.4s per mussel (n = 13).

At sites where man-derived food was abundant, but usually fairly large (e.g. fish offal), handling times also were short. At Hout Bay, mean handling time for fish and fish offal was 1.3 ± 0.7s, and handling time of domestic waste at Strandfontein averaged 1.7 ± 2.3s. When feeding on small and/or rapidly-handled prey, there were no obvious age-related differences in foraging efficiency (Table 2).

Table 2. Age related prey intake rates of Kelp Gulls feeding on small and/or rapidly handled prey. Sample sizes in parentheses.

Site	Prey	Intake rate (prey per minute)			
		AD	2Y	1Y	
CHILE					
Mehuin	M. donacium	9.7 ± 6.4(15)	13.8 ± 9.6(7)	10.0 ± 8.4(8)	
Porvenir	Gaimardia sp.	14.9 ± 9.7(14)	15.3 ± 3.8(7)	10.0 ± 3.8(7)	min*
		41.9 ± 12.1(31)	30.6 ± 8.7(5)	50.5 ± 9.0(6)	max*
SOUTH AFRICA					
Olifantsbos	Small invertebrates	5.8 ± 12.1(13)	3.1 ± 1.9(3)	23.7(2)	
		Handling time(s) per prey item: all age classes			
Hout Bay	Fish offal		1.3 ± 0.7 (33)		
Strandfontein	Domestic waste		1.7 ± 2.3 (18)		

* From Hockey 1988.

The more complex prey handling techniques required to break open large bivalves, by dropping them from the air, greatly increased handling times (Table 3), and these prey were dealt with more efficiently by adults than by younger gulls. Age-related differences in hunting and prey-handling efficiency (at Las Cruces) are illustrated in Fig. 1. Search times of successful hunts did not differ between age classes (Fig. 1a), but 1Y birds searched unsuccessfully for longer than other age classes before giving up (Mann-Whitney $U_{7,10} = 60.0$, $P<0.001$, Fig 1b). The proportion of successful plunges for mussels increased with age (Fig. 1c), adults requiring significantly fewer plunges than younger birds ($U_{14,66} = 238.0$, $P<0.05$, Fig. 1d). Young birds required more drops than older birds to break a mussel (Kruskal-Wallis one-way ANOVA, $H_{26,16,63} = 12.62$, $P<0.005$, Fig. 1e), resulting in significant age-related differences in prey handling times

($H_{14,12,37} = 19.0$, $P<0.002$, Fig. 1f). The contrastingly high foraging efficiency of 1Y birds at Mar Brava does not stand up to statistical testing: of all 1Y birds present, only one opened shells by dropping them; all others swallowed small whole shells - the data are thus pseudoreplicated.

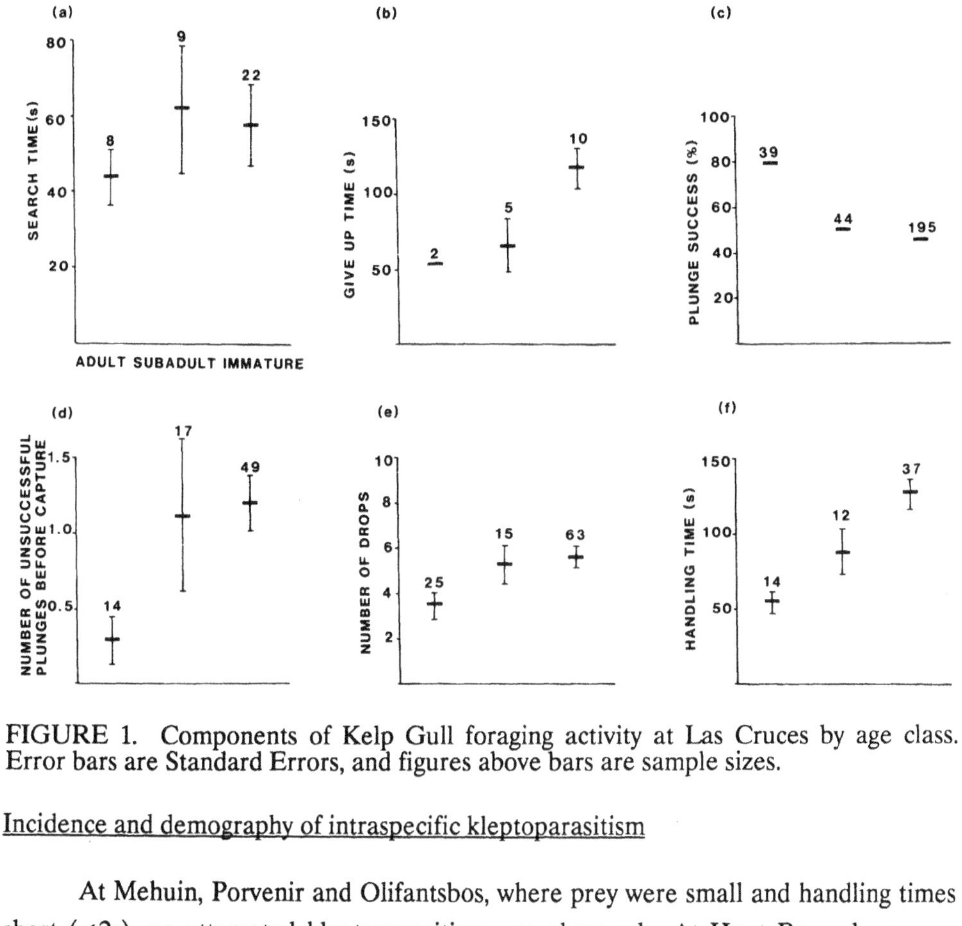

FIGURE 1. Components of Kelp Gull foraging activity at Las Cruces by age class. Error bars are Standard Errors, and figures above bars are sample sizes.

Incidence and demography of intraspecific kleptoparasitism

At Mehuin, Porvenir and Olifantsbos, where prey were small and handling times short (<2s), no attempted kleptoparasitism was observed. At Hout Bay, where prey were much larger, but were also handled rapidly (Table 2), only 2.7% of prey was lost to conspecifics, although kleptoparasitic attempts were 25% successful. At Strandfontein, where prey handling times were slightly greater, but prey were superabundant and generally of low quality, the incidence of attempted kleptoparasitism was only 33% of that at Hout Bay, although the success rate was slightly higher (35%). Kleptoparasitic attacks were only made over relatively large prey, even though these may not have been of high quality (e.g. plastic and orange peel). Such prey were taken infrequently and attempts to steal them usually involved aerial pursuits.

At sites where shell dropping was the major handling technique, and potential hosts were therefore out of contact with their prey for some of the handling time, kleptoparasitism was much more prevalent. At all such sites except Mehuin, between 10.1% and 30.8% of prey captured were subsequently stolen (Table 3). Comparing rates of prey loss with T_h across all sites and all age classes of gulls for different prey, the positive relationship predicted by Brockmann & Barnard (1979) between T_h and risk of prey loss is tenuous once T_h exceeds a few seconds (Fig. 2). Considering only data in Table 3, there is a strong relationship between the number of times a prey item is dropped and the risk of losing that prey item (Fig. 3). The shell dropping technique improves with age, and adults are less likely to lose prey handled this way than are younger birds (Fig. 1e, Table 3). At Mehuin, although all age classes of gulls were present, this prey handling technique was used almost exclusively by adults. The dominant prey dropped at this site was the bivalve *Mulinia* sp., which required fewer drops before breakage than did *Mesodesma* (Table 3).

Table 3. Age-related components of foraging activity for Kelp Gulls preying on bivalves which were dropped to break them open. Ts = search time for prey capture (s), d = number of times a bivalve was dropped before breakage, T_h = handling time (s). Values are means ± S.D. Sample sizes in parentheses.

Site	Gull age class	Ts	d	T_h	% stolen by kleptoparasites
CHILE					
Las Cruces	AD	44 ± 21(8)	3.4 ± 3.1(26)	54 ± 29(14)	15.4
	2Y	62 ± 53(9)	5.3 ± 3.7(15)	88 ± 55(12)	25.9
	1Y	58 ± 49(22)	5.6 ± 4.0(63)	127 ± 59(37)	29.6
Mehuin	AD	37 ± 14(15)	1.4 ± 0.8(32)	84 ± 24(25)	0
Mar Brava	AD	18 ± 16(74)	2.6 ± 1.6(62)	63 ± 29(56)	13.1
	2Y	31 ± 36(13)	3.8 ± 1.7(9)	72 ± 32(9)	30.8
	1Y	26 ± 21(14)	2.5 ± 1.2(13)	46 ± 26(13)	20.0
SOUTH AFRICA					
Yzerfontein	AD	212 ± 162(31)	3.1 ± 2.2(22)	146 ± 79(35)	10.1

The incidence of kleptoparasitism was not evenly distributed with respect to population demography (Table 4). At all sites adults attempted kleptoparasitism less frequently than predicted from their occurrence in the population, and the converse was true for 1Y birds. Mean ± S.D. for the ratios of % attack initiation:% of the population for AD, 2Y and 1Y birds respectively were 0.67 ± 0.26, 1.08 ± 0.89 and 2.23 ± 0.91.

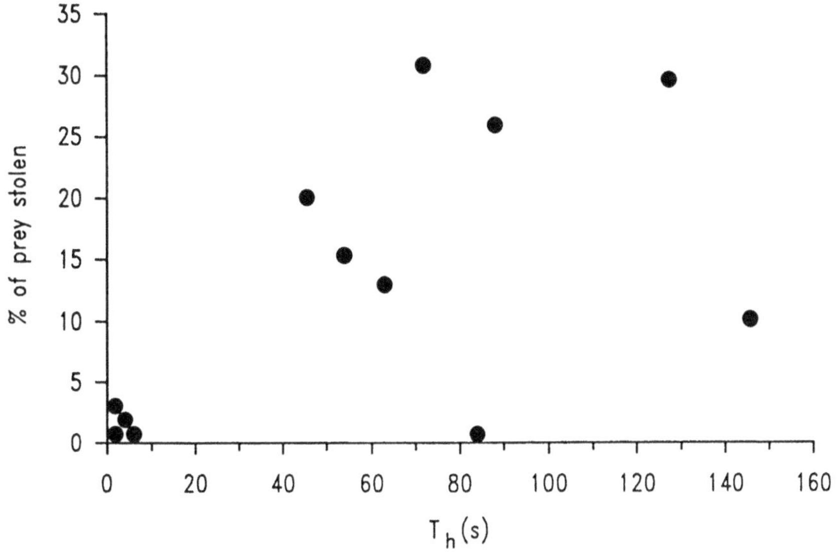

FIGURE 2. Relationship between prey handling time (T_h, in seconds) and the proportion of prey stolen by conspecifics. Data are mean values for different hunting and prey handling techniques at all study sites.

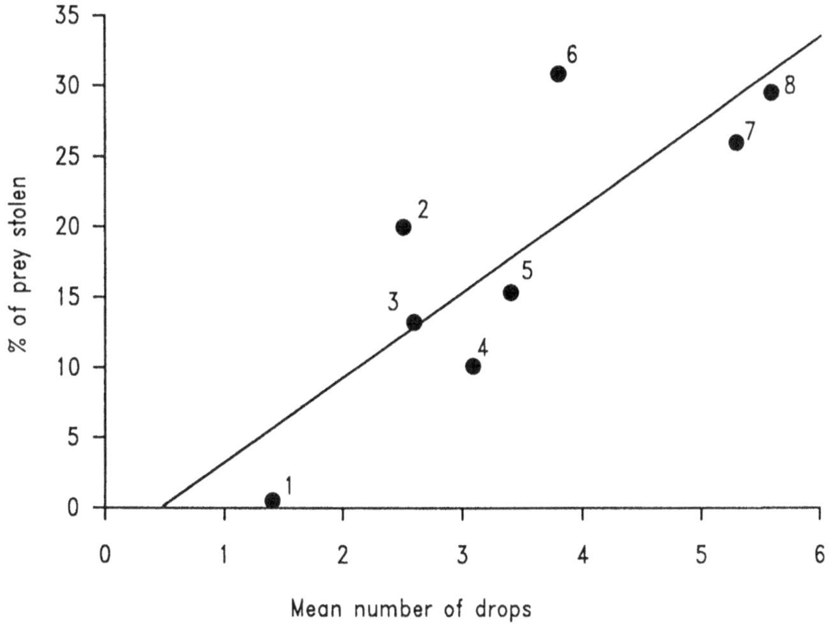

FIGURE 3. Relationship between the mean number of times a bivalve is dropped by a Kelp Gull before breakage and the proportion of prey stolen by conspecifics. 1 = Mehuin AD, 2 = Mar Brava 1Y, 3 = Mar Brava AD, 4 = Yzerfontein AD, 5 = Las Cruces AD, 6 = Mar Brava 2Y, 7 = Las Cruces 2Y, 8 = Las Cruces 1Y.

On average, therefore, when large prey with long handling times were involved, 1Y birds attempted kleptoparasitism 3.1 times as frequently as adults. Young birds (2Y and 1Y) are also favoured as hosts for attack (Table 4). The ratios of % host frequency: % of the population for AD, 2Y and 1Y birds were 0.89 ± 0.11, 1.65 ± 0.81 and 1.34 ± 0.68. Therefore, although 1Y birds are favoured over adults as hosts, they are selected as hosts less frequently than they attempt kleptoparasitism (ratio of 0.60, *cf* 1.53 and 1.33 for 2Y and AD birds respectively). 1Y birds gain more food through kleptoparasitism than they lose as a result of being parasitised, and the reverse is true for 2Y and adults, even though the latter are more efficient at handling their prey, and have a greater success rate when they do attempt kleptoparasitism (adult success = 45.5%, 2Y = 38.1%, 1Y = 35.0% - Las Cruces).

Table 4. Population demography, proportional initiation of kleptoparasitic attacks and proportional distribution of hosts by age class.

	% of population			% attack initiation				% host			
Site	AD	2Y	1Y	AD	2Y	1Y		AD	2Y	1Y	
CHILE											
Las Cruces	10	10	80	6	6	88	*	10	16	74	*
Mar Brava	88	3	9	83	0	17	ns #	72	9	19	**
SOUTH AFRICA											
Strandfontein	82	10	8	58	24	18	***	73	14	13	***
Hout Bay	51	23	26	14	26	60	**	38	19	43	**
Yzerfontein	88	7	5	73	9	18	**	88	10	2	n.s.#

#, 1Y and 2Y combined due to small sample size

* $p<0.05$

** $P<0.01$

*** $P<0.001$

The reason for adult losses through kleptoparasitism exceeding gains probably relates to their relatively high hunting efficiency. At Las Cruces, mean $T_s + T_h$ for prey successfully eaten were 98s, 150s and 185s for AD, 2Y and 1Y birds respectively. At this site, per unit active hunting time, adults caught 1.9 mussels per mussel caught by immatures. Individually, their potential availability as hosts is therefore higher.

Exercising options: stealing from other species

Interspecific kleptoparasitism at the Chilean study sites was rare. At Mar Brava, in addition to Kelp Gulls, American Pied Oystercatchers *Haematopus palliatus* and Chimango Caracaras *Milvago chimango* preyed extensively on *Mesodesma donacium*. However, although caracaras occasionally kleptoparasitised oystercatchers, and both species kleptoparasitised intraspecifically, neither was host to gulls. Rather, caracaras concentrated on removing small fragments of flesh from *Mesodesma* dropped by gulls after the shells had been abandoned. At Porvenir, Chilean Skuas *Catharacta chilensis* fed alongside Kelp Gulls, but no kleptoparasitic interactions were observed.

At all four South African study sites, the much smaller (*ca* 30% of body mass) endemic Hartlaub's Gull was present. At Olifantsbos, prey size and handling speed precluded kleptoparasitism even though the diets of the two species overlapped considerably. At Yzerfontein, Hartlaub's Gulls made up 9% of the gull population. Here, *Donax serra* dominated Kelp Gull diet, but Hartlaub's Gulls preyed mainly on the small isopod *Eurydice longicornis*, and dietary overlap was minimal. The only incidence of interspecific kleptoparasitism was one unsuccessful attempt by a Hartlaub's Gull to steal *Donax* from an adult Kelp Gull.

The only two sites where interspecific kleptoparasitism was prevalent were Hout Bay and Strandfontein, where abundant 'unnatural' food was available. At both these sites there was substantial dietary overlap between the two gull species.

Adult Kelp Gulls attempted interspecific kleptoparasitism of Hartlaub's Gulls approximately twice as frequently as did 1Y birds. The percentage of all attacks directed interspecifically at the two sites were : Hout Bay - AD = 78.7%, 2Y = 56.8%, 1Y = 41.1%; Strandfontein - AD = 34.4%, 2Y = 13.3%, 1Y = 15.9%. Thus, although adults are more efficient at kleptoparasitising interspecifically, the opportunities to use this advantage under natural conditions are few. Consequently, if risks or gains from kleptoparasitic behaviour influence either prey choice or the distribution of Kelp Gulls, such constraints will be determined through intraspecific rather than interspecific interactions.

Risk-sensitive foraging: does kleptoparasitism influence prey choice?

When evenly-spaced prey of three different sizes were offered to gulls simultaneously, the smallest prey item normally was selected first, and the largest last (Table 5). This pattern mirrors one of decreasing probability of successful prey retention after initial capture (Table 5). If the probability of successful retention and

Table 5. Prey choice by Kelp and Hartlaub's Gulls when presented with fish of three different sizes. A "benefit" ratio of small: medium: large fish of 1:2:3 was used in calculating gross energy return.

		Prey size	
Order of selection (%)	small	medium	large
1	46	42	10
2	30	46	16
3	24	8	56
Not taken	0	4	18
Percentage taken by:			
Kelp Gull AD	24	42	34
2Y	24	47	29
1Y	43	23	34
Hartlaub's Gull	75	20	5
Probability of retaining prey (Pr)			
Kelp Gull AD	1.00	0.75	0.22
2Y	1.00	0.50	0.33
1Y	0.87	0.87	0.67
Hartlaub's Gull	0.93	0.50	0.00
(Pr)x gross energetic return			
Kelp Gull AD	1.00	1.50	0.66
2Y	1.00	1.00	0.99
1Y	0.87	1.74	2.01
Hartlaub's Gull	0.93	1.00	0.00

consumption is multiplied by an index of the gross energy return (ratio of small:medium:large prey = 1:2:3), the 'benefit' of taking medium or large prey equals or exceeds the 'benefit' of taking small prey. Overall, adult gulls ate mainly medium-sized fish, which maximised their gross energy return. 2Y birds received equal return by preying on fish of any size, but also concentrated on medium-sized fish. 1Y birds preyed mainly on small fish, which gave the lowest gross return, but also had the lowest risk of being stolen (Table 5). However, such a simplistic calculation cannot take into account the energetic cost of attempting to avoid kleptoparasites.

Few sufficiently detailed field data exist to compare energy intake rates between age classes and feeding techniques. At Mehuin, intake rates of 2Y and 1Y birds, when swallowing small *Mesodesma*, exceeded that of adults feeding in the same way (Table 6). At this site most adults fed on large *Mulinia*, which they dropped. At Las Cruces, where no small mussels were available, energy intake rates of 2Y and 1Y birds were *ca* 50% of that of adults (Table 6).

When prey of equal size and differing quality were offered to gulls, selection was clearly for high quality prey. In two of 15 replicates, one of ten fish remained when gulls either abandoned the grid or were disturbed. All fish were eaten in the other replicates. On average, however, 5.8 $\pm$ 2.3 pieces of bread remained. Of the 148 fish taken, there were 11 attempts at kleptoparasitism (7.4%) of which 4 (2.7%) were successful. The proportion successsfully stolen was identical to the proportion of prey lost to conspecifics at Hout Bay fishing harbour where small fish and offal were the main prey. Of 63 pieces of bread taken, there was one (unsuccessful) attempt at kleptoparasitism Differences between rates of attempted kleptoparasitism of fish and bread are not significant.

Table 6. Intake rates (kJ min^{-1}) during active foraging by Kelp Gulls of different ages feeding on *Mesodesma donacium* at three localities in Chile.

Site	Handling technique	Ad	2Y	1Y
Mehuin	Swallow small mussels	135	183	182
Mar Brava	Swallow small mussels	164	-	-
Las Cruces	Drop large mussels	186	95	94

Does prey dispersion influence kleptoparasitism?

Experimental provisioning of Kelp Gulls with prey of equal size and quality at different densities indicated that the frequency of attempted kleptoparasitism increased as prey dispersion increased. When gulls fed on widely dispersed prey, the probability of kleptoparasitic attack was more than twice that when prey were highly clumped (Table 7). Although trends are consistent, differences between experimental treatments were not significant (X^2_2=2.81). This may be due to small sample size. The proportion of prey taken in each treatment was similar (X^2_2=0.38, NS).

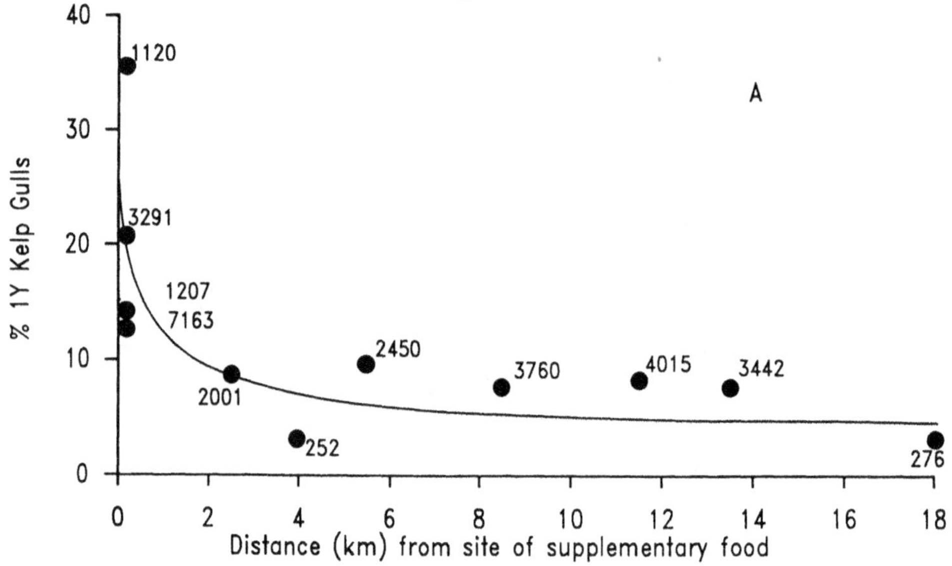

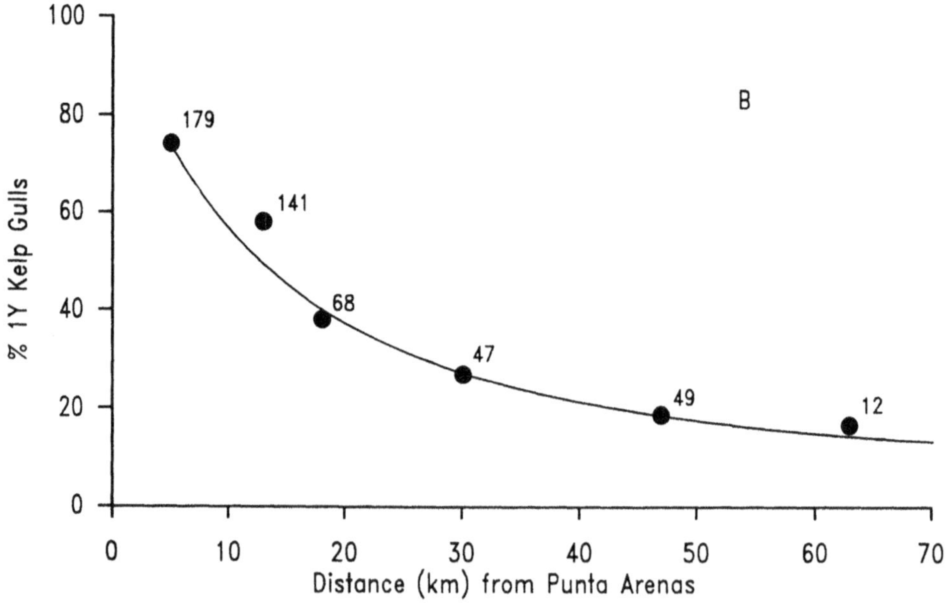

FIGURE 4. Proportion of 1Y birds in Kelp Gull populations in relation to the distance from a site of supplementary food in a) South Africa and b) Chile. N = sample sizes on which age ratios based.

Table 7. Influence of prey dispersion on kleptoparasitism. Twenty pilchards were offered to gulls either in one pile of 20 fish, five piles of four fish or as 20 individually-spaced fish.

No. of piles of fish	No of replicates	No of fish offered	No of fish taken	Attempts at kleptoparasitism
1	10	200	180 (90%)	7 (3.9%)
5	8	160	148 (93%)	9 (6.1%)
20	8	160	154 (96%)	13 (8.4%)

Table 8. Ratios of % frequency with which birds of different age classes initiated attacks *vs* % frequency with which they themselves were attacked.

	KELP GULL					HERRING GULL *Larus argentatus*	
Gull age	Las Cruces	Mar Brava	Strand.	Yzer.	Hout Bay	NE England[1]	NW. England[2]
AD	0.6	1.2	0.8	0.8	0.4	1.5	0.9
2Y	0.4	0	1.7	0.9	1.4	3.4	1.2
1Y	1.2	0.9	1.4	9.0	1.4	2.1	1.8

1. Greig *et al*. 1983 (gulls at rubbish tip)

2. Verbeek, 1977b (gulls eating starfish)

Population demography in relation to prey characteristics

Kelp Gulls were not evenly distributed with respect to age class (e.g. Table 4). In the southwestern Cape, the proportion of 1Y birds in the population (P_{1Y}) was inversely related to the distance (D, in km) from a site where supplementary food was available ($P_{1Y} = 10.01 - 1.54 \ln D$, $r_9 = 0.75$, $P<0.01$, Fig. 4a). A similar pattern was observed on the Brunswick Peninsula between the fishing port of Punta Arenas (53° 09'S, 70° 55'W) and San Juan to the south west ($P_{1Y} = 113.21 - 24.17 \ln D$, $r_4 = 0.98$, $P<0.001$, Fig. 4b).

Apart from sites where food was supplemented, only Las Cruces supported a high proportion of 1Y birds (80%). This may reflect early postfledging dispersal from nearby offshore islands where the birds breed. In the southwestern Cape, where year-round observations were made, initial dispersal of young birds from breeding islands is to sandy beaches, and subsequently towards urban areas. Yzerfontein is directly opposite Dassen Island, where some 2 900 pairs of Kelp Gulls breed (Crawford *et al.* 1982). Immediately following the breeding season, the proportion of 1Y birds on the beach was as high as 24%, far exceeding the annual average of 5%.

DISCUSSION

Prey type, abundance and handling technique

Small prey which are handled rapidly offer little opportunity to kleptoparasites. Even relatively large prey, such as *Mesodesma*, are not likely to be stolen if they are swallowed as soon as they are found. Once handling time exceeds a few seconds, the risk of prey loss increases markedly, as predicted by Brockmann and Barnard (1979). Data from this study do not suggest that absolute handling time, once in excess of a few seconds, is necessarily correlated with risk of losing prey (Fig. 1). However, handling technique influences risk strongly. The risk of losing bivalves which are dropped to break them open is directly related to the number of times a shell is dropped (Fig. 2).

Quantifying prey abundance, and more particularly prey availability, is difficult. At Las Cruces, where prey clearly were abundant, risk of prey loss was high (due to the handling technique). At Strandfontein, where prey also were abundant, but handled rapidly, risk of loss was low.

Within the range of prey dispersion studied experimentally, increased dispersion apparently led to an increased likelihood of kleptoparasitic attack. This parallels the Las Cruces/Strandfontein comparison, food at the beach being less clumped than food at the dump.

Hunting *vs* stealing: age-related differences

At sites where prey were small, there was no evidence of age-related differences in foraging efficiency (Table 2). At Las Cruces and Mar Brava, where birds hunted large bivalves, adults were more efficient in finding and handling prey than were younger birds. This pattern has been recorded in several gull species (Barash *et al.* 1975; Verbeek 1977b,c; Ingolfsson and Estrella 1978; Ulfstrand 1979; Monaghan 1980; Burger and Gochfeld 1981; Rockwell 1982; Greig *et al.* 1983). It was suggested by Lack (1967, 1968) and several subsequent workers that deferred maturity in seabirds occurs because young birds take time to master foraging techniques and identify foraging areas. In addition to being more efficient foragers, adult gulls are also more efficient kleptoparasites (Burger and Gochfeld 1981; this study). At Las Cruces, adults were 30% more successful than 1Y birds.

In general, flocks of gulls contain adult birds which are both efficient foragers and stealers, and young birds which are less efficent at both foraging and stealing. These young birds therefore capture less prey than adults and hence, individually, are available as hosts less often than adults, but are less efficient at handling their prey once caught. In this study, the differential between kleptoparasitic efficiency of old and young birds was less than the differential in hunting success. Therefore, given that host availability ultimately determines the opportunity for kleptoparasitism, young birds should attempt kleptoparasitism proportionately more than older birds. This is borne out by the ratios of % of attacks:% of hosts by age-class (Table 8). At four of five sites, adult Kelp Gulls attempted kleptoparasitism less frequently than they were selected as hosts. The reverse was true for 1Y birds. The same pattern has been recorded for Herring Gulls *Larus argentatus* in England feeding on both natural and man-derived food (Table 8). Burger and Gochfeld (1981) found the opposite pattern at a rubbish dump in New Jersey, where adults obtained more food by piracy than young birds and also lost less food to pirates. They also showed, indirectly, that young birds were less efficient foragers. At face value, this result is difficult to interpret in evolutionary terms, as the most efficient hunters were stealing food from the least efficient hunters. However, 80% of the piracy attempts in their study were made by birds not already feeding at the dump, and parasitism was therefore acting exogenously on, rather than endogenously within, the feeding group.

Kleptoparasitism and food quality

The evidence for prey quality affecting the incidence of kleptoparasitism is equivocal. When high and low quality prey of equal size and handling time (bread and fish) were equally available to gulls, there was clear selection for the high quality prey. However, the incidence of kleptoparasitic attacks on bread- and fish-carrying gulls were not significantly different (although this may be a function of sample size). At Strandfontein, the risk of attack was a function of prey size, and hence visibility. Gulls carrying worthless (e.g. plastic) but large prey were pursued, indicating little discrimination by parasites on the basis of prey quality.

Kushlan (1978) compared the foraging and kleptoparasitic behaviour of five sympatric piscivorous wading species (stork, ibis, egret and two herons). All five species kleptoparasitised intraspecifically more than they stole from any heterospecific. Interspecific kleptoparasitism was directed at hosts smaller than themselves, but prey stolen were larger than the average prey caught by the parasite. However, for the one species studied in detail (Great Egret *Casmerodius albus*), inclusion of kleptoparasitism within the foraging repertoire decreased the net rate of energy intake by the parasite.

The apparent lack of fine tuning in intraspecific kleptoparasitism contrasts with the tendency towards optimality displayed in interspecific interactions. At certain times of the year, some Blackheaded Gulls *Larus ridibundus* obtain 100% of their food by stealing from Lapwings *Vanellus vanellus* and Golden Plovers *Pluvialis apricaria* (Kallander 1977; Thompson 1986). Most attacks by Blackheaded Gulls, and the sizes of prey selected for attack, conformed closely to the predictions of two energy maximisation models (Thompson 1986). Guillemots *Uria aalge* carry larger prey than Atlantic Puffins *Fratercula arctica*, and on this basis should perhaps be favoured hosts of Arctic Skuas *Stercorarius parasiticus*. However, the skuas have a lower success rate when attacking Guillemots, and attack puffins preferentially (Furness 1987).

Lowering risk: prey choice and interspecific stealing

Potential hosts may reduce the likelihood of being attacked in several ways. They may move to areas where densities of conspecifics are low, or pre-empt attack by avoiding local concentrations of kleptoparasites (e.g. Grant 1971). They may select smaller prey which are handled rapidly, are less conspicuous and less profitable to steal (Fuchs 1977; Thompson 1983). Alternatively, they may resort to obtaining more of their own food by kleptoparasitism (Burger and Gochfeld 1981).

It appears that at least the latter two options are followed by Kelp Gulls. At Mehuin, a wide range of sizes of bivalves was available. All three age-classes of gulls swallowed small mussels, but 1Y and 2Y birds had a more rapid rate of prey intake than the few adults that fed in this way (Table 6). On the other hand, with one exception, only adults dropped large mussels at this site. Young birds avoid the risk of kleptoparasitism by selecting rapidly handled prey, but may incur a metabolic cost related to low absorption efficiency of prey with a high indigestible ash-equivalent content (cf Speakman 1987). There was no kleptoparasitism of adults at this site, but their prey, *Mulinia*, required an average of only 1.4 $\pm$ 0.8 drops to break. At other sites where *Mesodesma* and *Donax* were preyed on, mussels were dropped an average of 2.5 to 5.6 times, and losses to kleptoparasites ranged between 10% and 31%.

When prey of different sizes were offered to Kelp Gulls, the smallest item normally was selected first, and the largest last. Thus, although the potential energetic rewards from large prey were three times those of small prey, small prey were preferred (18% of large prey offered were not eaten at all - Table 5). The risk of losing small prey was negligible, but ranged between 33% and 78% for large prey (Table 5). Patterns of selection were therefore based on an assessment of risk rather than on the energy content of the prey *per se*.

Because of the greater foraging efficiency of adults, an individual adult potentially is available as a host more frequently than is a younger bird. At most sites, adults were hosts to attack more often than they were the attacker, the reverse being true for younger birds (Table 8). From these observations, one would predict that, in areas where heterospecific hosts were available, and gulls were able to steal from them, interspecific kleptoparasitism would be most prevalent among those sections of the gull population which had the highest cost:benefit ratio from intraspecific kleptoparasitism. At both sites where heterospecific hosts were parasitised regularly, adults attacked heterospecifics twice as often as did 1Y birds. At both these sites, unnatural food was abundant. We suggest that Hartlaub's Gulls take a larger range of prey sizes than under natural conditions, rendering them susceptible to attack by Kelp Gulls. This was demonstrated experimentally by Duffy *et al.* (1987).

Most gulls take two or more years to attain full adult plumage (Harrison, 1983). In addition to having a distinctive plumage and being inefficient foragers, young birds generally are subordinate to older birds (Ulfstrand 1979; Monaghan 1980). The signalling value of plumage (i.e. broadcasting dominance or subordinance) has been demonstrated experimentally (Rohwer 1975; Rohwer and Rohwer 1978). Evolutionary attempts to cheat on this apparent disadvantage would fail if there was a lack of congruence between status signalled and 'back-up' behaviour (Ulfstrand 1979). It has been suggested (Moyle 1966) that subordinate juvenile gulls may be able to suppress adult aggression by virtue of their coloration, thus making it easier for them to displace adults at a food source. However, there is little supporting evidence for this theory

(Monaghan 1980; Hockey *et al.* in press). It seems likely that increasing interspecific kleptoparasitism with age, as in Kelp Gulls, is related to learning and aptitude rather than to young birds being particularly successful when attempting to steal from conspecific adults rather than heterospecifics.

Intraspecific kleptoparasitism and Kelp Gull distribution

Dominance relationships within a population may have a profound effect on post-breeding dispersal of different age-classes and sexes. When food is in short supply, distribution may be influenced asymmetrically, with subordinate individuals being forced to move farther afield (Davis 1975; Gauthreaux 1978; Monaghan 1980). Monaghan (*op. cit.*) found that adult Herring Gulls dominated females and immatures at rubbish dumps during winter, forcing increased movements of the latter between feeding sites.

The movements of Kelp Gulls are poorly understood. In South Africa, where observations were made throughout the year, and large numbers of nestlings were ringed, some patterns emerge. Immediately after breeding, large numbers of gulls gather on mainland shores near the breeding sites. Most of the juvenile birds then disperse towards coastal urban centres. By the middle of the non-breeding season, sites where natural food only is available, irrespective of food type, contain few 1Y birds (Table 1, Fig. 3). Closer to urban centres, the proportion of 1Y birds increases: this pattern was detected both in South Africa and Chile (Figs 3a,b). The greatest concentrations of 1Y birds occurred at fishing harbours rather than rubbish dumps. It has been shown elsewhere (Verbeek 1977b,c) that young gulls feeding at dumps are less efficient than adults due to their relative inability to detect prey which are varied and 'unnatural'. At fishing harbours, prey is highly conspicuous, resembles or is natural prey, is easily handled and is of a relatively high quality.

Populations of many gull species worldwide are increasing: this has been attributed to increased food availability through man's activities (Verbeek 1977d; Horton *et al.* 1983). Although many Kelp Gulls in South Africa aggregate at dumps and harbours, the food fed to nestlings is largely of natural origin and of high quality (WKS unpubl. data). There is no reason to suppose that increasing numbers of Kelp Gulls are a result of increased breeding (fledging) success *per se*. Rather, juvenile survival is almost certainly enhanced by artificial food supplies which tide birds over the critical period of post-fledging independence when natural mortality is high (Burger 1980).

Intraspecific kleptoparasitism in an evolutionary context

Most reported instances of kleptoparasitism make little attempt to provide a functional or evolutionary explanation of the behaviour (Brockmann and Barnard 1979; Furness 1987). The term kleptoparasitism was originally coined in the context of interspecific kleptoparasitism (Rothschild and Clay 1952), but Brockmann and Barnard (1979) consider that there is no fundamental difference between inter- and intraspecific kleptoparasitism and that a direct analogy exists between parasitic individuals within a species and parasitic species within mixed groups. Furthermore they propose that there is no reason to doubt that natural selection works in the same way in both types of interaction. The assumption inherent in this argument, and in the model of Barnard and Sibly (1981), is that individuals within a species are either 'producers' or 'scroungers'; in other words they follow a single, not a mixed strategy. In the case of the House Sparrows *Passer domesticus* studied by Barnard and Sibly (*op. cit.*), this assumption may be true. In such a situation, the concept of a continuum between inter- and intraspecific kleptoparasitism is valid, because, with respect to food stealing, an individual is solving only one evolutionary problem. In species which follow a mixed strategy, the evolutionary problem is more complex as an individual may be fighting an evolutionary arms race with itself.

In specialised, interspecfic parasite-host relationships, hosts lose only a small proportion of their prey to kleptoparasites and, for this reason, Furness (1987) argues that these are highly evolved, stable relationships. Furness (*op. cit.* - Table 5.3) cites nine interactions among seabirds in which hosts lost 5% or more of their prey to kleptoparasites. Four of these (44%) involved intraspecific kleptoparasitism among larids. Combining data from Furness with data from Table 3, the average percentage of prey lost to conspecifics in such interactions is 20 ± 9%. (N = 11) The highest rate of prey loss recorded among seabirds is 35%, for intraspecific piracy by Glaucouswinged Gulls *Larus glaucescens* (Rockwell, 1982). In interactions between *Larus* gulls and other charadriiforms, hosts lost 13 ± 14% of prey (N = 6). Kleptoparasitism by gulls of phylogenetically distant hosts (puffins) resulted in 0-10% of prey being lost (N = 2) (Furness 1987). Hosts being parasitised by three species of skuas lost on average 1.5 ± 1.4% of prey (N = 9). In general, although opportunistic and specialist kleptoparasites have equal success in attacks, the highest rates of prey loss occur when attacks are directed against conspecifics, and this behaviour is most pronounced among opportunistic species (Furness 1987). An opportunistic scrounger involved in counter-adaptive arms races with several phylogenetically distant producers is unlikely to show fine-grained adaptation to any one of them because selection is likely to favour a range of competing adaptations (Barnard 1984). Exploitation of more closely related producers, however, may result in similar adaptations being selected for in each case and therefore allow scroungers to become more specialised with respect to each producer (Barnard 1984). In the case of interspecific kleptoparasitism by gulls, this pattern appears to be true: the highest success rates are achieved when stealing from closely related species. However, the highest success of all is achieved when stealing from conspecifics. Barnard (1984) considers that the intensity of selection will determine not only how far the arms race will proceed in any one case, but, because it is unlikely to be equal in the two lineages, it may also determine the winner of the race.

Can a species win an evolutionary arms race against itself if individuals follow mixed strategies by being both parasites and hosts? Because many gull populations have increased dramatically in recent years, high rates of interspecific kleptoparasitism by gulls (e.g. Nettleship 1972; Hulsman 1976; Verbeek 1977a) may not be evolutionarily stable. In other words gulls are stealing too much food from heterospecifics for host populations to withstand the loss in the long term. Perhaps true, but difficult to test or prove. There is, however, evidence that increasing numbers of Herring and Lesser Blackbacked Gulls *Larus fuscus* in the Finnish archipelago have led to an increase in interspecific kleptoparasitism (Bergman 1982). When food stealing occurs intraspecifically, without the option for interspecific kleptoparasitism, its incidence is unlikely to change dramatically with changes in gull population size because opportunity is constrained by host availability.

Monaghan (1980) found that different age-classes of Herring Gulls were unevenly distributed. Less efficient foragers (juveniles) and females were relegated to suboptimal feeding sites through the dominance behaviour of adult males. Age-classes among Kelp Gulls also are unevenly distributed, with young birds aggregating at sites where food is superabundant and easily obtained. If such sites were unavailable, and the existing adult:juvenile ratios at natural sites are stable with respect to the incidence of kleptoparasitism, one would predict an increased mortality of juvenile gulls. It appears that juveniles offset their poor hunting success either by selecting foraging sites where minimal hunting skills are required, or by stealing from conspecifics when prey handling techniques render potential hosts vulnerable to attack.

Kleptoparasitism is a complex behaviour and its incidence varies in relation to prey type, abundance, size and handling technique as well as host and parasite characteristics. Few or no paradigms exist, certainly in the case of intraspecific kleptoparasitism which has been little studied in an evolutionary context. The present study of kleptoparasitism in Kelp Gulls needs to be extended to populations which have not increased or been influenced by man's activities (e.g. at Subantarctic islands), and to large-scale manipulations of food supply (e.g. at fishing harbours). The real test of the evolutionary significance of the behaviour can only be made on populations which are severely food-stressed, but which cannot easily disperse to other sites when such conditions prevail. The massive mortalities of *Mesodesma donacium* which occur on the Peruvian coast during El Nino events may provide such an opportunity. In one study, densities of *Mesodesma* fell from *ca* 3 000 per m^2 to zero in three months without replacement by other species (Arntz *et al.* 1987).

CONCLUSIONS

1. Small and/or rapidly-handled prey are seldom stolen. Once prey handling time exceeds a few seconds the risk of prey loss increases, but not in a predictable manner. In populations where shell-dropping is an important prey-handling technique, the risk of prey loss is directly proportional to the number of times a shell is dropped before breakage. There is some evidence that increasing prey dispersion leads to an increased risk of attack from kleptoparasites.

2. Hunting and prey handling efficiency of Kelp Gulls increases with age except when prey are small and abundant. Young birds attempt kleptoparasitism proportionally more often than older birds. Adult gulls attempt kleptoparasitism less often than they are selected as hosts even though they are more successful than younger birds when they do attempt kleptoparasitism. The reverse is true for 1Y birds.

3. Kelp Gulls select prey on the basis of quality if high and low quality prey of equal handling time are equally available. However, there is little evidence that prey quality *per se* influences the likelihood of kleptoparasitic attack. Such attacks generally are directed at birds carrying large prey irrespective of prey quality.

4. Because of their greater hunting efficiency, individual adults are 'available' as hosts more often than individual young birds. At sites where heterospecific hosts were kleptoparasitised, adult Kelp Gulls attacked them twice as frequently as did 1Y gulls. However, interspecific kleptoparasitism occurred only at sites where unnatural food was available. Congeners (Hartlaub's Gulls) were present at all South African sites, but were not kleptoparasitised under natural conditions, presumably because Hartlaub's Gulls prey mainly on small, rapidly-handled invertebrates.

5. In prey choice experiments, the order of prey selection was determined by risk of loss rather than the energy content of the prey. The low prey handling efficiency of young gulls renders then particularly susceptible to prey loss through kleptoparasitism. When

offered prey with differing handling times, young gulls selected small prey (with a low risk of loss) more often than did older gulls.

6. Kelp Gulls are unevenly distributed geographically with respect to age class. Young birds are concentrated in areas of human activity where food is supplemented (e.g. fishing harbours and rubbish dumps). We suggest that the existence of such sites enhances the survival of young gulls by obviating the need to develop hunting skills rapidly, and reducing the need to obtain food by kleptoparasitsm of adults.

ACKNOWLEDGEMENTS

We are grateful to the Universidad Catolica de Chile, Santiago, and the Universidad Austral de Chile, Valdivia, for providing access to field station facilities at Las Cruces and Mehuin respectively. Alison Bosman, Peter Ryan and Jane Turpie assisted in the field. We thank the Cape Town City Council for allowing access to the Strandfontein dump. Funding for this study was provided by the South African Council for Scientific and Industrial Research's Foundation for Research Development, and by the University of Cape Town.

REFERENCES

Arntz WE, Brey T, Tarazona J, Robles A (1987) Changes in the structure of a shallow sandy-beach community in Peru during an El Nino event. In: Payne AIL, Gulland JA, Brink KH (eds) The Benguela and comparable ecosystems. S Afr J Mar Sci 5:645-658

Barash DP, Donovan P, Myrick R (1975) Clam dropping behaviour of the Glaucous-winged Gull (*Larus glaucescens*). Wilson Bull 87:60-64

Barnard CJ 1984 The evolution of food-scrounging strategies within and between species. In: Barnard CJ (ed) Producers and scroungers: strategies of exploitation and parasitism. Croom Helm, London, p 95

Barnard CJ, Sibly RM (1981) Producers and scroungers: a general model and its application to captive flocks of House Sparrows. Anim Behav 29:543-550

Bergman G (1982) Population dynamics, colony formation and competition in *Larus argentatus*, *fuscus* and *marinus* in the archipelago of Finland. Ann Zool Fennici 19:143-164

Brockmann HJ, Barnard CJ (1979) Kleptoparasitism in birds. Anim Behav 27:487-514

Burger J (1980) The transition to independence and post-fledging parental care in seabirds. In: Burger J, Olla B, Winn H (eds) Behaviour of marine animals, vol. 4: marine birds. Plenum Press, New York, p 367

Burger J, Gochfeld M (1981) Age-related differences in piracy behaviour of four species of gulls, *Larus*. Behaviour 77:242-267

Crawford RJM, Cooper J, Shelton PA (1982) Distribution, population size, breeding and conservation of the Kelp Gull in southern Africa. Ostrich 53:164-177

Davis JWF (1975) Specialization of feeding location by Herring Gulls. J Anim Ecol 44:795-804

Dawkins R, Krebs JR (1979) Arms races between and within species. Proc Roy Soc Lond B 205:489-511

Duffy DC, Heseltine S, La Cock GD (1987) Food size and aggressive interactions between two species of gulls: an experimental approach to resource partitioning. Ostrich 58:164-167

Fuchs E (1977) Kleptoparasitism of Sandwich Terns *Sterna sandvicensis* by Black-headed Gulls *Larus ridibundus*. Ibis 119:183-190

Furness RW (1987) Kleptoparasitism in seabirds. In: Croxall JP (ed) Seabirds: feeding ecology and role in marine ecosystems. Cambridge University Press, Cambridge, p77

Gauthreaux SA jr (1978) The ecological significance of behavioural dominance. In: Bateson PPG, Klopfer PH (eds) Perspectives in ethology, vol. 3. Plenum Press, New York, p 17

Grant PR (1971) Interactive behaviour of puffins *Fratercula arctica* L. and skuas *Stercorarius parasiticus* L. Behaviour 40:263-281

Greig S, Coulson JC, Monaghan P (1983) Age-related differences in foraging success in the Herring Gull *Larus argentatus*. Anim Behav 31:1237-1243

Harrison P (1983) Seabirds: an identification guide. Croom Helm, Beckenham

Hockey PAR (1988) Kelp Gulls *Larus dominicanus* as predators in kelp *Macrocystis pyrifera* beds. Oecologia 76:155-157

Hockey PAR, Ryan PG, Bosman AL (to be published) Age-related intraspecific kleptoparasitism and foraging success in Kelp Gulls *Larus dominicanus*. Ardea

Hopkins CD, Wiley RH (1972) Food parasitism and competition in two terns. Auk 89:583-594

Horton N, Brough T, Rochard JBA (1983) The importance of refuse tips to gulls wintering in an inland area of south-east England. J Appl Ecol 20:751-765

Hulsman K (1976) The robbing behaviour of terns and gulls. Emu 76:143-149

Ingolfsson A, Estrella JT (1978) The development of shell-cracking behaviour in Herring Gulls. Auk 95:577-579

Kallander H (1977) Piracy by Black-headed Gulls on Lapwings. Bird Study 24:186-194

Kushlan JA (1978) Non-rigorous foraging by robbing egrets. Ecology 59:649-653

Lack D (1967) Interrelationships in breeding adaptations as shown by marine birds. Proc Int Orn Congr 14:3-42

Lack D (1968) Ecological adaptations for breeding in birds. Methuen, London

Monaghan P (1980) Dominance and dispersal between feeding sites in the Herring Gull (*Larus argentatus*). Anim Behav 28:521-527

Moyle P (1966) Feeding behaviour of the Glaucous-winged Gull on an Alaskan salmon stream. Wilson Bull 78:175-190

Nettleship DN (1972) Breeding success of the Common Puffin *Fratercula arctica* L. on different habitats at Great Island, Newfoundland. Ecol Monogr 42:239-268

Rockwell ED (1982) Intraspecific food-robbing in Glaucous-winged Gulls. Wilson Bull 94:282-288

Rohwer S (1975) The social significance of avian winter plumage variability. Evolution 29:593-610

Rohwer S, Rohwer FC (1978) Status signalling in Harris Sparrows: experimental deceptions achieved. Anim Behav 26:1012-1022

Rothschild M, Clay T (1952) Fleas, flukes and cuckoos. Collins, London

Speakman JR (1987) Apparent absorption efficiencies for Redshank (*Tringa totanus* L.) and Oystercatcher (*Haematopus ostralegus* L.): implications for the predictions of optimal foraging models. Amer Natur 130:677-691

Tarifeno-Silva E (1980) Studies on the biology of the surf clam *Mesodesma donacium* (Lamarck, 1818) (Bivalvia: Mesodesmatidae) from Chilean sandy beaches. Unpubl. PhD thesis, University of California, Los Angeles

Thompson DBA (1983) Prey assessment by plovers (Charadriidae): net rate of energy intake and vulnerability to kleptoparasites. Anim Behav 31:1226-1236

Thompson DBA (1986) The economics of kleptoparasitism: optimal foraging, host and prey selection by gulls. Anim Behav 34:1189-1205

Ulfstrand S (1979) Age and plumage associated differences of behaviour among Blackheaded Gulls *Larus ridibundus*: foraging success, conflict victoriousness and reaction to disturbance. Oikos 33:1601-66

Verbeek NAM (1977a) Interactions between Herring and Lesser Black-backed Gulls feeding on refuse. Auk 94:726-735

Verbeek NAM (1977b) Comparative feeding behaviour of immature and adult Herring Gulls. Wilson Bull 89:415-421

Verbeek NAM (1977c) Age differences in the digging frequency of Herring Gulls on a dump. Condor 79:123-125

Verbeek NAM (1977d). Comparative feeding ecology of Herring Gulls *Larus argentatus* and Lesser Black-backed Gulls *Larus fuscus*. Ardea 65:25-42

ORAL DISCUSSION – P.A.R. Hockey

Q. (Ball) Are the clams found across all your sites ?
A. The clams are present in all the sandy beaches that we studied. The handling techniques are basically the same for all of the clams.

Q. (DeMott) You can test the importance of kleptoparasitism in your system by looking at the correlation between the height of flight (when dropping the clams) and kleptoparasitism.
A. There's a trade–off. Should a bird drop a clam several times from a low height or once from a greater height ? The higher the bird flies, the more likely is another bird to get the dropped clam. There is also the question of where to drop the prey, and young gulls are less efficient than adults in choosing areas of hard sand.

Q. (Hobbs) What is the incidence of kleptoparasitism compared with normal foraging ?
A. Young birds may lose 30 – 35% of their catch, but they obtain more than they lose by kleptoparasitism. An adult will satisfy its daily energy demand in about half the active foraging time that a juvenile will.

Q. (Vadas) Did birds come in contact and fight over a piece of food ?
A. Fighting occurs only rarely, when the parasitic bird misjudges its time to get to the dropped prey and meets the bird who dropped the prey on the beach.

Q. (Ansell) Do the birds get the clams out of the water ?
A. Gulls can obtain clams in the water column by plunge diving, or paddling when the water is shallow.

Q. (Ansell) Do the gulls take more than one species of bivalve and if so, do different bivalve species require different numbers of drops to be broken ?
A. 'Heavy' species such as *Mulinia* require fewer drops to break than *Mesodesma* or *Donax*, but *Mulinia* was preyed on at only one study site.

Foraging in the Black–Headed Gull: Compensatory site selection by immatures

L.R. Hesp,
University of Nottingham

Introduction

Studies of the foraging behaviour of gulls have indicated that there may be considerable differences in the feeding efficiencies of young and adults (Verbeek 1977a,b; Ingolfsson & Estrella 1978; Searcy 1978; Burger & Gochfield 1981,1983; MacLean 1986). It appears that young gulls are less adept at a variety of skills needed for successful foraging and that in some species these may take 3 – 4 years to acquire (Greig *et al* 1983; MacLean 1986). It has been suggested that foraging inefficiency in young and sub–adult gulls is one factor which has contributed to their delayed maturity, with most species not breeding until they are three years or older (Lack 1968; Burger 1987).

Gulls are an extremely adaptable group of birds, foraging in a wide range of habitats and employing a variety of feeding methods (Hunt & Hunt 1973; Mudge & Ferns 1982). Each feeding method is likely to require different foraging skills and give different returns in terms of capture rate (Burger & Gochfield 1983).

Immature gulls appear to compensate for their inefficient foraging behaviour in three main ways. Firstly, they may forage for longer periods of time than adults (van der Weghe 1971; Burger 1981; Porter & Sealy 1982; Chapman & Parker 1985). Secondly, they may make more attempts at prey capture than adults per unit time (Burger & Gochfield 1979; Hesp & Barnard 1989). Thirdly, they may adopt feeding methods which require little skill but give high capture rates (Verbeek 1977a; Chapman & Parker 1985), resulting in them foraging in different areas to adults (Burger & Gochfield 1983).

Several studies have found that adult and immature gulls are not equally distributed over the foraging area. Most of these have concluded that immature birds are being excluded from the best foraging areas by the adults and are thus congregating in low quality areas (Moyle 1966; van der Weghe 1971; Monaghan *et al* 1976; Hudson & furness 1989). Only one study has so far provided evidence for immature gulls

NATO ASI Series, Vol. G 20
Behavioural Mechanisms of Food Selection
Edited by R. N. Hughes

compensating for inefficient foraging techniques by congregating in areas where capture rate is high - Burger & Gochfeld (1983) found that the capture rate differed between feeding methods for foraging laughing gulls (Larus atricilla) and that there were proportionally more immatures in those situations where their capture rate was high. Capture rate also appeared to be an indicator of the degree of difficulty of the foraging task, as when capture rate was high, then the capture rate of immature gulls approached that of adults. Dunn (1972) also found that when food was abundant and capture rate high, the differences in capture rate between adult and immature sandwich terns was much reduced. It appears, then, when the foraging is easy, the capture rate of immatures may approach that of adults.

The black-headed gull (Larus ridibundus) is probably the most widespread gull species in Britain (Cramp & Simmons, 1983; Bowes et al, 1984). In winter it can be found foraging from coastal mudflats to inland arable fields, employing a wide variety of feeding techniques (Crook, 1953; Källander, 1977). It can be reliably aged as adult or 1st-year (immature) on the basis of plumage differences (Grant, 1982). These two age-classes have been shown to differ with respect to foraging efficiency (Ulfstrand, 1979; Burger, 1987; Hesp & Barnard, 1989). Adults and immatures also occur in different ratios in different foraging situations (Spencer & Welch, 1956; van de Weghe, 1971; Hesp & Barnard, 1989). Black-headed gulls thus provide a good opportunity to test a number of predictions based on age-related foraging behaviour and the use of different foraging methods.

When comparing black-headed gull foraging behaviour across a range of feeding techniques, we might expect that : 1) adults would have overall higher capture rates than immatures, 2) capture rates would differ for both age classes between different feeding methods, 3) there would be proportionally more immature gulls in those foraging situations where the feeding method employed resulted in high rates of capture for the immatures and 4) the differences between adult and immature foraging rates would be reduced when food was abundant and the foraging task was easy.

Immature birds of several species are suspected to move between foraging sites more often than adults (Davis, 1975; Draulens & van Vessem, 1975). One way for immature black-headed gulls to be over-represented at those foraging sites where capture rates are high would be to arrive at these sites at a faster rate than adults. This could be achieved by moving out of low quality flocks more often than adults. We might also expect then that 5) immature gulls move between foraging flocks more often than adults.

This study sets out to test these predictions.

METHODS

Foraging behaviour.

The foraging behaviour of black-headed gulls was studied at an inland site during the winters of 1986/87, 1987/88 and 1988/89 on an area of mixed farmland to the south-east of Nottingham. Foraging gulls were present in the area every day from October to February in all three years. Flocks of gulls were located opportunistically during regular circuits of the study area in a vehicle. Once located, an initial scan was made to determine the number and age-composition of the black-headed gulls present. Seven different feeding methods were identified according to behaviour and habitat and classified as:

Kleptoparasitism - theft of food from lapwings (*Vanellus vanellus*) on agricultural land. The gulls attacked and stole earthworms from lapwings (see Källander, 1977; Barnard & Thompson, 1985; Hesp & Barnard, 1989). This was further subdivided into kleptoparasitism on ploughed land and on pasture, as there were significant differences in gull capture rates between the two (Hesp, unpublished data).

Pasture - gulls foraged for earthworms and other surface invertebrates by walking over agricultural grassland.

Playing fields - gulls foraged for earthworms and other surface invertebrates by walking over playing fields.

Plough - gulls foraged for earthworms and other surface invertebrates by walking over land that had been ploughed at any time within the last year. This included both freshly ploughed fields and fields that had crops such as winter cereals growing in them.

Following tractor - gulls followed a tractor and foraged for earthworms as it ploughed up agricultural fields. This mostly occurred in the late autumn.

Flight feeding - gulls flew over ploughed fields at a height of 2-5m, dropping occasionally to the ground to capture prey.

In each foraging situation, focal gulls were chosen randomly and aged as adult or first winter (immature), on the basis of a pale brown carpal bar and black tail band on the immatures (Grant, 1982; Hesp & Barnard, 1989).

Capture rates.

Gulls were taking a variety of soil invertebrates, although the major prey item was earthworms (Hesp, 1989). Focal birds were observed for varying periods (mean period of observation $\pm$1 s.e. =130$\pm$2.2s, range =90-310s) through 10x50 binoculars or 22-47x telescope, during which time a record was made of the number of successful captures of prey items, as defined by obvious swallowing movements after a peck at the ground. A record was not used in the analysis if the observation time on the focal gull was less than 90s. A longer observation period was used for kleptoparasitic foraging because capture rates by gulls were very low. The mean observation time ($\pm$1 s.e.) for this feeding method was 518$\pm$9.4s, range 303-961s, with records rejected if the observation time was less then 300s.

The measure of foraging success used in this study was capture rate (food items taken per minute). A mean capture rate for each of the seven foraging options was calculated by combining the foraging data collected across years. All means are given ±1 standard error.

Surveys of the feeding distribution of gulls.

During the winters of 1987/88 & 1988/89, eight surveys of the study area were carried out per month to determine the number and age-composition of black-headed gulls employing each of the seven foraging options. This involved driving a set 36km route around the area at the same time of day and recording the activity, numbers and ages of all the gulls present.

Movements of gulls between flocks.

During these surveys, black-headed gulls were noted in flight in small groups. These groups were distinct from the large flight lines observed to and from the roosting site at dusk and dawn. The mean size of these groups flying over was 6.4±1.2 gulls, although 61% (47/77) of observations consisted of just 1 or 2 birds. Small numbers of black-headed gulls are often observed leaving or arriving at foraging flocks (Barnard & Thompson, 1985; Hesp, 1989) and so these birds observed in flight were considered to be gulls moving between foraging flocks. The ages of these birds were therefore recorded where possible, for the purpose of comparing the rates of movement of adults and immature gulls within the area.

Resting flocks of black-headed gulls also occasionally built up in the study area during the day, with gulls arriving or departing from the flock in small groups (mean group size arriving =1.7±0.24 gulls; mean group size departing = 2.0±0.39 gulls) at regular intervals (mean number of groups arriving or departing per minute = 1.41±0.12). These flocks were watched intensively on four occasions (mean size of the resting flock= 52.6±3.0 gulls), with the number and age of departing and arriving gulls recorded. This gave a second indicator of the rate of adult and immature gull movement between flocks.

RESULTS

Age-related success and feeding method.

Figure 1 shows the mean capture rates for adult and immature gulls over the seven feeding methods and demonstrates that immatures had lower rates of capture than adults in all cases but one (playing field). Capture rates also varied between different feeding methods.

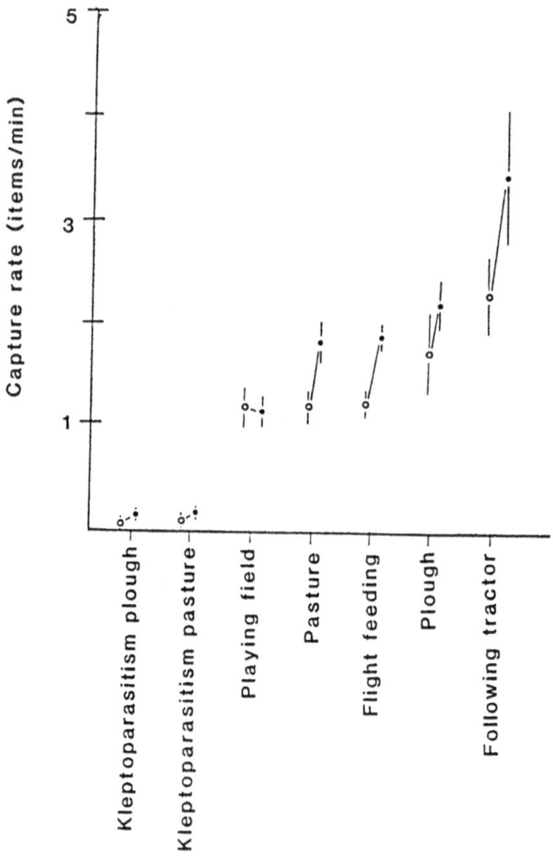

Figure 1. The mean (± se) capture rate in items/min for black-headed gulls using seven different foraging options. Solid points are adults, open points are immatures. Number of focal adult gulls = 247, number of focal immature gulls = 151.

A two-way analysis of variance on capture rate revealed a significant effect of both the age of the gull (F-ratio =9.98, df=1,396, P<0.01) and the feeding method employed (F-ratio =46.92, df=6,396, P<0.001). These results are thus consistent with predictions 1) and 2), the capture rate of black-headed gulls in this study was dependent on both the age of the gull and the feeding method used.

Distribution of immature gulls and capture rate.

It was predicted that the distribution of immature gulls across feeding methods would be related to the capture rate of prey by immatures. Figure 2 shows the mean percentage of immature gulls in foraging flocks for the seven feeding methods in relation to the mean capture rate by immatures in these flocks (items/min).

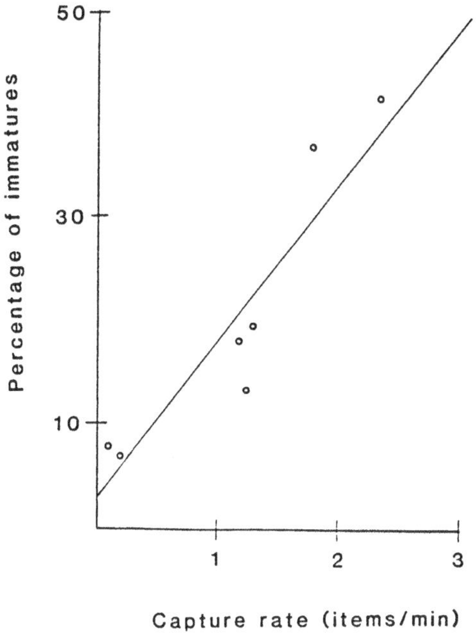

Figure 2. Regression analysis of the mean capture rate of prey for immature black-headed gulls (items/min) against the mean percentage of immature gulls in the foraging flock. Total number of focal immature capture rates = 151, total number of foraging flocks where the percentage of immature gulls was known = 406.

As expected, the mean immature capture rate correlated significantly with the mean percentage of immatures engaged in each method (F-ratio= 28.9, df=6, P<0.01). There were proportionally more immature gulls in those feeding situations where the rate of capture of prey was high.

Capture rate and efficiency index.

It was expected that differences in capture rates between adult and immature gulls would be lowest when capture rates were high, assuming that capture rate was related to the difficulty of the foraging task. An age efficiency index was therefore computed for each foraging method (see Burger & Gochfeld, 1983). This was simply adult capture rate divided by immature capture rate. An index of 2 meant that adult gulls had a capture rate twice that of immatures. It was expected that this index would have a negative correlation with capture rate; as capture rate increased then the difference in capture rate between adult and immature gulls should have decreased. In actual fact, there was no significant correlation between the efficiency index and immature capture rate (F-ratio= 1.45, df=6, NS). Prediction 4) therefore did not hold true, differences between adult and immature capture rate were not greatest when capture rate was low.

Age-related movement between flocks.

It was suggested that immature gulls may have been concentrated in areas offering high capture rates by moving out of low quality foraging flocks more often than adults. Of the 481 black-headed gulls recorded as moving between flocks, only 29 could be specifically aged. Of these, 10 (34.5%) were immature and 19 (65.5%) were adult. Of the 4838 gulls engaged in all activities in the study area that could be specifically aged, 17.8% were immatures. There were therefore significantly more immature gulls recorded flying between flocks than expected, based on the total number of immatures present in the area (X^2=5.40. df=1. P<0.05)

Table 1 shows the mean percentage of immatures in each of the resting flocks studied, the number of immatures observed either moving into or out of the flock during the observation period and the number of immatures we would have expected to see moving based on their percentage in the flock.

RESTING FLOCK	MEAN % IMMATURES	No. OF IMMS MOVING IN OR OUT	EXPECTED No.
1	11.1±0.24	11	5.4
2	7.8±1.05	7	1.7
3	10.8±0.74	10	4.4
4	4.6±0.32	7	2.0

TABLE 1 Data for 4 black-headed gull resting flocks showing mean (± se) percentage of immatures in the flock, number of immatures observed moving in and out of the flock during the period of observation, and number expected moving based on the percentage occurrence of immatures in the flock.

In all cases, the observed number of immatures was greater than the expected number. Due to small expected values, the data for the flocks could not be analysed separately. When the data are combined we find that immatures made significantly more movements into and out of resting flocks than expected (X^2=33.6, df=1. P<0.001). These two measures of immature movement rate between flocks therefore seem to be in agreement with prediction 5). Immatures moved between flocks of gulls more often than adults.

DISCUSSION

Differences in the percentage of immature gulls in different feeding flocks have been attributed previously to intraspecific competition keeping immatures away from best feeding areas (Moyle, 1966; van de Weghe 1971; Monaghan et al, 1986; Hudson & Furness; 1989). In this study however, the distribution of immature gulls appeared to be dependent on their capture rates within the flocks, with proportionally more immatures in those flocks where capture rate was highest. As capture rate varied between different feeding methods, this led to differences in the percentages of immature gulls employing the different options. The situations where immature gulls appear to be excluded from the best foraging areas occur where access to the foraging site is limited, thus giving rise to severe competition, for example along river banks (Moyle, 1966), on fresh areas of rubbish tips (Monaghan et al, 1986) and behind fishing boats (Hudson & Furness, 1989). Aggressive interactions between foraging gulls were rarely recorded in this study (Hesp, 1989). Foraging over open fields meant that none of the sites had limited access, competition for foraging places was low and immature gulls were not prone to competitive exclusion.

Age-related differences in foraging efficiency have been discovered in most species of gull studied so far (Ulfstrand, 1979; Greig et al, 1983; Chapman & Parker, 1985; MacLean, 1986; Monaghan et al, 1986; Burger, 1987). Immature black-headed gulls had lower capture rates than adults across a variety of feeding methods looked at in this study. It would appear that immature gulls were compensating for their relatively low capture rate by searching for and staying with those flocks that were feeding in areas offering high foraging returns. There was evidence from two measures of the rates of movements between flocks that immatures spent more time searching for foraging flocks than adults. Observations on radio-tagged grey herons (Ardea cinerea) (Draulens & van Vessem, 1985) and colour ringed herring gulls (Larus argentatus) (Davis, 1975) both suggest that foraging immatures may move from site to site more often than adults. Any gains made by immature black-headed gulls from foraging in these flocks with high capture rates may thus be partly offset by the extra costs involved in searching out the flocks.

If capture rate was a reflection of the difficulty of the foraging task, then we might have expected immatures to be relatively less efficient than adults when the foraging task was difficult and capture rates were low. It has been suggested for the laughing gull (Burger & Gochfeld, 1983) and the sandwich tern (Dunn, 1972) that the foraging efficiency of immatures most closely approaches that of adults when food is abundant, so that the foraging task is easy and requires less skill. That the same relationship was not found in the black-headed gull in this study suggests that the capture rate here may not have been a good indicator of the degree of difficulty of the foraging task.

SUMMARY

Immature black-headed gulls had a lower capture rate than adults across a range of feeding methods. The capture rate of gulls also varied between feeding methods. The percentage of immature gulls in a foraging flock was correlated with the mean capture rate by immatures in the flock. Immature gulls thus appeared to compensate for their relatively low capture rates by concentrating on feeding methods which resulted in high capture rates. The greater proportion of immature gulls in those areas of high capture rate was apparently due to immatures moving out of low quality foraging flocks at a faster rate than adults. Any gains made by immatures from foraging in flocks with high capture rates may thus be partly offset by the extra costs involved in searching out the flocks. The prediction that the foraging efficiency of immatures would most closely approach that of adults when capture rate was high did not hold true. It appeared that in this study at least, capture rate may not have been an accurate measure of the degree of difficulty of the foraging task.

ACKNOWLEDGEMENTS

I wish to thank Chris Barnard and Dave Grewcock for constructive comments on this manuscript. Sara Houghton and my parents provided moral and financial support, as ever. The work was supported by a University of Nottingham studentship.

REFERENCES

Barnard CJ, Thompson, DBA (1985) Gulls and Plovers. The ecology and behaviour of mixed-species feeding groups. Croom-Helm, Beckenham.

Bowes A, Lack PC, Fletcher, MR (1984) Wintering gulls in Britain, January 1983. Bird Study 31:161-170.

Burger J (1981) Feeding competition between laughing gulls and herring gulls at a sanitory landfill. Condor 83:328-335.

Burger J (1987) Foraging efficiency in gulls : A congeneric comparison of age differences in efficiency and age of maturity. In: Hand J, Southern W, Vermeer K (eds) Ecology and behaviour of gulls. Studies in avian biology no. 10. Cooper Ornithological Society, Los Angeles.

Burger J, Gochfeld M (1979) Age differences in ring billed gull kleptoparasitism on starlings. Auk 96:806-808.

Burger J, Gochfeld M (1981) Age related differences in piracy behaviour of four species of gulls. 77:242-267.

Burger J, Gochfeld M (1983) Feeding behaviour in laughing gulls : compensatory site selection by young. Condor 85:467-473.

Chapman B, Parker JW (1985) Foraging areas, techniques and schedules of wintering gulls on southeastern Lake Erie. Colonial Waterbirds 8:135-141.

Cramp S, Simmons KEL (eds) (1983) The birds of the Western Palearctic Vol 3. Oxford University Press, Oxford.

Crook JH (1953) An observational study of the gulls of Southampton water. British Birds 56:385-397.

Davis JWF (1975) Specialization in feeding location by herring gulls. J Anim Ecol 44:795-804.

Draulens D, van Vessem J (1985) Age-related differences in the use of time and space by radio-tagged grey herons (Ardea cinerea) in winter. J Anim Ecol 54:771-780.

Dunn K (1972) Effect of age on the fishing ability of sandwich terns. Ibis 114:360-366.

Grant PJ (1982) Gulls : A guide to identification. T & AD Poyser, Calton.

Greig S, Coulson JC, Monaghan P (1983) Age-related differences in foraging success in the herring gull (Larus argentatus). Anim Beh 31:1237-1243.

Hesp L (1989) Age-related foraging behaviour in black-headed gulls (Larus ridibundus). PhD thesis, University of Nottingham.

Hesp L, Barnard CJ (1989) Gulls and Plovers: age-related differences in kleptoparasitism among black-headed gulls (Larus ridibundus). Behav Ecol Sociobiol 24:297-304.

Hudson AV, Furness RW (1989) The behaviour of seabirds foraging at fishing boats around Shetland. Ibis 131:225-237.

Hunt GL, Hunt MW (1973) Habitat partitioning by foraging gulls in Maine and NW Europe. Auk 90:827-839.

Ingolfsson A, Estrella BT (1978) The development of shell-cracking behaviour in herring gulls. Auk 95:557-579.

Källander H (1977) Piracy by black-headed gulls on lapwings. Bird Study 16:45-52.

Lack D (1967) Interrelationships in breeding adaption as shown by marine birds. Proc. 14th Int Ornth Congr 32.

MacLean AAE (1986) Age-specific foraging ability and the evolution of deferred breeding in three species of gulls. Wils Bull 98:267-279.

Monaghan P, Metcalfe NB, Hansell MH (1986) The influence of food availibilty and competition on the use of a feeding area by herring gulls (Larus argentatus). Bird Study 33:87-90.

Moyle P (1966) Feeding behaviour of the glaucous-winged gull on an Alaskan salmon stream. Wils. Bull. 78:175-190.

Mudge GP, Ferns PN (1982) The feeding ecology of 5 species of gulls in the inner Bristol Channel. J Zool Lond 197:497-510.

Porter JM, Sealy SG (1982) Dynamics of seabird multispecies feeding flocks : age-related feeding behaviour. Beh 81:91-109.

Searcy WA (1978) Foraging success in three age classes of glaucous-winged gulls. Auk 95:586-588.

Spencer KG, Welch A (1956) Immature-adult proportions and behaviour in black-headed gull flocks. The Naturalist, July-Sept, p114.

Ulfstrand S (1979) Age and plumage associated differences of behaviour among black-headed gulls (Larus ridibundus) : foraging success, conflict victoriousness and reaction to disturbance. Oikos 33:160-166.

Van de Weghe JP (1971) Relations entre adultes et juveniles chez la mouette rieuse (Larus ridibundus) et le goeland cendre (Larus canus) en hivernage. Gerfaut 61:111-124.

Verbeek NAM (1977a) Comparative feeding behaviour of immature and adult herring gulls. Wils Bull 89:415-421.

Verbeek NAM (1977b) Age differences in the digging frequency of herring gulls on a dump. Condor 79:123-125.

ORAL DISCUSSION – L. Hesp

Q. (Sih): Why do adults and juveniles have such different foraging strategies ?
A. I think it is a trade–off. The places where foraging rate is high occur very infrequently. Juveniles go out and search, whereas adults stay where they are.

Q. (Sih): But if it is a good strategy for juveniles to incur high costs to get high benefits, why don't adults do it too ? Are immatures more risk prone ?
A. I think so.

Q. (Sih): Do adults have some other reason for not moving about, like staying around for aggressive interactions to achieve a dominant status, later important in the breeding season ?
A. There are no aggressive interactions to speak of at his time of the year, so far away from the breeding colonies. I think immatures are more prepared to take bigger risks because they have lower foraging efficiencies.

Q. (Ball): Is there any difference in body mass between adults and juveniles ?
A. From the literature, there seems to be little morphological difference at all.

Q. (Ball): So presumably the energetic cost of flight is equal and you would expect the energy–maximizing strategy to be the same for both adults and juveniles ? Hesp: Yes

Q. (Milinski): Did you say that competitive weights stayed the same across different patch qualities ?
A. No – the birds aren't competing with one another. In a huge field there may be relatively few birds.

Q. (Milinski): There was no interference ?
A. No. There are very low rates of aggression in these flocks at this time. Milinski: Interference does not necessarily mean aggression, it could be exploitative competition.

Q. (Hockey): Your data suggest that juveniles are moving about because they perceive themselves to be under food stress. In your poorest habitats, both adults and juveniles exhibit much kleptoparasitism, suggesting that adults in these areas should also be food stressed. Is there any evidence of a greater turnover of birds that are feeding by kleptoparasitism than by other techniques ?
A. I haven't looked at rates–of–leaving associated with different foraging techniques, so I can't compare leaving–rates across flocks. I think that flocks where kleptoparasitism occurs are unstable; perhaps birds only stay an hour or so before moving on to a different foraging method.

INFORMATION OVERLOAD AND FOOD SELECTION

Manfred Milinski
Abteilung Verhaltensökologie
Zoologisches Institut
Universität Bern
Wohlenstrasse 50 a
3038 Hinterkappelen
Switzerland

It is said that Julius Caesar was able to dictate two different letters to two secretaries almost simultaneously, switching between letters sentence after sentence. The fact that this anecdote has been passed on to our time suggests that Caesar's ability is appreciated as an exceptional one. Given that the anecdote is true, modern experimental psychology would tell us that Julius Caesar's brain must have had a supernormal channel capacity. Normally the nervous system is limited in its ability to process sensory data (= channel capacity) which makes it increasingly difficult to fulfil two different tasks at the same time, as each single task gets harder (Broadbent 1965, Blough 1979).

Finding food in the most economic way is only one of the many sub-goals for maximizing fitness. Another one is avoiding to be preyed upon. Unfortunately, an animal is mostly not able to fulfil the tasks sequentially, for example foraging when hungry and avoiding predators when satiated (Milinski 1986). Since a predator can strike any time, an animal has to scan the environment almost continuously for signs of a pending predatory attack. When the forager has to pay attention to potential predators, some fraction of its channel capacity is utilized. If the foraging task is so difficult that it requires more than the remaining channel capacity, the animal has to decide which task to fulfil less efficiently. We would expect that the animal gives priority to the task which has higher fitness consequences, for example, being very vigilant but foraging less efficiently when nearly satiated and/or the risk of predation is very high.

NATO ASI Series, Vol. G 20
Behavioural Mechanisms of Food Selection
Edited by R. N. Hughes

Fish foraging on swarming prey.

Aggregations of tiny prey animals in the zooplankton are a valuable food source, especially for small fish for which tiny items are relatively large. Small fish can feed on prey only one at a time. Such particulate feeders (Janssen 1976) can potentially increase their feeding rate when they choose a swarm of high density because the distances between prey items are shorter there. Many predators, however, suffer from a so-called confusion effect (Miller 1922) when they attack dense swarms of prey. Therefore, there should be an optimal swarm density which allows for a maximum feeding rate.

When waterfleas (*Daphnia magna*) were offered to a stickleback (*Gasterosteus aculeatus*), significantly more strays were caught near the end of the fish's feeding sequence (Milinski 1977a). This suggests that there is no optimal prey density, but the one preferred by the fish is dependent on its hunger. To test this hypothesis the sticklebacks were given singly the choice between two test tubes, one containing a swarm of 40 waterfleas, the other one only two waterfleas.

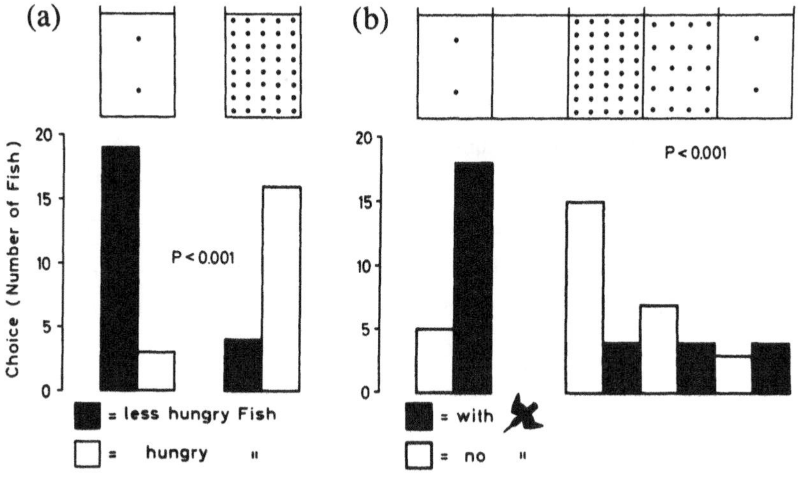

Fig. 1. (a) Choice of either a low or a high density of prey by less hungry sticklebacks and by hungry ones. P from Chi-square test. (modified after Milinski 1977a). (b) Choice of different swarm densities by sticklebacks which were either undisturbed or frightened by a model kingfisher. P from Freeman-Halton test. (after Milinski & Heller 1978).

One group of fish was hungry whereas another one had become nearly satiated before the trial by feeding upon *Tubifex* worms (Milinski 1977a). Most of the hungry fish attacked the swarm, whereas most less-hungry fish chose the singletons. (Fig. 1a). Further experiments confirmed that the hungry fish did not simply overlook the singletons but chose to attack the swarm (Milinski 1977b). Does this mean that hunting in a high density is most profitable for hungry fish, and attacking a low-density swarm is more rewarding for more satiated fish? This was tested in further experiments in which hungry sticklebacks and more-satiated ones were allowed to feed upon either a high or a low prey density kept constant during a trial (Heller & Milinski 1979). The hungry fish fed significantly faster in the high prey density whereas the less hungry ones achieved a higher feeding rate in the low prey density. Therefore, it pays a hungry fish to choose a high-density swarm and a less-hungry one to attack a lower density (see also Morgan & Ritz 1984, Godin & Smith 1988). If hunting in place A is more profitable than in B, there is no reason for a less-hungry predator to hunt in B. Thus, apart from the risk of starvation, there must be other costs at stake which increase with the density of the prey swarm attacked.

If sticklebacks suffer from the confusion effect of swarming prey, hungry sticklebacks must be able and willing to overcome the confusion imposed by the high density swarm, otherwise they could not achieve their highest feeding rate there. This implies that the less-hungry fish were not *willing* to pay the costs for overcoming the confusion. These costs may consist in utilizing a large proportion of the channel capacity when the fish has to concentrate on the difficult task of tracking one out of many similarly looking targets of a dense swarm. When feeding in a high prey density, the hungry fish probably cannot pay sufficient attention to a suddenly approaching predator of its own. Therefore, it must have weighed the fitness gain of feeding higher than that of avoiding predation. When the sticklebacks' expected risk of predation was increased by moving a model of a European kingfisher (*Alcedo atthis*) over the tank before the stickleback was allowed to choose between different prey densities (Milinski & Heller 1978), frightened fish attacked the lowest prey density whereas unfrightened ones preferred the highest density (Fig. 1b). This result implies that a hungry stickleback suffers from information overload when it attacks a high density swarm at a high rate as compared to a similarly hungry fish attacking a low density swarm.

Do the costs of the information overload consist in being less likely to detect an approaching predator? If we find that a stickleback feeding in a high prey density does not react to a predator, we have the problem to distinguish experimentally between *not willing* and *not able to react*. We have to decide whether the fish is not *able* to recognize the predator. In a further experiment hungry sticklebacks were allowed either to prey upon a high prey density of waterfleas or a low one. After they had consumed the fourth waterflea, a conspicuous model of a kingfisher, or a cryptic model or no model, was flown over the tank (Milinski 1984). If the fish detected the predator, they should hesitate before attacking the fifth waterflea. If they do not react differently to the cryptic model and to the conspicuous one when hunting in the *low* prey density, then they rank both predators as similar threat. If they react to the conspicuous model but not to the cryptic one in the *high* prey density, they must have overlooked the cryptic one.

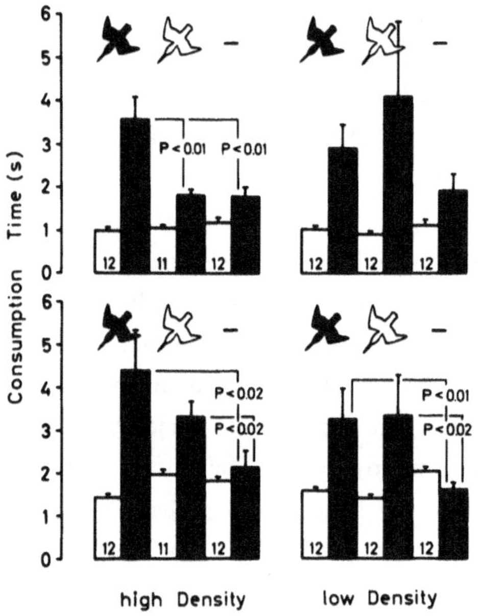

Fig. 2. Mean (+SE) consumption time for the fourth waterflea (white columns) and for the fifth one (black columns) of fast fish (upper part) and of slow fish (lower part) feeding either upon a high or a low prey density. After the fourth waterflea had been caught, sticklebacks were shown either a conspicuous model predator or a cryptic one or no predator. Number of fish (trials) given in white columns. P from Mann-Whitney U-test. (modified after Milinski 1984).

Figure 2 shows that both quick and slow fish (as determined by the time they needed for catching the fourth waterflea) reacted similarly to the cryptic model and the conspicuous one in the low prey density. In the high prey density, however, the quick fish reacted only to the conspicuous model. Therefore they must have overlooked the cryptic model. Some of the slower fish must have seen the cryptic model and some must have overlooked it in the high prey density. Hence, hungry sticklebacks hunting in a high prey density at a high rate suffer from information overload and cannot pay sufficient attention to other stimuli, e.g. an approaching predator.

A similar result was obtained by Godin & Smith (1988) in a very similar experiment using guppies (*Poecilia reticulata*) instead of sticklebacks and a live fish predator (*Hemichromis bimaculatus*) instead of a model kingfisher.

Humans foraging on swarming prey.

To decide whether an animal is unable or unwilling to fulfil a task requires a complicated experimental procedure (see above). This is different when the experiment is done with first semester students who can be assumed to do their best. They can be told exactly what to do and asked after the trial whether they have detected what was shown to them during the trial. Since Gillett, Hogarth & Noble (1979) found that human subjects suffered from confusion in a very similar way as a lizard predator when they hunted for swarming locusts, it seemed promising to investigate whether humans fail to pay attention to sudden events when attacking a swarm of prey. Such a study was done with the help of 25 female and 25 male students at the Ruhr-University Bochum during the summer semester 1983 (Milinski, Kosakowski & Krause unpublished).

The prey were white paper dots (2 mm diameter) made by the tape punch of a computer. They were glued on black paper (35 cm x 26.5 cm) covering an area of 28 cm x 18 cm. Swarms of three different densities were used, a low density swarm (0.5 dots/cm^2), a medium density swarm (2.1 dots/cm^2) and a high density swarm (8 dots/cm^2). As a control, one sheet of paper contained only one single dot in its centre. Each sheet had a bigger white dot (5 mm diameter) placed centrally 2.5 cm above the edge of the swarm. The student who sat in front of the swarm had to alternate with the blunt tip of a crochet-needle between the big dot and a central one (any dot) of the swarm (Fig. 3). A trial consisted of alternating twenty times between the big and a central dot as quickly as possible but as slowly as necessary to hit each dot precisely. In order to distract the student's visual attention there were four white lamps attached to the periphery of the swarm (Fig. 3). Any of the lamps could be switched on for 1/10 s when the student left the big dot. There were 7, 8 or 9 light presentations during the student's twenty attacks. The student had to count the light presentations irrespective of their position and to tell the number after the trial. She or he was told before a trial whether lights would be presented. Each student had three training trials of twenty attacks each with the one-dot pattern, during the third training trial lights were presented.

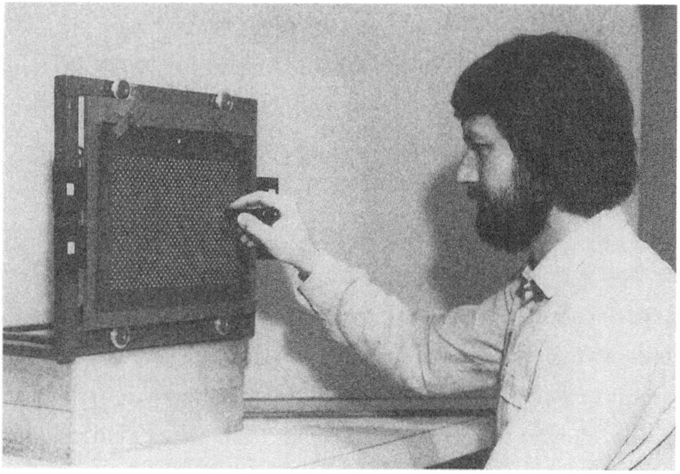

Fig. 3. Human subject hits a dot of the high density swarm. (Milinski, Kosakowski & Krause unpublished).

Without visual distraction, there was no significant difference between the time needed to attack prey items in the one-dot pattern, in the low density swarm and in the medium density swarm (Fig. 4). Only when attacking the high density swarm the students needed significantly more time than with lower prey densities ($p<0.001$, 2-tailed Wilcoxon test). When there was visual distraction by light presentations, the students needed increasingly more time with increasing swarm density. It is not surprising that the additional task cost time. It is, however, important to note that the detection of the same number of light presentations increased the time to attack twenty dots the more strongly the higher the prey density was ($p<0.01$, difference between additional time in low and in medium density, $p<0.01$, difference between additional time in medium and in high density, 2-tailed Wilcoxon tests).

As in the sticklebacks, human subjects suffered from information overload when they attacked high density swarms. Contrary to the sticklebacks, the students were told before a trial whether they would have to pay attention to additional stimuli. To be able to fulfil both tasks precisely the students slowed down their attack rate when the foraging task got harder with increasing prey density. The sticklebacks did not know about the predator. They made more mistakes in detecting the predator when hunting in the high prey density as compared to hunting in the low prey density. Only after they had detected the predator they decreased their feeding rate. Since there were no obvious fitness consequences for the students of doing one task less efficiently than the other, it is not clear how they decided to which one to give priority. Obviously, they gave priority to detecting the lights at the expense of a high "feeding" rate. Only the 17 quickest students made slightly more mistakes in detecting the light presentations than the 16 medium quick students ($p<0.05$, 2-tailed Binomial test).

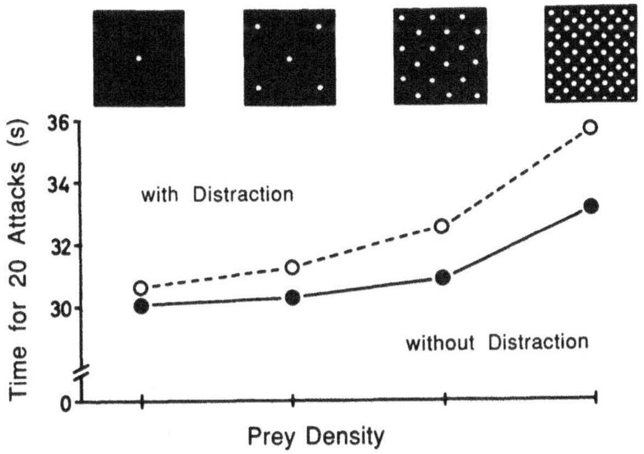

Fig. 4. Mean Time 50 students needed for 20 attacks dependent on swarm density either with or without visual distraction by light presentations. (Milinski, Kosakowski & Krause unpublished).

When attacking swarms of prey, sticklebacks and humans suffer from information overload. When they have to pay attention to additional events, e.g. a potential predator, they either become less efficient foragers or make mistakes in detecting the simulated predator.

The vegetarian stickleback.

In his investigation of the mechanisms of the confusion effect, Ohguchi (1981) presented sticklebacks with isolated events from a swarm of waterfleas. With the help of his "confusion machine" (Fig. 5) he made, e.g., one waterflea move back and forth in a vertical glass tube and a second one in a horizontal glass tube which was attached to the vertical one. The tubes were connected by pipes and therefore had a common water column which was moved back and forth by an excenter pressing and releasing a rubberball at one end of the tube-pipe system at regular intervals. The two "cross-moved" waterfleas met exactly in the centre of the cross.

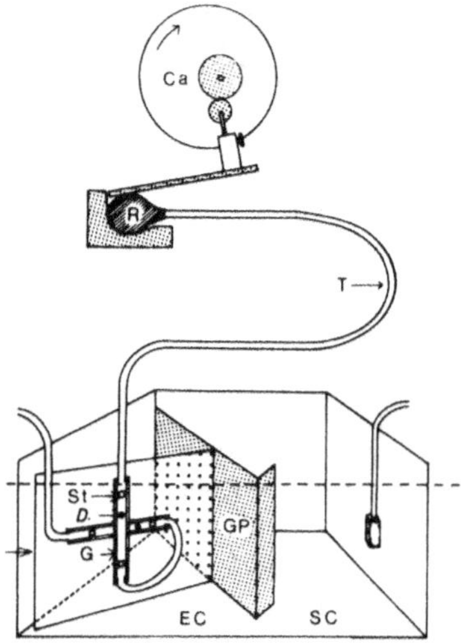

Fig. 5. Ohguchi's "confusion machine". Ca: cam, Co: cover, D: *Daphnia*, EC: experimental compartment, G: glass tube, GP: grey partition, St: stopper of plastic tube, R: rubber ball, SC: start compartment, T: plastic tube. Further explanations in text. (After Ohguchi 1981).

The sticklebacks attacked two cross-moved waterfleas of different colour more often than two waterfleas of similar colour (Ohguchi 1981). This showed that confusion is enhanced by uniformity of the prey's coloration already in swarms of two prey items. In order to test whether uniformity of shape has a similar effect,

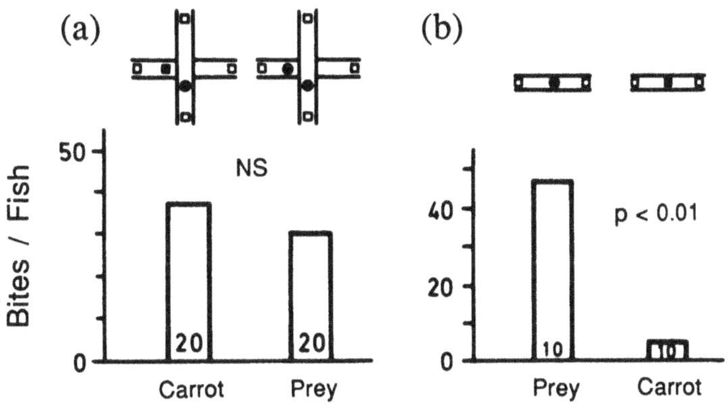

Fig. 6. (a) Mean number of bites of sticklebacks to a vertically moved red *Daphnia* which displays cross-movement with a horizontally moved piece of carrot and to two cross-moved red *Daphnia*. (b) Responses to a horizontally moved red *Daphnia* and to a horizontally moved piece of carrot. Number of fish (trials) in the columns. P from Mann-Whitney U-test. (After Ohguchi 1981).

Ohguchi made a angular piece of red carrot "cross-move" with a red waterflea. There was no significant difference between the number of attacks delivered to the carrot-waterflea system and the two-waterflea system (Fig. 6a). Ohguchi was surprised to find that the piece of carrot received almost the same percentage of attacks as the red waterflea in the same position in the two-waterflea system. He wrote (page 30): "The piece of carrot probably appeared to be an edible prey object". He said (personal communication, my translation): "The sticklebacks are completely stupid because carrot is a vegetable and sticklebacks are not vegetarians". Since a carrot normally does not have the size of a waterflea and does not move like a waterflea, Ohguchi's carrot seemed to be a very good mimic of a waterflea. However, in a further experiment, Ohguchi (1981) presented the fish either with a piece of carrot moved horizontally or with a red waterflea moved in the same way. This time the sticklebacks avoided to attack the carrot almost completely (Fig. 6b). This shows that sticklebacks are not vegetarians and that they cannot distinguish an inedible object from an edible one when both are moved simulaneously in such a way that they produce information overload in the predator. Information overload therefore leads to less efficient diet selection.

The wary blackbird.

When information overload results from two different tasks which have to be fulfilled simultaniously, the animal can choose which task to do less precisely. Stickleback and human predators foraged less efficiently when they knew that a predator or its simulation would probably appear. Lawrence (1985) investigated whether blackbirds (*Turdus merula*) searching for cryptic and conspicuous food alter their vigilance for predators depending on the difficulty of the foraging task.

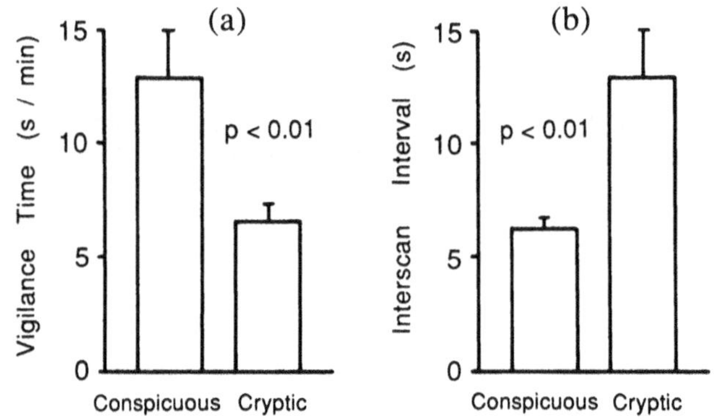

Fig. 7. (a) Mean (+SE) vigilance time of eight blackbirds. Longer periods of vigilance occurred during tests with conspicuous prey compared with tests with cryptic prey. (b) The mean interval between scans was shorter during tests with conspicuous prey compared with cryptic prey. P from t-test. (After Lawrence 1985).

In the field Lawrence offered individually colourringed birds 200 small pastry baits scattered on the surface of metal trays (70 cm x 80 cm) filled with small pebbles. The baits were coloured green. They were placed either on a background matching in colour (green painted), or contrasting in colour (brown), i.e. prey items were either cryptic or conspicuous. Each of eight adult blackbirds received a random sequence of eight tests: four tests with conspicuous prey and four tests with cryptic prey. Since the birds were always exposed to potential predation from sparrowhawks (*Accipiter nisus*) and kestrels (*Falco tinnunculus*), they seemed to be very alert when approaching the food patch. At the feeding site, they fed rapidly for up to 70 s, and then departed abruptly straight for cover.

The birds' vigilance behaviour varied signficantly between the two feeding conditions. When feeding on conspicuous prey, the blackbirds spent almost double the proportion of time keeping their heads up and scanning the surroundings (Fig. 7a) compared with feeding on cryptic food. Moreover, when feeding on cryptic prey, twice as much time elapsed between scans compared with feeding on conspicuous baits (Fig. 7b). Hence, the birds were much less vigilant when searching for cryptic baits than when feeding on conspicuous food.

These results suggest that blackbirds faced with the difficult foraging task of searching for cryptic prey would have been less likely to detect predators than birds feeding on conspicuous food. When more information has to be processed for fulfilling a difficult foraging task, a predator is more likely to be overlooked. Metcalfe (1984) found some indication that purple sandpipers (*Calidris maritima*) preferred easier but less profitable prey when in habitats which did not permit a clear view of their surroundings. Contrary to the sandpipers, Lawrence's blackbirds did not have a choice between prey types and therefore reduced their vigilance behaviour when offered difficult prey.

The fallible salmon.

Metcalfe, Huntingford & Thorpe (1987) examined how an increase in predation risk affects the attention devoted to food by juvenile salmon, *Salmo salar*. These fish are sit-and-wait predators, orienting from a resting position to food items passing in the water current and then darting out to intercept them. Since current velocities are often high, the salmon must discriminate quickly between suitable and unsuitable items.

In the experiment, a salmon was offered three small and three large food pellets in an unpredictable order, one pellet introduced into the current every 10 minutes. The smaller sized pellet was that recommended to produce optimal feeding responses and growth for the size of fish tested (Wankowski & Thorpe 1979), whereas the larger pellets were too big for the fish to swallow. Therefore large pellets were inedible. There were high and low predation risk trials. Predation risk was varied using a fibreglass model of a brown trout, *Salmo trutta*, placed in an adjacent compartment and moved to within 25 cm of the resting salmon, which could see it through a transparent partition. In low risk trials the predator was not visible. Predation risk was increased for each high risk trial by moving the model trout into view for 30 s immediately before the trial began.

Under low risk conditions only a quarter of the attacks attempted by the salmon were on the inedible pellets, i.e. were "mistaken" attacks (Fig. 8). This abil-

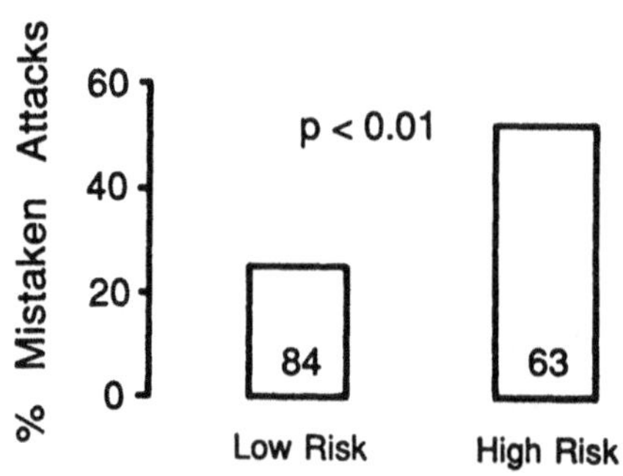

Fig. 8. Attacks on the inedible pellets ("mistaken attacks") as a percentage of all attacks made by the salmon. The proportion of mistaken attacks increased significantly with an increase in predation risk. P from chi-square test. Sample sizes shown in columns. (After Metcalfe, Huntingford & Thorpe 1987).

ity to discriminate between edible and inedible items was absent in the high predation risk trials. Both pellet sizes were attacked with equal frequency, i.e. 50%. Furthermore, there was a significant reduction in the number of orientations to edible pellets with increasing predation risk, but no change in the number of orientations to inedible pellets. This indicates that the smaller pellets were more often overlooked when predation risk was high (Metcalfe, Huntingford & Thorpe 1987). Again, information overload led to less efficient diet selection.

INFORMATION OVERLOAD IN THE PREY

A prey animal which is fleeing from an attacking predator has to monitor the predator's movements continuously to be able to escape safely from it. Since the prey has to process a lot of sensory information while fleeing, it may pay the predator to provide additional but irrelevant information. By sending additional information which is of no use for the prey, the predator may exhaust the prey's channel capacity for processing the relevant information. Wilson, Ryan, James & Wilson (1987) investigated whether complicated colour patterns of some piscivores may enhance prey capture.

Although marine vertebrates are typically dark dorsally and pale ventrally, a pattern that may confer low detectability for both predators and prey (Cott 1985), some dolphins and penguins are highly conspicuous due to black and white patterned flanks (Fig. 9). To see whether this might be involved in prey capture, Wilson et al. (1987) examined coloration and diet of marine odontocedes from published data. They found that only 3 out of 15 uniform or countershaded species were known to eat pelagic school fish whilst 13 out of 14 of the highly conspicuous species did so. This suggests that a complex coloration may be advantageous for a predator attacking swarming prey.

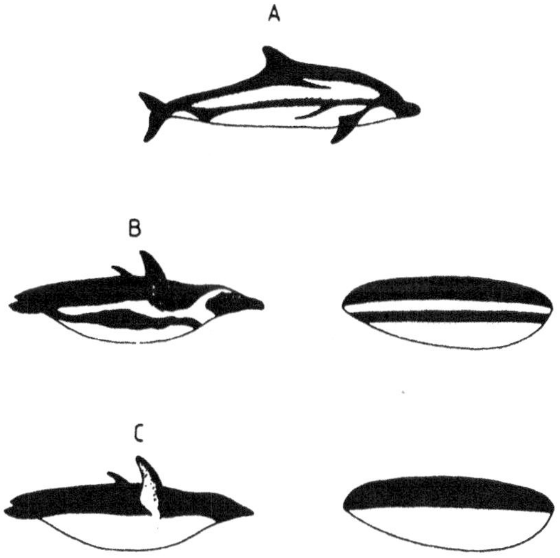

Fig. 9. Plumage patterns of swimming penguins and the experimental designs to represent them. (B) Adult *Spheniscus penguin* (African). (C) Typical non-*Spheniscus* penguin (Adelie *Pygoscelis adelie*). (A) Dolphin coloration pattern shown for comparison. Striped dolphin *Stenella coeruleoalba* (a pelagic fish feeder). (After Wilson, Ryan, James & Wilson 1987).

Unlike most other penguin species, the four *Spheniscus* species feed on pelagic school fish which react to predators by becoming highly polarized. Since polarization is regarded as an effective anti-predator strategy in schooling fish (e. g. Pitcher 1986), Wilson et al. (1987) tested the reaction of pelagic school fish to different pigmentation patterns by using penguin colour models. They exposed a schoal of 31 anchovy (*Engraulis capensis*) to three different penguin models: a transparent perspex control, a black and white "typical" penguin model, and a black and white *Spheniscus* model with a black lateral stripe on the white area (Fig. 9). The models were of the appropriate size of an African penguin. They were suspended from a rotating arm which moved through the tank at 2.3 m/s, similar to the foraging speed of African penguins. From film records it was measured when the fish moved as polarized group and when they were depolarized (defined as complete breakdown of the school where all fish acted independently for longer than 0.1 s).

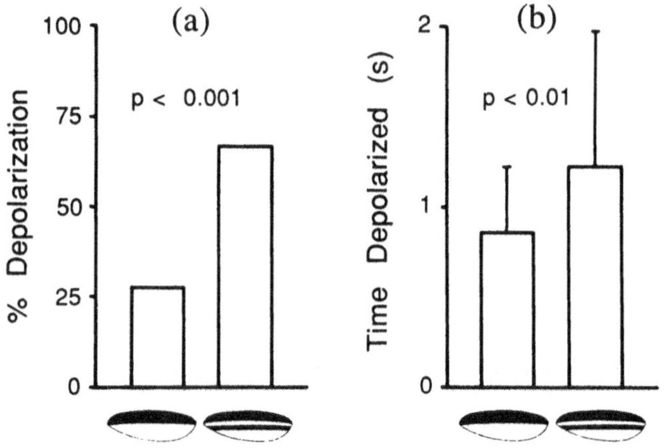

Fig. 10. (a) The frequency of depolarization of a shoal of 31 anchovy as a percentage of 135 passes of a typical penguin model and as a percentage of 95 passes of a *Spheniscus* model. P from chi-square test. (b) Mean (+SD) duration of depolarization. The *Spheniscus* model induced depolarization for longer than the typical penguin model. P from t-test. (After Wilson, Ryan, James & Wilson 1987).

The fish did not react to the transparent model. Although they reacted to the typical penguin model with a coordinated escape response, the school was depolarized 38 times in 135 passes of the model for a mean duration of 0.86 s (Fig. 10). When the *Spheniscus* model was used, the school depolarized 64 times in 95 passes of the model for a mean duration of 1.22 s (Fig. 10). Thus, the *Spheniscus* model induced depolarization significantly more frequently and for longer than the typical penguin model. Since it is well known that depolarized fish are more susceptible to predation than fish in polarized schools (e.g. Neill & Cullen 1974, Major 1978), the complex coloration of *Spheniscus* penguins appears to enhance predatory effeciency. One possible explanation for the observed depolarization of the fish is that the complex visual stimulus of the *Spheniscus* coloration overrides the processing of sensory information on the relative positions of each individual fish's neighbour.

Thus, predators may forage more efficiently when the prey suffers from information overload.

SUMMARY

Because of the limited channel capacity of the nervous system in processing sensory information an animal has increasingly difficulties in fulfilling two different tasks at the same time, as each single task gets harder. Recently, there has been an accumulation of studies showing that animals become less efficient in foraging or in avoiding to be preyed upon when they have to pay attention to potential predators at the same time. Also predators may forage more efficiently when the prey suffers from information overload. In this paper I review several of these investigations.

ACKNOWLEDGEMENTS

I thank Lennart Persson and Les Real for comments on the manuscript.

REFERENCES

Blough DS (1979) Effects of the number and form of stimuli on visual search in the pigeon. J exp Psychol Anim Behav Process 5:211-223

Broadbent DE (1965) Information processing in the nervoussystem. Science 150:457-462

Cott HB (1985) Coloration, adaptive. In: Campbell E, Lack E (eds) A dictionary of birds. Poyser, Calton, p 97

Gillett SD, Hogarth PJ, Noble FEJ (1979) The response of predators to varying densities of *Gregaria* locust nymphs. Anim Behav 27:592-596

Godin J-G, Smith SA (1988) A fitness cost of foraging in the guppy. Nature Lond 333:69-71

Heller R, Milinski M (1979) Optimal foraging of sticklebacks on swarming prey. Anim Behav 27:1127-1141

Janssen J (1976) Feeding modes and prey size selection in the alewife (*Alosa pseudoharengus*). J Fish Res Board Can 33:1972-1975

Lawrence ES (1985) Vigilance during 'easy' and 'difficult' foraging tasks. Anim Behav 33:1373-1374

Major PF (1978) Predator-prey interactions in two schooling fishes, *Caranx ignobilis* and *Stolephorus purpureus*. Anim Behav 26:760-777

Metcalfe NB (1984) The effects of habitat on the vigilance of shorebirds: is visibility important? Anim.Behav 32:981-985

Metcalfe NB, Huntingford FA, Thorpe JE (1987) Predation risk impairs diet selection in juvenile salmon. Anim Behav 35:931-933

Milinski M (1977) Experiments on the selection by predators against spatial oddity of their prey. Z Tierpsychol 43:311-325

Milinski M (1977) Do all members of a swarm suffer the same predation? Z Tierpsychol 45:373-388

Milinski M (1984) A predator's costs of overcoming the confusion-effect of swarming prey. Anim Behav 32:1157-1162

Milinski M (1986) Constraints placed by predators on feeding behaviour. In: Pitcher TJ (ed) The behaviour of teleost fishes. Croom Helm Ltd, London, p 236

Milinski M, Heller R (1978) Influence of a predator on the optimal foraging behaviour of sticklebacks (*Gasterosteus aculeatus* L.). Nature Lond 275:642-644

Miller RC (1922) The significance of the gregarious habit. Ecology 3:122-126

Morgan WL, Ritz DA (1984) Effect of prey density and hunger state on capture of krill, *Nyctiphanes australis* Sars, by Australian salmon, *Arripis trutta* (Bloch & Schneider). J Fish Biol 24:51-58

Neill SRSJ , Cullen JM (1974) Experiments on whether schooling by their prey affects the hunting success of cephalopod and fish predators. J Zool Lond 172:549-569

Ohguchi O (1981) Prey density and selection against oddity by three-spined sticklebacks. Adv Ethol 23:1-79

Pitcher TJ (1986) The functions of shoaling behaviour. In: Pitcher TJ (ed) The behaviour of teleost fishes. Croom Helm, London, p 294

Wankowski JWJ, Thorpe JE (1979) The role of food particle size in the growth of juvenile Atlantic salmon (*Salmo salar* L.). J Fish Biol 14:351-370

Wilson RP, Ryan PG, James A, Wilson M-PT (1987) Conspicuous coloration may enhance prey capture in some piscivores. Anim Behav 35:1558-1560

ORAL DISCUSSION – M. Milinski

Q. (Okamura) Are there instances when it will be advantageous to individual water fleas not to aggregate in the presence of their predators ?
A. It may be advantageous for water fleas always to aggregate since individuals in the centre of aggregations experience on average a smaller risk of being eaten, especially when a stickleback's predator is present.

Q. (Sih) Is there a difficulty in separating the effects of differential responses from differential detection ? That is, at high densities of waterfleas, predators are detecting prey as efficiently as at low densities, but are not responding to the prey, for alternative reasons ?
A. Since the delay in feeding was the same after either the cryptic or conspicuous model had been shown in the low prey density, the two models were regarded as equally dangerous by the sticklebacks. This implies that the cryptic predator model was not detected by fish feeding upon the high prey density, because their feeding response was delayed only after the conspicuous model had been shown.

DIET SELECTION UNDER THE RISK OF PREDATION

Jean-Guy J. Godin
Department of Biology
Mount Allison University
Sackville
New Brunswick
Canada E0A 3C0

INTRODUCTION

Animals continuously make choices or "decisions" (sensu Dill 1987) about when to feed, where to feed, which search and attack tactics to use and which prey types to eat, in addition to other behavioural decisions which have fitness consequences (Krebs et al. 1983; Stephens and Krebs 1986; Dill 1987; Lima and Dill 1990). In making such foraging decisions, an individual is expected to behave in ways that maximize its fitness (= lifetime production of offspring) (Stephens and Krebs 1986; Dill 1987; Lima and Dill 1990). Classical foraging theory assumes that selection favours foragers that maximize their long-term net rate of energy intake, which is presumed to be positively correlated with fitness (Krebs et al. 1983; Pyke 1984; Stephens and Krebs 1986). This fitness currency seems appropriate for animals that forage on food items differing mainly in body size (Pyke 1984) or that can make use of surplus energy to enhance their growth rate, fecundity and survivorship, such as fishes (Ware 1975; Wootton 1979), but it may not be for others that have specific nutrient requirements, such as certain herbivores and granivores (Pyke 1984; Stephens and Krebs 1986).

Animals in nature, however, may rarely be able to maximize their feeding rate since other important activities, such as vigilance for predators and defence against territorial intruders, commonly conflict with foraging behaviour for expression in their time budgets. Because behavioural alternatives have different benefits and costs, individual animals should choose the behavioural option that maximizes their fitness when faced with such conflicts (Milinski 1986; Dill 1987; Mangel and Clark 1988; Lima and Dill 1990). Therefore, under certain circumstances, the forager must trade off foraging against some other activity at the cost of reduced energy intake rate (Werner and Gilliam 1984; Milinski 1986; Dill 1987; Lima and Dill 1990). The optimal trade-off or optimal "decision rule" will depend mainly on the animal's current physiological state (e.g. hunger) and current environmental states (e.g. food availability,

NATO ASI Series, Vol. G 20
Behavioural Mechanisms of Food Selection
Edited by R. N. Hughes

perceived risk of predation) (Krebs et al. 1983; Milinski 1986; Dill 1987; Mangel and Clark 1988; Lima and Dill 1990).

Foraging theory has been mainly concerned with the foraging decisions made by animals on two spatial scales; namely, between habitats or food patches (between-patch decisions) and within habitats or food patches (within-patch decisions). The particular food patch currently exploited by a forager, its residency time in that patch and the prey types it eats within the patch should depend not only on the forager's ability to assess the relative energetic profitabilities of the available patches and prey types in the environment (Krebs et al. 1983; Pyke 1984; Stephens and Krebs 1986), but also on variation in the risk of predation between and within food patches or habitats (Werner and Gilliam 1984; Milinski 1986; Dill 1987; Lima and Dill 1990). There is mounting evidence that animals can assess variations in both energetic profitability and risk of predation among available habitats and trade off expected foraging gains and perceived risk of mortality to predation in deciding where to feed (reviewed in Werner and Gilliam 1984; Milinski 1986; Dill 1987; Lima and Dill 1990). Much less attention has been given, however, to foraging decisions on a smaller spatial scale, i.e. within food patches, under the risk of predation (cf. Lima and Dill 1990). If by feeding in different locations in the food patch, on different parts of the prey aggregation or on different prey types the forager incurs different risks of mortality to predation, then the risk of predation may also be an important determinant of the forager's feeding rate and diet choice.

In this paper I review the evidence for the potential conflict between vigilance for predators and feeding, and thus for an increased risk of predation while foraging, and the evidence for predator-mediated changes in foraging behaviour within prey patches, in particular diet selection, that are consistent with trade-offs between food intake rate and predator avoidance. A dynamic model of diet selection under the risk of predation is then presented and used to functionally explain a diet shift observed in the guppy (*Poecilia reticulata*) foraging under predation hazard. However, before proceeding, a brief examination of the classical diet or prey model and certain of its limitations will help emphasize at the outset the need to develop new models and experimental approaches to further understand diet selection in animals.

THE CLASSICAL PREY CHOICE MODEL AND PARTIAL PREFERENCES

The first generation of optimal diet models, now referred to as the classical prey or diet choice model (Stephens and Krebs 1986), predict the types of prey eaten by animals foraging with a fixed search strategy within a homogeneous food patch. The model assumes that the forager was designed (by natural selection) to maximize the amount of food or energy intake (E) over some finite time period (T). The forager's overall rate of intake is therefore

$$\frac{E}{T} = \frac{\Sigma\, p_i\, \lambda_i\, e_i}{1 + \Sigma\, p_i\, \lambda_i\, h_i} - s \quad , \qquad (1)$$

where λ_i is the rate of encounter with prey items of type i, e_i is the expected net energy gained from an individual prey item of type i, h_i is the expected time spent handling an individual prey item of type i, s is the rate of energy expenditure during search (a constant), and p_i is the probability that a given prey item of type i will be attacked and eaten upon encounter. The model further assumes that (i) the forager is capable of unambiguously and instantaneously recognizing prey types and assessing their respective energetic profitability (= e_i / h_i), (ii) prey types are encountered sequentially and randomly, (iii) the rate of energy expenditure during prey handling is similar for all prey types, (iv) handling time for each prey type is a fixed constant and is mutually exclusive of searching for other prey, and (v) there are no trade-offs with other behavioural activities during foraging. In short, the model predicts that foragers should never exhibit "partial preferences"; that is, prey types should either always be attacked and consumed upon encounter (p_i = 1) or always ignored upon encounter (p_i = 0), and that they should always specialize upon the highest ranking prey (in terms of energetic profitability) and ignore the low-ranking prey if the encounter rate with the next higher-ranking prey exceeds a certain (acceptance) threshold (Krebs et al. 1983; Pyke 1984; Stephens and Krebs 1986).

Previous reviews of tests of the classical diet model (Krebs et al. 1983; Pyke 1984; Stephens and Krebs 1986) revealed that animals commonly show partial preferences; that is, their prey selectivity is not absolute (= step function) but follows a sigmoid pattern with changes in prey encounter rate. This observation is at variance with the prediction of the above diet model. Several factors are known or hypothesized to generate partial preferences in foragers (Table 1). Most of these factors violate the explicit assumptions of the classical diet model, and it is therefore not surprising that they consequently may result in the forager exhibiting partial preferences. Some of these factors represent constraints on the animal's foraging behaviour, either imposed by certain physical features of the environment (e.g. foraging time constraints such as daylength and tidal cycle), the forager's physiology (e.g. nutrient requirements) and its perceptual and learning capabilities (e.g. minimum prey recognition time, variance in estimating

Table 1. Factors that may result in foragers showing partial preferences when choosing among different prey types.

Factor	References
Constraints on foraging time	Lucas (1985, 1987)
Forager hunger level	Charnov (1976), Snyderman (1983)
Nutrient constraints	Belovsky (1978), Pyke (1984)
Learning-dependent foraging rate or handling time	Hughes (1979), McNair (1981), Croy and Hughes (this volume)
Sampling of available prey types	Krebs et al. (1977)
Non-instantaneous prey recognition time	Hughes (1979)
Prey discrimination errors	Rechten et al. (1983), Getty and Krebs (1985)
Simultaneous prey encounters	O'Brien et al. (1976), Engen and Stenseth (1984)
Low encounter rate with profitable prey	Krebs et al. (1977)
Stochasticity in threshold prey encounter for inclusion into the diet	Krebs et al. (1983), Stephens (1985)
Stochasticity in net reward value of prey	Stephens and Charnov (1982), McNamara and Houston (1987)
Discounting of future prey rewards	Lea (1979), Kagel et al. (1986)
Forager maximizing a currency other than energy	McNamara and Houston (1987)
Runs of successful encounters with particular prey types	Milinski (1982), McNamara and Houston (1987)
Averaging diet choice over different conditions	McNamara and Houston (1987)
Threat of competition	Barnard and Brown (1981), Milinski (1982), Jaeger et al. (1983)
Risk of predation	This paper

prey encounter rates), or features of the prey types (e.g. crypsis, distribution pattern, variance in energetic value). Other factors, such as the risk of predation, have also been viewed as contraints on maximization of energy intake rate (e.g. Krebs et al. 1983; Lucas 1985; Milinski 1986), based on the premise that the forager must sacrifice some foraging time or foraging efficiency to avoid or defend against predators. Alternatively, predator-mediated reductions in an animal's food intake rate may result from active decision making, whereby the forager "chooses" a behavioural option which lies somewhere along a continuum between energy maximization (at the total expense of predator avoidance) and minimization of the risk of mortality to predation (at the total expense of foraging). The optimal behavioural trade-off will thus depend on the animal's current perceived threat of predation and the expected benefits gained from foraging, rather than on any particular factor "constraining" it from choosing one of the extreme behavioural options (cf. Dill 1987; Lima and Dill 1990).

The classical prey choice model assumes that the forager does not trade off its rate of energy intake for greater safety from predation when "deciding" to attack or ignore encountered prey items within a patch. However, one might expect this decision to be influenced by the risk of predation, particularly if the forager incurs a differential risk of mortality to predation when attacking and eating different prey types. Therefore, under such circumstances, the forager may alter its prey preferences to decrease the risk of predation, at the cost of a reduced rate of energy intake, and thus exhibit partial preference. Because the classical prey choice model is a static and deterministic model (Krebs et al. 1983; Pyke 1984; Stephens and Krebs 1986), it is not designed to address changes in diets that result from behavioural trade-offs. The latter can now be better understood with the aid of a new generation of dynamic, stochastic models, which take into account the changing internal state of the animal and use a common fitness currency to express the benefits and costs of the different behavioural options available to the animal (reviewed in Houston et al. 1988; Mangel and Clark 1988).

Because of the diverse factors (Table 1) potentially determining the diet composition of animals in nature, in addition to the energetic profitabilities and densities of available prey types in the environment, there is an obvious need for more realistic diet choice models that incorporate the effects of known constraints on the animal's foraging efficiency (e.g. Hughes 1979; Rechten et al. 1983), its changing internal state (e.g. Houston and McNamara 1985) and its potential trade-offs with other activities (e.g. Lima 1987, 1988a; Gilliam, this volume). Such models will help identify and focus research on important ecological and physiological determinants of diet choice and ultimately contribute to a greater understanding of the diverse mechanisms underlying prey selection in animals.

THE RISK OF MORTALITY TO PREDATION AND FORAGING

An animal's risk of predation can be defined as the probability of being eaten by a predator during some discrete time interval, which in turn depends mainly on its encounter rate with predators (*r*), the time it spends vulnerable to an encounter or attack (*T*), and the probability of death given an encounter (*d*). The latter three parameters are referred to as the basic components of predation risk (Lima and Dill 1990). An individual may incur a greater risk of predation when foraging because (i) its level of vigilance for predators may be compromised when feeding at high rates (*d* and*T* increased), (ii) it is rendered more conspicuous by its movements (*r* and *d* increased), (iii) feeding may produce products (e.g. chemicals, sound) that attract predators (*r* and *T* increased), and (iv) predators may aggregate in habitats where food is most abundant (*r* and *T* increased) (Milinski 1986; Dill 1987; Lima and Dill 1990).

Evidence for an increased risk of predation associated with foraging is largely circumstantial, based mainly on observed changes in the vigilance pattern of animals when foraging (refs. in Godin and Smith 1988). However, more direct evidence has recently been obtained by Godin and Smith (1988). They demonstrated empirically an increased probability of mortality, on a given attack by an ambush predatory fish, in female guppies foraging on waterfleas (*Daphnia magna*). Guppies feeding at the highest food density presented (20 *Daphnia* per litre) suffered a 120-140% increased risk of predation, depending on their food deficit (hunger level), compared with non-feeding conspecifics (Fig. 1). Moreover, the pre-attack instantaneous feeding rate of guppies which were captured by the predator was higher on average than that of survivors. Since the guppy's single encounter with the cichlid fish predator was staged in this study, the observed increase in it's probability of death during the encounter (= *d*) with increasing feeding rate is most attributable to a reduced likelihood of detecting the attacking predator in time while foraging at high rates. This conclusion is supported by Milinski's (1984) previous observation that sticklebacks (*Gasterosteus aculeatus*) feeding at high rates on dense prey are less likely to detect an overhead cryptic avian predator model than fish feeding more slowly on less dense prey. Therefore, foragers which attempt to maximize their rate of food intake will likely incur a greater risk of predation. Because of this fitness cost of foraging, animals are expected to trade off foraging gains for greater safety from predators in such a way as to maximize fitness. There exists considerable evidence that animals can apparently assess their current risk of predation and modify accordingly their behaviour in ways that are consistent with a foraging gain – predation risk trade-off (reviewed in Milinski 1986; Dill 1987; Lima and Dill 1990). Evidence for such a trade-off involving diet selection within a patch/habitat is reviewed below.

WITHIN-PATCH FORAGING DECISIONS UNDER THE RISK OF PREDATION

Animals may move to another, safer habitat or shift their activity pattern to safer diel periods in response to a perceived increase in the risk of predation in the currently occupied habitat or at certain times, respectively. Such shifts in habitat and diel time period use are often made at the expense of food intake rate (reviewed in Werner and Gilliam 1984; Milinski 1986; Dill 1987; Lima and Dill 1990). Changes in the diet composition of the prey animal may coincide with these predator-mediated shifts in habitat (e.g. Vance and Schmitt 1979; Edwards 1983; Werner et al. 1983) or diel activity pattern (e.g. Hall et al. 1979; Caldwell 1986). These dietary changes, however, may simply be owing to the forager encountering a different assemblage of prey types in the safer habitat or diel foraging time window, or experiencing increased food competition there (then), compared with the previously exploited riskier habitat or time period, for example. Therefore, they do not represent modification in the forager's prey selectivity to avoid predation *per se*, but are rather indirect, second-order effects of predator avoidance and will thus not be considered further here.

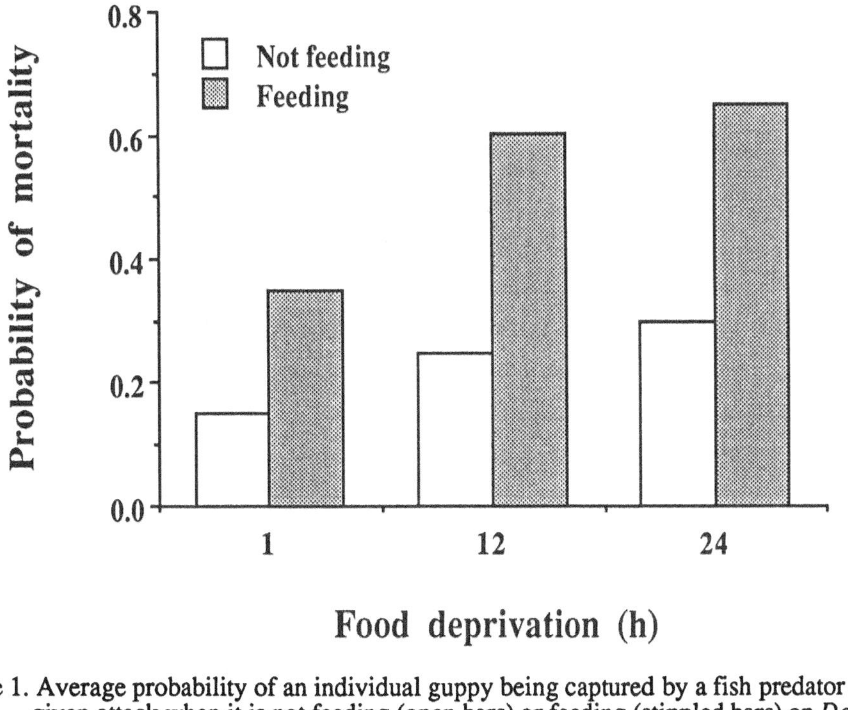

Figure 1. Average probability of an individual guppy being captured by a fish predator on a given attack when it is not feeding (open bars) or feeding (stippled bars) on *Daphnia* at a high density (20 individuals / litre). Guppies were tested after being deprived of food for one of three pre-determined periods shown. N = 20 for each bar. Adapted from Godin and Smith (1988) by permission from Macmillan Magazines Ltd.

It is not uncommon for prey animals to be in close proximity to predators in nature (Foster 1980; Pitcher 1980) and to remain in the currently occupied patch/habitat while being concurrently within the attack range of predators (Potts 1983; Pitcher et al. 1986; Magurran and Pitcher 1987; Milinski 1987; Dugatkin 1988). In these circumstances, there will likely exist a conflict between foraging and predator avoidance behaviours. In resolving this conflict, the prey animal may reduce its risk of being detected and attacked by a predator by seeking refuge, becoming more cryptic, or mobbing (attacking) the predator (reviewed in Sih 1987; Lima and Dill 1990). These antipredator strategies are usually associated with complete cessation of feeding. Alternatively, the animal may continue foraging in the face of predation hazard by trading off some foraging gains for increased safety from predation (e.g. Lima et al. 1985; Fraser and Huntingford 1986; Godin 1986; Fraser and Gilliam 1987) or by accepting a greater risk of predation to maintain a high feeding rate (e.g. Gilliam and Fraser 1987; Godin and Sproul 1988; Pitcher et al. 1988; Abrahams and Dill 1989). The latter strategy is beyond the scope of this paper. In the former situation, however, the forager may be able to reduce its risk of predation within the food patch by being less active, joining a social group, approaching and "inspecting" the predator, altering its use of space within the patch, exploiting different parts of the prey aggregation, adjusting the time spent handling individual prey types, and(or) altering its choice of prey types to eat (reviewed in Milinski 1986; Dill 1987; Lima and Dill 1990). Of particular interest here are direct predator-mediated changes in the diet selection of foragers over the short term and in prey handling time, inasmuch as the latter may alter the profitability of the prey item and the forager's decision to include or exclude it from its diet. I summarize below a number of case studies, some of which were previously reviewed by Lima and Dill (1990), reporting changes in the diet selection of animals that appear to be directly predator-mediated and, for the most part, consistent with an energy intake rate – predator avoidance trade-off.

The types of seed harvested within a particular microhabitat (patch) by desert heteromyid rodents would appear to be sensitive to their perceived vulnerability to predation in that patch. In one study, Hay and Fuller (1981) reported that *Perognathus fallax* and *Dipodomys merriami* increased by three-fold their selectivity for preferred seed types over equally available non-preferred seeds when foraging in open, riskier areas compared to when they foraged in safer areas under vegetation canopies. Similarly, in another study (Bowers 1988), *D. mirriami* exhibited a greater preference for milo seeds over equally abundant millet seeds when foraging in open microhabitats compared to when they foraged within the protective cover of bushes. Unfortunately, it is not possible to assess from the data presented in either study whether the observed change in seed selectivity within the microhabitats was based on a trade-off between foraging efficiency and predator avoidance.

When an animal chooses to leave the safety of a refuge to forage in a non-structured habitat/patch, its risk of predation may be largely dependent on the time spent vulnerable to attack (*T*) in the latter habitat/patch. A component of this time is the total time spent handling prey items. Therefore, if available prey types vary in their respective handling times, then one might expect a forager to alter its prey selectivity to minimize *T* ; this may, however, entail a cost of reduced energy intake rate. Once a grey squirrel (*Sciurus carolinensis*) has located a food item in an open area away from the protective cover of trees and shrubs, it can either carry the food back to cover for consumption and thus minimize *T*, but at the cost of reduced feeding efficiency because of the added time and energy costs of returning to cover, or stay and eat the food where located which maximizes feeding efficiency, but at the cost of increasing *T* (Lima et al. 1985). Maximizing feeding efficiency and minimizing the risk of predation are conflicting demands in this situation. Lima and Valone (1986) found that grey squirrels resolved this conflict by altering their diet selection when locating food items in a patch as a function of both distance from cover and the size of the food items. More specifically, the squirrels increased their tendency to reject the more energetically profitable food items in favour of locating larger, less profitable items, which are then carried to cover for consumption, as their distance to cover decreased and as the size of the less profitable food type increased. The squirrels therefore exhibited partial preference in their diet selection, reflecting a trade-off between foraging efficiency and risk of predation.

Similarly, parental birds may be faced with a conflict between maximizing the rate at which they deliver food (energy) to their nestlings and minimizing the time offspring are left unguarded, and presumably more vulnerable to predators, when they are foraging at some distance from the nest. At least two studies report changes in the foraging decisions, including diet choice, of parental birds in response to an experimentally increased threat of predation to nestlings. Freed (1981) observed that, following the short presentation of a tethered snake predator at their nest compared to before its presence, parental house wrens (*Troglodytes aedon*) of both sexes significantly reduced their time searching for insect prey in a field near the nest, increased their predator surveillance time at the nest (where they mobbed the snake), foraged closer to the nest and increased the frequency of smaller prey brought back to the nestlings. The shift in the bird's diet selection (Fig. 2) is reflected mainly in an increase in the frequency of smaller insects ($\leq$ 10 mg) at the expense of the less abundant and larger (> 80 mg) prey. This shift in selectivity persisted regardless of the duration of the bird's foraging trips. Selectivity for intermediate sized insects remained unchanged (Fig. 2). Therefore, in response to an apparent increased predation threat to their nestlings, the wrens changed their foraging decisions (i.e. search time and prey selection) within the patch (field) to minimize their time away from the nest, thereby reducing the risk of predation to their offspring, but at the expense of maximizing

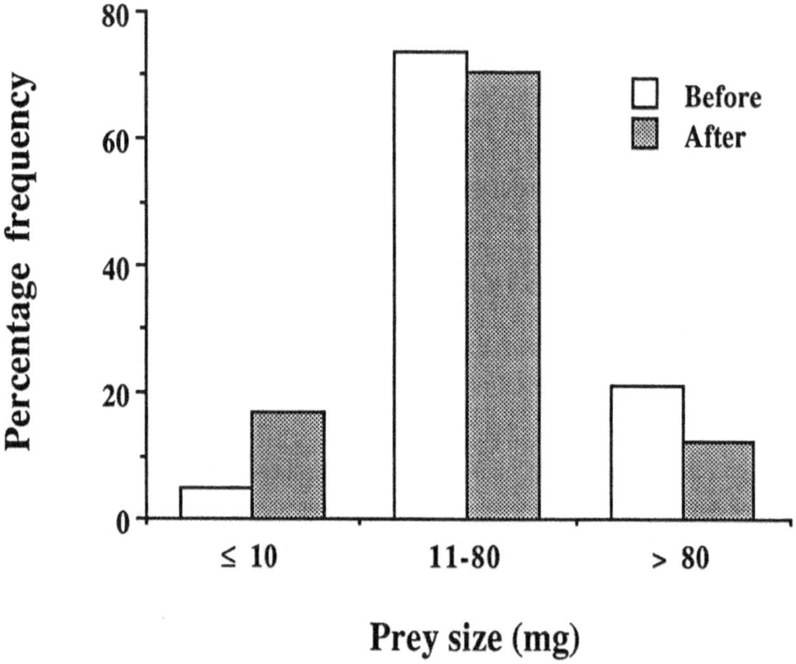

Figure 2. The percentage frequency of invertebrate prey items, falling into three size classes, that were fed to nestlings by foraging parental house wrens before (open bars) and after (stippled bars) the appearance of a snake predator near the nest. Adapted from Freed (1981) by permission from the author.

the rate of energy delivery to the young. By adaptively altering their diet to include more small insects, which presumably required shorter search times because of their greater abundance and shorter handling times, the parents reduced the time their nest was left undefended during a period of high predation risk. Male, but not female, parental Gila woodpeckers (*Melanerpes uropygialis*) showed a similar response to the presence of a potential predator (stuffed Gila woodpecker) at the nest (Martindale 1982). The diet selection of the above parental birds foraging under the risk of predation is at variance with the expectations of the classical diet model, but it is consistent with a foraging efficiency – predation risk trade-off. Alternatively, Lucas (1985) has interpreted the decreased preference for larger insects by Martindale's (1982) woodpeckers as being caused by shorter foraging bouts (i.e. a time constraint on foraging), resulting from the bird's continuous foraging time being interrupted by the presence of a predator at the nest, rather than by an active decision making process of the kind described above.

A reduction in the distance travelled to capture food items in response to an increased risk of predation has also been reported for juvenile stream-dwelling salmon, which are typically sit-and-wait predators that swim varying distances from their feeding "station" near or on the substrate to intercept individual insects drifting in the current (see Godin and Rangeley (1989) for further details). The salmon are likely more visible to visual predators when moving than when stationary near or on the substrate due to their cryptic colouration. Therefore, the further they travel to feed on drifting prey, the longer they remain vulnerable to predatory attacks from birds and larger fish. Juvenile coho, *Oncorhynchus kisutch* (Dill and Fraser 1984) and Atlantic salmon, *Salmo salar* (Metcalfe et al. 1987a) have been shown to reduce their tendency to move to intercept drifting prey, and thus exhibited shorter attack distances, especially for the largest, most profitable prey in the case of the coho (Dill and Fraser 1984), after sighting a model of a predatory trout. Such a behavioural response to an apparent increase in predation hazard presumably reduces the salmon's likelihood of being detected by a predator while feeding, but at the cost of a reduction in feeding rate, and furthermore may have consequences for its diet selection. For example, since the coho salmon's predator-mediated reduction in attack distance is greater for the larger prey items encountered (Dill and Fraser 1984), its diet should therefore comprise a smaller proportion of large prey items when feeding during periods of increased predation risk compared with periods of lower perceived risk. This prediction, however, remains to be tested.

We have already noted that an animal feeding at a high rate is less likely to detect an approaching predator (Milinski 1984) and is consequently more likely to be captured on a given predatory attack (Godin and Smith 1988; Fig. 1) than one that is feeding more slowly or not at all. Searching for, assessing and capturing prey items and being vigilant for predators are apparently difficult visual tasks to carry out simultaneously for any animal (Milinski, this volume). A conflict therefore arises between these two tasks when the forager requires a period of time to assess or handle an encountered prey item during which its ability to detect predators is reduced. How the animal resolves this conflict under the risk of predation has implications for its diet selection. In situations where the degree to which food handling and vigilance for predators are mutually exclusive varies positively with prey profitability, the optimal diet may be non-classical, comprising a mixture of prey types rather than the most profitable prey type exclusively. This was illustrated by Lima (1988b), who provided dark-eyed juncos (*Junco hyemalis*) with a simple two-prey environment, a single patch consisting of equally available small food items of high profitability, but which could only be eaten with the head down (vigilance and handling thus mutually exclusive), and larger, less profitable items that could be eaten with the head up (vigilance and handling could occur simultaneously). The junco's selectivity was not absolute (Fig. 3); it included an increasing proportion of the larger, less profitable food type in its diet with both a decrease in the absolute profitability of the more profitable food type and a decrease in flock size, that is, as the apparent individual risk of predation, and thus the need to be vigilant, increased. Such changes in diet selection are consistent with a trade-off between maximization of energy intake rate and minimization of the risk of predation.

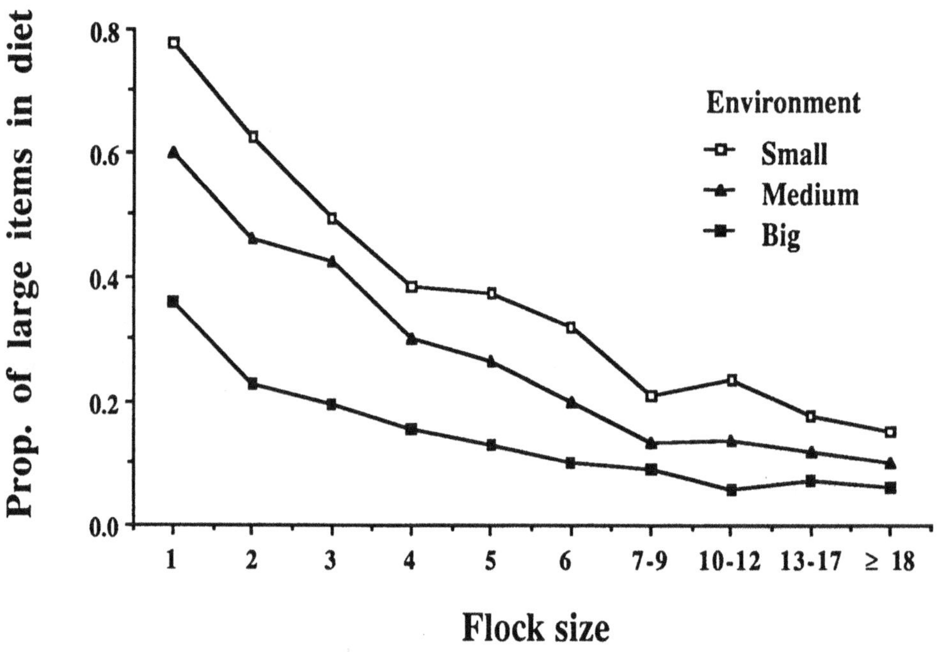

Figure 3. Proportion of larger, less profitable food items in the diet of dark-eyed juncos as a function of flock size and the size of the more profitable, smaller items ("bits") available in the environment. Variance bars around the means were omitted for simplicity. Adapted from Lima (1988b) by permission from the National Research Council of Canada.

Animals that feed on moving prey require a discrete period of time to visually track, pursue and successfully capture an individual prey (Lanchester and Mark 1975; Curio 1976; Ohguchi 1981). Any sudden interruption of this visual task, such as a brief visual scan of the environment for predators or competitors, will likely reduce the animals' foraging efficiency and may also have implications for its diet selection (e.g. Ohguchi 1981; Milinski, this volume). For example, if the forager gives priority to vigilance when it perceives its risk of predation to be high, its accuracy at discriminating among encountered potential prey types may decrease, and this may lead to non-classical partial preferences. A case in point is the observed increase in the frequency of attacks on inedible food items (= discrimination errors) by juvenile Atlantic salmon after sighting a model predatory trout (Metcalfe et al. 1987b).

For many animals, larger prey items require more time to handle (= total time elapsed between the onset of a stalk or attack to ingestion) than smaller ones. The degree to which prey handling time reduces vigilance for predators (i.e. increases *d*) may be positively correlated with the size of the food item in certain situations. In such cases, when the forager perceives its risk of predation to have increased, one would expect it to alter its diet by choosing preferentially smaller prey with associated shorter handling times, at a potential cost of reduced net rate of energy intake. This hypothesis is currently being tested in my laboratory with female guppies foraging in a "patch" consisting of three equally abundant size classes of waterfleas (*Daphnia magna*) in the absence and presence of a potential fish predator. Female guppies have previously been shown to be sensitive to the risk of predation when choosing where to forage (Abrahams and Dill 1989) and how fast to feed (Fraser and Gilliam 1987). They would therefore appear to be suitable subjects for investigating diet selection under the risk of predation. In our study, guppies showed a significant shift in their diet selectivity, from preferring the largest prey items to preferring the smallest ones, when foraging under apparent predation hazard, as predicted. Since this study is ongoing, only preliminary results are summarized here (next section); the results of the completed study (J.-G.J. Godin, J. Miller and M. Mangel, unpubl. data) will be published elsewhere.

DIET CHOICE IN THE GUPPY UNDER THE RISK OF PREDATION

We presented individual guppies with 90 *Daphnia*, 30 of each of three distinct size classes (Table 2), in an aquarium. Each *Daphnia* size class was equally abundant (1.17 prey/ litre). The fish were allowed to eat only 10 out of the total number of prey available to avoid significant prey depletion effects on diet choice. Such a foraging trial was carried out separately with each of 30 similar-sized guppies in both the absence and presence of a potential predator, a cichlid fish (*Astronotus ocellatus*). The predator was located behind a clear Plexiglas partition at one end of the aquarium in full view of the guppy (during predator-present trials only). These two trials were carried out in random order on separate days for each guppy, whose initial hunger level was standardized between trials. We noted the size class of each *Daphnia* attacked, whether it was successfully ingested or not, and the behavioural responses of the guppy towards the fish predator.

In the absence of the predator, the guppies fed selectively and preferred the largest *Daphnia* over the medium and small ones (Fig. 4). Because guppies are gape-limited foragers, they required significantly more time, on average, to successfully attack, pursue and ingest (= total handling time) a large *Daphnia* than a small one (Table 2). Large *Daphnia* were therefore

Table 2. Characteristics of the three *Daphnia* size classes presented to the guppies and other parameters used in the dynamic prey choice model (equation 3). Parameter values are shown for guppies foraging in both the absence and presence (in brackets) of a predator where appropriate. Only mean values are shown for simplicity. Total handling time is the sum of the prey attack, pursuit and mandibulation times; for brevity, the latter are not reported here. Fish reaction distance is the distance travelled by the fish to "strike" a prey once detected. All time and distance measurements were obtained from video films of foraging guppies. Fish encounter rate with prey was estimated using known mean values of search speed and reaction distance, following Confer and Blades (1975) and Eggers (1977). The net energy value and profitability of the prey were calculated following the procedure of Bence and Murdoch (1986).

Parameter	Prey size class		
	Small	Medium	Large
Body length (mm)	2.02	2.84	3.51
Body weight (μg)	52.2	111.7	180.6
Gross energy value (J) = Y_i	1.23	2.63	4.26
Net energy value (J) = e_i	0.858	1.832	2.967
Handling time (s) = τ_i	2.1	5.2	8.6
	(2.7)	(5.1)	(9.3)
Profitability (J/s)	0. 409	0.352	0.345
	(0.317)	(0.359)	(0.319)
Encounter rate (prey/s) = λ_i	0.015	0.030	0.050
	(0.016)	(0.018)	(0.030)
Reaction distance (cm)	1.7	2.4	3.1
	(2.1)	(2.3)	(2.9)
Attack success (proportion) = φ_i	0.922	0.808	0.650
	(0.959)	(0.719)	(0.353)

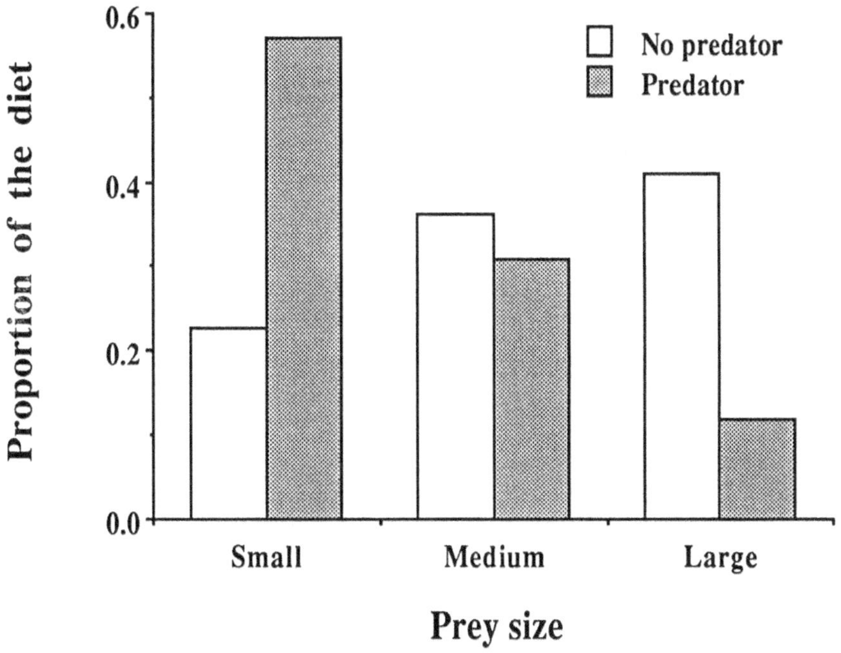

Figure 4. Mean proportions of three *Daphnia* size classes eaten by individual guppies (N = 30) in the absence (open bars) and presence (stippled bars) of a cichlid fish predator. Standard error bars around the means were omitted for simplicity.

slightly less profitable, in terms of net energy return per unit handling time, and were more difficult to capture and ingest than either the medium or small *Daphnia* (Table 2). However, based on the assumptions and the threshold prey acceptance rule of the classical diet model (equation 2.11 in Stephens and Krebs 1986), all three *Daphnia* size classes fall within the optimal prey set that should maximize the fish's expected net rate of energy intake. All three prey types are therefore expected to be consumed upon encounter in proportion to their relative encounter rates. Given the estimated fish encounter rates with the three prey types and their energy values and handling times (Table 2) as parameters, the classical diet model predicts qualitatively the observed guppy preference for the largest *Daphnia* (Fig. 5A). However, this model overestimates the frequency of large prey and underestimates the frequencies of the medium and small prey in the diet ($G = 18.0$, $df = 2$, $P < 0.001$; Fig. 5A).

The reactive visual field model (RVFM; Eggers 1982, Wetterer and Bishop 1985), an alternative sequential prey encounter model, makes a dietary prediction similar to that of the classical diet model; it also overestimates the proportion of large prey and underestimate the proportions of medium and small prey in the guppy's diet ($G = 18.4$, $df = 2$, $P < 0.001$; Fig.

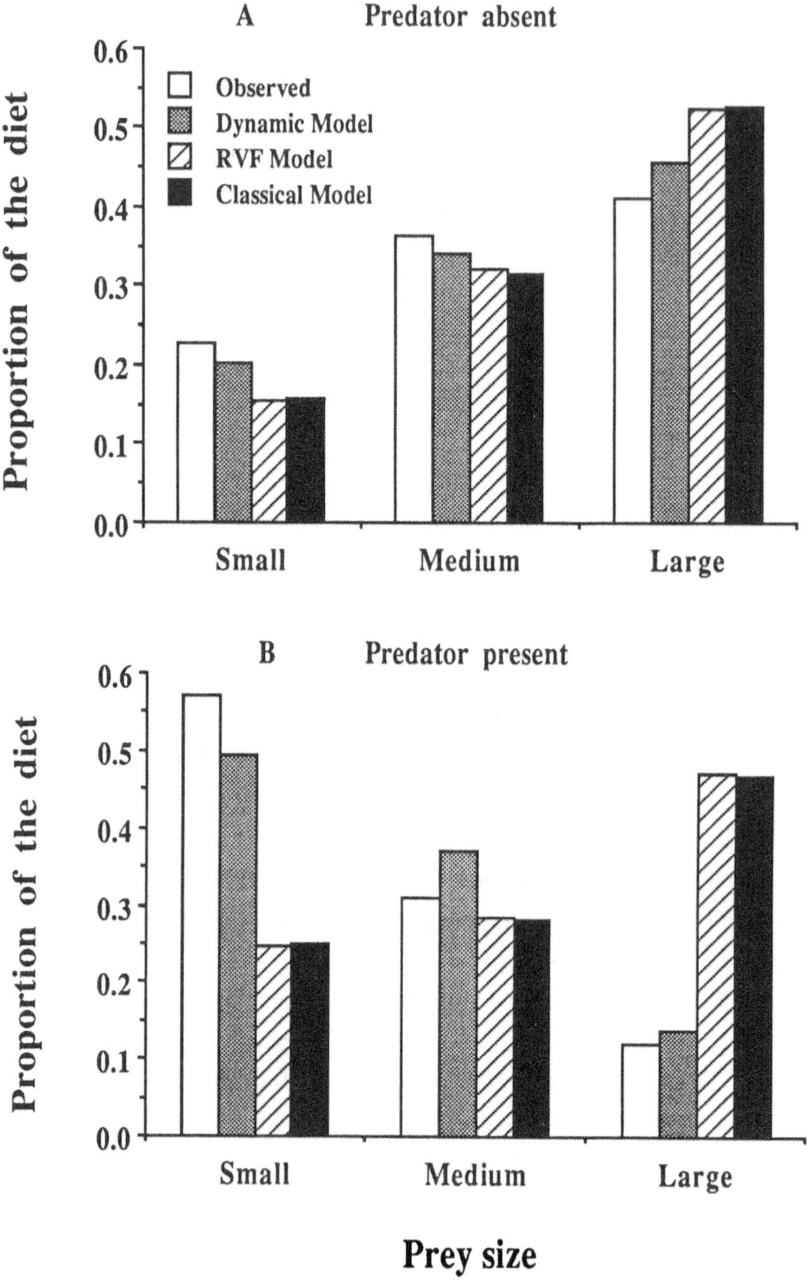

Figure 5. Comparisons of the observed mean proportions of three *Daphnia* size classes eaten by guppies (open bars), both in the absence (upper panel A) and presence (lower panel B) of a fish predator, with the diets predicted by three different prey choice models. The models are described in detail in the text.

5A). Assuming that the guppy possesses a cylindrical visual field (an assumption commonly made for planktivorous fish; Eggers 1977, Confer et al. 1978) whose volume is proportional to its reaction distance (RD_i) towards prey of a particular size *i*, the RVFM predicts that the forager will consume prey types as encountered in proportion to their relative reaction distances squared (RD_i^2) times their respective density (D_i) in the forager's visual field. Because the reaction distance of planktivorous fish towards prey is positively correlated with prey length (Eggers 1977, 1982; Confer et al. 1978), large prey items should be encountered more frequently than smaller ones, even when their densities are similar. Therefore, the expected frequencies (F_i) of *n* prey types in the diet is

$$F_i = \frac{RD_i^2 \cdot D_i}{\sum_{j=1}^{n} RD_j^2 \cdot D_j} \qquad (2)$$

When predation risk was apparently low (i.e. predator not in view), the guppy's prey selectivity did not accurately conform with that predicted by either of the above two static prey choice models. It would therefore seem that the guppy was not foraging in a manner consistent with maximization of net rate of energy intake only, nor was it simply eating *Daphnia* as randomly encountered. Alternatively, the fish may have been actively choosing among prey types in a way that reflects a compromise between energy maximization and minimization of the probability of death to predation while foraging. Three lines of evidence support this hypothesis.

First, the guppy's selectivity for *Daphnia* was not static during the course of a foraging trial (Fig. 6A). Data averaged over all fish show that they initially preferred the larger *Daphnia*, but this preference waned over consecutive prey ingestions and the three prey types became equally likely to be eaten once seven prey items had been consumed (Fig. 6A). If foraging on large *Daphnia* is associated with a greater risk mortality to predation (because of their longer handling time) compared with the other two smaller prey types, then the guppy's reduced preference for the larger *Daphnia* with increasing satiation might reflect a shift in its optimization criterion from energy maximization, when hungry early in a foraging period, to minimization of its risk of predation (by choosing prey with shorter handling times), when less hungry later in

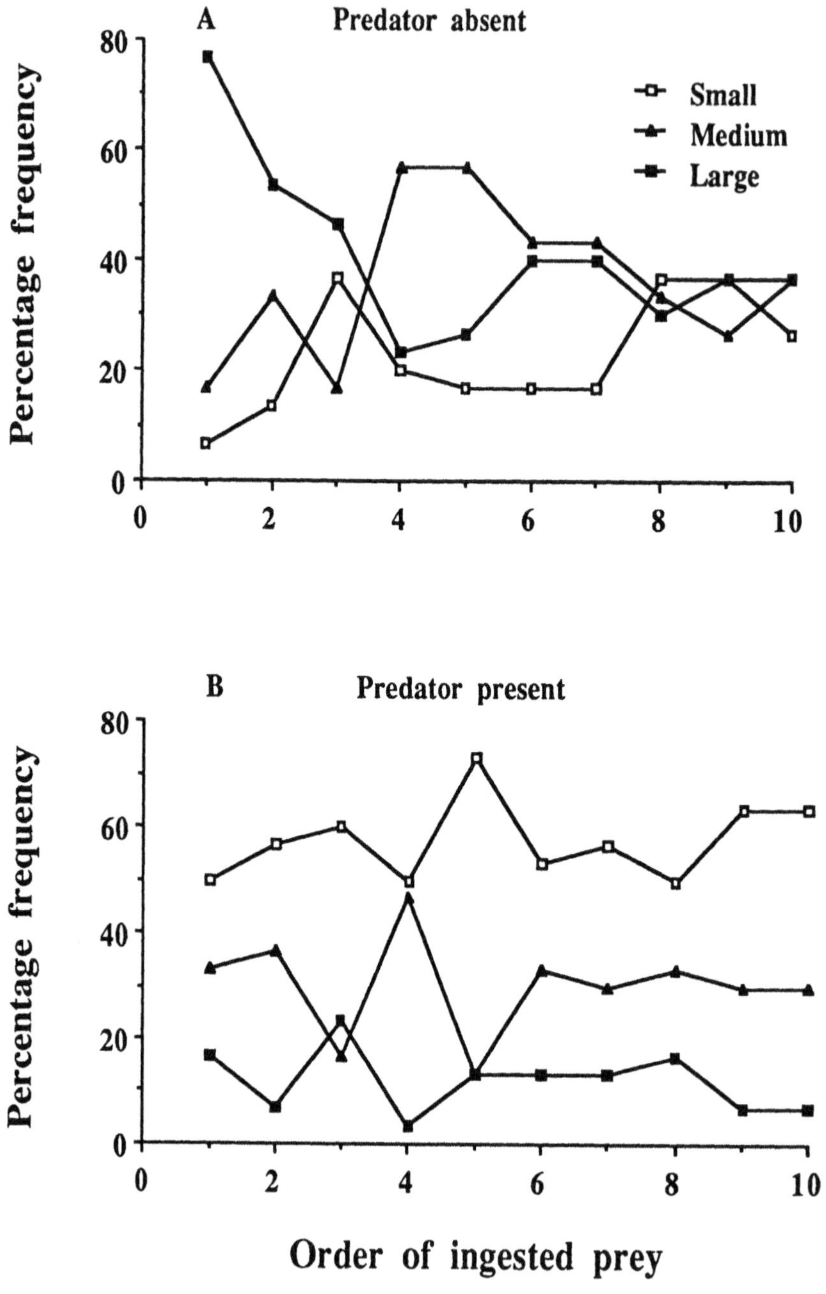

Figure 6. Sequential rank order in which *Daphnia* of three different size classes were eaten by guppies, both in the absence (upper panel A) and presence (lower panel B) of a fish predator. Each value represents the mean for 30 guppies, each having eaten 10 *Daphnia*.

the foraging period. Similar hunger-related trade-offs between foraging and the risk of predation have been previously reported by Milinski and Heller (1978), Heller and Milinski (1979) and Dill and Fraser (1984), for example.

The second line of evidence is based on the close concordance ($G = 2.8$, $df = 2$, $P > 0.1$) obtained between the observed guppy diet and one simulated by a stochastic, dynamic diet selection model which we developed (Fig. 5A). This model, described in more detail in the next section, predicts foraging decisions (i.e. attack or ignore encountered prey) as a function of the amount of food currently in the forager's stomach (i.e. its hunger level), the time remaining until the end of the foraging period, and the probability of mortality to predation while searching for food and while handling a particular prey type.

Third, if the guppies were indeed adjusting their prey selectivity to increase vigilance for predators, and thereby reduce their risk of predation, with decreasing hunger level when predation risk was apparently low, we would therefore expect them to shift their preference from the largest to the smallest *Daphnia*, whose associated short handling time would conflict less with vigilance, in the presence of a predator. This is precisely what we observed (Fig. 4). The guppy's selectivity for the largest and medium *Daphnia* decreased and that for the smallest *Daphnia* increased under apparently higher predation risk, as expected if prey handling detracts from being vigilant for predators. Neither of the two static prey choice models tested predicted the observed predator-mediated shift in the guppy's diet selection (Fig. 5B). Our dynamic prey choice model, however, did predict qualitatively such a dietary shift, but its simulated dietary composition did not quite accurately conform ($G = 24.8$, $df = 2$, $P < 0.001$) with the observed fish diet under the risk of predation (Fig. 5B). Nonetheless, the dietary shift predicted by the dynamic model suggests that the guppies altered their prey choice decisions to minimize the probabilty of death to predation in response to a perceived increase in predation risk.

Moreover, the greater selectivity for the smallest *Daphnia*, and the lower one for the larger *Daphnia*, remained relatively constant with increasing satiation (Fig. 6B), contrary to the pattern noted in the absence of the predator (Fig. 6A). This suggests that the changing energetic state of the fish did not affect its prey selectivity when the risk of predation was apparently high. In the presence of the predator, the guppy was less successful in attacking medium and large *Daphnia* , which normally require longer pursuits for successful capture, compared with small *Daphnia* (Table 2). This reduction in attack success for the larger two prey size classes can be attributed at least in part to the fish's apparent decision to delay attacking these prey, with resulting shorter reaction distances (Table 2), and to abort their pursuit earlier in the presence of the predator than in its absence. Pursuit times for medium and large *Daphnia* thus decreased by

50% (from 1.8 to 0.9 s) and 24% (from 3.4 to 2.6 s), respectively, although total handling time for the large prey type (and for the smallest one also) increased slightly (Table 2) owing to longer times taken to mandibulate them before ingestion under predation hazard. However, *Daphnia* mandibulation and predator vigilance are presumed not to be mutually exclusive activities for the guppy, and therefore the fish's longer prey mandibulation time under predation hazard is not expected to increase its risk of predation, but a longer pursuit time presumably would. Prey handling time has usually been treated as a fixed time constraint on the forager's feeding efficiency by classical diet models (Krebs et al. 1983, Pyke 1984, Stephens and Krebs 1986). Our results, and others reviewed by Lima and Dill (1990), show that handling time, or some component of it, can be flexible in certain cases and that it may be adaptively altered by the forager in response to the perceived risk of predation.

The observed decrease in pursuit time and attack success for the medium and large *Daphnia* in our study can thus be interpreted as a cost of foraging on these two prey types whilst simultaneously attempting to be vigilant for predators. By shifting its diet in favour of small *Daphnia* in response to a threat of predation, the guppy presumably increased the time available for vigilance, but incurred an estimated 6% reduction in its net rate of energy intake (from 0.344 to 0.324 J/s, calculated using equation 1 above) in doing so.

A STOCHASTIC, DYNAMIC DIET SELECTION MODEL

We have developed a discrete-time, state-variable model which generates optimal prey choice decision rules, and thus predicts an optimal diet composition that maximizes the fitness of an animal foraging under the risk of predation. Using stochastic dynamic programming (described by Houston et al. 1988, Mangel and Clark 1988), optimal prey selection strategies were derived for the forager as a function of its current energetic state and the current time (t) over the course of a longer finite foraging period (T). These sets of optimal prey choice decision rules were then used to simulate an optimal diet for guppies foraging under the risk of predation. Since the details of this model and its simulated results will be published elsewhere (J.-G.J. Godin, J. Miller and M. Mangel, unpubl. data), only certain of their salient features are summarized here.

The Model

To begin, imagine that an individual female guppy may encounter n prey types within a patch. Each prey type can be characterized by a handling time (τ_i), an energy value (Y_i), an energetic handling cost per unit time (α_i), a probability of being encountered (λ_i) and a probability of being eaten given an attack (φ_i). The fish is therefore faced with a decision to attack or ignore any prey item of type *i* upon encounter. If the fish attacks and handles a particular prey item of type *i*, it incurs a certain probability of mortality to predation (β_H) over the interval τ_i. Alternatively, if the forager either ignores an encountered prey item or does not encounter any prey item during a small discrete time interval *t*, it incurs a lower probability of mortality (β_S), as observed by Godin and Smith (1988) for the guppy. In the model $\beta_H > \beta_S$, since the forager's vigilance for predators is presumably compromised more by handling a prey item for a given period than by either continued search for prey or ignoring an encountered prey item over the same period. Thus, in the absence of predators in the forager's visual field (a presumed low predation hazard situation), we arbitrarily set $\beta_S = 1.0 \times 10^{-6}$ and $\beta_H = 2.0 \times 10^{-6}$ in the model. Both these values were increased to $\beta_S = 1.0 \times 10^{-3}$ and $\beta_H = 2.0 \times 10^{-2}$ in the presence of a potential predator, reflecting a presumed increase in predation hazard. Note that the magnitude of this increase in the probability of mortality to predation is greater for a fish handling a food item than for one not handling prey.

Let's further assume that the guppy's lifetime fitness is positively correlated with the amount of food (energy) it consumes over the long term. This is a reasonable assumption for female fish since they can make use of surplus energy to enhance their growth rate, survivorship and fecundity (Ware 1975, Wootton 1979). Let the state variable $X(t)$ characterize the amount of food (energy) in the guppy's stomach at time *t* of a longer foraging period *T*. Next, we define a fitness function $F(x,t,T)$ as the maximum expected amount of food consumed (net energy gained) by the forager from period *t* to period *T*, given that at period *t* the forager is alive and the value of the state variable $X(t)$ is *x*. The guppy does not die if $X(t)$ falls below 0. At the start of the foraging period *T*, the terminal fitness function is $F(x,T,T)$. By definition the optimal prey choices are those that lead to this maximum amount of energy intake, assuming that the guppy chooses the optimal prey type at each period $t, t+1 \ldots\ldots T$. The model determines the optimal decision rules when it calculates $F(x,t,T)$ for each discrete time period. It is also assumed that the forager has a stomach capacity constraint *C*. Thus

$$X(t) \leq C \qquad \text{for all } t$$

We set $C = 50$ Joules in the model; this value is close to the observed maximum (satiation) amount of *Daphnia* energy equivalents that individual guppies are capable of consuming in a continuous foraging bout. Therefore, the one value of $F(x,t,T)$ that is known at the outset is the terminal one, $F(x,T,T) = x$ J. The end condition $F(x,T,T)$ is 50 J only for a fully satiated guppy.

The dynamic programming equation defining the fitness function is

$$F(x,t,T) = \lambda_0 (1 - \beta_S) F(x - \alpha_0, t+1, T) + \Sigma \lambda_i \max [(1 - \beta_S) F(x - \alpha_0, t+1, T)] ; \quad (3)$$
$$\varphi_i (1 - \beta_H)^{\tau i} F(x - \alpha_i + Y_i, t + \tau_i, T) + (1 - \varphi_i)(1 - \beta_H)^{\tau ius} F(x - \alpha_{ius}, t + \tau_{ius}, T))$$

The dynamics of the state variable x are as follows. The forager is assumed to encounter prey items sequentially and at random, and no significant prey depletion occurs over the foraging period T. At any given discrete time period t, the forager either encounters a prey item of type i with probability λ_i or does not encounter any prey, with probability $\lambda_0 = 1 - \Sigma \lambda_i$. The prey encounter rate parameter λ_i approximates the forager's probability of encountering prey type i during a small discrete time interval (i.e. as $dt \rightarrow 0$; cf. Mangel and Clark 1988). As expressed by equation 3 for $t < T$, changes in the value of the fitness function between successive time periods depends on stochastic events associated with predation, encountering prey of different types, and the forager's success at attacking encountered prey.

The first expression on the right-hand side of the equality sign in equation 3 specifies the change in the fitness function if no prey is encountered (with probability λ_0) during time period t. The state variable $X(t)$, that is the current amount of energy (x) in the guppy's stomach, is reduced by its metabolic expenditure during that period. Thus, $X(t) = x - \alpha_0$, where α_0 = RMR x t. RMR is the fish's routine metabolic rate, estimated at 0.0016 J/s (after Sage 1971), and t = 1 s. If the forager survives predation, with probability $(1 - \beta_S)$, it then enters the next time period $t + 1$ with the state variable $X(t + 1)$. The next term in the equation defines the change in the fitness function if a particular prey item of type i is encountered during period t (with probability λ_i). The forager either ignores or attacks the encountered prey item under a certain risk of mortality to predation. The optimal foraging decision is the one which maximizes the value of the particular fitness function on one side or the other of the semicolon in equation 3.

The expression on the left-hand side of the semicolon specifies the change in the value of the fitness function if the encountered prey of type i is ignored. In this case, the state variable x is decreased by the metabolic cost α_0. Therefore, if the forager survives predation in period t (with probability $1 - \beta_S$), it then enters the next time period $t + 1$ with the state variable $X(t + 1)$. The lengthy expression on the right-hand side of the semicolon defines the change in the state variable x when the encountered prey of type i is attacked. An attack has two stochastic outcomes, either a successful ingestion with probability φ_i or a miss with probability $1 - \varphi_i$, which have different metabolic costs. First, let's consider the chance event that the prey item is successfully attacked. In this case, the forager incurs the metabolic cost (α_i) of attacking, pursuing and mandibulating (ingesting) the prey item. This takes a total handling time of τ_i. The forager, however, gains Y_i energy units from ingesting the prey. Since the forager's prey handling is assumed to detract from vigilance for predators, its risk of predation is thus modified by handling time τ_i, and its probability of surviving predation during period τ_i becomes $(1 - \beta_H)^{\tau_i}$. If the forager survives predation during this period, it then enters the next period $t + \tau_i$ with the state variable $X(t + \tau_i)$. Alternatively, with probability $1 - \varphi_i$, the encountered prey item of type i is attacked and pursued, but not captured and ingested. The forager therefore incurs the metabolic cost of an unsuccessful attack (α_{ius}) during the elapsed handling period τ_{ius}, without of course gaining any energy from foraging. If the forager survives this time period, with probability $(1 - \beta_H)^{\tau_{ius}}$, it then enters the next period $t + \tau_{ius}$ with the state variable $X(t + \tau_{ius})$.

The metabolic cost α_i is the sum of the metabolic costs of attack (C_A), pursuit (C_P) and mandibulation (C_M). Since the guppy's swimming speed was greatest during the attack phase, intermediate during pursuit and lowest during prey mandibulation in our study, the metabolic costs of these different phases of the predation cycle are expected to be dissimilar. Similarly, the total handling time elapsed during a successful attack (τ_i) is the sum of the attack time (h_{Ai}), pursuit time (h_{Pi}) and mandibulation time (h_{Mi}) for prey type *i*, whereas the handling time for an unsuccessful attack is $\tau_{ius} = h_{Ai} + h_{Pi}$. The guppy's metabolic rate during attack (AMR) is estimated to be 22% higher than its routine metabolic rate (RMR), following Puckett and Dill's (1984) estimates of the metabolic cost of attacking prey items in salmon. Unfortunately, no such data exist for the guppy. Thus, AMR = 1.22 x RMR and C_A = AMR x h_{Ai}. Given that the guppy's swimming speed during prey pursuit was only about one quarter of that during the attack phase, its metabolic rate during pursuit (PMR) was estimated to be only 5% higher than its RMR. Thus, PMR = 1.05 x RMR and C_P = PMR x h_{Pi}. The metabolic cost of prey mandibulation is C_M = RMR x h_{Mi}. Therefore, the estimated total metabolic cost of a successful attack on prey *i* is $\alpha_i = C_A + C_P + C_M$, and that for an unsuccessful attack is $\alpha_{ius} = C_A + C_P$. The values of the other parameters in equation 3 are summarized in Table 2. Although total prey handling times are shown, their component attack, pursuit and mandibulation times are not for brevity; they will be reported elsewhere when the study is completed.

There are at least three important features of equation 3. First, the optimal prey choice decision will depend on the current value of the state variable *x* and the current time *t*. Second, the forager's fitness in the current period *t* will affect its fitness in the next period, until the terminal period *T* is reached. Third, the forager's probability of surviving predation during any given discrete time period is a function of its decision to ignore or attack any encountered prey item of type *i* and of the handling time associated with this prey.

The model essentially determines the optimal prey choice decision rule (i.e. ignore or attack) for each prey type encountered when it computes the forager's fitness function $F(x,t,T)$, defined by equation 3, for each value of the state variable *x* and each time period *t*. Since the value of the terminal fitness function is given as $F(x,T,T) = x$ J, the expression on the right-hand side of the equality sign in equation 3 is known, and therefore the fitness function for the penultimate period $t = T - 1$ can be computed as $F(x,T\text{-}1,T)$. The model then continues to derive sets of optimal prey choice decisions as it similarly computes the fitness function for each discrete time period from T to t = 1 by a backward iterative process. Details of the algorithm for dynamic programming equations like equation 3 are given in Mangel and Clark (1988).

Simulated Diet

For the sake of brevity, I will not present here the numerous sets of optimal prey choice decisions generated by the dynamic model described above. Suffice it to mention that the optimal prey choice decisions depended on the risk of predation, the current value of the forager's state x and the current time t. However, as time to go $(T - t)$ in the foraging bout increased (i.e. as $T \rightarrow t = 1$), the prey choice decision rules became relatively insensitive to the value of t and depended essentially only on the value of the forager's state (x). This phenomenon is referred to as stationarity (cf. Mangel and Clark 1988). Consequently, we assumed that the guppy begins a foraging period of duration T with a set of fixed prey choice decision rules for each value of its state x, as determined by the model for $t = 1$. Using this particular set of decision rules (see Fig. 7) and the parameter values shown in Table 2 and described above, we simulated an average diet for a sample of 100 guppies, each foraging separately on three *Daphnia* size classes (Table 2) within a patch for an arbitrary period of $T =$ 200 s under both the absence and presence of a predator. Each simulated fish had a small amount of food (= 2 J) in its stomach at the start of the foraging period.

Fig. 7 summarizes the optimal prey choice decision rules as a function of the guppy's current energetic state (x) for both the predator absent and predator present situations. In the former situation (Fig. 7, upper panel), the optimal policy is for the guppy to attack any one of the three *Daphnia* types when encountered until it is about 90% satiated (= 45 J in stomach), at which point it should ignore the largest *Daphnia* if encountered. When the fish is very close to complete satiation, it should only attack the smallest prey and ignore the other two prey types if encountered. Using this set of decision rules, the model simulated a diet biased towards the largest *Daphnia* and which closely matched the observed guppy diet in the absence of the cichlid predator (Fig. 5A). In the presence of a predator (Fig. 7, lower panel), the optimal prey choice decisions differed markedly from those described above. The guppy should attack all three prey types upon encounter when it is hungry. However, it should ignore the largest *Daphnia* when only about 20% satiated (= 10 J) and continue to attack the small and medium *Daphnia* upon encounter until it reaches 90% satiation, at which point the medium-sized prey is dropped in turn from the optimal diet. Finally, the guppy should ignore all prey types when it is almost completely satiated. Using this particular set of decision rules, the model simulated a diet biased towards the smallest *Daphnia* and which also closely matched the observed guppy diet in the presence of a cichlid predator (Fig. 5B).

The similarity between the observed (Fig. 4) and simulated (Fig. 5A) guppy diets in the absence of a predator and the observed predator-mediated shift in prey selectivity (Fig. 4), which was predicted by the dynamic model (Figs. 5B and 7) but not by the static diet models

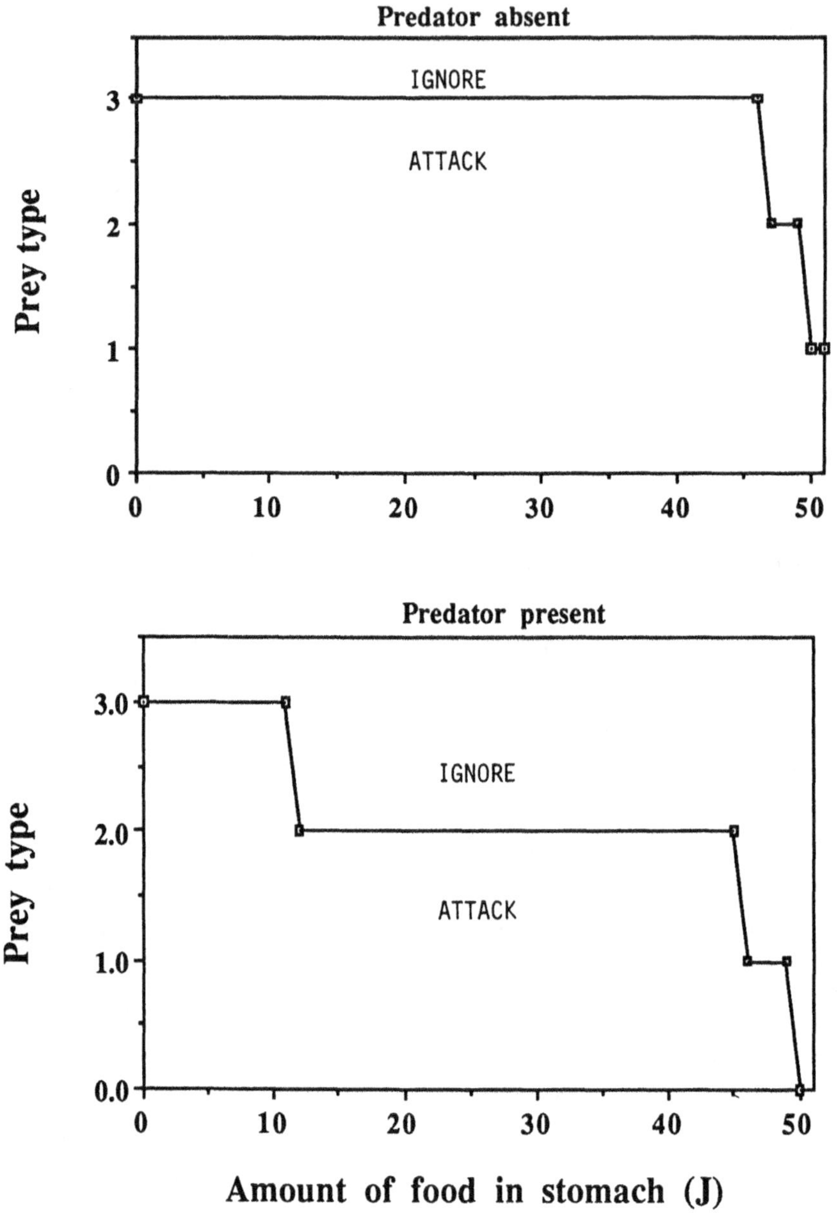

Figure 7. Optimal prey choice decision rules (i.e. attack or ignore prey), as a function of the current amount of food (energy) in the guppy's stomach, when foraging in the absence (upper panel) and presence (lower panel) of a predator. A completely satiated fish has 50 J of food in its stomach. Prey types above the plotted line should be ignored, whereas those below the line should be attacked. Prey types 1, 2 and 3 denote small, medium and large *Daphnia*, respectively. The values of the parameters used in the model are stated in the text.

tested (Fig. 5B), strongly suggest that selection by the guppy for prey types that differ apparently only in size reflects a trade-off between the rate of net energy intake and the risk of mortality to predation. The guppy would thus appear to have used prey choice decision rules in our study that are expected to maximize fitness under the risk of predation.

SUMMARY

Animals can expect to encounter predators by chance during their lifetime. An individual's risk of predation will depend on the abundance and behaviour of its predators and may also be determined to some extent by its own behaviour. For example, foraging may increase the animal's probability of mortality to predation, perhaps because searching for and handling prey items detract from being vigilant for predators. Animals are therefore expected to be sensitive to the risk of predation when making foraging decisions and to trade off foraging gains for greater safety from predation in a manner that maximizes fitness. Considerable evidence indicates that animals can indeed assess their risk of predation and can modify various aspects of their foraging behaviour in ways that reflect such a trade-off. However, the effects of predators on the within-patch (habitat) prey choice decisions of foragers are largely unknown.

In this paper I review the evidence for the presumed greater probability of death to predation associated with foraging and for predator-mediated modifications in the within-patch diet selectivity of animals that are consistent with a trade-off between foraging and the risk of predation. Several studies have documented changes in the foraging behaviour of animals, in response to an increased risk of predation within the currently exploited food patch (habitat), that have implications for their diet selectivity. In some of these studies, the foraging animal altered its preference for available prey types when it perceived the risk of predation to have increased. In situations where the risk of predation associated with eating different prey types varies, predator-mediated shifts in prey selectivity would apparently reduce the time the forager is vulnerable to predation and(or) enhance its vigilance for predators, thereby decreasing the probability of death to predation but at a cost of reduced energy intake rate. Moreover, the forager may also adjust the time it takes to handle accepted individual prey items to reduce its probability of mortality to predation. A recent example of this strategy is the ongoing work on guppies in my laboratory. These fish shifted their preference from large *Daphnia* to the smallest ones, which require shorter handling times, and reduced the time taken to pursue individual prey in the presence of a fish predator.

Such predator-mediated shifts in prey selectivity, and resulting partial preferences, and flexibility in prey handling time cannot be accounted for by classical models of foraging behaviour. Most of these models are static and therefore cannot adequately address behavioural trade-offs. However, more recent dynamic models, which incorporate the changing internal state of the animal and the risk of predation, are playing an increasingly important role in functionally explaining predation-risk – related trade-offs in animal decision making and in focusing research on the underlying mechanisms of such phenomena. I describe in this paper a dynamic prey selection model that predicted, with relatively high accurary, the observed dietary composition of guppies foraging in both the absence and presence of a predator.

Based on the evidence reviewed, it appears that the risk of predation does not simply act as a constraint on foraging behaviour, but rather animals have evolved the ability to make foraging decisions under the risk of predation, such as altering their prey choice and prey handling time, that maximize fitness. More attention should therefore be directed in the future on the effects of predation risk on the underlying behavioural mechanisms of food selection in animals.

ACKNOWLEDGMENTS

The research on guppies partially reported in this paper was carried out in collaboration with J. Miller and was supported by research grants from NSERC Canada and the Donner Canadian Foundation awarded to me. The dynamic diet selection model described herein was largely developed in collaboration with M. Mangel while we were both on sabbatical leave in the Edward Grey Institute at the University of Oxford. J.R. Krebs and C.M. Perrins kindly provided us with space and facilities there, and my leave was supported by NSERC Canada and the Royal Society (London) under their bilateral scientific exchange program and the M.Y. Bell Travel Fund at Mount Allison University. I thank R.L. Caldwell, J.F. Gilliam, V. Gotceitas, M. Mangel and A. Sih for their helpful comments on parts of or the entire manuscript in an earlier form.

REFERENCES

Abrahams MV, Dill LM (1989) A determination of the energetic equivalence of the risk of predation. Ecology (In press)

Barnard CJ, Brown CAJ (1981) Prey size selection and competition in the common shrew (*Sorex araneus* L.). Behav Ecol Sociobiol 8: 239-243

Belovsky GE (1978) Diet optimization in a generalist herbivore: the moose. Theor Popul Biol 14: 105-134

Bowers MA (1988) Seed removal experiments on desert rodents: the microhabitat by moonlight effect. J Mamm 69: 201-204

Bence JR, Murdoch WW (1986) Prey size selection by the mosquitofish: relation to optimal diet theory. Ecology 67: 324-336
Caldwell GS (1986) Predation as a selective force on foraging herons: effects of plumage color and flocking. Auk 103: 494-505
Charnov EL (1976) Optimal foraging: attack strategy of a mantid. Am Nat 110: 141-151
Confer JL, Blades PI (1975) Omnivorous zooplankton and planktivorous fish. Limnol Oceanogr 20: 571-579
Confer JL, Howick GL, Corzette MH, Kramer SL, Fitzgibbon S, Landesberg R (1978) Visual predation by planktivores. Oikos 31: 27-37
Curio E (1976) The ethology of predation. Springer-Verlag, Berlin Heidelberg New York
Dill LM (1987) Animal decision making and its ecological consequences: the future of aquatic ecology and behaviour. Can J. Zool 65: 803-811
Dill LM, Fraser AHG (1984) Risk of predation and the feeding behavior of juvenile coho salmon (*Oncorhynchus kisutch*). Behav Ecol Sociobiol 16: 65-71
Dugatkin LA (1988) Do guppies play TIT FOR TAT during predator inspection visits? Behav Ecol Sociobiol 23: 395-399
Edwards J (1983) Diet shifts in moose due to predator avoidance. Oecologia (Berlin) 60: 185-189
Eggers DM (1977) The nature of prey selection by planktivorous fish. Ecology 58: 46-59
Eggers DM (1982) Planktivore preference by prey size. Ecology 63: 381-390
Engen S, Stenseth NC (1975) A general version of optimal foraging theory: the effect of simultaneous encounters. Theor Popul Biol 26: 192-204
Foster JR (1980) Factors influencing the predator-prey relations of a small esocid, the grass pickerel *Esox americanus vermiculatus*. Unpubl. Ph.D. thesis, University of Toronto, Toronto
Fraser DF, Gilliam JF (1987) Feeding under predation hazard: response of the guppy and Hart's rivulus from sites with contrasting predation hazard. Behav Ecol Sociobiol 21: 203-209
Fraser DF, Huntingford FA (1986) Feeding and avoiding predation hazard: the behavioral response of the prey. Ethology 73: 56-68
Freed LA (1981) Breeding biology of House Wrens: new views of avian life history phenomena. Unpubl. Ph.D. thesis. University of Iowa
Getty T, Krebs JR (1985) Lagging partial preferences for cryptic prey: a signal detection analysis of great tit foraging. Am Nat 125: 39-60
Gilliam JF, Fraser DF (1987) Habitat selection under predation hazard: test of a model with foraging minnows. Ecology 68: 1856-1862
Godin J-GJ (1986) Risk of predation and foraging behaviour in shoaling banded killifish (*Fundulus diaphanus*). Can J Zool 64: 1675-1678
Godin J-GJ, Rangeley RW (1989) Living in the fast lane: effects of cost of locomotion on foraging behaviour in juvenile Atlantic salmon. Anim Behav 37: 943-954
Godin J-GJ, Smith SA (1988) A fitness cost of foraging in the guppy. Nature (London) 333: 69-71
Godin J-GJ, Sproul CD (1988) Risk taking in parasitized sticklebacks under threat of predation: effects of energetic need and food availability. Can J Zool 66: 2360-2367
Hall DJ, Werner EE, Gilliam JF, Mittelbach GG, Howard D, Doner CG, Dickerman JA, Stewart AJ (1979) Diel foraging behavior and prey selection in the golden shiner (*Notemigonus crysoleucas*). J Fish Res Board Can 36: 1029-1039
Hay ME, Fuller PJ (1981) Seed escape from heteromyid rodents: the importance of microhabitat and seed preference. Ecology 62: 1395-1399
Heller R, Milinski M (1979) Optimal foraging of sticklebacks on swarming prey. Anim Behav 27: 1127-1141
Houston A, Clark C, McNamara J, Mangel M (1988) Dynamic models in behavioural and evolutionary ecology. Nature (London) 332: 29-34
Houston A, McNamara J (1985) The choice of two prey types that minimises the probability of starvation. Behav Ecol Sociobiol 17: 135-141
Hughes RN (1979) Optimal diets under the energy maximization premise: the effects of

recognition time and learning. Am Nat 113: 209-221
Jaeger RG, Nishikawa KCB, Barnard DE (1983) Foraging tactics of a territorial salamander: cost of territorial defence. Anim Behav 31: 191-198
Kagel JH, Green L, Caraco T (1986) When foragers discount the future: constraint or adaptation? Anim Behav 34: 271-283
Krebs JR, Erichsen JT, Webber MI, Charnov EL (1977) Optimal prey-selection by the great tit (*Parus major*). Anim Behav 25: 30-38
Krebs JR, Stephens DW, Sutherland WJ (1983) Perspectives in optimal foraging. In: Brush AH, Clark GA Jr (eds) Perspectives in ornithology. Cambridge University Press, Cambridge pp. 165-216
Lanchester BS, Mark RF (1975) Pursuit and prediction in the tracking of moving food by a teleost fish (*Acanthaluteres spilomelanurus*). J Exp Biol 63: 627-645
Lea SEG (1979) Foraging and reinforcement schedules in the pigeon: optimal and non-optimal aspects of choice. Anim Behav 27: 875-886
Lima SL (1987) Vigilance while feeding and its relation to the risk of predation. J Theor Biol 124: 303-316
Lima SL (1988a) Maximizing feeding efficiency and minimizing time exposed to predators: a trade-off in the black-capped chickadee. Oecologia (Berlin) 66: 60-67
Lima SL (1988b) Vigilance and diet selection: a simple example in the dark-eyed junco. Can J Zool 66: 593-596
Lima SL, Valone TJ, Caraco T (1985) Foraging-efficiency – predation-risk trade-off in the grey squirrel. Anim Behav 33: 155-165
Lima SL, Valone TJ (1986) Influence of predation risk on diet selection: a simple example in the grey squirrel. Anim Behav 34: 536-544
Lima SL, Dill LM (1990) Behavioral decisions made under the risk of predation: a review and prospectus. Can J Zool (In press)
Lucas JR (1985) Time constraints and diet choice: different predictions from different constraints. Am Nat 126: 680-705
Lucas JR (1987) The influence of time constraints on diet choice of the great tit, *Parus major*. Anim Behav 35: 1538-1548
Magurran AE, Pitcher TJ (1987) Provenance, shoal size and the sociobiology of predator-evasion behaviour in minnow shoals. Proc R Soc Lond B 229: 439-465
Mangel M, Clark CW (1988) Dynamic modeling in behavioral ecology. Princeton University Press, Princeton
Martindale S (1982) Nest defense and central place foraging: a model and experiment. Behav Ecol Sociobiol 10: 85-89
McNair JN (1981) A stochastic foraging model with predator training effects. II. Optimal diets. Theor Popul Biol 19: 147-162
McNamara JM, Houston AI (1987) Partial preferences and foraging. Anim Behav 35: 1084-1099
Metcalfe NB, Huntingford FA, Thorpe JE (1987a) The influence of predation risk on the feeding motivation and foraging strategy of juvenile Atlantic salmon. Anim Behav 35: 901-911
Metcalfe NB, Huntingford FA, Thorpe JE (1987b) Predation risk impairs diet selection in juvenile salmon. Anim Behav 35: 931-933
Milinski M (1982) Optimal foraging: the influence of intraspecific competition on diet selection. Behav Ecol Sociobiol 11: 109-115
Milinski M (1984) A predator's costs of overcoming the confusion-effect of swarming prey. Anim Behav 32: 1157-1162
Milinski M (1986) Constraints placed by predators on feeding behaviour. In: Pitcher TJ (ed). The behaviour of teleost fishes. Croom Helm, London pp. 236-252
Milinski M (1987) TIT FOR TAT in sticklebacks and the evolution of cooperation. Nature (London) 325: 433-435
Milinski M, Heller R (1978) Influence of a predator on the optimal foraging behaviour of sticklebacks (*Gasterosteus aculeatus* L.). Nature (London) 275: 642-644
O'Brien WJ, Slade NA, Vinyard GL (1976) Apparent size as the determinant of prey selection

by bluegill sunfish (*Lepomis macrochirus*). Ecology 57: 1304-1310.
Ohguchi O (1981) Prey density and selection against oddity by three-spined sticklebacks. Z Tierpsychol Suppl 23: 1-79
Pitcher TJ (1980) Some ecological consequences of fish school volumes. Freshwat Biol 10: 539-544
Pitcher TJ, Green DA, Magurran AE (1986) Dicing with death: predator inspection behaviour in minnow shoals. J Fish Biol 28: 439-448
Pitcher TJ, Lang SH, Turner JA (1988) A risk-balancing trade off between foraging rewards and predation hazard in a shoaling fish. Behav Ecol Sociobiol 22: 225-228
Potts GW (1983) The predatory tactics of *Caranx melampygus* and the response of its prey. In: Noakes DLG, Lindquist DG, Helfman GS, Ward JA (eds). Predator and prey in fishes. Dr. W. Junk Publ., The Hague Boston London, pp. 181-191
Puckett KJ, Dill LM (1984) Cost of sustained and burst swimming to juvenile coho salmon (*Oncorhynchus kisutch*). Can J Fish Aquat Sci 41: 1546-1551
Pyke GH (1984) Optimal foraging theory: a critical review. Ann Rev Ecol Syst 15: 523-575
Rechten C, Avery M, Stevens A (1983) Optimal prey selection: why do great tits show partial preferences. Anim Behav 31: 576-584
Sage M (1971) Respiratory response of the teleost *Poecilia reticulata* to rising temperature. Comp Biochem Physiol 40A: 851-853
Sih A (1987) Predators and prey lifestyles: an evolutionary and ecological overview. In: Kerfoot WC, Sih A (eds). Predation: direct and indirect impacts on aquatic communities. University Press of New England, Hanover, NH. pp. 203-224
Stephens DW (1985) How important are partial preferences? Anim Behav 33: 667-669
Stephens DW, Charnov EL (1982) Optimal foraging: some simple stochastic models. Behav Ecol Sociobiol 10: 251-263
Stephens DW, Krebs JR (1986) Foraging theory. Princeton University Press, Princeton
Snyderman M (1983) Optimal prey selection: the effects of food deprivation. Behav Anal Letters 3: 359-369
Vance RR, Schmitt RJ (1979) The effect of the predator-avoidance behavior of the sea urchin, *Centrostephanus coronatus*, on the breadth of its diet. Oecologia (Berlin) 44: 21-25
Ware DM (1975) Relation between egg size, growth, and natural mortality of larval fish. J Fish Res Board Can 32: 2503-2512
Werner EE, Gilliam JF (1984) The ontogenetic niche and species interactions in size-structured populations. Ann Rev Ecol Syst 15: 393-425
Werner EE, Gilliam JF, Hall DJ, Mittlebach GG (1983) An experimental test of the effects of predation risk on habitat use in fish. Ecology 64: 1540-1548
Wetterer JK, Bishop CJ (1985) Planktivore prey selection: the reactive field volume model vs. the apparent size model. Ecology 66: 457-464
Wootton RJ (1979) Energy costs of egg production and environmental determinants of fecundity in teleost fishes. Symp Zool Soc London 44: 133-159

ORAL DISCUSSION – J.-G. Godin

Q. (Milinski) As Jakobsenson and Johnson have shown for *Bosmina*, and I have seen for *Daphnia magna*, larger animals flee from a disturbance caused by the predators more than do small ones. I think your guppies, as well as my sticklebacks in a similar set–up, will prefer the back part of a tank. That means the bigger *Daphnia* will flee this part of the tank more quickly than the smaller ones.
A. We have actually done some measurements on this. The guppy reduces its activity. It often does go the to back of the tank and it does do a lot of predator inspection.
Milinski: Is it more active in the back or the front ?
A. It is more active in the back half of the tank than the front half. *Daphnia* came right through its visual field very often and the guppy did not go after them. The frequencies of encounter with larger *Daphnia* were still higher than with small *Daphnia*. The guppy would often visually fixate large *Daphnia* within 3 – 4 cm of its mouth but was likely to ignore them in the presence of a predator. But as soon as a small *Daphnia* was encountered in the presence of a predator, the guppy quickly went after it.

Q. (Sih) Is it feasible for guppies that are not moving around much to actually observe a number of encounters and measure all those components. Can you justify your use of an indirect measure of encounter rate ?
A. I used two measures that are both commonly reported in the literature and obtained similar relative encounter rates. One method used reaction distances and the other actual number of attacks on the prey, divided by total search time minus the handling time, to estimate encounter rates. With these two independent measures, I derived encounter rates that were almost identical.

INTERACTING EFFECTS OF PREDATOR AND PREY BEHAVIOR IN DETERMINING DIETS

Andrew Sih and Robert D. Moore
T.H. Morgan School of Biological Sciences
University of Kentucky
Lexington, Kentucky 40506
U.S.A.

A predator's diet is the array of food types eaten by that predator. From an ecological and evolutionary standpoint, diets are particularly interesting when predators show non-random preferences; that is, when the relative abundances of different prey types in the diet differ from a random sampling of the environment (Chesson 1983). Non-random diets are ecologically important, because the effects of predators on prey communities often appear to be mediated by non-random predation (Paine 1966; Zaret 1980; Sih et al. 1985). Given that predator preferences are ecologically important, ecologists have been interested in the determinants of preference. In behavioral ecology, the dominant theme has been to interpret behavior in terms of its adaptive significance (Krebs and Davies 1984). To explain non-random diets, behavioral ecologists have used optimal diet theory (ODT) which, in its simpler forms predicts diets based on the notion that predators choose prey types that result in maximum net energy intake (Stephens and Krebs 1986).

There is a crucial distinction, however, between predator preference and what is addressed by ODT. ODT makes predictions on active predator choice; i.e., given an encounter, what should predators choose to attack? This decision is only one component of the overall predation process that determines preference. Preference can also be influenced by non-random encounter rates, and unequal capture and consumption probabilities. These components can be heavily influenced by both predator and prey behaviors. Here, we will: 1) Discuss prey and predator behaviors that can influence components of preference; 2) Note that the diet

NATO ASI Series, Vol. G 20
Behavioural Mechanisms of Food Selection
Edited by R. N. Hughes

patterns predicted by ODT can be generated by adaptive prey behavior in the absence of active predator choice; alternatively, prey behavior can generate diets that differ from those predicted by ODT; 3) Re-analyze some existing literature to suggest that if prey are mobile (relative to predators), then observed diets do not fit ODT predictions very well; 4) Survey studies on the diets of aquatic invertebrate predators; the overall theme will be that in this group, prey behavior appears to be more important than predator behavior in determining diet preferences; and 5) Present our own new data on variation in the preference of salamander larvae given a choice between two invertebrate prey, isopods and amphipods; we will show how path analysis can be used to quantify the relative importance of different pathways in determining patterns of preference.

COMPONENTS OF THE PREDATOR-PREY INTERACTION

A predator-prey interaction can be divided into stages.

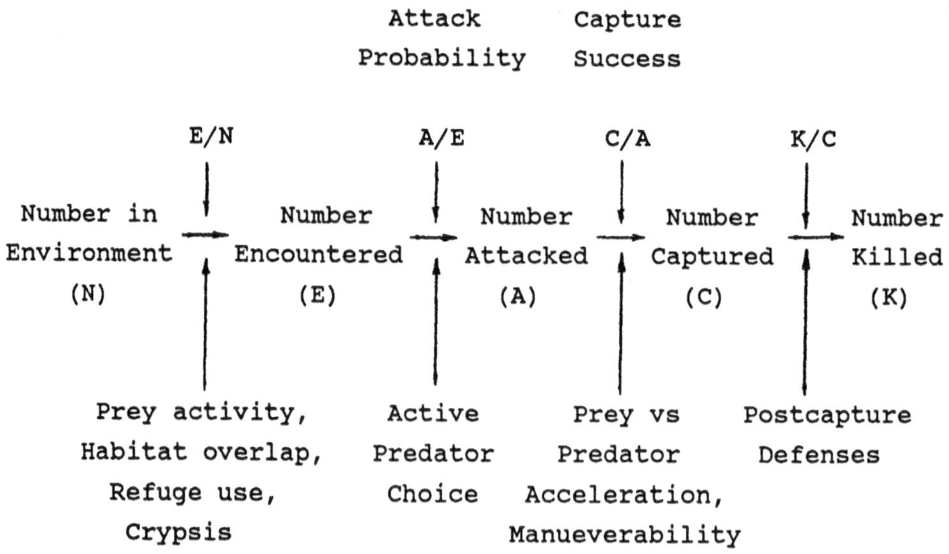

Figure 1: Components of a predator-prey interaction.

The key point is that active predator choice is only one of several factors that contribute to preference. Preference is proportional to K/N, where K is the number of prey eaten per unit time and N is the number of prey in the environment. For example, with two prey types (1 and 2), Chesson's (1983) measure of preference for type 1 is: (K1/N1)/(K1/N1 + K2/N2). Four factors affect K/N: 1) relative encounter rate, E/N; 2) relative attack probability, A/E; 3) relative capture success, C/A, and relative consumption probability, K/C (E is the number of encounters/time, A is the number of attacks/time, and C is the number of captures/time.

ODT makes predictions about attack probabilities. Theory predicts that the decision to attack or not should hinge on the prey item's relative value (as compared to the value of other items) and on the predator's encounter rate with prey of higher value. Although prey behavior can indirectly influence attack probability by altering prey value, the decision to attack or not would seem to be ultimately, largely the predator's decision.

In contrast, the other stages of the predation act are clearly influenced heavily by both prey and predator behaviors. Encounter rates depend on habitat overlap between predators and prey and on the movement velocities of both. Predators attempt to increase habitat overlap with prey by aggregating in patches with more prey (Hassell 1978; Stephens and Krebs 1986); in turn, prey reduce habitat overlap with predators by using refuges, areas that predators do not use, or areas where predators are ineffective at capturing prey (Sih 1987a). In a similar vein, predators can increase encounter rates with prey by increasing foraging speed; while prey often respond to predators by decreasing prey activity and thus encounter rates with predators (Sih 1987a,1987b). Predator diets can obviously be affected by the relative responses of different prey types to predators. For example, imagine an ambush predator faced with two prey types. If one prey type is largely inactive, while the other is very active, then even if both prey are equally abundant, and if both prey are equally likely to be attacked or captured (given an encounter), the prey type that is more active will be preferred simply because it encounters the predator more often; that is, if all else is the same, preference can be determined by prey activity.

Similarly, capture success (or to use the prey view, escape success) depends on the relative abilities of predators and prey. For example, in a pursuit-flee scenario, capture success will depend on the relative velocity, acceleration and manueverability of prey versus predator. Again, if all else is the same, predators will tend to capture (and thus prefer) prey that have relatively poor escape ability. The same can be said for postcapture events.

The above points seem obvious, but have often been overlooked by empiricists testing ODT. Although ODT addresses active predator choice, many tests of the theory actually examine predator preferences. In some, often highly artificial systems (e.g., birds feeding on mealworms on a conveyor belt), the investigator determines encounter rates and capture success is essentially 100%. These studies directly study active predator choice. However, in many other more natural systems, investigators test ODT predictions by comparing prey relative abundances in the environment versus in the diet. In experimental studies, investigators control prey abundance in the environment, while in observational studies, prey abundance is not manipulated. In either case, non-random diets are often interpreted as being due to active predator choice, even in the absence of data on the relative roles of the 4 components (figure 1) that can influence preference. Our simple suggestion is that empiricists should consider alternative hypotheses including both prey and predator behaviors when studying mechanisms that determine preference.

PREY BEHAVIOR AND THE PREDICTIONS OF OPTIMAL DIET THEORY

This section explores the possibility that variation in prey behavior can result in patterns of preference that concur with the predictions of ODT. In that case, studies that corroborate ODT, but do not directly quantify active predator choice, might find the "right answer for the wrong reason". Alternatively, adaptive prey behavior might result in diet patterns that do not fit ODT; this condition is particularly interesting if it helps us to understand existing studies where observed patterns do not fit ODT. Here, we present verbal arguments; details will be provided in a later paper.

The standard predictions of ODT are: 1) Predators should prefer more profitable prey; 2) Preference for more profitable prey should increase with increasing abundance of these prey; that is when higher value prey are more abundant, lower value prey should be dropped from the optimal diet; and 3) Preference for less profitable prey should not be affected by their own abundance. Again, the theory actually addresses active predator choice, not preference. We state the predictions in terms of preference because empirical tests of ODT often quantify preference rather than active predator choice.

Prey value can be defined as: $(Ec-(1-c)X)/H$ where E is net energy gain if prey are captured, c is capture success (C/A), X is energy cost if prey are attacked but not captured and H is handling time. Some empiricists have expressed prey value as simply E/H; that is, they have left out capture success. This is incorrect; obviously, foragers only gain energy from prey if prey are successfully captured. The fact that capture success is part of prey value results in an automatic positive relationship between prey value and preference. We thus have two hypotheses that yield prediction 1 via very different cause-and-effect pathways. The conventional ODT view is that an increase in the value of a given prey type should alter predator behavior (predators should increase their attack probability on that prey type) such that preference for that prey type increases. An alternative view is that capture success is, by definition, positively related to both prey value and preference. A positive correlation between prey value and preference can then be due simply to cross-correlations with capture success. Prediction 1 should thus be upheld even in the absence of active predator choice.

Prey behavior can also generate an alternative view on prediction 2. Imagine a system where a prey type has a limited number of refuges (R) and all refuges are filled. As prey abundance (N) increases, the proportion of prey that are in refuge (R/N) decreases. Assuming that prey that are out of refuge are more likely than prey in refuge to encounter predators and to be captured given an attack, the result should be an increase in mean capture success and a disproportionate increase in mean encounter rate. Both should tend to increase preference for that prey type even if predators do not alter their patterns of active predator choice. Thus prediction 2 can also be upheld even in the absence of active predator choice.

Note, however, that the above mechanism should also work for less profitable prey; i.e., for any prey type with a limited number of refuges, an increase in prey abundance should result in an increase in preference. Stephens and Krebs' (1986) literature review found several cases where an increase in the relative abundance of poor prey resulted in an increased tendency for predators to attack that prey type. Second generation optimal diet models have offered several mechanisms that can explain this discrepancy; to these we add the limited refuges mechanism discussed above.

Adaptive prey behavior can also generate other diet patterns that do not fit ODT. For example, prey "safety in numbers" can result in a negative relationship between prey abundance and preference. Many studies have noted that prey enjoy "safety in numbers" due, for example, to earlier prey detection of predators or increased predator confusion (Bertram 1978; Pulliam and Caraco 1984). Both mechanisms should decrease capture success and thus decrease prey value. Decreased capture success should decrease preference even in the absence of active predator choice. If predators obey optimal diet criteria, then a decrease in prey value should result in reduced attack probability, thus further decreasing preference.

Our point is not that predators do not often exercise active choice, nor is it that most optimal diet studies that corroborate ODT predictions have done so for the wrong reason. Our point is simply that alternative hypotheses based on prey behavior should be considered. Variations in prey behavior can produce the diet patterns that fit the standard predictions of ODT in the absence of active predator choice. Given alternative hypotheses that yield the same predictions, empiricists should strive to gather data on all components of the predator-prey interaction to more fully understand mechanisms explaining preference.

MOBILE PREY AND OPTIMAL DIETS

Most of our comments about the effects of prey behavior on relative encounter rates or capture success do not apply if prey are immobile. Thus for predators that feed on immobile prey, preference might indeed be determined primarily by active predator choice; this suggests that ODT predictions are most likely to be upheld if prey are immobile. If prey are mobile, then prey behavior can determine preference and the resulting diet patterns might be less likely to fit ODT predictions. To test the notion that relative prey mobility (relative to predator mobility) is inversely related to the ability of ODT to explain dietary patterns, we re-analyzed a literature survey found in Stephens and Krebs (1986, Table 9.1). Stephens and Krebs' Table 9.1 provides the following information on each of 60 studies: 1) the types of forager and prey; 2) the quality of fit to ODT predictions (1 indicates a quantitative fit, 4 indicates a major discrepancy with ODT); 3) whether the study was a laboratory or field study; 4) whether the study was experimental or observational; 5) whether each of 4 predictions was tested or not; and 6) whether each of 4 assumptions of ODT was satisfied or not. See Stephens and Krebs (1986) for details. We did a multiple regression with "fit to ODT predictions" as the dependent variable, and each of 11 other factors as independent variables (relative prey mobility, lab vs. field, experimental vs. observational, and the 8 predictions and assumptions).

Except for relative prey mobility, the other independent variables were assigned values of 0 or 1. We devised 4 levels of relative prey mobility: 1) completely immobile (flowers, leaves, seeds, dead mealworms); 2) essentially immobile (bivalves); 3) slightly mobile (zooplankton vs. fish, insects or worms vs. birds) and 4) mobile (aquatic insect prey vs. aquatic insect predators, voles vs. stoats). Of the 60 studies that had full information, 28 (47%) involved completely immobile prey, 8 (13%) involved essentially immobile prey, 16 (27%) used slightly mobile prey, and 8 (13%) used mobile prey.

Partial correlation analysis showed a significant positive partial correlation coefficient ($r = 0.29$, $p < .05$) between relative prey mobility and quality of fit of results to ODT theory; that is, after all other factors are accounted for, relative prey mobility explains a significant portion of the variation in fit of results to theory. The positive coefficient indicates that when prey were more mobile (type 4), the study was more likely to yield results that do not fit ODT (fit=4). This analysis affirms the value of considering the effects of prey behavior on diets, whenever prey are capable of responding behaviorally to predators.

No other factor yielded a partial correlation that was significant at the .05 level. Surprisingly, the prediction that high value prey should be preferred was marginally significantly less likely than were other predictions to fit ODT ($p = .06$), while the prediction that high abundance of high value prey results in greater selectivity was more likely than other predictions to fit ODT ($p = .07$). Laboratory experiments showed a non-significant tendency to yield a poorer fit than field studies to ODT predictions ($p = .23$), and experimental studies showed a non-significant tendency to produce a better fit than observational studies to ODT predictions ($p = .23$).

Specific case studies where predators foraging for mobile prey did not fit ODT yielded interesting insights on how dominant theory can guide interpretations of results (e.g. Stein 1977; Erlinge 1981). Due to space constraints, we will not discuss these or other examples in detail; instead, we next survey recent literature on the important role of prey behavior in determining the diets of a particular group of predators, aquatic invertebrate predators.

DETERMINANTS OF PREFERENCE FOR AQUATIC INVERTEBRATE PREDATORS

Predator-prey ecologists working in aquatic systems, in particular in freshwater, have had an admirable record of taking data on several stages of the predation act. Various workers have noted that preference can be due either to active predator choice or to some combination of relative encounter rates and capture success. Pastorok (1981) coined the term "vulnerability" which he defined as the product of encounter rate x capture success. He suggested that preference patterns of a predatory midge (Chaoborus trivitattus) feeding on zooplankton could be equally well explained by prey value or by prey vulnerability. Midges rarely ate small prey that were relatively immobile and thus rarely encountered, or large prey that were too mobile and thus difficult to capture; medium-sized prey were most vulnerable and most often eaten. Medium-sized prey were also most profitable; small prey contained too little energy while large prey took too long to handle. Note, however, that the fact that predators preferred the most profitable prey does not demonstrate that they actively chose those prey. Active choice can only be demonstrated convincingly by having direct data on variation in attack probabilities.

To address the relative roles of various factors in determining preference, Sih (1987b) reviewed studies of the diets of 20 freshwater insect predators feeding on mobile prey (zooplankton, insect nymphs, tadpoles). We updated the survey to include 13 more aquatic predator-prey interactions (Riessen et al. 1984; Cooper et al. 1985; Greene and Landry 1985; Anholt 1986; Greene et al. 1986; Allan et al. 1987; Allan and Flecker 1988). For each study, we noted whether preference appeared to be explained by: prey movement, prey refuge use (both of these affect encounter rates between predators and different prey), active predator choice or capture success. In many cases, preference was affected by more than one factor. The results of our survey are summarized in Table 1.

Table 1: Determinants of preference for aquatic invertebrate predators feeding on mobile prey. Shown are the number of studies in which each of four factors had a signficant effect on preference.

Predators and Prey	# of Studies	Prey Movement	Refuge Use	Active Choice	Capture Success
Dragonfly and Damselfly Nymphs					
Zooplankton	6	6	0	0	0
Aquatic insects	2	0	2	0	1
Amphibian larvae	1	1	0	0	0
Aquatic Bugs and Beetles					
Zooplankton	9	8	3	2	4
Amphibian larvae	2	2	0	0	0
Stoneflies					
Mayfly, blackfly	4	3	0	2	4
Phantom Midges					
Zoolplankton	6	5	0	2	5
Leech, Copepod, Ctenophore					
Flies, Copepods	3	3	0	0	3
TOTALS	33	28	5	6	17
% of total		(85)	(15)	(18)	(52)

The factors that were most often important in determining diet preferences were relative prey movement and relative ease of capture; that is, predators tended to prefer either more active prey that were therefore encountered more often, or prey that were easy to catch. Using Pastorok's (1981) terminology, preference was determined primarily by prey vulnerability. Active predator choice only appeared to be important in 18% of the cases. In several instances, patterns of active choice (or lack of it) appeared to go against the predictions of ODT (Scott and Murdoch 1983; Spitze 1985; Allan and Flecker 1988). Refuge use was rarely implicated; however, most studies were done in simple laboratory containers that lacked refuges. In natural conditions where refuge use is often a crucial component of antipredator behavior, relative refuge use might often have important effects on diets.

MULTIVARIATE ANALYSES OF THE COMPONENTS OF PREFERENCE

The above survey included studies that varied considerably in the quality of data gathered and analyzed in support of conclusions. Some studies were designed to assess the role of active predator choice in determining preference. These studies often got direct evidence that active predator choice was not important, but collected little, if any, data on prey behavior. The authors' conclusion that prey behavior was important was based largely on anecdotal evidence. Other studies included detailed data on: 1) preference; 2) components of the predator-prey interaction (encounter rates, attack probabilities and capture success); and 3) prey and predator movement. These studies then based conclusions on comparisons of patterns across prey types. Often, they noted which prey were more or less preferred and then apparently simply eyeballed patterns of relative movement, encounter, attack or capture to deduce which factor explained variation in preference. In some cases, explicit statistical analyses were done to test the notion that a particular factor was indeed closely related to preference; for example, Cooper et al. (1985) used Spearman rank correlations to show that encounter rate and preference were significantly positively correlated. To our knowledge, no previous study used statistical tests to quantitatively assess the relative effects of various factors in determining preference.

Several standard multivariate tests should be relevant. The dependent variable is preference, and predictor variables could include components of predation (encounter rates, attack probabilities, and capture success) and mechanisms influencing these components (refuge use, prey and predator activity). The goal is to isolate the effects of each component on preference. Two related techniques that do this are partial correlation analysis and path analysis. Partial correlation coefficients measure the relationship between the dependent variable and a given predictor variable with all other predictor variables held constant and accounted for. If the usual assumptions of multiple regression are met, then standard methods exist for doing tests of significance on partial correlation coefficients.

Multiple regression and partial correlation assume that the predictor variables are independent. However, in many cases, we know that the predictor variables are related via a particular set of causal pathways. For example, even the simplest view of the components of preference yields the following set of relationships:

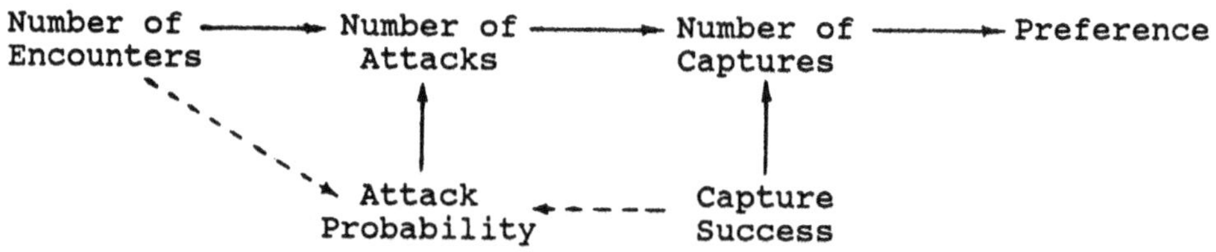

Figure 2: Relationships among components of predation that influence preference. If the number of prey in the environment is the same for all prey types, then preference is proportional to the number of captures. The number of captures is determined by the number of attacks and capture success given an attack; in turn, the number of attacks is determined by the number of encounters and attack probability.

Figure 2 is a simple path diagram. Each of the three components is linked to preference by a single path. In a more complex view, we can account for the fact that ODT predicts that encounter rates and capture success should causally affect attack probabilities. This would then yield additional paths (dashed lines) linking encounter rate and capture success to preference through effects on attack probability.

Path analysis is appropriate for assessing the relative importance of different routes linking dependent and predictor variables. It is particularly valuable where two points are connected by more than one pathway. Thus in the above example, path analysis would partition the overall effects of encounter rate on preference into a "direct" path through direct effects on the number of attacks and an "indirect" path through effects on attack probability. This partitioning could provide interesting insights that have not been previously addressed. For example, many studies have noted that an increase in the absolute abundance of a high value prey type results in an increase in preference for that type. ODT predicts that this should occur because the increased encounter rate with high value prey should cause predators to reduce their attack probability on low value prey. However, it is conceivable that the increase in preference occurs via other pathways; an increase in the abundance of high value prey might result in a disproportionate increase in encounter rate with that type (e.g., due to limited number of refuges) or it might increase attack probability on high value prey (as opposed to decreasing attack probability on low value prey). Path analysis can provide a measure of the relative importance of each of these causal pathways.

To account for differences in the magnitudes of different variables, all variables are standardized to a mean of zero and standard deviation of one. Path coefficients are standardized partial regression coefficients; path coefficients and partial correlation coefficients bear a simple relationship to each other (see Li 1976). The importance of a path ("path value") involving several steps is quantified by the product of the path coefficients along that path. Although there exist standard tests of significance for path coefficients, we know of no standard tests for the significance of path values. For a more detailed description of path analysis see Li (1976). For discussions of the use of path analysis in evolutionary biology see Lande and Arnold 1983, Arnold and Wade 1984, and Mitchell-Olds and Shaw 1987.

In the remainder of this paper, we will explore inferences gained from the use of multiple regression techniques (partial correlation analysis and path analysis) to examine the relative effects of encounter rates, attack probabilities and capture success on the diet preferences of insect predators (Cooper et al. 1985) and salamander larvae (our own new data).

CASE STUDY: THREE PREDATORY INSECTS - COOPER ET AL. 1985

Cooper et al. (1985) documented preferences, encounter

rates, attack probabilities and capture success for five aquatic predators (four insects and one fish) offered a choice of up to 14 prey types. By comparing differences among prey in each of these variables, they concluded that encounter rates were the most important determinant of predator preference, with capture success also playing a key role in some situations. We gleaned data from their figures and tables to do a quantitative analysis on the relative importance of different components on preference for three of the predators (a beetle, Acilius semiscultatus; a bug, Buenoa scimitra, and an odonate, Pachydiplax longipennis). Our goal was not to show that their conclusions were incorrect; in fact, our analysis yielded conclusions that corroborated theirs. Instead, our goal was to explore quantitative techniques for drawing inferences on components of preference.

For each of the three predator species, we ran partial correlation analyses and path analyses (using the path diagram in figure 2) to isolate the effects of three components of the predator-prey interaction (encounter rate, attack probability and capture success) on preference. Data points were means for each prey type. For example, for each of 8 prey types simultaneously offered to the predator Buenoa scimitra, we used the prey type's mean value (pooling replicates of one trial) for preference, encounter rate, attack probability and capture success.

Partial correlation analyses and path analyses yielded qualitatively similar conclusions (Table 2); both concurred with the assessment of Cooper et al. (1985). For all three predators, differences among prey in encounter rates were the most important factor explaining variation among prey in preference. For all three predators, capture success was of intermediate importance; for two of the three, the partial correlation between preference and capture success was significant or close to significant. Finally, for all three predators, attack probability played relatively little role in determining preference.

Table 2: For each of three predators, the relative importance of different components (encounter rate, attack probability, capture success) in determining preference. Shown are path values that are the product of path coefficients along a given path, and p-values for partial correlation coefficients.

	Predator					
	Acilius		Anax		Buenoa	
Component	Path Value	P-Value	Path Value	P-Value	Path Value	P-Value
Encounter	.813	.10	.916	.02	.802	.002
Attack	.062	.87	.056	.11	.214	.16
Capture	.207	.43	.091	.06	.242	.04

Cooper et al. (1985) also collected information on prey movement speeds. Movement speed for a given prey type differed substantially in different trials; we thus only did analyses in the one trial (with Buenoa as the predator) where they provided data on movement speed and on the other predictor variables and preference. A simple regression with encounter rate as the dependent variable and prey movement speed as the independent variable showed that the two are highly correlated ($r=0.96$, $p<.01$, encounter rate = -0.35 + 5.36 (prey movement speed). Path analysis indicates a strong link between prey movement speed and preference (path value = 0.766).

Overall, our statistical analyses confirmed Cooper et al.'s (1985) conclusions: 1) Prey vulnerability explains preference; 2) Relative encounter rate is the more important component of prey vulnerability; 3) Prey movement speed explains variation in encounter rate; 4) Capture success also influences vulnerability and preference; and 5) Active predator choice has little, if any, effect on preference.

CASE STUDY - SALAMANDERS FEEDING ON ISOPODS AND AMPHIPODS

We quantified preference and components of preference for small-mouthed salamander larvae, Ambystoma barbouri, feeding on isopods, Lirceus fontinalis and amphipods, Gammarus spp.

25 individuals (about 2.5 mm) of each type were placed in circular, glass dishes (14 cm dia) filled to a depth of 1.5 cm with ionically-balanced freshwater. Thirty minutes later, we added one *A. barbouri* (17-22 mm snout-vent length). To standardize hunger level, salamanders were fed *ad libitum* for one day and then starved 16-20 hr before the experimental trial. A run began either 10 minutes after the salamander was added, or when the salamander ate its first prey (in most cases, within 5 minutes). Trials lasted for 10 minutes. During this time, predators typically ate 5-10 prey: enough to quantify preference, but not enough to significantly deplete prey. To quantify predation events and prey and predator behaviors, we videotaped entire trials with commentary added on to the soundtrack. To date, we have viewed the tapes to generate data on encounter, attack and capture rates. A more complete analysis including data on prey and predator activity will be presented in a later paper. To assess the relative importance of components of preference, we used the following path diagram:

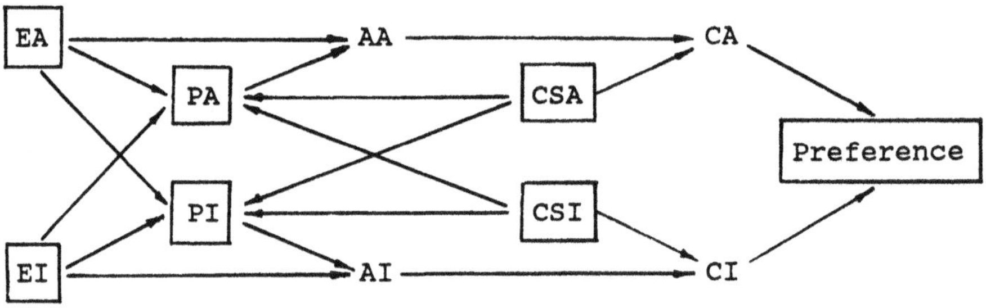

Figure 3: Path diagram for the relative importance of different components in determining preference of salamander larvae for amphipods over isopods.

EA and EI are encounter rates (number of encounters/10 min) with amphipods and isopods; PA and PI are attack probabilities on the two prey; CSA and CSI are capture successes; AA and AI are the numbers of attacks/10 min; CA and CI are the numbers of captures/ 10 min.

Figure 3 gives 14 paths linking preference and components

of preference (boxed). Six paths are "direct"; i.e., they do not involve one of the other components. Eight paths run from encounter rates and capture success to preference through attack probabilities. We included these paths because ODT makes predictions about them. ODT predicts that high value prey should always be attacked; thus, in general, changes in encounter rates or capture success on either prey type should not affect attack probabilities on high value prey. The exception is if relative capture success changes enough to switch the ranking of prey types. In contrast, various components should affect the probability of attacking low value prey. An increase in encounter rate with high value prey should result in a decrease in attack probability on low value prey. Encounter rates with low value prey should not affect its own attack probability. An increase in capture success of high value prey should further decrease attack probabilities on low value prey, whereas an increase in capture success of low value prey should tend to increase attack probabilities on low value prey. The path analyses will assess the relative importance of the 14 direct and indirect paths and test the above optimality hypotheses.

We ran 10 trials in control freshwater, and 8 trials using water taken from a 38 liter tank holding one green sunfish (Lepomis cyanellus, 100 mm SL). Earlier studies showed that small-mouthed salamanders and both prey species detect and respond to the smell of fish by increasing refuge use and decreasing activity (Petranka et al. 1987; J.R. Holomukzi; Sih (personal observation)). Thus we expected fish smell to decrease encounter rates with both prey types. No animals were used in more than one trial.

Table 3 shows the effects of fish smell on various aspects of the predator-prey interaction between salamanders, isopods and amphipods. In the absence of fish smell, salamanders encountered amphipods at a much higher rate than isopods; relative encounter rates with amphipods (number of encounters with amphipods/ total encounters) significantly differed from random (t-test for relative encounter > 0.5: $t=19.6$, $p<.001$). As expected, the presence of fish smell resulted in a decrease in encounter rates between predators and both prey types; the decrease was more dramatic for amphipods than it was for isopods. With fish smell, relative encounter rates did not differ from random ($t=1.03$, $p>.10$). Our unanalyzed observations suggest that the encounter patterns could be explained by the relative activity of the two prey types; that is: 1) amphipods appeared to be generally more active than isopods; 2) both prey types decreased their activity with fish smell; 3) amphipods decreased their activity more than did isopods.

Table 3: Effects of fish smell on salamander foraging. Shown are means and standard errors. * indicates a significant difference between treatments at the .05 level.

	No Fish Smell		Fish Smell Present
Number of encounters/10 min			
Amphipods	158.3 (14.2)	*	35.6 (4.2)
Isopods	45.9 (5.3)	*	23.6 (1.8)
Probability of attack			
Amphipods	.059 (.013)	*	.327 (.038)
Isopods	.018 (.012)	*	.237 (.040)
Capture success			
Amphipods	.729 (.052)	*	.612 (.025)
Isopods	.667 (.033)	*	.873 (.044)
Number killed			
Amphipods	6.0 (1.0)	NS	6.6 (0.7)
Isopods	0.5 (0.3)	*	4.6 (0.7)
Total	6.5 (1.2)	*	11.3 (1.3)
Relative			
Encounter rates	.774 (.014)	*	.592 (.089)
Attack probabilities	.878 (.067)	*	.596 (.028)
Preference	.940 (.043)	*	.601 (.034)

Perhaps in response to the decrease in encounter rates, with fish smell, salamanders had much higher probabilities of attack given an encounter. Relative attack probabilities for amphipods ($PA/(PA + PI)$), which reflect active predator choice for amphipods were also influenced by fish smell. In both the presence and absence of fish smell, salamanders chose amphipods (fish smell absent: t= 5.64, p<.001; fish smell present: t=3.43, p<.01); however, their tendency to actively choose amphipods was decreased by fish smell (t=3.88, p<.01). Capture success was not affected by fish smell. Preference showed patterns that were similar to those seen for active predator choice; in both situations, salamanders preferred amphipods (fish smell absent: t=10.23, p<.001; fish smell present: t= 2.97, p=.01); preference for amphipods was reduced by the presence of fish smell (t=6.18, p<.001). Interestingly, the number of amphipods killed was not affected by fish smell, while the number of isopods killed was actually increased by fish smell; thus salamander larvae ate significantly more prey in the presence of fish smell.

Table 4 shows path values for all 14 direct and indirect paths, for all runs pooled, for "no fish" runs only and for "fish" runs only. Pooled data address determinants of variation in preference across treatments; these are the more interesting analyses as far as tests of ODT are concerned. For example, Table 3 shows that preference for amphipods is generally lower with fish smell present. This pattern fits ODT predictions: fish smell results in decreased encounter rates with preferred amphipods which should cause an increase in attack probability on less preferred isopods. Path values from pooled data tell us whether this pathway is actually important in explaining the difference in preference between treatments. Path analyses within treatments address determinants of individual variation in preference for amphipods in a given situation.

For pooled data, the main determinant of preference was a direct negative effect of PI; this is not surprising, it says that an increase in the attack probability on isopods results in a decrease in preference for amphipods. In contrast, the direct positive effect of PA on preference was not nearly as strong. The main factor affecting PI was EA (encounter rates with amphipods). The overall pattern fit ODT predictions: variation in preference is primarily driven by changes in attack probability on less preferred prey (PI) that, in turn, are associated with changes in encounter rates with preferred prey (EA). Other direct and indirect pathways had relatively weak effects on variation in preference.

Table 4: Path values measuring the relative importance of different paths that can influence the preference of salamander larvae for amphipods over isopods. Shown are path values for both treatments pooled, for runs with no fish, and for runs with fish smell present.

	Pooled	No Fish Smell	Fish Smell
Direct Paths			
Amphipods			
Encounter rate	.083	.224	.862
Attack probability	.201	.426	.949
Capture success	.114	.197	.330
Isopods			
Encounter rate	-.006	-.072	-.931
Attack probability	-.870	-.845	-1.544
Capture success	-.138	-.312	-.303
Indirect Paths Via Attack Probabilities			
Attack probability on amphipods			
Encounter with amphipods	-.195	-.321	-.721
Encounter with isopods	.033	<-.001	.187
Capture success on amphipods	.004	.127	-.102
Capture success on isopods	.015	-.004	.293
Attack probability on isopods			
Encounter with amphipods	.703	.570	.653
Encounter with isopods	-.037	-.161	.753
Capture success on amphipods	-.064	-.306	.110
Capture success on isopods	-.119	-.279	-.272

Results from runs with no fish smell mirrored those generated by pooled data. Thus even with a narrower range of variation in PI and EA, PI had the largest direct effect, and the indirect path linking EA to preference via effects on PI was also important.

Path values in the presence of fish smell were very different from those found in the other two analyses. Here, both PI and PA had large effects on preference. In addition, encounter rates with both prey types had large, conflicting direct and indirect effects on preference. An increase in encounter rate with amphipods was associated with a decrease in attack probabilities on both prey types; these two trends largely cancelled one another. EA, however, also had a large direct positive effect on preference; with all else equal, some salamanders encountered more amphipods and as a result had higher preference for amphipods. Encounter rates with isopods had a large, direct, negative effect on preference for amphipods that was cancelled by indirect, positive effects of EI on P via effects on PA and PI. The large, direct effects of encounter rates on preference may be associated with the low encounter rates observed in the presence of fish smell; perhaps encounter rates *per se* were a limiting factor in determining the number of each prey type captured.

The value of the path analyses lies in their ability to partition the importance of different paths that link two variables. For example, in the pooled analysis, EA has a major effect on preference. ODT would predict this via the scenario outlined above. However, EA could also affect P via two other pathways. The path analysis suggests that in this case, the path predicted by ODT was indeed the important one. In contrast, in trials with fish smell, the overall, strong effect of EA is largely due to a direct, positive pathway. For another example, in the trials with fish smell, the overall effect of EI is close to zero; however, the path analysis suggests that EI actually had strong, but opposing effects on preference via different pathways.

Path analyses also allow us to more directly test the predictions of ODT. Strictly speaking, ODT makes predictions on attack probabilities, not on preference. ODT predicts that an increase in encounter rate with amphipods (EA) should cause a decrease in attack probability on isopods (PI). Other than that, changes in EA and EI should not affect PA or PI. Changes in capture success (CSA and CSI) should affect prey value. Changes in prey value, however, should not affect attack probabilities on more valuable prey (here, assumed to be amphipods); except for in the extreme case where changes in CSA or CSI actually switch the rankings of the two prey types. We did not observe any instance where salamanders actually switched to prefer isopods; thus we predict that CSA and CSI should not affect PA. In contrast, changes in the value of the less profitable prey (here, isopods) should affect its inclusion in the diet. PI should be positively related to CSI and negatively related to CSA.

Table 5 lists the predictions of ODT and observed patterns for effects of encounter rates and capture success on attack probabilities. Several trends fit the predictions of ODT: 1) EI, CSA and CSI had weak and conflicting effects on PA; 2) PI was strongly, negatively related to EA; when encounter rates with amphipods were high, salamanders attacked a relatively low proportion of isopods encountered; and 3) The effect of CSI on PI was generally (though weakly) positive.

Table 5: The effects of encounter rates and capture success on attack probabilities. Shown are the predicted signs of the relationships based on optimal diet theory, and path coefficients for pooled data and for trials with fish smell absent and present. See the text for a description of the abbreviations.

	Optimal Diet Predictions	Pooled Data	No Fish Smell	Fish Smell
Effects on PA				
EA	zero	-.97	-.75	-.76
EI	zero	.16	-.001	.20
CSA	zero	.02	.30	-.11
CSI	zero	.07	-.01	.31
Effects on PI				
EA	negative	-.81	-.68	-.42
EI	zero	.04	.19	-.49
CSA	negative	.07	.36	-.07
CSI	positive	.14	.33	.18

Contrary to predictions: 1) PA was strongly, negatively influenced by EA; when salamanders had high encounter rates with amphipods, they attacked a relatively low proportion of amphipods; 2) PI was not negatively influenced by CSA; and 3) With fish smell, EI had a negative effect on PI.

CLOSING COMMENTS

Over the past 25 years, The Grand Idea in evolutionary

ecology has been the promise that optimality theory might explain or even predict behaviors that have major effects on community dynamics. Non-random predation (predator preference) has often been shown to be important in determining the structure of prey communities. Optimality theory, however, has focused too narrowly on active predator choice, only one of several components that can determine predator preferences. Predator preference can also be influenced by relative encounter rates and capture success. Both of these are probably determined as much by prey behavior as by predator behavior. Here, we explore the use of path analysis, a novel approach (or actually, an old statistical technique used for a new purpose) for partitioning paths that determine preference. In our example, we only went part way, we used path analysis to examine direct and indirect paths that link encounter rates, attack probabilities and capture success to preference. We did not explicitly address effects of prey and predator behaviors such as relative activity or microhabitat use. Our closing suggestion is that future empirical and theoretical studies should address all of the behaviors and components outlined in figure 1. Empiricists should use multivariate statistics (e.g., path analysis) to examine the importance of various factors in determining preference. Further experiments would then directly assess the importance of the strong pathways. Similar multivariate approaches could be used to examine components of active predator choice (the subject of optimal diet theory). A more complete, quantitative approach to understanding diet preferences should yield extremely valuable insights.

SUMMARY

1. Optimal diet theory makes predictions about active predator choice which is only one component of preference. Empirical tests of ODT often examine preference. Preference is of inherent interest to community ecologists because predator preferences can have important effects on prey community structure.

2. Other components of preference (encounter rates, capture

success) are likely to be influenced by both prey and predator behaviors.

3. The standard predictions of ODT can also be generated by adaptive prey behavior in the absence of active predator choice. Prey behavior can also yield diet patterns that do not fit ODT.

4. A re-analysis of a literature survey from Stephens and Krebs (1986) suggests that when prey are mobile, ODT predictions do not work as well as when prey are less mobile.

5. A review of studies on aquatic invertebrate predators foraging on mobile prey showed that prey behaviors that influence encounter rates are the main determinant of preference in this group; relative capture success is also important.

6. Finally, we presented the results of our new study on the effects of predation risk on salamander larval preference for amphipods versus isopods. We used path analysis to assess the relative importance of different components in determining variation in preference. In 2 out of 3 analyses, the strong pathways were the ones predicted by ODT. In the presence of fish smell, the direct effects of encounter rates were also important.

ACKNOWLEDGMENTS

We thank Karen Thompson for her helping hands on the salamander experiment, and Marie-Sylvie Baltus, Bob Jaeger, Stephanie Porter and Loric Sih for thought-provoking comments. This work was supported by a grant from the National Science Foundation.

LITERATURE CITED

Allan JD, Flecker AS (1988) Prey preference in stoneflies: a comparative analysis of prey vulnerability. Oecologia 76:496-503

Allan JD, Flecker AS, McClintock NL (1987) Prey preference of stoneflies: sedentary vs. mobile prey. Oikos 49:323-331

Anholt B (1986) Prey selection by the predatory leech Nephelopsis obscura in relation to three alternative models of foraging. Can J Zool 64:649-655

Arnold SJ, Wade MJ (1984) On the measurement of natural and sexual selection: Theory. Evolution 38:709-719

Bertram BC (1978) Living in groups: predators and prey. In: Krebs JR, Davies NB (eds) Behavioural ecology: an

evolutionary approach. Blackwell, Oxford pp 64-96
Chesson J (1983) The estimation and analysis of preference and its relationship to foraging models. Ecology 64:1297-1304
Cooper SD, Smith DW, Bence JR (1985) Prey selection by freshwater predators with different foraging strategies. Can J Fish Aquat Sci 42:1720-1732
Erlinge S (1981) Food preference, optimal diet and reproductive output in stoats Mustela erminea in Sweden. Oikos 36:303-315
Greene CH, Landry MR (1985) Patterns of prey selection in the cruising calanoid predator Enchaeta elongata. Ecology 66:1408-1416
Greene, CH, Landry MR, Monger BC (1986) Foraging behavior and prey selection by the ambush entangling predator Pleurobrachia bachei. Ecology 67:1493-1501
Hassell MP (1978) The dynamics of arthropod predation. Princeton University Press, Princeton
Krebs JR, Davies NB (1984) Behavioural ecology: an evolutionary approach. Sinauer, Sunderland
Lande R, Arnold SJ (1983) The measurement of selection on correlated characters. Evolution 37:1210-1226
Li CC (1976) First course in population genetics. Boxwood, Pacific Grove
Mitchell-Olds T, Shaw RG (1987) Regression analysis of natural selection: statistical inference and biological interpretation. Evolution 41:1149-1161
Paine RT (1966) Food web complexity and species diversity. Am Nat 100:65-75
Pastorok RA (1981) Prey vulnerability and size selection by Chaoborus larvae. Ecology 62:1311-1324
Petranka JW, Kats LB, Sih A (1987) Predator-prey interactions among fish and larval amphibians: use of chemical cues to detect predatory fish. Anim Behav 35:426-434
Pulliam HR, Caraco T (1984) Living in groups: is there an optimal group size? In: Krebs, JR, Davies NB (eds) Behavioural ecology: an evolutionary approach. Sinauer, Sunderland pp 122-147
Riessen HP, O'Brien WJ, Loveless B (1984) An analysis of the components of Chaoborus predation on zooplankton and the calculation of relative prey vulnerabilities. Ecology 65:514-522
Scott MA, Murdoch WW (1983) Selective predation by the backswimmer, Notonecta. Limnol Oceanogr 28:352-366
Sih A, Crowley PH, McPeek MA, Petranka JW, Strohmeier K (1985) Predation, competition, and prey communities: A review of field experiments. Ann Rev Ecol Syst 16:269-311
Sih A (1987a) Predators and prey lifestyles: an evolutionary and ecological overview. In: Kerfoot WS, Sih A (eds) Predation: direct and indirect impacts on aquatic communities. University Press of New England, Hanover pp 203-224
Sih A (1987b) Nutritional ecology of aquatic insect predators. In: Slansky F, Rodriguez, JG (eds) Nutritional ecology of insects, mites and spiders. Wiley and Sons
Spitze K (1985) Functional response of an ambush predator: Chaoborus americanus predation on Daphnia pulex. Ecology 66:938-949

Stein RA (1977) Selective predation, optimal foraging, and the predator-prey interaction between fish and crayfish. Ecology 58:1237-1253
Stephens DW, Krebs JR (1986) Foraging theory. Princeton University Press, Princeton
Zaret TM (1980) Predation and freshwater communities. Yale University Press, New Haven

ORAL DISCUSSION – A. Sih

Q. (Schuler): Might not prey be killed and not eaten ? This could have consequences for behaviour.
A. It certainly would make a difference to the predator: if it killed prey but did not eat it, the predator would not gain energy. Concerning compenents of preference, what I have done is to simplify the predator–prey interaction by considering only four components: encounter, attack, capture and consumption. One could recognise other stages: detection, recognition and so on.

Q. (Kacelnik) What do gut contents represent ? If there is active choice to the extent that the animal is rejecting certain items altogether, then these will leave no trace in the gut.
A. The process of quantifying preference, particularly in field conditions is full of problems. Measuring prey availability in the field is difficult, quantifying diets in the field is difficult, defining an encounter can by very difficult and so on. My point is that, even if you could do all that, there are still some clarifying points to be made on interpretations of that information.

Q. (Ball) The very first step in the flowchart – the number of prey available in the environment – seems very subjective. Unless you are viewing things from the point of view of the animal, you will get a biased view of preference. Animals encounter prey that are patchily distributed in space and time – its environment is not the same as your environment.
A. The only time you could probably have an accurate measure of prey availability is in simple laboratory conditions where you have provided the prey yourself, and you are confident that all of them are available. I would agree that assessing prey availability in the field it is a major problem.

Q. (Lucas) It is not easy to figure out what prey the animal encounters: that could be an important step in the problem. The prey could be unavailable because they are in specialized micro–habitats.
A. I agree. Refuge use could be a major determinent of preference that has great ecological importance. I think path analysis, or perhaps some other multivariate technique, could be very useful for assessing the relative importance of different mechanisms that affect preference.

Q. (Peterson) Distastefulness has important effects because it decreases prey value, and thus results in a potential arrow back to attack probability.
A. Yes, path analysis forces you to make you assumptions explicit. To write out a path diagram you must clearly think out your assumptions on what the relationships are among various components of preference. In that respect it is similar to dynamic programming.

HUNTING BY THE HUNTED: OPTIMAL PREY SELECTION BY FORAGERS UNDER PREDATION HAZARD

James F. Gilliam
Behavioral Ecology Group
Department of Biological Sciences
State University of New York at Albany
Albany, New York 12222
U.S.A.

In nature, foraging animals are members of food webs, in which a given forager seeks to consume other organisms, and might itself be eaten by other predators. Studies of diet choice usually focus on a forager and "lower" levels in the web, i.e. the forager and its potential prey. This focus on the forager and its prey has indeed been a very effective line of research. In particular, the classic model based on the premise of maximizing the long-term rate of energetic gain (Equation 1, below) has been especially useful in empirical work (Stephens and Krebs 1987), and energy maximization remains a cornerstone of heuristic and practical approaches to predicting habitat and diet choice.

Nonetheless, my own experiences in natural systems lead me to believe that where (habitat choice) and what (diet choice) a given forager eats often depends as much on the "upper" parts of the food web, i.e. relationships between the forager and its predators, as it depends on the relationships between the forager and its prey. For example, in the absence of predators, small bluegill sunfish consumed invertebrate prey from open areas of a pond in accordance with an energy-maximization premise (Werner et al. 1983b), but in the presence of a predator (largemouth bass), the bluegill's diet shifted to different invertebrate taxa collected from vegetated areas at the pond's edge; the bluegills sacrificed energetic intake rate for safety (Werner et al. 1983a). However, safety does not necessarily override energetic gain in determining habitat choice; stream-dwelling cyprinid fish (creek chubs) forage in safer areas if resource levels are similar in alternative sites, but can be induced to prefer more hazardous sites if resources are enriched enough there (Gilliam and Fraser 1987). Predator-influenced spatial shifts have been shown

NATO ASI Series, Vol. G 20
Behavioural Mechanisms of Food Selection
Edited by R. N. Hughes

in many other recent experimental studies, including some field experiments, in which predators or models of predators have been manipulated (e.g., among others, Stein and Magnuson 1976, Fraser and Cerri 1982, Sih 1982, Cooper 1984, Kotler 1984, Lima 1985, Power et al. 1985, Schmitt and Holbrook 1985, Holomuzki 1986, Milinski 1986, Kneib 1987, Carpenter et al. 1987, Brown et al. 1988, Gilliam and Fraser 1988, Harvey et al. 1988, McIvor and Odum 1988, Rahel and Stein 1988, Schlosser 1988, Abrahams and Dill 1989). None of this is especially surprising, as both gathering resources and avoiding predators are salient features of fitness, but the work focuses attention on the need to consider both the "lower" and "upper" portions of the food web surrounding a particular forager, if we are to understand use of space in natural communities.

Spatial shifts are perhaps the most obvious and dramatic way in which predators influence a forager's diet, and understanding habitat selection under predation hazard deserves special attention, especially if one is interested in linking individual behavior and community structure. However, we might also expect predators to influence a forager's diet *within* a given habitat. For example, a forager searching from a protective location might hesitate to attack a prey item seen in an adjacent exposed location (e.g. Dill and Fraser 1984), even though the item is in the optimal prey set based on a energy-maximizing model. Relative to studies of habitat choice, there is less experimental work on choice of diet within a site in the presence of predators, but Godin (this volume) provides a cogent review of the available literature, summarizing several empirical studies in which the frequency of different prey types in a forager's diet depends upon the presence or absence of predators.

The purpose of this paper is to address the problem of optimal prey choice, within a habitat, by foragers under a chronic threat of being killed while foraging. The idea is to take an approach which has proved useful in experimental studies of habitat selection (Gilliam and Fraser 1987), and explore how optimal prey sets under predation hazard differ from those predicted by the classical energy-maximizing model. In the habitat selection study (Gilliam and Fraser 1987), we asked how an animal should distribute its time among different foraging sites, if it were to minimize the probability of being killed over some time period (e.g. a day), subject to the constraint of achieving at least some specified rate of energetic gain for the period (the foraging rate constraint). The case we tested experimentally allowed foragers to use a foodless refuge when not foraging, and to choose between feeding sites which differed in the risk of being killed (i.e. number of predators) and resource level (oligochaete worms in natural sediments). For this case, the model yielded an especially simple result, which we called the "minimize μ/f " rule. The "minimize μ/f " rule predicts that an animal will, when feeding, adopt feeding tactics so as to achieve the lowest ratio of mortality rate (μ, probability of death per unit time) to gross foraging rate (f, joules assimilated per unit time). Simply put, the rule just says that the forager should pick up each bit of food at the lowest mortality cost, thereby accumulating a given total amount of food with the least total exposure to predation. This result assumes (1) the metabolic rates are similar in alternative sites (or different activities within sites) (2) the forager spends some time not foraging, during which time its mortality rate is zero (e.g. retreats to a foodless refuge), and (3) the foraging rate achieved by following the "μ/f " rule at least equals the foraging rate constraint; otherwise a richer site must be chosen. Altering these assumptions of course alters the predictions of the model in some interesting ways (Gilliam and Fraser 1987). However, we were encouraged that an initially difficult problem could be reduced to such a simple form, and, especially, that the approach proved useful in predicting habitat preferences by fish asked to choose among sites differing in resource levels and number of predators (Fig. 6 in Gilliam and Fraser 1987).

Given this initial experimental support for a "minimize μ/f " criterion as a simple, workable basis for field work, my first purpose is to apply the "minimize μ/f " criterion (equivalently, for long-term rates, "maximize f/μ") to the problem of prey choice within a habitat, and ask how optimal prey sets by that model differ from optimal prey sets predicted by the classical energetic-rate-maximizing model ("maximize f "). I then consider graphically a broader class of problems of this sort, in which a "refuge" may or may not be present.

THE MODEL, AND PREY RANKINGS

First consider the classical prey choice model (Schoener 1971, Emlen 1973, Maynard Smith 1974, Pulliam 1974, Werner and Hall 1974, Charnov 1976a) in its simplest form. A forager encounters prey singly by a Poisson process. Let

λ_i = encounter rate with prey type i,

e_i = assimilable energetic content of type i

h_i = handling time on type i,

p_i = probability type i is accepted, given encounter.

The long-term rate of energetic intake (the gross foraging rate) while foraging is

$$f = \frac{\Sigma p_i \lambda_i e_i}{1 + \Sigma p_i \lambda_i h_i}, \qquad (1)$$

where the summations are taken over i=1...n. The optimization problem in the classical case is to find decision variables $p_1 ... p_n$ such that f is maximized. The familiar algorithm is to rank prey types by e_i/h_i, and add prey until f is maximized.

Now suppose the forager suffers mortality rate μ_s (probability of death per time) while searching for prey. If the forager chooses to attack an item of prey type i, the forager not only gains e_i, but also is killed with probability $\mu_i h_i$. Hence μ_i is an average mortality rate (probability of death per time) while handling prey type i . Assuming μ_s and the μ_i's are small, the mortality rate of the forager, while foraging, is

$$\mu = \frac{\mu_s + \Sigma p_i \lambda_i \mu_i h_i}{1 + \Sigma p_i \lambda_i h_i} \qquad (2)$$

It is assumed that μ_s and the μ_i's are positive, but nothing is assumed about the relative magnitudes of μ_s and the μ_i's. For example, a stream-dwelling fish hunting from an undercut bank or a bird hunting from a tree might be relatively safe while searching, but incur a larger chance of being killed by an aerial predator while taking an excursion to eat a prey item detected at some given distance. Similarly, an insect might be safer while flying between flowers than when landing and risking attack by an ambush bug, or a spider might increase hazard if it attacks a prey item in its web. Alternatively, searching might be more dangerous than handling, e.g. if the forager can watch for predators while husking a seed, or duck under cover while eating an item, or if the main threat to the insect mentioned above comes from a predator which catches it flying between flowers.

Finally, suppose for now that the forager has the option of spending time in a "Refuge," during which time

μ_R = mortality rate while in Refuge,

f_R = intake rate while in the Refuge (usually zero).

For a foodless absolute refuge, $\mu_R = f_R = 0$. The "refuge" need not be a physical refuge, but could just as well be a "behavioral" refuge in the sense that the animal can adopt some behavior which renders it somewhat safer than when actively foraging (e.g. extreme vigilance, or an inconspicuous posture or microhabitat location). The forager controls

q = proportion of time foraging.

The overall mortality rate (call it U) and the overall foraging rate (F) are then

$$U = q\mu + (1-q)\mu_R \ , \qquad (3)$$

$$F = qf + (1-q)f_R \ . \qquad (4)$$

The objective is to choose q and the p_i's so as to

$$\left.\begin{array}{l} \text{minimize } U \\ \text{subject to } F \geq F' \\ \qquad 0 \leq q \leq 1 \\ \qquad 0 \leq p_i \leq 1 \text{ for all } i \end{array}\right) \qquad (5)$$

where F' is the foraging rate constraint. Optimal values of the p_i's specify the optimal diet, and the optimal value of q specifies the optimal proportion of time to spend foraging.

We can begin the analysis by considering the "minimize μ/f " approach, a special case of the model defined in (5) above. That is, assume the forager has the option of spending some time in a foodless absolute refuge ($\mu_R = f_R = 0$), and, following Gilliam and Fraser (1987), ask what diet would minimize μ/f while out of the refuge. The result will be more obvious if we express this problem as the equivalent "maximize f/μ" (assuming μ is positive). Dividing Eq. 1 by Eq. 2, the denominators cancel, yielding

$$\text{maximize } f/\mu = \frac{\Sigma p_i \lambda_i e_i}{\mu_s + \Sigma p_i \lambda_i \mu_i h_i}. \qquad (6)$$

The solution (optimal p_i's) is unchanged if we divide the denominator in Eq. 6 by a positive constant, in this case μ_s, so an equivalent problem is

$$\text{maximize} \quad \frac{\Sigma p_i \lambda_i e_i}{1 + \Sigma p_i \lambda_i (\mu_i/\mu_s) h_i}. \qquad (7)$$

Equation 7 is of exactly the same form as Eq. 1, with $(\mu_i/\mu_s)h_i$ substituted for h_i. Recalling that e_i/h_i is the natural ranking of prey in maximizing Eq. 1, by exact analogy we conclude that the natural ranking for maximizing f/μ (Eq. 6) is $e_i/(\mu_i/\mu_s)h_i$. This is the same ranking as $e_i/\mu_i h_i$, since μ_s is a constant. To maximize f/μ (i.e. minimize μ/f), prey are added according to this ranking, exactly analogously to the maximization of Eq. 1. As in the classical model, an "all-or-none" rule applies: for a given i, the optimal value of p_i is either 0 or 1, or possibly the range 0 to 1.

The first result gleaned from the model, then, is how prey rankings can differ from the classical model. Iwasa et al. (1984) produced an analogous ranking of potential oviposition sites, if a parasitoid wasp is to maximize its fecundity per death, or, equivalently, its lifetime fecundity. Pearson (1976) also produced this ranking for prey items, starting from the premise that maximizing energy gained per death incurred would maximize the animal's expected energy intake over its lifetime. The premise of the present model is quite different, i.e. how to go about obtaining so arbitrary energetic gain over a day at the least mortality cost, but it is not surprising that the three approaches yield analogous rankings. Charnov and Stephens (1988), like Iwasa et al. (1984) considering optimal oviposition by parasitiods, obtained a different ranking, because they considered a mortality cost of egg production rather than different mortality costs while searching and handling.

It is important, and perhaps comforting, to note that the prey ranking in the present model ($e_i/\mu_i h_i$) yields the same ranking of prey as in the classical model (e_i/h_i) if the μ_i's are equal for all prey types, which is to say that the probability of being killed while handling a given prey is proportional to its handling time. If not, then differences in prey ranking can occur. For example, a forager searching from an ambush site and venturing some distance to capture encountered prey should show the same prey rankings as the classical model if handling time (including pursuit) and probability of being killed are both proportional to distance to the prey, but may not if handling time is proportional to distance, but risk of being killed increases as the square of distance. Even if rankings do not differ for a given case, diet breadth, i.e. how far down the rankings we "dip," can vary between the present model and the classical model.

DIET BREADTH

The "minimize μ/f " case, and the more general "minimize U subject to $F \geq F'$ " model, can both be solved graphically by constructing graphs like Figs. 1, 2, and 3. The graphical solution takes three steps. First, possible values of μ and f are graphed. Second, a line of least slope is drawn between the point representing the refuge and the set of possible values of μ and f (for the "minimize μ/f " case the refuge is represented by the origin). Third, the solution is read directly from the

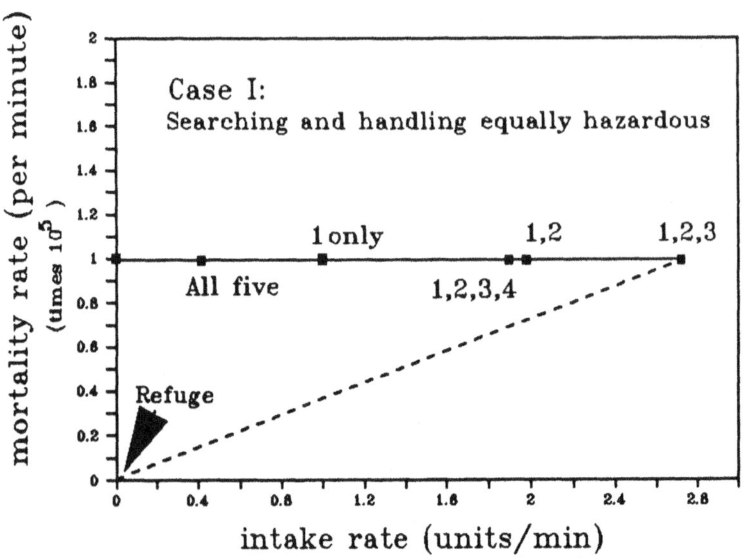

Figure 1. Case I (searching and handling equally hazardous). All diets lie on a horizontal line depicting μ and f. For this example, taking prey types 1, 2, and 3 maximizes f and also minimizes μ/f. $\mu_s = \mu_h = 10^{-5}$/min.

graph. Here, I initially consider the "minimize μ/f " case, and then turn to the more general case.

Three cases are considered in Figs. 1, 2 and 3: i.e. whether mortality hazard while handling is equal to (Case I), greater than (Case II), or less than (Case III) hazard while searching. For this analysis, I simplify the problem somewhat, by partitioning mortality rate into just two components: a mortality rate while searching (μ_s, as before), and a mortality rate while handling, μ_h, so $\mu_i = \mu_h$ for all i. This simplification assures that the prey rankings are the same by the "minimize μ/f " criterion (ranking by $e_i/\mu_i h_i$) and the classical model (ranking by e_i/h_i), so the analysis can concentrate on issue of diet breadth; i.e. how far down the prey list the forager should go. The simplification is also pragmatic; in nature, it may be rather easy to identify cases in which handling is more hazardous than searching, or vice versa, but much more difficult to associate different μ_i's with different prey types. To illustrate results, a numerical example is shown in Figs. 1, 2, and 3. The example assumes n = 5 prey types, and in each figure identical values of μ_s and the e_i's, h_i's, and λ_i's are assumed ($\mu_s = 10^{-5}$/min, $e_1 = 100$ units, $e_2 = 10$, $e_3 = 1.0$, $e_4=0.1$, $e_5=0.01$, $h_i=6$ sec for all i, $\lambda_1 = 0.01$/min, $\lambda_2 = 0.1$, $\lambda_3 = 1.0$, $\lambda_4 = 10$, λ_5

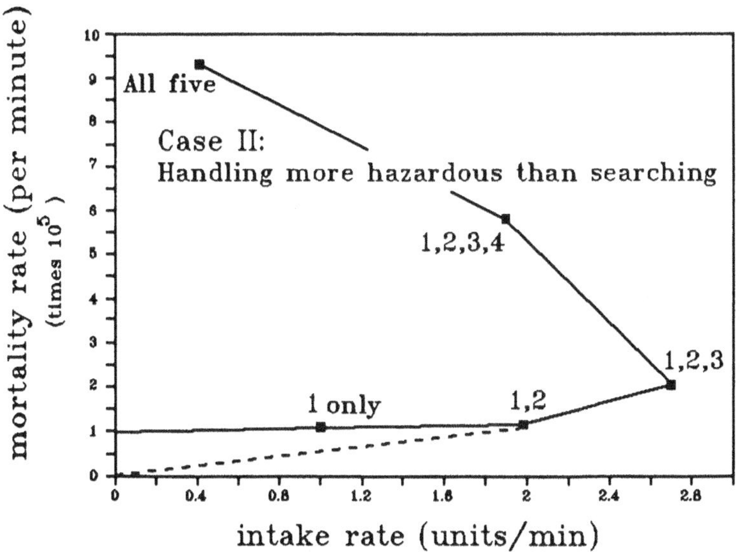

Figure 2. Case II (handling more hazardous than searching). Successively broader diets yield higher values of μ. Optimal diets tend to be *narrower* than the classical diet. $\mu_s = 10^{-5} < \mu_h = 10^{-4}$.

= 100). Prey ranking is type 1 (highest) to type 5 (lowest). Only μ_h is varied between figures.

The first use of Figs. 1, 2, and 3 is to compare the breadth of the optimal diet in the classical model (maximize f) versus the diet which minimizes μ/f.

In Case I, searching and handling are equally hazardous, so μ does not depend on the diet chosen, as can be immediately confirmed by substituting μ_s for μ_i in Eq. 2. This means that the values of μ and f for different diets lie on the horizontal solid line in Fig. 1. The diet which maximizes f is the one on the right end of the horizontal line, in this case acceptance of prey types 1, 2, and 3, while narrower (type 1 only, or 1 and 2) or broader diets yield a lower f. The diet which minimizes μ/f is found by "anchoring" the dashed line at the origin, and raising the slope of the line until it just touches the solid line. In Case I, obviously, the diet which minimizes μ/f is the same diet as maximizes f, because μ is independent of prey choice.

Case II (Fig. 2) and Case III (Fig. 3) are more interesting. In Case II, handling prey requires a greater exposure to predators than does searching, so successively broader diets result in higher values of μ. Unlike Case I, the solid line is not horizontal, but rather traces a counterclockwise path as diet breadth is increased.

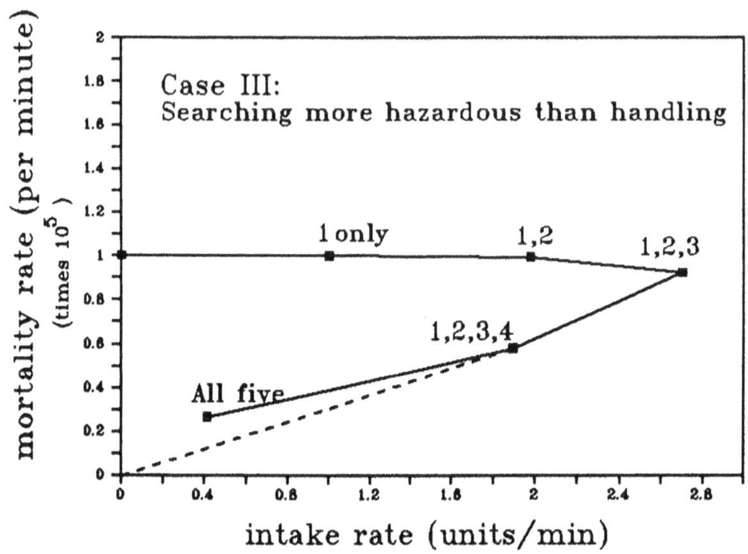

Figure 3. Case III (searching more hazardous than handling). Successively broader diets yield lower values of μ. Optimal diets tend to be *broader* than the classical diet. $\mu_s = 10^{-5} > \mu_h = 0.2X10^{-5}$.

Relative to the diet which maximizes f (here, prey types 1, 2, and 3), narrower or broader diets all yield lower values of f, but only narrower diets yield lower values of μ. This means that the diet which minimizes μ/f will always be *narrower* than, or possibly the same as, the diet which maximizes f, but never wider (this statement still assumes the μ_i's are equal). For the example in Fig. 2, the diet which minimizes μ/f is types 1 and 2 only.

In Case III (Fig. 3), the situation reverses. Now successively broader diets result in lower values of μ, and the solid line traces a clockwise path as diet breadth is increased. The diet which minimizes μ/f is always *wider* than or the same as the diet which maximizes f.

This analysis suggests that the classical model (maximize f), when applied to free-ranging foragers under chronic predation hazard, will tend to predict broader diets than observed if handling is more hazardous than searching, but narrower diets than observed if searching is more hazardous than handling. This statement of course assumes that the biology of a given case approximates the assumptions of the models treated here. As a cautionary note, it should be mentioned that just comparing diets in the presence and absence of visible predators may not form a suitable test, because the forager might adjust its λ_i's, e_i's, and/or h_i's when the predator appears. The present model deals with optimal prey selection, given values of these parameters. Thus a proper test is to measure the parameters in the presence of predators, and ask whether the classical model produces systematic errors in prediction, and whether the "minimize μ/f " diet or an appropriate variant produces a better fit.

VARIATIONS ON THE THEME

Dependence of diet on the energetic constraint

One distinct advantage of constructing graphs like Figs. 1, 2, and 3 is that it allows easy assessment of any dependency of the optimal prey set on the foraging rate to be achieved (the foraging constraint F'). For example, consider Fig. 2. Assume that across a day an animal is required to accumulate a maintenance ration, and the maintenance ration requires a mean intake rate across the day of F' = 0.8 units/min. The solution to "minimize U subject to F $\geq$ 0.8" can be read directly from the graph. The solution is characterized by the point, on the dashed line, which corresponds to the intake rate of 0.8 on the horizontal axis (i.e. starting from the point marked 0.8 on the intake axis, move vertically upward to the dashed line). This point denotes a behavior in which the animal splits its time between the refuge and foraging activity, consuming types 1 and 2 while foraging. The proportion of time foraging, q, is given by how far along the dashed line the optimal behavior is located, in this case about 0.4 of the distance between the point denoting the refuge and the point denoting the optimal diet. The complete solution of the problem is thus these optimal values of the decision variables: $p_1^* = p_2^* = 1$, $p_3^* = p_4^* = p_5^* = 0$, $q^* = 0.4$. Resultant optimal values of μ, f, and U can also be read directly from the graph.

In Case II (Fig. 2), increasing F' beyond 0.8 changes the optimal time to spend out of the refuge, but not the optimal diet, unless F' is increased beyond about 2.0. Beyond 2.0, the animal should abandon the refuge; it should spend all of its time

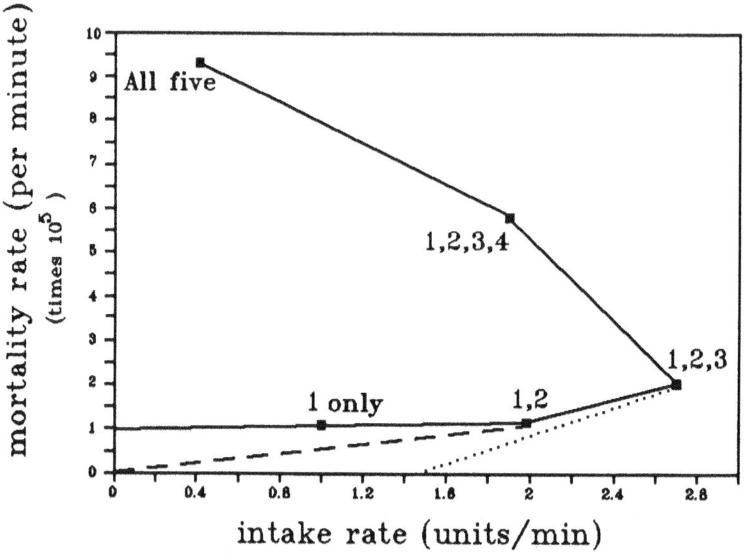

Figure 4. Enriching the refuge (i.e. $f_R > 0$, dotted line) tends to result in less time spent out of the refuge, and a lower overall mortality rate, but also "less cautious" foraging (higher μ and f) when out of the refuge.

foraging, spending some time consuming just types 1 and 2, and some time consuming types 1, 2, and 3. This amounts to, in effect, a partial preference for type 3. That is, at F' > 2.0, the optimal solution is to abandon the refuge ($q^* = 0$), always consume types 1 and 2, and consume just enough of type 3 to meet the foraging constraint. Notably, the solution for F' > 2.0 is *not* to spend some time in the refuge and accept types 1, 2, and 3 (maximize f) when foraging. This is because a line connecting the refuge and the square depicting the diet of types 1, 2, and 3 lies above the optimal options.

If the refuge were absent (i.e. no dashed line on Fig. 2), then for very small F' the solution is to take only just enough of type 1 to meet the constraint; as F' is increased, the forager should take a greater proportion of the type 1's encountered, then eventually add some of type 2 in the diet, and finally both types 1, 2, and some type 3. Only if F' takes the highest value for which there is a solution should the animal maximize f.

In general, for Case II (handling more hazardous than searching), increases in F' tend to *expand* the diet, with the important proviso that, if a refuge is present and used, increases in F' over some range does not change the optimal diet, but only the time spent foraging. For Case III (Fig. 3, searching more hazardous than handling), the situation is in one way reversed; increases in F' tend to *contract* the diet, but again the diet is unchanged over some range if a refuge is used.

Dependency of the optimal prey choice on conditions in the refuge

An interesting property of the present model is that the optimal prey choice in a given site can depend on conditions elsewhere in the environment. Figure 4 is an adaption of Fig. 2. For the case considered before (refuge depicted by the origin), recall that an animal using the refuge part of the time should consume only prey types 1 and 2 when out of the refuge. However, now consider the dotted line. Here, the refuge has now been enriched, in the sense that a positive intake rate can be gained in the refuge; for example, some food might occasionally drift into a fish's hiding site, or a bird hunting from a perch might occasionally find food on the branches. Curiously, perhaps, enrichment of the refuge shifts the optimal behavior while out of the refuge to a "*less* cautious" behavior; i.e. the forager should accept a higher μ for a higher f, in this case consuming prey types 1, 2, *and* 3.

It is interesting that the optimal diet can be shifted to more hazardous cases by either enriching the refuge, in some sense "helping" the animal, or by increasing F', in some sense "stressing" the animal. One practical implication is that shifts towards more hazardous diets cannot by themselves be taken as signs of energetic stress.

A COMMENT ON THE PATCH CASE

The classical patch problem is how long to stay in a patch (Charnov 1976b). The simplest case is as follows. A forager encounters patches at some expected inter-patch time T_s, and stays in patches for time T_P, with a cumulative energetic intake given by a function $e(T_P)$. The problem is to choose T_P so as to maximize intake rate, i.e.

$$\text{maximize } f = \frac{e(T_P)}{T_s + T_P} \qquad (8)$$

Now suppose travelling between patches entails different risks than does feeding in a patch. Letting μ_s be the mortality rate while searching for patches, and μ_P be the mortality rate while in a patch, the mortality rate while foraging is

$$\mu = \frac{\mu_s T_s + \mu_P T_P}{T_s + T_P} \qquad (9)$$

How does the "minimize μ/f " solution differ from the classical "maximize f " solution? To minimize μ/f, we want to find the T_P so as to

$$\text{maximize } f/\mu = \frac{e(T_P)}{\mu_s T_s + \mu_P T_P} \qquad (10)$$

The optimal T_P is unchanged if the denominator in Eq. 10 is divided by a constant, μ_P, so the same answer is given by

$$\text{maximize } \frac{e(T_P)}{(\mu_s/\mu_P)T_s + T_P} \qquad (11)$$

Equation 11 is the key to an especially straightforward comparison of the classical model and the variant including mortality hazard. Figure 5 shows the classical solution, which yields an optimal T_P shown with the dotted line. The solution for the "maximize f/μ" criterion is shown with the dashed line, for the case of searching between patches being more hazardous than within patches. Unsurprisingly, the solution is to stay longer in patches than specified by the classical model, and the graphical solution says how much longer. The situation is reversed

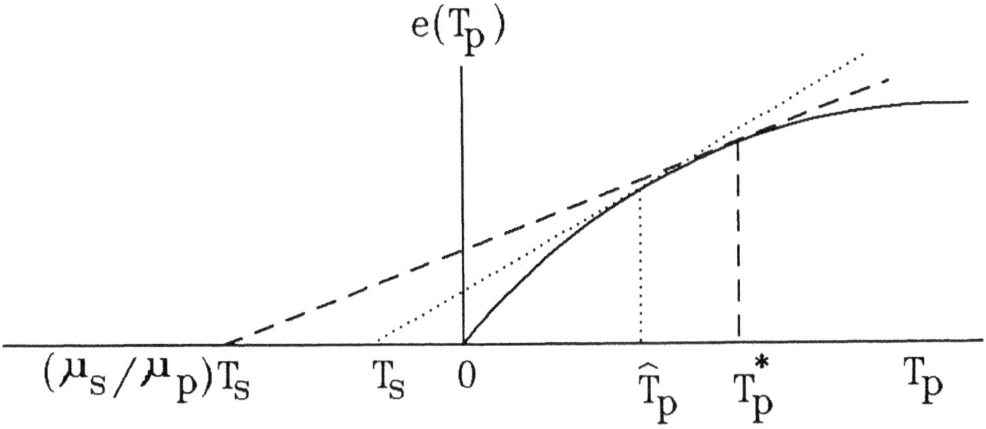

Figure 5. The patch case. If searching for patches is more hazardous than being in patches, the "maximize f/μ" model (dashed line) predicts how much longer the foragers will stay in patches, compared to the energy-maximizing solution (dotted line).

for the case that searching for patches is safer than being in a patch; the optimal patch residence times shorten. In a case with multiple patch types, by direct analogy to the classical model, each patch should be left at the same marginal rate of intake to mortality rate in the patch. For example, if patch A exposes the forager to twice the hazard as patch B (i.e. μ_p twice as large in A than B), then the intake rate in A at an optimal departure time is twice the intake rate in B at departure, rather than the equal intake rates at departure in the classical model. If foragers behave in this way, then spatial variation in predation hazard can "filter down," via forager behavior, to structure remaining resource levels on the same spatial scale (cf. Gilliam and Fraser 1988 for the habitat case).

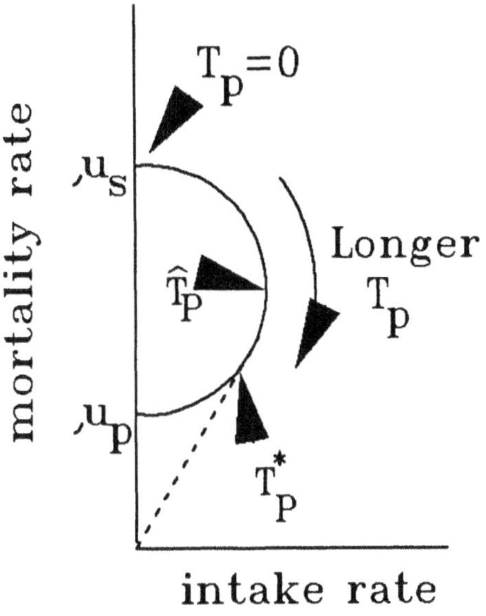

Figure 6. The patch case, plotted on the same axes as Figs. 1-4.

The patch problem can also be graphed on the same axes as was done in Figs. 1-4. This graphing is done in Fig. 6, for the case of one patch type. Successively longer patch residence times (T_P) trace a clockwise arc through the μ-f plane. The point which minimizes μ/f is given as in the prey choice problem: anchoring the dashed line at the origin, and increasing the slope until the dashed line just touches the solid line. For the case shown, that point corresponds to a patch residence time greater than the one which would maximize intake rate. If, instead, mortality rate were higher in patches than when searching for patches, increases in T_P would trace a counterclockwise arc. Figure 6 would also allow assessment of variations on the problem, such as the dependency on F', and effects of changed conditions in the refuge. As for the prey choice case, enriching the refuge (shifting the end of the dashed line to the right of the origin) would tend to shift the solution to "less cautious" (higher μ) behavior while out of the refuge. If F' is sufficiently high, the optimal T_P can, of course, shift towards that which maximizes f.

CONCLUSION

In this essay, I have explored some consequences of taking the simplest form of the classical energetic-rate-maximizing model, and extending it in a straightforward way to include mortality hazard. The approach has preserved much of the analytical tractability of the classical model, and has yielded some rather broad-brushed predictions. The main results are most conveniently summarized by contrasting cases in which handling (pursuing, capturing, processing) prey is more hazardous than searching (Case II, Fig. 2), versus cases where searching is more hazardous than handling (Case III, Fig. 3). Regarding diet breadth, the model predicts that diets of animals falling into Case II will tend to have narrower diets than would be predicted by the classical model, but animals falling into Case III will tend to have broader diets than predicted by the classical model. For both cases, if a refuge is used part of the time, increases in overall foraging rate (F) are purchased by expanding the time spent foraging, rather than altering prey selection while foraging.

However, if a refuge is not present, or F is made high enough to force abandonment of the refuge, increases in foraging rate are purchased by expanding the diet for Case II, but narrowing the diet in Case III. Finally, as was noted earlier for the habitat selection problem (Gilliam and Fraser 1987), optimal feeding behavior while out of the refuge is not independent of conditions in the refuge. Perhaps the most curious prediction is that enriching the refuge shifts the optimal behavior while out of the refuge towards increasing foraging rate at the cost of greater hazard, as if the animal were energetically stressed.

This analysis of the problem has benefited from thinking of foraging problems in two steps. The first step is to characterize foraging decisions which would minimize mortality rate, given a foraging rate constraint ($F \geq F'$). This is the step considered here (diets and patches), and in the habitat selection model (Gilliam and Fraser 1987). The approach can be used to study other foraging problems with tradeoffs; one just plots the options in a plane in which the abscissa is the objective function and the ordinate is the constraint, and reads off effects of varying the constraint. I find this procedure valuable, not only because it simplifies the problem and provides graphical insight, but also because it makes explicit the nature of the tradeoff between two components of fitness, in this case mortality rate and foraging rate.

The second step is to allow the "constraint" also to be a decision variable subject to optimization. That is, in the present model, what should F' (or just F) be? For some purposes, the second step may not be needed. If a reasonable energetic "requirement" can be stated *a priori* (e.g. a maintenance ration for a homeotherm), then F can be set as the rate which "points" at or just above the "target" required intake at the end of the day; an animal might adjust its F if the target moves, or the forager has runs of luck encountering prey during the day. In other cases, the second step may not be needed if precise knowledge of an optimal F is not essential for the question asked. For example, in Gilliam and Fraser (1987), the predictions of *where* to forage did not depend on precise knowledge of F' (H', net foraging rate, in that paper), although predicting *how long* to forage would have required knowledge of F'. We did not make the latter prediction, in part because we did not expect the fish to behave as if there were a fixed "requirement." Empirically, the fish did show flexibility in the total amount of food consumed in a bout, spending less time in the refuge in treatments with higher resource levels and/or less overall hazard (Fig. 7 in that paper), and dismissing the existence of a fixed "requirement."

In some studies, however, determination of an optimal F will be necessary to determine the optimal diet, and sometimes F and the time spent foraging may themselves be of primary interest. For example, Abrams (1984, 1989) has emphasized how the proportion of time spent foraging affects food webs and functional responses. Since life history theory deals with tradeoffs among components of fitness, one effective way to combine foraging rate and mortality rate is to construct a foraging model which uses some form of dynamic optimization to maximize some measure of fitness over the animal's lifetime or some segment thereof (e.g. Katz 1974, Craig et al. 1979, Werner and Gilliam 1984, Sibly et al. 1985, McNamara and Houston 1986, Mangel and Clark 1986, Clark and Levy 1988, Godin in this volume). The present model provides raw material for dynamic optimization, by making explicit the tradeoff between hazard and feeding. Below, I suggest three closely related ways we might proceed to combine foraging and life history theory, to produce a workable integration of feeding rate and mortality rate. Analytical methods might be most appealing to experimentalists, but I feel numerical solutions can also play an important role.

The first suggestion applies to purely size-structured populations, in which properties of individuals depend only on their size rather than their age. Under certain assumptions (Werner and Gilliam 1984, Gilliam and Fraser 1987, and Gilliam manuscript in review), animals would maximize fitness by maximizing the probability of reaching each size (not age), which is accomplished by minimizing the ratio of mortality rate to growth rate at each size. Step One, above, tells us the minimal mortality rate for a given intake, i.e. $U^*(F)$, and supposing growth rate is a function of intake, $G(F)$, then we just choose F so as to minimize $U^*(F)/G(F)$.

The second suggestion follows from an approach to life histories dating at least to Williams (1966), Shaffer (1974), Pianka and Parker (1975), and Leon (1976). Under certain assumptions, an animal maximizes fitness by maximizing, at each time t, the sum $b_t + P_tV_{t+1}$, where b_t = fecundity at time t, P_t = probability of surviving from t to t+1, and V_{t+1} = reproductive value at time t + 1. For a non-reproductive period (e.g. a given day in a non-reproductive season), the problem reduces to maximizing P_tV_{t+1}. If both P_t and V_{t+1} can be written as functions of intake rate achieved, then this expression becomes $P_t(F)][V_{t+1}(F)]$, and the optimal F is taken to maximize that product. The problem, of course, is constructing a realistic "final function" V_{t+1}, but taking reasonable guesses as to its form, and perhaps trying different forms, some good insight can often be obtained into possible behaviors of foragers (e.g. McNamara and Houston 1986, Mangel and Clark 1986, Clark and Levy 1988, Godin in this volume).

A final option is to abandon, or de-emphasize, attempts at analytical solutions, in favor of numerical solutions retaining the full richness of stochastic dynamic optimization, using stochastic optimal control theory or stochastic dynamic programming (McNamara and Houston 1986, Mangel and Clark 1986, Godin this volume). This approach might be criticized for loss of analytical insight, but nonetheless is nearly foolproof in producing some answers to a wide range of problems. Mangel and Clark (1988) have produced a lucid tutorial on one such approach (stochastic dynamic programming), which has a great advantage of not requiring special knowledge of either mathematics or computer programming. However, I do not agree with the strength of the statements in Mangel and Clark (1988) which claim superiority of this method over other methods of addressing optimal behavior under tradeoffs. Rather, I feel a pluralistic, eclectic approach to methodology will be more useful, bringing to bear on the problem techniques of both static and dynamic optimization, in both deterministic and stochastic forms, and both analytical and numerical results. Notably, if early workers had used only a single numerical methodology, we would never have had the classical analytical rate-maximizing models which have fueled the growth of the field. Given the deep importance of resource use in evolution, behavior, and ecology, continued efforts to produce practical, simple, and heuristic approaches to choices of diets and feeding sites are sure to be rewarding.

ACKNOWLEDGEMENTS

Roger Hughes accidentally forced crystallization of the analysis here, by asking that this volume address hard-core "*diet*" selection, rather than habitat selection. I have discussed this approach to the problem with several people, and especially thank Doug Fraser, Jerram Brown, Alberto Sabat, Jeff Tyler, and Gregg Recer. The work was supported by grants from the National Science Foundation (USA).

LITERATURE CITED

Abrahams MV, Dill LM (1989) A determination of the energetic equivalence of the risk of predation. Ecology 70:999-1007

Abrams PA (1984) Foraging time optimization and interactions in food webs. American Naturalist 124:80-96

Abrams PA (1989) Decreasing functional responses as a result of adaptive consumer behavior. Evolutionary Ecology 3:95-114

Brown JS, Kotler BP, Smith RJ, Wirtz WO (1988) The effects of owl predation on the foraging behavior of desert rodents. Oecologia (Berlin) 76:408-415

Carpenter SR, Kitchell JF, Hodgson JR, Cochran PA, Elser JJ, El ser MM, Lodge DM, Kretchmer D, He X, von Ende CN (1987) Regulation of lake primary productivity by food web structure. Ecology 68:1863-1876

Charnov EL (1976a) Optimal foraging: attack strategy of a mantid. American Naturalist 110:141-151

Charnov EL (1976b) Optimal foraging: the marginal value theorem. Theoretical Population Biology 9:129-136

Charnov EL, Stephens DW (1988) On the evolution of host selection in solitary parasitoids. American Naturalist 132:707-722

Clark CW, Levy DA (1988) Diel vertical migration by juvenile sockeye salmon. American Naturalist 131:271-290

Cooper SD (1984) The effects of trout on water striders in stream pools. Oecologia 63:376-378

Craig RB, DeAngelis DL, Dixon KR (1979) Long- and short-term dynamic optimization models with application to the feeding strategy of the loggerhead shrike. American Naturalist 113:31-51

Dill LM, Fraser AHG (1984) Risk of predation and the feeding behavior of juvenile coho salmon (*Oncorhynchus kisutch*). Behavioral Ecology and Sociobiology 16:65-71

Emlen JM (1973) *Ecology: An Evolutionary Approach*. Addison-Wesley New York

Fraser DF, Cerri RD (1982) Experimental evaluation of predator-prey relationships in a patchy environment: consequences for habitat use patterns in minnows. Ecology 63:307-313

Gilliam JF, Fraser DF (1987) Habitat selection under predation hazard: test of a model with foraging minnows. Ecology 68:1856-1862

Gilliam JF. Fraser DF (1988) Resource depletion and habitat segregation by competitors under predation hazard. Pages 173-184 in Ebenman B, Persson L (eds) *Size-structured populations: ecology and evolution*. Springer Berlin Heidelberg New York

Harvey BC, Cashner RC, Matthews WJ (1988) Differential effects of largemouth and smallmouth bass on habitat use by stoneroller minnows in stream pools. Journal of Fish Biology 33:481-487

Holomuzki JR (1986) Predator avoidance and diel patterns of microhabitat use by larval tiger salamanders. Ecology 67:737-406

Iwasa Y, Suzuki Y, Matsuda H (1984) Theory of oviposition strategy of parasitoids. I. Effect of mortality and limited egg number. Theoretical Population Biology 26:205-227.

Katz PL (1974) A long-term approach to foraging optimization.American Naturalist 108:758-782

Kneib RT (1987) Predation risk and use of intertidal habitats by young fishes and shrimp. Ecology 68:379-386

Kotler BP (1984) Risk of predation and the structure of desert rodent communities. Ecology 65:689-701

Leon JA (1976) Life histories as adaptive strategies. Journal of Theoretical Biology 60: 301-335

Lima SL (1985) Maximizing feeding efficiency and minimizing time exposed to predators: a trade-off in the black-capped chickadee. Oecologia 66:60-67

Mangel M, Clark CC (1986) Toward a unified foraging theory. Ecology 67:1127-1138

Mangel M, Clark CC (1988) *Dynamic modelling in behavioral ecology*. Princeton University Press Princeton New Jersey USA

McNamara JM, Houston AI (1986) The common currency for behavioral decisions. American Naturalist 127:358-378

Maynard Smith J (1974) *Models in Ecology*. Cambridge University Press Cambridge UK

McIvor CC, Odum WE (1988) Food, predation risk, and microhabitat selection in a marsh fish assemblage. Ecology 69:1341-1351

Milinski M (1986) Constraints placed by foragers on feeding behavior. Pages 236-252 in Pitcher TJ (ed) *The Behavior of Teleost Fishes.* Johns Hopkins University Press Baltimore MD USA

Pearson NE (1976) Optimal foraging: some theoretical consequences of different feeding strategies. Ph.D. dissertation, University of Washington

Pianka ER, Parker WS (1975) Age-specific reproductive tactics. American Naturalist 109:453-464

Power WE, Matthews WJ, Stewart AJ (1985) Grazing minnows, piscivorous bass, and stream algae: dynamics of a strong interaction. Ecology 66:1448-1456

Pulliam HR (1974) On the theory of optimal diets. American Naturalist 108:108:59-75

Rahel FJ, Stein RA (1988) Complex predator-prey interactions and predator intimidation among crayfish, piscivorous fish and small benthic fish. Oecologia (Berlin) 75:94-98

Schlosser IJ (1988) Predation rates and the behavioral response of adult brassy minnows (*Hybognathus hankinsoni*) to creek chub and smallmouth bass predators. Copeia 1988:691-687

Schoener TW (1971) Theory of feeding strategies. Annual Review of Ecology and Systematics 2:369-404

Schmitt RJ, Holbrook SJ (1985) Patch selection by juvenile black surfperch (Embiotocidae) under variable risk: interactive influence of food quality and structural complexity. Journal of Experimental Marine Biology and Ecology 85:269-285

Schaffer WM (1974) Selection for optimal life histories: Effects of age structure. Ecology 55: 291-303

Sibly R, Calow P, Nichols N (1985) Are patterns of growth adaptive? Journal of Theoretical Biology 112:553-574

Sih A (1982) Foraging strategies and the avoidance of predation by an aquatic insect, *Notonecta hoffmanni*. Ecology 63:786-796

Stein RA, Magnuson JH (1976) Behavioral response of crayfish to a fish predator. Ecology 57:751-761

Stephens DW, Krebs JR (1987) *Foraging Theory*. Princeton University Press Princeton New Jersey USA

Werner EE, Gilliam JF (1984) The ontogenetic niche and species interactions in size-structured populations. Annual Review of Ecology and Systematics 15:393-425

Werner EE, Hall DJ (1974) Optimal foraging and the size selection of prey by the bluegill sunfish (*Lepomis macrochirus*). Ecology 55:1042-1052

Werner EE, Gilliam JF, Hall DJ, Mittelbach GG (1983a) An experimental test of the effects of predation risk on habitat use in fish. Ecology 64:1540-1548

Werner EE, Mittelbach GG, Hall DJ, Gilliam JF (1983b) Experimental tests of optimal habitat use in fish: the role of relative habitat profitability. Ecology 64:1525-1539

Williams GC (1966) Natural selection, the costs of reproduction, and a refinement of Lack's principle. American Naturalist 100:687-690

ORAL DISCUSSION – J. Gilliam

Q. Hobbs: Just how successful were the predators in the habitat selection experiment with creek chubs ?
A. We chose the species and size of the predator so as to have measurable hazard, but without instant killing of all our experimental subjects. With 8 foragers and 1 predator present, the predator took about 1.2 foragers per day. With 2 predators, about 1.8 foragers were killed per day.

Q. Hobbs: How did the foragers estimate predation rate ?
A. First, I should describe a bit of how the foragers behaved. The foragers would leave the refuge, approach a food dish, and then upon an approach of a predator, either go back to the refuge and perhaps try the other feeding area, or, circle around the predator and come back to feed in the same area. Qualitatively, the fish tended to circle around the predators in treatments with high food levels, but leave upon approach if food was low (we have no quantitative data on this, however). Thus, a very proximate cue could be the frequency of approaches by predators, and the probability of leaving seems to reflect frequency of approach, conditions on current intake rate. However, I feel that assessment of hazard is almost surely multifaceted, involving such factors as frequency of approaches, visual sightings, perhaps odor, perhaps the lateral line sense. In nature, physical features such as potential ambush sites could also be important. In any event, the "u/f" hypothesis has the feature that the animal does not need to estimate either absolute hazard or absolute intake rates available. Rather, a fish might effect the "rule" by just following the rule of thumb: "if site A appears twice as dangerous, don't go there unless twice as much food can be obtained there". Whatever the information and mechanism the fish use to accomplish such decisions, we can expect strong selection for abilities to discriminate hazard and intake levels, and to somehow balance their values.

Q. (Belovsky): Are your models of habitat switching and diet choice constrained by the need to get enough food to avoid starvation ?
A. Yes, in that the constraint F' is assumed to be at least sufficient to avoid starvation, but can also reflect further benefits of energy beyond the starvation level. However, the models assume the animals do avoid starvation, and are concerned with how to avoid predatory death while doing so. The models are not really intended to address cases in which animals are actually threatened with dropping over dead upon dipping below some critical value of energy reserves due to stochastic events. In some cases such as small birds in winter, or perhaps larval fish, starvation might be a substantial component relative to predatory death, and a stochastic version such as addressed here by McNamara and Houston is more appropriate.

ON THE ROLE OF ECOLOGICAL EXPERIMENTATION IN RESOURCE MANAGEMENT: MANAGING FISHERIES THROUGH MECHANISTIC UNDERSTANDING OF PREDATOR FEEDING BEHAVIOUR

Charles H. Peterson
University of North Carolina at Chapel Hill
Institute of Marine Sciences
Morehead City, North Carolina 28557
USA

The field of fisheries management truly diverged from its academic parent discipline of ecology at the time of the revolutionary development of quantitative fisheries models. With emergence and elaboration of the yield-per-unit-effort and yield-per-recruit models of Beverton and Holt (1957), Ricker (1954, 1958) and others more recently (e.g., Cushing 1981), fisheries resource management attained status as a recognizably separate, academic discipline. Over the past three decades (1960-1990), a different sort of revolution has been occurring in the parent field of ecology. Ecology in this period has matured from an observational to an experimental and process-exploring science (e.g., Connell 1972). I will argue here that the fields of fisheries resource management and ecology are now suffering from their separation academically and could profit from enhanced integration.

The discipline of fisheries resource management could benefit enormously in at least four different ways from application of recent advances in ecology. First, fisheries regulations, which represent to some often large degree an educated guess, should be formulated using modern principles of experimental design to allow rigorous testing of their influences on key population and system variables. There is no reason why experimental hypothesis testing often on a landscape scale cannot be built into the structure of the regulations. Such science is even inexpensive because the fishing industry provides the treatment in the process of conducting business. Second, marine fisheries management could benefit from greater allocation of resources to research programs in what is now called fisheries oceanography (e.g., Sinclair 1988). Fisheries oceanography is the interdisciplinary study of ecological processes in the context of the physical, chemical, and various biological features of the marine environment. For example,

NATO ASI Series, Vol. G 20
Behavioural Mechanisms of Food Selection
Edited by R. N. Hughes

the long-standing problem of what controls recruitment variation in marine fish stocks may be best addressed by collaboration of physical oceanographers, biological oceanographers, and fisheries biologists using modern oceanographic technology to evaluate from first principles how features of the sea alter feeding, predation, reproduction, and physical mortality of larval life stages (Rothschild 1986). Third, fisheries management may benefit from enhanced application of recent theory in optimal foraging with the fisherman as the forager. Surely, human foraging should incorporate some aspects of rationality, which is a fundamental tenet of optimal foraging theory. Fourth, the powerful experimental and analytical methodologies developed in part in seashore and mudflat ecology (e.g., Underwood 1981, 1986) should be applied to testing explicit hypotheses involving foraging ecology that are of significance to proper management of fisheries resources. These questions of foraging will often combine approaches from the fields of ecology and feeding physiology. It is this fourth application of ecology to fisheries resource management that I address here.

Hypotheses concerning foraging behaviour are especially amenable to direct test using modern methodologies of ecological and physiological experimentation because these processes can be examined at scales feasible for experimental study. Many other important questions in management of fisheries resources, such as what controls recruitment of species with planktonic larvae, involve spatial and/or temporal scales that render experimentation difficult unless achieved on a landscape scope by the structure of fisheries regulations. The foraging questions most often relevant to management of fisheries resources are issues of how to manage the fishery to minimize losses of stock to natural predators. These are the issues that I will discuss and illustrate here, although I acknowledge that analogous questions often of equal importance can be formulated around the foraging and growth of the fisheries species themselves. Experimental tests of physiological constraints to foraging, density dependence, prey preference, size selectivity, and habitat relationships of important natural consumers of fishery resources should be utilized by resource managers. Here, I shall describe some invertebrate fisheries with which I am familiar to illustrate the role that experimental tests of predator foraging processes do or could play in resource management. Finally, I will address the role that optimal foraging theory may play in implementing my recommendations that managers of fisheries resources utilize empirical experimental tests of the factors that influence natural rates of predation on fishery stocks.

The bay scallop fishery in North Carolina

The bay scallop, Argopecten irradians, has a life span of about 18-20 months (Belding 1910, Gutsell 1930). Geographically, it is found along the east coast of North America, stretching from the west coast of Florida to Massachusetts (Belding 1910, Gutsell 1930). In North Carolina, settlement of larval bay scallops occurs in autumn: for example in 1988 virtually all the year's successful settlement occurred during September (Peterson, unpub.). New recruits settle on emergent hard substrate, almost exclusively seagrass blades in North Carolina, where they remain attached by byssal threads for their first 4 or so months of life. This obligate, epibiotic attachment serves to elevate the juvenile scallops above the sediment surface, which may help eliminate stressfully high levels of turbidity (Duggan 1973) or alternatively expose the scallop to more favorable current flows and food fluxes (Eckman et al. 1989). Even after detachment from the seagrass blades, bay scallops tend to remain closely associated with seagrass beds (Thayer and Stuart 1974); when experimentally transplanted both into unvegetated areas and into nearby seagrass patches, adult scallops emigrate by swimming (clapping their valves and jetting forward - Peterson et al. 1982a) away from unvegetated bottom and into seagrass beds (Peterson et al. 1989).

The bay scallop fishery in the state of North Carolina is managed through a combination of biological, economic, and social considerations. Because the bay scallop is an annual crop, the key issue in management is when to open the fishing season on adult bay scallops, not how many adults to conserve for replenishment of the stock (Table 1). The opening of the fishing season in North Carolina is set to occur after autumn spawning is complete. That is the key biological consideration. On that ground alone, bay scallop harvest could begin as early as October perhaps. In practice, however, the fishery is not opened until December for three reasons: one biological, one economic, and one social. After spawning, the meat weights (adductor muscle) are greatly depressed because of physiological reallocation of resources to reproduction during the spawning season (Kellogg and Spitsbergen 1983). Consequently, harvest immediately after spawning would produce

TABLE 1. FACTORS AFFECTING THE TIMING OF OPENING OF THE BAY SCALLOP FISHING SEASON IN NORTH CAROLINA

Date	Biological			Economic	Social
	Life history	Meat wt	Mortality rate	Price per unit wt	Fishermen's wishes
Aug-Nov	larval life and settlement	-----	high but unknown	-----	-----
Oct-Jan	attached by byssus to seagrass	-----	unknown	-----	-----
Feb-June	growth to adult size reclining unattached on the bottom	negligible	unknown	-----	-----
July	adult size	relatively high	unknown	relatively high before market is flooded with Calico scallop crop	to continue shrimping, not to scallop
Aug-Nov	spawning	low because of diversion of energy to spawning	unknown	low because market is flooded by Calicos	to continue shrimping and fall fin-fishing
Dec	post-spawning	still low from prior spawning	unknown	still low	to begin scal-loping because of lack of alternative fisheries and need for Christmas money
Jan-May	final months of life	assymptotically increasing	unknown	relatively high	to continue scalloping

a low yield per recruit as measured in weights of meats per scallop. Furthermore, the market price per unit weight of scallop meat is ordinarily depressed in autumn because of the influx of competing scallop meats at the time of the autumn opening of the Calico scallop fishery off Florida. Finally, small-boat fishermen in North Carolina are still committed to shrimp and various finfish pursuits in autumn so there is social pressure to postpone the opening of the bay scallop fishery until a time of inactivity in other fisheries (Table 1).

A bioeconomic model (Kellogg and Spitsbergen 1983) was developed to assess when the fishing season for bay scallops should be set to maximize the economic returns to the fishermen. This model demonstrates that delaying opening of the bay scallop fishing season until much later in winter would be expected to enhance greatly the value of the bay scallop harvest. These projections, however, are based on an assumption of negligible mortality of adult bay scallops from natural causes during the potential fishing season from about the first of December until early May. At the time of conducting this analysis, there were no data available on natural rates of adult bay scallop mortality in North Carolina estuaries. In 1987, the North Carolina Marine Fisheries Commission, which is responsible for setting all fisheries management regulations in state waters, considered the projections of the bioeconomic model and decided not to follow its implicit recommendation to delay opening of the bay scallop fishing season later into the winter. Social considerations were one contributing factor in that some fishermen argued the need for pre-Christmas income, even if economic returns would be far greater by postponing harvest. The major factor inhibiting the commission from delaying the fishing season for bay scallops, however, was a concern over the untested biological assumption of negligible natural mortality of adult bay scallops during winter. Clearly, the model predictions are sensitive to assumptions of natural mortality rate: as natural mortality rates increase, the optimal fishing season will move progressively earlier in the winter to minimize losses to natural sources of mortality.

Mortality of adult bay scallops from natural causes clearly does occur during winter months. Recently, adult bay scallop mortality from the beginning of December until the end of January was estimated at 21%, averaged across three important scallop grounds (Summerson and Peterson 1990). This estimate, however, is almost certainly inflated over normal winters by the influence of the first recorded red tide outbreak in North Carolina waters in that winter of 1987-1988 (Summerson and Peterson 1990). One abundant type of predator clearly consumes adult bay scallops during winter, overwintering gulls, both herring gulls and ringbilled gulls (Peterson et al. 1989, Prescott 1989). Because gull foraging occurs out of water, gulls and large shorebirds represent ideal candidates for feasible study of foraging behavior (e.g., O'Connor and Brown 1977, Zwarts and Drent 1981, Sutherland 1982). Foraging studies that document how the intensity of gull predation on bay scallops varies in time and space would greatly enhance the wise management of the bay scallop resource.

Factors influencing foraging by consumers of fisheries species

Physiological constraints to foraging

Gradients in physiological stress are present in all natural environments and often influence the foraging pattern of predators. Along intertidal shores, tidal elevation is one fundamental feature of the environment that implies a particular intensity of physiological stress. Marine predators are limited in their upward distribution by increasing amounts of aerial exposure (Connell 1972), whereas terrestrial consumers have limited access to lower elevations on the shore (Frank 1982, Quammen 1984, Irons et al. 1986, Marsh 1986). Controlled experiments placing equal numbers of adult bay scallops at different elevations on tidal flats have shown that herring and ringbilled gulls will rarely forage on scallops covered by as little as 3 cm of water (Peterson et al. 1989, Prescott 1989). Gull predation in these experiments was intense on all scallops exposed to air (Fig. 1). Consequently, physiological constraints on gull foraging impose an elevation pattern to their predation on bay scallops and probably set a limit to the importance of gull predation because the seagrass beds, where most bay scallops occur (Thayer and Stuart 1974), are almost always covered by water.

Although bay scallops tend to occur in deeper waters below the depths of low-tide susceptibility to gull predation, gulls can be seen eating scallops in large numbers during winter low tides (e.g., Gutsell 1930, Prescott 1989). It seems likely that managers of the bay scallop fishery could utilize the known physiological constraint of the overwintering gulls to

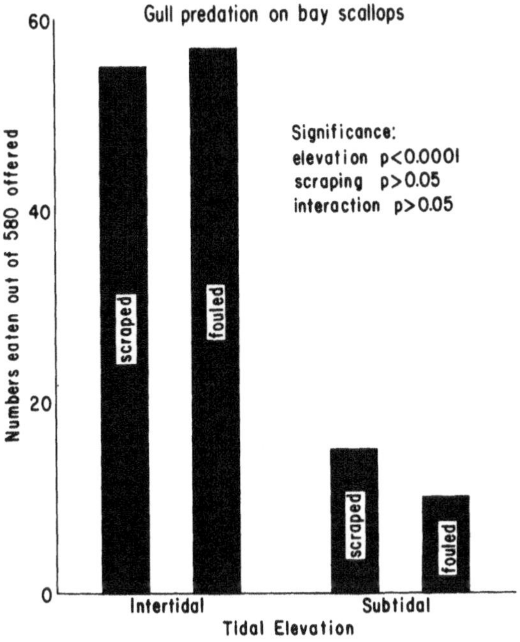

Fig. 1. Effects of tidal elevation and shell scraping on the rate of gull predation on adult bay scallops (Argopecten irradians) during a single low tide. Data bars pool experimental results from twelve winter dates, two sites in North Carolina, and two habitats (seagrass and sandflat), combining data in both Peterson et al. (1989) and Prescott (1989). The intertidal sites became exposed to air for 1.5 hours at low tide, whereas the subtidal sites were always covered by at least 3 cm of water. Scraping removed all fouling by epibiota. Statistical results come from a 3-way contingency test using log-linear models.

foraging on exposed bottom together with estimates of bay scallop abundance by water depth and tide tables to model the temporally varying impact of gull predation on adult bay scallop survivorship during winter. For example, there is a seasonal pattern to the occurrence of extreme low tides with winter being the season of greatest frequency in North Carolina (Fig. 2). Such a model of time-dependent predation rate by gulls on adult bay scallops would serve management by permitting an assessment of the degree of cumulative scallop mortality expected under different scenarios of openings of the bay scallop fishery season. This would fill a major void in

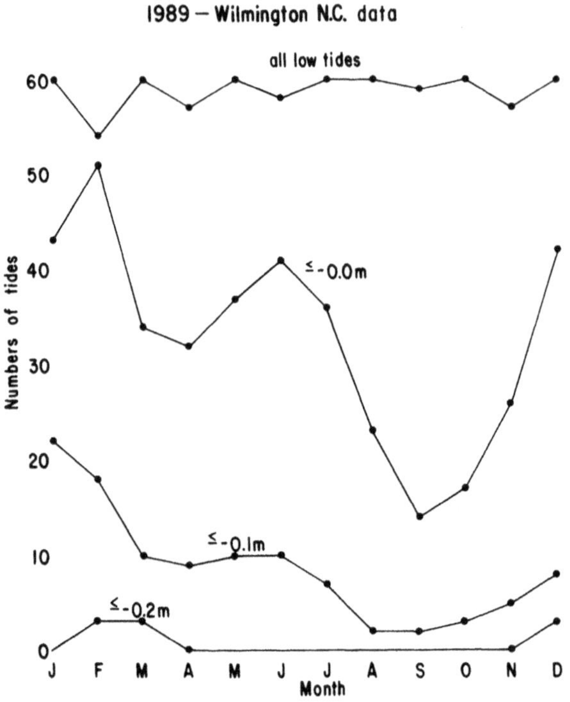

Fig. 2. Frequency of extreme low tides by month at the Wilmington, North Carolina tide station (data from published NOAA, U. S. Dept. of Commerce).

our present understanding of how much natural mortality occurs among adult bay scallops during winter and allow a rational application of the Kellogg and Spitsbergen (1983) bioeconomic model.

Such a model could be profitably elaborated and improved by further application of field experimentation. Peterson et al. (1989) hypothesized that gull predation rate on bay scallops depends on the recent history of strong windstorm events. Even if wind-driven waves do not transport adult scallops preferentially onto shallow flats, where exposure to air and gull predation is increased, because scallops tend to occur in seagrass beds in deeper waters, the mere introduction of entropy would suffice to increase their abundances on shallow flats. If this interaction between windstorm events and gull predation were shown to occur, then windstorms would add a large stochastic variance to the prediction of how gull predation acts on bay scallops over winter months. Because weather histories are recorded, this factor could be incorporated into the model of expected predation but would add variance to the expectation. Such models should probably be constructed specific to sites because the geography of each scallop bed would determine the potential susceptibility of scallops to stranding as a function of wind direction and velocity.

If deemed necessary, even greater fine tuning of such a model of how gull predation affects bay scallop population size could be achieved by performing field and flume experiments on how a scallop's epibiota influences its susceptibility to transport by currents and waves (Orensanz 1986). Adult bay scallops can have extensive epibiotic coverage, including large algae such as _Codium_ and _Sargassum_ and large tunicates such as _Styela_ (Peterson et al. 1989). Furthermore, the incidence and kinds of large epibiotic species on bay scallops varies predictably among sites. Since the presence of large epibiota increases the drag forces acting on the host when flow is applied (Witman and Suchanek 1984, Black and Peterson 1987), it is conceivable that scallops carrying large epibiotic loads will be more readily transported and stranded on flats (as hypothesized by Peterson et al. 1989). Gulls do not discriminate in their predation between scallops with and without epibiota (Fig. 1). This hypothesis, that the degree of gull predation varies with both the recent history of windstorm events and epibiotic coverage, could be tested and could serve as a further elaboration of the foraging model needed to evaluate expected site-specific contributions of gull predation to mortality of bay scallops during winter.

Density dependence and prey preference

Predation rate on a particular prey type is not independent of prey density, the abundance of alternative prey, and the relative preference rankings of available prey types. Because of switching behavior (Murdoch 1969), prey preferences themselves may often be dependent upon the frequencies of available prey types. Such relationships are of obvious relevance to fisheries managers who need to estimate the predation rate upon specific fishery stocks. Again, the use of experiments is almost unavoidable if one wishes to construct a rigorous foraging model of how predation rate affects stock abundance. To address how predation by overwintering gulls on adult bay scallops varies with abundance of scallops and of alternative prey types, a combination of experiments and field observations may be necessary because of the spatial scale of foraging by gulls.

Data from replicated experimental strandings of adult bay scallops on intertidal flats at low tide during different winter months and at several sites described in Peterson et al. (1989) and Prescott (1989) imply high predation rates on exposed scallops. For example, during individual periods of low tide emersion averaging 1.5 hours, approximately 10-15% on average of all exposed bay scallops on unvegetated sand flats were consumed by herring and ringbilled gulls. If extrapolated over longer times of exposure and multiple low tides, this rate suggests that gulls are likely to consume a large fraction of those bay scallops that occur high enough on tidal flats to become exposed at low tides. Nevertheless, these experimental data come from a year when adult bay scallop abundances were unusually low at the study sites because of the influence of a red tide outbreak (Summerson and Peterson 1990). One might reasonably hypothesize that predation rate (as opposed to numbers eaten) by gulls on adult bay scallops would be negatively density dependent: in years with abundant bay scallops the relatively fixed number of overwintering gulls might be expected to consume a smaller percentage of exposed scallops than in years of scallop rarity.

A test of this hypothesis, achieved by comparing results of identical experimental strandings in years differing in bay scallop abundance, is clearly important to resource managers to allow prediction of the predation rate by gulls on adult scallops. If this hypothesis were supported by such data and if gull predation were to prove a significant source of mortality at some levels of bay scallop abundance, then an appropriate strategy for resource management may be to vary the opening date of the bay scallop fishing season. In years of plenty, the harvest could be delayed to take advantage of improving meat weights and improving prices later in winter without appreciable stock losses to gulls. In years of scarcity, the season could be opened earlier to avoid proportionately large losses to gulls.

This management strategy presumes that the few scallops lost to gulls in a poor scallop year are more important than the larger numbers, but lower percentages lost in a good scallop year. Such an assumption is based upon the social importance of even the small scalloping income in a poor year and the possible economic effect of supply on unit price. If these assumptions are unjustified, then the strategy of delaying the opening of the bay scallop fishing season in proportion to annual stock abundance may not be appropriate. Nevertheless, in the absence of rigorous experimentation to test elements of plausible foraging models, the role of bay scallop abundance in affecting predation by and stock losses to overwintering gulls would go unassessed and not contribute to management of the resource.

A complete foraging model for gulls consuming bay scallops must include not only bay scallop abundance but also the abundance of alternative prey and the relative preferences for different prey types. In hypothesizing a tight negative relationship between predation rate by gulls and bay scallop abundance, I assume that at some realistic abundance of bay scallops the numbers of scallop strandings at low tide can exceed the capacity or the interest of gulls to consume scallops. If the bay scallop were a highly preferred prey item for gulls in winter, then the relationship between bay scallop abundance and predation rate from gulls would be relatively insensitive to the abundance of alternative prey. Prey preferences for the various prey types available to gulls in winter are not known but could be assessed by rigorous experimentation. Fish and crabs are important food items in the winter diets of herring and ringbilled gulls (Prescott 1989). Since the availability of both of these food types declines as the weather grows colder into January and February, it is possible that the intensity of predation on stranded bay scallops increases from about November until February in part because of declining abundance of alternative prey. This relationship would be a particularly likely one if bay scallops proved not to be highly preferred over fish and crabs. Proper management of the bay scallop resource would involve experimental assessment of the preferences of gulls for bay scallops relative to major alternative foods and the relationship between predation rate on scallops and the abundance of alternative foods. Such a model would permit fine tuning of management of bay scallops by allowing the opening of fishing season to be delayed when alternative winter prey are abundant.

Size selectivity of predators

An excellent example of how models of size selectivity

play a role in fisheries management comes from the management of hard clam, Mercenaria mercenaria L., aquaculture. The aquaculture of hard clams involves three phases, the first two on land and the final one in natural waters. The initial phase, the hatchery stage, constitutes spawning of adults and nuturing the feeding planktonic larvae to the time of settlement and metamorphosis into the benthic life form. The second phase, the nursery stage, is also usually conducted on shore and involves feeding the juvenile clams in raceways, upwellers, or downwellers (alternative devices to achieve a flux of suspended foods past the feeding clams). This nursery stage continues until the juvenile clams have reached a size at which they can be introduced into plots on the estuarine bottom with expectation of high survivorship. Growth to market size does not continue on shore because as the small clams grow their food requirements also grow to the point where marginal expenses of pumping, food augmentation, and expanded facilities render the land-based culture uneconomical at present market prices. Thus at some size, the juvenile cultured clams must be introduced into the estuary for the final grow-out stage.

The field grow-out stage is the most problematic in the entire process of hard clam aquaculture because of the need to prevent natural predators from decimating the crop. Although a variety of locally important natural predators on juvenile hard clams exists, including seastars (Asterias forbesi), moon snails (Polinices duplicatus), whelks (Busycon spp.), oyster drills (Urosalpinx cinerea), various rays, oystercatchers (Haematopus spp.), and other consumers, the most universally significant consumers are crabs. In North Carolina, blue crabs (Callinectes sapidus) are the primary concern, although mud crabs (Panopeus herbstii) can also be pests. Predation from crabs represents the biggest single impediment to development of an extensive industry in hard clam aquaculture.

Numerous research projects have tested imaginative solutions to the problem of crab predation. Application of gravel or shell hash to the bottom to interfere with foraging success of the crabs, use of mesh cages to exclude crabs, and confining the clams with toadfish (Opsanus tau) or other predators on juvenile crabs have all proven to be of some value (Castagna and Kraeuter 1977, Gibbons and Castagna 1985). In North Carolina, however, none of these protective measures is predictably successful if the hard clams are introduced into the field plots at too small a size. If introduced at sizes less than 1 cm in length, clams are at serious risk from snapping shrimp (Alpheus spp.) and perhaps even penaeid shrimps

(Beal 1983). At sizes less than about 2 cm in length, clams are subject to intense predation by mud crabs (Whetstone and Eversole 1978). Risk of predation for blue crabs also declines with increasing clam size (Arnold 1984).

The crab-bivalve system has proven to be a popular one for testing a variety of hypotheses in optimal foraging theory (e.g., Elner and Hughes 1978, Seed 1980, Hughes and Seed 1981). These studies generally reveal that predators tend to select smaller bivalves over larger ones, which is often shown to be consistent with predictions based upon the expected energetic gain per unit of foraging cost (usually measured by foraging time only). This pattern appears to hold well for the blue crab's size selection of hard clams (Arnold 1984). I, too, have performed several laboratory trials testing the feeding responses of blue crabs to mixtures of different sizes of juvenile hard clams (Peterson, unpub.). These results (Fig. 3) show a clear pattern of declining risk of predation with increasing clam size. Furthermore, there appear to be two major breaks in the curve relating predation rate to hard clam size: below about 1.5 cm in length predation rate is extremely intense, but above about 2.5 cm clams gain some substantial, although not absolute, protection in size. These laboratory experiments were conducted in large (0.24 m^2) wooden aquaria without sediments for convenience, but a control experiment suggested that the absence of sediments did not bias the tests of size selectivity (Fig. 4). Field experiments on a bottom lease confirm that in summer when crabs are active, a clam of 2.5 cm in length experiences substantially less predation than smaller clams (Table 2).

Based on the observed size selectivity of blue crabs while foraging on juvenile hard clams, I propose a management strategy to optimize returns to the aquaculture fishery. The goal is to introduce the clams into the field plots at as small a size as possible, to avoid the escalating costs of reaching larger and larger sizes in the land-based nursery system, yet large enough to insure adequate survivorship in the face of potentially decimating predation, largely by blue crabs. During winter in North Carolina and other southeastern states, blue crabs become inactive and remain buried in mud in deeper waters for several months. Consequently, there exists a temporal refuge for hard clams from predation (Table 3). During these months, hard clam growth in the southeast is still substantial (Peterson and Fegley 1986), as opposed to areas north of Chesapeake Bay where hard clam growth ceases in winter. Consequently, an optimal strategy for management of

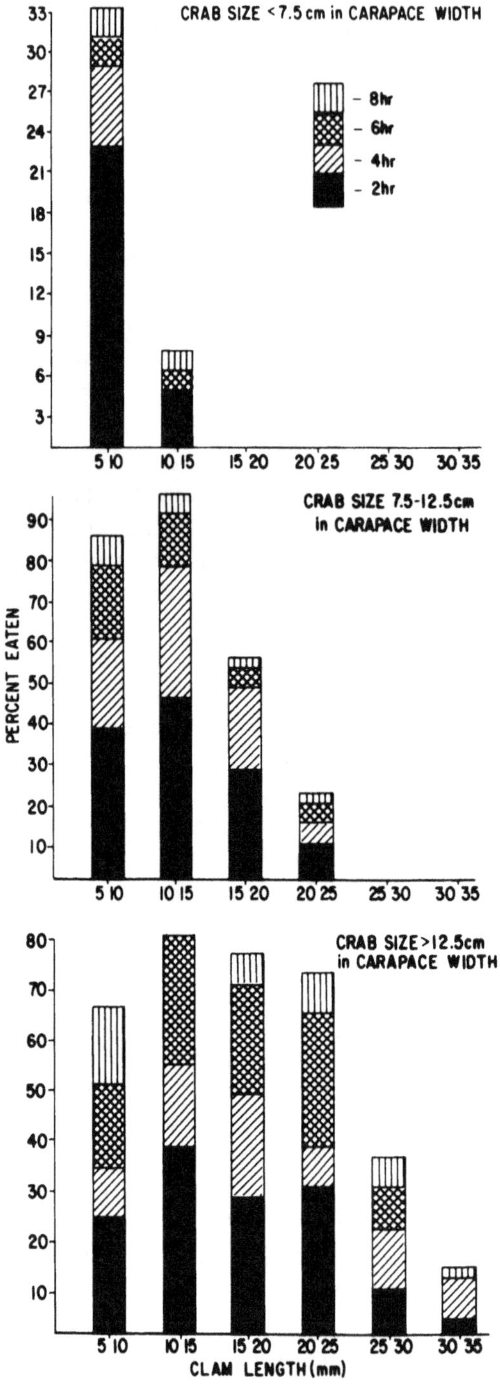

Fig. 3. Size selectivity of three sizes of blue crabs (Callinectes sapidus) feeding upon six sizes of juvenile hard clams (Mercenaria mercenaria). Data represent pooled totals of 7 replicate crabs of each predator size class. Experiments were conducted during the day under natural and fluorescent lighting in 0.24 m^2 wooden aquaria without sediments. Aquaria were supplied with 28-30°C unfiltered seawater, the ambient temperature in August 1984 when the experiments were conducted. Crabs used were males starved for three days prior to use. Ten clams of each size class were used in each replicate trial.

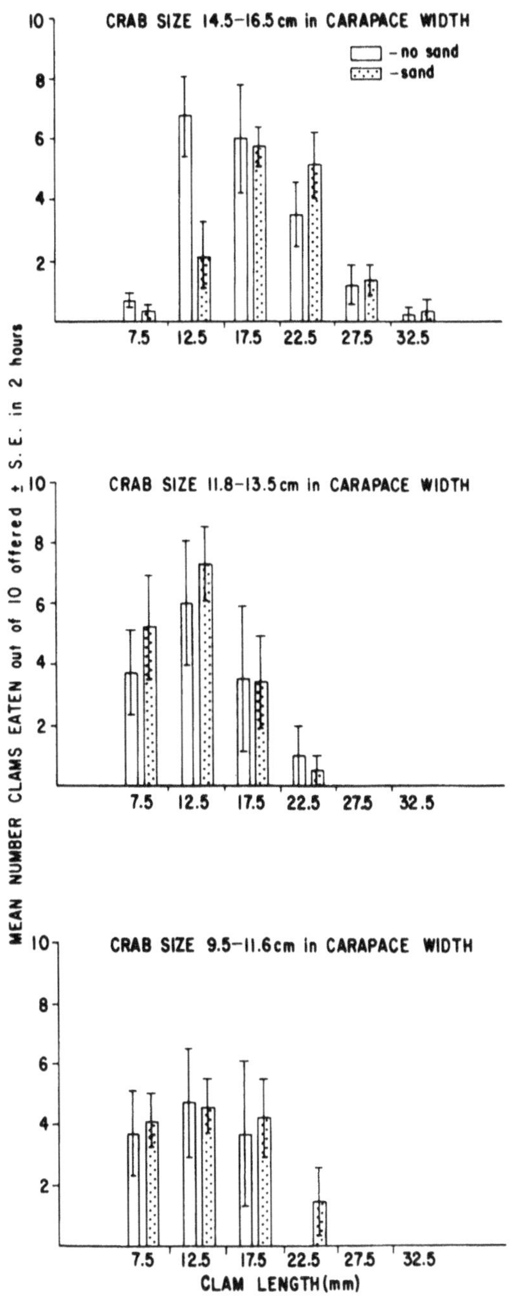

Fig. 4. Size selectivity and consumption rate for 2 hours of feeding by blue crabs (*Callinectes sapidus*) feeding upon six size classes of juvenile hard clams (*Mercenaria mercenaria*). Treatments contrast exposure on the bare aquarium floor to burial in 6 cm of muddy sand sediments. Data represent pooled totals of 4 replicate crabs of each predator size class that were exposed sequentially to both sediment treatments. Experiments were conducted during the day under natural and fluorescent lighting in 0.24 m^2 wooden aquaria. Aquaria received flowing 28-30°C unfiltered seawater, the ambient temperature in August 1984. Crabs used were males starved for three days prior to use. Ten clams of each size class were used in each replicate trial. A contingency test using log-linear models showed no significant difference ($p>0.10$) in size frequency of diet and a paired t-test no significant difference in total number eaten as a function of presence/absence of sediments for any size class of predators.

TABLE 2. SURVIVAL OF SEED ON NEWPORT RIVER LEASE (JUNE 15-OCT 24)*

Protection Treatment	Seed Size: Small-9 mm (1/8-1/4 inch)	Medium-12 mm (1/4-3/8 inch)	Large-20 mm (3/8-3/4 inch)
Unprotected-200	3%	8%	50%
Tent-50	43%	50%	68%
Tent-100	42%	60%	76%
Tent-200	48%	54%	78%
Cage-200	38%	63%	79%
Shell-200	40%	58%	76%

*Average 4-month survivorship of juvenile hard clams (<u>Mercenaria mercenaria</u>) of three different shell lengths under six combinations of density and caging in a Newport River, North Carolina bottom lease. Means average four replicate 1-m^2 plots except for the cage-200 treatment, which had eight replicates. Density code refers to total clam numbers per m^2. Clams of all three sizes were enclosed together in each plot in a numerical ratio of 1 large:1 medium:3 small. Vertical lines connect means that were not significantly different at α=0.05 in a Duncan's test after 1-factor ANOVA's showed significant treatment effects on the small and medium sizes. Tents were 6 mm polypropylene mesh covers anchored by metal bars around the perimeter. Cages were 6 mm polyethelene mesh, fabricated into vertical walls penetrating 10 cm into the sediments and projecting 5 cm up above the bottom, to which a mesh top was fastened, propped up by three internal supports. Shell was about 1 bushel per m^2 of coarsely crushed oyster shell spread evenly on the bottom. Unprotected controls were marked by corner posts only.

the hard clam resource in aquaculture systems would be to place clams into field plots in late autumn (around November in North Carolina), after blue crabs have ceased foraging, using clams of a size that predicted growth over winter will be sufficient to enable them to attain the much less vulnerable sizes >2.5 cm by the time in spring (around May) when blue crabs return to activity (Table 3). My experiments on winter growth of juvenile hard clams in one area (Middle Marsh) of North Carolina suggest that juvenile clams should be introduced into the field plots at a size of 1.0 cm in length in November (Fig. 5). These clams will exhibit tremendous variation in individual growth rate but the majority will attain the relatively safer size of 2.5 cm by May. Even at this size, field grow-out requires protection from caging of around 6 mm mesh to reduce losses to foraging crabs. Nevertheless, this management

TABLE 3. MONTHLY MORTALITY ON 13-16 MM SEED CLAMS IN BACK SOUND LOW INTERTIDAL MUD*

Period	Monthly Mortality Rate (1982)
January	0%
February	3%
March	6%
April	6%
May	10%
June	31%
July	31%
August	6%
September	11%
October	4%
November	14%
December	4%

*Average monthly mortality rates of a cohort of 360 13-16 mm long hard clams (Mercenaria mercenaria) followed in natural field plots in Middle Marsh, Back Sound, North Carolina. Each month begins on the 19th of the previous month and ends on the 19th of the listed month. Data come from Peterson and Fegley (1986) and were taken in 1982.

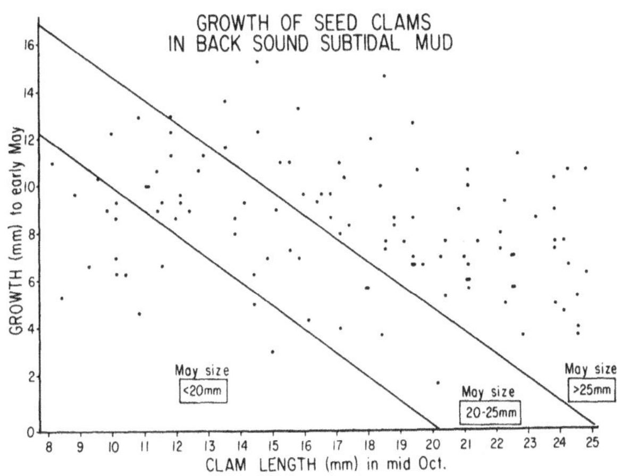

Fig. 5. Growth of individually marked juvenile hard clams (Mercenaria mercenaria) under natural field conditions as a function of initial shell length. Unpublished data come from experiments in Middle Marsh, Back Sound, North Carolina done during four successive years (1978-79, 1979-80, 1980-81, 1981-82).

recommendation illustrates how results of foraging experiments that evaluate size selectivity of predators can be profitably applied to management of a fisheries resource.

Predator habitat relationships

The influence of habitat variation on predator foraging behavior has received a substantial amount of attention by terrestrial, aquatic, and marine ecologists. One common generalization arising from a wide range of studies of predation-habitat relationships is that the addition of structural complexity to a habitat tends to reduce the effectiveness of foraging by the predators (e.g., Crowder and Cooper 1982). In simple laboratory systems, some structural complexity is usually required even to permit coexistence of predator and prey populations (e.g., Gause 1934, Huffaker 1958) because in the absence of refugia provided by the habitat complexity the predator's foraging efficiency is too great to allow prey survival. Other sorts of spatial or temporal refuges can play a similar role in stabilizing predator-prey interactions (e.g., Caswell 1978).

Habitat structure has a strong influence on the foraging success of several consumers of both bay scallops and hard clams in estuarine environments. The presence of seagrass acts to reduce predation on both bay scallops and hard clams. For example, Peterson (1982) showed that densities of hard clams in seagrass beds were about thirty times the densities in adjacent unvegetated sand-bottom habitats and that experimental removal of the seagrass allowed predatory whelks (Busycon spp.) to decimate the clam population. The proximate mechanism for the differential impact of whelk predation on clams in different benthic habitats may well be habitat selection by foraging whelks, although the ultimate cause for lower rates of clam predation by whelks is probably the difficulty of burrowing and foraging through sediments that are compacted by the binding of below-ground roots and rhizomes (Peterson 1982). Blue crab predation on buried molluscs is also apparently inhibited by the presence of seagrasses (Orth 1977). An alternative explanation for the higher hard clam densities inside seagrass beds, namely that recruitment of clams is enhanced by the structure provided by emergent seagrasses, was shown incapable of explaining the majority of the observed density enhancement (Peterson 1986).

Bay scallops in North Carolina are virtually confined to

seagrass habitats, even though or perhaps because the adult scallops possess the ability to move among benthic habitats by swimming (Peterson et al. 1989). Controlled field experiments have demonstrated that the presence of seagrass reduces the rate of predation by gulls on scallops exposed at low tide (Fig. 6). Furthermore, predation by whelks (Busycon spp.) upon experimentally tethered bay scallops is greatly reduced by the presence of seagrass (Fig. 7). The use of tethers in the whelk experiments almost certainly enhances the rate of whelk predation by restricting the escape response of the scallops (Peterson et al. 1982a), but whelks do naturally prey upon bay

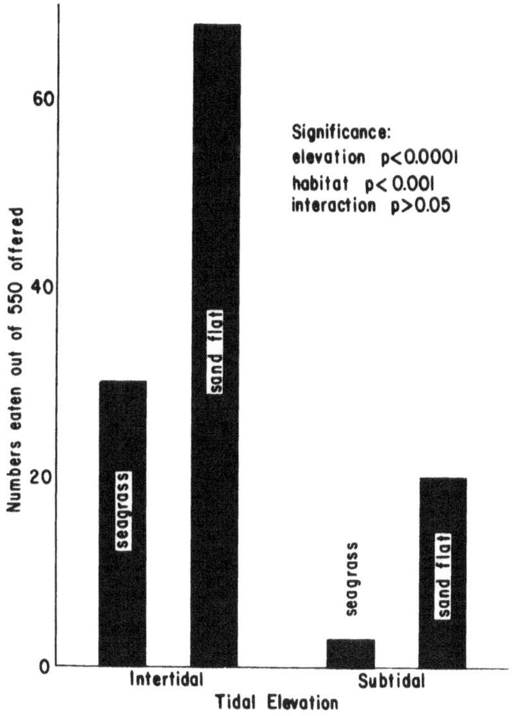

Fig. 6. Effects of tidal elevation and habitat on the rate of gull predation on adult bay scallops (Argopecten irradians) during a single low tide. Data bars pool experimental results from eleven winter dates, two sites in North Carolina, and two treatments of shell epibiota (scraped and fouled). The intertidal sites became exposed to air for 1.5 hours at low tides whereas the subtidal sites were always covered by at least 3 cm of water. Data come from Prescott (1989). Statistical results come from a 3-way contingency test using log-linear models.

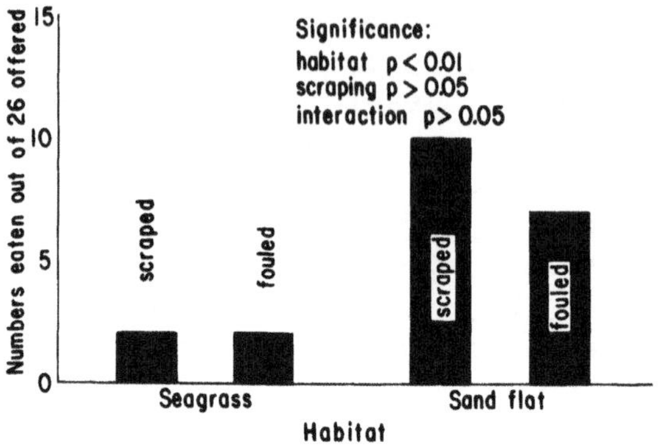

Fig. 7. Effects of habitat and shell scraping on the rate of whelk (Busycon spp.) predation on tethered adult bay scallops (Argopecten irradians). Data came from a 5-day experiment in November 1986 at Middle Marsh, Back Sound, North Carolina (from Peterson et al. 1989). Statistical results come from a 3-way contingency test using log-linear models.

scallops (Peterson et al. 1989, Prescott 1989) so the habitat difference is probably a real one.

Experiments showing that seagrass beds represent a partial refuge from natural predators for hard clams and bay scallops has been incorporated into resource management for several years. Seagrass habitat is preferentially protected from destruction by development activities, channel dredging, and destructive fishing practices. For example, fishermen are allowed to use only those hand implements that do not greatly uproot seagrass when they fish for clams in seagrass beds. More destructive "bull rakes" (Peterson et al. 1983) and mechanical harvest gear such as "clam kickers" and hydraulic dredges (Peterson et al. 1987) are prohibited by regulation from use in North Carolina's seagrass beds. These regulations are justified largely on the basis of observational and experimental evidence that the seagrass beds represent a habitat that acts as an important refuge where many fisheries species are less vulnerable to foraging by natural enemies.

Role of optimal foraging theory

I have advocated and illustrated the use of largely

empirical experimentation by fisheries resource managers as a means of incorporating more accurate and complete information on foraging behavior of consumers of those resources into management schemes. Following this recommendation does not necessarily involve the use of optimal foraging theory. Nevertheless, optimal foraging theory may enter into this process in two important ways. First, the null hypotheses and specific factors examined in experiments designed to improve management of fisheries resources may often be generated by knowledge of how similar consumers under analogous conditions optimize their foraging behavior. In particular, in the absence of appreciation of the formalisms of optimal foraging theory, experiments constructed to understand foraging behavior may fail to include critically important factors, which may even confound interpretation of results. Second, to the degree that optimal foraging considerations provide a rational explanation for observed foraging behavior, one might have more credence in the results of otherwise empirical experimentation and more confidence in their generality. Nevertheless, this use of optimal foraging theory can easily become abused if resource managers too readily and unrigorously fall victims to the Lorelai appeal of a unifying theory that allows rational explanation for all observed foraging behaviors. Among many resource managers, the existence of optimal foraging theories may be more dangerous than helpful if the desire to generalize and especially extrapolate outstrips rigor.

One fundamental concern that may be raised with my suggestion that fisheries resource managers implement manipulative ecological experiments to assess the role in resource dynamics of foraging by consumers, is the issue of cost. Such experiments are labor-intensive. This characteristic, however, makes the approach especially suitable to adoption in many developing countries where labor costs are minimal. Increased training of fisheries resource managers from the "third world" in the developed countries promises a supply of technically competent managers to organize, supervise, and conduct such experiments. Furthermore, in such experiments with natural and managed ecosystems there is no direct cost of applying the treatment. In foraging experiments, the natural consumers react to manipulation and through their behavior apply the treatment. In managed systems, the fishing industry applies a treatment in the conduct of business. Such treatments do not have direct cost to the resource manager unlike treatments addressing the differential consequences of alternative manufacturing processes, for example. Consequently, many experimental tests of consumer behavior should be affordable to resource managers and should serve fisheries resource management by exploiting rigorous methodologies of modern ecological hypothesis testing.

Summary. Management of fisheries can often be improved by substituting models based upon direct experimental tests of explicit hypotheses for traditional yield-per-unit-effort or yield-per-recruit models. Largely sessile shellfishes are particularly amenable to application of the experimental methodologies developed by rocky-shore and tidal-flat ecologists. Some problems, such as questions involving controls on recruitment of species with planktonic larvae, incorporate spatial and temporal scales that render experimentation difficult. The factors that influence predator behaviour and predation rate, however, can be readily assessed by experiment. Tests of physiological constraints, size selectivity, prey preferences, density dependence, and habitat selection in predation processes can be and should be incorporated directly into resource management. Optimal foraging theory plays two roles in this process of applying experimental hypothesis testing to fisheries management. First, optimal foraging theory generates reasonable hypotheses to test. Second, and more importantly, optimal foraging theory, if applied rigorously, can provide a rational explanation for specific predation patterns, which lends credence to conclusions based on otherwise purely empirical evidence and supports generalization from those conclusions. The utility of experimental tests of foraging hypotheses is well illustrated by an evaluation of aspects of how the bay scallop resource and hard clam aquaculture could be managed along the southeastern Atlantic coast of North America.

Acknowledgements. The research described here on bay scallops and hard clams was supported by the U. S. Department of Commerce, NOAA, under the National Sea Grant College Program, the University of North Carolina Sea Grant College Program, and the State of North Carolina.

References

Arnold WS (1984) The effects of prey size, predator size, and sediment composition on the rate of predation of the blue crab, Callinectes sapidus Rathbun, on the hard clam, Mercenaria mercenaria (Linne). J Exp Mar Biol Ecol 80:207-219

Beal BF (1983) Predation of juveniles of the hard clam Mercenaria mercenaria (Linne) by the snapping shrimp Alpheus heterochaelis Say and Alpheus normanni Kingsley. J Shellfish Res 3:1-9

Belding DL (1910) A report on the scallop fishery of Massachusetts. Comm. of Massachusetts, Boston

Beverton RJH, Holt SJ (1957) On the dynamics of exploited fish populations. Fish Invest Lond, Ser 2, 19

Black R, Peterson CH (1987) Biological vs. physical explanations for the non-random pattern of host occupation by a macroalga attaching to infaunal bivalve molluscs. Oecologia 73:213-221

Boulding EG, Hay TK (1984) Crab response to prey density can result in density-dependent mortality of clams. Can J Fish Aquat Sci 41:521-525

Castagna M, Kraeuter JN (1985) Mercenaria culture using stone aggregate for predator protection. Proc Natl Shellfish Assoc 67:1-6

Caswell H (1978) Predator-mediated coexistence: a nonequilibrium model. Am Nat 112:127-154

Connell JH (1972) Community interactions on marine rocky intertidal shores. Annu Rev Ecol Syst 3:169-192

Crowder LB, Cooper WE (1982) Habitat structural complexity and the interaction between bluegills and their prey. Ecol 63:1802-1813

Cushing DH (1981) Fisheries biology: a study in population dynamics, 2nd edn. Univ Wisconsin Press, Madison

Duggan WP (1973) Growth and survival of the bay scallop, Argopecten irradians, at various locations in the water column and at various densities. Proc Natl Shellfish Assoc 63:68-71

Eckman JE, Peterson CH, Cahalan JA (1989) Effects of flow speed, turbulence, and orientation on growth of juvenile bay scallops (Argopecten irradians concentricus Say). J Exp Mar Biol Ecol: in press

Elner RW, Hughes RN (1978) Energy maximization in the diet of the shore crab, Carcinus maenas. J Anim Ecol 47:103-116

Frank PW (1982) Effects of winter feeding on limpets by black oystercatchers, Haematopus bachmani. Ecol 63:1352-1362

Gause GF (1934) The struggle for existence. Hafner, New York

Gibbons MC, Castagna M (1985) Biological control of production by crabs in bottom cultures of hard clams using a combination of crushed stone aggregate, toadfish, and cages. Aquaculture 47:101-104

Gutsell JS (1930) Natural history of the bay scallop. Bull US Bur Fish 46:569-632

Huffaker CB (1958) Experimental studies on predation: dispersion factors and predator-prey oscillations.

Hilgardia 27:343-383
Hughes RN, Seed R (1981) Size selection of mussels by the blue crab Callinectes sapidus: energy maximizer or time minimizer? Mar Ecol Prog Ser 6:83-89
Irons DB, Anthony RG, Estes JA (1986) Foraging strategies of glaucous-winged gulls in a rocky intertidal community. Ecol 67:1460-1470
Kellogg R, Spitsbergen D (1983) Predictive growth model for the meat weight (adductor muscle) of bay scallops in North Carolina. UNC Sea Grant Publ UNC-SG-WP-83-6, Raleigh
Marsh CP (1986) Rocky intertidal community organization: the impact of avian predators on mussel recruitment. Ecol 67:771-786
Murdoch WW (1969) Switching in general predators: experiments on predator specificity and stability of prey populations. Ecol Monogr 39:335-354
O'Connor RJ, Brown RA (1977) Prey depletion and foraging strategy in the oystercatcher Haematopus ostralegus. Oecologia 27:75-92
Orensanz JM (1986) Size, environment, and density: the regulation of a scallop stock and its management implications. Can Spec Publ Fish Aquat Sci 92:195-227
Orth RJ (1977) The importance of sediment stability in seagrass communities. In: Coull BC (ed) Ecology of marine benthos. Univ South Carolina Press, Columbia, pp 182-300
Peterson CH (1982) Clam predation by whelks (Busycon spp.): experimental tests of the importance of prey size, prey density, and seagrass cover. Mar Biol 66:159-170
Peterson CH (1986) Enhancement of Mercenaria mercenaria densities in seagrass beds: is pattern fixed during settlement season or altered by subsequent differential survival? Limnol Oceanogr 31:200-205
Peterson CH, Ambrose WG, Hunt JH (1982a) A field test of the swimming response of the bay scallop (Argopecten irradians) to changing biological factors. Bull Mar Sci 32:939-944
Peterson CH, Fegley SR (1986) Seasonal allocation of resources to growth of shell, soma, and gonads in Mercenaria mercenaria. Biol Bull 171:597-610
Peterson CH, Summerson HC, Fegley SR (1982b) The relative efficiency of two clam rakes and their contrasting impacts on seagrass biomass. Fish Bull 81:429-434
Peterson CH, Summerson HC, Fegley SR (1987) Ecological consequences of mechanical harvesting of clams. Fish Bull 85:281-298
Peterson, CH, Summerson HC, Fegley SR, Prescott RC (1989) Timing, intensity, and sources of autumn mortality of adult bay scallops, Argopecten irradians concentricus Say. J Exp Mar Biol Ecol 127:121-140
Prescott RC (1989) Sources of predatory mortality in the bay scallop Argopecten irradians in North Carolina: interactions with seagrass and epibiotic coverage. J Exp Mar Biol Ecol: in press
Quammen ML (1984) Predation by shorebirds, fish, and crabs on invertebrates in intertidal mud flats: an experimental test. Ecol 65:529-539
Ricker WE (1954) Stock and recruitment. J Fish Res Board Can

11:559-623
Ricker WE (1958) Handbook of computations for biological statistics of fish populations. Bull Fish Res Board Can 119:1-300
Rothschild BJ (1986) Dynamics of marine fish populations, Harvard Univ Press, Cambridge
Seed R (1980) Predator-prey relationships between the mud crab Panopeus herbstii, the blue crab, Callinectes sapidus and the Atlantic ribbed mussel Geukensia (=Modiolus) demissa. Estuarine Coastal Shelf Sci 11:445-458
Sinclair M (1988) Marine populations: an essay on population regulation and speciation, Univ Washington Press, Seattle
Summerson HC, Peterson CH (1990) Patterns of recruitment failure of the bay scallop Argopecten irradians concentricus during the first red tide Ptychodiscus brevis outbreak recorded in North Carolina. Estuaries: in review
Sutherland WJ (1982) Spatial variation in the predation of cockles by oystercatchers at Traeth Melynog Anglesey. I. The cockle population. J Anim Ecol 51:481-490
Thayer GW, Stuart HH (1974) The scallop makes its bed of eelgrass. U S Nat Mar Fish Serv Mar Fish Rev 36:27-39
Underwood AJ (1981) Techniques of analysis of variance in experimental marine biology and ecology. Oceanogr Mar Biol Annu Rev 19:513-605
Underwood AJ (1986) The analysis of competition by field experiments. In: Kikkawa J, Anderson DJ (eds) Community ecology: pattern and process. Blackwell, London, pp 240-268
Whetstone JM, Eversole AG (1978) Predation on hard clams Mercenaria mercenaria, by mud crabs Panopeus herbstii. Proc Natl Shellfish Assoc 68:42-48
Witman JD, Suchanek TH (1984) Mussels in flow: drag and dislodgement of epizoans. Mar Ecol Prog Ser 16:259-268
Zwarts L, Drent RH (1981) Prey depletion and the regulation of predator density: oystercatchers (Haematopus ostralegus) feeding on mussels (Mytilus edulis). In: Jones NV, Wolf WJ (eds) Feeding and survival strategies of estuarine organisms. Plenum, New York, pp 193-216

ORAL DISCUSSION – C. H. Peterson

Q. (Real) In a hierarchical Optimal Foraging problem, foraging in the biological sense is related to what caused the mortality of scallops, but there is also the OF problem of fishing for scallops. It might be interesting to analyse the first problem assuming fixed mortality rates, but the fisherman's behaviour is going to be very much influenced by the amount of uncertainty, so that estimated variance (in mortality) is going to be critical.

A. The manager looking at natural mortality must quantify the penalty for delaying the opening of the season from December to February. You need to predict the income as you go along, then you have fishing mortality on top of the natural mortality. We had a year when fishing was closed because of a red–tide outbreak, enabling us to estimate natural mortality in the absence of fishing mortality. Natural mortality rate was 21% over December and January, so that would have been the price paid in a bioeconomic model. However, the red–tide may have caused an unusual level of mortality in either direction. It could have killed the scallops, or it could have made them distasteful to gulls and therefore reduced the rate of pilfering. So how do you treat that number ? That is why we need a mechanistic (for example Optimal Foraging) model, because the more you can explain at that level, the better.

Q. (DeMott) Your data show tremendous differences between some years, but if you are going to delay the season, you must be able to tell the fishermen whether the year is going to be a good or a bad one for fishing.
A. We can do that to some degree, because the numbers of recruits correlate well with the numbers of adults one year later at the beginning of December. So under the current opening scheme we can give fishermen a year's warning, based on recruitment monitored in the previous autumn.

Q. (Ansell) I can't understand why the bay scallop spawns, then goes on living for another 6 months, during which time it develops its adductor muscle. In other scallops all energy would be used to develop the gonads for spawning earlier in the year.
A. Actually, there is a second spawning in the spring, for those animals that survive that long. But I don't think this spawning yields successful recruits in North Carolina, because of intense predation by juvenile fishes, prawns and crabs. The first autumn spawning yields recruits when their predators are absent or dormant. A very small number of recruits from the second spawning may survive and it would be useful to model this influence on life history selection.

Q. (Gilliam) Do you have any information on whether the gulls switched to or from scallops in response to the relative abundance of major alternative foods ?
A. There is seasonal variation in the diet of gulls: they eat crabs and fish, but not many scallops very early in the autumn. They begin to eat scallops in about December and increase their predation on scallops as winter progresses. This is the same time period when fish and crabs become rive in the estuarine waters.

THE ROLE OF THE OPTIMAL DIET PREDATOR IN MULTISPECIES FISHERY ASSESSMENT

T. J. Pitcher & G. F. Turner
School of Biological Sciences
University of Wales
BANGOR
Gwynedd LL57 2UW
United Kingdom

Introduction

This paper asks whether detailed knowledge of the optimal diet selection behaviour of a predator would be likely to assist in making an assessment of a commercial fishery.

In the management of fisheries, there is a current attempt to move away from treating each species as an isolated independent unit (Brander, 1988). Multispecies models vary from complex ecosystem compartment simulations (Laevastu *et al.* 1982), *ad hoc* aggregated compartment models (Larkin & Gazey 1982), *ad hoc* empirical and simplified ecological relationships (eg Polovina 1984), multispecies versions of the classical surplus yield fishery model (e.g. Kirkwood 1982), or multispecies dynamic pool models (e.g. Doubleday & Beacham 1982) with various degrees of species interactions built in. Only models restricted in complexity to one or two predator/prey or competitive interactions can be considered to have been in any way successful. In relatively few cases have the models been successful enough to be considered in setting fishery quotas (an exception is Brander & Bennet 1987), but a number have had heuristic utility.

It is very difficult to predict with confidence the response of a community to changes in the exploitation of some of its component species: this depends on the realism of the model (Sainsbury 1988). The predictive power of models at first increases with their complexity and realism, but soon decreases as unmeasured (and often unmeasurable) parameters introduce uncertainties and errors (Sugihara *et al.* 1984). For the purposes of management, the optimal model complexity for the spatially and temporally aggregated data of commercial fisheries is low (Sainsbury 1988), precluding the investigation of detailed community responses. In our case, however, we wish only to compare the overall consequences of two alternative styles of predation for fishery management, and can

NATO ASI Series, Vol. G 20
Behavioural Mechanisms of Food Selection
Edited by R. N. Hughes

therefore justify employing a simplified two-species steady state model.

A general framework for steady-state multispecies fishery models has been devised (Shepherd 1988) and has been followed here. In a two-species predator-prey model Shepherd considers an unselective predator with consumption based on the weight of fish tuned to real preferences calculated from North Sea cod and haddock stomach data. It is suggested that the yield of fish is relatively insensitive to the specific predator preference rules. This paper tests Shepherd's hypothesis for the specific case of optimal diet versus an unselective predator.

Based on the energy balance of feeding benefits to the costs of handling and capture, optimal diet theory makes some very strong predictions about the way that predators should select prey of different values (Stephens & Krebs 1986). In the case of fish, which exhibit indeterminate growth, the diet of a given size of predator should expand and contract to include or exclude a range of sizes of prey fish (Hart 1986). Optimal diet theory has accumulated an impressive body of laboratory evidence, and even some field data (Hughes & Burrows 1989), that predators obey its general rules.

If optimal diet theory shows the precise way that predators select their diet, then at first sight it would seem perverse not to use it in when formulating the 'predator preference' sub-model in a multispecies predator-prey fishery model. If we make the assumption that our predator feeds in a fine-grained environment and cannot easily feed in different habitats, the inclusion of less rewarding prey in the diet depends only on the encounter rate with prey already in the diet, and so the optimal diet dynamics are relatively easy to model. There is, however, a considerable cost in obtaining data to characterise the optimal diet dynamics of a particular predator-prey interaction. This paper investigates the need for obtaining such data in relation to the questions generally posed in fishery assessment.

Methods

The model is built around the age-structured dynamic pool paradigm with explicit submodels for growth, mortality and recruitment (Pitcher & Hart 1982). The core of the model is the standard fisheries age-structured catch equation:

$$Y = \sum_{i=1}^{i=tl} [F_i/Z_i\ (1-\exp(-Z_i))\ N_i\ W_i]$$

where Z_i is the total instantaneous mortality rate of age class i, F_i is the instantaneous mortality rate due to fishing, N_i is the number alive at the start of age class i, W_i is the mean weight of individual of age class i and tl is the oldest age group in the stock. Separate single-species catch equations were incorporated for predator and for prey.

Our deterministic model provides a long-term steady-state assessment aimed at providing goals for management and therefore contains no stochastic element in the predation sub-model or in recruitment. The only difference this would make to a long-term assessment is in the estimation of variability and risk (Beddington & May 1977; Pitcher & Hart 1982), in contrast to the short-term forecasting used to set fishery quotas.

Vectors of mean weight simulated the growth of predator and prey (See Table 1). We assume that growth and fishing mortality of the predator and prey are known: our predator and prey were loosely based on cod and haddock respectively.

	Mass (g)										Mortality	Recruitment parameters		
Age	1	2	3	4	5	6	7	8	9	10	M	a	S'	beta
Predator	25	400	950	2100	4200	6500	8300	9600	10400	11000	0.10	0.002	10^7	1 or 2
Prey	3	45	100	230	400	700	900	1050	1200	1300	0.05	0.020	$5x10^6$	1 or 2

Table 1: Mass-with-age for the predator and prey together with natural instantaneous mortality rate (other than predation) and the recruitment curve parameters; the prey is approximately 10% of the weight of the predator. Values are loosely based on those in Shepherd (1988) for cod and haddock.

To simplify the model, fisheries for our predator and prey were deemed independent, entailing different fishing gear for example. Management options for the fishery were simulated by varying F from 0 to 1.1 independently for both predator and prey. F was equivalent across all ages as there seemed little point in modelling the subtleties of gear selection for the purposes of comparing the effect of the different types of predator.

Both predator and prey suffered natural mortality other than from predation, M (See Table 1), which was added to the fishing mortality rate to make up total mortality:

$$Z_i = F_i + M_i$$

The total mortality for the prey also included mortality due to the predator:

$$Z_i = F_i + M_i + PM_i$$

where PM is mortality from the predator, modelled as set out below.

The conventional use of yield-per-recruit can be misleading in fishery assessments, and so recruitment to the fishery was explicitly modelled using Shepherd's flexible stock-recruitment relationship (SRR) (Shepherd 1982).

$$R = (aS)/(1 + (S/S')^{beta})$$

where R is recruits, a is a proportionality coefficient, S is spawning stock biomass, S' is the stock biomass at which density dependence reduces recruits below a linear model, beta is a power varying between 0 and 2.

Because of the great variability and uncertainty concerning appropriate stock recruitment models, two extreme versions of the SRR were compared. First, recruitment was set to be mildly density dependent, and the consequent number of recruits asymptotic: under the Shepherd model this is quasi-Beverton & Holt recruitment (Pitcher & Hart 1982) with beta set to 1. Secondly, strongly density dependent recruitment was simulated with beta set to 2: this is a quasi-Ricker curve. (The parameters were chosen to be within the subset of values with a stable age distribution). Age of maturity to produce the spawning stock for the recruitment curve was set to 3 years old for both predator and prey.

Mortality from the predator, PM_i, was age-structured, and for each age class of prey was proportional to the numbers of the predators in each age class which were allowed to eat it.

Prey selection matrices were constructed with a cell for each age of predator eating each age of prey. Using the weight-at-age vectors, a matrix of predator to prey weight ratios (prey weight as a percentage of predator weight) for each age class interaction of predator and prey was drawn up (Table 2).

	Predator ages									
Prey	1	2	3	4	5	6	7	8	9	10
ages 1	12.0	0.8	0.3	0.1	0.1	0.0	0.0	0.0	0.0	0.0
2	180.0	11.3	4.7	2.1	1.1	0.7	0.5	0.5	0.4	0.4
3	400.0	25.0	10.5	4.8	2.4	1.5	1.2	1.0	1.0	0.9
4	920.0	57.5	24.2	11.0	5.5	3.5	2.8	2.4	2.2	2.1
5	1600.0	100.0	42.1	19.0	9.5	6.2	4.8	4.2	3.8	3.6
6	2800.0	175.0	73.7	33.3	16.7	10.8	8.4	7.3	6.7	6.4
7	3600.0	225.0	94.7	42.9	21.4	13.8	10.8	9.4	8.7	8.2
8	4200.0	262.5	110.5	50.0	25.0	16.2	12.7	10.9	10.1	9.5
9	4800.0	300.0	126.3	57.1	28.6	18.5	14.5	12.5	11.5	10.9
10	5200.0	325.0	136.8	61.9	31.0	20.0	15.7	13.5	12.5	11.8

Table 2: Potential prey weights as percentages of predator weight for ten age class interactions.

Alternative consumption matrices for the predator were drawn up using the table of relative prey weights (Table 2).

In Table 3 a blank in a cell indicates no predation whereas a block indicates that this age of predator eats this age of prey. In the model (matrix PS), these are represented by zero and one respectively. If more than one age class of prey was to be eaten by a given age of predator, the prey selection coefficients were divided equally among them, i.e. for three age classes of prey, the coefficient was one third in each case. So the diets were compared on the basis of equivalent numbers, not mass, within each age class.

The following rules were devised for the model to reasonably simulate reality: the precise values have been varied, but make relatively little difference to the results.

UNSELECTIVE PREDATOR

prey age \ predator age	1	2	3	4	5	6	7	8	9	10
1	■	■	■	■	■	■	■	■	■	■
2		■	■	■	■	■	■	■	■	■
3		■	■	■	■	■	■	■	■	■
4		■	■	■	■	■	■	■	■	■
5			■	■	■	■	■	■	■	■
6			■	■	■	■	■	■	■	■
7				■	■	■	■	■	■	■
8				■	■	■	■	■	■	■
9				■	■	■	■	■	■	■
10				■	■	■	■	■	■	■

OPTIMAL DIET PREDATOR above threshold

	1	2	3	4	5	6	7	8	9	10
1	■									
2		■								
3			■							
4				■						
5					■					
6						■				
7							■			
8								■		
9									■	
10										■

OPTIMAL DIET PREDATOR below threshold

	1	2	3	4	5	6	7	8	9	10
1	■									
2		■	■							
3		■	■	■						
4			■	■	■	■				
5				■	■	■	■	■	■	■
6					■	■	■	■	■	■
7					■	■	■	■	■	■
8						■	■	■	■	■
9						■	■	■	■	■
10						■	■	■	■	■

Table 3: Alternative prey selection matrices (PS) for the predator. ■ = consumption allowed; blank = consumption disallowed.

The unselective predator was allowed to eat any prey less than 75% of its own weight (see Table 2). When prey numbers were above a diet switch threshold, the 'optimal diet' predator ate only prey of the same age as itself, in each case approximately 10% (9.5% - 12%) of its own weight (Table 3). When prey numbers were below the threshold, the optimal diet predator broadened its diet to include all prey between 3% and 25% of its own weight (see Tables 2 and 3).

The switch threshold at which the optimal diet predator would broaden its diet was set to half the virgin stock prey numbers for each age class. The predator therefore switched at half the encounter rate with prey that it experienced in the unfished stock: in different runs of the model this value was altered between zero and 0.9.

In each case, predators consumed prey in proportion to their numbers (coefficient, cp = 3×10^{-5}). The instantaneous rate of mortality due to predation for prey age i, PM_i, was therefore given by

$$PM_i = \sum_{j=i}^{j=tl} [cp \times ps_{ij} \times n_j]$$

where cp is a coefficient of proportionality, ps_{ij} is the prey selection coefficient for prey age i by predator age j, n_j is the numbers of predators aged j. For the chosen values of the coefficients PM_i values varied between 0.1 and 1.2.

For each run of the model, the starting population age structure was based on mortality in the virgin stock. The model was iterated for each combination of fishing mortalities until a stable age distribution was reached (Shepherd 1988). Output from the model was contoured plots of the total numbers of prey eaten, prey fishery yield and total joint yield from the fishery, prey biomass and prey recruitment.

Results

In absence of predation, the simulated fishery produced a strongly peaked type of yield curve, as might be expected where the age of entry is equivalent to the age of recruitment and there is no lag between spawning and recruits entering the fishery (Pitcher & Hart 1982). The shape of this region of the dynamic pool yield surface might be expected to be most sensitive to changes in natural mortality from our predator-prey interaction.

Figure 1 shows the numbers of prey consumed by the predator at each combination of fishing mortalities on the predator and prey. Values for the selective predator are in the right hand column; those for the unselective predator in the left. Strongly density-dependent Ricker recruitment is at the top of the figure, while the asymptotic Beverton and Holt recruitment is at the bottom. Contours are drawn at convenient levels; maxima are indicated by the stippled area.

Figure 2 illustrates the recruitment of the prey population, contoured as percentages of the maximum values, and shown for each type of predator and recruitment curve as before. Figure 3 shows the biomass of the prey population, again contoured as percentages of the maximum values. Figure 4 contours the yield in weight of the prey population. Figure 5 shows the total joint yield (in weight) of the fishery for the predator and the prey.

In each case there are only minor differences in the shape of the parameter surfaces between selective and

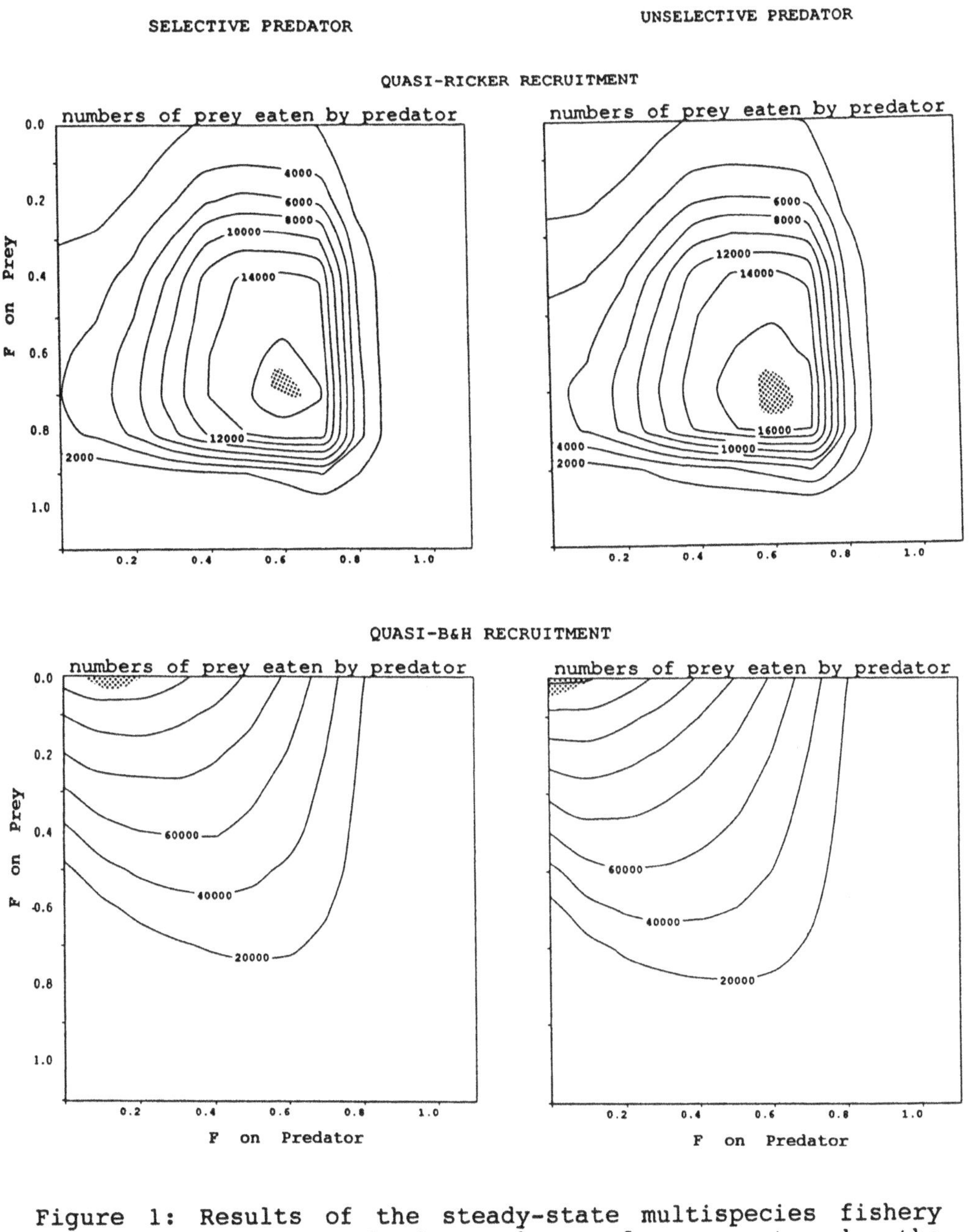

Figure 1: Results of the steady-state multispecies fishery model. Contours of the numbers of prey eaten by the predator population under different levels of fishing mortality, F, on the predator and the prey. The optimal diet predator is on the left and the unselective predator on the right. The top half of the diagram refers to quasi-Ricker recruitment and the bottom half to quasi-Beverton and Holt recruitment. Peaks are shown by shading.

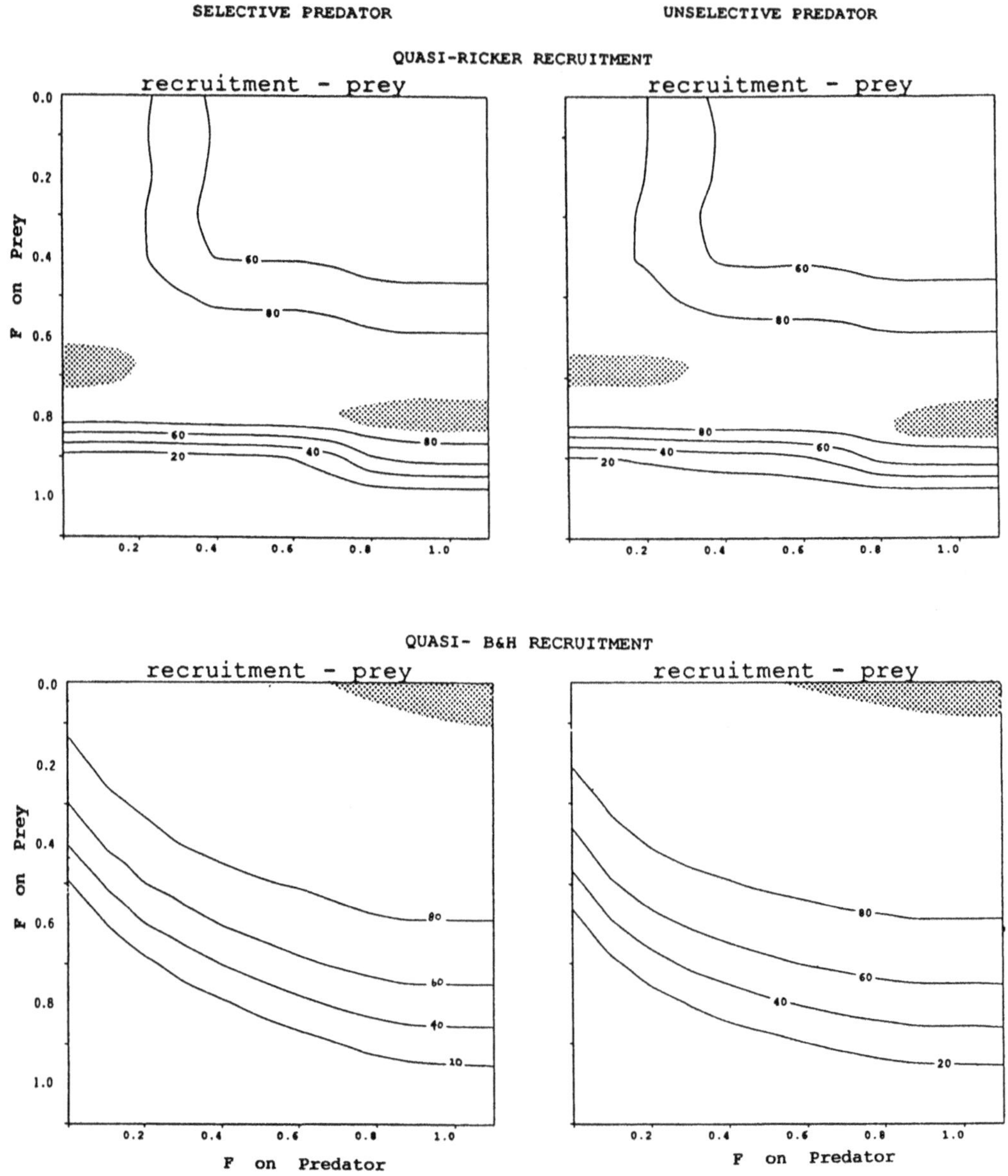

Figure 2: Contours of the recruitment of prey under different levels of fishing mortality, F, on the predator and the prey. The optimal diet predator is on the left and the unselective predator on the right. The top half of the diagram refers to quasi-Ricker recruitment and the bottom half to quasi-Beverton and Holt recruitment. Contours are at percentages of the maximum value; peaks are shown by shading. Note that the effect of the predator is to produce two maxima for Ricker recruitment.

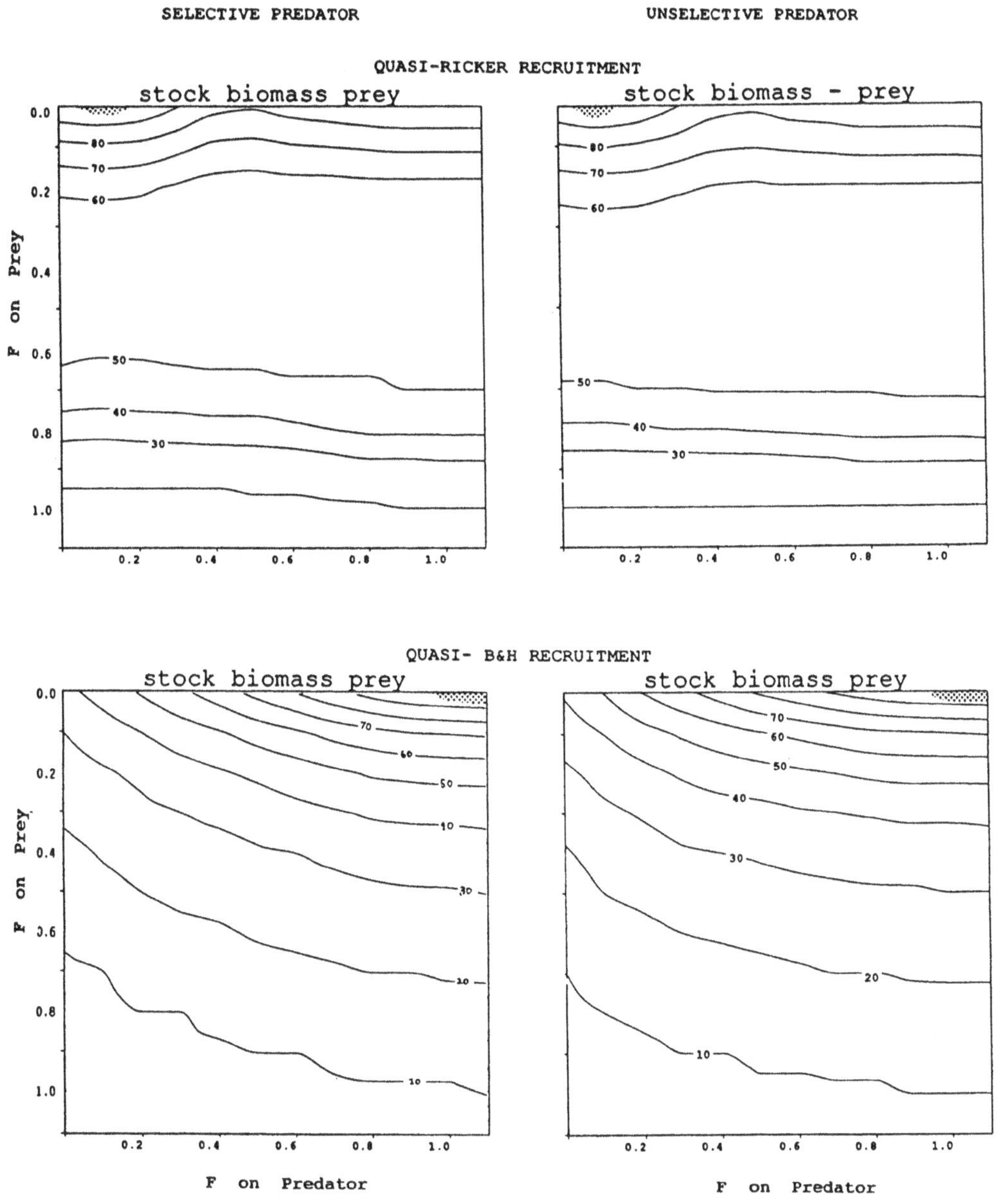

Figure 3: Contours of prey biomass under different levels of fishing mortality, F, on the predator and the prey. The optimal diet predator is on the left and the unselective predator on the right. The top half of the diagram refers to quasi-Ricker recruitment and the bottom half to quasi-Beverton and Holt recruitment. Contours are at percentages of the maximum value; peaks are shown by shading.

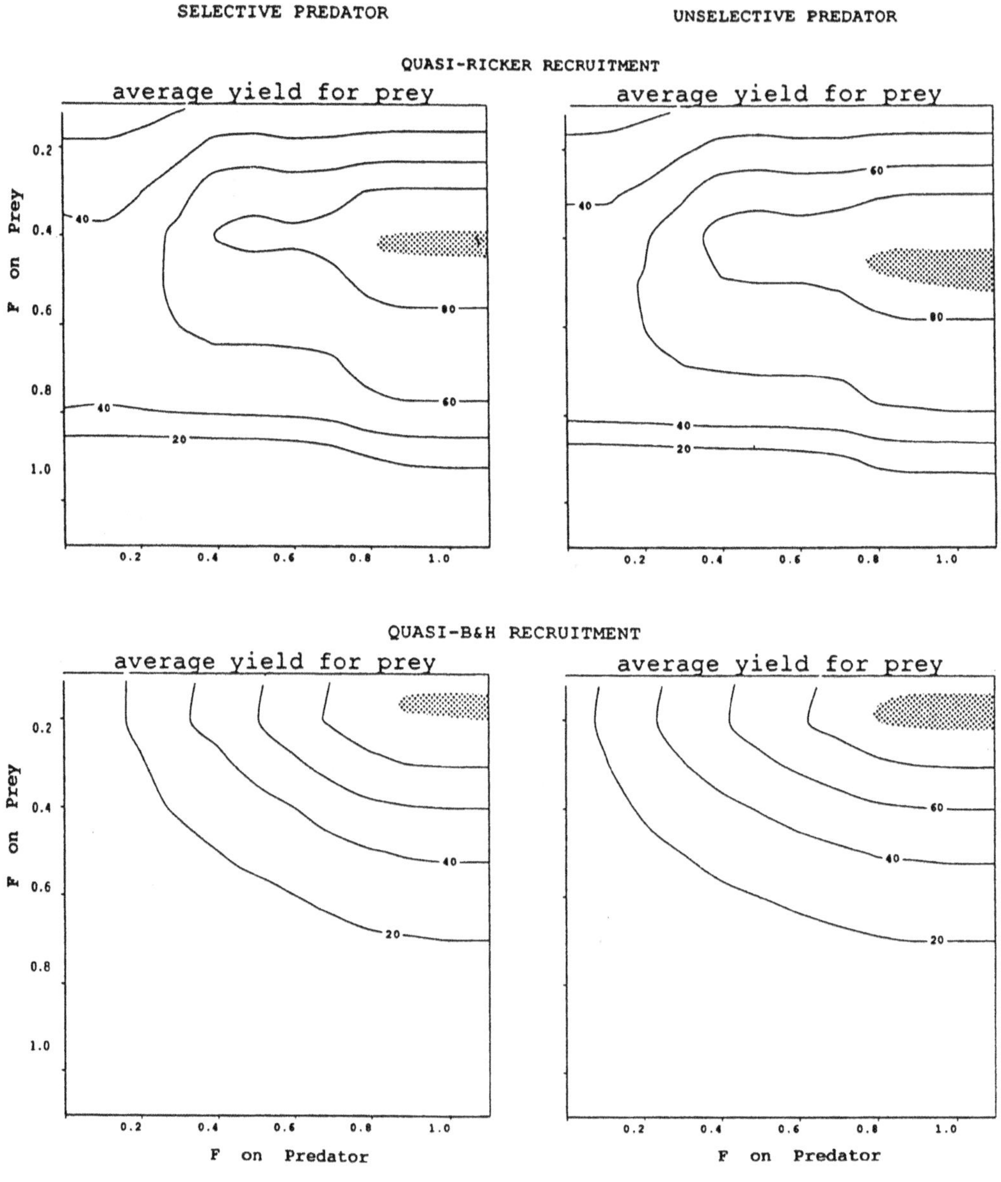

Figure 4: Contours of the yield in weight in the prey fishery under different levels of fishing mortality, F, on the predator and the prey. The optimal diet predator is on the left and the unselective predator on the right. The top half of the diagram refers to quasi-Ricker recruitment and the bottom half to quasi-Beverton and Holt recruitment. Contours are at percentages of the maximum value; peaks are shown by shading.

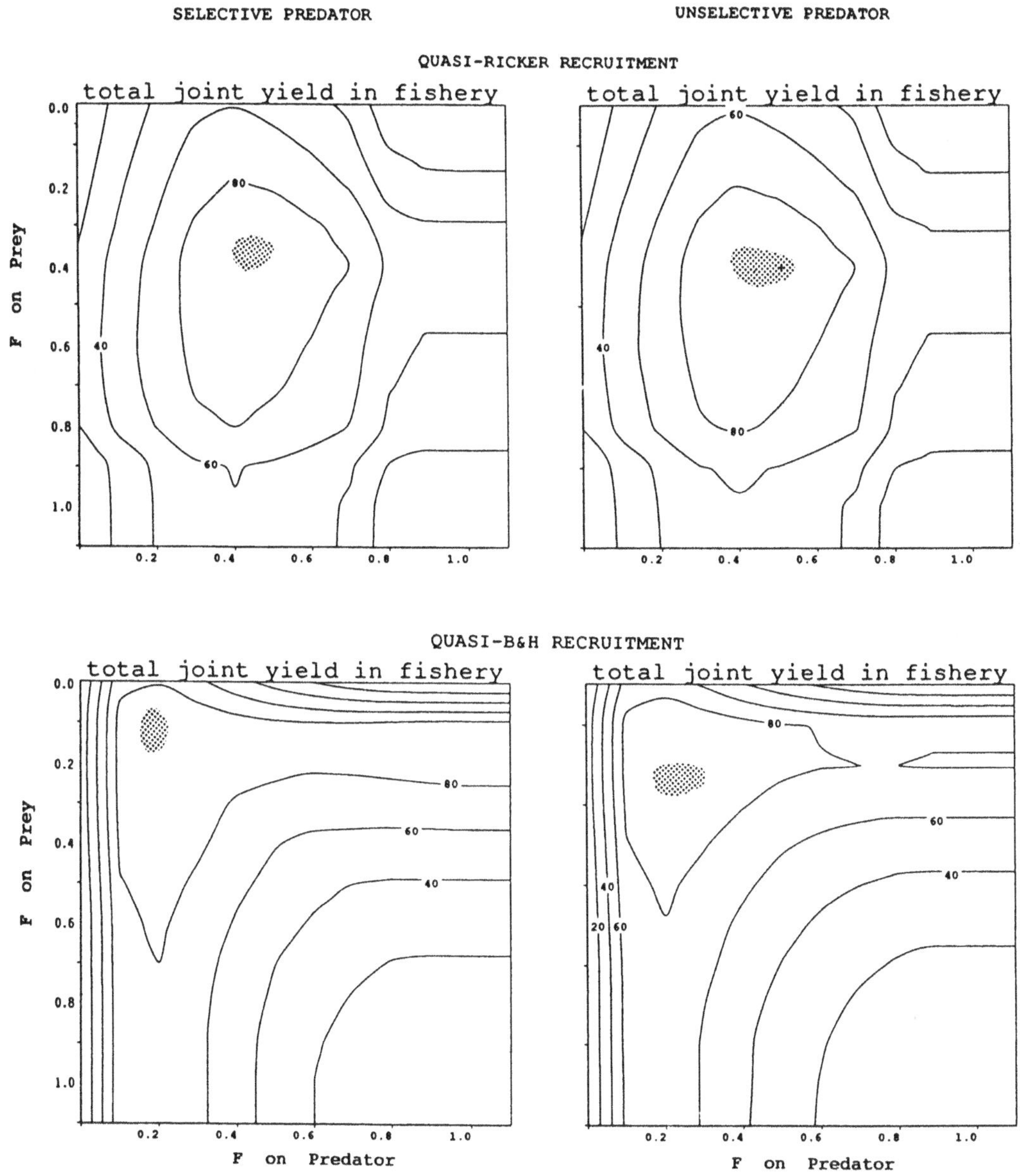

Figure 5: Contours of the joint yield in weight in the prey and predator fisheries under different levels of fishing mortality, F, on the predator and the prey. The optimal diet predator is on the left and the unselective predator on the right. The top half of the diagram refers to quasi-Ricker recruitment and the bottom half to quasi-Beverton and Holt recruitment. Contours are at percentages of the maximum value; peaks are shown by shading.

unselective predators. There are some differences in the exact location of the peaks, but these are small compared to the overall precision of this type of model.

We looked at the effects of altering the threshold at which the optimal diet predator broadened its diet (*0 to 0.9*), the proportionality coefficient for prey consumption by the predator(± *50%*), the growth rate (*linear versus sigmoid*) and age of maturity (*1 to 3*) of the fish. As Shepherd (1988) predicted, the difference between the selective and unselective predator was insensitive to quite large changes in these parameters.

Contour maps like those in Figures 1 to 5 are generally employed to target long-term desired yields or stock biomass in relation to the current location of the fishery. The current location itself may not be known with any great precision, and so the maps can indicate only a general direction for management actions. In particular, the type of recruitment in the fishery makes a very large difference to the stock dynamics, as mimicked by our model. Since recruitment data is characteristically of very poor quality, in practice fisheries often have to be assessed by comparing management strategies which are robust against alternative recruitment scenarios (e.g. Wilson & Pitcher 1985). Faced with such practical difficulties, the small differences generated by optimal diet as opposed to unselective predators seem insignificant.

Discussion

In order to use a model like this in management of an actual fishery, in addition to the usual data on growth, recruitment and mortality, we would have to know the predator's diet selection rules in some detail, including the density threshold at which the predator broadened its diet. This information would be expensive and time-consuming to obtain. Our model suggests that the gains in terms of increased precision in making a long-term assessment of the fishery would be trivial. Errors in estimating mortality parameters would doubtless be larger than this. Our conclusions support Shepherd's hypothesis that the type of preferences shown by predators for prey make only small differences to the aggregated outputs of yield and biomass of an age-structured fish population. Moreover, optimal diet theory seems to have little help to offer in making a practical long-term assessment of a commercial fishery.

It is possible that our conclusion might not be fully general. It would be worth investigating any effects on age structure in relation to stability and the dynamics of short-term forecasting. In many fisheries only fish of a certain range of sizes are marketable or efficiently processed by on-shore machinery.

A more ambitious model might include a link between predator growth or recruitment and the amount of prey consumed. This could affect the stability properties of the predator-prey/harvesting system. We have not included this type of effect since our predator simulates a fish which can easily switch to alternative prey if the modelled prey is not available. Predators more closely-coupled to a restricted prey resource might produce different results.

Summary

An age-structured predator-prey dynamic pool model for the assessment of multispecies commercial fisheries is described. Unselective predator behaviour is compared with a predator which selects prey according to the rules of optimal diet theory. As hypothesised from previous work, there were very few important differences between the two types of predator for the aggregated outputs of yield or stock biomass which are generally used in fishery management. Our conclusion is that the considerable expense and effort required to produce convincing field data on optimal diet dynamics is not necessary for managing age-structured fish populations.

Acknowledgements We would like to thank Qun Liu for the contour plotting routine.

References

Beddington J, May RM (1977) Harvesting populations in a randomly fluctuating environment. **Science 197**:463-465

Brander K (1988) Multispecies fisheries of the Irish Sea. In: Gulland JA (ed) Fish population dynamics, 2nd edition. Wiley, Chichester, UK. 303-328

Brander K, Bennet DB (1986) Interactions between Norway lobster (*Nephrops norvegicus*) and cod (*Gadus morhua*) and their fisheries in the Irish Sea. **Can Spec Publ Fish Aquat Sci 92**:269-281

Doubleday WG, Beacham TD (1982) Southern Gulf of St. Lawrence cod: a review of multispecies management models and management advice. **Can Spec Publ Fish Aquat Sci 59**:133-140

Hart PJB (1986) Foraging in Teleost Fishes. In: Pitcher TJ (ed) The Behaviour of Teleost Fishes. Croom Helm. p211-235

Hughes RN, Burrows MT (1989) The role of energy maximisation in the natural foraging behaviour of dogwhelks, *Nucella lapillus*. In: 24th Symposium on Marine Biology, Aberdeen University Press. *(in press)*

Kirkwood GP (1982) Simple models of multispecies fisheries. In: Pauly D, Murphy GI (eds) Theory and Management of Tropical Fisheries. ICLARM, Manila, Philippines. p 83-98

Laevastu T, Favorite F, Larkins HA (1982) Resource assessment and evaluation of the dynamics of the fisheries resources in the northeastern Pacific with numerical ecosystem models. **Can Spec Publ Fish Aquat Sci 59**:81

Larkin PA, Gazey W (1982) Applications of ecological simulation models to management of tropical multispecies fisheries. In Pauly D, Murphy GI (eds) Theory and Management of Tropical Fisheries. ICLARM, Manila, Philippines. p123-140

Pitcher TJ, Hart PJB (1982) Fisheries Ecology. Croom Helm, London, p414

Polovina JJ (1984) Model of a coral reef ecosystem 1. The ECOPATH model and its application to French Frigate Shoals. **Coral Reefs 3**:1-11

Sainsbury KJ (1988) The ecological basis of multispecies fisheries, and management of a demersal fishery in tropical Australia. In: Gulland JA (ed) Fish population dynamics, 2nd edition. Wiley, Chichester, UK. p349-382

Shepherd JG (1982) A versatile new stock and recruitment relationship for fisheries. **J Cons int Explor Mer 40**:67-75

Shepherd JG (1988) A exploratory method for the assessment of multispecies fisheries. **J Cons int Explor Mer 44**:189-199

Stephens DW, Krebs JR (1986) Foraging Theory. Princeton University Press, USA. p247

Sugihara G (1984) Ecosystem Dynamics. In: May RM (ed) Exploitation of Marine Communities. Springer-Verlag, Berlin. p366

Wilson JPF, Pitcher TJ (1985) Predictive modelling of a stock of Lough Neagh pollan, *Coregonus autumnalis pollan* by a Monte Carlo method. **Aquaculture and Fisheries Management 1**:37-43

ORAL DISCUSSION - G.Turner.

Q. Did you investigate what would happen if the predators retained a fixed preference and didn't switch ?
A. Yes, in the sensitivity analysis, we varied the switching threshold from proportions of 0 – 0.9 of the virgin stock biomass for each prey age class. At zero this meant that there was no predator switching. It made very little difference to the results.

Q. (Sih): Perhaps the predators are being selective only at very low predation pressure, when it doesn't really have all that much effect, but at high predation pressure thay are non–selective because of the way the model is built. Perhaps because of the size structure, the effects of the different predator size classes cancel each other out, and the aggregate mass of predators is non–selective.
A. Under the model, the predation pressure was often quite small compared with the fishing mortality, when fishing pressure on both the predators and prey was high. In these cases the risk of elimination of the preferred size class by the predators was probably quite small. It would help to look at the numbers of each size class of prey eaten by the predators throughout the time before stable age distibrution is reached, but we haven't done this yet. The fact that disallowing switching from the preferred prey class made no difference to the results suggests that they are robust. Another possible cancelling–out effect could be due to the symmetry of the selection matrix, whereby predators of all age classes prefer to feed on prey of the same age class.

Q. (Gilliam): The predators are age selective rather than size selective ?
A. We assumed a good correlation between age and size, and only looked at mean size values.

Q. (Gilliam): So all ages of predator consume prey and all ages of prey are consumed by predators. I agree with Andy Sih that this would tend to produce uniformity, with a whole mass of predators taking a whole mass of prey. One other point. Not many age–structured or size–structured models of interacting predator and prey have been made, but those that are available suggest a very rich dynamics of cycles and waves. These can be stabilised or de–stabilised by changing the mortality rates. If predator recruitment were related to prey intake, then you might expect some of these changes. Then the behaviour of the predator could produce stable equilibria, prey cycles or chaos. This might be very sensitive to predator behaviour. I wouldn't say the model was wrong, but it might be premature to say that predator foraging behaviour is irrelevant to fisheries management.
A. I appreciate the point that some feedback from predator intake to growth and recruitment would be more realistic, but perhaps not more useful. For example, if the predator were a generalist species which only included the prey species in question as one of many in its diet,

then of course, the optimal foraging models of switching between different prey species would apply, and make for even more complications. When you increase the complexity of the model to include feedback to predator recruitment and cyclic phenomena, predator behaviour might become more important, but would it then be simple enough to have much practical value for fisheries management ?

Gilliam: I understand that problem. But, I think it was Albert Einstein who said that everything should be made as simple as possible, but not simpler.

Round–Table Discussion

"Can there be a general theory of diet selection ?"

Discussion leaders: G. Taghon, A. I.Houston, G.E.Belovsky

Rapporteur: J.M. McNamara

Although members of different trophic groups feed differently, behave differently and are subject to different morphological, physiological and ecological constraints, they all potentially maintain fitness by maximizing the gain per unit cost of foraging. This raises questions of feasibility and utility of developing a theoretical approach to foraging behaviour applicable across trophic groups.

Discussion began contentiously: on the one hand, it was felt that models must be built to reflect underlying biology, in which case there could be no general theory of optimal diet choice or, if there were a general theory, it would have to be very complex. Although it seemed to be the majority view that there could be no single theory, certain classes of models were thought to have wide applicability. Alimentary– constraint models, developed for herbivores, seemed applicable to any animals that have to cope with food items ranging widely in digestibility, for example salamanders foraging on chitinous and non– chitinous prey, or polychaetes feeding on mixtures of and nutrient–poor particles. The matching law, used extensively by psychologists to explain behavioural choice, was thought to have potential generality, although there are specific problems in its application, for example defining reinforcement.

There was considerable discussion on the relationship between empirical and theoretical work and on how one chose a model applicable to a particular situation. Sih argued that it would be useful to have general guidelines indicating which model was appropriate in a particular situation. These guidelines could be in the form of a flow diagram, which if followed would lead to a specific model. Several contributors did not think that this approach would be useful, and that it might devalue the position of empirical research relative to theory. Sih also argued that there should be guidelines indicating which technique, such as dynamic programming or linear programming, was the best and simplest in a given situation. This view was opposed by Gilliam, who pointed out that there were many ways to mathematically analyse the standard prey choice model. The approach adopted would inevitably reflect personal background, with different approachesgiving different perspectives.

Several people emphasized that in model building, the biology must come first and that modelling could not be a substitute for biological understanding. Theoreticians had made simple models, such as the prey choice model, which then became the victim of a bandwagon effect. The models had been misapplied because people had not thought out whether they were appropriate for the particular biological situation under consideration. Gilliam thought that the

NATO ASI Series, Vol. G 20
Behavioural Mechanisms of Food Selection
Edited by R. N. Hughes

best solution was for empiricists to build their own models. In that way they would include what was biologically important. Jørgensen urged theoreticians to drop anthropomorphic terms such as "trade–off" in favour of a more neutral terminology.

The most fundamental aspect of modelling foraging behaviour in an evolutionary ecological context is that of defining fitness. While many regarded fitness as being essentially survival and reproduction, others objected that there was no unified theory of what constitutes fitness and that genetic constraints were problematic. To arrive at a meaningful definition one might have to look at a problem in its population–ecological setting, although in many systems this may not be feasible.

The testability of models was another major point of discussion. Concern was expressed that Optimal Foraging Theory could not be falsified, but Maynard Smith's view was reiterated, that we are testing particular models based on optimization, not optimization itself. Kacelnik pointed out that one does not test, say, the Marginal Value Theorem, but only whether the consequences of the theorem give a good description of the situation under study. One can reject a model based on data, but it is a different matter to reject OFT as a whole. OFT is tightly linked to natural selection. One would only abandon explanations of behaviour in terms of natural selection under general and repeated failure of such explanations.

Discussion was concluded in general agreement that the diverse models comprising Optimal Diet Theory continue to be useful. The consensus was in favour of specific applications of ODT rather than the quest for a general theory, which many felt to be an impossible goal. Nevertheless, some thought it would be revealing to apply specific models, such as the type developed for herbivores, to a wider range of trophic groups.

Round-Table Discussion

"The role and importance of optimal foraging theory in ecology"

Discussion leaders: C. Barnard, M. Milinski, C.H. Peterson

Rapporteur: J.-G.J. Godin

Ecologists have traditionally been most interested in understanding patterns of species abundance and diversity and the factors that structure populations and communities. On the other hand, behavioural ecologists have concerned themselves mainly with functional explanations of the behaviour of individual animals, the role that ecology plays in determining behaviour and the fitness consequences of behavioural decisions made by animals. Many have argued in the past that an understanding of the factors controlling dynamic processes at the population and community levels cannot be complete without some knowledge of the behaviour of the individuals that comprise the population. Perhaps fewer have championed an opposite, top-down approach to investigating ecological phenomena, with studies of population and community processes preceeding and forming the framework within which functional questions about animal behaviour can be more effectively addressed. Of course, both approaches have their merits as well as limitations.

This round-table discussion was focused mainly at the interface between behavioural ecology and population and community ecology and on assessing whether foraging theory can lead to an increased understanding of the complex, dynamic processes that occur at the population and community levels. There was general agreement among the discussion leaders and participants that an understanding of the latter phenomena cannot be complete without knowledge of how individual animals respond behaviourally to a wide range of ecological factors, such as gradients of food resources, competitors and predation pressure for example, which may in turn have implications for population and community structures. That is, a bottum-up approach to such inquiry would appear to be most appropriate and fruitful. In this context, foraging theory provides a functional framework to investigate the foraging behaviour of individuals and the dynamics of predator-prey interactions. In contrast, population and community ecologists have typically viewed the population as a unitary black box, with little or no attention given to the components of the population, namely the individuals.

Foraging theory, optimization models and the experimental approach have proved useful in certain cases (particularly in aquatic systems) in predicting the utilization of space and food by individuals across food, competitor and predator gradients. Understanding the behavioural responses of individuals to such ecological factors may explain observed patterns of spatial distribution, food resource utilization and mortality due to predation of animals within a population at

NATO ASI Series, Vol. G 20
Behavioural Mechanisms of Food Selection
Edited by R.N. Hughes

a given time, which may in turn lead to insights on how populations and communities are structured and which ecological forces regulate species abundance and diversity over the long term. Furthermore, there already exists a few conceptual and mechanistic models (e.g. Holling's functional response) that link the foraging behaviour of individuals to population consequences. It was recognized, however, that more realistic models interfacing individual behaviour and dynamic population processes are needed in cases where it is clear that an understanding of particular population phemonena requires detailed knowledge of the behaviour of individual animals.

In conclusion, if contemporary foraging theory is to make significant contributions to population and community ecology in the future, further progress will have to be made in characterizing (i) the fitness consequences of individual foraging decisions, (ii) the functional links between the exploitation of food resources by individuals and their dispersal, and thus immigration and emigration, and recruitment into the population, and (iii) the impact of the foraging (predatory) behaviour of individuals at one trophic level on the population demographics of other species (prey or predator) at other trophic levels. Although there will always be ecologists who see little or no value in using knowledge of the behaviour of individuals to understanding population and community processes, there was general concensus among the participants at this round-table discussion that foraging theory, used as a functional and conceptual tool, has and will continue to play a role in linking individual behaviour to ecological processes at the population and community levels.

LIST OF PARTICIPANTS

ANSELL, Alan, Dunstaffnage Marine Research Laboratory, P.O. Box 3, Oban, Argyll PA34 4AD, Scotland

BALL, John P. Delta Waterfowl and Wetlands Research Station, R.R. 1 Portage la Prairie, Manitoba R1N 3A1, Canada

BARNARD, Christopher J. Animal Behaviour Research Group, University of Nottingham, University Park, Nottingham NG7 2RD, UK

BAZELY, Dawn R. Edward Grey Institute for Field Ornithology, Department of Zoology, South Parks Road, Oxford OX1 3PS, UK

BELOVSKY, Gary E. Department of Biology and School of Natural Resources, University of Michigan, Ann Arbor, Michigan 48 109– 1115, USA

BURROWS, Michael T. School of Biological Sciences, UCNW Bangor, Gwynedd LL57 2UW

CALDWELL, Roy L. Department of Integrative Biology, University of California at Berkeley, Berkeley CA 94720, USA

CROY, Marion I. School of Biological Sciences, UCNW, Bangor, Gwynedd LL57 2UW, UK

DeMOTT, William R. Department of Biological Sciences and Crooked Lake Biological Station, Indiana University–Purdue University at Fort Wayne, Fort Wayne, Indiana 46815, USA

GALIS, Frietson, Division of Ecological Morphology and Department of Population Biology, University of Leiden, P.O. Box 9516, 2300 RA Leiden, Netherlands

GILLIAM, James F. Department of Biological Sciences, State University of New York, Albany, NY 12222, USA

GODIN, Jean–Guy J. Department of Biology, Mount Allison University, Sackville, New Brunswick EOA 3CO, Canada

GORDON, Iain J. Macaulay Land Use Research Institute, Pentlandfield, Roslin, Midlothian EH25 9RF, Scotland

GOSS, Simon, Unit of Behavioural Ecology, CP 231, Universite Libre de Bruxelles, 1050 Bruxelles, Belgium

HART, Paul J.B. Department of Zoology, University of Leicester, University Road, Leicester LE1 7RH, UK

HESP, Lee H. Department of Zoology, University of Nottingham, University Park, Nottingham NG7 2RD, UK

HOBBS, N.Thompson, Research Center, Colorado Division of Wildlife, 317 West Prospect Street, Fort Collins, CO 80526, USA

HOCKEY, Philip A.R. Percy FitzPatrick Institute of African Ornithology, University of Cape Town, Rondebosch 7700, RSA

HOUSTON, Alasdair I. NERC Unit of Behavioural Ecology, Department of Zoology, University of Oxford, Oxford OX1 3PS, UK

HUGHES, Roger, N. School of Biological Sciences, UCNW, Bangor, Gwynedd LL57 2UW, UK

ILLIUS, Andrew W. Department of Agriculture, University of Edinburgh, West Mains Road, Edinburgh EH9 3JG, Scotland

JAEGER, Robert G. Department of Biology, University of Southwestern Louisiana, Lafayette, Louisiana 70504, USA

JØRGENSEN, C. Barker, Zoophysiological Laboratory A, University of Copenhagen, Universitetsparken 13, 2100 Copenhagen Ø, Denmark

JUMARS, Peter A. School of Oceanography, WB–10, University of Washington, Seattle, WA 98195, USA

KACELNIK, Alejandro, King's College Research Centre, King's College, Cambridge CB2 1ST, UK

LUCAS, Jeffrey, Department of Biological Sciences, Purdue University, West Lafayette, Indiana 47907, USA

McNAMARA, John M. Department of Mathematics, University of Bristol, University Walk, Bristol BS8 1TW, UK

MILINSKI, Manfred, Abteilung Verhaltensokologie, Zoologisches Institut, Universitat Bern, Wohlenstrasse 50 a, 3038 Hinterkappelen, Switzerland

NORTON, Trevor A. University of Liverpool, Port Erin Marine Laboratory, Isle of Man, UK

OKAMURA, Beth, Department of Zoology, South Parks Rd.,Oxford OX1 3PS

PERSSON, Lennart, Department of Animal Ecology, University of Umea, S–901 87 Umea, Sweden

PETERSON, Charles H. University of North Carolina at Chapel Hill, Institute of Marine Sciences, Morehead City, North Carolina 28557, USA

PITCHER, Tony J. School of Biological Sciences, UCNW, Bangor, Gwynedd LL57 2UW, UK

PROVENZA Fred D. Department of Range Science, Utah State University, Logan, Utah USA 84322–5230

REAL, Leslie A. Department of Zoology, North Carolina State University, Raleigh, NC 27695–7617, USA

SAETHER, Bernt–Eric, Norwegian Institute for Nature Research, N–7004 Trondheim, Norway

SCHULER, Werner, Zoological Institute II, University of Gottingen, Berliner Strasse 28, D–3400 Gottingen, Federal Republik of Germany

SIH, Andrew, T.H. Morgan School of Biological Sciences, University of Kentucky, Lexington, Kentucky 40506, USA

TAGHON, Gary L. College of Oceanography, Hatfield Marine Science Center, Oregon State University, Newport, Oregon 97365, USA

TURNER, George F. School of Biological Sciences, UCNW Bangor, Gwynedd LL57 2UW, UK

VADAS, Robert L. Department of Botany and Plant Pathology, University of Maine, Orono, Maine 04469, USA

VANDERPLOEG, Henry A. US Department of Commerce, NOAA, Great Lakes Environmental Research Laboratory, 2205 Commonwealth Blvd. Ann Arbor, Michigan, USA 48105–1593

ZIMMER–FAUST, Richard K. Marine Environmental Sciences Consortium and Aquatic Biology Program, University of Alabama, P.O. Box 369–370, Dauphin Island, Alabama 36528, USA

INDEX

NATO ASI Series G

Vol. 1: **Numerical Taxonomy.** Edited by J. Felsenstein. 644 pages. 1983. (out of print)

Vol. 2: **Immunotoxicology.** Edited by P.W. Mullen. 161 pages. 1984.

Vol. 3: **In Vitro Effects of Mineral Dusts.** Edited by E.G. Beck and J. Bignon. 548 pages. 1985.

Vol. 4: **Environmental Impact Assessment, Technology Assessment, and Risk Analysis.** Edited by V.T. Covello, J.L. Mumpower, P.J.M. Stallen, and V.R.R. Uppuluri. 1068 pages. 1985.

Vol. 5: **Genetic Differentiation and Dispersal in Plants.** Edited by P. Jacquard, G. Heim, and J. Antonovics. 452 pages. 1985.

Vol. 6: **Chemistry of Multiphase Atmospheric Systems.** Edited by W. Jaeschke. 773 pages. 1986.

Vol. 7: **The Role of Freshwater Outflow in Coastal Marine Ecosystems.** Edited by S. Skreslet. 453 pages. 1986.

Vol. 8: **Stratospheric Ozone Reduction, Solar Ultraviolet Radiation and Plant Life.** Edited by R.C. Worrest and M.M. Caldwell. 374 pages. 1986.

Vol. 9: **Strategies and Advanced Techniques for Marine Pollution Studies: Mediterranean Sea.** Edited by C.S. Giam and H.J.-M. Dou. 475 pages. 1986.

Vol. 10: **Urban Runoff Pollution.** Edited by H.C. Torno, J. Marsalek, and M. Desbordes. 893 pages. 1986.

Vol. 11: **Pest Control: Operations and Systems Analysis in Fruit Fly Management.** Edited by M. Mangel, J.R. Carey, and R.E. Plant. 465 pages. 1986.

Vol. 12: **Mediterranean Marine Avifauna: Population Studies and Conservation.** Edited by MEDMARAVIS and X. Monbailliu. 535 pages. 1986.

Vol. 13: **Taxonomy of Porifera from the N.E. Atlantic and Mediterranean Sea.** Edited by J. Vacelet and N. Boury-Esnault. 332 pages. 1987.

Vol. 14: **Developments in Numerical Ecology.** Edited by P. Legendre and L. Legendre. 585 pages. 1987.

Vol. 15: **Plant Response to Stress. Functional Analysis in Mediterranean Ecosystems.** Edited by J.D. Tenhunen, F.M. Catarino, O.L. Lange, and W.C. Oechel. 668 pages. 1987.

Vol. 16: **Effects of Atmospheric Pollutants on Forests, Wetlands and Agricultural Ecosystems.** Edited by T.C. Hutchinson and K.M. Meema. 652 pages. 1987.

Vol. 17: **Intelligence and Evolutionary Biology.** Edited by H.J. Jerison and I. Jerison. 481 pages. 1988.

Vol. 18: **Safety Assurance for Environmental Introductions of Genetically-Engineered Organisms.** Edited by J. Fiksel and V.T. Covello. 282 pages. 1988.

Vol. 19: **Environmental Stress in Plants. Biochemical and Physiological Mechanisms.** Edited by J.H. Cherry. 369 pages. 1989.

Vol. 20: **Behavioural Mechanisms of Food Selection.** Edited by R.N. Hughes. 886 pages. 1990.

GPSR Compliance

The European Union's (EU) General Product Safety Regulation (GPSR) is a set of rules that requires consumer products to be safe and our obligations to ensure this.

If you have any concerns about our products, you can contact us on ProductSafety@springernature.com

In case Publisher is established outside the EU, the EU authorized representative is:

Springer Nature Customer Service Center GmbH
Europaplatz 3
69115 Heidelberg, Germany

Batch number: 10761693

Printed by Printforce, the Netherlands